Exploring Biology in the Laboratory

Murray P. Pendarvis
Southeastern Louisiana University

John L. Crawley

MORTON
PUBLISHING

925 W. Kenyon Avenue, Unit 12
Englewood, CO 80110

www.morton-pub.com

Book Team

Publisher	Douglas Morton
Biology Editor	David Ferguson
Project Manager	Desireé Coscia
Cover	Bob Schram, Bookends, Inc.
Illustrations	Imagineering Inc.
Copyeditor	Carolyn Acheson
Proofreader	Patricia Billiot
Interior Design and Composition	Focus Design

**This book is dedicated to the memory
of Shaun Paton Pendarvis.**

About the Cover

The green lynx spider, *Peucetia viridans*, is the largest of the North American lynx spiders. It is a non venomous spider that is beneficial to agriculture due to the fact that it preys on many pest species.

Printed in the United States of America

10 9 8 7 6 5 4 3 2 1

ISBN–13: 978-0-89582-799-9

Library of Congress Control Number: 2010943081

Preface

WHY I TEACH

"Ya'll BIOLOGIZE out there" are the final words of wisdom that I often shout to my biology classes as they bolt for the door after class. Many of my students listen and return to my next class with photographs, fossils, leaves, and critters, as well as a myriad of questions and stories related to their exploration of the world of life. I enjoy and look forward to sharing their enthusiasm and excitement about biology.

For me, Introductory Biology is about more than fulfilling a lab requirement for graduation. This class might be my one opportunity to inspire a student to pursue a career in biology research or education. Falling short of that, it is my 14-week chance to create enthusiasm for students to explore some of life's mysteries. When the light bulb goes off for future math or business majors, I hope the excitement of discovery will lead them to whichever path life takes them.

Thinking back over my teaching career at both the high school and the university levels, I have immersed myself in biology. It has become a passion, a hobby, and a way of life. Many of my fondest memories stem from my experiences with students in the classroom, in the laboratory, and in the field. I have the honor of teaching the most intriguing subject in academia and working with young people. I have dedicated my life to quality biology education.

HOW I TEACH

This laboratory manual is based upon my vision of biology education and the way I teach biology. The activities have been designed to be safe, interesting, and meaningful. I have arranged the chapters in an order that builds on previous chapters. In addition, I have organized the chapters to reflect variations on a theme. For example, Chapter 7, Histology: Understanding Plant and Animal Tissues emphasizes the study of tissues as fundamental to understanding plants and animals. This subtle but important idea—that all living things share a commonality of purpose and a similarity in design—is vital to understanding the unifying concepts of biology. Throughout the manual, many activities emphasize the unity of all living things and evolutionary forces that have resulted in (and continue to act on) the diversity that we see before us today.

Biology education doesn't happen only in the classroom or the laboratory. Students can enhance their appreciation for and understanding of biology by using ubiquitous camera phones, digital cameras, and iPods. Several activities encourage students to use their cameras in the biology lab and in the field. These activities include photographing procedures and specimens, photographic scavenger hunts, and specific photographic assignments. Thanks to the technological advances in today's smartphones, students can photograph living organisms in their natural environments without compromising either the observer or the observed. In addition, these biophotography activities help students develop the ability to impart their knowledge to others. Perhaps as the result of their participating in these activities, students may pursue photography as a hobby for a lifetime.

Sprinkled throughout the manual are interesting facts and figures designed to capture students' imagination and stimulate thought beyond the laboratory. Also included are inspirational and insightful quotes from influential scientists, statesmen, and other well-known historical figures. Whether they are profound or humorous, they all emphasize the pivotal role of scientific study in shaping humankind's understanding of the world around us.

If you have any questions or comments about the manual, or suggestions for how it might be improved in future editions, please contact us at exploringbiology@morton-pub.com

—**Murray P. "Pat" Pendarvis**

Acknowledgments

Many professionals have assisted in the preparation of *Exploring Biology in the Laboratory*, and have shared our enthusiasm of its value for students of biology. Thank you to Sarah Jean Rayner for her contributions to chapters 8 through 19. We are appreciative of Dr. Barry Ferguson, Dr. Byron Adams at Brigham Young University, Dr. Samuel R. Rushforth and Dr. Robert Robbins at Utah Valley University, and James Loyd at Southeastern Louisiana University. We appreciate all of the feedback we received from biology laboratory professionals during the creation of the book. We are particularly indebted to Thomas Pitzer, at Florida International University and Warner Bair, at Lone Star College—Cy Fair for their detailed suggestions for improvement. Any outstanding errors are the sole responsibility of the authors. We welcome any feedback you might have after using this book. We would also like to thank Patricia Billiot who was instrumental in proofing the lab manual.

We gratefully acknowledge the assistance of Imagineering Media Services, Inc. for the art throughout the book. We also appreciate Focus Design and Malina Nielson for the layout and organization of this lab manual. We are indebted to Douglas Morton, David Ferguson and the personnel at Morton Publishing Company for the opportunity, encouragement, and support to prepare this lab manual.

Many of the photographs of living plants and animals were made possible because of the cooperation and generosity of the San Diego Zoo, San Diego Wild Animal Park, Sea World (San Diego, CA), Hogle Zoo (Salt Lake City, UT), San Francisco Zoo, Aquatica (Orem, UT), Avery Island, La., and Tickfaw State Park (Springfield, La.). We are especially appreciative to the professional biologists at these fine institutions. Thanks to Satsuma Seafood (Satsuma, La) for letting us photograph crayfish (crawfish) specimens.

The following students served as models in the manual: Soo Ahn, Kristen Hilliard, Scott Nation, Laurie Beth McCoy, Bryan Pendarvis, Pramir K.C., Ariel Ellis, and Dayo Felix. In addition Cynthia M. Williams, Whitney Curry, and Allison Lanclos were instrumental in organizing components of this manual.

The authors would like to thank their families and friends for their support during this endeavor as well as countless students through the years. A special thanks to Sheree Foster and her family for her support and patience during the writing of this manual.

A heartfelt thanks to Shaun Paton Pendarvis (1982–2009) for his strength, courage, and inspiration while battling cystic fibrosis.

Photo Credits

About the Authors

Murray P. "Pat" Pendarvis a resident of Walker, Louisiana, is currently an assistant professor of biological sciences at Southeastern Louisiana University. He teaches general biology for majors and nonmajors, honor's biology, medical terminology, anatomy and physiology, the history of biology, and biophotography. In addition, he is an adjunct professor of biology at Our Lady of the Lake College in Baton Rouge, Louisiana teaching general biology, environmental science, paleontology, evolution, medical genetics, and the history of medicine. Prior to teaching at the university level, he taught science at Doyle and Walker High Schools in Livingston Parish, Louisiana.

As a result of his dedication to quality science education, he has received a number of awards including The Presidential Award for Excellence in Mathematics and Science Education and The National Association of Biology Teachers Outstanding Teacher Award. In addition to teaching, Pat is an author writing biology texts for McGraw-Hill and Morton Publishing.

One of Pat's passions is the science and art of photography. He has published many photographs in several biology and ecology texts. Pat has earned a B.S. degree in biology education and a M.S. degree in zoology from Southeastern Louisiana University and a Ph.D. in science education from the University of Southern Mississippi.

John L. Crawley currently resides in Provo, Utah. He received his degree in Zoology from Brigham Young University in 1988. While working as a researcher for the National Forest Service and Utah Division of Wildlife Resources in the early 1990s, John was invited to work on his first project for Morton Publishing, *A Photographic Atlas for the Anatomy and Physiology Laboratory. Exploring Biology in the Laboratory* is John's fifth title with Morton Publishing.

John has spent much of his life taking pictures. His photography has allowed him to travel widely and his photos have appeared in national ads, magazines, and publications. He has worked for groups such as Delta Airlines, National Geographic, Bureau of Land Management, and many others. His projects with Morton Publishing have been a great fit for his passion for photography and the biological sciences.

Contents

Safety in the Laboratory

The laboratory should provide students and instructors alike with an environment conducive to accomplishing specific scientific tasks. It is imperative that everyone involved in the laboratory recognize the importance of safety in the laboratory. In addition, everyone using the laboratory must be aware of potential safety hazards such as faulty electrical outlets, frayed wires, broken glassware and slides, chemical spills, and potentially dangerous organisms. Students must follow proper safety practices in the laboratory and immediately report any safety hazards or accidents to the instructor.

Basic Rules for the Laboratory

1. Follow your laboratory instructor's directions.
2. Be familiar with the location of safety equipment (first aid kit, eyewash, gas shut-off, fire blanket, fire extinguisher), emergency telephone numbers, and exits.
3. Be familiar with the activity of the day and potential safety issues. Read labels carefully.
4. Treat all laboratory equipment such as microscopes with care, and store the equipment as instructed.
5. Treat all living things with respect. Avoid causing unnecessary stress or discomfort to living animals.
6. Do not open specimen jars unless instructed.
7. Avoid horseplay in the laboratory.
8. Do not eat, drink, or smoke in the laboratory.
9. Always keep your work area clean and uncluttered. Thoroughly clean your laboratory station before and after each activity.
10. Always wash your hands with soap and water before and after the laboratory experience. Keep your hands away from your face.
11. Properly dispose of broken glass, slides, and disposable laboratory equipment.
12. Wear closed-toed shoes, eye protection, gloves, and laboratory coats when instructed.
13. When dissecting specimens, always wash them thoroughly before you begin the dissection. At the completion of the laboratory activity, properly dispose of the specimen as instructed.
14. Use caution when employing sharp instruments such as scalpels and dissecting pins. Immediately report any cuts or punctures to your instructor.
15. Place a stopper in any chemical bottle when it is not in use. Follow your instructor's directions when carrying bottles and pouring chemicals. Report any spills immediately to your instructor. Do not taste any chemicals.
16. Keep flammable chemicals away from open flames and take precaution when handling hot items. Roll up your sleeves when working around open flames.
17. If you have long hair, tie it back.
18. During outdoor activities, be aware of poisonous plants, venomous animals, and potentially dangerous environments such as cliffs and water. Work in teams.

I have read and understand the basic safety rules for the laboratory.

Name _____

Class _____

Date _____

Chapter 1
Introduction: The Nature of Science

Student Outcome Objectives

At the completion of this exercise, the student will be able to:

1. Define science and discuss the characteristics of science.
2. Compare and contrast physical, earth, and life science.
3. Compare and contrast science and technology.
4. Name the three aspects of science.
5. Discuss the steps of the scientific method.
6. Discuss the basic biology, ecology, and behavior of termites.
7. Conduct and record observations about termite behavior.
8. Construct a hypothesis regarding termite behavior.
9. Discuss the scientific method and process skills used in this activity.

Overview

The wonders of science surround us every day, from the rising of the sun to each beat of our heart. A descriptive definition of science, however, is elusive. Classically, **science** is defined as an organized body of knowledge that attempts to explain natural phenomena. This general definition does not recognize that science functions as a dynamic process, encompassing exploration, experimentation, and discovery. Perhaps science is best described by several fundamental characteristics.

- Science is based upon observations that incorporate our senses or instruments that extend our senses to interpret natural phenomena. Science does not allow any room for mysticism, superstition, or thought that is contrary to observations.
- Science is a search for regularities. These regularities may include easily observed phenomena or patterns in nature.
- After observations have been recorded about regularities in nature, scientists must process information. Through qualitative and quantitative techniques such as computer analysis, scientists are better able to understand natural processes.
- Science is a self-correcting process in which previously existing concepts can be expanded, modified, or replaced if necessary. In some instances, changes to existing ways of thought are not easily accepted. Ideas that

redefine our place in the universe conflict with centuries of existing dogma. Scientific study is an ongoing, active process, and the scientific body of knowledge is growing exponentially each year.

While recognizing that science is a tremendous endeavor, it can be divided into three major fields of study: physical science, earth science, and life science.

1. **Physical science** attempts to describe the laws that govern the universe and the composition of the universe. Physics and chemistry are two of the fundamental disciplines of physical science.
2. **Earth science** attempts to describe our place in the universe and includes many disciplines, such as astronomy, geology, oceanography, and meteorology.
3. **Life science**, or biology, attempts to describe living things and consists of many disciplines from anatomy to zoology.

Scientific study is a unified endeavor. The methods, reasoning, and zest for science are universal. Investigating a complex biological phenomenon such as photosynthesis requires knowledge of the physics of light, chemical reactions, the atmosphere, soils, cell and plant anatomy, and evolution. To fully understand a process, we must use science as a way of thought, not simply as a storehouse of specific information.

Science is a great game. It is inspiring and refreshing. The playing field is the universe itself.

—**Isidor Isaac Rabi (1898–1988)**

Figure 1.1 Students studying in a classroom.

Some Representative Disciplines of Biology

cytology: the study of cells
histology: the study of tissues
anatomy: the study of structure
physiology: the study of function
botany: the study of plants
zoology: the study of animals
microbiology: the study of microorganisms
mycology: the study of fungi
phycology: the study of algae
entomology: the study of insects
ichthyology: the study of fishes
ornithology: the study of birds
taxonomy: the identification and naming of organisms
genetics: the study of heredity
ecology: the study of the interrelationship between organisms and their environment
hematology: the study of blood
paleontology: the study of fossils
ethology: the study of animal behavior

Figure 1.2 Darwin as a young man.

Sir Isaac Newton once stated, "If I have seen a little further, it is by standing on the shoulders of giants." Today in biology, we stand upon the shoulders of giants such as Anthony von Leeuwenhoek, Carolus Linnaeus, Charles Darwin (Fig. 1.2), Louis Pasteur, and countless other contributors. Our ancestors have left us with a legacy and wealth of knowledge about our world. Through their discoveries, triumphs, and toils, we better understand our universe and life itself. Thus, we celebrate the history of science and biology as we march into the future.

Although the terms *science* and *technology* often are used interchangeably, we must distinguish between these terms. **Science** is a tool that allows us to comprehend natural phenomena. It is neither good nor bad; it simply provides us with an understanding. **Technology** is the application of science. Science feeds technology, as evidenced by the recent growth in genetic engineering after understanding the molecular configuration of DNA. Depending upon the user, technology-related fields such as nuclear physics, drilling for oil, and genetics can be highly beneficial or diabolically harmful. A literate society is responsible for ensuring that technology is applied in beneficial ways.

Scientific responsibility begins with a solid science education for everyone. As a result of this education, the general public, as well as those who govern our world, will be able to make better decisions. Presently we are facing global environmental issues such as loss of habitat, overpopulation, pollution, and global climate change. In addition, we are facing scientifically based ethical and moral dilemmas and the threat of bioterrorism. To address these issues and ensure a better future, quality science education must be present at all levels of society (Fig. 1.3).

Check Your Understanding

1. Define *science*.

_____

_____

2. Provide an example from the world of science that addresses each of the characteristics of science.

_____

_____

3. Name and define five disciplines within the biological sciences.

_____

_____

_____

_____

_____

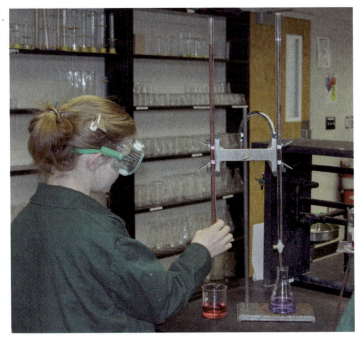

Figure 1.3 A scientist at work in a lab.

4. Based upon a biological phenomenon such as photosynthesis, discuss the relationship between the physical, earth, and life sciences.

_____

_____

_____

_____

5. Compare and contrast science and technology, and provide an example of each.

_____

_____

_____

_____

The scientist does not study nature because it is useful;
he studies it because he delights in it, and he delights
in it because it is beautiful. If nature were not beautiful,
it would not be worth knowing, and if nature were not
worth knowing, life would not be worth living.

—Jules Henri Poincaré (1854–1912)

THE THREE ASPECTS OF SCIENCE

Over the past five decades, science has been characterized by ever-accelerating changes, including a flood of new information, ideas, and unifying concepts. Students today must be able to balance classical and contemporary science in a meaningful way. To be successful, students who are studying and participating in a science program must understand the three aspects of science—content, process, and attitude.

Content

The content of science is represented by a vast database of factual and theoretical information that has been accumulated through the centuries. Today, more than any other time in history, scientific content is reaching new heights through exploration, experimentation, and discovery. As a result, the public, students, and instructors, are being exposed to an information explosion. To be scientifically literate, we must comprehend content material that will facilitate an understanding of the nature of scientific laws and principles, enhance an understanding of the environment, enrich an understanding of mankind, and facilitate proper decision-making based upon the understanding of scientific content. The fundamental core of content knowledge consists of facts, concepts, and generalizations.

Facts

Facts are the fundamental unit of content that relate directly to observations such as in these examples:

• The numbers of eosinophils (eosin-staining white blood cells) are elevated during periods of parasitic infections and allergies
• Cytosine binds with guanine in DNA
• Mammals have a four-chambered heart

Without a sufficient number of facts (including vocabulary), students will not be able to develop concepts and generalizations.

Concepts

Concepts are abstractions of ideas that are used to simplify understanding of a phenomenon. Concepts consist of a variety of facts that generally bring about a mental image, for example:

• mammal
• camouflage
• leaf

Concepts serve to link facts together to form categories and require higher-level thought than just memorizing facts.

Generalizations

Generalizations represent patterns in science by relating concepts to each other. In addition, generalizations allow us to predict events based upon concepts. Examples of generalizations are the following:

- Many marine fishes have poorly developed kidneys.
- Offspring of individuals who are closely related have an increased risk for recessive genetic disorders.
- The higher the elevation above sea level, the lower the boiling temperature of water.

Biology is full of exceptions. For example, not all mammals have seven cervical vertebrae. Therefore, we must be careful in making broad statements concerning facts, concepts, and generalizations.

Process

All of the content in the universe is not relevant unless it can be applied in science and technology. Science is an active process, not merely a collection of facts and unrelated information. Knowledge of science based upon facts alone will merely produce a "robot." Processes are the active components of science and refer to the way that science works. If processes are the basis for a program in which they are not interwoven with factual information, however, frustration and misunderstanding may result. Thus, a quality science program must balance content and processes.

The American Association for the Advancement of Science and the National Science Foundation under the guidance of Robert Gagne, developed "Science: A Process Approach." This program suggested that, to be scientifically functional, science students should master various skills. These skills, known as the basic and integrated process skills of science, represent a cumulative hierarchy of performance skills that all students of science should master.

Basic Process Skills

The basic process skills represent fundamental skills that should be acquired early in a scientific endeavor and the integrated skills will be acquired with experience.

1. *Observing:* Senses or extensions of the senses are used to gather information about the natural world (Fig. 1.4).
2. *Classifying:* Observable properties are used to sort and categorize objects. Some schemes of classification are simple, such as describing whether an organism possesses certain traits. Other schemes, such as those used in taxonomy, are highly complex.
3. *Using numbers:* The investigator finds the quantitative relationships among data. Mathematics is the backbone of science.
4. *Measuring:* Instruments are used to quantify observations. The metric system is used in all scientific measurements.
5. *Using space-time relationships:* The investigator states the locations and shapes of objects or describes the position and changes in position of moving objects. This relationship also can include time-based activities.
6. *Communicating:* Compiled information is used to describe a system consisting of objects and/or interactions. These systems can be represented in oral, written, graphic, or pictorial form.
7. *Predicting:* Future events are forecast from trends found within a solid base of evidence. These events can fall within collected data (interpolation) or go beyond the scope of the data (extrapolation).
8. *Inferring:* Speculations are developed to account for observations. Inferences go beyond the data.

Integrated Process Skills

1. *Defining operationally:* Operational definitions describe a system in terms of what one can observe. Many times, the operational definition serves as an accurate and practical working definition. As examples, the operational definition may define a specific physiological process as it occurs in the experiment, a dosage as it is given in an experiment, or a population to be used in an experiment.
2. *Identifying and controlling variables:* The experimenter identifies and controls variables while experimenting so a single variable is manipulated and a single variable responds to manipulation. A **variable** is any factor that can cause changes within a system. In setting up an experiment, three types of variables are recognized.
   a. The **control variable** is held constant and is used as a baseline for comparison.
   b. The **independent variable** is the variable being manipulated in the experiment.
   c. The **dependent variable** is also known as the responding variable.
3. *Formulating hypotheses:* The experimenter predicts relationships between the independent and dependent variables. The function of the **hypothesis** is to provide direction for gathering data. The hypothesis must be testable and practical. Keep in mind that a rejected hypothesis is not considered a failure. Several formats are used in constructing hypotheses. Initially, hypotheses are written in an "if–then" form. A **null hypothesis** states that there is no relationship between the independent and dependent variables.

4. *Interpreting data:* Interpreting data is a composite skill consisting of communicating, predicting, and inferring. In interpreting data, statistical methods, as well as charts and graphs, should be used to make the results understandable.

5. *Experimenting:* This is the cumulative process skill that encompasses all of the basic and integrated process skills.

Attitude

The third aspect of science, attitude, is overlooked many times but perhaps is the most important. The term *attitude* has two different connotations in science: one's attitude toward science and one's scientific worldview. Scientific attitudes begin to develop at an early age and often are well formed by the time a student reaches the university level.

Today, many students do not possess a healthy scientific view of the world because they do not understand the role of science in society. As a result, they do not respect the foundations of science. To build a realistic scientific worldview, the student must approach science with curiosity, open-mindedness, logical thought, math skills, resolution of superstitions, problem-solving techniques, patience, and persistence. Also, students must suspend judgment until the facts are known, and record the data honestly. The Hollywood stereotypical scientist has distorted many students' views regarding scientists. Keep in mind that scientists are people too!

Science is a synthesis of content, process, and attitude. Without content, the student cannot understand the grandeur of the universe. Without process, the student cannot experience the excitement of discovery. Without a healthy scientific attitude, the student cannot value content and cannot appreciate the excitement of discovery.

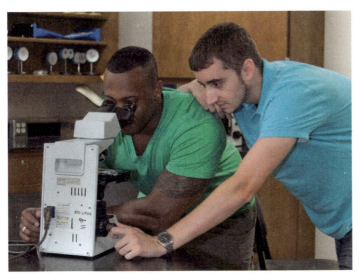

Figure 1.4 Students working with a microscope.

The mere formulation of a problem is far more often essential than its solution, which may be merely a matter of mathematical or experimental skill. To raise new questions, new possibilities, to regard old problems from a new angle, requires creative imagination and marks real advances in science.

—Albert Einstein (1879–1955)

 ## Check Your Understanding

1. List three biological facts, three concepts, and three generalizations.

_____

_____

_____

_____

_____

_____

_____

_____

2. Describe several variables that may impact learning in this laboratory.

_____

_____

_____

3. In an experiment designed to test the effect of salinity upon the development of brine shrimp, identify the control and the independent and dependent variables.

_____

_____

_____

_____

4. Write a valid hypothesis based upon the variables established in the above question.

_____

_____

_____

5. Describe the importance of understanding the three aspects of science.

_____

_____

_____

6. Research reveals that many college freshmen, as well as the general public, possess high anxiety levels related to science. Suggest some common causes of scientific anxiety.

_____

_____

_____

7. Research indicates many misconceptions regarding science and scientists. Identify some common misconceptions.

_____

_____

_____

THE SCIENTIFIC METHOD

Through the ages, the scientific method typically has been used to describe the way that science works. The scientific method, however, is just a guide to solve problems and initiate investigations. Not all scientific discoveries rigidly follow the scientific method. Many discoveries involve serendipity and even a little luck. Representatively, the scientific method consists of the following steps (Fig. 1.5):

1. A phenomenon sparks the interest of the potential investigator, who makes observations and reviews literature about the phenomenon.
2. Once the investigator has an understanding of the problem, a prediction is made and a hypothesis (or hypotheses) is constructed.
3. After the hypothesis is stated, the investigator performs experiments and gathers data about the problem.
4. Based upon the data, the hypothesis is rejected or accepted and a conclusion is drawn.

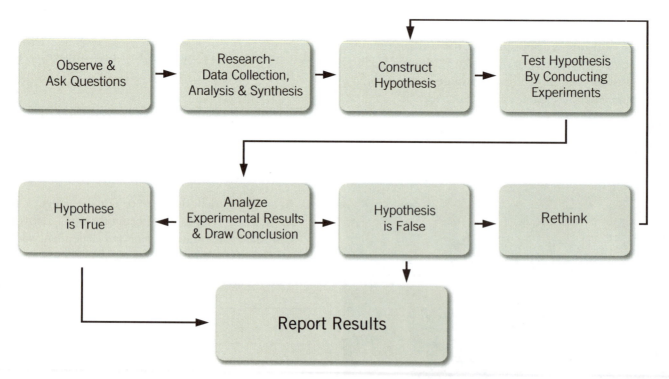

Figure 1.5 The scientific method.

SCIENTIFIC THEORIES

The term **theory** is often misused. In science, the statement, "Oh, it's just a theory," has no place because it suggests that a theory is nothing more than a guess. Theories, however, stand on their own accord and represent the current well-supported explanation about some aspect of the natural world. Examples of well-known theories in the biological sciences are cell theory, evolutionary theory, and germ theory of disease.

Termites

The term *termite* is derived from the Latin word *termes*, which literally means woodworm. Nearly 2000 species of termites comprise the insect order Isoptera (Greek, *isos*—the same; *ptera*—wing). The isopterans possess equal-sized paired wings in the reproductive form. In the environment, termites serve as decomposers, helping to break down and recycle dead wood. They become pests when they consume cellulose-based materials such as wood and paper. In addition, termites are responsible for the destruction of valuable forest resources. To digest these cellulose-based sources, termites exist in a symbiotic relationship with either bacteria or ciliated protists.

Termites are soft-bodied insects that rarely exceed 10 mm in length. They are highly social animals living in subterranean colonies or dead wood. Colonies can be composed of several dozen members or millions of members depending upon the species and age of the colony. With the exception of the immature nymphs, a termite colony consists of three basic functional groups called castes: workers, soldiers, and reproductives.

The vast majority of members in a termite colony are wingless, sterile, and blind workers. They are milky white in color and possess hard chewing mouth parts. *Workers* look after the eggs and nymphs, feed the soldiers and reproductive forms, build and maintain the colony, and forage for food. The workers are responsible for the telltale signs of termite damage. *Soldiers* are also wingless, sterile, and blind. This caste has a milky white body with a yellowish-brown head and prominent mandibles. Soldiers defend the colony mostly from ants. The *reproductives*, also called the *royal caste*, consist of the king and queen. Their only function is to reproduce. The royal caste is dark brown in color and has functional eyes. The queen seems to be striped because the segments of her abdomen are distended. Swarming termites are called *alates* or *swarmers*. They are winged reproductives that eventually establish new nests (Fig. 1.6).

Because termites are highly social insects, communication is essential. Termites use chemicals known as **pheromones** for communication. Specific pheromones are used in mating, producing an alarm, keeping nymphs from forming reproductive castes, and establishing a trail. Termites use the trail hormones when they are attempting to lure other termites to follow them to a food source or other region of interest.

It has been discovered that the solvents in Papermate™ and Bic™ ballpoint pens are similar to the trail hormones used by termites. In this activity, the student will use the scientific method and process skills of science to investigate the relationship between various inks and termite trailing behavior.

Figure 1.6 Termites are in the order Isoptera.
1. Worker termites   2. Soldier termites

Student Activity—Ink and Termites

Materials
- worker termites
- Petri dish lined with a moist paper towel to house the termites
- typing paper
- artist's brush
- paperclip
- black, blue, and red Papermate™ or Bic™ ballpoint pens

Procedure 1.1
Termites

1. Divide into teams as directed by the instructor.
2. Procure the termites and supplies for the activity. Ensure that the termites are housed in a covered Petri dish lined with a moist paper towel. When they are not being used, cover the termites with a moist paper towel.
3. Place the typing paper on the desktop, and carefully use the brush to move and orient the termites as they are fragile.

Q. What variables may influence termite movement?

_____

_____

_____

Q. Describe the movement of the termites. Does their movement exhibit any pattern?

_____

_____

_____

4. Using one pen at a time draw a 2 cm straight line on the typing paper. Using the artist's brush or paperclip, gently guide the termite toward the line. Record and describe the response of the termites to each line. Do they prefer a one pen over the others?

_____

_____

5. Using fresh paper and the same pens, draw circles that are approximately 2 cm in diameter. Using the artist's brush or paperclip, gently guide the termite toward the center of the circle. Record and describe the response of the termites to each circle. Do they prefer one pen over the others?

_____

_____

6. Using fresh paper and the pen that the termites preferred, draw several geometrical patterns with sharp turns and a figure 8. Using the artist's brush or paperclip, gently guide the termite toward the patterns, testing one design at a time. Record and describe the response of the termites to each geometrical pattern and the figure 8.

_____

_____

7. Using fresh paper and the pen that the termites preferred, draw several broken straight-line patterns. Using the artist's brush or paperclip, gently guide the termite toward the line.

Q. Do the termites respond differently to broken lines?

_____

_____

Q. Is there any relationship between termite movement and the gaps between the lines?

_____

_____

8. Based upon your observations using black ink, design an investigation that tests other ink colors or types of pens.

   a. Identify the variable that you are going to test.

   b. Construct a hypothesis for your investigation and state it below.

   _____

   _____

   _____

   c. Briefly describe your procedure.

   _____

   _____

   d. Discuss your findings. Did you accept or reject your hypothesis?

   _____

   _____

   e. Describe several other variations of this basic activity that would be good topics for an investigation.

   _____

   _____

   _____

9. Place the termites back in the container, and return them and the supplies to their distribution area.

Check Your Understanding

Q. How did you incorporate the scientific method in this experiment?

_____

_____

_____

_____

_____

_____

_____

_____

_____

Q. List and describe the basic and integrated process skills used in this activity.

_____

_____

_____

_____

_____

_____

_____

_____

_____

_____

_____

_____

_____

_____

_____

NOTES

Name: _____   Date: _____   Section: _____

?? Review Questions

1. Describe the characteristics of science.

   _____

   _____

2. Why is science considered a unified endeavor?

   _____

   _____

3. Compare and contrast science and technology.

   _____

   _____

4. What did Sir Isaac Newton mean when he stated, "If I have seen a little farther, it is by standing on the shoulders of giants"?

   _____

   _____

5. Why are the process skills considered life skills?

   _____

   _____

6. Why must all three aspects of science be addressed in a successful science education program?

   _____

   _____

7. Relate the classical scientific method to the process skills of science.

   _____

   _____

8. What can be done to improve scientific attitudes and education in our nation?

   _____

   _____

9. Justify spending money on seemingly trivial scientific endeavors.

_____

_____

10. Describe several characteristics of a good experiment.

_____

_____

11. In an experiment designed to test the effects of outboard motor oil on the growth of algae, identify the control, the independent variable, the dependent variables, and construct a simple hypothesis.

_____

_____

_____

12. Why is the term "theory" often misused?

_____

_____

13. Why is there no room for superstition and mysticism in science?

_____

_____

14. Why is a fundamental knowledge of biology necessary for everyone?

_____

_____

15. Interpret "Science, an Epilogue."

"Science, an Epilogue"
The task of science is to simplify, not to glorify.
The task of science is to seek truth and understanding, not to mystify.
The task of science is to benefit humans, not to destroy them.
The task of humans is to know enough science to see that this happens.

—**Anonymous**

_____

_____

_____

Chapter 2
Experimentation: The Process Skills in Action

Student Outcome Objectives

At the completion of this exercise, the student will be able to:

1. Design and conduct a basic scientific experiment.
2. Identify and control variables.
3. Write a formal hypothesis.
4. Interpret and present data in a meaningful manner.
5. Write a formal laboratory report as outlined by the instructor.
6. Identify and discuss the basic and integrated process skills of science that are used in this experiment.
7. Discuss the effects of temperature on goldfish respiration.
8. Discuss extensions on the theme of the experiment.

Overview

This exercise is designed to facilitate acquisition of the basic and integrated process skills of science by having the students design and conduct an experiment that tests the effect of temperature on the respiration rate of goldfish (*Carrasius auratus*) (Fig. 2.1). The student will identify variables, construct hypotheses, make observations, collect, record, and interpret data, and draw conclusions based upon the data. The laboratory instructor and students will treat the goldfish in a humane manner at all times.

In aerobic respiration, an organism takes in oxygen from its environment and releases carbon dioxide as a waste product. Organisms have specialized structures to carry out respiration. Many aquatic animals utilize gills for respiration. In fish, the gills can be found beneath a protective covering called the **operculum**. The **gills** are made of **gill filaments**, which serve to increase the surface area. This allows maximum exposure to the oxygen-laden water environment.

When a fish "breathes," its operculum closes and its mouth opens. To allow water to pass over the gill filaments, the mouth closes and the pharynx contracts. In turn, oxygen diffuses into the capillary circulatory network and is distributed throughout the fish's body. Carbon dioxide diffuses from the capillary network and enters the environment. The process of opening and closing the mouth or opening and closing the operculum constitutes one breath in fish.

Several variables affect the respiration rate of fish. In this experiment, student teams will discover the effects of temperature on goldfish respiration and report their results in a scientific manner. Students should make every effort to ensure the survival of the experimental goldfish.

Student Activity—Goldfish Respiration

Materials

- goldfish
- aquarium net
- thermometer
- crushed ice
- aquarium water
- 250 ml beaker
- 500 ml beaker
- stirring rod
- stopwatch
- graph paper

Procedure 2.1
Goldfish

Read all of the instructions before beginning the experiment! In preparation, divide into working teams of four students. In this activity, one student should serve as timer, one student as recorder, one student as counter, and one student as worker.

1. Describe variables that may influence goldfish respiration. Discuss and identify the control, independent, and dependent variables found in this experiment.

_____

_____

_____

Figure 2.1 A goldfish, *Carrasius auratus*.

2. Construct a valid hypothesis pertaining to the experiment.

_____

_____

_____

3. Add approximately 150 ml of aquarium water to the 250 ml beaker. Place the thermometer in the beaker, and take the temperature. Practice adding small amounts of crushed ice to the water until you can easily lower the temperature of the water approximately 2 degrees Celsius.

4. Empty the beaker and refill it with 150 ml of aquarium water. This will provide the starting temperature and allow the goldfish enough water to swim.

5. Carefully capture one goldfish, and place it gently into the beaker of water. Measure and record the temperature of the water in degrees Celsius. Once the goldfish has adjusted to the new environment for 3 minutes, count its breaths. Develop a consistent procedure for counting the breaths of the fish for a full minute and record the data.

6. Add enough ice to lower the temperature of the water approximately 2°C, stir gently with the glass rod, and wait 1 minute for the fish to adjust. Count and record the number of breaths the fish takes within a 1-minute period at each temperature.

7. Each time the temperature is lowered approximately 2 degrees, record the number of times the fish breathes in 1 minute. If the goldfish shows signs of stress, abort the experiment and record the temperature and the number of breaths at this point. The laboratory instructor will discuss the symptoms of stress at the beginning of the class.

8. Continue lowering the temperature of the water approximately 2°C at a time. Allow the fish time to adjust until the water reaches a temperature of 4°C.

9. At the conclusion of the experiment, gradually replace the cold water with aquarium water until the fish has recovered and the aquarium water temperature is attained. Return the goldfish to a recovery container.

10. Enter your group's data on the chart provided on the board. Complete the chart in your manual using the data from each group. Record your group's results on the chart, then graph your group's data versus the class average from the data on the board.

11. For the next class meeting, complete a laboratory report in the format your laboratory instructor instructs.

Sample Chart

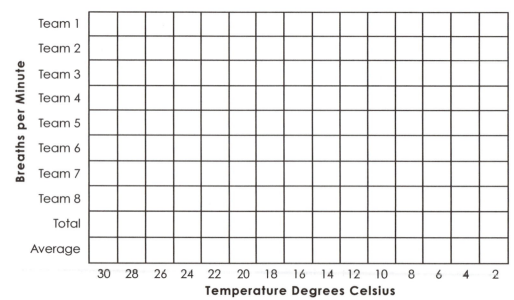

NOTES

Name: _____   Date: _____   Section: _____

Review Questions

1. List and discuss the basic and integrated process skills used in this experiment.

_____

_____

_____

_____

_____

2. State the hypothesis. Was it rejected or accepted? Why?

_____

_____

_____

_____

_____

3. Identify the control, the independent variable, and the dependent variables incorporated in this experiment.

_____

_____

_____

_____

_____

4. List other variables that may affect the respiration rate of fish.

_____

_____

_____

_____

_____

5. Discuss how your group's results compared to the results of the entire class.

_____

_____

_____

_____

6. Discuss how the theme of this experiment can apply to the natural world and the care of aquatic animals.

_____

_____

_____

_____

7. List some possible sources of error in this experiment.

_____

_____

_____

_____

8. Discuss some extensions of this experiment that may be worth investigating.

_____

_____

_____

_____

_____

_____

_____

Chapter 3
Measuring: Using Scientific Notation and the Metric System

Student Outcome Objectives

At the completion of this exercise, the student will be able to:

1. Discuss the importance of precise measurements in science.
2. Discuss the importance and use of scientific notation.
3. Convert between standard notation and scientific notation.
4. Add, subtract, multiply, and divide using scientific notation.
5. Describe the basic units and prefixes used in the metric system.
6. Convert from a given metric unit to another unit.
7. Convert from English units of measure to metric units.
8. Convert from metric units of measure to English units.
9. Measure length, mass, volume, and temperature using metric units.

Figure 3.1 Precise measurements are essential in science.

The Universe is a grand book of philosophy. The book lies continually open to man's gaze, yet none can hope to comprehend it who has not first mastered the language and characters in which it has been written. This language is mathematics.

—**Galileo (1564–1642)**

Overview

What is the price of oil per barrel? How many milliliters are in a tablespoon? What is your respiratory rate? How many grams of fat are in that cookie? What does this job pay per hour? Numbers are an integral part of our everyday lives from checking gas mileage to determining a batting average. Seldom does a day pass in which we do not use mathematics or measure something.

It is said that mathematics is the backbone of science. To succeed in scientific endeavors, we have to acquire sound mathematics and measurement skills (Fig. 3.1). Occasionally, scientists use extremely small or large numbers such as measuring the width of the cell membrane or the distance to another galaxy. In using scientific notation, these intimidating and awkward numbers become easier to manage. In addition, standard measurements are important in science. The metric system provides science with a logical, precise, and easy-to-use universal system.

SCIENTIFIC NOTATION

The distance to the bright star Alpha Centauri is approximately 38,000,000,000,000,000 km, and the mass of a mitochondrion in a nerve cell of a 21-day-old rat is approximately 0.00000000000308 g. Scientists constantly work with extremely large and small numbers. In standard notation these numbers can become awkward and confusing. Scientific notation provides scientists with a numbering system that is much easier to use and interpret (Fig. 3.2).

In scientific notation, numbers are composed of three components: the coefficient, the base, and the exponent. Thus, the number 93,000,000 can be expressed as 9.3×10^7. In this number, the coefficient is 9.3, the base is 10, and the exponent is 7. Specific rules have been developed to express a number in scientific notation properly. The base always has to be 10, the coefficient has to be greater than or equal to 1 but less than 10, and the exponent has to reflect the number of places that the decimal has to be moved to change the number to its standard notation. In writing scientific notation, numbers greater than 1 should be expressed in positive exponents and numbers less than 1 should be expressed in negative exponents.

Example numbers:

9,000 can be expressed as 9×10^3

229,000,000 can be expressed as 2.29×10^8

7.63×10^5 can be expressed as 763,000

0.0008 can be expressed as 8×10^{-4}

0.000000000452 can be expressed as 4.52×10^{-10}

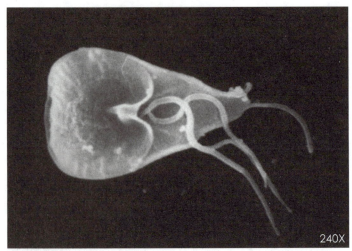

Figure 3.2 In 1681, the "father of the microscope," Anthony van Leeuwenhoek, described *Giardia lamblia* living in his own feces. The organism is approximately 1.4×10^{-7} meters long.

Performing Basic Operations

In multiplying values expressed in scientific notation, multiply the coefficients and add the exponents. Always convert the answer to properly written scientific notation. In dividing in scientific notation, divide the coefficients and subtract the exponent of the divisor by the exponent of the dividend. Convert the resulting answer to a properly written scientific notation.

Example calculations:

$(3.0 \times 10^{12}) \times (6.0 \times 10^{3}) = 1.8 \times 10^{16}$

$(8.1 \times 10^{7}) \times (3.5 \times 10^{-3}) = 2.8 \times 10^{5}$

$(6.4 \times 10^{-5}) \times (2.4 \times 10^{-3}) = 1.5 \times 10^{-7}$

$(8.0 \times 10^{6}) \div (4.0 \times 10^{2}) = 2.0 \times 10^{4}$

$(5.25 \times 10^{8}) \div (2.75 \times 10^{-3}) = 1.91 \times 10^{11}$

$(7.64 \times 10^{-7}) \div (3.22 \times 10^{-4}) = 2.37 \times 10^{-3}$

When adding and subtracting values expressed in scientific notation, always convert the exponent to the same value before beginning the calculation. In many cases, this requires changing the decimal place of the coefficient as well. Add or subtract the coefficients, and leave the base and exponent the same. Convert the resulting answer to properly written scientific notation.

Example calculations:

$4.0 \times 10^{6} + 2.0 \times 10^{8} = 2.04 \times 10^{8}$

$7.25 \times 10^{8} + 9.63 \times 10^{6} = 7.35 \times 10^{8}$

$5.4 \times 10^{4} + 3.1 \times 10^{-3} = 5.4 \times 10^{4}$

$8.0 \times 10^{9} - 4.0 \times 10^{9} = 4.0 \times 10^{9}$

$6.4 \times 10^{3} - 2.2 \times 10^{-3} = 6.4 \times 10^{3}$

$9.2 \times 10^{-5} - 6.1 \times 10^{-4} = -5.2 \times 10^{-5}$

 Student Activity—Scientific Notation and the Metric System

 **Procedure 3.1
Scientific Notation**

1. Convert 124.95000000 to scientific notation.

_____

2. Convert 0.000000000567 to scientific notation.

_____

3. Convert 2.94×10^{7} to standard notation.

_____

4. Convert 5.43×10^{-8} to standard notation.

_____

5. Multiply $(3.6 \times 10^{5}) \times (2.1 \times 10^{6})$ _____

6. Multiply $(2.6 \times 10^{-4}) \times (1.5 \times 10^{-4})$ _____

7. Multiply $(5.8 \times 10^{8}) \times (3.3 \times 10^{-5})$ _____

8. Divide $4 \times 10^{10} \div 1.9 \times 10^{5}$ _____

9. Divide $6.8 \times 10^{-4} \div 3.4 \times 10^{-2}$ _____

10. Divide $8.8 \times 10^{12} \div 4.2 \times 10^{-2}$ _____

11. Add $5.2 \times 10^{9} + 3.2 \times 10^{5}$ _____

12. Subtract $7.3 \times 10^{5} - 2.1 \times 10^{2}$ _____

13. An average human has 125 trillion cells; convert this number to scientific notation.

_____

14. A blue whale has 1.893×10^{7} ml of blood; convert this value to standard notation.

_____

15. A nanometer is 0.000000001 m; convert this to scientific notation.

_____

16. Convert Avogadro's number (6.02×10^{23}) to standard notation.

_____

17. Approximately 4.3×10^9 kg of matter is converted to energy by the sun each second; convert this number to standard notation.

_____

18. The approximate number of stars in the Milky Way is 2×10^{11}; convert this value to standard notation.

_____

19. If an average person produces 2 million red blood cells per second, how many red blood cells will be produced in a 24-hour period? Express the value in both standard and scientific notation.

_____

_____

20. If the sun is 1.5×10^8 km from Earth, how long does it take light to strike the Earth when traveling at 3.0×10^5 km/s?

_____

 Beyond the Lab—Searching for Scientific Notation Values

Q. What is the average thickness of a human hair?

_____

Q. What is the mass of an electron?

_____

Q. What is the mass of the Earth?

_____

Q. What is the diameter of a typical mitochondrion?

_____

Q. What is the speed of sound?

_____

Keep in Mind

1. In the United States, the _meter_ and _liter_ spellings are commonly used. In other nations, the spellings _metre_ and _litre_ are used more frequently.
2. Do not follow unit symbols by a period except when they are the last word in a sentence. For example, 75 m. is written incorrectly.
3. Unit symbols are case-sensitive. For example, the abbreviation for meter is m and liter is L.
4. Unit symbols are singular not plural forms. Do not write 62 ms. It is correct to use the plural form in writing the metric units. For example, 27 grams is written correctly.
5. A space must separate digits from unit symbols. For example, 4.3 m is correct, not 4.2m. In measuring temperature, do not use the space. For example, use 37°C.
6. In writing a quotient of two units, do not use a p to represent per. For example 88 km/h is correct, not 88 kph.
7. Always use a zero before the decimal point when the number is less than one. For example, use 0.61 instead of .61.
8. Never mix unit symbols such as 7.2 m 14 cm.

Figure 3.3 When diving, a Peregrine falcon can attain a speed of 200 miles per hour, or 322 km/hr.

To myself I seem to have been only like a boy playing on the seashore, and diverting myself in now and then finding a smoother pebble or prettier shell than ordinary, whilst the great ocean of truth lay all undiscovered before me.

—Isaac Newton (1642–1727)

THE METRIC SYSTEM

The metric system had its origin in France in 1790. During the same year, Thomas Jefferson proposed a decimal system of measurement for the United States, but the United States continues to use the awkward and antiquated English system based upon units such as inches, pounds, and gallons. The modernized version of the metric system is called *Le Système International d'Unités*, or the SI system.

Today, in science and industry, the metric system serves as the universal system of measurement. The metric system is based on units of 10. Conversions between the units of the metric can be performed by simply shifting the decimal place. The basic units of measurement in the metric system are the meter (length), gram (mass), and liter (volume) (Table 3.2). In addition, temperature in the metric system is measured in degrees Celsius (Table 3.3). In the metric system, Greek or Latin prefixes are placed before the base unit to denote powers of 10. In the SI system, the meter is the unit of length and the kilogram is the unit of mass.

Table 3.1 Commonly Used Metric Prefixes

Prefix	Symbol	Factor	Equivalent	Name
pico	p	10^{-12}	0.000000000001	trillionth
nano	n	10^{-9}	0.000000001	billionth
micro	μ	10^{-6}	0.000001	millionth
milli	m	10^{-3}	0.001	thousandth
centi	c	10^{-2}	0.01	hundredth
deci	d	10^{-1}	0.1	tenth
base unit		10^{0}	1	one
deka	da	10^{1}	10	ten
hecto	h	10^{2}	100	hundred
kilo	k	10^{3}	1,000	thousand
mega	M	10^{6}	1,000,000	million
giga	G	10^{9}	1,000,000,000	billion
tera	T	10^{12}	1,000,000,000,000	trillion

Table 3.2 *Le Système International d'Unités* Base Units

Quantity	Unit	Symbol
Length	meter	m
Mass	kilogram	kg
Time	second	s
Electric current	ampere	A
Thermodynamic temperature	kelvin	K
Amount of substance	mole	mol
Luminous intensity	candela	cd

Table 3.3 Commonly Used Metric Units

Measurement	Unit	Symbol
Length	millimeter	mm
	centimeter	cm
	meter	m
	kilometer	km
Mass	milligram	mg
	gram	g
	kilogram	kg
Volume	milliliter	mL
	cubic centimeter	cm³
	liter	L
	cubic meter	m³
Temperature	degrees Celsius	°C
Time	second	s
Area	square meter	m²
	hectare	ha
	square kilometer	km²
Velocity/Speed	meter per second	m/s
Acceleration	meter per second squared	m/s²
Force	Newton	N
Density	kilogram per cubic meter	kg/m²
Power	watt	W
	kilowatt	kW
Energy	joule	J
	kilojoule	kJ
	kilowatt hour	kW·h
Illumination	lux	lx
Frequency	hertz	Hz
Pressure	pascal	Pa

Procedure 3.2 Conversions

Metric-to-Metric Conversions

3.5 meters = _____ centimeters

9.7 kilograms = _____ grams

2 liters = _____ milliliters

59.04 grams = _____ kilograms

767 nanometers = _____ kilometers

Adult humans average 5,000 milliliters of blood; this equates to _____ liters.

If the average human heart has a mass of about 300 grams, what is its mass in milligrams? _____

Some individuals with diabetes insipidus can produce as much as 15 liters of urine a day. This value equates to _____ milliliters a day.

A ciliate can be 10 micrometers in length. What is its length in meters? _____

A blue whale has a mass of 130,000 kilograms. What is its mass in milligrams? _____

English/Metric Relationships

Because the English system is still commonly used in the United States (Fig. 3.3), we have to perform conversions between the metric system and the English system. Several conversion factors are included in the lists below. We suggest that you learn just one conversion factor per unit. The common values are denoted with an asterisk.

Length
*1 inch = 2.54 cm
1 mile = 1.6 km
1 foot = 30.35 cm
1 yard = .91 m
39.37 inches = 1 m
1 angstrom = 0.0001 μm

Mass
1 ounce = 28.3 g
*1 pound = 454 g
2.2 pounds = 1 kg

Volume
1 teaspoon = 5 mL
*1 quart = 0.946 L
1 gallon = 3.79 L
1 fluid ounce = 30 mL

Temperature
°F = °C × 9/5 + 32
°C = 5/9 × (°F−32)

English/Metric Conversions

70 miles = _____ kilometers

35.6 gallons = _____ liters

37°C = _____ °F

150 pounds = _____ kilograms.

.65 milliliters = _____ teaspoons

During the Carboniferous, some dragonflies had a wingspan of 70 centimeters. Convert this value to inches: _____.

If an average woman's brain has a mass of 1450 grams, what is its equivalent in pounds? _____

If an average cow produces 60 liters of saliva daily, how many gallons does this equal? _____

Some redwood trees have a diameter of 11 meters. What is their diameter in feet? _____

−40 degrees Fahrenheit equals _____ degrees Celsius.

Beyond the Lab—Searching for Metric Values

Q. What is the diameter of a human erythrocyte?

_____

Q. What kind of tree is the General Sherman, and what is its height?

_____

Q. Compare the mass of a flea and a blue whale.

_____

Q. Where is the world's largest fungus located? Describe it, using metric values.

_____

Q. How deep is the Marianas trench?

_____

Q. What was the wingspan of the flying reptile *Quetzalcoatlus*? How does that value compare to the wingspan of a F-18?

_____

Q. Describe the size of the dust mite.

_____

Q. Compare and contrast the size of a typical human sperm cell and ovum.

_____

Q. Describe the size of *Bacillus anthracis*.

_____

Q. What is the estimated length of the giant squid?

_____

Student Activity—Measuring

Measuring Length

In the metric system, the basic unit of linear measurement is the **meter**, defined as the distance traveled in 1/299792458 of a second by plane electromagnetic waves in a vacuum. A meter is 39.37 inches long, a little longer than a yard. In the laboratory, nanometers (1×10^{-9} m) and micrometers (1×10^{-6} m) are used to measure microscopic objects. Millimeters (1×10^{-3} m) and centimeters (1×10^{-2} m) are commonly used to measure the size of macroscopic organisms and kilometers (1×10^{3} m) are used to measure longer distances (Fig. 3.4).

Procedure 3.3
Measuring Length

1. Obtain a meter stick and a metric ruler from your instructor.
2. Examine the meter stick and the metric ruler.
   a. How many centimeters are found in each instrument?

   _____

b. How many millimeters are found in each instrument?

_____

c. Approximately how many centimeters are there in 1 yard?

_____

d. Approximately how many centimeters are there in 1 foot?

_____

e. Approximately how many centimeters are there in 1 inch?

_____

3. Measure the following using the appropriate metric unit of length.
   a. Length and width of the room:

   _____

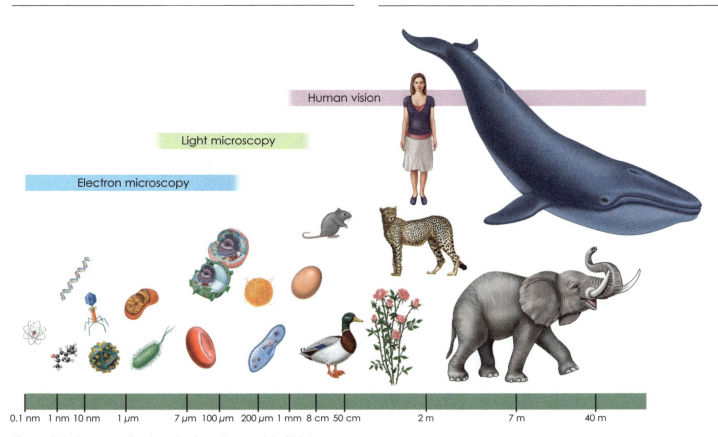

Human vision

Light microscopy

Electron microscopy

0.1 nm   1 nm   10 nm   1 μm   7 μm 100 μm 200 μm 1 mm 8 cm 50 cm   2 m   7 m   40 m

Figure 3.4 Comparative lengths from the world of biology.

b. Length and width of the laboratory table:

_____

c. The height of the group members working at your laboratory table:

_____

d. The length and width of this page:

_____

e. The diameter of a penny:

_____

4. The laboratory instructor will provide five objects to be measured. Record their measurements below.

    a. _____

    b. _____

    c. _____

    d. _____

    e. _____

5. Return the material to the designated location.

Measuring Mass

The terms mass and weight are sometimes used interchangeably. The use of weight as a synonym for mass should be completely avoided. **Mass** is defined as the measure of the quantity of matter. The basic unit of mass in the metric system is the gram. In the SI system, the basic unit of mass is the kilogram. There are 1,000 grams in a kilogram. The standard for the kilogram is a cylinder composed of platinum-iridium housed in Sèvres, France. One English pound is equal to 454 grams and 2.2 pounds is equal to 1 kilogram (Fig. 3.5).

The majority of laboratories use either a triple-beam balance or an electronic balance to measure the mass of an object. Both the triple-beam balance and the electronic balance use a thin piece of paper or weighing pan (boat) to hold the sample. The weighing pan must be *dry and clean* before using it in an experiment. Before working with the sample, both the triple-beam balance and the electronic balance must be set to compensate for the weight of the pan or paper. This process is called **zeroing** or **taring** the balance. To zero the balance, place the paper or pan on the balance and set the weight to 0. The laboratory instructor will provide directions for zeroing the specific balance used in this laboratory.

The triple-beam balance is used to measure mass down to one-tenth of a gram. The balance uses three

$1.63 \times 10^{-24}\,g$ $1.5 \times 10^{-3}\,g$ $2.8 \times 10^{-2}\,g$ 5 kg   68 kg   240 kg   6000 kg $1 \times 10^{5}\,kg$ $6 \times 10^{24}\,kg$

Figure 3.5 Comparative masses from the world of biology.

separate horizontal beams and associated masses to measure an object's mass. Depending upon the type of balance, one beam measures in 100 gram increments, another beam measures in 0–10 gram increments, and the third in 0.1 gram increments (Fig. 3.6).

A variety of electronic balances exist. Electronic balances usually can determine mass in hundredths or thousandths of a gram. The laboratory instructor will have specific instructions for the electronic balance used in this laboratory.

Procedure 3.4
Measuring Mass

Materials
- balance
- piece of gravel
- quarter
- penny
- paperclip
- cell phone
- ring
- 100 ml cylinder
- water

1. Using the laboratory instructors directions, zero your balance with either the paper or the weighing pan. This should be done each time a specimen is measured.
2. Place a piece of gravel in the center of the paper or the weighing pan.
3. If an electronic balance is used, follow the instructor's directions for using this device. If a triple-beam balance is used, slide the masses all the way to the left. Move one mass at a time toward the right, starting with the largest unit, until the balance arm lines up with the zero mark. The mass of the object is the sum of the masses on the beams of the balance.

What is the mass of the piece of gravel? _____

4. Determine the mass of the following:

   a. A quarter _____

   b. A penny _____

   c. A small paperclip _____

   d. A cell phone _____

   e. A ring _____

   f. Your choice _____

   g. Your choice _____

   h. Your choice _____

   i. Empty 100 mL graduated cylinder _____

   j. 100 mL graduated cylinder filled with 100 ml of water _____

   k. 100 mL graduated cylinder filled with 50 ml of water _____

   l. 100 mL graduated cylinder filled with 30 ml of water _____

5. Thoroughly clean your glassware and laboratory station, and return the material to the designated location.

Figure 3.6 Triple-beam and electronic balances are commonly used in the biology laboratory.

Measuring Volume

Volume is a measure of the space occupied by an object. The liter is the basic unit of volume in the metric system. One liter is equal to 1.06 quarts, and a typical 2-liter soft drink bottle is equal to 0.5283 gallons. Milliliters (1×10^{-3} L) and microliters (1×10^{-6} L) are used to measure small volumes. Often, volume is measured in cubic centimeters (cm^3); 1 mL is equal to 1 cm^3. In the laboratory, several devices are used to measure volume, including flasks, beakers, graduated cylinders, and pipets (Fig. 3.7).

Figure 3.7 Several different devices are used to measure volume.

 ## Procedure 3.5
Measuring Volume

1. From the laboratory instructor, obtain one 250 mL beaker, one 100 mL beaker, one stirring rod, one dropper pipet, one 100 mL flask or beaker, one 100 mL graduated cylinder, and one 10 mL pump or bulb pipet.

2. Fill the 250 mL beaker with tap water to the designated line, and add five drops of red food coloring. Stir the solution gently.

3. Most beakers and flasks have volume markings. This marking is a close approximation and not as accurate as other measuring devices. Beakers and flasks should be used when measuring approximate volumes.

   a. Carefully pour 100 milliliters of the red solution into a 100 mL flask or beaker.

   b. Add or remove the solution with a dropper pipet to obtain 100 mL.

4. Graduated cylinders range in volume capacity. In this activity, a 100 mL graduated cylinder is to be used. This graduated cylinder is marked at 1 mL

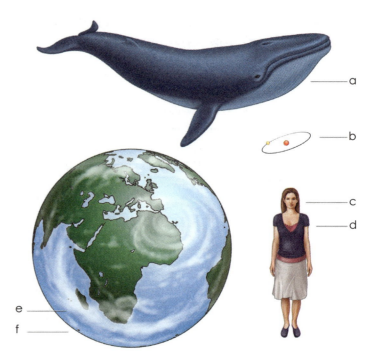

Figure 3.8 Comparative volumes from the world of biology. (a) Volume of blood in a blue whale: 7.098×10^3 L, (b) Volume of a proton: 1×10^{-12} L, (c) Inspiration and expiration of an adult human: 0.5 L of air, (d) Cardiac output of an adult human heart at rest: 5.25 L/minute, (e) Water in Earth's Oceans: 1.3×10^{21} L, and (f) Planet Earth: 1×10^{24} L

increments, and the total volume is labeled every 10 milliliters. Carefully pour the solution from the 100 mL beaker into a 100 mL graduated cylinder. To read a glass-graduated cylinder, hold the cylinder at eye level and read the bottom of the curve, this is known as the **meniscus**. The meniscus results from the surface tension of the solution and adhesion of the solution to the sides of the glass. No meniscus is formed in plastic graduated cylinders (Fig. 3.9).

   a. How many mL of solution did the 100 mL beaker contain? _____ mL

   b. How close did you come to pouring 100 mL in the beaker or flask? _____

5. Carefully remove 25 mL of the solution from the graduated cylinder and place it in the 250 mL beaker. To ensure accuracy, add or remove small amounts of solution with a dropper pipet.

6. Pipets are used to distribute and extract liquid. Several sizes of pipets are available, each measuring different volumes. The maximum and minimum volumes are denoted on the side of the pipet, along with the measurement increments. The 0-mark serves as the initial point of measure. *Never use your mouth to extract liquid with a pipet.* One common type of pipet is a bulb

pipet. It utilizes a bulb to draw liquid into the pipet. Pump or pi-pump pipets have a thumbwheel and plunger that are used in extracting and distributing liquids. Dropper pipettes are used in extracting and distributing small amounts of liquid. Dropper pipettes provide only approximate values.

a. State the range and increments of the 10 mL pipet.

Range _____

Increments _____

b. If it is not assembled already, place the bulb or pump on the pipet as directed by your laboratory instructor.

c. Practice extracting various amounts of liquid from the 250 mL beaker. Several attempts with a bulb pipet may be needed to obtain the desired amount of liquid. In pump pipettes, the thumbwheel is used to obtain and distribute liquids. The plunger is used to flush the pipet (Fig. 3.10). Moving the thumbwheel clockwise extracts the liquid, and moving the thumbwheel counterclockwise distributes the liquid. Always remember to use the plunger to expel the residual liquid.

d. Using the 10 mL pipet, extract the following amount of liquid from the graduated cylinder. Discard the liquid back into the 250 mL beaker. 10 mL, 7 mL, 1 mL, 4.5 mL, 3.7 mL, and 2.2 mL

e. Extract 10 mL of liquid from the graduated cylinder. Release the following amounts of liquid back into the beaker: 2 mL, 3.5 mL, 2.6 mL

Q. How many milliliters did you place into the beaker?

_____

Q. How many millimeters are left in the pipet?

_____

Q. How do your values compare to the instructed amount? Account for possible discrepancies between the two values.

_____

_____

7. Dispose of the liquid as indicated by your laboratory instructor.

8. Thoroughly clean your glassware and laboratory station and return the material to the designated location.

Figure 3.9 For accuracy when measuring the volume of a liquid in a glass graduated cylinder, the bottom of the meniscus is viewed.
1. Bottom of meniscus

Figure 3.10 A pump pipet.
1. Thumbwheel   2. Plunger

Measuring Temperature

Temperature measures the average kinetic energy of molecules. In the metric system, temperature is measured in degrees Celsius, named in honor of the Swiss astronomer Anders Celsius. In his scale, the freezing point of water is 0°C and the boiling point of water is 100°C. Thermometers are used to measure temperature. A variety of thermometers and probes are available in science (Fig. 3.11).

Procedure 3.6
Measuring Temperature

1. Obtain a Celsius thermometer from your laboratory instructor.

2. What is the range of your thermometer?

_____

3. Conduct the following measurements:

a. Record the room temperature.

_____

b. After holding the bulb of the thermometer tightly in your hand for 3 minutes, record the temperature.

_____

c. What is the temperature of the tap water at your station?

_____

d. What is the temperature of the hot water at your station?

_____

e. Place four ice cubes in 250 mL of water and record the following:

initial reading _____

after 1 minute _____

after 2 minutes _____

after 3 minutes _____

f. What is the temperature inside the laboratory refrigerator?

_____

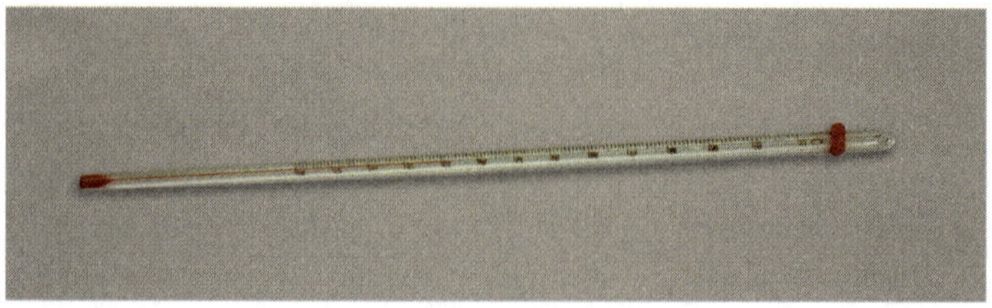

Figure 3.11 Thermometers are used to measure temperature.

g. What is the temperature inside the laboratory freezer?

_____

h. Your choice:

_____

i. Your choice:

_____

j. Your choice:

_____

4. Thoroughly clean your glassware and laboratory station, and return the material to the designated location.

NOTES

_____

_____

_____

_____

_____

_____

_____

_____

_____

Name: _____ Date: _____ Section: _____

Review Questions

1. What is the advantage of using scientific notation?

_____

_____

_____

_____

2. Why is a standard unit of measurement necessary in science and industry?

_____

_____

_____

3. Describe the advantage of using the metric system over the English system.

_____

_____

_____

_____

4. Indicate the standard unit for measuring length, mass, volume, and temperature in the metric system.

_____

_____

_____

_____

5. What is the SI system?

_____

_____

_____

_____

6. Convert the following to standard notation:

$5.36 \times 10^{-7} =$ _____

$7.62 \times 10^{5} =$ _____

$3.63 \times 10^{8} =$ _____

$9.452 \times 10^{-6} =$ _____

7. Convert the following to scientific notation:

$0.00000061 =$ _____

$0.0256 =$ _____

$29420000000000 =$ _____

$47000 =$ _____

8. Perform the following calculations:

$(6.02 \times 10^{23}) \times (5.1 \times 10^{3}) =$ _____

$(8.36 \times 10^{-4}) \times (2.23 \times 10^{3}) =$ _____

$3.6 \times 10^{8}/2.2 \times 10^{4} =$ _____

$6.4 \times 10^{6}/3.2 \times 10^{-2} =$ _____

$7.2 \times 10^{8} + 3.3 \times 10^{7} =$ _____

$8.4 \times 10^{6} - 4.2 \times 10^{5} =$ _____

9. Perform the following conversions:

7.03 cm = _____ nm

88 mph = _____ km/h

88 μm = _____ mm

16 miles = _____ mm

1.7 mg = _____ kg

30 gallons = _____ L

72°F = _____ °C

−13°C = _____ °F

3.3×10^{5} g = _____ kg

8.33×10^{-3} mL = _____ gallons

4°C = _____ °F

2.6 g = _____ pounds

3.62×10^{-4} L = _____ mL

100 yards = _____ m

15 nm = _____ m

25 μg = _____ ounces

Note: For more information on the metric system and the SI system, refer to US Metric Association, http://lamar.colostate.edu/~hillger

Chapter 4
Microscopy: Using the Tools of Biology

Student Outcome Objectives

At the completion of this exercise, the student will be able to:

1. Discuss the importance of the microscope in biology.
2. Label the parts of a compound microscope and a stereomicroscope.
3. Describe the function of the parts of a compound microscope and a stereomicroscope.
4. Properly handle and care for a compound microscope and stereomicroscope.
5. Exhibit the proper technique when using and focusing a compound microscope and a stereomicroscope.
6. Determine the total magnification of a compound microscope using different objectives.
7. Properly prepare a wet mount.
8. Explain the direction of movement of objects while using a compound microscope.
9. Calculate the field of view of low and high power.
10. Estimate the size of objects seen under the compound microscope.
11. Define and display an understanding of the following: magnification, resolving power, working distance, parfocal, illumination, depth of field.
12. Compare and contrast compound microscopes, fluorescence microscopes, phase contrast microscopes, Nomarski microscopes, transmission electron microscopes, scanning electron microscopes, scanning transmsission electron microscopes, and stereomicroscopes.

Overview

The microscope is one of the most important and frequently used tools in the biological sciences. It allows the user to peer into the world of the cell, as well as discover the fascinating world of microscopic organisms. A typical compound microscope, similar to the one that we will use in today's activity, is capable of extending the vision of the observer more than a thousand times. Other microscopes, such as the transmission electron microscope, can magnify objects up to one million times. Since its invention more than 300 years ago, the microscope has greatly improved our understanding of the cell, tissues, disease, and ecology.

Figure 4.1 The compound microscope is an essential instrument in the biology laboratory.

Student Activity—The Microscope

Materials (Procedures 4.1–4.9)
- compound microscope
- lens paper
- prepared slides (instructor's choice)
- blank slides and coverslips
- tap water and pond water
- eyedropper
- plastic ruler
- paper towels
- stereomicroscope

Materials (cont.)
- stereomicroscope
- macro objects (instructor's choice)
- forceps
- protoslo
- oil immersion

Microscope Comparisons

- A microscope that uses any kind of light to view a specimen is called a **light microscope**. A simple

Figure 4.2 "Father of the Microscope" Anton van Leeuwenhoek.

light microscope can have a single lens, similar to the early microscopes made by Anton von Leeuwenhoek (Fig. 4.2). A compound microscope uses two sets of lenses to magnify an object.

- The compound **bright field microscope** is capable of a magnification range of 10–2,000× and resolution of 300 nanometers. Light is transmitted directly through the specimen, and the specimen generally appears as a dark object against a light background. This microscope is used to examine various types of cells, microscopic organisms, and tissues.

- The compound **dark field microscope** is similar to the bright field microscope except that a special condenser causes light rays to reflect off the specimen at an angle, making the specimen appear bright against a dark background. This type of microscope is particularly useful when viewing specimens that lack contrast with a bright field microscope.

- The **fluorescence microscope** uses ultraviolet light and fluorescent dyes to study specimens. This microscope is capable of magnifications from 10–3,000× and has a resolution of 200 nanometers. It is used in advanced biological laboratories and medical laboratories to study cells, antibodies, microscopic organisms, and tissues.

- The **phase contrast microscope** uses regular light for illumination but possesses a special condenser to accent minute differences in the refractive index of structures within a specimen. As a result, it is useful in studying cellular components and microscopic organisms. It is capable of magnifications from 10–1,500× and has a resolution of 200 nanometers.

- The **Nomarski microscope** uses differences in the refractive index of a specimen to study structures. This microscope has better resolution than a phase contrast

The motion of most of these little animacules in the water was so swift and varied—upward, downward, and round about—that 'twas wonderful to see! I judge that some of these little creatures had a volume less than one thousandth that of the smallest ones that I have ever seen upon the rind of cheese, in wheaten flour, and the like.

—**Anton van Leeuwenhoek (1632–1723)**

microscope and produces nearly three-dimensional images. It is used to study the finer details of the internal structure of organisms.

- **Electron microscopes** utilize beams of electrons to magnify a specimen. They are capable of greater magnification than compound microscopes.

  - **The transmission electron microscope** (TEM) uses extremely thin sections of specimens treated with heavy metal salts and is capable of magnifications of 200–1,000,000×. The TEM has a resolution of 0.1 μm because of the shorter wavelength of the electron beam. This instrument is used to study the ultrastructure of cells and certain biochemicals.

  - The **scanning electron microscope** (SEM) provides three-dimensional views of objects and has a greater depth of focus. It is capable of magnifications from 10–500,000× and a resolution of 5.0 to 10.0 nanometers. The SEM is useful in studying the surface features of specimens.

  - The **scanning transmission electron microscope** (STEM) is a combination of the TEM and SEM. It is used in the analysis of specimens and various chemicals (Figs. 4.3 and 4.4).

THE ANATOMY OF A COMPOUND MICROSCOPE

Microscopes are expensive and delicate scientific instruments, so proper care and handling procedures must be followed when working with a microscope. Carefully remove the microscope from the case according to the instructor's directions, and place it on the laboratory table. Identify the following parts of the compound microscope (Fig. 4.5).

- **Ocular** (eyepiece): The uppermost lens or series of lenses through which a specimen is viewed. Most oculars have a magnification of 10×. A microscope with one ocular is called a **monocular microscope,** and a microscope with two oculars is called a **binocular microscope.** The distance between the oculars can be adjusted to match different distances between pupils. Many times, a trinocular head is added to a microscope

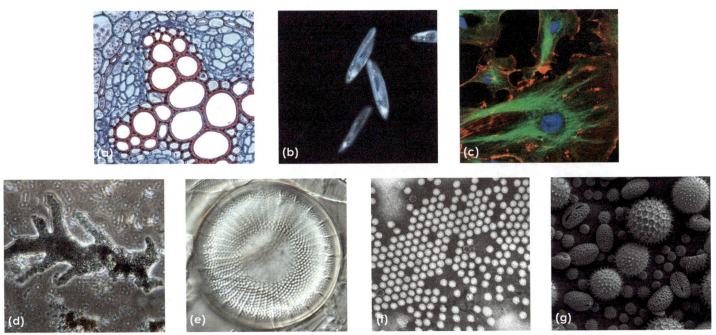

Figure 4.3 Example images taken with several kinds of microscopes, (a) bright light field microscope, (b) dark field microscope, (c) fluorescence microscope, (d) phase contrast microscope, (e) Normarski microscope, (f) transmission electron microscope, and (g) scanning electron microscope.

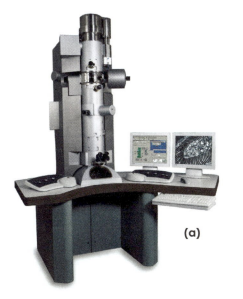

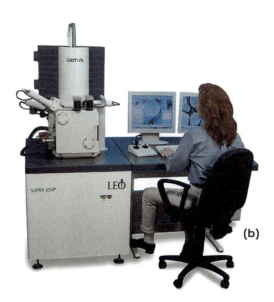

Figure 4.4 Types of electron microscopes: (a) a scanning electron microscope (SEM); (b) a transmission electron microscope (TEM).

for teaching and photography purposes. Some oculars contain a **pointer** or a micrometer disc that is used to determine the size of an object. *Never touch the ocular lens with your fingers.*

- **Draw tube:** Connects the ocular to the body tube.
- **Body:** Holds the nosepiece at one end and includes the draw tube.
- **Arm:** Serves as a handle.
- **Nosepiece:** Revolves and holds the objectives. When changing objectives, always turn the nosepiece instead

of using the objectives themselves. The objectives will click into place when properly aligned.

- **Objectives:** Lower lenses attached to the nosepiece. The magnification of each objective is stamped on the housing of the objective. The magnification may vary with different brands. *Never touch the objective lenses.*
  - **Scanning objective:** Used for viewing larger specimens or searching for a specimen; the shortest objective and usually magnifies an object 4× or 5×.

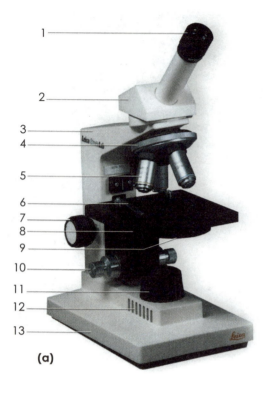

(a)

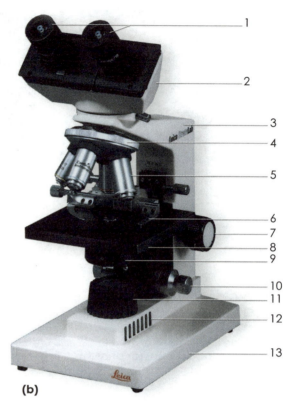

(b)

Figure 4.5 Light microscopes: (a) a compound monocular microscope, and (b) a compound binocular microscope.

1. Eyepiece (ocular)
2. Body
3. Arm
4. Nosepiece
5. Objective
6. Stage clip
7. Focus adjustment knob
8. Stage
9. Condenser
10. Fine focus adjustment knob
11. Collector lens with field diaphragm
12. Illuminator (inside)
13. Base

- **Low-power objective:** Used for coarse and preliminary focusing; magnifies an object approximately 10×.
- **High-power objective:** Used for final and fine focusing, magnifies an object approximately 43× or 45×.
- **Oil immersion objective:** Uses the optical properties of immersion oil to help magnify a specimen. Oil immersion objectives are capable of magnifications of 93×, 95× or 100×. Care should be taken with oil immersion objectives in that the oil should not be allowed to build up on the objective or contaminate other objectives. After use, the oil immersion objective and slide should be thoroughly wiped off and cleaned.
- **Stage:** Platform on which slides are placed. Some microscopes have a mechanical stage to accurately control the movement of slides. Stage clips secure the slide.
- **Light source** (illuminator): Serves as the source of illumination for the microscope.
- **Iris diaphragm:** Regulates light entering the microscope; usually is controlled by a mechanical lever or rotating disc.
- **Condenser:** A lens system found beneath the stage; used to focus the light on the specimen.

- **Coarse adjustment knob:** Used to adjust the microscope on scanning and low power only.
- **Fine adjustment knob:** Used to adjust the specimen into final focus.
- **Base:** The supportive portion of the microscope that rests on the laboratory table.

 Procedure 4.1
Using the Microscope

1. Ensure that your work area is clean and uncluttered.
2. After procuring your microscope from the cabinet, carry the microscope close to your body in an upright position with one hand placed under the base and the other hand holding the arm. Place the microscope directly in front of you on the table. Once in place, do not drag the microscope across the table for others to view. Doing so can damage the intricate optics and mechanisms of the microscope.
3. Examine the anatomy of the microscope.
4. Carefully clean the ocular and objectives with *lens paper only*. If a smudge or scratch persists, consult the laboratory instructor.

5. Make sure that the microscope was stored with the scanning or low power objective in place. Never begin a session with the high power objective in place. If necessary, use the revolving nosepiece ring to click the low power objective in place.

6. Inspect the electric cord, making sure that it is not frayed or damaged. Plug in the electric cord as instructed so it will not get in your way, trip other students, or damage the microscope. Turn the switch to the "on" position.

7. A good quality clean slide will be provided for observation. Carefully place your slide on the stage, and use the stage clips or the mechanical stage to hold it in place. Center the specimen under the objective.

8. If using a binocular microscope, adjust the distance between the oculars to match the distance between your pupils. One of the oculars on some binocular microscopes can be focused individually for precision. If using a monocular microscope, keep both eyes open to view the object, as closing one eye will result in eyestrain and a headache. If you are having difficulty keeping just one eye open, consult your laboratory instructor.

9. Use the coarse adjustment knob to focus the specimen. This knob is to be used to view specimens under scanning and low power only. Depending upon the brand of the microscope, either the nosepiece will move toward the stage or the stage will move toward the nosepiece. Practice your microscopy skills by viewing various parts of the slide. While viewing the slide, do not rest your hand on the stage.

10. Reposition the slide to attain the desired view. Use the iris diaphragm and condenser to focus and regulate the light entering the microscope. On scanning and low power, the fine adjustment knob may be used to fine-tune the specimen. Sketch and record the name and magnification of your specimen below.

11. Using the revolving nosepiece ring, rotate the high power objective into position until you feel it click into place.

12. Many microscopes are **parfocal** (once the image is focused with one objective, it should be in focus with others) and require only minor adjustments in focusing. Using the fine adjustment knob only, focus your specimen. You may have to reposition your specimen carefully and adjust the iris diaphragm and condenser. Sketch and record the name and magnification of your specimen below.

DEVELOPING THE SKILLS OF MICROSCOPY

To become proficient with the microscope, you must understand fundamental principles of microscopy and acquire basic skills. The key to becoming skillful with a microscope is *practice*. Using your laboratory microscope, perform the following tasks.

Understanding Magnification

Microscopes are designed to magnify objects. The **magnification** of a specimen is the product of the power of the ocular and the power of the objective. Microscopes generally are equipped with a 10× ocular, a 4× or 5× scanning objective, a 10× low power objective, a 43× or 45× high power objective, and a 93×, 95×, or 100× oil immersion objective. Thus, the total magnification of a microscope on low power is the 10× ocular times the 10× low power objective, or 100×.

 ## Check Your Understanding

Q. What is the total magnification of a typical microscope using the high power objective?

_____

Q. If a scanning objective is used, what is the total magnification of a microscope?

_____

Q. What is the total magnification of a microscope using an oil immersion objective (100×)?

_____

Understanding Resolving Power

The ability to resolve objects is an important characteristic of microscopes and a measure of lens quality. The **resolving power** of a microscope depends upon the design and quality of the objective lenses. Quality lenses have a high resolving power, which is the ability to deliver a clear image in detail. In comparing microscopes, lower values represent better resolving power.

When using the microscope, magnifying a specimen is not always accompanied by an increase in the clarity of the detail within the specimen. At a certain point, greater magnification of the specimen yields progressively fuzzy images. Oil immersion objectives can increase the resolving power of a microscope.

Understanding Image Orientation

Procedure 4.2
Orientation

To understand image orientation:

1. Place a prepared slide of the letter "e" on the stage of your microscope. Using low power, observe the slide. Sketch and describe your slide.

```



```

Q. What is the difference between the orientation of the letter between the unaided eye and the microscope?

2. Move the slide to the left and to the right. Describe the direction of the movement of the image.

Understanding Illumination

When using a microscope, proper illumination is essential as improper illumination can result in poor images and eyestrain. Some microscopes are equipped with a mirror to gather and concentrate light. Generally, the mirror has a smooth side and a concave side. The smooth side should be used in most applications, and the concave side should be used when it is necessary to concentrate light on the specimen. If your microscope uses a mirror, adjust the mirror to deliver maximum light on the specimen, and use the iris diaphragm to regulate the intensity of the light.

Most microscopes are equipped with an illuminator, attached to the base. The iris diaphragm is used to adjust the amount of light entering the objective lens by rotating a disc or moving an aperture adjustment control. The iris diaphragm can be used to regulate the light passing through an object, as well as the contrast of the object. Although closing the diaphragm results in poorer resolution, it can be useful in viewing certain specimens.

Some microscopes have a **substage condenser.** For an introductory laboratory, the condenser should be left in its uppermost position. Many microscopes have a light intensity control; use this control as directed by the instructor.

Procedure 4.3
Illumination

To understand illumination:

1. Carefully place a slide (instructor's choice) on the stage, and focus using low power.
2. Use the iris diaphragm to attain the best illumination for the specimen. Open and close the iris diaphragm and notice changes in the contrast of the specimen.
3. Follow the same procedure using the high power objective.
4. After you finish, return the slide to the slide tray.
5. Briefly describe the changes in illumination and contrast that you observed.

Understanding Working Distance and Diameter of Field

Working distance is the distance between the objective lens and the specimen. As the working distance decreases, magnification increases. More light is needed when the working distance decreases. Thus, more light must be made available when using the high power and oil immersion objectives.

The circular field that you see when you look through the ocular is known as the field of view. The field of view changes at different magnifications. The size of a specimen can be estimated if one knows the diameter of the field of view at each magnification. To do this, one must first determine the diameter of the field of view for both low and high power.

Procedure 4.4
Distance and Diameter

To determine the field of view:

1. Place a translucent plastic millimeter ruler on the microscope stage as if it were a slide. If no plastic ruler is available, mount a paper millimeter ruler. If your microscope has an ocular micrometer use it instead of the ruler.

2. Using the low power objective, record the number of millimeters (to tenths) that you see along the field.

   Your value _____

3. Convert the figure that you attained to micrometers (1 millimeter = 1000 micrometers). This is the diameter of the field of view for the low power objective (LPD).

   Your value (LPD) _____

4. The diameter of the field of view for the high power objective will be difficult to attain using this method. However, a simple inverse proportion can be developed in to determine the diameter of the field of view for the high power objective.
   HPD = high power objective diameter of field of view
   LPD = low power objective diameter of field of view
   LPM = low power magnification
   HPM = high power magnification
   HPD = LPD/1 (×) LPM/HPM

5. Based on the formula, find the high power objective field of view for your microscope.

   LPD = _____

   LPM = _____

HPM = _____

HPD = _____

6. Using the above numbers, approximate the size of the following specimens (instructor's choice).

   _____

   _____

   _____

7. A more precise method for measuring the size of a microscopic object is to use a standardized ocular micrometer. These devices are found on more advanced microscopes. In some microscopes, the pointer that is attached to the inside of the ocular and appears in the field of view can be used for measuring objects.

Understanding Planes of Focus and Depth of Field

Microscope lenses have a restrictive **plane of focus**, a specific distance from the lens where the specimen can be sharply focused. It is possible to focus through different planes of the specimen by changing the plane of focus. The plane of focus has some depth called the **depth of field**, the thickness of the specimen that is in focus at any one time. When using lens systems, including cameras, the greater the magnification, the less is the depth of field.

Procedure 4.5
Focus and Depth

To understand planes of focus and depth of field:

1. Procure or prepare a slide containing three different-colored threads, and place it on the stage of your microscope.

2. Locate the point on the slide where the threads intersect, and center it under the objective.

3. Using the low power objective, slowly focus until the first thread comes into focus. The fine adjustment can be used in addition to the coarse adjustment.

Q. What is the color of the first thread?

_____

4. Continue focusing through the threads, noting the order of the colored threads.

Q. What is the color of the second thread?

_____

Q. What is the color of the third thread?

_____

Draw the levels of the threads.

┌─────────────────────────────────────┐
│   │
│   │
│   │
│   │
│   │
│   │
└─────────────────────────────────────┘

5. Switch to the high power objective and notice that the depth of field is shallower. If you constantly use the fine adjustment, the three-dimensional form of the threads can be determined.
6. The laboratory instructor will provide a slide of a tissue such as columnar epithelium or an organism such as a rotifer. Focus through your specimen and record your observations.

Q. When will an understanding of planes of focus and depth of field be valuable in the laboratory?

_____

_____

_____

Q. Discuss planes of focus and depth of field based upon your observations of both the low power and high power objectives.

_____

_____

_____

Using Oil Immersion

Many microscopes have an oil immersion objective for observing small organisms and the fine detail of specimens (Fig. 4.6). It appears longer than the other objectives and is clearly marked. Oil immersion objectives are capable of magnifications of 93×, 95×, or 100×.

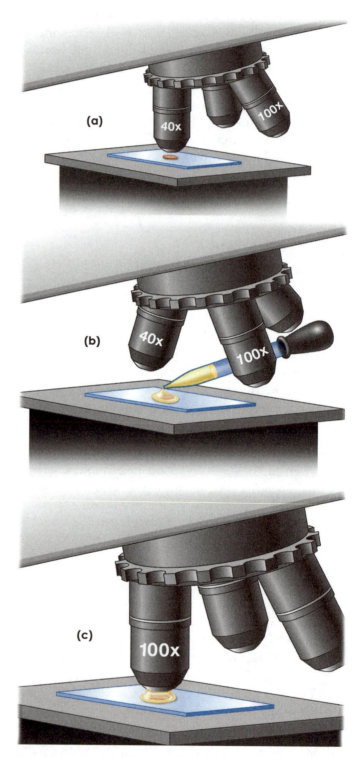

Figure 4.6 Using an oil immersion objective. (a) Focus on specimen using a lower power objective. (b) Swing the lower power objective away from specimen, place a drop of oil immersion oil on slide covering the specimen. (c) Swing the oil immersion objective over the specimen so that the tip of the objective is in contact with the oil and adjust focus.

Procedure 4.6
Oil Immersion

To practice using oil immersion:

1. Procure a prepared slide of human blood and place it on the stage of the microscope. The red blood cells, or **erythrocytes**, will appear as lightly stained biconcave discs with a lighter colored center. White blood cells, **leukocytes,** will appear large with a dark-stained nucleus.

2. Focus and view the specimen under low power. Click the high power objective into place, focus, and view the specimen.

3. Sketch and record the name and magnification of your observations below.

Oil immersion

4. Carefully swing the high power objective away from the slide. Place one drop of oil immersion oil (confirm that you are using oil immersion oil) in the center of the viewing area of the slide. Do not use any other objectives with oil immersion oil, and be careful not to get the oil on anything. Click the oil immersion objective into place. Focus on the specimen using the fine adjustment knob only. The oil immersion objective will dip into the oil.

5. Sketch and record the name and magnification of the specimen below.

Low power

High power

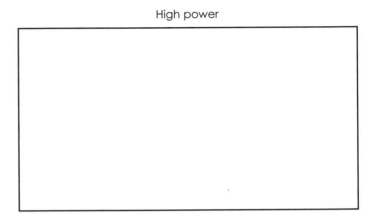

6. Once the exercise is complete, clean the slide and objective thoroughly with lens paper. Lens paper with a drop of xylene is often used to clean the objective and slide. Return the lowest power objective into place.

MICROSCOPE SLIDES

Prepared slides are commonly used in the biology laboratory. These slides are the products of tedious work. Many of the specimens on prepared slides have been stained for better observation. The slides must be carried by the edge or the end, never placing your fingers over the viewing area. The label of the slide may indicate that the specimen is a whole mount (w.m.), a cross-section (c.s. or x.s.), or a longitudinal section (l.s.).

When viewing slides, always note the type of mount that you are viewing. In addition, do not become dependent on dye colors when learning specimens. Prepared slides are expensive, so treat them carefully and report any broken or damaged slides to your laboratory instructor.

Wet Mounts

Wet mounts are often made in the laboratory to view fresh specimens (Fig. 4.7).

Procedure 4.7
Wet Mount

To prepare a wet mount:

1. Clean and dry a slide and coverslip thoroughly.

2. When mounting bits of tissue, place a small piece of the tissue to be studied in the center of the slide. Using a teasing needle, tease the tissue out flat. Add a drop of water, stain, or other reagent, and lower the coverslip on the tissue as discussed next. Your instructor will tell you what to use specifically. For pond culture studies (Fig. 4.8), place a drop of the culture liquid in the center of the slide and lower a coverslip. Always place a coverslip over wet mounts.

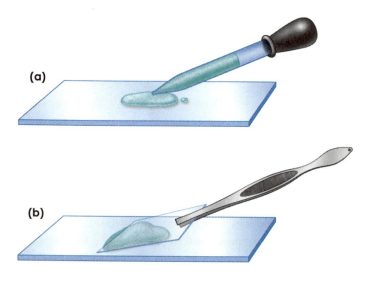

Figure 4.7 Preparing a wet mount. (a) Add a drop of pond water or a solution over the center of the slide. (b) Place the coverslip next to the droplet along one edge as shown in the illustration above. The side resting against the glass will act as the pivot point as you lower the coverslip over the sample. (c) Lower the coverslip into place. As you do, the drop will spread outward and suspend the sample between the slide and coverslip.

3. Holding the coverslip at an angle, place its edge in the margin of the drop of liquid on the slide and release the coverslip slowly. A pair of forceps will help you lower the coverslip. The procedure will minimize air bubbles from forming beneath the coverslip.

4. If the coverslip floats on the liquid, it will be caused by excess water on the slide. The excess water may be siphoned off with the edge of a paper towel. Too much liquid (water or stain) should be removed. If some of the reagents get on the microscope, they may corrode it and cause serious damage. You may

In the history of science, we often find that the study of some natural phenomenon has been the starting point in the development of a new branch of knowledge.

—**Chandrasekhara V. Raman (1888–1970)**

have to add water at the edge of the coverslip to compensate for evaporation in wet mounts such as in pond cultures.

5. At certain times it will be necessary to support the coverslip to prevent crushing the organisms that are being observed. A small piece of lens tissue at each corner of the coverslip or small pieces of broken coverslips may be placed beneath the top coverslip.

6. Many times when observing living specimens, they appear to move rapidly. To slow them down, place a drop of a prepared slowing solution, methyl cellulose or dishwashing liquid, in the drop of water. The result will be equivalent to a person swimming in a pool of molasses.

7. Prepare and observe a wet mount of an *Elodea* leaf and pond water. Use the scanning, low power, and high power objectives in this activity.

8. Sketch your observations below.

Elodea

Pond water

USING THE STEREOMICROSCOPE

A **stereomicroscope** (Fig. 4.9) also is known as a dissecting microscope or binocular microscope. The stereomicroscope has two oculars and is capable of magnifications of 4× to 50×. Stereomicroscopes provide a significantly greater field of view and depth of field than compound microscopes. This type of microscope is advantageous when viewing larger objects and dissecting.

The anatomy of a stereomicroscope is similar to that of the compound microscope. The two oculars can be adjusted for the individual viewer by moving the oculars or using an interpupillary adjustment. The specimen can be lighted from above (reflection) with an external or built-in illuminator, or from below (transmission) with a substage light or both. When viewing a specimen, determine which type of light is best for your specimen. Some stereomicroscopes have fixed magnifications, while others have dial type or zoom magnification ability. Stereomicroscopes have a single focusing knob. Treat the stereomicroscope with the same respect afforded the compound microscope.

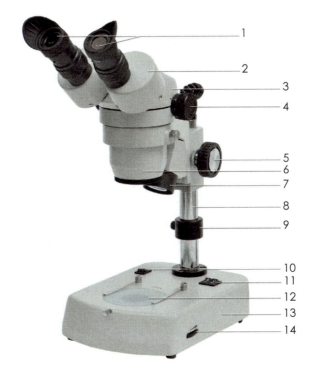

Figure 4.9 The stereomicroscope.

1. Eyepiece (ocular)
2. Eyepiece tube housing
3. Head
4. Magnification knob
5. Focus adjustment knob
6. Objective
7. Incidental light
8. Post
9. Locking support collar
10. Main power switch
11. Light switches
12. Stage plate
13. Base
14. Rheostat light control

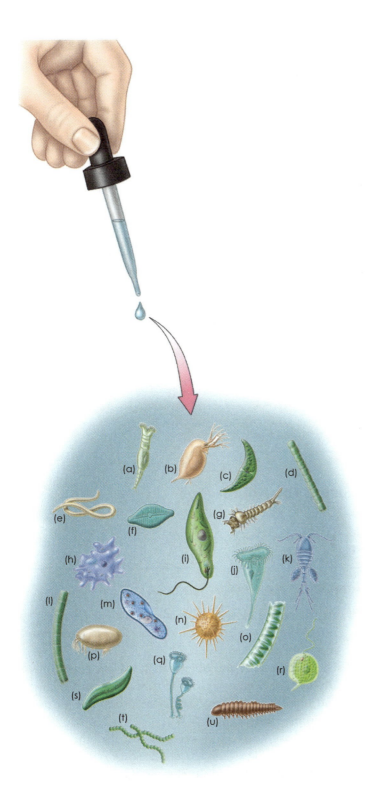

Figure 4.8 Life in a drop of pond water. Common pond organisms include, (a) rotifer, (b) *Daphnia*, (c) desmid, (d) cyanobacteria, (e) nematode, (f) diatom, (g) larva, (h) amoeba, (i) *Euglena*, (j) *Stentor*, (k) copepod, (l) cyanobacteria, (m) paramecium, (n) radiolarian, (0) green algae, (p) tardigrade, (q) *Vorticella*, (r) *Phacus*, (s) diatom, (t) cyanobacteria, and (u) oligochaete.

Procedure 4.8
Stereomicroscope

To become familiar with the stereomicroscope:

Materials
- ring
- coin
- dollar bill
- fossil
- feather
- leaf
- pond water
- other specimens, instructor's choice

1. View the following objects with the stereomicroscope: your fingernail, the surface of a stone on a ring, a coin, a dollar bill, a fossil, a feather, a leaf, several biological specimens provided by your instructor, and some pond water in a Petri dish.
2. Try using various light and magnification settings with your stereomicroscope.
3. Sketch several of your observations below.

Storing the Microscope

Procedure 4.9
Storage

To properly store a microscope:

1. At the completion of each laboratory experience, rotate the lowest power objective into place and re-move the slide.
2. Clean the ocular and objectives with lens paper only. If you have been using oil immersion, clean the slide and objectives as instructed in the procedure for using oil immersion.
3. If you have been using wet mounts, clean the stage with a cleaning tissue or a clean cloth.
4. Turn off the light and unplug the microscope. Wrap the cord around the base as directed by the instructor. If a plastic dust cover is available, cover the microscope and return it to the cabinet.

Name: _____   Date: _____   Section: _____

Review Questions

1. What is the difference between a simple microscope and a compound microscope?

_____

_____

2. What are some useful applications of a stereomicroscope?

_____

_____

3. What are some of the advantages and disadvantages of using electron microscopy?

_____

_____

_____

4. When should an oil immersion objective be used?

_____

_____

5. Why do many biologists prefer a phase contrast microscope?

_____

_____

6. What is the advantage of parfocal objectives?

_____

_____

7. Discuss resolving power.

_____

_____

8. Discuss the direction of movement of the protozoans in pond water in relation to movement of the slide.

_____

_____

9. Discuss the relationships between plane of focus, depth of field, illumination, and magnification.

_____

_____

_____

_____

10. What is the magnification of a microscope with a 10× ocular and a 95× oil immersion objective?

_____

_____

_____

11. List three rules to remember when focusing a microscope.

_____

_____

_____

12. Label the parts of the microscope.

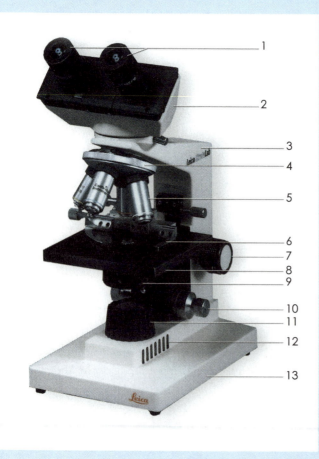

1. _____

2. _____

3. _____

4. _____

5. _____

6. _____

7. _____

8. _____

9. _____

10. _____

11. _____

12. _____

13. _____

Chapter 5
Chemistry:
Understanding the Composition of Living Things

Student Outcome Objectives

At the completion of this exercise, the student will be able to:

1. Discuss the role of chemistry in the biological sciences.
2. Define elements, atoms, atomic number, atomic mass, and isotopes.
3. Define compounds and molecules.
4. Differentiate among ionic, covalent, and hydrogen bonds.
5. Distinguish between inorganic, organic, and biochemistry.
6. Define solution, aqueous solution, solvent, and solute.
7. Define and describe the properties of an acid and a base.
8. Determine the pH of a substance.
9. Define and describe the characteristics of carbohydrates.
10. Discuss the composition and functions of monosaccharides, disaccharides, and polysaccharides.
11. Perform the Benedict's Test for the presence of a reducing sugar.
12. Perform the Iodine Test for the presence of a starch.
13. Define and describe the characteristics of lipids.
14. Perform the Grease-Spot Test for the presence of a lipid.
15. Perform the Sudan Test for the presence of a lipid.
16. Define and describe the characteristics of proteins.
17. Perform the Biuret's Test for the presence of a protein.
18. Define and describe the characteristics of nucleic acids.

Overview

To develop a meaningful understanding of biology, a fundamental knowledge of chemistry is necessary (Fig. 5.1). **Chemistry** is the study of the composition, structure, properties, and interaction of matter. **Matter** is anything that occupies space and has mass. Matter, whether it is this page, a meteorite, or a gallstone, is composed of elements. An **element** is a substance that cannot be broken down into other substances by ordinary chemical means. Although more than 112 elements have been described,

only 92 elements are naturally occurring; the others are "human-made."

Of the naturally occurring elements, approximately 25 are common to living systems. A typical human is composed of 65% oxygen (O), 18.5% carbon (C), 9.5% hydrogen (H), and 3.2% nitrogen (N). Other important elements in humans and other living systems include phosphorus (P), sulfur (S), calcium (Ca), potassium (K), sodium (Na), chlorine (Cl), magnesium (Mg), iron (Fe), and silicon (Si).

Elements are composed of **atoms**. An atom is the smallest part of an element that retains the properties of that element. Atoms are composed of a variety of subatomic particles, the best known of which are protons, neutrons, and electrons. Positively charged protons and uncharged neutrons can be found in the nucleus, and negatively charged electrons exist in orbitals surrounding the nucleus.

The **atomic number** of an element reflects the number of protons in the nucleus. For example, copper (Cu) has an atomic number of 29. The **atomic mass** of an element indicates the number of protons plus the number of neutrons in the nucleus. The atomic mass of copper is 63.546. The atomic mass reflects that the number of neutrons can vary in the nucleus. Atoms that have the same number of protons and a varied number of neutrons are called **isotopes**. For example, the element tin (Sn) has 10 isotopes. Many isotopes are radioactive.

A **molecule** results from the chemical union of two or more elements. Some molecules are simple, such as a molecule of oxygen (O_2) and a molecule of water (H_2O). Others are quite large, such as a molecule of chlorophyll ($C_{55}H_{72}O_5N_4Mg$). **Compounds** are molecules

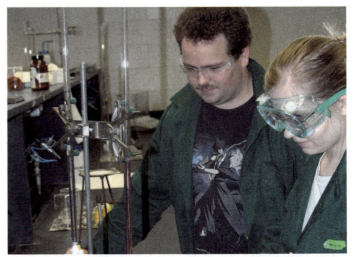

Figure 5.1 A fundamental knowledge of chemistry is important to all students of biology.

composed of different elements. Water and chlorophyll are compounds. Other examples of compounds are glucose ($C_6H_{12}O_6$), hydrogen peroxide (H_2O_2), baking soda ($NaHCO_3$) and ethyl alcohol (C_2H_5OH).

Atoms combine with other atoms by means of chemical bonds. Ionic compounds such as salt (NaCl) are held together by an attraction between positive and negatively charged ions. This loss or gain of electrons forms an **ionic bond**. When placed in water, ionic compounds dissociate (dissolve).

- **Covalent bonds** result from sharing of electrons. Covalent bonds are strong and are common in living systems.
- **Nonpolar covalent bonds** involve an *equal sharing* of electrons. Nonpolar covalent bonds between carbon and hydrogen form a stable framework for building larger molecules such as octane (C_8H_{10}).
- **Polar covalent bonds** involve an *unequal sharing* of electrons.

The most important polar molecule to life on earth is water (H_2O). **Hydrogen bonds** are weak bonds between the positively charged region of a hydrogen atom of a polar covalent molecule and the negatively charged region of oxygen or nitrogen of another polar covalent molecule. Hydrogen bonds give shape and three-dimensional structure to complex molecules such as proteins. Hydrogen bonds also provide water with several special properties.

In this chapter we divide the discussion of chemistry into inorganic chemistry and organic chemistry. **Inorganic chemistry** addresses chemical principles excluding the special properties of carbon. **Organic chemistry** addresses chemistry involving the special properties of carbon. **Biochemistry**, the study of the chemistry of life, will be discussed in later chapters.

An Expensive Meal

Perhaps Cleopatra consumed the most expensive drink in history. She dropped one of her massive pearl earrings into a goblet filled with vinegar. The vinegar fizzed, and the pearl—worth as much as 2 million ounces of silver—eventually dissolved. Once the pearl dissolved, she consumed the unusual concoction. Pearls are composed mostly of calcium carbonate ($CaCO_3$), the main ingredient in calcium supplements.

Suppose we liken individual chemical reactions to so many simple musical themes each played on a different instrument, then the operation of a living cell depends upon the combination of all these elements into a symphony. Knowing something of the theory of the simple elements can we find something about the rules of composition of the symphony?

—**Cybril N. Hinshelwood (1897–1967)**

EXERCISES IN INORGANIC CHEMISTRY

Understanding Acids and Bases

A **solution** is a liquid composed of a uniform mixture of two or more substances. In a solution, the dissolving medium is the **solvent**, and the dissolved substance is the **solute**. One of the unique properties of water is that it is an excellent solvent. An **aqueous solution** uses water as the solvent. Some inorganic molecules are held together by ionic bonds. In an aqueous solution, these substances undergo dissociation and produce positive and negative ions. For example, sodium chloride (NaCl) dissociates into sodium ions (Na^+) and chloride ions (Cl^-). Substances that release ions in an aqueous solution are called **electrolytes**. Other biologically important electrolytes include potassium chloride (KCl), calcium chloride ($CaCl_2$), and sodium bicarbonate ($NaHCO_3$).

Although many people think that acids and bases are chemicals that are best left in the laboratory, many common substances are classified as an acid or a base. These substances can be important and, in some instances, potentially dangerous to living systems.

Both inorganic and organic acids are common in nature. An **acid** is a substance that yields (donates) a hydrogen ion in solution. Acids share a number of structural characteristics and properties.

- Acids contribute one or more hydrogen atoms to a solution when they dissociate in water.
- Acids have a sour taste. (*Note: Never taste unknown chemicals!*)
- Acids may be corrosive or poisonous.
- Acids react with certain metals to liberate hydrogen.
- Acids neutralize bases.
- Acids affect the color of certain indicators.

Several common inorganic acids are:

sulfuric acid (H_2SO_4)

nitric acid (HNO_3)

phosphoric acid (H_3PO_4)

hydrochloric acid (HCl)

Some common organic acids are:

citric acid (lemon juice)

acetic acid (vinegar)

carbonic acid (carbonated water)

malic acid (apple juice)

formic acid (bee stings)

lactic acid (sour milk)

Bases are commonly known as **alkalines**. These chemicals share several structural characteristics and properties.

- Bases decrease the hydrogen ion concentration of its aqueous solution or release hydroxide ions (OH^-) in solution.
- Bases have a bitter taste.
- Bases feel slippery.
- Bases may be corrosive or poisonous.
- Bases neutralize acids.
- Bases affect the color of certain indicators.

Some common bases are sodium hydroxide ($NaOH$), calcium hydroxide ($Ca(OH)_2$), and magnesium hydroxide ($Mg(OH)_2$). Examples of industrial bases are potassium hydroxide, barium hydroxide, and strontium hydroxide.

Chemists use a variety of indicators to determine if a substance is an acid or a base.

- Acids turn blue litmus indicators red, are colorless in phenolphthalein, and turn methyl orange-red.
- Bases turn red litmus indicators blue, turn phenolphthalein pink, and turn methyl orange-yellow.

Why are Fresh Eggs Harder to Peel?

Fresh eggs have a higher carbon dioxide content than older eggs. The carbon dioxide dissolves in the eggs' natural moisture and forms carbonic acid (H_2CO_3). As the egg ages, the carbon dioxide diffuses out of the egg and the egg becomes less acidic. This condition weakens the inner membrane, preventing it from sticking to the white and making it easier to peel. *Be careful. You do not want an egg to be too old!*

Using Natural pH Indicators

One of the most interesting properties of acids and bases is their ability to change the color of some plant materials. Purple cabbage and elderberry extracts respond in an amazing manner to an acidic or basic solution. In this activity, purple (red) cabbage is used to develop an acid or a base scale and determine the approximate pH value of various substances.

Q. Why do some plant materials change color in the presence of acidic or basic substances?

_____

_____

_____

Student Activity—pH Level and CO_2

Carbon dioxide (CO_2) is a gaseous byproduct released during expiration. When released into water, it dissociates, forming carbonic acid

(H_2CO_3): $H_2O + CO_2 \wedge H_2CO_3$

This change can be observed by blowing CO_2 through a straw into a basic solution containing the indicator phenolphthalein (Fig. 5.2). The indicator turns bright pink in a basic solution, and in an acidic solution the phenolphthalein is colorless.

Materials
- distilled water
- 500 ml beaker
- phenolphthalein
- baking soda
- clean straw
- stirring rod

Figure 5.2 The carbon dioxide that is exhaled forms carbonic acid in the water. In turn, the carbonic acid neutralizes the basic solution.

 Procedure 5.1
Carbon Dioxide

1. Pour 250 ml of distilled water into a 500 beaker. Add a small amount of baking soda ($NaHCO_3$) to the water and stir with a stirring rod.
2. Add a small amount of phenolphthalein to the solution, and stir.

Q. What color is the solution?

_____

3. Place a clean straw into the solution and begin blowing steadily into the straw. Observe the color changes in the solution. This may take a few minutes.

Q. Describe the color changes in the solution.

_____

_____

4. Discuss the results of the demonstration and what happened as CO_2 from your breath was blown into the basic solution.

_____

_____

_____

The **pH** (potential of hydrogen) **scale** is the most commonly used method to measure the acidic or basic nature of a substance. The pH of a solution is the negative logarithm of the hydrogen ion (H^+) concentration, in moles per liter. Because the scale is based on logarithms, a substance with a pH of 2 such as lemon juice has 10 times more H^+ ions than a substance such as grapefruit juice with a pH of 3, and 10,000 times more H^+ ions than human urine with a pH of 6. In a solution, as the H^+ increases the OH^- decreases and as the H^+ decreases, the OH^- increases.

Q. If tomato juice has a pH of 3, how many more H^+ ions are present in lemon juice with a pH of 2?

_____

Q. If hydrochloric acid that comprises gastric juice has a pH of 1, how many more H^+ ions are present than in grapes with a pH of 4?

_____

The pH scale has a range from 0 to 14. A solution with a pH less than 7 is considered acidic. A substance with a pH of 7 is considered neutral, and a substance with a pH greater than 7 is considered basic. Ethyl alcohol and some substances do not dissociate in water and do not have a pH. The pH value of these substances reflects the water from which they were made, or other ingredients in the product (Fig. 5.4).

Figure 5.3 A pH meter.

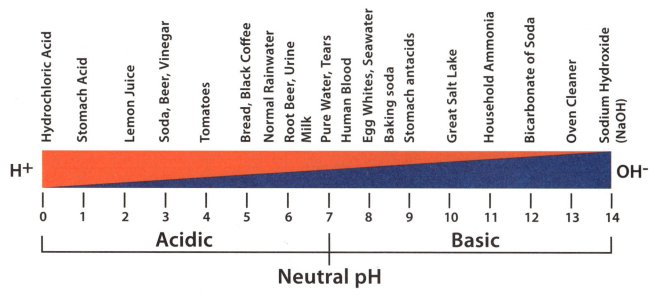

Figure 5.4 pH scale and the pH of several common substances.

 Student Activity—pH Levels

DETERMINING THE PH OF COMMON SUBSTANCES

The pH of a substance can be measured using a variety of methods. Using pH paper is an inexpensive and easy method to measure the pH of a substance. The pH paper has been treated with an indicator that is sensitive to certain pH values. A color chart provided with the pH paper relates the color of the pH paper to a specific pH. Many labs have a pH meter (Fig. 5.3) that records the pH. If a pH meter is available, follow your instructor's directions for its use in this activity.

Materials
- pH paper
- pH meter
- substances for investigation to be determined by the instructor

 Procedure 5.2
pH

1. Complete the following chart.

Substance	pH paper value	pH meter value	Actual pH
1.			
2.			
3.			
4.			
5.			
6.			
7.			
8.			
9.			
10.			

Q. Was the pH paper or the meter more accurate? _____

Student Activity—Using Natural pH Indicators

Materials

- purple cabbage extracts
- substances for investigation (to be determined by the instructor)
- pipets
- test tubes
- reaction wells
- toothpicks
- colored pencils
- beakers
- paper towels

 ## Procedure 5.3
Purple Cabbage Extract

1. Divide into working groups of at least two students.
2. Procure a beaker of prepared purple cabbage extract, 8 test tubes, a reaction well plate, pipettes, toothpicks, substances to test, and several paper towels.
3. To establish a standard for comparison, place 7 test tubes in a rack and label them according to pH 2, 4, 6, 7, 8, 10, and 12.
4. Pipette 5 drops of substance of known pH into the appropriate test tube.
5. Pipette 5 drops of purple cabbage juice extract into each test tube and gently swirl the test tube mixing the substances.
6. Record the color of each test tube in Table 5.1. Let the substance sit in pH 12 for a few minutes. The color will shift because of the instability of the pigments at a high pH. Record the initial and final colors at pH 12. Once completed, Table 5.1 will serve as the standard for comparison.

Table 5.1 The Color of Standard Solutions in Purple Cabbage Juice Extract

pH	Substance	Color
2		
4		
6		
7		
8		
10		
12		

7. Place 5 drops of each of the 11 substances to be tested into an individual well on the reaction well plate, and record its position. If no well plates are available, test tubes can be used.

_____

_____

8. Place 5 drops of the red cabbage extract into each reaction well, and stir with a toothpick.
9. Record the color changes and color of the given template. You may want to color the appropriate well in Table 5.2.

Table 5.2 Color Changes in Purple Cabbage Juice

Well Number	Substance	Color Change
1.		
2.		
3.		
4.		
5.		
6.		
7.		
8.		
9.		
10.		
11.		

10. Look up the pH values for the substances tested in the experiment and record them on Table 5.2.

Q. How did the cabbage juice pH values compare to the actual values?

_____

_____

11. Procure an unknown substance from the instructor and follow the instructions in steps 7, 8, and 9. Place the unknown in the empty 12th well.
12. Determine the approximate pH of your unknown from the scale that you have developed.

   a. What was the color of the unknown?

_____

b. What is the estimated pH of the unknown?

_____

c. If permitted, use pH paper or a pH meter to determine the pH of your unknown.

_____

d. How did the two values compare?

_____

_____

 Student Activity—Antacids

Over-the-counter antacids are used to neutralize "acid-stomach" or "heartburn." These products should absorb excess hydrogen ions from stomach acid and relieve the sufferer. A number of antacids claim that they are the best. The following exercise is designed to test which antacid neutralizes acid the best. Keep in mind that this activity does not promote any product over another.

Materials
- distilled water
- mortar and pestle
- 250-ml beaker
- pipette
- test tubes
- 0.1 M hydrochloric acid solution (provided by the instructor)
- over-the-counter antacids (instructor's choice)
- Phenolphthalein indicator

 Procedure 5.4 Antacids

1. Procure the equipment and supplies for the lab table.
2. If you have an antacid in the solid form, use a mortar and pestle to pulverize it into a fine powder.
3. Place the pulverized antacid into a beaker, add 100 ml of distilled water and stir vigorously.
4. Once the powder is dissolved in the solution, add 10 ml of the solution into a test tube. Gently swirl the solution.
5. Add 5 ml of the indicator phenolphthalein and gently swirl. Note the color of the solution.
6. Carefully add the prepared 0.1 M hydrochloric acid (HCl) solution one ml (drop) at a time, using a pipette. Swirl the solution gently. Continue adding the HCl into the test tube one ml at a time and counting each drop until the solution turns clear. When the solution turns clear, it has become neutralized.

Q. How many drops of 0.1 M HCl were used to neutralize the antacids recorded in Table 5.3?

_____

Table 5.3 Drops of 0.1 m HCl Required to Neutralize an Antacid	
Antacid	**Drops Of 0.1 m HCl**
1.	
2.	
3.	
4.	
5.	

Q. Which antacid is potentially the strongest?

_____

EXERCISES IN ORGANIC CHEMISTRY

Organisms are composed primarily of organic compounds. Originally the term "organic" referred to the belief that such compounds could be made by only living things. Today, **organic chemistry** is a branch of chemistry that studies the compounds of carbon.

Carbon is a unique element. Because of its tetravalent nature, carbon has the ability to combine with itself and other elements to form an almost limitless number of compounds. More than 2 million organic compounds have been cataloged, and approximately 100,000 new organic compounds are described yearly. Examples of common organic products are plastics, octane, isopropyl alcohol, formaldehyde, aspirin, and detergents. Organic molecules commonly found in living things include carbohydrates, lipids, and proteins, each of which is explored in this chapter.

Organic chemistry just now is enough to drive one mad. It gives me the impression of a primeval forest full of the most remarkable things, a monstrous and boundless thicket with no way to escape, and into which one may dread to enter.

—**Baron Jon Jakob Berzelius (1779–1848)**

Carbohydrates

Carbohydrates comprise a class of organic compounds that contain carbon, hydrogen, and oxygen in a 1:2:1 ratio. The number of carbons in carbohydrates varies from 3 to more than 1,000. In nature, carbohydrates may store energy, serve as fuels, and provide support and protection. Carbohydrates are classified as monosaccharides, disaccharides, and polysaccharides (Fig. 5.5)

Monosaccharides

The simplest forms of carbohydrates are known as single sugars, simple sugars, or **monosaccharides**. These molecules are composed of three to seven carbon atoms and their appropriate hydrogen and oxygen atoms. Primarily, monosaccharides serve as fuels for organisms and building blocks for larger molecules. Examples of simple carbohydrates are ribose $C_5H_{10}O_5$ and glucose $C_6H_{12}O_6$. Ribose and deoxyribose are five-carbon sugars called pentoses (*pent* = five) that are important components of the nucleic acids RNA and DNA. Glucose is a six-carbon sugar known as a hexose (*hex* = six). Glucose serves as an important source of cellular fuel. Fructose and galactose are also hexoses, but they have a different molecular architecture than glucose and are known as **isomers** of glucose.

Disaccharides

Two monosaccharides combine in a process known as **dehydration synthesis**, or condensation, to form a double sugar, or **disaccharide**. Dehydration synthesis involves the removal of an OH^- from one sugar and an H^+ from another sugar. As water (H_2O) is removed, the simple sugars link, forming a disaccharide.

- **Maltose** is a common disaccharide composed of two glucose molecules. It is used in making beer, malts, and malted milk balls.
- **Sucrose** is a common disaccharide composed of glucose and fructose. Sucrose is found in plants. Table sugar is sucrose that is harvested from the stems of sugarcane or the roots of sugarbeets.
- **Lactose**, or milk sugar, is formed from a dehydration synthesis reaction between glucose and galactose.

Disaccharides can be broken down into their simpler sugars through a process called **hydrolysis**.

Polysaccharides

Polysaccharides are complex carbohydrates built from simple carbohydrates, linked through dehydration synthesis. Polysaccharides form large molecules that can be linear or highly branched. Some polysaccharides serve as energy storage molecules. **Starch** is a storage polysaccharide found in plants that consists of glucose molecules. Potatoes, corn, and rice are sources of starch in the diet of humans. **Glycogen**, or animal starch, is a highly-branched glucose-rich polysaccharide that is stored in the liver and skeletal muscle of animals.

Polysaccharides also can serve as structural molecules. The structural polysaccharide **cellulose**, found in the cell wall of plants, is the most abundant carbohydrate on Earth. Cotton consists of more than 90% cellulose, and some woods are about 50% cellulose. Herbivores have modified digestive systems that allow them to digest cellulose. In the human diet, cellulose serves as a fiber, which is important in the proper functioning of the digestive

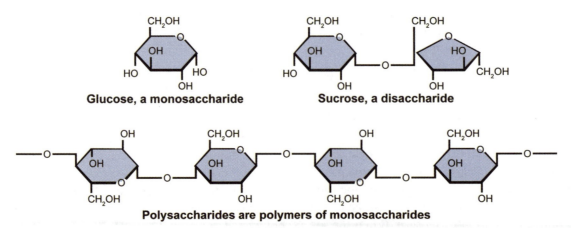

Glucose, a monosaccharide

Sucrose, a disaccharide

Polysaccharides are polymers of monosaccharides

Figure 5.5 Carbohydrates are classified as monosaccharides, disaccharides, and polysaccharides.

Figure 5.6 (a) Structural polysaccharides such as cellulose comprise the cell wall of plants such as aloe, and (b) chitin is found in the exoskeleton of arthropods like the crayfish.

tract. **Chitin** is a modified polysaccharide that is the main component in the cell walls of some fungi and the exoskeleton of insects and other arthropods. Some chitin-based exoskeletons, such as those of crayfish (Fig. 5.6), are reinforced and hardened by calcium carbonate.

Performing Benedict's Test for Reducing Sugars

Benedict's reagent, containing sodium citrate, sodium bicarbonate, and copper sulfate, is used to identify **reducing sugars.** These sugars can be monosaccharides or some disaccharides that possess a free aldehyde or ketone group. Reducing sugars have the ability to reduce (remove oxygen or add hydrogen to) oxidizing agents such as the copper compound in Benedict's solution. Polysaccharides and some disaccharides such as sucrose are not reducing sugars because they do not possess a free aldehyde or ketone group and therefore do not donate protons to oxidizing agents.

Benedict's reagent is blue because of the cupric or copper ions. When the reagent is mixed with a solution containing reducing sugars and is heated, a colorful precipitate will form in the bottom of the test tube. If no reducing sugar is present, the solution will remain blue and not form a precipitate. In the presence of a very small amount of reducing sugar, it will produce a green precipitate; in a low amount it will produce a yellow precipitate; in a moderate amount it will produce a yellowish-orange precipitate; in a high amount it will produce an orange precipitate; and in an extremely high amount it will produce a brick red precipitate.

 Student Activity—Benedict's Test

Materials
- hot plate
- 500 ml beaker
- test tube rack and test tube holder
- 10 test tubes
- wax pencils
- metric ruler
- dropper
- Benedict's reagent★
- test solutions: distilled water, clear diet soda, clear non–diet soda, pineapple juice, onion juice, potato juice, milk, glucose solution, sucrose solution, corn syrup, and tap water.

★ *Benedict's reagent is corrosive. If it splashes on to your skin, wash the area immediately with soap and water.*

 **Procedure 5.5
Benedict's Test**

1. Place approximately 250 ml of tap water into a 500 ml beaker.
2. Place the beaker on a hotplate and begin to heat the water. When the water begins to boil, set the temperature to medium.
3. Procure 10 clean test tubes and wax pencils. Label the test tubes 1–10.
4. Starting at the bottom of the test tube, place a 0.5 cm mark and a 1.0 cm mark on each test tube.
5. Using the order of the test materials in Table 5.4, fill each test tube with 0.5 cm of the solution to be tested.
6. Your instructor may assign additional substances.

7. Carefully add Benedict's reagent to the 1.0 cm mark of each test tube, and gently swirl each test tube.
8. Place the test tubes in the beaker, and let them heat for 3 minutes.

9. Remove the test tubes from the beaker with a test tube holder, and let them cool for 2 minutes.
10. Record the color changes in Table 5.4.
11. Discard the solutions according to the instructor's directions, and return the test tubes to the collection area.

Table 5.4 Benedict's Reagent Test for Reducing Sugar			
Test Tube	**Solution**	**Color**	**Conclusion**
1	Distilled water		
2	Diet soda		
3	Non-diet soda		
4	Pineapple juice		
5	Onion juice		
6	Potato juice		
7	Milk		
8	Glucose solution		
9	Sucrose solution		
10	Corn syrup		
11			
12			
13			
14			
15			

Q. In the above test, rank the solutions from non–reducing sugar to strongest reducing sugar.

1. _____   6. _____   11. _____

2. _____   7. _____   12. _____

3. _____   8. _____   13. _____

4. _____   9. _____   14. _____

5. _____   10. _____   15. _____

Q. Why was Benedict's reagent used to test the urine of diabetics before more sophisticated methods were developed?

_____

_____

_____

_____

Student Activity—Iodine Test

IODINE TEST FOR STARCH

The **Iodine Test** using iodine-potassium iodide (I_2KI) has been developed to distinguish starch from other carbohydrates. Because starch is a coiled glucose polymer, the I_2KI solution interacts with the starch, producing a bluish-black color. The I_2KI does not react with non-coiled carbohydrates and remains a yellowish-brown color. Some carbohydrates, such as dextrin and glycogen, will produce an intermediate color reaction.

Materials
- I_2KI solution
- test tube rack
- 3–10 test tubes
- wax pencils
- metric ruler
- dropper
- test solutions and materials: distilled water, clear diet soda, clear non-diet soda, onion juice, potato juice, milk, glucose solution, sucrose solution, paper, and cotton.

Procedure 5.6
Iodine Test

1. Procure 10 clean test tubes and wax pencils. Label the test tubes 1–10.

2. Starting at the bottom of each test tube, place a 0.5 cm mark on each one.

3. Using the order of the test materials in Table 5.5, fill test tubes 1–8 with 0.5 cm of the solution to be tested. Place a small amount of paper and cotton in test tubes 9 and 10.

4. Your instructor may assign additional substances.

5. Carefully add 5 drops of I_2KI to each test tube, and swirl.

6. Record the color changes in Table 5.5.

7. Discard the solutions according to the instructor's directions, and return the test tubes to the collection area.

Table 5.5 Iodine Test for Starch

Test Tube	Solution	Color	Conclusion
1	Distilled water		
2	Diet soda		
3	Non-diet soda		
4	Onion juice		
5	Potato juice		
6	Milk		
7	Glucose solution		
8	Sucrose solution		
9	Paper		
10	Cotton		
11			
12			
13			
14			
15			

Q. Which substances contained starch?

_____

_____

Q. Compare and contrast the reaction of potato juice and onion juice.

_____

_____

LIPIDS

Lipids comprise a diverse group of organic compounds that includes fats, waxes, phospholipids, and steroids. These compounds are characterized by being insoluble in water and soluble in nonpolar compounds such as ether. Lipids consist mostly of carbon and hydrogen atoms with few oxygen atoms. Lipids also may contain small amounts of other atoms such as phosphorus and sulfur.

Fats, or triglycerides, are the most abundant lipids in living organisms. In the body, fat serves as an energy reserve. One gram of fat stores significantly more energy than one gram of starch or one gram of glycogen. In addition, fats are important in insulation and cushioning.

A fat molecule is composed of a glycerol molecule attached to three fatty acid molecules. Fatty acid chains may vary in their length and the way their carbon atoms join. In **saturated fatty acids**, all of the carbon atoms are linked by single bonds, and **unsaturated fatty acids** have one or more double bonds between the carbon atoms. Saturated fatty acids contain the maximum number of hydrogen atoms, and unsaturated fatty acids do

Is It Counterfeit?

One safeguard taken when producing paper currency is to remove all traces of starch from the paper. Fortunately, counterfeiters have not mastered this technique. A common starch known as amylose turns blue in the presence of iodine. If iodine is applied to the money with a special pen, the money will turn blue if it is counterfeit, because of the presence of starch.

not have the maximum number of hydrogen atoms. If an unsaturated fatty acid has one double bond, it is known as **monounsaturated**. If more than one double bond is present, the fatty acid is known as **polyunsaturated**.

Saturated fats consist of saturated fatty acids and include animal fats and vegetable shortenings. These substances, such as butter and lard, are solid at room temperature. Diets rich in saturated fats are associated with cardiovascular disease. **Unsaturated fats** consist of unsaturated fatty acids and consist primarily of plant and fish oils. The unsaturated fats tend to be liquid at room temperature. Unsaturated fats are considered safer for the heart (Fig. 5.7).

Waxes consist of an alcohol bonded with a long-chain fatty acid. Waxes are solid at room temperature and repel water. The cuticle found on the surface of leaves and fruits consists of waxes and a lipid, cutin. The cuticle helps to conserve water and repel insects. Birds preen, using waxes, fatty acids, and fats to waterproof their feathers (Fig. 5.8). **Cerumin**, a waxy secretion better known as earwax, traps foreign substances that potentially may enter the middle ear.

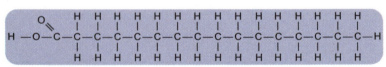

Saturated fatty acids

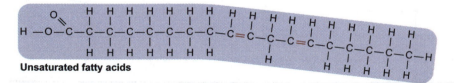

Unsaturated fatty acids

Figure 5.7 Saturated and unsaturated fats.

Figure 5.8 Water beading on a duck's head.

Student Activity—Grease-Spot Test

Grease-Spot Test for Lipids

You probably have noticed that lipids are greasy, especially after eating a bag of potato chips. The Grease-Spot Test is a simple test used to identify the lipid nature of substances.

Materials

- brown paper bag or brown wrapping paper
- scissors
- ruler
- dropper
- test substances: water, vegetable oil, egg white, soda, potato chip, honey, salad dressing, butter, rubbing alcohol, beef jerky.

 ## Procedure 5.7
Grease-Spot

1. Cut a brown paper bag or brown wrapping paper into 2.54 cm squares.
2. On the back of the paper, write the substance to be tested.
3. Following the order in Table 5.6, either gently rub a solid substance or place a drop of liquid on the brown paper.
4. Let the substance and paper stand for approximately 15 minutes, then hold the brown paper up to a light source. The presence of a translucent spot indicates a lipid.
5. Your instructor may assign additional substances.
6. Clean the lab station and discard the paper.

Table 5.6 Grease-Spot Test for a Lipid

Substance	Result	Conclusion
1. Water		
2. Vegetable oil		
3. Egg white		
4. Soda		
5. Potato chip		
6. Honey		
7. Salad dressing		
8. Butter		
9. Rubbing alcohol		
10. Beef jerky		
11.		
12.		
13.		
14.		
15.		

Q. If you rub some nose grease (oil from the bridge of the nose) on the paper, what will happen?

Student Activity—Sudan Test

SUDAN TEST FOR A LIPID

A stain known as **Sudan III** combines with lipid molecules to produce a brilliant orange color. Sudan works by forming a hydrophobic interaction with nonpolar molecules. The more vivid the orange color, the greater the intensity of the interaction.

Materials
- Sudan III stain
- filter paper
- ruler
- pencil
- forceps
- test substances: distilled water, whole milk, low-fat milk, skim milk, egg white, egg yolk, butter, margarine, corn syrup, and pineapple juice

 Procedure 5.8
Sudan Test

1. Procure several sheets of filter paper.

2. On the perimeter of the paper, draw several circles of 2 cm diameter, spaced equally apart. You may need more than one sheet of filter paper to make 10 circles.
3. Number each circle in accordance with the numbering in Table 5.7.
4. Your instructor may assign additional substances.
5. Place 2 drops of each substance in the appropriate circle. Blot off any excess liquid and allow the paper to dry completely.
6. Procure a bowl containing the Sudan III stain.
7. Immerse the filter papers in the bowl and let them sit for 5 minutes.
8. Remove the filter paper with forceps.
9. Wash the filter paper in a pan of water for 1 minute.
10. Examine the color in each circle and record the color in Table 5.7.
11. Discard the solutions according to the instructor's directions, and return the material to the collection area.

Q. Which substances interacted the greatest with Sudan?

Q. Which substances did not react with Sudan?

Table 5.7 Sudan Test for a Lipid		
Substance	**Color**	**Conclusion**
1. Distilled water		
2. Whole milk		
3. Low-fat milk		
4. Skim milk		
5. Egg white		
6. Egg yolk		
7. Butter		
8. Margarine		
9. Corn syrup		
10. Pineapple juice		
11.		
12.		
13.		
14.		
15.		

Proteins

Proteins are the most numerous and complex molecules found in living organisms. More than 100,000 proteins have been identified. Proteins provide support, movement, storage, defense, and regulation. In addition, hormones and enzymes are composed of proteins. Common proteins include keratin (hair, fingernails), hemoglobin (blood), ovalbumin (egg white), actin and myosin (muscle), insulin and glucogon (pancreas), lysozyme (tears), and collagen (connective tissues).

Proteins are composed of building blocks known as **amino acids**. Amino acids are compounds that have an amino group ($-NH_2$) and a carboxyl group ($-COOH$) attached to the same asymmetrical alpha carbon atom. A third group known as the *R group* varies tremendously and influences the characteristic of the molecule. Twenty amino acids serve as the building blocks of proteins (Fig. 5.9).

Organisms can synthesize the majority of amino acids. Amino acids that animals cannot synthesize are called **essential amino acids**. In humans, the essential amino acids are:

histidine	isoleucine
leucine	threonine
lysine	tryptophan
methionine	valine
phenylalanine	arginine (in children)

Amino acids are linked by **peptide bonds**. A peptide bond is a covalent bond that forms between the amino group of one amino acid and the carboxyl group of another amino acid. Additional amino acids can be joined to the chain, forming **polypeptides**. Some complex proteins can be made of more than 100,000 amino acids.

A protein is a polypeptide composed of more than 100 amino acids. The **conformation** of a protein refers to the final folded arrangement of the polypeptide chain of a protein. Proteins can vary in size and three-dimensional structure. Four levels of protein structural complexity are found in nature: primary structure, secondary structure, tertiary structure, and quaternary structure.

 Student Activity—Biuret's Reagent

BIURET'S TEST FOR PROTEIN

The **Biuret Test** is commonly used to detect the presence of a protein. The Biuret reagent is a strong blue-green colored solution containing 1% copper sulfate ($CuSO_4$), and sodium hydroxide (NaOH) or potassium hydroxide (KOH). The reagent changes color from blue-green to violet in the presence of proteins. The change in color results from the interaction between the copper ions and the peptide bonds of the protein. The more the peptide bonds, the darker is the resulting color.

Materials
- Biuret's reagent [*Caution:* Biuret's reagent is extremely corrosive. Handle this chemical with great care. If you get some on your skin, wash the area with mild soap and water and notify your laboratory instructor. Carefully follow your instructor's directions regarding the use and disposal of Biuret's reagent.]
- 10 test tubes
- test tube rack
- wax pencil
- dropper
- test substances: distilled water, sucrose solution, whole milk, bread, ground peanuts, chicken broth, vegetable oil, egg white, egg yolk, and albumin

 **Procedure 5.9
Biuret's Reagent**

1. Procure 10 test tubes, a wax pencil, a dropper, and a test tube rack.
2. Label the test tubes 1–10. Place a 1.0 cm mark from the bottom on each test tube.
3. Using the order of the test materials in Table 5.8, fill the test tubes 1–10 up to the 1.0 cm mark with the substance to be tested. Place the test tubes in the rack.
4. Your instructor may assign additional substances.
5. Add 3 drops of the Biuret's reagent to each test tube. Gently tap each test tube with your finger to mix the solution.
6. Allow 2 minutes for the colors to develop, and record the observed color changes in Table 5.8.
7. Discard the solutions according to the instructor's directions, and return the test tubes to the collection area.

Q. Which substances were proteinaceous?

_____

_____

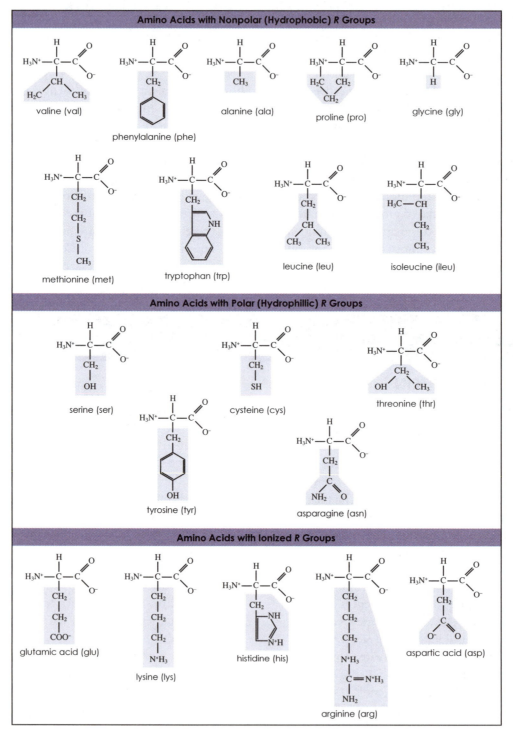

Figure 5.9 The 20 amino acids. The *R* Groups are light purple.

Table 5.8 Biuret's Test for a Protein		
Substance	**Color**	**Conclusion**
1. Distilled water		
2. Sucrose solution		
3. Whole milk		
4. Bread		
5. Ground peanuts		
6. Chicken broth		
7. Vegetable oil		
8. Egg white		
9. Egg yolk		
10. Albumin		
11.		
12.		
13.		
14.		
15.		

Why Do Shrimp Turn Pink?

When shrimp are boiled, they mysteriously turn an orangish-pink color. This color change is the result of protein chemistry. Shrimp consume planktonic organisms that contain pigments made from carotenoids. Prior to boiling, the carotenoids are bound to protein molecules in their shells and appear dark green. When the shrimp are boiled, the proteins break down (denature) and the carotenoid becomes visible.

NUCLEIC ACIDS

Nucleic acids are polymers composed of nucleotides and are responsible for the transmission of hereditary information, as well as the production of proteins by cells. The two kinds of nucleic acids are **deoxyribonucleic acid (DNA)** and **ribonucleic acid (RNA)**. DNA makes up genes and possesses instructions for making proteins and RNA. RNA serves primarily to build proteins from amino acids.

Chemically, a nucleic acid consists of a pentose sugar (deoxiribose or ribose), a phosphate group, and a nitrogenous base. In DNA, nitrogenous bases are pyrimidines (cytosine and thymine) and purines (guanine and adenine) (Fig. 5.10). In RNA, the pyrimidine uracil replaces thymine.

The DNA molecule is a double helix with as many as 3 billion nucleotides. Hydrogen bonds between pyrimidines, and purines hold the double helix together. Complementary base pairing occurs in DNA, with cytosine pairing with guanine and thymine pairing with adenine. DNA and RNA will be discussed in detail in Chapter 16.

ATP AND OTHER NUCLEOTIDES

Adenosine triphosphate (ATP) is a nucleotide that serves as the energy currency of cells. Nicotinomide adenine dinucleotide (NAD^+) and flavin adenine dinucleotide (FAD^+) are coenzymes in metabolic processes. A **coenzyme** aids the action of an enzyme.

TESTING AN UNKNOWN

The instructor will provide you with an unknown substance. Using the procedures incorporated in previous activities, determine if the substance is a reducing sugar, a starch, a lipid, or a protein.

Q. Based upon the tests below, the unknown was a:

_____

_____

Test	Result	Conclusion + / –
Benedict's Test		
Iodine Test		
Sudan's Test		
Biuret's Test		

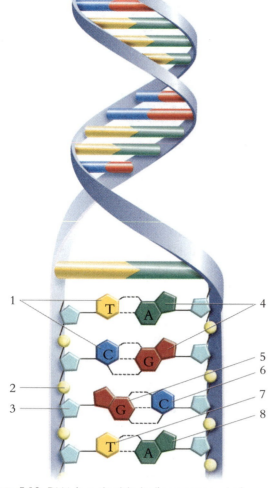

Figure 5.10 DNA is a double helix composed of a pentose sugar, a phosphate group and the pyrimidines (cytosine and thymine) and the purines (guanine and adenine).

1. Pyrimidines
2. Phosphate
3. Pentose sugar
4. Purines
5. Guanine
6. Cytosine
7. Thymine
8. Adenine

… If (and oh, what a big if!) we could conceive in some warm little pond, with all sorts of ammonia and phosphoric salts, light, heat, electricity, etc…, present that a compound was chemically formed ready to undergo still more complex changes…

—Charles Darwin (1809–1882)

NOTES

_____

_____

_____

_____

_____

_____

_____

_____

_____

_____

_____

_____

Name: _____ Date: _____ Section: _____

Review Questions

1. Compare and contrast an acid and a base, and provide examples of each.

_____

_____

_____

_____

2. Distinguish between inorganic and organic molecule.

_____

_____

_____

_____

3. Why was water or distilled water tested in each activity?

_____

_____

4. Compare and contrast ionic and covalent bonds, and give an example of each.

_____

_____

_____

_____

5. Describe what happens when a polar covalent and a nonpolar covalent substance are combined. Provide an example of a mixture with these components.

_____

_____

_____

_____

6. Define monosaccharides, disaccharides, and polysaccharides, and provide two examples of each.

_____

_____

_____

_____

_____

_____

_____

7. What is a peptide bond?

_____

_____

_____

8. What are phospholipids, and where are they found?

_____

_____

_____

9. What is an essential amino acid?

_____

_____

_____

10. What is acid rain? Describe the impact of acid rain upon the environment.

_____

_____

_____

_____

Chapter 6
The Structure and Function of Cells: Understanding the Building Blocks of Life

Student Outcome Objectives

At the completion of this exercise, the student will be able to:

1. State the cell theory.
2. Discuss the factor that limits cell size.
3. Define and give examples of unicellular, colonial, and multicellular organisms.
4. Compare and contrast prokaryotic and eukaryotic cells.
5. Describe the basic anatomy of a typical prokaryotic cell.
6. Compare and contrast plant and animal cells.
7. Describe the basic anatomy of a plant cell and an animal cell.
8. Perform microscopic observations of bacteria, protist, fungi, plant, and animal cells.

Overview

From tiny bacteria to the great blue whale, the cell serves as the fundamental building block of all living things (Fig. 6.1). A student reading this paragraph consists of nearly 125 trillion cells working together to maintain a state of biological balance, or **homeostasis**. Even the hamburger and fries sitting next to this manual are made up of a multitude of plant and animal cells. Yes, that burger and fries also have their share of bacterial cells.

The **cell** is the smallest unit of biological organization that can undergo the activities associated with life, such as metabolism, response, and reproduction. The British scientist Robert Hooke first described the cell in 1665. In the late 1830s, two German scientists—the botanist Matthias Schleiden and the zoologist Theodor Schwann—provided a powerful understanding of the structure and function of plant and animal cells through the cell theory. Basically the **cell theory** states that all living things are composed of cells and that the cell is the basic unit of structure and function of all living things.

In the 1850s, the German physician Rudolph Virchow added to the cell theory that cells come only from preexisting cells. Virchow also pointed out that the cell is the fundamental link in the cellular level of organization that includes tissues, organs, systems, and ultimately the complete organism. Today, the cellular level of organization has been expanded to include populations, communities, ecosystems, and the biosphere.

Although cells vary in size from a bacterium 1 to 10 micrometers in diameter to a chicken egg larger than 1 centimeter in diameter, most cells are microscopic.

The inclusions and organelles within the cell are much smaller and are measured in nanometers. The reason for the absence of giant cells is a matter of the surface area-to-volume ratio. If the surface area of a cell increases, the volume does not increase in a direct proportion; the volume increases proportionally faster. Thus, the surface area could not support the metabolic needs of the increased volume. Some cells, such as frog eggs, chicken eggs, and ostrich eggs, can become large because they are not metabolically active until they begin to divide. Other cells, such as nerve cells, can possess extensions of more than a meter, but the extensions are narrow and have little volume.

Bacteria and many protists, such as the green alga *Spirogyra* and the protozoan paramecium, are composed of one cell and are called **unicellular**. Despite having just one cell, these organisms carry on all of the life processes efficiently. Several species of protists exist as colonies that are loosely connected groups or aggregates of cells. Examples of **colonial organisms** are the alga *Volvox* and *Scenedesmus*. Organisms such as an azalea, a mushroom, and a walrus, which are composed of many cells, are called **multicellular**. These organisms exhibit a division of labor and have a variety of specialized tissues.

Although innumerable forms of cells exist in nature, only two basic types of cells comprise life on Earth: prokaryotic cells and eukaryotic cells.

1. **Prokaryotic cells** lack a membrane-bound nucleus and organelles and are much smaller than eukaryotic cells. The cytoplasm of prokaryotes is surrounded by a plasma membrane, and the majority of prokaryotes are encased in a protective cell wall. Prokaryotic organisms are placed within the kingdoms Archaebacteria and Eubacteria.
2. **Eukaryotic cells** are more structurally complex and larger than prokaryotic cells and have a membrane-bound nucleus and organelles. Members of the kingdoms Protista, Plantae, Fungi, and Animalia possess eukaryotic cells.

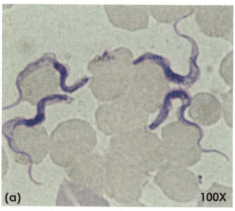
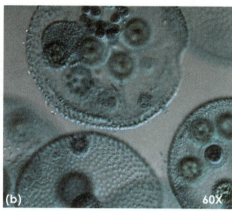

Figure 6.1 The cell is the basic unit of structure and function of all living things: (a) *Trypanosoma* sp. is a unicellular organism, (b) *Volvox* sp. is a colonial organism, and (c) *Diceros bicornis Black Rhinoceros* is a multicellular organism.

PROKARYOTIC CELLS

The most cosmopolitan organisms on Earth today are the prokaryotes. They exist in every possible environment, even those that do not seem conducive to life. Although the prokaryotes are small in body (1 to 50 micrometers in width and diameter), they are economically, ecologically, and medically important. Two distinct groups of prokaryotic organisms are archaebacteria and bacteria. Bacterial fossils have been dated at older than 3.5 billion years.

1. The **archaebacteria**, or ancient bacteria (Fig. 6.2) can be found living in extreme environments such as exceedingly salty habitats (extreme halophiles), exceptionally hot environments (extreme thermophiles), the anaerobic mud of swamps, and the gut of termites and many mammals (methanogens).

2. The **eubacteria**, or true bacteria, are better known to the general public. Although the majority of eubacteria are harmless or helpful, such as *Lactobacillus acidophilus*, which is placed in yogurt, there are several medically important species. Examples of these organisms are *Yersinia pestis* (black plague), *Clostridium perfringins* (gangrene), *Helicobacter pylori* (ulcers), *Vibrio cholerae* (cholera), *Staphylococcus aureus* (boils), and *Bacillus anthracis* (anthrax).

Q. Identify the bacterial species responsible for five diseases in humans that are not mentioned above?

_____

_____

_____

_____

_____

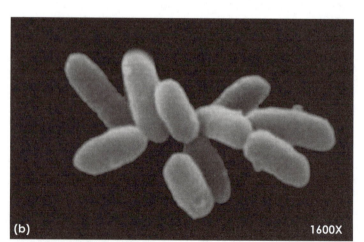

Figure 6.2 (a) Extreme thermophiles are archaebacteria that can thrive in extremely hot water such as these in Yellowstone National Park; (b) SEM of the archaebacterium *Halobacteria*.

Observing Cyanobacteria

The **cyanobacteria**, once classified as the blue-green algae, are photosynthetic eubacteria. These rather large prokaryotes do not possess chloroplasts; the chlorophyll *a* is located in the thylakoid membranes. The cyanobacteria have a number of accessory pigments that can mask the green color of chlorophyll. As a result, species of cyanobacteria appear red, yellow, brown, or blue-green. The cyanobacteria are common and can be found in a number of environments, including in the soil, on sidewalks, on the sides of buildings, on trees, and in bodies of water such as ditches (Fig. 6.3).

Q. What are stromatolites, and what is their scientific significance?

_____

_____

_____

 Student Activity—Cyanobacteria

Materials
- microscope
- prepared slides of cyanobacteria (*Gloeocapsa* and *Nostoc*)
- blank slides and coverslips
- living specimens of cyanobacteria (*Oscillatoria* and *Anabaena*)
- forceps

 Procedure 6.1
Observing Cyanobacteria

1. Procure a microscope, prepared slides, blank slides, and coverslips.
2. Using proper microscopy techniques, observe the prepared slides of *Gleocapsa* and *Nostoc*.
3. Sketch and describe *Gleocapsa*.

_____

_____

4. Sketch and describe *Nostoc*.

_____

_____

5. Properly prepare a wet mount of *Oscillatoria* and *Anabaena*.

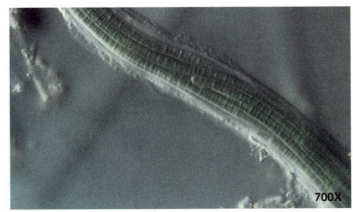

Figure 6.3 *Microcoleus* sp., one of the most common cyanobacteria in and on soils throughout the world. It is characterized by several filaments in a common sheath.

6. Using proper microscopy techniques, observe *Oscillatoria* and *Anabaena*.

7. Sketch and describe *Oscillatoria*.

8. Sketch and describe *Anabaena*.

_____

_____

_____

_____

_____

_____

9. After completion of the activity, clean up your work area and return or dispose of the material as instructed.

Student Activity—Bacteria

Observing Bacteria

Most bacteria are significantly smaller than the cyanobacteria. The bacteria are simple in form and anatomy and exhibit three basic shapes: **bacillus** (rod-shaped), **coccus** (spherical-shaped), and **spirillum** (spiral-shaped). An electron microscope is used to observe the anatomical detail of a typical bacterium (Table 6.1 and Fig. 6.4).

Materials
- microscope
- immersion oil
- prepared slides of *Escherichia coli* and *Helicobacter pylori*
- blank slides and coverslips
- toothpick
- pipette
- plain yogurt

Procedure 6.2
Observing Bacteria

1. Procure a microscope, immersion oil, prepared slides, blank slides, coverslips and toothpicks.

2. Using proper microscopy techniques, observe the prepared slides of *Escherichia coli* and *Helicobacter pylori*. To view the specimens properly, use an oil immersion objective, if available. *After use, ensure that all oil is removed from the stage, oil immersion objective, and slide.*

3. Sketch and describe *Escherichia coli*.

_____

_____

_____

Table 6.1 Common Anatomical Features of a Generalized Bacterium	
Structure	**Function**
Cell wall	In eubacteria, a peptidoglycan envelope that provides protection and shape.
Plasma membrane	A phospholipid bilayer that provides support and regulates the movement of substances into and out of the cell.
Cytoplasm	Semifluid medium within the cell.
Nucleoid	Region that houses the bacterial DNA in a single chromosome. Some bacteria possess small circular fragments of DNA called plasmids.
Ribosome	Site of protein synthesis.
Fimbriae	Short hair-like structures that aid in attachment.
Pilus	Rigid hair-like structures that are important for attachment and the exchange of genetic information.
Flagellum	An elongated structure used for locomotion. The number of flagella and their location are important in determining the species of bacteria.
Capsule	A protective slime-like area lying outside the cell wall. It helps the bacterium adhere to certain surfaces, keeps it from drying out, and protects the bacterium from phagocytosis by other organisms or cells.

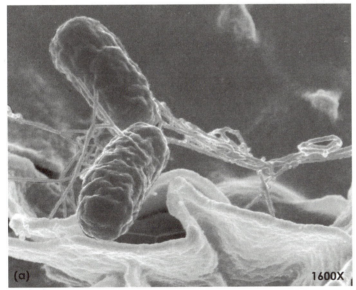 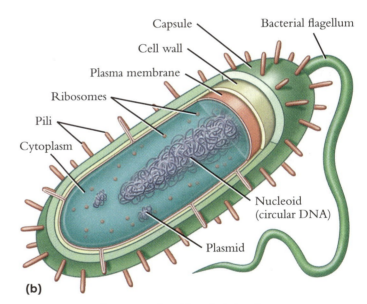

Figure 6.4 (a) SEM showing Salmonella typhimurium and (b) the basic structure of a generalized bacterial cell.

In science, as in life, learning and knowledge are distant, and the study of things, and not of books, is the source of the later.

—**Thomas Henry Huxley (1825–1895)**

4. Sketch and describe *Helicobacter pylori*.

[sketch box]

_____

_____

_____

5. Procure a small amount of plain yogurt on the tip of a toothpick. Rub the yogurt onto the central portion of a blank slide. Place one drop of water on the yogurt with a pipette, and mix with a toothpick. Gently place the coverslip on the water/yogurt mixture.
6. Observe the bacteria in the yogurt under high power. (The majority of bacterial cells in yogurt are *Lactobacillus acidophilus*.)
7. Sketch and describe *Lactobacillus*.

[sketch box]

_____

_____

8. After completing the activity, clean up your work area and return or dispose of the material as instructed.

Check Your Understanding

Q. Why is *Lactobacillus* used in yogurt?

_____

_____

_____

Q. Why is the presence of *Lactobacillus* in yogurt considered beneficial to your health?

_____

_____

_____

_____

EUKARYOTIC CELLS

Eukaryotic cells had their origins nearly 2 billion years ago. Eukaryotes include the protists, fungi, plants, and animals. The cells of eukaryotes possess a membrane-bound nucleus and a variety of membrane-bound organelles.

Observing Protists

Protists include a diverse group of organisms (Fig. 6.5). In fact, the former kingdom Protista is undergoing reorganization and one day will consist of several new kingdoms. Presently, the protists can be separated into the plant-like protists (algae), fungi-like protists (slime and water molds), and animal-like protists (protozoans). The protists are discussed in more depth in Chapter 21.

Within every cell there are eloquently evolved molecular machines, nucleic acids, enzymes, the cell architecture; every cell a triumph of natural selection and we are made of trillions of cells. We are each of us a multitude. Within us is a little universe.

—Carl Sagan (1934–1996)

Student Activity—Protists

Materials
- microscope
- immersion oil
- prepared slide of *Volvox* and *Amoeba proteus*
- blank slides and coverslips
- toothpicks
- pipette
- culture of *Spirogyra* and *Paramecium* sp.
- Protoslo®

Procedure 6.3
Observing Protists

1. Procure a microscope, prepared slides, blank slides, coverslips, and toothpicks.
2. Using proper microscopy techniques, observe the prepared slides of the colonial alga *Volvox* sp. and the protozoan *Amoeba proteus*.
3. Sketch and describe *Volvox* sp.

4. Sketch and describe *Amoeba proteus*.

5. Carefully prepare a wet mount of *Spirogyra* and *Paramecium* sp. and observe the living protists. Protoslo may have to be added to the slide with paramecia to slow them down for observational purposes.

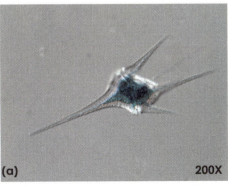
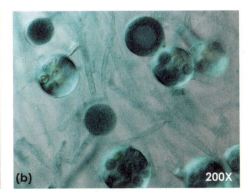
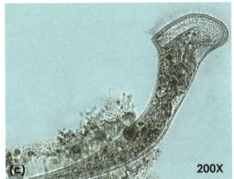

Figure 6.5 Example protists include (a) *Ceratium* sp., a dinoflagellate, (b) *Saprolegnia* sp., a water mold, and (c) *Stentor* sp., a protozoan.

6. Sketch and describe *Spirogyra*.

7. Sketch and describe *Paramecium* sp.

_____

_____

_____

_____

_____

_____

8. After completing the activity, clean up your work area and return or dispose of the material as instructed.

Student Activity—Fungi

Observing Fungi

Kingdom **Fungi** includes a diverse group of mostly multicellular heterotrophic organisms. Examples of fungi are mushrooms, truffles, morels, rusts, bread mold, ring-worm, and yeast (Fig. 6.6). The fungi are discussed in more depth in Chapter 28.

Materials (Procedures 6.4–6.5)
- microscope
- blank slides and coverslips
- culture of Baker's yeast (*Saccaromyces cerevisiae*)
- culture of *Paramecium* sp.
- pipettes
- methylene blue
- Protoslo
- yeast cells stained with Congo red dye

Procedure 6.4
Observing Fungi

1. Procure a microscope, blank slides, coverslips, and a pipette.
2. Carefully prepare a wet mount of Baker's yeast.

3. Observe the slide. If the yeast is difficult to see, carefully add one drop of methylene blue stain to the wet mount with a pipette. *Avoid inhalation and skin contact with methylene blue. Immediately rinse it off the skin with mild soap and water as methylene blue will stain clothing.*
4. Describe and sketch Baker's yeast (*Saccaromyces cerevisiae*).

_____

_____

_____

Figure 6.6 Examples of kingdom Fungi are (a) *Aspergillus* and (b) *Amantia* or death angel mushroom.

5. After completion of the activity, clean up your work area and return or dispose of the material as instructed.

Q. Describe the smell of the yeast culture.

_____

_____

_____

Q. Why does the culture have a characteristic smell?

_____

_____

_____

_____

 Student Activity—Paramecia and Yeast

Extension: Paramecia and Yeasts

Baker's yeast is an excellent food source for paramecia. If the yeast is heated and stained with Congo red dye, these organisms are highly visible. They can be easily observed within the food vacuole of a hungry paramecium.

 **Procedure 6.5
Observing Paramecia and Yeast**

1. Procure a microscope, blank slides, coverslips, a pipette, toothpicks, and Protoslo®.
2. Place a drop of yeast cells stained with Congo red dye on a blank slide. Add a drop of paramecia from the culture. Then add a drop of Protoslo® with a toothpick to the slide. Then add a coverslip.
3. Let the slide sit for at least 10 minutes to allow the paramecia to begin feeding.

4. Describe and sketch the interaction between the yeast and the paramecia.

_____

_____

5. After completion of the activity, clean up your work area and return or dispose of the material as instructed.

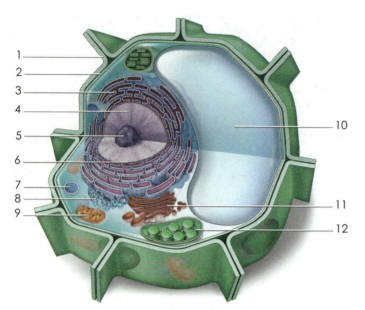

Figure 6.7 A typical eukaryotic plant cell.

1. Cell wall
2. Cell (plasma) membrane
3. Rough endoplasmic reticulum
4. Nucleus Mitochondrion
5. Nucleolus
6. Nuclear membrane (envelope)
7. Peroxisome
8. Smooth endoplasmic reticulum
9. Mitochondrion
10. Vacuole
11. Golgi complex
12. Chloroplast

PLANT AND ANIMAL CELLS

Kingdoms **Plantae** and **Animalia** include the most conspicuous organisms on Earth. The plant kingdom contains approximately 280,000 species of multicellular, photosynthetic authotrophs. Plants vary in size and complexity from the minute duckweed to the giant redwood tree. Kingdom Animalia encompasses more than 1.5 million species of multicellular heterotrophs. Members of the animal kingdom vary tremendously, from simple sponges to humans. The plants and animals are discussed in more depth in later chapters.

Observing Plant Cells

Elodea is a common plant that lives in freshwater habitats such as ponds and lakes (Fig. 6.9). It provides an excellent example for studying basic plant cell anatomy. The leaves of *Elodea* are only a few cells thick and allow light to pass through the leaf without special preparation techniques. Refer to Figure 6.7 and Table 6.2 for references to plant cell anatomy.

Every fact that is learned becomes a key to other facts.

—**Edward Livingston Youmans (1821–1887)**

 Student Activity—Plant Cells

Materials
- microscope
- living specimen of *Elodea*
- blank slides and coverslips
- pipette

 **Procedure 6.6
Observing Plant Cells**

1. Procure a microscope, a blank slide, coverslips, and a pipette.
2. Carefully remove a single healthy leaf from the *Elodea*.
3. Place the leaf in a drop of water on the blank slide with the top surface facing upward. (The cells on the upper surface are much larger and easier to observe.) Place a coverslip over the *Elodea*. Periodically check the leaf, making sure that it does not dry out. If the leaf begins to dry, add a drop of water with a pipette.

4. Examine the leaf surface with the scanning and low power objectives. Focus through the cell layers of the *Elodea*.
5. Describe and sketch *Elodea* using scanning and low power objectives.

Table 6.2 Common Anatomical Features of Eukaryotic Cells

Structure	Function
Cell wall	In plant cells, a cellulose envelope that provides protection and shape.
Plasma membrane	A phospholipid bilayer that provides support and regulates the movement of substances into and out of the cell.
Cytoplasm	A semifluid medium located between the plasma membrane and nucleus. Inclusions and organelles are found in the cytoplasm.
Nucleus	The control center of the cell.
Nuclear envelope	Membrane surrounding the nucleus; possesses numerous nuclear pores.
Nucleoplasm	Cytoplasm within the nucleus.
Nucleolus	Chromatin-rich region that serves to combine proteins and RNA to make ribosomial subunits. Many cells possess numerous nucleoli.
Chromatin	Diffuse thread-like strands composed of DNA and proteins.
Mitochondria	Site of aerobic cellular respiration.
Endoplasmic reticulum (ER)	Network of membranes throughout the cytoplasm. Synthesis of protein and non-protein products.
Rough ER	Lined with ribosomes. Involved in the synthesis and assembly of a variety of proteins, and production of membranes.
Smooth ER	Not associated with ribosomes. Main site of steroid, fatty acid, and phospholipid synthesis. Site of detoxification.
Golgi apparatus	Stacks of flattened membranous sacs or cisternae. Receives, packages, stores, and ships protein products. Produces lysosomes and other vesicles.
Peroxisome	Vesicles containing enzymes that help in breaking down fatty acids and neutralizing hydrogen peroxide.
Lysosome	In animal cells, vesicles containing hydrolytic digestive enzymes that are used in destroying cellular debris and worn-out organelles. They also are important in programmed cell death.
Centrioles	Found in animal cells with the exception of roundworms (nematodes). Appear as a pair of cylindrical structures made of microtubules. Form the spindle apparatus in cell division.
Ribosomes	Sites of protein synthesis.
Cytoskeleton	Structures that help the cell maintain its shape, anchor organelles, and move. Three kinds of cytoskeletal elements are recognized: microtubules, microfilaments, and intermediate fibers.
Chloroplasts	In plant cells, they are sites of photosynthesis. Contains grana that are stacks comprised of chlorophyll rich thylakoids.
Central vacuole	In plant cells, it is a large fluid-filled sac that helps maintain the shape of the cell and stores metabolites.
Middle lamellae	Region between adjacent plant cells that cements the cell walls together.

6. Using the high power objective, examine a single cell of *Elodea*. Attempt to locate the structures indicated in Figure 6.8. The gray-colored nucleus may be difficult to locate. The nucleus may become more evident if a drop of iodine is placed upon the leaf. In a good preparation, the nucleolus may be evident. Carefully notice if the cytoplasm and chloroplast are moving. This process is called **cytoplasmic streaming.**

7. Describe and sketch *Elodea* on high power.

_____

_____

_____

8. After completion of the activity, clean up your work area and return or dispose of the material as instructed.

Check Your Understanding

Q. Describe the function of the components that you viewed in *Elodea*.

_____

_____

_____

Q. Describe the shape and size of the central vacuole.

_____

_____

Q. Describe the number and shape of the chloroplasts.

_____

_____

Q. Describe the location of the nucleus and the majority of the chloroplast.

_____

_____

_____

_____

Q. Describe and indicate the function of cytoplasmic streaming.

_____

_____

_____

Q. Describe the color of any water-soluble accessory pigments that may have appeared. What are the accessory pigments composed of, and what is their function?

_____

_____

_____

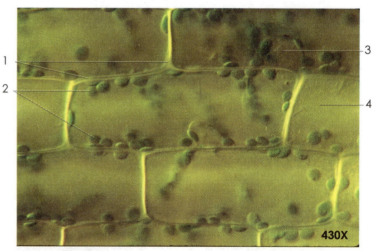

Figure 6.8 *Elodea* is a common plant found in freshwater ponds and lakes.
1. Cell wall   3. Nucleus
2. Chloroplasts   4 Vacuole

Student Activity—Animal Cells

Observing Animal Cells

A typical animal cell can be collected from the lining of your mouth (Fig. 6.9). These simple cells, known as squamous epithelial cells, are flat and thin and possess an obvious nucleus. Epithelial cells appear in regions of wear and tear and are constantly being sloughed away. In this specimen, only the cell membrane, cytoplasm, and nucleus will be easily observed. Refer to Figure 6.10 and Table 6.2 for references to animal cell anatomy.

Materials
- microscope
- blank slides and coverslips
- methylene blue
- clean toothpicks
- pipette

 Procedure 6.7
Observing Animal Cells

1. Procure a microscope, a blank slide, coverslips, clean toothpicks, and a pipette.

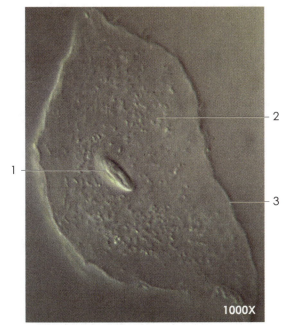

Figure 6.9 Epithelial cells can be collected from the lining of the mouth.

1. Nucleus
2. Cytoplasm
3. Cell membrane

2. Using a clean toothpick, gently scrape the inside of your cheek.
3. Place a small drop of water on a blank slide. Gently roll and swirl the end of the toothpick with the epithelial scrapings into the drop of water. Discard the used toothpick into the designated container.
4. Carefully place a drop of methylene blue on the drop of water. *Avoid inhalation and skin contact with methylene blue. Rinse it off the skin with mild soap and water immediately, as methylene blue will stain clothing.*
5. Place a coverslip over the specimen and make your observations.
6. Describe and sketch the epithelial tissue.

7. After completion of the activity, clean up your work area and return or dispose of the material as instructed.

I traveled among cells, watched their functioning… and realized that within myself was a grand assemblage of living organisms, all of which added up to be me.

—**John C. Lilly (1915–2001)**

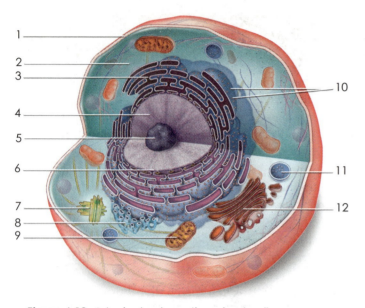

1
2
3
10
4
5
6
11
7
8
9
12

Figure 6.10 A typical eukaryotic animal cell.

1. Cell membrane
2. Cytoplasm
3. Rough endoplasmic reticulum
4. Nucleoplasm
5. Nucleolus
6. Nuclear membrane (envelope)
7. Centrosomes
8. Smooth endoplasmic reticulum
9. Mitochondrion
10. Ribosomes
11. Lysosome
12. Golgi complex

Check Your Understanding

Q. Describe the function of the components that you viewed in the epithelial tissue.

_____

_____

_____

Q. Describe the shape and size of the nucleus. Did you see a nucleolus? If so, describe the nucleolus.

_____

_____

_____

_____

NOTES

_____

_____

_____

_____

_____

_____

_____

_____

_____

Name: _____ Date: _____ Section: _____

Review Questions

1. Compare and contrast prokaryotic and eukaryotic cells.

_____

_____

_____

2. Describe and give an example of unicellular, colonial, and multicellular organisms.

_____

_____

_____

3. What factor prevents cells from becoming the size of blobs that consume city blocks in science fiction movies?

_____

_____

4. Compare and contrast plant and animal cells.

_____

_____

_____

5. Label the anatomical features of the bacterium below.

1. _____

2. _____

3. _____

4. _____

5. _____

6. _____

7. _____

8. _____

9. _____

6. Describe the function of the following features of a prokaryotic cell:

a. Cell wall _____

b. Pili _____

c. Fimbriae _____

d. Nucleoid _____

e. Flagellum _____

7. Describe the function of the following features of a plant cell:

a. Cell wall _____

b. Chloroplast _____

c. Central vacuole _____

d. Middle lamellae _____

e. Cytoplasm _____

8. Describe the function of the following structures:

a. Mitochondrion _____

b. Plasma membrane _____

c. Nucleus _____

d. Golgi apparatus _____

e. Lysosome _____

f. Chromatin _____

g. Golgi apparatus _____

h. Peroxisomes _____

i. Rough ER _____

j. Smooth ER _____

Chapter 7
Histology: Understanding Basic Plant and Animal Tissues

Student Outcome Objectives

At the completion of this exercise, the student will be able to:

1. Define the terms *tissue* and *histology*.
2. Describe the levels of biological organization.
3. Discuss the classification and basic characteristics of common plant tissues and human tissues.
4. State the location and function of the tissues used in the exercises in this chapter.
5. Identify the tissues shown through microscopic examination.
6. Draw the tissues used in this exercise, and label specified anatomical features.
7. Demonstrate proficiency with the microscope.

Overview

The cell is the basic unit of structure and function of all living things. A cell serves as the exclusive functional unit in unicellular organisms. In multicellular organisms ranging from trees to mushrooms to chimps, however, cells differentiate to perform a variety of specialized functions. Multicellular organisms involve a **division of labor,** with certain groups of cells becoming highly specialized to perform duties that benefit the entire organism. Groups of cells and their intercellular substances that are similar in structure and function are called **tissues.** The study of tissues is termed **histology.**

Tissues are a fundamental part of the **biological level of organization** (Fig. 7.1), which begins with **atoms**, which comprise **molecules**, which eventually form **cells**. Cells, in turn, form tissues (Fig. 7.2). To perform specific functions, tissues are organized into **organs**, such as the heart and the brain. Organs may contain several representative tissues, and the arrangement of these tissues determines the organ's structure and function. In turn, organs comprise **organ systems**, composed of several organs working together, such as the circulatory system and the nervous system.

The complete **organism** consists of an individual containing several systems working together. Organisms that are members of the same species make up a **population**. Populations of organisms that interact in a defined region comprise a **community**. Communities of organisms and their physical environment constitute an

ecosystem. The Earth's **biosphere** encompasses all of the regions inhabited by living things.

This chapter develops the concept of the biological levels of organization and introduces students to the diversity of tissues found in plants and humans. Histology complements the study of gross anatomy and provides the structural basis for studying organ physiology. Students are advised to spend quality time observing and drawing the representative tissues discussed in this chapter.

Atom

Molecule

Cell

Tissue

Organ

Organ System

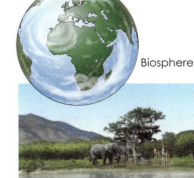
Biosphere

Ecosystem

Population

Community

Organism

Figure 7.1 The biological levels of organization.

Student Activity—Tissues in Plants and Animals

Materials (Procedures 7.1–7.21)
- microscope
- prepared slides
- blank slides and coverslips

Procedure 7.1
Tissues

1. Procure the slides of the tissue indicated in each section of the activity.
2. Carefully observe the specimens under the powers indicated in each section.
3. Sketch the tissues in the space provided in each section.
4. Return the slides.

Histology Study Hints

- Read the description of the tissue, and study the pictures thoroughly.
- Do not become dependent on the color of the tissue.
- View the tissue using the microscope powers suggested by your instructor.
- In viewing, scan different parts of the slide and use different zones of focus.
- Accurately draw and color what you view.
- Spend quality time viewing the specimens. *Don't rush!*

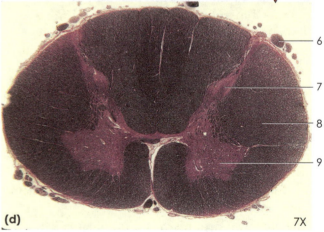

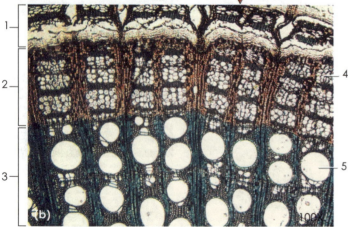

Figure 7.2 Plants and animals are composed of tissues: (a) The stem from a grape, *Vitis* sp. and (b) a transverse section (c) A chimpanzee, *Pan troglodytes*, and (d) a transverse section of the spinal cord.

1. Bark
2. Secondary phloem
3. Secondary xylem
4. Sieve tube elements
5. Vessel

6. Posterior (dorsal) root of spinal nerve
7. Posterior (dorsal) horn (gray matter)
8. Spinal cord tract (white matter)
9. Anterior (ventral) horn (gray matter)

VASCULAR PLANT TISSUES

To colonize the terrestrial environment successfully, plants developed many special adaptations. One of the main adaptations was the development of vascular tissues to carry water and nutrients throughout the body of the plant. The vascular plants of today possess a **vascular tissue system** composed of conducting tissues. **Xylem** is a specialized tissue that carries water, and **phloem** is a specialized conducting tissue that carries nutrients. Examples of seedless vascular plants are horsetails and ferns. More advanced vascular plants consist of conifers such as pine trees and ginkgo trees, and flowering plants such as maple trees and roses. The tissues discussed here are common to the more advanced vascular plants.

Meristematic Tissues

Plants have permanent regions of growth composed of **meristematic tissues**. In these tissues, cells are actively undergoing cell division. The new cells resulting from cell division usually are small and six-sided with a prominent nucleus. As the newly formed cells mature, they begin to take on their characteristic size, shape, and function (Fig. 7.3).

Meristematic tissues that are found at or near the tips of roots and stems comprise the **apical meristem** (Fig. 7.4). Growth of the apical meristem, known as primary growth,

involves increasing the length of the root or stem. The apical meristem gives rise to three distinct regions:

1. the **protoderm**, which gives rise to the epidermis;
2. the **ground meristem**, which gives rise to building block tissue called parenchyma, which usually exists between the epidermis and the vascular tissue, and
3. the **procambium**, which gives rise to vascular tissue such as xylem and phloem.

The **lateral meristem** provides the plant growth in girth or secondary growth. Two derivatives of the lateral meristems are:

1. the **vascular cambium,** or simply **cambium**, that gives rise to tissues that are important in support and protection.
2. the **cork cambium** in woody plants gives rise to cork tissue, which makes up the protective bark. The cork is impregnated with the waxy substance **suberin**, which makes the cells impenetrable to water.

Grasses do not possess a vascular cambium or cork cambium, but they do have apical meristematic tissue called **intercalary meristems** near **nodes** (regions of leaf attachment) at intervals throughout the plant. Intercalary meristems allow grass to grow back quickly after being grazed by a cow or cut by a lawn mower (Fig.7.5).

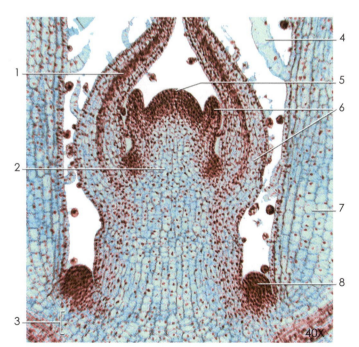

Figure 7.3 A longitudinal section of the stem tip of the houseplant *Coleus* sp. showing the apical meristem.
1. Procambium
2. Ground meristem
3. Leaf gap
4. Trichome
5. Apical meristem
6. Developing leaf primordia
7. Leaf primordium
8. Axillary bud

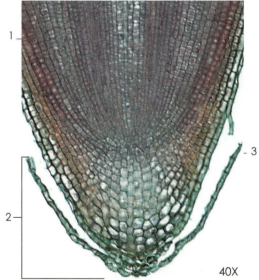

Figure 7.4 A longitudinal section of the root tip of a pear, *Pyrus* sp., showing the apical meristem.
1. Elongation region
2. Root cap
3. Apical meristem

Figure 7.5 Intercalary meristems allow grass to recover quickly from a grazing cow or a lawn mower.

Procedure 7.2
Meristematic Tissues

1. Acquire a slide of a stem tip and a root tip longitudinal section.
2. Record and label your observations in the space provided below.

SIMPLE PLANT TISSUES

Parenchyma tissue is composed primarily of parenchyma cells, the most abundant and diverse type of cell in plants. Parenchyma cells vary in size and shape and tend to have large vacuoles (Fig. 7.6). Parenchyma cells are involved in storage, photosynthesis, support, secretion, repair, and the movement of water and food in plants.

The soft, edible parts of apples and other fruits consist mostly of parenchyma cells. In potatoes, parenchyma cells store starch. Parenchyma cells with numerous **chloroplasts** are sites of **photosynthesis**. These cells, termed **chlorenchyma**, are abundant in leaves and the stems of herbaceous plants.

Parenchyma with extensive air spaces found in water plants are known as **aerenchyma**. This tissue helps to support the plant and, when squeezed, is crunchy. Some mature parenchyma cells can divide when stimulated. When a plant is damaged, parenchyma cells are important in repair. When gardeners make cutting, they take advantage of the growth of parenchyma cells.

Procedure 7.3
Parenchyma Tissues

1. Acquire a microscope slide of parenchyma tissue, as well as an example from a living specimen.
2. Sketch the tissue in the space provided below.

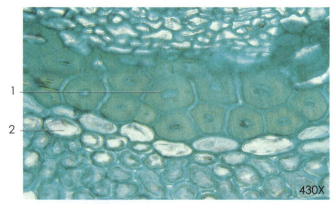

Figure 7.6 A transverse section through the stem of flax, *Linum*. Note the thick-walled fibers as compared to the thin-walled parenchyma cells.
1. Fibers   2. Parenchyma cell

Collenchyma tissue is composed of elongated collenchyma cells. These cells develop thick, flexible walls that support young plants and specific plant structures such as leaves and flower parts. Collenchyma resides beneath the epidermis in stems. In a fresh specimen, collenchyma tissue glistens. This tissue borders the veins of leaves and makes up the "strings" in celery (Fig. 7.7).

Procedure 7.4
Collenchyma Tissues

1. Acquire a microscope slide of collenchyma tissue, as well as a sample of celery.
2. Remove and observe a string from the celery.
3. Sketch the tissue in the space provided below.

Figure 7.7 (a) Celery strings and (b) a micrograph of celery strings showing collenchyma tissue.

Science bestowed immense new powers on man, and, at the same time, created conditions which were largely beyond his comprehension.

—Winston Churchill (1874–1965)

Schlerenchyma tissue is composed of schlerenchyma cells, which are thick and often impregnated with the plant polymer **lignin**. Unlike parenchyma and collenchyma cells, schlerenchyma cells are dead at maturity and primarily provide support. The two types of scherenchyma are:

1. **fibers**, long, slender cells that occur in strands. They are commonly found in roots, stems, leaves, and fruits. Fibers are used in the manufacture of ropes, string, and canvas.
2. **sclereids**, or stone cells (Fig. 7.8), responsible for the gritty texture of pear tissue, in which they may occur singly or in groups throughout the plant; also, a major component of the shell of various nuts and the pit of a peach.

Procedure 7.5
Schlerenchyma Tissues

1. Acquire a microscope slide of schlerenchyma tissue, as well as a section of pear tissue.
2. Sketch the tissue in the space provided below.

Figure 7.8 A section through the stem of a wax plant, *Hoya carnosa*. Thick-walled sclereids (stone cells) are evident.
1. Parenchyma cell containing starch grains   2. Sclereid (stone cell)

Complex Plant Tissues

Complex tissues consist of two or more types of cell. Complex tissues can be divided into:

1. **dermal tissue**, consisting of the epidermis and the periderm, and
2. **vascular tissues** consisting of xylem and phloem.

Dermal Tissues

The **epidermis** constitutes the outermost layer of cells in plant structures, such as roots, stems, leaves, floral parts, fruits, and seeds. The epidermis generally is one cell layer thick and does not undergo photosynthesis. Because epidermal cells are in direct contact with the environment, they vary in form and function. The walls of many epidermal cells are covered with a waxy cuticle, minimizing water loss and protecting the plant against pathogens. The waxy **cuticle** can be easily observed on magnolia leaves. Epidermal cells also can form **root and leaf hairs** that increase the surface area. Numerous small, pore-like structures are found primarily on the underside of leaves. The structures are called **stomata**, and each stoma is bordered by a pair of guard cells. Stomata allow for gas exchange in plants.

In the roots and stems of woody plants, the epidermis is sloughed off and replaced by the **periderm**, which makes up the outer bark composed of box-shaped cork cells. Mature cork cells are dead. A fatty substance, **suberin**, is found in the walls of cork cells, providing protection from mechanical injury, desiccation, and extreme temperatures.

Procedure 7.6
Dermal Tissue

1. Acquire a microscope slide of epidermal tissue, as well as a piece of onion skin (Fig. 7.9 and Fig. 7.10).
2. Sketch the tissues in the space provided below.
3. Observe the underside of a leaf and sketch the stomata and guard cells.

Stomata tissue

Epidermal tissue

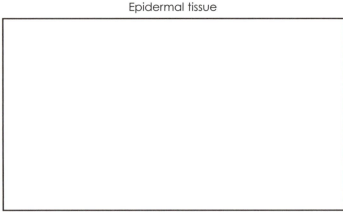

Peridermal tissue

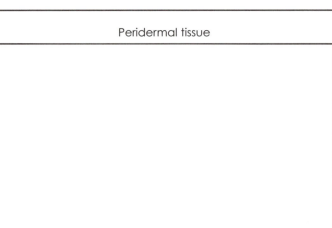

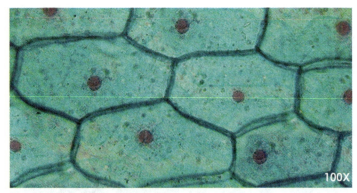

Figure 7.9 Epidermal cells from onion skin.

Figure 7.10 Stoma on the underside of a *Tradescantia* leaf.
1. Stoma

Vascular Tissues

Vascular tissue makes up the **vascular bundle** (Fig. 7.11). In plants, **xylem tissue** is the primary water-conducting tissue and also serves in support, food storage, and the conduction of minerals. Xylem is composed of five basic types of cells:

- **Parenchyma cells** serve in storage.
- **Tracheids** and **vessel elements** are the major conducting cells of the xylem.
- **Ray cells** serve in lateral conduction and storage.
- **Fibers** also can occur in xylem, adding support and storage.

The role of xylem and its structure will be discussed in detail in Chapter 26.

　Phloem tissue conducts dissolved food materials throughout the plant body. The food is composed primarily of sugars produced through photosynthesis (Fig. 7.12). Phloem is composed of

1. parenchyma cells, which provides storage
2. sclerenchyma, which provides support
3. sieve tube members, which provide conduction
4. companion cells, which also provide conduction.

　The sieve tube members and companion cells provide conduction. The role of phloem and its structure will be discussed in detail in Chapter 26.

Procedure 7.7
Complex Tissue

1. Acquire microscope slides of the complex tissues, vascular, xylem, and phloem.
2. Sketch the tissues in the space provided.

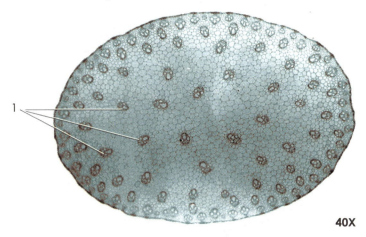

40X

Figure 7.11 A transverse section from the stem of a monocot, *Zea mays* (corn) showing the vascular bundles. The pattern of vascular bundles in a monocot, is known as an atactostele.
1. Vascular bundles

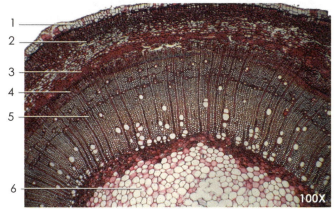

100X

Figure 7.12 A transverse section through one-year-old *Fraxinus* sp. stem showing secondary growth.
1. Periderm   4. Secondary phloem
2. Cortex   5. Secondary xylem
3. Phloem fibers 6. Pith

Vascular tissue

Xylem tissue

Phloem tissue

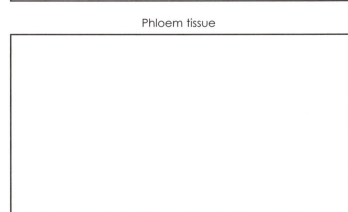

Complete the following table.

Tissue	Description	Location	Function
Meristematic tissue			
Simple Tissue			
Parenchyma			
Collenchyma			
Sclerenchyma			
Complex Tissue			
Epidermis			
Periderm			
Xylem			
Phloem			

ANIMAL TISSUES

Primary Tissues

Tissues are generally classified into one of four primary types: epithelial, connective, muscular, and nervous. These tissues resemble each other only to the extent that they are composed of cells and intercellular substances. There are many different types of primary tissues, each with morphological and functional modifications (Fig. 7.13).

Epithelial Tissue

Epithelial tissue, or epithelium, refers to tissue that covers body surfaces, lines body cavities, and forms glands. Usually this tissue is attached to connective tissue by a basement membrane and has one surface exposed to the environment. Epithelial tissues prevent most objects and substances from the outside environment from entering the body, and they keep most of the internal material within the cell. Epithelial tissue is highly modified for absorption, excretion, and secretion.

Fundamental characteristics of epithelial tissue are the following:

1. The cells that make up epithelial tissue are relatively regular in shape.

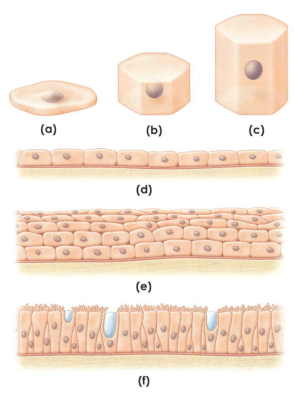

Figure 7.13 Some common arrangements and shapes of tissues: a) squamous, b) cuboidal, c) columnar, d) simple, e) stratified, and f) pseudostratified.

2. Because this tissue is tightly packed, there is little or no intercellular material between adjacent cells.

3. Junctional complexes hold cells together and allow them to function as a unit.

4. Epithelial tissue is arranged in single or multi-layered sheets.

5. Epithelial tissue is anchored to the underlying connective tissue by a basement membrane; this membrane serves as support and as a partial barrier for diffusion and filtration.

6. This tissue reproduces readily through cell division.

7. Epithelial tissue is primarily avascular, as blood vessels are located in underlying vascularized tissue.

8. Some secretory epithelial tissue is modified into glands that secrete their substances through ducts or directly into the blood.

9. Some epithelial tissue is involved in excretion of waste, water, and dissolved substances.

10. Epithelial tissues usually have several types of surface specializations. Although some epithelial tissues have smooth surfaces, most tissues have many complex folds, or **microvilli.** The microvilli on the free surface of epithelial cells is termed **brush border.** The primary function of microvilli is to increase surface

Tissue Terminology

- **Simple:** a single layer.
- **Stratified:** two or more layers.
- **Pseudostratified:** appears to have many layers because the cells vary in height and shape. The nuclei appearing at different heights give an impression of false layers.
- **Squamous:** flat or scale-like cells forming a mosaic pattern.
- **Cuboidal:** cells appear to be cube-like in cross section.
- **Columnar:** cells are long and cylindrical like a column.
- **Transitional:** stratified epithelium, usually with no distinct basement membrane. The surface cells cannot be classified according to shape because it changes as the surface is distended. This tissue can be found in the urinary tract and kidney calyxes. (This tissue is not viewed in this laboratory).
- **Glandular:** epithelial cells are modified to perform excretion. The glands are found throughout the body and include: sweat glands, mammary glands, salivary glands, and endocrine glands such as the thyroid gland. (This tissue is not viewed in this laboratory.)

area. **Cilia** occur in some epithelial cells, such as those lining the trachea.

11. Epithelial tissue is generally classified according to arrangement of the cells, number of layers of cells, and shape of the cells in the superficial layer.

COMMON TISSUES

The common epithelial tissues are discussed below.

Simple Squamous Epithelium

Simple squamous epithelium occurs as a single layer of flattened cells that are tightly held together (Fig. 7.14). The nuclei appear broad and thin and are parallel to the surface. This tissue is thin and highly adapted for osmosis, diffusion, and filtration. Simple squamous epithelium is found in regions of little wear and tear. The two kinds of simple squamous epithelia are:

1. **endothelium**, which lines the heart, blood vessels, and lymph vessels; and

2. **mesothelium**, found in serous membranes such as those lining the thoracic and abdominopelvic cavities.

Figure 7.14 Simple squamous epithelium.
1. Single layer of flattened cells

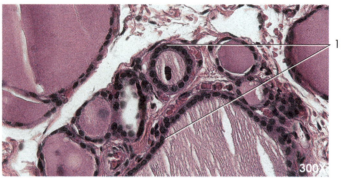

Figure 7.15 Simple cuboidal epithelium.
1. Single layer of cells with round nuclei

Procedure 7.8
Squamous Epithelium

1. View the slide labeled "Simple Squamous
   Epithelium" under both low and high power.
2. Sketch the tissue in the space provided on page 94.

Simple Cuboidal Epithelium
Simple cuboidal epithelium consists of a single layer
of cube-shaped cells. When viewed from above, the cells
appear as polygons. The large nuclei are centrally located
and are rounded in shape (Fig. 7.15). This tissue is found
in the lining of many glands and their ducts, the surface
of the ovaries, the inner surface of the lens of the eye, the
pigmented epithelia of the retina of the eye, and some
kidney tubules. Simple cuboidal epithelium is active in
absorption and secretion.

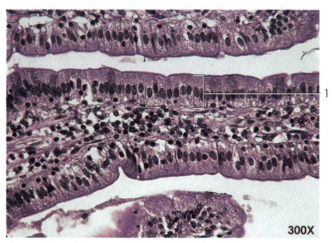

Figure 7.16 Simple columnar epithelium.
1. Single layer of cells with oval nuclei

Procedure 7.9
Cuboidal Epithelium

1. View the slide labeled "Cuboidal Epithelium, or
   "Kidney Section," or "Thyroid Gland" under both
   low and high power.
2. Sketch the tissue in the space provided on page 94.

Simple Columnar Epithelium
Simple columnar epithelium consists of a single layer
of long, column-shaped cells. The nuclei are large and
oval-shaped, usually located near the base of the cell (Fig.
7.16). Simple columnar epithelium serves in protection,
secretion, absorption, and initiating movement. This tissue
can be ciliated or non-ciliated depending on its location
and function. Ciliated tissue can be found in the oviduct,
where it helps to sweep the egg cell toward the uterus after it leaves the ovary. Non-ciliated cells can be found in
the stomach, digestive glands, and gallbladder, where they
protect the delicate linings and function in absorption

and secretion. In the intestines, modified cells called
goblet cells are interspersed in the columnar cells and
secrete mucus that protects and lubricates the walls of the
digestive tract.

Procedure 7.10
Columnar Epithelium

1. View the slide labeled "Columnar Epithelium" or
   "Frog Intestine" under both low and high power.
2. In the slide, note that the columnar epithelium and
   goblet cells can be found in the **villi** (finger-like
   projections) of the small intestine. In addition, note
   the presence of smooth muscle in this slide.
3. Sketch the tissue in the space provided on page 94.

Pseudostratified Ciliated
Columnar Epithelium
Because of the irregularities in cells, the nuclei appear
at several different levels and give the false impression of
multi-layered stratification. When cilia are associated with
the tissue, it is called **pseudostratified ciliated co-**
lumnar epithelium. It consists of a single layer of cells

of varying height and shape, all of which are attached to a basement membrane but not all reaching the surface (Fig. 7.17). These cells can be found in the trachea and in the male reproductive tract. Mucous-producing goblet cells also can be associated with this tissue.

Procedure 7.11
Pseudostratified Ciliated

1. View the slide labeled pseudostratified ciliated columnar epithelium or trachea under both low and high power.
2. In addition, note the presence of adipose tissue and hyaline cartilage in this slide.
3. Sketch the tissue in the space provided on page 94.

Stratified Squamous Epithelium

Stratified squamous epithelium usually consists of several layers of cells, but only the superficial layer consists of squamous cells (Fig. 7.18). The underlying basal cells are modified columnar and cuboidal cells. The basal cells

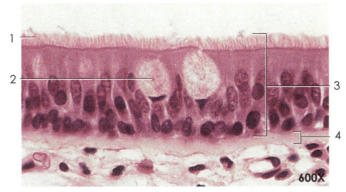

Figure 7.17 Pseudostratified columnar epithelium.
1. Cilia
2. Goblet cell
3. Pseudostratified columnar epithelium
4. Basement membrane

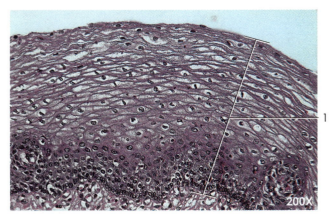

Figure 7.18 Stratified squamous epithelium.
1. Multiple layers of cells, which are flattened at the upper layer

have the ability to replace the superficial squamous cells as they become damaged or worn. Because of its regenerative powers, stratified squamous epithelium appears in areas of drying, wear, injury, and friction. Some stratified squamous epithelium is associated with a protective protein called **keratin**. Keratinized tissue can be found in the epidermis. Non-keratinized tissue can be found in the mouth, vagina, and esophagus.

Procedure 7.12
Stratified Squamous

1. View the slide labeled "Stratified Squamous" under both low and high power.
2. Sketch the tissue in the space provided on page 94.

Connective Tissue

Connective tissue is the most widely dispersed and abundant type of tissue in the body. As a general rule, connective tissues have abundant extracellular fibrous material, which supports the cells of other tissues. Connective tissues perform a variety of functions, but they primarily protect, support, and bind together other tissues. Other functions include: insulation and cushioning, storage of fat, repair of body tissues, and **hematopoiesis** (production of red blood cells). These tissues vary in their morphology and anatomy.

In this laboratory, several representative types of connective tissue will be studied. The tissues explored next are adipose tissue, hyaline cartilage, and bone. Blood cells and blood-forming tissues are included because they have the same embryonic origin (mesenchyme) as the connective tissues.

Characteristics of connective tissue are as follows.

1. Connective tissue is well vascularized, with the exception of cartilage, tendons, and ligaments.
2. The extracellular fibers and ground substances comprise the non-living **matrix**. The arrangement, composition, and function of the substances in the matrix vary with the different kinds of connective tissue. The matrix is responsible for the strength associated with connective tissue. Bone, ligament, tendon, and cartilage have few cells and a large amount of matrix (dense tissue); adipose tissues are composed mostly of cells (loose tissue).
3. Connective tissues are classified according to the arrangement of cells and of the extracellular fibers, as well as the consistency of the ground substance in which the fibers are embedded.
4. The three types of fibers found in connective tissue are:
   a. **collagenous fibers**, white in color, the most common type of connective tissue fiber.

DRAWINGS

Simple squamous epithelium

Simple cuboidal epithelium

Simple columnar epithelium

Pseudostratified ciliated columnar epithelium

Stratified squamous epithelium

Collagenous fibers are composed primarily of the protein collagen and are sturdy and flexible. These fibers support and protect organs, connect muscles to bone (tendon), and bone to bone (ligament).

b. **reticular fibers**, delicately branched networks of inelastic fibrils having the same chemical composition and molecular structure as collagenous fibers but are thinner. Reticular fibers support fat cells, capillaries, nerves, muscle fibers, and secretory liver cells. In addition, they form the reticular framework of the spleen, lymph nodes and bone marrow.

c. **elastic fibers**, yellow in color, appearing singly rather than in bundles, but branch to form networks. Elastic fibers contain the protein elastin. These fibers give organs, such as the skin, the ability to move, stretch, and contract.

5. The ground substance is a homogeneous, extracellular material that ranges from a semifluid to a thick gel in consistency. The ground substance provides a suitable medium for the passage of nutrients and waste products between the cells and the bloodstream.

6. The cells of the various kinds of connective tissues are specialized to help produce the extracellular matrix. The cells' names end in suffixes such as –blasts, –cytes, and –clasts.

a. **Blasts** are responsible for creating matrix.

b. **Cytes** are responsible for maintaining matrix.

c. **Clasts** are responsible for the breakdown and remodeling of matrix.

7. Connective tissue cells are classified as fixed or wandering.

a. **Fixed cells** have a permanent site and are concerned primarily with long-term functions such as storage, maintenance, and synthesis.

b. **Wandering cells** usually are involved in short-term activities such as repair and defense. Examples of wandering cells are: leukocytes, plasma cells, and mast cells.

(1) **Leukocytes** include the five types of white blood cells.

(2) **Plasma cells** produce antibodies to destroy antigens.

(3) **Mast cells** produce histamine, a chemical that dilates small blood vessels during inflammation in addition to the anticoagulant heparin.

Adipose Tissue

Adipose tissue consists of numerous adipocytes, or fat cells, and a small amount of reticular matrix. Adipose tissue may appear singly or in clusters. A single adipocyte appears as a large, clear sphere containing lipids with a large nucleus flattened against the side. This tissue resembles signet rings or fish net. Adipose tissue functions in insulation, cushioning, and the storage of energy. This tissue is found beneath the epidermis and surrounding organs, and throughout the body. (Fig. 7.19)

Procedure 7.13
Adipose

1. To find adipose tissue, view the slide labeled "Pseudostratified Ciliated Columnar Epithelium" (or "Trachea") under both low and high power.

2. In addition, note the presence of pseudostratified ciliated columnar epithelium tissue and hyaline cartilage.

3. Sketch the tissue in the space provided on page 99.

Cartilage/Hyaline Cartilage

Cartilage is a type of connective tissue that provides support and aids in movement. Cartilage is avascular. Oxygen, nutrients, and waste products diffuse through the cartilage matrix. The three types of cartilage are:

1. **hyaline** cartilage (studied here)

2. **fibrocartilage**, somewhat flexible and capable of withstanding pressure; found in the pubic symphysis, intervertebral discs, knee joints, and the temporomandibular joints.

3. elastic **cartilage**, provides rigidity and great flexibility; found in the epiglottis, external ear, and Eustachian tubes.

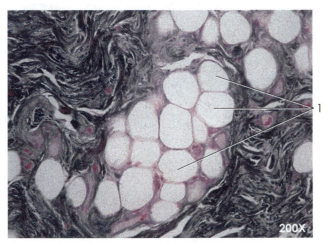

Figure 7.19 Adipose connective tissue.
1. Adipocytes (adipose cells)

Cartilage cells are called **chondrocytes** and are embedded in small cavities within the **matrix**. The small cavities are known as **lacunae**.

Hyaline cartilage is the most common and rigid type of cartilage. The collagenous fibers are scattered in a network that is completely filled with ground substance. It usually is enclosed in a fibrous covering called the **perichondrium**. Hyaline cartilage forms a major part of the skeleton of embryos and reinforces respiratory passageways in the trachea, larynx, and bronchi. At the ends of long bones it is called **articular cartilage**, and at the distal ends of ribs it is called **costal cartilage** (Fig. 7.20).

Procedure 7.14
Hyaline Cartilage

1. View the slide labeled "Hyaline Cartilage," "Pseudostratified Ciliated Columnar Epithelium," or "Trachea" under both low and high power.
2. Sketch the tissue in the space provided on page 99.
3. Label the matrix, lacunae, and chondrocytes.

Bone or Osseous Tissue

Bone tissue, or **osseous tissue**, is a hard, connective tissue that consists of living cells dispersed in an organic and mineralized matrix (Fig. 7.21). The organic portion of the matrix contains collagen fibers and other organic molecules. The mineral part of the matrix contains tricalcium phosphate crystals called hydroxyapatite and calcium carbonate. The human body is about 62% water, but bone tissue contains about 20% water. As a result of the minerals and lack of water, bone tissue is stronger and more durable than other tissues. Bone serves in protection, support, movement, **hematopoiesis** (production of blood cells), mineral homeostasis, storage of energy, and storage of calcium.

The two types of bone are spongy bone and compact bone.

1. **Spongy bone** has spaces between the plates (trabeculae) of bone. The spaces between the trabeculae of some bones are filled with red bone marrow. Spongy bone tissue in the sternum, vertebrae, ribs, hipbones, and near the ends of long bones is involved in hematopoiesis. Spongy bone makes up most of the bone tissue in short, flat, and irregular bones. It also is found at the ends of long bones.

2. **Compact bone** contains few spaces and also is known as dense bone. It comprises the external layer of all bones and makes up the bulk of the shaft of long bones. Compact bone tissue provides support and protection and helps the long bones resist stress.

Compact bone contains cylinders of calcified bone known as **osteons** or **Haversian systems**. These cylinders consist of four to 20 concentric rings of bone called **lamellae**. The lamellae contain numerous **lacunae**, each housing a bone cell or **osteocyte**. Radiating out from each lacunae are the **canaliculi**, which channel nutrients and wastes by diffusion into and out of the blood vessels in the **Haversian canal** (central canal), the most prominent portion of the osteon. These longitudinal channels contain nerves and blood vessels.

Connected to and running at a right angle to the Haversian canals are **Volkman's canals** (perforating canals). The Volkman's canals extend the nerves and vessels outward to the **periosteum** (outer covering) and **endosteum** (inner lining) of the bone marrow cavity. The osteon complex provides the strength necessary to resist everyday stress.

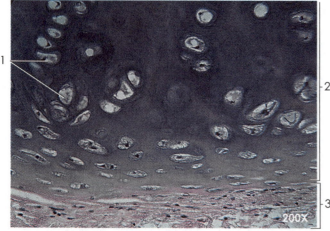

Figure 7.20 Hyaline cartilage.
1. Chondrocytes within lacunae   3. Perichondrium
2. Hyaline cartilage

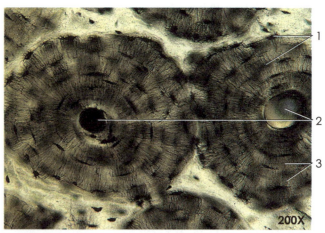

Figure 7.21 Cross section of two osteons in bone tissue.
1. Lacunae   3. Lamellae
2. Central (haversian)
   canals

Procedure 7.15
Bone

1. View the slide labeled "Ground Bone" under both low and high power.
2. Sketch the tissue in the space provided on page 99.
3. Label the matrix, Haversian canal, lamellae, lacunae, and canaliculi.

The Blood

Blood is classified as a specialized kind of fluid connective tissue. Blood contains a fluid matrix called **plasma** and **formed elements** called **erythrocytes**, **leukocytes**, and **platelets**. The functions of blood include: transportation of respiratory gases, nutrients, enzymes, hormones, and waste products; regulation of acid-base balance; regulation of body temperature; regulation of electrolytes; and defense against toxins and harmful microorganisms (Fig. 7.22).

Blood has the following characteristics:

1. An average man has 5 to 6 liters of blood, and an average woman has 4 to 5 liters of blood. Blood makes up about 7%–9% of the total body weight.
2. The straw-colored liquid portion of the blood, called **plasma**, makes up about 55% of the total blood volume. Plasma is 92% water and 8% dissolved or suspended molecules such as plasma proteins, gases, cellular waste products, hormones, and ions.
3. **Red blood cells (RBCs)** are termed **erythrocytes**. These cells make up 99% of the formed elements of blood; the human body has about 25 trillion erythrocytes. Typically, six RBCs placed in a row will line up across a period (like the one following this sentence). An RBC contains approximately 280 million molecules of the oxygen-carrying red pigment **hemoglobin**. Mature red blood cells in humans are **anucleated** and appear as a biconcave disc, which has more surface area for diffusion and is flexible so it can more readily pass through blood vessels. The surface antigens of RBCs are responsible for the various blood groups such as the ABO group and Rh group.
4. **White blood cells (WBCs)** are termed **leukocytes**. Adults have about 1,000 erythrocytes for every leukocyte. The five basic kinds of leukocytes are neutrophils, eosinophils, basophils, lymphocytes, and monocytes. WBCs are classified into two major groups based on their nuclei:
   a. **granular leukocytes (granulocytes)**: have conspicuous granules in the cytoplasm and a lobed nucleus. The granulocytes include **neutrophils**, **eosinophils**, and **basophils**.
   b. **agranular leukocytes (agranulocytes)**: no cytoplasmic granules can been seen under the light microscope. The two kinds of agranulocytes are the **lymphocytes** and **monocytes**.
5. **Platelets** are disc-shaped cell fragments of large multinucleate cells (megakaryocytes) formed in the bone marrow. When a blood vessel is injured, platelets move to the site and begin to clump together, attaching themselves to the damaged area. If the injury is slight, the platelets form a platelet plug to stop bleeding. If the damage is more intense, platelets begin the process of clotting.

Think About It!

Tough meat contains a high percentage of collagenous fibers. To make the meat tender, it usually is simmered in a soup or stew. The simmering turns the collagen into a gelatinous material, which is softer. In contrast, leather is tanned or toughened by adding tannic acid to the material. The tannic acid converts the collagen into a firm, insoluble material.

Now it's your turn!

1. When you eat a steak, what is the gristle?

2. What causes wrinkles?

3. Why should sun worshipers beware?

4. What is Wharton's jelly?

5. What is the difference between a carcinoma and a sarcoma?

6. Relate connective tissues to scars.

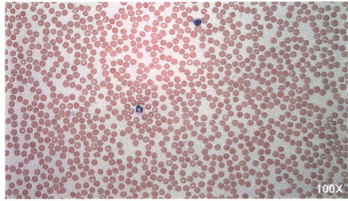

100X

Figure 7.22 Human blood.

Procedure 7.16
Blood

1. View the slide labeled "Human Blood" under high power. Note that the mature human RBCs are anucleated.
2. Identify erythrocytes, leukocytes, and platelets.
3. Your instructor may want you to identify the different kinds of leukocytes.
4. Sketch the slide in the space provided on page 99.

The Leukocytes (optional section)

Granulocytes

1. **Neutrophils:** The neutrophils, the most common leukocyte in the blood, comprise 60%–70% of the total white blood count. Neutrophils have nuclei with two to six lobes (polymorphonucleated). The granules appear pale and lilac-colored. They destroy microorganisms and foreign particles and are a major component of pus. High neutrophil counts occur in bacterial infections, burns, and inflammation. Low numbers may occur in systemic lupus erythematosus, and vitamin B_{12} deficiency.

2. **Eosinophils:** The eosinophils comprise approximately 2%–4% of the total WBC count. The nucleus of an eosinophil is usually bilobed. The granules in eosinophils range from red to red-orange in acid stain. Common in allergic reactions and parasitic infections, eosinophils can phagocytize antigen-antibody complexes formed during the allergic reaction. Low numbers may occur in Cushing's syndrome.

3. **Basophils:** The basophils make up less than 1% of the total white blood count. The nucleus is bilobed or irregular in shape, often in the form of an S. The granules are large, stain blue-black to red-purple, and often obscure the nucleus. These cells contain the anticoagulant heparin. In addition, they liberate histamine and serotonin to intensify the inflammatory response. High numbers may occur in some allergic responses and cancers. Low numbers may occur during ovulation and some pregnancies.

Agranulocytes

4. **Lymphocytes:** Lymphocytes, the smallest of the leukocytes, comprise approximately 20%–30% of the total white blood count and encompass a number of different types. Lymphocytes have a round, dark-stained nucleus, and the cytoplasm appears as a thin ring around the nucleus. These cells are abundant in lymphoid tissue and play a major role in immunity

and antibody production. High counts can occur in viral infections, immune diseases, and some leukemias. Low numbers appear in severe illness and immunosuppression.

5. **Monocytes:** The largest of the leukocytes, monocytes comprise 2%–8% of the total white count. They have a dark-stained, large, kidney-shaped nucleus, and the cytoplasm appears bluish and foamy. Monocytes generally remain in circulation for 3 days, leave the circulation, and become migrating macrophages. They phagocytize bacteria, dead cells, cell fragments, and other debris. An increase in monocytes is associated with chronic infections (viral infections, fungal infections, infectious mononucleosis, some leukemias, and tuberculosis).

Procedure 7.17
Amphibian Blood

1. View the slide labeled "Amphibian Blood," "Frog Blood," or "*Amphihuma* Blood" under both low and high power (Fig. 7.23). Note that amphibian RBCs are nucleated. In addition, the *Amphihuma* has the largest erythrocytes in the animal kingdom.
2. Identify erythrocytes, leukocytes, and platelets.
3. Your instructor may want you to identify the different types of leukocytes.
4. Sketch the slide in the space provided on page 99.

Muscle Tissue

Muscle tissue is specialized to generate force, perform work, generate heat, maintain posture, and provide movement. The three major types of muscle tissue are:

1. smooth muscle,
2. skeletal muscle, and
3. cardiac muscle.

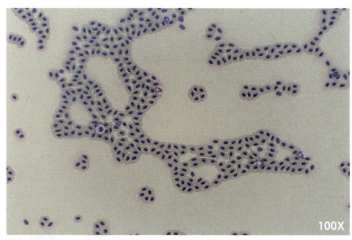

Figure 7.23 *Amphihuma* blood.

DRAWINGS

Adipose tissue

Cartilage

Bone tissue

Blood

Amphihuma blood

Leukocyte (optional)

These three types of muscle differ from each other in their microscopic anatomy, location, and control by the nervous and endocrine systems.

Muscle tissue has the following characteristics:

1. Muscle tissue exhibits **contractility**. As muscle tissue contracts, it generates force to do work.
2. Muscle tissue exhibits **excitability**, the ability to receive and respond to stimuli. Muscle tissue responds to neurotransmitters released by neurons or hormones distributed by the blood.
3. Muscle tissue exhibits **extensibility**, the ability to be stretched.
4. Muscle tissue exhibits **elasticity**, the ability to return to its original shape after constriction or extension.

Smooth Muscle

Smooth muscle is considered to be involuntary muscle because it is controlled by the autonomic (involuntary) division of the nervous system. Under the microscope, the muscle appears to lack the striations characteristic of skeletal and cardiac muscle. Instead, this muscle exists in fibers that bulge in the center and are tapered at both ends (Fig. 7.24). A single oval nucleus is located within each fiber of this muscle. Smooth muscle is not connected to bone. Smooth muscle is distributed throughout the body and is more variable in function than other types of muscle. The two types of involuntary muscle are:

1. **visceral** (single-unit) smooth muscle, the most common type, found in wraparound sheets in the walls of small blood vessels and the hollow viscera such as the urinary bladder, uterus, and small intestines.
2. **multiunit** smooth muscle, found in the walls of the larger blood vessels, the bronchioles of the lungs, and in the capsules of the spleen; in addition, it can be found in arrector pili that attach to hair follicles and in the muscles of the iris that adjust the diameter of the eye's pupil.

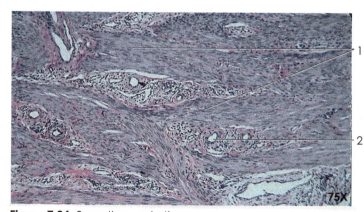

Figure 7.24 Smooth muscle tissue.
1. Smooth muscle   2. Blood vessel

Did You Know?

The term *muscle* is derived from the Latin word *musculus*, which literally means "little mouse." Early observers thought the visible movements of muscle under the skin looked like little mice running around!

The contraction and relaxation periods of smooth muscle are slower than in striated and cardiac muscle. Contractions can last longer than 30 seconds without the muscle tiring. In addition, the action of smooth muscle contractions is rhythmical. These characteristics make possible processes such as peristalsis (propelling food through portions of the digestive tract).

Procedure 7.18
Smooth Muscle

1. View the slide labeled "Smooth Muscle" or "Small Intestines" (c.s.) under both low and high power.
2. Sketch the tissue in the space provided on page 102.

Skeletal Muscle

Skeletal muscle, or **striated muscle**, is commonly called voluntary muscle because most of these muscles are associated with movement. Skeletal muscle makes locomotion possible, generates heat, and guards the entrances and exits of the respiratory, digestive, and urinary tracts. Skeletal muscle is described as striated because of the alternating light and dark bands of the proteins **actin** and **myosin** (Fig. 7.25).

Striated muscle is composed of long, cylindrical multinucleate cells called muscle fibers. A thin membrane, the **sarcolemma**, encloses each muscle fiber. The fiber consists of protoplasm called **sarcoplasm**, several nuclei, numerous mitochondria, and thread-like fibers, or **myofibrils**. Skeletal muscle is capable of hard work for short times.

Procedure 7.19
Skeletal Muscle

1. View the slide labeled "Skeletal Muscle," "Striated Muscle," or "Voluntary Muscle" under both low and high power.
2. In the slide, note the striations and prominent nuclei.
3. Sketch the tissue in the space provided on page 102.

Cardiac Muscle

As the name indicates, cardiac muscle is found exclusively in the heart. Cardiac muscle pumps blood throughout the body. It contains the same type of myofibril and protein composition as skeletal muscle. Although this muscle is closely packed, each cell is separate and has its own nucleus. This hard-working muscle contains many mitochondria.

Cardiac muscle is characterized by having **intercalated discs,** which appear as dark bands running perpendicular to the cardiac muscle. The functions of intercalated discs are to help impulses pass quickly from one cell to the next, strengthen the junction between cells, and separate the cells within a muscle fiber.

Procedure 7.20 Cardiac Muscle

1. View the slide labeled cardiac muscle under both low and high power.
2. On the slide, note the striations, prominent nuclei, and intercalated discs.
3. Sketch the tissue in the space provided on page 102.

Nervous Tissue

The nervous system is responsible for integrating and coordinating the other systems of the body, sensing stimuli, and continuously monitoring the external and internal environment. The nervous system is made up of more than 100 billion nerve cells called **neurons** and other cells called neuroglia ("nerve-glue") that serve as supportive cells.

Characteristics of nervous tissue are the following:

1. Neurons are one of the most specialized types of cells. These cells display great diversity in size and shape. The cell body of some neurons range in diameter from 5 microns (smaller than a RBC) to 135 microns (large enough to be seen by the unaided eye). Some

neurons extend more than 3 feet. Neurons vary in shape from star-shaped to oval.

2. A typical neuron has three parts: a cell body, dendrites, and axons (Fig. 7.27).

   a. The cell body, or **soma**, consists of varying amounts of cytoplasm with a prominent nucleus and nucleolus. A variety of **organelles** reside within the cytoplasm. **Neurofibrils** are numerous in the cell body and provide internal support.

   b. **Dendrites** are neuron processes that conduct impulses toward the nerve cell body. The dendrites usually are unmyelinated, tapered, and branched. These structures are generally extensions of the cell body and share organelles with the cell body. Some neurons have more than 200 dendrites.

   c. **Axons** are myelinated (myelin is a lipid sheath used for insulation), long, thin, cylindrical neuron processes that carry impulses away from the cell body. Generally, neurons possess one axon that may branch into collaterals.

3. **Neurons** also can be classified according to their functions. Neurons carrying impulses from the sensory receptors in the internal organs and skin are called **sensory** or **afferent neurons**. The nerve cell bodies of sensory neurons are always located outside of the central nervous system. Neurons that carry activating signals from the central nervous system to the body muscles and glands are called **motor** or **efferent neurons**. The cell bodies of motor neurons are located within the central nervous system.

4. Neurons may be classified by the number of processes attached to the nerve cell body. There are three basic types of neurons.

Figure 7.25 Longitudinal section of skeletal muscle tissue.
1. Skeletal muscle cells, note striations
2. Multiple nuclei in periphery of cell

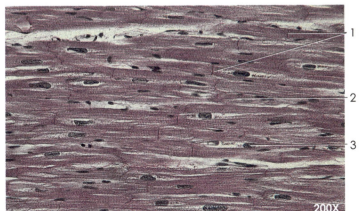

Figure 7.26 Cardiac muscle tissue.
1. Intercalated discs
2. Light-staining perinuclear sarcoplasm
3. Nucleus in center of cell

DRAWINGS

Skeletal muscle

Cardiac muscle

Smooth muscle

Nervous tissue

a. **Unipolar neurons** are sensory neurons in which the dendrites and axons are continuous and the cell body is off to one side. Some of these neurons exceed a meter in length.

b. **Bipolar neurons** have two distinct processes—one axon and one dendrite attached to the nerve cell body. These neurons are relatively rare and can be found in the sense organs.

c. **Multipolar neurons** possess several dendrites and only one axon. This type of neuron can be found in the brain and spinal cord.

Procedure 7.21
Neuron

1. View the slide labeled "Multipolar Neurons," or "Ox Spinal Smear" under both low and high power.
2. Sketch the tissue in the space provided above.

Just Think!

Considering that an average heart beats about 100,000 times a day, how many times does an average heart beat in a lifetime of 70 years?

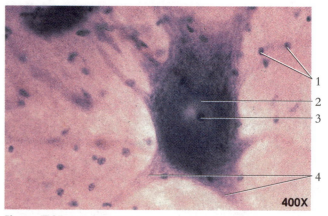

400X

Figure 7.27 Neuron smear.
1. Nuclei of surrounding neuroglial cells
2. Nucleus of neuron
3. Nucleolus of neuron
4. Dendrites of neuron

Complete the following tables.

Tissue	Description	Location	Function
Epithelial Tissue			
A. Simple squamous			
B. Cuboidal			
C. Columnar			
D. Pseudostratified			
E. Stratified			
Connective Tissue			
A. Adipose			
B. Hyaline cartilage			
C. Bone			
Blood			
Connective/Vascular Tissue			
A. Erythrocytes			
B. Leukocytes			
Neutrophils			
Eosinophils			
Basophils			
Lymphocytes			
Monocytes			

Tissue	Description	Location	Function
Connective/Vascular Tissue (continued)			
C. Platelets			
Muscle Tissue			
A. Smooth			
B. Skeletal/Striated			
C. Cardiac			
Nervous Tissue			
A. Unipolar			
B. Bipolar			
C. Multipolar			

NOTES

Name: _____ Date: _____ Section: _____

Review Questions

1. Sketch the following tissues in the boxes below: parenchyma cells, collenchyma cells, sclereids, epidermal cells, stomata and guard cells, periderm, vascular bundle with xylem, and vascular bundle with phloem.

Parenchyma cells

Collenchyma cells

Sclereids

Epidermal cells

Stomata cells

Guard cells

Periderm

Vascular bundle with Xylem

Vascular bundle with Phloem

2. Compare and contrast xylem and phloem.

_____

_____

_____

_____

3. What is the major function of the periderm?

_____

_____

_____

4. What are intercalary meristems?

_____

_____

_____

5. What is the function of a tracheid?

_____

_____

_____

6. What is the function of the cuticle?

_____

_____

_____

7. What is the function of stoma?

_____

_____

_____

Name: _____ Date: _____ Section: _____

8. Why was the evolution of vascular tissue important to the conquering of the land by plants?

_____

_____

_____

9. Label the following (cross-section of the corn stem):

1. _____   4. _____

2. _____   5. _____

3. _____   6. _____

10. Label the following (underside of a leaf):

1. _____

11. What is the function of a goblet cell?

_____

_____

_____

12. What is the function of cilia in pseudostratified ciliated columnar epithelium?

_____

_____

_____

13. Compare and contrast human blood and amphibian blood.

_____

_____

_____

14. Compare and contrast smooth, skeletal, and cardiac muscle.

_____

_____

_____

_____

15. Describe the location and function of intercalated discs.

_____

_____

_____

16. Why does simple squamous tissue appear in thin layers?

_____

_____

_____

17. What are the components of the striations in striated and cardiac muscle?

_____

_____

_____

Name: _____ Date: _____ Section: _____

18. In the boxes provided below, sketch the following:

Bone

Cuboidal Epithelium

Cardiac Muscle

Motor Neuron

Simple Squamous

Human Blood

Hyaline Cartilage

Columnar Epithelium

NOTES

Chapter 8
Enzymes:
Understanding How Enzymes Work*

Overview

Within an organism, the sum total of chemical processes is called **metabolism**. Some processes break down substances and are called **catabolic** (degradation). Other processes build new substances and are called **anabolic** (synthesis). Most of the chemical reactions within living systems are controlled by specialized proteins called enzymes. An **enzyme** is a biological catalyst that accelerates a chemical reaction without itself being affected by the reaction.

In recent years, non-enzyme substances such as RNA catalysts, or **ribozymes**, have been described as speeding up the rate of certain reactions as well. Within an organism, many metabolic pathways are involved in breaking down and forming products. Enzymes ensure that these pathways do not slow down and become congested by lowering the activation energy required for a reaction to take place.

The **activation energy** is the original input of energy necessary to initiate a reaction. In Figure 8.1, the activation energy can be described as the amount of energy necessary to push the reactants over a barrier so the reaction can begin. Two physical means of attaining the activation energy are heat and agitation, which increase the number of collisions between reactant and speed up the reaction. These means, however, may damage living systems. Thus, in living systems, enzymes serve as catalysts to lower the activation energy. Notice in Figure 8.1 that less energy is needed to overcome the barrier in the presence of an enzyme.

In an enzymatic reaction, the reactant that the enzyme acts upon is the **substrate**. Enzymes are substrate-specific. The specificity of an enzyme results from the enzyme's unique three-dimensional molecular shape. While the enzyme and the substrate form to make an **enzyme-substrate complex**, catalytic actions of the enzyme converts the complex into one or more products. The enzyme then exits the reaction unaltered and ready to proceed to the next reaction.

Portions of the enzyme called **active sites** bind to the substrate. Usually, weak ionic or hydrogen bonds link the substrate and the active site. In some cases, an enzyme reacts with a specific substrate thousands of times per second. The **induced fit model** explains how an active site on an enzyme changes its shape slightly to accommodate the substrate. In addition, many enzyme reactions are reversible; the same enzyme can catalyze a reaction in either the forward or the reverse direction.

More than a thousand different enzymes have been cataloged. Enzymes may be specific to a certain type of cell and a certain species of organism. Enzyme names usually end in *ase*, such as sucrase and amylase. An example of a metabolically important enzyme found in vertebrate red blood cells is carbonic anhydrase. Without the presence of carbonic anhydrase, the reversible chemical reaction carbon dioxide plus water yields carbonic acid ($CO_2 + H_2O \rightarrow H_2CO_3$) occurs slowly. In a red blood cell, it is estimated

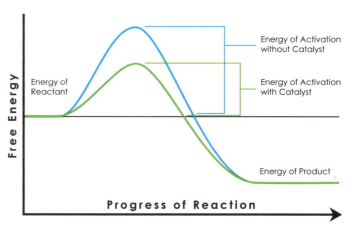

Figure 8.1 Less activation energy is needed in the presence of a catalyst.

* Thank you to Sarah Jean Rayner for her contributions to chapters 8 through 19.

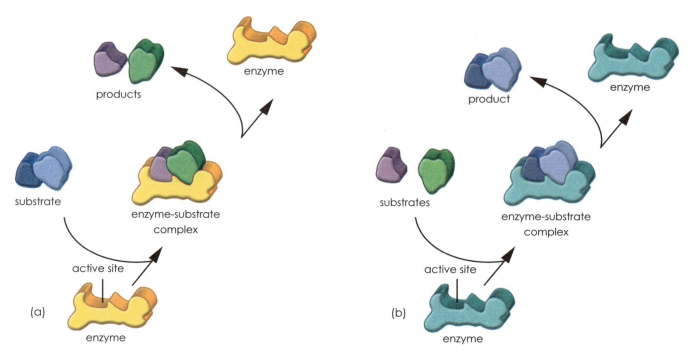

Figure 8.2 Enzymatic action: (a) catabolism or degradation; (b) anabolism or synthesis.

that as few as 200 carbonic acid molecules form per hour without the aid of carbonic anhydrase. In the presence of carbonic anhydrase, nearly 600,000 molecules of carbonic acid are generated per second. Without enzymes, the chemistry of life would be too slow.

Enzyme action often is assisted by other molecules (Fig. 8.2). **Cofactors**, which usually are non-organic metal ions such as copper, zinc, and manganese, aid the action of an enzyme. **Coenzymes** are nonprotein organic molecules that aid enzymatic action. An important coenzyme involved in energy relationships is nicotinamide adenine dinucleotide (NAD). Also, vitamins can serve as coenzymes.

Enzymatic action can be affected by a number of external factors (e.g., concentration of substrate, concentration of enzymes, temperature, pH, salinity). An enzyme functions best within certain parameters called **optimal conditions**. Any factor that may alter the unique shape of an enzyme may affect its ability to serve as a catalyst. If the shape of an enzyme is altered and loses its function, the enzyme is said to be **denatured**. For example, when egg white solidifies when it is cooked, it has been denatured. **Inhibitors** bind to an enzyme and decrease its activity. Usually, the end product of a given reaction inhibits the action of an enzyme. Conversely, **activators** increase enzyme activity.

A number of variables, including temperature and pH, can affect the activity of enzymes (Fig. 8.3). All enzymes have an optimal temperature at which they react more rapidly. Most human enzymes are more efficient near 37° Celsius. Increasing the temperature of an enzyme-driven reaction speeds the reaction to a certain point. If the temperature is below or above the optimum temperature, the reaction of the enzyme and the substrate is impeded. The optimum pH for most enzymes is between 6 and 8. Exceptions include the digestive enzyme pepsin, which digests proteins in the stomach at pH 2.

BROMELIAN ACTIVITIES

The following laboratory activities have been designed to investigate an enzyme that occurs in pineapples (*Ananas cosmosus*) called **bromelian**. Pineapple plants, along with orchids, are classified as bromeliads (Fig. 8.4). Although native to South America, pineapples are grown in many tropical areas including Hawaii, Brazil, and Thailand. The enzyme bromelian can be found in the leaves, stems, and fruit of pineapples. Bromelian is a **proteolytic enzyme (protease),** which breaks down proteins into their amino acids by **hydrolysis**.

The directions of many gelatins including Jello™ recommend against the user placing certain fruits, including fresh pineapple, in Jello. Gelatin is a protein obtained from collagen (a structural protein component of fibrous connective tissues such as animal hooves). The proteins essentially trap and absorb water, allowing the gelatin to set. Bromelian and other proteolytic enzymes degrade the gelatin proteins and prevent the gelatin from setting. A number of variables, including temperature and pH, can affect how bromelian reacts with gelatin.

Figure 8.3 Enzymes determine the color pattern in Siamese cats. A heat-sensitive enzyme that helps to control melanin production is less active in the warmer (lighter colored) regions of the body. The cooler extremities are darker in color.

Figure 8.4 The stems, leaves, and fruit of pineapples contain the enzyme bromelian.

Student Activity—Bromelian

Procedure 8.1
Bromelian and Jello

The Effect of Bromelian on the Formation of Jello™

This simple activity illustrates the effect of bromelian on the formation of gelatin (Jello™). Based upon your knowledge of enzymes, you will construct a simple hypothesis for this activity and identify the control and the independent and dependent variables.

1. Obtain a test tube rack containing 12 test tubes. In this activity, three test tubes will be used.
2. To each of the three test tubes, add 3 ml of warm gelatin.
3. Label the test tubes "water," "fresh," and "canned" accordingly.
4. With separate syringes, obtain 2 ml of water, 2 ml of fresh pineapple juice, and 2 ml of canned pineapple juice.
5. Rapidly push the water, fresh pineapple juice, and canned pineapple juice into the appropriate test tube to mix the gelatin with the solution.
6. Place the test tubes on ice. Carefully watch the test tube containing the water; it should solidify within 5 to 10 minutes. After it solidifies, observe the test tubes containing fresh and canned pineapple juice.

7. Record your observations in the following chart.

TEST TUBE	RESULT
Water	
Fresh pineapple juice	
Canned pineapple juice	

8. Discuss the results of this activity.

_____

_____

_____

_____

Materials (Procedures 8.1–8.3)

- test tube rack
- test tube clamp
- 12 test tubes
- pipettes
- syringes
- ice container
- hot plate
- water
- canned pineapple juice
- fresh pineapple juice
- ice
- HCL
- NaOH
- eye protection
- gelatin

Procedure 8.2
Bromelian and Temperature

The Effect of Temperature on the Rate of Bromelian Activity

This activity illustrates the effect of temperature on the rate of bromelian activity. *Eye protection is required.* Based upon your knowledge of enzymes and temperature, construct a simple hypothesis for this activity and identify the control, independent, and dependent variables.

1. Procure three test tubes. Label the test tubes: water cold, and hot.
2. Add 2 ml of water to the test tube labeled water and place it in the test tube rack.
3. Add 2 ml of fresh, not canned, pineapple juice to the test tubes labeled cold and hot.
4. Leave the test tubes labeled water and cold at room temperature. Place the test tube labeled hot into a 70°C water bath for 5 minutes.
5. After 5 minutes, carefully remove the test tube labeled hot from the water bath. Add 3 ml of warm gelatin to each test tube.
6. Place all three test tubes in ice until the tube containing the water and gelatin begins to solidify.

Every living thing is a sort of imperialist, seeking to transform as much as possible of its environment into itself and its seed.

—**Bertrand Russell (1872–1970)**

7. Observe the test tubes labeled cold and hot and record your results in the following chart.

TEST TUBE	RESULT
Water	
Cold water	
Hot water	

8. Discuss the results of this activity.

_____

_____

_____

Procedure 8.3
Bromelian and pH

The Effect of pH on the Rate of Bromelian Activity

This simple activity illustrates the effect of pH on the rate of bromelian activity. *Eye protection is required.* Based upon your knowledge of enzymes and pH, construct a simple hypothesis for this activity. In the activity, identify the control, independent, and dependent variables.

1. Procure six test tubes. Label the test tubes 1–6.
2. Carefully add the following to the test tubes with a pipette:

Test tube number 1:	2 ml of water
Test tube number 2:	2 ml of fresh pineapple juice
Test tube number 3:	1 ml HCl and 1 ml fresh pineapple juice
Test tube number 4:	1 ml HCL and 1 ml water
Test tube number 5:	1 ml NaOH and 1 ml fresh pineapple juice
Test tube number 6:	1 ml NaOH and 1 ml water

3. Mix the components of the tube by gently and carefully swirling the tubes and letting them sit for 3 minutes.

4. Add 3 ml of gelatin to each test tube.
5. Place all six test tubes in ice until the tube containing the water and gelatin begins to solidify.
6. Observe the test tubes and record your results in the chart below.

TEST TUBE	RESULT
2 ml of water	
2 ml of fresh pineapple juice	
1 ml HCl and 1 ml fresh pineapple juice	
1 ml HCl and 1 ml water	

TEST TUBE	RESULT
1 ml NaOH and 1 ml fresh pineapple juice	
1 ml NaOH and 1 ml water	

7. Discuss the results of this activity.

_____

_____

_____

_____

8. At the completion of the activity, return the supplies to the main table and dispose of the material as outlined by the laboratory instructor.

 ## Student Activity—Bromelian and Film

Getting the Picture! Activities with Bromelian and Film

In the age of digital cameras, we often forget about film (especially black-and-white film). Black-and-white photographic film is composed of a layer of thin plastic coated with gelatin impregnated with light-sensitive silver salts known as the emulsion. The gelatin holds the silver salts in place on the plastic (Fig. 8.5). If the gelatin layer is destroyed by bromelian, the silver will be released, leaving a clear plastic sheet.

Figure 8.5 Black-and-white photographic films consist of a plastic layer and the emulsion composed of gelatin and silver salts.

The following experiment illustrates the effect of bromelian upon the gelatin layer in exposed black-and-white photographic film. The time required for the film to become clear indicates the rate at which bromelian breaks down the photographic gelatin. _This experiment uses silver salts, heat, acids, and bases, so eye protection and gloves are required. In addition, follow the laboratory instructor's directions for proper disposal of all chemicals._

Materials
• test tube rack
• test tube clamp
• 6 test tubes
• pipettes
• gloves
• eye protection
• fresh pineapple juice
• hot plate
• water baths
• developed black and white strips of film
• tape
• wax pencil

Procedure 8.4
Bromelian and Temperature

The Effect of Temperature on the Rate of Bromelian

In this experiment, strips of developed black and white film will be placed in pineapple juice (bromelian) at different temperatures. The amount of time required for the film to become clear will be an indicator of the rate at which the enzyme bromelian breaks down the gelatin on the film. Based upon your knowledge of enzymes and pH, construct a simple hypothesis for this activity.

1. Procure 6 test tubes and place them in a test tube rack. With a wax pencil, label each test tube as: 70°C, 60°C, 50°C, 40°C, 30°C, and 20°C.
2. Procure six strips of exposed black and white film from the instructor. Place a piece of tape on one end for handling. Handle the film by the edges and by the tape to ensure that finger grease does not get on the film.
3. Fill each of the test tubes with 7 ml of fresh pineapple juice.
4. Be sure that the water baths are available at each of the designated test temperatures. The water baths must maintain the proper temperature throughout the activity.
5. Place the strip of film into the test tube, with the taped portion at the top of the test tube.
6. Place the labeled test tube and film into the designated water bath. Accurately record the time, or begin a stopwatch.
7. Every 5 minutes, vigorously agitate the test tube for 5 seconds. Examine the film for clearing.
8. Record how long the film takes to clear at each temperature. If the film does not clear after 60 minutes, record "no reaction."
9. Construct a table relating the amount of time it takes the film to clear.

10. Complete the line graph using temperature as the x-axis and the clearing time as the y-axis.

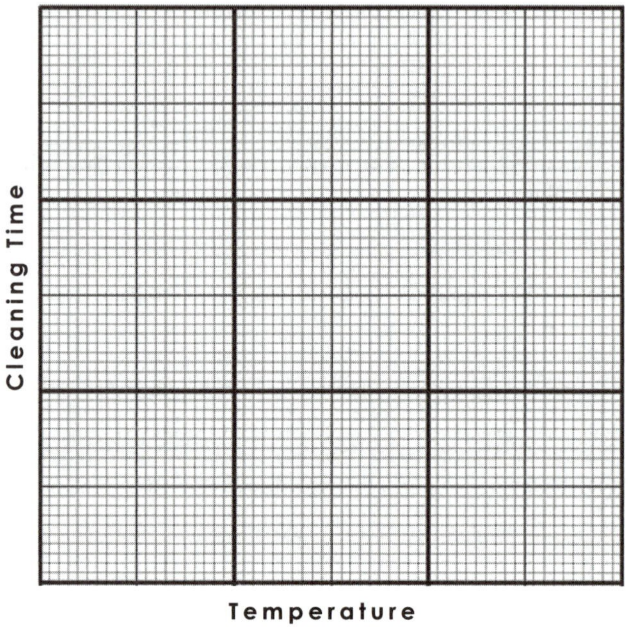

Temperature

(y-axis label: Cleaning Time)

11. Dispose of the pineapple juice as indicated by the instructor, thoroughly clean the test tubes, dispose of the film in the designated container, and return the test tube rack.

Check Your Understanding

Q. Express your conclusions regarding the effect of temperature upon the action of enzymes.

_____

_____

_____

Q. Did you accept or reject your hypothesis?

_____

Q. Based upon your data, what was the optimum temperature for bromelian to break down the gelatin in the film?

_____

_____

ENZYMES AND CHEESE

Various kinds of cheese, as well as many other dairy products, are produced through the breakdown of proteins in milk (Fig. 8.6). The enzyme rennin converts milk protein (casein) to insoluble paracasein. The paracasein, in turn, precipitates out of solution, forming curd (the cheese) and whey (a watery residue). In the first part of this activity, milk will be converted into curd and whey.

Cheese is composed of protein, which in turn is made up of amino acids. Cheese can be broken down into its component amino acids by the enzyme bacterial protease. Using the indicator ninhydrin, this activity illustrates how bacterial protease breaks down the curd (cheese) produced in the first activity into its component amino acids.

Figure 8.6 Cheese is produced by the breakdown of proteins in milk.

 Student Activity—Enzymes and Cheese

Part 1: Making Cheese from Whole Milk

This activity is designed to make curd (cheese) and whey from whole milk. The curd will be used in Part 2 of the activity.

Materials
- whole milk
- hot plate
- two 400 ml beakers
- rennilase
- cheesecloth
- 1 N HCL
- pipette
- glass rod

Did You Know?

The stomach glands of newborn infants produce rennin, also known as chymosin. Rennin is not produced in adults.

 ### Procedure 8.5
Making Cheese

1. Pour 250 ml (1 cup) of whole milk into a 400 ml beaker.
2. Add 5 drops of 1 N HCL to the milk.
3. Heat the milk to 32°C while stirring continuously.
4. While stirring, add three drops of rennilase to the milk, then remove the beaker from the heat source.
5. Allow the milk to stand undisturbed for approximately 15 minutes or until the milk coagulates.
6. After the curd has formed, break up the curd with a glass rod, and filter the curd from the whey, using the cheesecloth and another 400 ml beaker.
7. Dispose of the whey as indicated by the instructor.
8. Let the cheese sit for approximately 10 minutes.
9. Discuss the results.

_____

_____

_____

Part 2: Converting Cheese into Amino Acids

This activity is designed to convert the curd (cheese) produced in the previous experiment into its component amino acids.

Materials
- curd (cheese)
- bacterial protease
- two 250 ml beakers
- stirring rods
- ninhydrin
- water
- cheesecloth
- 2 test tubes
- hot plate and water bath
- 2 funnels

Procedure 8.6
Converting Cheese

1. Procure two 250 ml beakers. Label one beaker "Enzyme" and the other beaker "No Enzyme."
2. To each beaker add one-half of the curd produced in the previous activity.
3. Add 100 ml of water to each beaker. Stir vigorously to break up the curd.
4. Add 1 gram of the enzyme bacterial protease to the beaker labeled "Enzyme." Do not add bacterial protease to the beaker labeled "No Enzyme." Briefly stir both beakers and allow them to sit undisturbed for 5 minutes.
5. Procure two test tubes and two funnels. Label two test tubes "Enzyme" and "No Enzyme." Place cheesecloth over the funnels and filter 5 ml of the liquid from the beakers into the two separately labeled test tubes.
6. Add 1.0 ml of the indicator ninhydrin to each test tube, and gently mix by tapping the sides of the test tube. *Ninhydrin is an indicator that turns purple in the presence of amino acids.*
7. Place the test tubes in a boiling water bath for 10 minutes.
8. Remove the test tubes and record the color changes.
9. Dispose of the curd and contents of the test tubes as indicated by your instructor.
10. Clean the glassware, and return it to the proper place.

Q. Describe the color difference in the test tubes.

_____

_____

_____

Q. Explain what happened in the test tubes.

_____

_____

_____

_____

NOTES

Name: _____ Date: _____ Section: _____

Review Questions

1. What is an enzyme?

_____

_____

_____

2. Explain the induced-fit model of enzyme action.

_____

_____

_____

3. Define catabolic and anabolic enzymes.

_____

_____

_____

_____

4. What is activation energy?

_____

_____

_____

5. Name several variables that may affect enzyme action.

_____

_____

_____

_____

_____

6. What happens when an enzyme is denatured?

_____

_____

_____

7. What was the optimum temperature for bromelian activities?

_____

_____

_____

8. Did canned pineapple juice and fresh pineapple juice induce different reactions?

_____

_____

_____

_____

9. Can various foods affect the nature of digestive enzymes?

_____

_____

_____

_____

10. How is cheese made?

_____

_____

_____

_____

_____

_____

_____

Chapter 9
Cellular Transport Mechanisms: Understanding Diffusion and Osmosis

Student Outcome Objectives

At the completion of this exercise, the student will be able to:

1. Compare and contrast passive and active transport mechanisms.
2. Describe several means of passive and active transport.
3. Provide specific examples of passive and active transport.
4. Compare and contrast endocytosis and exocytosis.
5. Define equilibrium.
6. Define solution, solvent, and solute.
7. Compare and contrast hypotonic, hypertonic, and isotonic solutions.
8. Describe cytolysis and plasmolysis.

Overview

One of the major functions of the plasma membrane is to regulate the movement of substances into and out of the cell (Fig. 9.1). This process is essential in maintaining the homeostatic state of the cell. If you recall, the plasma membrane is composed primarily of a phospholipid bilayer and specialized proteins. The unique structure of the plasma membrane allows it to be **selectively permeable** to certain substances. The permeability of the membrane is regulated by several variables including size of the molecule, polarity of the molecule, and external conditions such as concentration, temperature, and pressure.

For molecules to enter or exit a cell, they must overcome the **concentration gradient**, which involves the movement of molecules from regions of greater concentration to regions of lesser concentration. Living systems have two primary mechanisms for moving substances in and out of the cell—passive and active transport. In **passive transport** the cell uses no energy (ATP) as essential substances are moved across the plasma membrane. Examples of molecules moved by the various means of passive transport are oxygen, water, and glucose.

The most fundamental means of passive transport is **diffusion**, the random movement of molecules from regions of greater concentration to regions of lesser concentration (Fig. 9.2). This random movement also is known as **Brownian motion**. A state of equilibrium is attained when an equal distribution of molecules exists throughout the system. The movement of water across the plasma membrane in living systems is called **osmosis**.

Other mechanisms of passive transport include facilitative diffusion and filtration.

1. In **facilitative diffusion**, carrier proteins along the cell membrane are required to ferry specific molecules such as glucose across the membrane into the cell.
2. **Filtration** involves hydrostatic pressure (water pressure) forcing molecules through a cell membrane. Filtration is an essential mechanism that takes place in the kidneys in the formation of urine.

In **active transport**, energy in the form of ATP is required for the movement of substances across the concentration gradient. Biological pumps essential in this process are the following.

1. The **proton** or **hydrogen pump** is necessary to maintain the normal pH of the stomach.
2. The **calcium pump** is important in nerve and muscle function, and
3. The **sodium-potassium** pump is integral in cellular metabolism.

Macromolecules such as polypeptides and polysaccharides are too large to traverse the cell membrane by either passive or pump systems. Instead, they must be transported

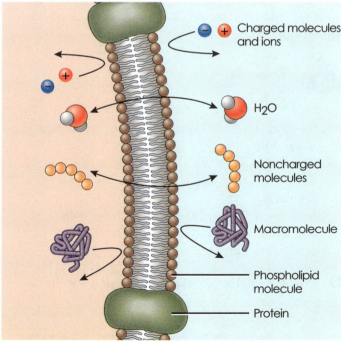

Figure 9.1 The cell membrane serves as the active interface between the cell and its environment.

Figure 9.2 The diffusion of food coloring in water from regions of greater concentration to regions of lesser concentration.

Figure 9.3 An amoeba capturing its food through phagocytosis.

into the cell by **endocytosis** and out of the cell by **exocytosis**. Both endocytosis and exocytosis employ the formation of vesicles in transporting substances.

In endocytosis, solid substances can be engulfed by a cell through **phagocytosis**. Everyday examples in biological systems include the ingestion of food particles by an amoeba (Fig. 9.3) and the engulfing of foreign particles by a macrophage. Another example of endocytosis is **pinocytosis**, or cell drinking, in which liquid droplets that may contain salts and other molecules are taken into the cell through vesicle formation. Specialized cells in the roots of plants use pinocytosis to ingest a variety of nutrients.

 Student Activity—Diffusion

Take a Whiff

The instructor will spray perfume or room deodorizer in a front corner of the classroom. The molecules will diffuse throughout the room in an attempt to attain a state of equilibrium. Timing devices are used to determine the rate of diffusion of the molecules across the room (Fig. 9.4).

 ### Procedure 9.1
Smell

1. Students should be equally dispersed throughout the classroom.
2. The instructor sprays a small amount of a scent in the front corner of the room and this will serve as time zero.
3. Students will record the time and raise their hand when they detect the odor.
4. Students discuss their results.

Q. How long did it take for the scent to reach the farthest point?

_____

Q. What variables can affect the rate of diffusion of the scent?

_____

_____

_____

Figure 9.4 Diffusion is the random movement of particles from regions of greater concentrations to regions of lesser concentrations.

What's in That Bag?

In this activity, dialysis tubing will serve as the selectively permeable membrane. After making a bag with the dialysis tubing and filling it with colorless cornstarch solution, the bag will be immersed in a beaker containing an iodine solution that is caramel in color. Movement of the iodine molecules across the membrane can be detected by a change in the cornstarch solution to a purplish-brown color.

Materials
- 250 ml beaker
- 25 ml graduated cylinder
- cornstarch solution
- iodine solution
- water
- dialysis tubing
- string
- timing device
- ruler
- paper towels

Procedure 9.2
Membrane

1. Procure glassware and needed accessory materials.
2. Measure and cut a 15 cm length of dialysis tubing.
3. Place the tubing in water until it becomes soft and pliable.
4. Using string, form a bag with the dialysis tubing closing one end.
5. Fill the dialysis tubing bag halfway with the cornstarch solution.

6. Using string, tie off the top end of the bag.
7. Immerse the bag containing the cornstarch solution into a beaker containing 200 ml of iodine solution and record the time and the color of the solutions.

Q. What is the color of the iodine solution?

_____

Q. What is the color of the cornstarch solution?

_____

8. Leave the apparatus undisturbed for 15 minutes.
9. Remove the bag from the solution and place it on a paper towel.
10. Observe the color changes in the dialysis bag.

Q. What color changes did you observe in the bag and in the solution?

_____

_____

Q. Explain the color changes in the solution and in the bag.

_____

_____

_____

Q. In the space provided below, graphically illustrate the results of this activity.

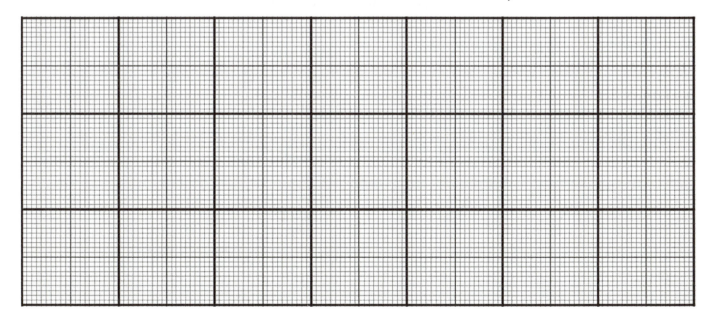

OSMOSIS ACTIVITIES

Recall that osmosis is the diffusion of water across a selectively permeable membrane, and diffusion always occurs from regions of greater concentration to regions of lesser concentrations. A typical **solution** consists of two components—the **solvent** as the dissolving medium and the **solute** as the substance dissolved in the solvent. In a saltwater solution, the water serves as a solvent and the salt as the solute.

Tonicity refers to the concentration of solute in the solvent.

1. In a **hypotonic** solution, there is a lower concentration of solute relative to the inside of the cell. If a cell such as a red blood cell or a potato cell is placed in a hypotonic solution, water will rush into the cell in attempt to reach a state of equilibrium. An ideal hypotonic solution is distilled water, because it devoid of solutes. As the cell begins to fill with solvent, the cell will swell, and perhaps burst. The bursting of cells in a hypotonic solution is called **cytolysis**.

2. In a **hypertonic** solution, there is a higher concentration of solute relative to the inside of the cell. If a cell such as a red blood cell or potato cell is placed in a hypertonic solution, water will be drawn out of the cell into the outside solution in an attempt to reach a state of equilibrium. This is called **plasmolysis** (Fig. 9.5).

In red blood cells, plasmolysis is distinctly seen as **crenation**. In plants, the swelling of cells placed in a hypotonic solution results in **turgor pressure**. The framework cell wall protects the cell from bursting. Turgor pressure keeps the plant erect. If the turgor pressure is lost in a plant, the plant will wilt. Just think of the plants in your yard on a hot summer's day.

Many cells exist in an **isotonic** solution, which has the same concentration of solute outside and inside of the cell. Normally, red blood cells exist in an isotonic state in plasma. Many organisms, such as some species of sharks, can undergo **osmoregulation**, in which they can alter their internal solutions to equal that of the environment to maintain homeostasis. This explains why some species of sharks can be found living in freshwater rivers near coasts.

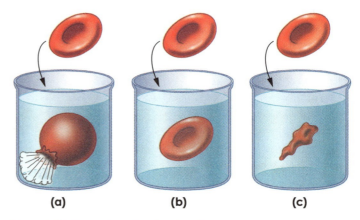

Figure 9.5 Osmosis and animal cells. When the outer solution is hypotonic (a) in comparison to the cell, the solution will move into the cell and the cell will lyse. In an isotonic solution (b), a homeostasis is achieved. When the solution is hypertonic (c) in comparison to the cell, the solution will move out of the cell and the cell will shrink, or become crenate.

 Student Activity—Plant Tissue: Hypotonic vs Hypertonic

Testing Tissues in Hypotonic and Hypertonic Solutions

Plant Tissues

In this activity, strips of potato tissue will be placed in hypotonic and hypertonic solutions, and observations will be made and recorded.

Materials
- potato strips
- 3 test tubes
- test tube rack
- wax pencil
- water
- distilled water
- 10% sodium chloride solution
- paper towels
- forceps
- scalpel
- microscope slides and coverslips
- microscope

 **Procedure 9.3
Plant Tissue**

1. Procure the needed equipment and bring it to your lab station.
2. With a wax pencil, label the tubes "Control," "Hypotonic Solution, "and "Hypertonic Solution."
3. Cut three strips of potato, each 2 cm in length and 0.5 cm wide.
4. Place each strip in a labeled test tube.
5. Add tap water to the test tube labeled "Control," totally immersing the potato strip.
6. Add distilled water to the test tube labeled "Hypotonic Solution," totally immersing the potato strip.
7. Add 10% sodium chloride solution to the test tube labeled "Hypertonic Solution," totally immersing the potato strip.
8. Let the solutions sit quietly for 30 minutes before making your observations.
9. Remove the potato strips and place them on a paper towel. Record your observations.
10. Using the scalpel, carefully cut a thin piece of tissue from each strip.
11. Prepare a wet mount of each strip.
12. Using the microscope, observe each slide, and record your results.
13. Clean your lab station.

Q. Which strip was most limp? Why?

_____

_____

_____

Q. Which strip was most stiff? Why?

_____

_____

_____

Sketch your microscopic observations.

Control

Hypotonic Solution

Hypotonic Solution

What we observe is not nature itself, but nature exposed to our method of questioning.

—Werner Karl Heisenberg (1901–1976)

Animal Tissues

In this activity, sheep red blood cells will be placed in hypotonic and hypertonic solutions, and observations made and recorded.

Materials

- whole sheep blood
- 3 test tubes
- test tube rack
- wax pencil
- water
- distilled water
- 10% NaCl
- 0.9% NaCl
- paper towels
- eyedropper
- microscope slides and coverslips
- microscope

 ### Procedure 9.4
Animal Tissue

1. Procure the needed equipment and bring it to your lab station.
2. With a wax pencil, label the tests "0.9% NaCl," "10% NaCl," and "Distilled Water."
3. Add 5 ml of 0.9% NaCl solution to the appropriate test tube.
4. Add 5 ml of 10% NaCl solution to the appropriate test tube.
5. Add 5 ml of distilled water to the appropriate test tube.
6. Using the eyedropper, add 5 drops of whole sheep blood to each test tube.
7. Let the solutions sit quietly for 1 minute.
8. Hold each test tube in front of this printed page to determine which one is more clear.
9. Prepare a wet mount from each test tube.
10. Using the microscope, observe each slide and record your results.
11. Clean your lab station and return the equipment.

Q. Which solution was the clearest? Why?

_____

_____

_____

Q. Which solution was less clear? Why?

_____

_____

Sketch and label your microscopic observations.

0.9% NaCl

[blank sketch box]

10% NaCl

[blank sketch box]

Distilled Water

[blank sketch box]

Q. Which solution more closely resembles the tonicity of plasma? Why?

_____

_____

_____

Student Activity—Chicken Egg: Hypotonic vs Hypertonic

Green Eggs, but Where is the Ham?

This 2-day activity is designed for you to do in your home or dormitory. Before engaging in this activity, review the definitions of hypertonic and hypotonic solutions. Acquire the materials and follow the procedure below.

Materials (Procedures 9.5–9.6)
- raw egg
- white vinegar
- green food coloring
- light corn syrup
- 2-quart mason jars or similar containers

 ## Procedure 9.5
Vinegar

Day 1

1. Procure the needed materials.
2. Fill a Mason jar or container approximately halfway with white vinegar.
3. Add 3–5 drops of green food coloring to the vinegar.
4. Carefully place the raw egg into the vinegar solution.
5. Set the experiment aside at least 8 hours.
6. Observe the egg.
7. Using a digital camera, video camera, or phone camera, record the egg.

Q. Describe the egg.

_____

_____

Q. Using your knowledge of solutions, describe what has happened to the egg.

_____

_____

Procedure 9.6
Corn Syrup

Day 2

1. Procure the needed materials.
2. Fill a Mason jar or container approximately halfway with light corn syrup.

3. Carefully place the egg into the corn syrup.
4. Set the experiment aside for at least 8 hours.
5. Observe the egg.
6. Using a digital camera, video camera, or phone camera, record the egg.

Q. Describe the egg.

_____

_____

Q. Using your knowledge of solutions, describe what has happened to the egg.

_____

_____

_____

Why Do Sea Turtles Cry?

Have you ever watched a nature show and seen sea turtles crying as they lay their eggs on some lonesome beach? Why do turtles cry? Is it because they are anticipating the future of their young, or perhaps they are sentimental, or maybe it hurts. No, it's not any of these reasons. Marine turtles have glands in the regions of their eyes that help them remove excess salt from the hypertonic solution in which they live. That is why these turtles appear to cry!

NOTES

Name: _____   Date: _____   Section: _____

??? Review Questions

1. Why does the cell membrane have to be selectively permeable?

_____

_____

_____

2. Compare and contrast active and passive transport.

_____

_____

_____

3. What variables affect the rate of diffusion in biological systems?

_____

_____

_____

4. Describe osmosis in your own words.

_____

_____

_____

6. What is the difference between pinocytosis and phagocytosis?

_____

_____

_____

7. Explain why a sailor set adrift cannot drink seawater?

_____

_____

_____

8. If a medical technologist notes that many red blood cells are crenated, what could have caused this phenomenon?

_____

_____

_____

9. Describe three active transport pumps that function in biological systems.

_____

_____

_____

10. Describe and draw what happens to a red blood cell when placed in a hypotonic solution, a hypertonic solution, and an isotonic solution.

_____

_____

_____

Going Beyond

1. Discuss specific adaptations of freshwater fishes that allow them to survive in freshwater environments.

_____

_____

_____

2. Discuss specific adaptations of marine fishes that allow them to survive in saline environments. Why are marine organisms said to literally live in a desert?

_____

_____

_____

3. What is the difference between an osmoregulator and an osmoconformer? Give several examples of each.

_____

_____

Chapter 10
Photosynthesis:
Understanding Photosynthesis

Student Outcome Objectives

At the completion of this exercise, the student will be able to:

1. Describe the role of photosynthetic organisms.
2. Discuss the contributions of von Helmont, Priestly, and Ingehousz on the history of understanding photosynthesis.
3. Compare and contrast the visible and electromagnetic spectrum.
4. Discuss the role of pigments in photosynthesis.
5. Discuss the internal anatomy of a leaf.
6. Describe the role of the stomata in a leaf.
7. Describe the function and organization of chloroplasts.
8. Discuss the location and products of the light-dependent and light-independent reactions.

Figure 10.1 Ultimately, plants convert the radiant energy from a solitary star known as the sun, more than 150 million kilometers away, into products essential to life on Earth.

Overview

It is time that plants and other photosynthetic organisms receive a much-deserved "thank you." Without them we would not exist. On the early earth, primitive photosynthetic bacteria and algae produced copious amounts of oxygen, changing the atmosphere forever and making it conducive to the development of more complex life forms. It is hard to believe, but the algae that inhabit primarily aquatic environments produce nearly 80% of the oxygen in the atmosphere today and serve as the basis of many food chains. Plants such as grasses, shrubs, and trees dominate terrestrial landscapes, providing habitat, food, and commercial products. Plants are eloquent green machines that harness sunlight and produce carbohydrates and oxygen necessary for the plants themselves, as well as other living things (Fig. 10.1). This remarkable process is termed **photosynthesis**. Keep in mind that photosynthesis is not the opposite counterpart to respiration.

Early philosophers and scientists were intrigued by the mystery of plant growth. Noted Greek scientists such as Aristotle and Theophrastus thought that plants obtained their essential nutrients exclusively from the soil. Although partially true, it took several centuries for the mystery of plant growth to unfold. The first step in understanding the mystery was taken by a Flemish physician Jan Baptista van Helmont, in the mid-1660s. He grew a willow tree weighing 2.5 kg in a pot filled with 91 kg of rich soil.

Van Helmont provided the tree with water periodically. After 5 years, the tree weighed nearly 75 kg and the soil lost only 57 g of weight. He concluded that the tree gained weight from the water, not the soil. Although he did not understand the role of sunlight and carbon dioxide in plant growth, he initiated the scientific study of plant biology.

In the 1770s, the English chemist Joseph Priestly performed a series of unique experiments emphasizing the importance of atmospheric gases in plant growth. Eventually, Priestly was credited with the discovery of oxygen. Building upon the works of Priestly and the "founder of modern chemistry" Antoine Lavoisier, the Dutch physician Jan Ingenhousz hypothesized that plants utilize sunlight to split carbon dioxide and use the carbon for growth. He also concluded that plants expelled oxygen as a waste product. As the result of his work, Ingenhousz is credited with describing photosynthesis. In the mid 1800s, the German botanist Julius van Sachs discovered chloroplasts and their role in photosynthesis. The light reaction, or light-dependent reaction, of photosynthesis was described by Theodor Engelmann in the 1880s. In 1940, Melvin Calvin traced the route of carbon throughout the entire process of photosynthesis.

Plants, instead of affecting the air in the same manner as animal respiration, reverse the effect of breathing and tend to keep the atmosphere sweet and wholesome.

—**Joseph Priestly (1733–1804)**

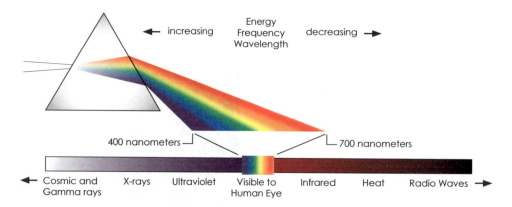

increasing ← Energy
Frequency
Wavelength → decreasing

400 nanometers ⌐   ⌐ 700 nanometers

← Cosmic and   X-rays   Ultraviolet   Visible to   Infrared   Heat   Radio Waves →
  Gamma rays   Human Eye

Figure 10.2 The electromagnetic spectrum.

Sunlight powers photosynthesis. Using a prism, the English physicist Sir Isaac Newton demonstrated that white light consists of a variety of colors ranging from red at one end of the **visible spectrum** to violet at the other end. In the mid 1800s, James Clerk Maxwell illustrated that the visible spectrum was a minute portion of a continuous spectrum or **electromagnetic spectrum** that includes, radio waves, visible light, x-rays, and cosmic rays (Fig. 10.2). Radiations of the spectrum travel in waves measured in nanometers (1 nm = 10^{-9} m). Radiations with longer wavelengths (radio waves) have less energy, and those with shorter wavelengths (x-rays) have more energy.

Q. Observe Figure 10.2. What has more energy, a microwave signal or a gamma ray?

_____

Q. Observe Figure 10.2. What has more energy, green light or purple light?

_____

Q. If you go scuba diving, how do the colors shift with depth? Why?

_____

_____

For an organism to utilize light energy, it has to be absorbed. In living systems, **pigments** absorb light energy. Some pigments, such as melanin, absorb all wavelengths of light, and they appear black. At the other end of the spectrum, many pigments absorb only certain wavelengths of light and reflect the other wavelengths. The light **absorption spectrum** of a pigment illustrates the wavelengths that are absorbed. For example, green

leaves contain the pigment **chlorophyll**, which reflects the green portion of the spectrum.

Chlorophyll is the most important pigment in photosynthesis. Several types of chlorophyll exist in nature. Chlorophyll *a* is the main photosynthetic pigment in some cyanobacteria and in plants. Other pigments important in plants but not involved directly in photosynthesis are called **accessory pigments**. Xanthophyll is a yellowish pigment (e.g., fall leaves), and carotene is an orange pigment (e.g., carrots). Chlorophyll *b* is considered an accessory pigment in plants, broadening the spectrum that can be used in photosynthesis.

Leaves are the most conspicuous part of a plant. They vary tremendously in shape and size, and some large trees have more than 100,000 leaves. One of the major functions of a leaf is as a photosynthesis factory. The internal anatomy of a typical leaf is complex (Fig. 10.3). A waxy **cuticle** covers the upper side of the leaf, and an **epidermis** completes the upper and lower layers of a typical leaf. Scattered primarily throughout the lower epidermis are **stomata** (*sing,* stoma), tiny openings regulated by **guard cells**. The stomata allow the carbon dioxide from the atmosphere to enter the leaf.

The center of the leaf consists of the **mesophyll**, composed of **palisade parenchyma** and **spongy parenchyma**. The palisade parenchyma is columnar in shape and usually appears beneath the upper epidermis. The spongy parenchyma is loosely packed and surrounded by numerous air spaces. Most of the photosynthetic activity occurs in the cells of the spongy parenchyma.

Chloroplasts reside within plant cells and serve as the organelle of photosynthesis. A chloroplast consists of two outer membranes that surround a semifluid matrix called the **stroma**. A third membrane system forms a series of flattened sacs called **thylakoids**. In some chloroplasts, the thylakoids become stacked, forming a

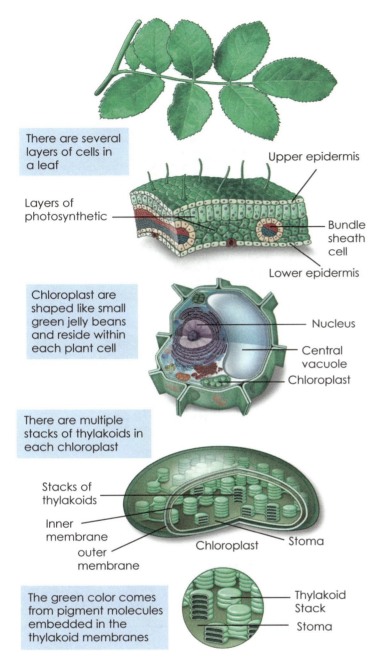

There are several layers of cells in a leaf

Layers of photosynthetic

Upper epidermis

Bundle sheath cell

Lower epidermis

Chloroplast are shaped like small green jelly beans and reside within each plant cell

Nucleus

Central vacuole

Chloroplast

There are multiple stacks of thylakoids in each chloroplast

Stacks of thylakoids

Inner membrane

outer membrane

Chloroplast

Stoma

The green color comes from pigment molecules embedded in the thylakoid membranes

Thylakoid Stack

Stoma

Figure 10.3 Leaf hierarchy.

granum. Pigment molecules embedded in the membranes of the thylakoids initiate photosynthesis. Sugars are synthesized in the stroma.

The overall reaction for photosynthesis is:

$$6CO_2 + 6H_2O \rightarrow C_6H_{12}O_6 + 6O_2$$

This reaction is the result of a series of chemical reactions that are controlled and carried out by specific enzymes. These reactions of photosynthesis are divided into two distinct metabolic pathways:

1. In the **light reaction**, or **light-dependent reactions**, the pigment chlorophyll absorbs light energy from sunlight and produces ATP, the coenzyme NADPH, and oxygen. The light-independent reaction takes place in the thylkoid membrane of the chloroplasts.
2. The **dark reaction**, also known as the **light-independent reaction** or **Calvin cycle**, takes place in the stroma of the chloroplasts. It is responsible for the fixing of a carbohydrate (glucose).

Photosynthetic organisms are estimated to produce as much as 200 billion metric tons of carbohydrate yearly.

During aerobic respiration, plants and other organisms use the oxygen produced as a byproduct of photosynthesis. The carbohydrates produced during photosynthesis are converted into many plant products such as fiber, wood, and other structural materials. In addition, the simple sugars the plant produces can be converted into disaccharides and polysaccharides, such as starch for energy storage. Sugars also are utilized in the synthesis of amino acids to form proteins and other cellular components. Life on Earth depends on the ability of photosynthetic organisms to convert the radiant energy of the sun into ATP, oxygen, and other nutrients.

EXTERNAL STRUCTURE OF A LEAF

Photosynthetic reactions usually take place in the leaves. Leaves generally consist of a blade and a petiole. The petiole attaches the flattened blade to the stem. In this activity, you will examine a typical dicot leaf.

Figure 10.4 Generalized leaf structure.
1. Blade
2. Petiole
3. Margin
4. Midrib

The Light-Dependent Reaction   The Light-Independent Reaction

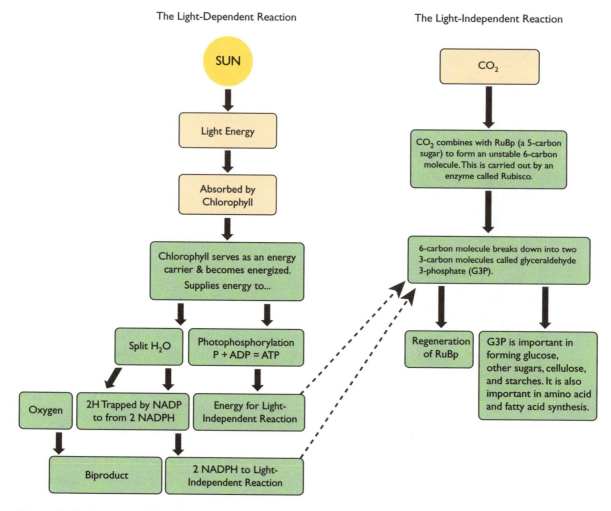

Figure 10.5 Overview of the light-dependent and light–independent reactions of photosynthesis.

 ## Student Activity—External Leaf Structure

Materials
- dicot leaf
- dissecting microscope or hand lens
- colored pencils

 ## Procedure 10.1
External Leaf Structure

1. Obtain a leaf from your teacher.
2. Sketch the leaf and label the blade and petiole.

Student Activity—Internal Leaf Anatomy

Internal Anatomy of a Leaf

In this activity, you will observe the internal structures of a leaf. Leaves have a waxy covering called a **cuticle**, which keeps water in the leaf. Leaves are covered with epidermal tissue on the top and bottom of the leaf. In between the epidermal tissue is **mesophyll** (palisade parenchyma and spongy parenchyma). Along the underside of the leaf are small openings called stomata that allow carbon dioxide to enter the plant and oxygen to exit the plant. Guard cells control the opening and closing of the stomata to minimize loss of water by the plant.

Materials
- prepared slide of a leaf
- compound microscope
- colored pencils

 ## Procedure 10.2
Internal Leaf Anatomy

1. Obtain a prepared slide of a leaf from your teacher.
2. Using a compound microscope, observe the leaf.
3. Sketch the leaf, and label the following structures: cuticle, epidermis, mesophyll, palisade parenchyma, spongy parenchyma stomata, guard cells.

USING PAPER CHROMATOGRAPHY TO SEPARATE PLANT PIGMENTS

The following laboratory activity has been designed to study plant pigments. Before energy is produced, light must be absorbed. Plant pigments function in the absorption of light. The principal pigments found in the thylakoids of plants are chlorophylls *a* and *b*, both of which absorb red and blue light and reflect green light. In addition to chlorophyll, accessory pigments such as **xanthophylls** and **carotenes** absorb light and transfer energy to chlorophyll *a*. Xanthophylls reflect yellow light, and carotenes reflect orange light. Chlorophyll can mask the xanthophylls and carotenes in the leaves, but in autumn, when the chlorophyll begins to break down, the other pigments can be seen.

Paper chromatography is a method used to separate the plant pigments. The pigments are separated based upon absorption and the size of the molecules. The separated components, or **chromatogram**, will show on the chromatography paper as colored streaks. (Fig. 10.6) The relative rate of migration (the R_f value) for each pigment can be determined from the chromatogram. The R_f value for each pigment can be determined by calculating the ratio of the distance of a pigment moved on the chromatogram to the distance the solvent front moved.

$$R_f = \text{distance moved by the pigment}/ \text{distance moved by solvent.}$$

Figure 10.6 A simple paper chromatogram showing the different distances that plant pigments have travelled.

Student Activity—Plant Pigments

Materials
- safety goggles
- scissors
- chromatography paper strip
- wax pencil
- capillary tube
- artificial plant pigment extract
- test tube
- cork stopper
- graduated cylinder
- chromatography solvent (alternate isopropyl alcohol)
- metric ruler
- stopwatch or clock with a secondhand
- hook or fashioned paperclip
- paper towels
- test tube rack
- mortar and pestle

Procedure 10.3
Chromatogram

1. Obtain a strip of chromatography paper, and cut it so it fits into a test tube and barely touches the bottom of the tube. (When handling the strip, touch the top of the paper only.)
2. Securely attach the top of the strip to a hook or fashioned paperclip at the bottom of a cork stopper. Test for fit. Remove the cork and the strip from the test tube.
3. Using a wax pencil, draw a faint line across the strip about 2 cm from the bottom tip of the strip.

4. Put the cork and strip in place, and with a wax pencil, mark the test tube 1 cm below the top of the stopper.
5. Place the strip of chromatography paper on a paper towel. Dip a capillary tube into the plant pigment extract provided by the instructor. The tube will fill on its own.
6. Apply the extract to the pencil line on the paper. Blow the strip dry, and repeat this application process three or four more times.
7. Using a graduated cylinder, carefully measure 5 mL of chromatography solvent and carefully pour into the test tube.
8. Place the chromatography strip in the test tube, and position it so the tip of the strip just touches the solvent. Be careful not to let the plant pigment extract touch the solvent.
9. Keeping the test tube capped, place the test tube in a test tube rack. Record your observations as the solvent rises up the paper.
10. When the solvent has moved up to the wax pencil line drawn on the test tube, remove the strip of paper. Set aside the paper to dry.
11. Identify the pigment bands. The chlorophyll *a* will be blue-green, the chlorophyll *b* will be olive-green, the xanthophylls will be yellow, and the beta-carotene will be a bright orange-yellow band.
12. Calculate the R_f (ratio factor) for each pigment by measuring the distance of the solvent from its origin to the highest point it traveled on the paper. Then measure the distance the different pigments travel from the origin (pencil mark) to the center of each pigment band. Record your measurements in the Table 10.1.

Color Band	Pigment	Color	Migration (mm)	R_f Value
1				
2				
3				
4				
The Solvent				

Table 10.1 Artificial Plant Pigments

Check Your Understanding

Q. Which of the pigments migrated the farthest from the point of origin? Why?

_____

_____

_____

PIGMENTS FROM NATIVE LEAVES

The laboratory instructor will provide several leaf specimens (at least three) collected from a nearby source, or with permission, you can collect leaves of interest. During the fall, try to collect different-colored leaves; during the remainder of the year, try to collect leaves of different colors. If you collect your leaves, be sure to identify them properly. Using a mortar and pestle, crush the leaves individually and collect a sample of extract. Follow the procedure outlined on page 136.

Describe the color of the leaves and their samples.

_____

_____

_____

_____

Compare and contrast the data collected from the leaf specimens.

_____

_____

_____

Follow-up Activities

Through the centuries, plant pigments have been used to dye textiles. For example, walnuts and pecans have been used to attain various shades of brown; indigo has been used to develop bluish dyes; safarrin has been used to develop yellow dyes; *Tradescantia* has been used to develop purple dyes. The American physician Alexander Garden used prickly pear cactus to develop red and pink dyes in the 1800s, and eventually the gardenia was named in his honor.

Procedure 10.4
Plant Dye

1. Determine the color of the dye obtained from the following plants.

• sassafras _____

• pomegranate _____

• hickory _____

• grenadine _____

• monkey grass _____

Procedure 10.5
Leaf Collection

1. Take a walk across campus and collect several leaves and flowers. Bring them back to the lab.
2. Identify the plant specimens, and smear them onto a piece of white paper.
3. Record the name of the plants and the colors obtained.

_____

_____

_____

_____

_____

Procedure 10.6
Extracting Pigments

1. Using the plants from Procedure 10.4, develop a method of extracting the pigments and dyeing swatches from an old T-shirt.
2. Record your results.

_____

_____

_____

_____

Student Activity—*Elodea*

The Uptake of Carbon Dioxide by *Elodea* During Photosynthesis

This activity illustrates the plant's uptake of carbon dioxide from the environment during the light-independent reaction, or Calvin cycle of photosynthesis.

Materials
- large, leafy stem of *Elodea*
- test tube
- test tube rack
- medicine dropper
- 1% solution of phenol red (pH indicator)
- stopper
- straw
- water

Procedure 10.7
Uptake of Carbon Dioxide

1. Fill two-thirds of a test tube with water.
2. Place the *Elodea* in the tube.
3. Add four or five drops of the phenol red to the test tube.
4. Insert a straw into the test tube and blow gently to release carbon dioxide. The water will become orange-yellow in color (more acidic), and carbonic acid is formed.
5. Immediately place the stopper on the test tube.
6. Place the test tube in a well-lit area for 10–20 minutes.

Check Your Understanding

Q. Describe the color change that has occurred in the test tube.

_____

_____

Where flowers degenerate, man cannot live.

—**Napoleon Bonaparte (1769–1821)**

Q. What is responsible for the color change of the solution?

_____

_____

_____

Q. If you were to wrap the test tube in aluminum foil, do you think the pH level would increase or decrease?

_____

_____

_____

Oxygen Production in *Elodea*

This activity illustrates the production of oxygen by a plant during the light-dependent reaction.

Materials
- large, leafy stem of *Elodea*
- test tube and fitted stopper
- test tube holder
- beaker
- glass funnel
- water
- safety glasses
- matches

Procedure 10.8
Oxygen Production

1. Procure a healthy piece of *Elodea* from the laboratory instructor.
2. Place the *Elodea* in a 500 ml beaker containing 350 ml of water.
3. Place the glass funnel into the beaker, and completely cover the *Elodea* so any oxygen the *Elodea* produces will pass through the funnel. Ensure that the stem of the funnel is under the water.
4. Completely fill a test tube with water. Place your thumb over the open end of the test tube.

5. Keeping your thumb over the open end of the test tube, invert the test tube and place it over the stem of the funnel, dipping down into the water.

6. Place the apparatus in direct sunlight, or expose it to lights provided by the laboratory instructor.

7. Observe tiny gas bubbles being liberated from the *Elodea*. These bubbles are oxygen and will slowly displace the water in the inverted test tube.

8. When the test tube is nearly filled with oxygen and still in the beaker, place a stopper over the end of the test tube. Carefully remove the test tube from the beaker.

9. Place the test tube in a test tube clamp. Put on your safety glasses.

10. Holding the test tube in the clamp, strike a match and bring it close to the test tube.

11. Remove the cork, bring the burning match a little closer, and make your observations.

12. Record your observations in the space below.

_____

_____

_____

Check Your Understanding

Q. How do you know that oxygen was produced, and where did it originate?

_____

_____

_____

_____

Procedure 10.9
Oxygen Production Rate

1. Design a simple experiment that will measure the rate of oxygen production during photosynthesis in *Elodea*.

2. Incorporating the process skills learned in previous chapters, collect qualitative and quantitative data, and interpret the data.

3. Develop a graph to represent the data below.

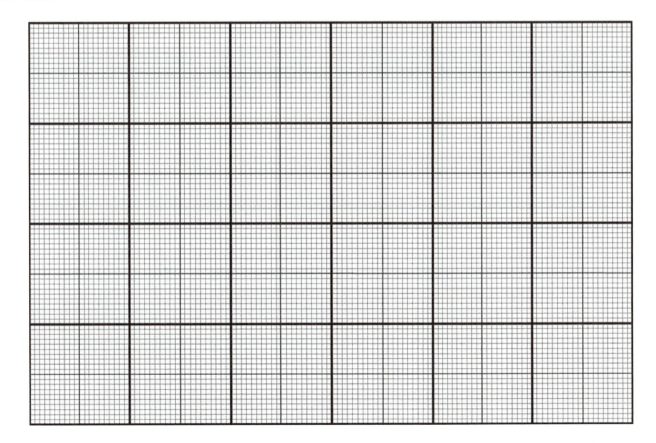

Variations in Photosynthesis

The vast majority of plants (perhaps as much as 90%) undergo typical photosynthesis as described above. These plants, known as C_3 plants, include azaleas, roses, oak trees, Kentucky bluegrass, and zinnias. Other plants have evolved alternative photosynthetic pathways to account for variations in CO_2 level, water availability, and light intensity. Some plants, especially those that thrive in the intense heat of summer, utilize another pathway and are known as C_4 plants. Examples of C_4 plants are sugar cane, corn, crabgrass, and rice. Desert plants such as cacti undergo the CAM pathway (Crassulacean Acid Metabolism), which have the ability to fix carbon dioxide in the dark. Other plants that utilize the CAM pathway include Spanish moss, pineapple, quillwort, and *Welwitschia*.

Identify five plants that undergo the C_3 pathway:

1. _____

2. _____

3. _____

4. _____

5. _____

Identify five plants that undergo the C_4 pathway:

1. _____

2. _____

3. _____

4. _____

5. _____

Identify five plants that undergo the CAM pathway:

1. _____

2. _____

3. _____

4. _____

5. _____

 Student Activity—Photosynthetic Photography

In 1914, the Austrian plant physiologist Hans Molisch developed techniques of photosynthetic photography. The exercise below is one of many activities that illustrate photosynthetic photography. The student is highly encouraged to experiment with different plant species, different amounts of light, and different frequencies of light.

The plant uses the carbohydrates it makes as an energy source to go about its planty business. And we animals, who are ultimately parasites on the plants, steal the carbohydrates so we can go about our business.

—Carl Sagan (1934–1996)

Materials
- geranium leaf
- slide projector★
- black cloth
- baking soda solution
- two sheets of glass
- negative
- slide housing
- 86% alcohol solution
- water
- iodine
- potassium iodide
- cup
- glass holder
- shallow dish

★ If a slide projector is unavailable, tape a negative to a geranium leaf and place a bright lamp near the leaf. Be careful not to burn the leaf.

Procedure 10.10
Photosynthetic Photography

1. Procure a healthy geranium plant and place it in a dark room for 3 days.
2. Remove a short stalk that has a healthy leaf on it.
3. Moisten a black cloth with a baking soda solution. Place the leaf on the cloth.
4. Sandwich the leaf and the cloth backing between two pieces of glass. Position the leaf and glass in an upright position with the stalk leading into cup of water. Face the leaf toward a slide projector.
5. Place your negative in a slide housing, and place the housing in the projector. Turn on the projector and project, and focus the image onto the leaf. Leave the projected image on the leaf for 1 hour.
6. Turn off the projector and remove the leaf from the glass. Leach out the leaf pigments by dipping the leaf briefly in a boiling solution of 86% alcohol.
7. Place the blanched leaf in a shallow dish containing iodine and potassium iodide.
8. Record your observations in the space provided below.

_____

_____

_____

_____

9. Using the techniques outlined above, develop experiments that investigate how plant species, types of leaves, negative density, light intensity, light frequency, and exposure time effects photosynthetic photography.
10. Record your experimental results.

_____

_____

_____

_____

Check Your Understanding

Q. Describe the leaf at the end of the activity.

_____

_____

_____

Q. Why was the plant placed in the dark for 3 days?

_____

_____

_____

Q. Why was the leaf boiled in alcohol?

_____

_____

_____

Q. Why was the leaf placed in a solution of iodine and potassium iodide?

_____

_____

_____

Q. Discuss how the image formed on the leaf.

_____

_____

_____

Q. What would happen if you were to project a regular slide on the target leaf?

_____

_____

In the space below, conduct an extension of this experiment as suggested in the 9th step of the procedure.

_____

_____

_____

_____

_____

_____

_____

_____

_____

_____

_____

_____

_____

_____

_____

_____

_____

_____

Name: _____ Date: _____ Section: _____

Review Questions

1. Write the equation for photosynthesis.

   _____

2. Why is photosynthesis considered the most important chemical reaction on Earth?

   _____

   _____

3. Compare and contrast the visible and electromagnetic spectrum.

   _____

   _____

   _____

4. Name and describe three pigments found in the labs in this chapter.

   _____

   _____

   _____

5. Sketch and label the basic internal anatomy of a leaf.

6. Describe the role of the stoma in leaves.

   _____

   _____

   _____

7. Sketch and label a chloroplast, including the membranes, stroma, thylakoids, and grana.

8. Describe the light-dependent and light-independent reactions of photosynthesis.

_____

_____

_____

_____

9. How is chromatography used to study plant pigments?

_____

_____

_____

10. Discuss the succession of C_3 and C_4 plants in your yard or an open field through the four seasons.

_____

_____

_____

11. Describe several products of photosynthesis that are important to humans.

_____

_____

Chapter 11
Cellular Respiration: Understanding Cellular Respiration

Student Outcome Objectives

At the completion of this exercise, the student will be able to:

1. Define cellular respiration.
2. Recite the equation for cellular respiration.
3. Describe glycolysis, and tell where in the cell it occurs.
4. Define aerobic respiration; include where in the cell the biochemical pathway occurs and the products formed.
5. Define anaerobic respiration.
6. Describe the two common fermentation pathways, and tell what products are formed in each pathway.
7. Explain the function of the electron transport chain.
8. Compare the amount of ATP formed in anaerobic and aerobic respiration.
9. Describe the interrelationship between photosynthesis and cellular respiration.

Overview

Life is a conquest of energy that all begins with the sun. The radiant energy of sunlight is converted into chemical energy (primarily glucose) through **photosynthesis**. In turn, the glucose is used to produce **adenosine triphosphate (ATP)**, the energy currency of living systems, through the process of **respiration**. Plants and animals have a unique evolutionary relationship based upon each using the other's products. As the result of photosynthesis, some bacteria, algae, and plants produce oxygen as a waste product. In turn, the oxygen is a vital component in aerobic cellular respiration. The plants then use the CO_2 produced in respiration to ultimately build carbohydrates.

In nature, two major types of respiration have evolved.

1. Some bacteria and fungi undergo **anaerobic respiration**, which occurs in the absence of oxygen. The ancestors of anaerobic organisms first appeared on Earth approximately 3.5 billion years ago. Today, many species of anaerobic organisms abound. Anaerobic bacteria can be found in certain soils, sediments in bodies of water, and the gut of some animals. Several species of anaerobic bacteria are responsible for diseases such as gangrene, tetanus, and botulism.

2. **Aerobic respiration** takes place in the presence of oxygen. The evolution of aerobic respiration began approximately 2.7 billion years ago, and the vast majority of organisms on Earth today are aerobic. The overall reaction for aerobic respiration is the reverse of photosynthesis:

$$C_6H_{12}O_6 + 6O_2 \rightarrow 6CO_2 + 6H_2O$$

Anaerobic and aerobic respiration both begin with a molecule of glucose produced through photosynthesis. With a few exceptions, the majority of living things undergo glucose metabolism to release the energy from the sun locked within a molecule of glucose. The first step in unlocking the energy within glucose is **glycolysis**. This complex chemical pathway occurs anaerobically in the cytoplasm of a cell. Because glycolysis is common to life on Earth, it arose early in the evolution of life.

The end products of glycolysis are four molecules of ATP (two of which are recycled in further glycolysis), two molecules of NADH (nicotinamide adenine dinucleotide phosphate, a coenzyme), and two molecules of pyruvate. If pyruvate is metabolized in the cytoplasm anaerobically, the process is known as **fermentation**. If the pyruvate is shuttled into the mitochondrion, the process continues aerobically and is known as **aerobic cellular respiration**.

Two significant types of fermentation reactions occur anaerobically: the **alcoholic fermentation reaction** and the **lactate fermentation reaction**. In both reactions, pyruvate is reduced by NADH to form either ethyl alcohol (C_2H_5OH) or lactate ($C_3H_5O_3$). Both reactions produce two molecules of water and only two molecules of ATP. Combined with the two ATPs produced from glycolysis, the net yield of ATP during each reaction is only four molecules of ATP. Thus, anaerobic organisms have no "energy to spare." That explains why anaerobic life forms cannot engage in a game of tennis or even "putt" around like a paramecium. Other types of commercially important microbial fermentation processes yield acetone and methanol.

AEROBIC RESPIRATION

One of the most significant events in the history of life was the evolution of aerobic respiration. In aerobic

Stage 1:
Breakdown of large macro-molecules to simple subunits

Stage 2:
Breakdown of simple subunits to acetyl CoA accompanied by production of limited ATP and NADH

Stage 3:
Complete oxidation of acetyl CoA to H_2O and CO_2 involves production of much NADH, which yields much ATP via electron transport

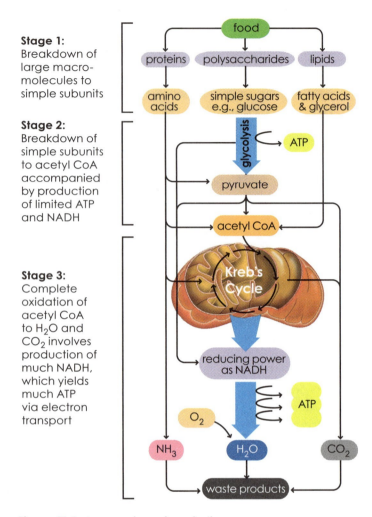

Figure 11.1 An overview of respiration.

transport chain. At the completion of the electron transport chain, typically 32 ATP molecules and water form. In the electron transport chain, oxygen serves as the final electron acceptor by combining with two hydrogen molecules to form water. The role of oxygen is responsible for the term **aerobic respiration** (Fig. 11.1).

ALCOHOLIC FERMENTATION

Yeasts are unicellular organisms in Kingdom Fungi that produce energy (ATP) anaerobically in a two-step pathway called the **alcoholic fermentation reaction** (Fig. 11.2). In the first step of the reaction, yeast chemically breaks down glucose into pyruvate in a series of metabolic reactions called **glycolysis**. A molecule of CO_2 then is removed from the pyruvic acid. This leaves a two-carbon compound.

In the second step of the reaction, two hydrogen atoms from NADH and H^+ are added to the two-carbon compound to form ethyl alcohol. The NADH is oxidized to form NAD^+, an essential molecule that allows the glycolysis pathway to continue.

Alcoholic fermentation is essential in making wine, beer, and bread. In making bread, the CO_2 produced causes the bread dough to rise. The ethyl alcohol evaporates during baking. Have you ever eaten a slice of pizza or a piece of bread that smells a little of beer? The beer odor is a bit of residual alcohol that has not completely evaporated!

respiration, the two molecules of pyruvate that result from glycolysis are converted further to two molecules of **acetyl coenzyme A** and two molecules of CO_2 in a mitochondrion. This is the CO_2 that you exhale. NAD^+ facilitates the removal of electrons from pyruvate and forms two molecules of NADH. The acetyl coenzyme (CoA) carries the acetyl group.

The acetyl group enters the **citric acid cycle**, or **Krebs cycle**, named after the biochemist Hans Krebs. The citric acid cycle is a complex series of metabolic reactions occurring in the matrix of the mitochondrion. These reactions are enzyme-driven, and the products in one step of the reaction become the reactants in the next. The end products of the citric acid cycle are two molecules of ATP, four molecules of CO_2, six molecules of NADH, and two molecules of $FADH_2$ (flavin adenine dinucleotide).

The final step of aerobic respiration occurs in the cristae of the mitochondrion and the plasma membrane of prokaryotes. This process is known as the **electron**

A Tidbit from Biohistory

In the mid 1800s, a French winery that produced alcohol from the fermentation of sugar beets ran into a big problem: Its prized wine tasted like vinegar. The owner, Monsieur Bigo, contacted the renowned Louis Pasteur to come up with a solution to the problem. At that time, no one including Pasteur really understood how sugar ferments into alcohol.

Pasteur sampled a healthy vat of fermenting sugar beets and, to his surprise, noted healthy yeasts under the microscope. Next he examined a vat that contained poor-tasting wine. In this vat, instead of finding yeasts, he observed rod-shaped bacteria. These bacteria were collected and grown in a nutritive media. Pasteur determined that instead of producing alcohol as a byproduct, the bacteria produced lactic acid. Thus, with this discovery, Pasteur was able to save the French wine industry by ensuring that yeasts and not bacteria are introduced to wine-making vats.

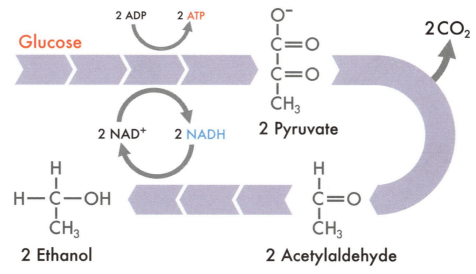

Figure 11.2 An overview alcoholic fermentation.

  Student Activity—Alcoholic Fermentation

Part 1: Demonstrating Alcoholic Fermentation in Yeasts

This activity has been designed to demonstrate alcoholic fermentation in yeasts (Fig. 11.3).

Materials
- package of Baker's yeast
- 1 tablespoon of granulated sugar
- water
- 250 ml beaker
- empty glass bottle (soda bottle)
- stirring rod
- glass dropper
- microscope
- microscope slides
- coverslips
- balloons
- hot plate
- thermometer
- rubber band

Procedure 11.1 Demonstrating Fermentation

1. Procure the materials to be used in this activity, and bring them to your lab station.
2. Add 200 ml of tap water to the beaker, and heat the water, using a hot plate, to approximately 35° C.

3. Carefully add the yeast and the sugar to the water, and stir. Wash your hands after handling the yeast.
4. Carefully pour the mixture into the glass bottle to approximately the halfway mark.
5. Cover the lip of the bottle with a balloon, and secure the base of the balloon to the bottle with a rubber band.
6. Observe and record what happens to the balloon over the next 30 minutes.
7. Using a glass dropper, take a drop of the mixture from the beaker.
8. Make a wet mount of the mixture.
9. Observe the mixture using a microscope on low and high power.
10. At the end of 30 minutes, remove the balloon from the glass bottle, waft your hand over the bottle, and smell the contents.
11. Clean your laboratory station, glassware, and hands.
12. Return your materials to the designated area.
13. Sketch your observations in the space below.

(a)

(b)

Figure 11.3 Bread making depends on the fermentation reaction by yeast. The CO_2 produced in the reaction forms bubbles in the bread dough, causing it to rise. (a) At the beginning of the process, and (b) four hours later.

Check Your Understanding

Q. What happened to the balloon? Why?

_____

_____

_____

Q. Describe the odor of the mixture.

_____

_____

_____

In eating the plants, we combine the carbohydrates with oxygen dissolved in our blood because of our penchant for breathing air, and so extract the energy that makes us go. In the process we exhale carbon dioxide, which the plants then recycle to make more carbohydrates. What a marvelous cooperative arrangement—plants and animals each inhaling the other's exhalations, a kind of plant-wide mutual mouth-to-stoma resuscitation, the entire elegant cycle powered by a star 150 million kilometers away.

—**Carl Sagan (1934–1996)**

Part 2: Observing Fermentation in Yeasts

This is a second activity to observe and record fermentation in yeasts.

Materials
- five 10-ml test tubes
- test tube rack
- six graduated pipettes
- pipette bulb
- five Pasteur pipettes
- five pieces of parafilm
- an actively fermenting yeast culture
- 10% glucose solution
- 10% sucrose solution
- 10% galactose solution
- 10% starch solution
- distilled water
- 37° C warm water bath
- wax pencil

Procedure 11.2
Observing Fermentation

1. Label the test tubes 1–5, using a wax pencil.
2. Place the test tubes in the test tube rack (Fig. 11.4).
3. Obtain a sample of the actively fermenting yeast culture from the instructor.
4. Using a graduated pipette, add 2 ml of yeast solution to each of the numbered test tubes.

5. Procure the distilled water and four sugar solutions from the instructor.
6. Using a clean pipette, add 2 ml of sugar solution to test tubes marked 1–4 and 2 ml of the distilled water to the fifth test tube. Leave the pipettes in the test tubes.
7. Gently swirl/shake each test tube.
8. Using a pipette bulb, draw up the yeast solution in test tube #1 into the graduated pipette, and place your finger over the top of the pipette to keep the solution in the pipette.
9. Carefully invert the pipette and seal the bottom end of the graduated pipette with a piece of parafilm.
10. Return the graduated test tube to its upright position.
11. Using a Pasteur pipette, fill the graduated pipette to the top (until it overflows) with more of the yeast solution from test tube #1.
12. Invert the graduated pipette again, and place it carefully into test tube #1.
13. Repeat steps 8–12 for the remaining four test tubes.
14. Place tubes in a warm water bath.
15. Record the gas level in each graduated pipette every 2 minutes for 20 minutes in the table below.

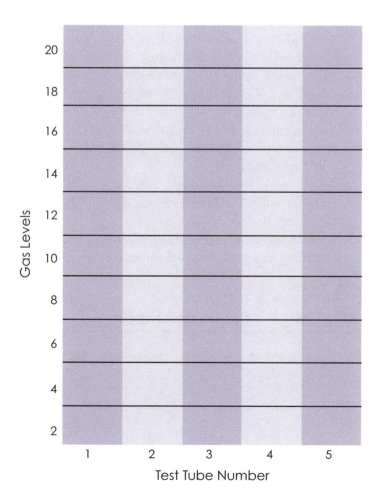

 Check Your Understanding

Q. In which tube (carbohydrate) did fermentation occur the fastest? Why?

_____

_____

Q. What is the function of the test tube containing the water and yeast solution?

_____

_____

_____

LACTATE FERMENTATION

Humans share with certain bacteria, fungi, and other animals a second alternative anaerobic pathway, the fermentation pathway. In the **lactate fermentation reaction**, glucose is broken down anaerobically in the cytoplasm of the cell to produce lactate, ATP, and water. As in alcoholic fermentation, this is a two-step reaction.

1. Glucose is broken down into pyruvate.
2. Two hydrogen atoms from NADH and H^+ are added to the 2-carbon pyruvate compound to form lactate.

Figure 11.4 Test tubes.

Water and ATP are also formed. The NADH is oxidized to form NAD+, an essential molecule to allow the glycolysis pathway to continue.

The equation for lactate fermentation is:

$$\text{Glucose} + 2\,\text{ADP} + P_i \rightarrow 2\,\text{Lactate} + 2\,\text{ATP} + 2\,H_2O$$

The lactate produced by the bacteria *Lactobacillus* and *Streptococcus* enhances the flavors of some cheeses and breads such as rye bread. The reaction also is important in the formation of soy sauce, kimchee, and sauerkraut. In the muscles, lactate can build up and cause muscle soreness and fatigue. The lactate is formed during strenuous exercise when the oxygen needed for aerobic respiration is not adequate. When cellular oxygen levels are low, pyruvate cannot enter the Krebs cycle, where large amounts of ATP are made. While cells store ATP in the form of **creatine phosphate**, these stores are rapidly used up for muscle contraction and relaxation. Once depleted, the cells will produce ATP anaerobically. The lactate produced is toxic to the cells. It is transported to the liver and eventually converted to glucose.

Student Activity—Lactate Fermentation

Making Kimchee
A popular dish in Korea is kimchee, or pickled cabbage, produced through lactate fermentation. Several types of bacteria play a role in the fermentation process. The succession of bacteria is pH-dependent, in which the most acid-tolerant bacteria are the last resident. Variables influencing the fermentation process include pH balance, temperature, oxygen availability, amount of salt used, and additional ingredients. The final texture and taste of kimchee is characterized as a unique combination of sour, sweet, and spicy. This exercise has been designed to demonstrate lactate fermentation by the making of kimchee.

Materials
- Chinese cabbage
- salt, pepper, garlic, and chili peppers
- clean 2-liter plastic bottle with cap
- plastic Petri dish and lid
- pH paper
- knife and cutting board
- bowl

Procedure 11.3
Making Kimchee

1. Procure a clean 2-liter plastic soda bottle, and cut the bottle into two pieces 15 cm from the top. Save the top portion that you removed, for later use.
2. Carefully cut the cabbage into 1-inch square pieces to fill approximately three-quarters of the plastic soda bottle.
3. In the bowl, thoroughly mix the cabbage with a small amount of salt, pepper, and garlic. If you like it super spicy, add chili peppers.
4. Place your mixture into the plastic soda bottle.
5. Clean your equipment and return it to the proper place.
6. Place the lid for a plastic Petri dish flat side down atop the cabbage mixture.
7. Press the cabbage mixture down with your fingers on the Petri dish, compacting the mixture.
8. Over the next hour, continue to press down on the Petri dish, further compacting the cabbage mixture.
9. Using pH paper or another pH indicator, measure and record the pH of the mixture.
10. Take the top of the bottle that you had cut and removed, and place it atop the Petri dish and mixture, forming a tight seal.
11. Each day for a week, take a pH reading of the mixture, and press on the sealed top portion of the soda bottle to remove the gas.
12. After a week, remove the seal, take a pH reading, remove the Petri dish, and—if you dare—taste your kimchee.
13. Disassemble your apparatus and discard the material according to your instructor's directions.

Check Your Understanding

Q. Describe the change in pH over the week. Why did this occur?

_____

_____

_____

Q. What gas was produced in the making of kimchee?

_____

_____

_____

Q. Describe the taste of kimchee.

_____

_____

_____

 ## Student Activity—Aerobic Respiration

AEROBIC RESPIRATION

Many people do not realize that plants carry out aerobic respiration. In aerobic respiration, the plant uses atmospheric oxygen (a byproduct of photosynthesis) to terminate the electron transport chain. As oxygen combines with two hydrogen molecules at the end of the electron transport chain, it forms water. In aerobic respiration, oxygen is used to release energy from glucose. Carbon dioxide is formed in the process.

Measuring Oxygen Consumption During Aerobic Respiration

In this experiment, oxygen consumption will be compared in germinating and nongerminating green peas. The peas contain a plant embryo that will germinate when conditions are favorable. During germination, the carbohydrates stored in the peas serve as fuel for growth of the embryonic plant. Your laboratory instructor will initiate the germination process by soaking the peas in water in a dark environment for 3 days. The nongerminated peas will be heat-killed by roasting the peas in a 200° F oven for 20 minutes.

To measure the amount of oxygen consumed by the green peas, potassium hydroxide (KOH) will be used to remove the carbon dioxide that is produced in the reaction. This reaction is:

$$CO_2 + 2KOH \rightarrow K_2CO_3 + H_2O$$

Materials
- germinating green peas
- nongerminating green peas
- glass beads
- absorbent cotton
- KOH pellets
- two test tubes
- two 2-hole rubber stoppers to fit test tubes

Materials (cont.)
- germinating green peas
- nongerminating green peas
- glass beads
- absorbent cotton
- KOH pellets
- two test tubes
- two 2-hole rubber stoppers to fit test tubes
- two 500 ml beakers
- two test tube clamps
- two ring stands
- water
- rubber tubing
- two graduated pipettes
- two tubing clamps
- black cloth
- Brodie manometer fluid
- Pasteur pipette
- wax pencil
- watch
- forceps

 ## Procedure 11.4 Aerobic Respiration

1. Procure the supplies needed for this experiment from your instructor.
2. To one test tube add 30 germinating peas, and to the second test tube add 30 nongerminating peas (Fig. 11.5).
3. Place a small wad of absorbent cotton on top of the peas in each test tube.
4. Using forceps, carefully place six pellets of KOH on the cotton in the test tube with the germinating peas. (*Caution: Do not handle the KOH with your bare fingers, as it is a caustic base.*)

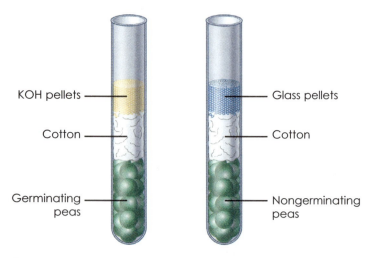

KOH pellets

Cotton

Germinating peas

Glass pellets

Cotton

Nongerminating peas

Figure 11.5 Test tubes with germinating and nongerminating peas.

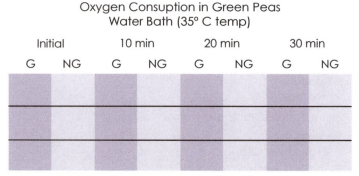

Rubber Tubing

Clamp

Graduated pipet

Test Tube Clamp

Dye

Ring Stand

Test Tube

KOH or Glass pellets

Cotton

Beaker

Room-temp water-bath

Either germinating or nongerminating peas

Figure 11.6 Apparatus housing room temperature water bath.

5. In the test tube containing the nongerminating peas, add 10 glass beads on top of the cotton.

6. Carefully insert the graduated pipette and rubber tubing into the rubber stoppers, as shown in Figure 11.6.

7. Insert the rubber stopper apparatus into the test tubes containing the peas, as shown in Figure 11.6.

8. Cover each test tube with a black cloth to block the light.

9. Attach test tube clamps to each ring stand as shown in Figure 11.6.

10. Prepare two room-temperature water baths by adding approximately 350 ml of tap water to each beaker.

11. Attach the test tubes containing the apparatus to each test tube clamp.

12. Gently submerge the test tubes into the water bath to the level just below the cotton.

13. Using a Pasteur pipette, carefully add a drop of Brodie's manometer fluid to the opened end of the graduated pipette.

14. Let the apparatus sit for 3 minutes for equilibration.

15. Pinch-clamp the rubber tubing, and mark the location of the dye droplet in the graduated pipette with the wax pencil. Record the location on the graduated pipette as 0 in the table below on the left.

16. At 10-minute intervals over the next 30 minutes, mark the position of the dye movement on the graduated pipette.

17. After marking the location of the dye movement, measure and record the distance the dye has traveled, and record it in the table below on the right.

18. Remove the pinch clamp from the rubber tubing, and pipette the dye back into the Pasteur pipette.

19. Disassemble your apparatus. Clean and return your equipment as directed by your laboratory instructor. (*Take precautions in the handling and disposal of KOH.*) Clean your laboratory stations.

20. Prepare a detailed laboratory report of this activity (see next page).

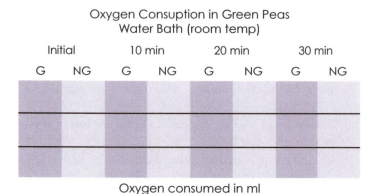

Oxygen Consuption in Green Peas
Water Bath (room temp)

Initial		10 min		20 min		30 min	
G	NG	G	NG	G	NG	G	NG

Oxygen consumed in ml

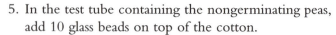

Oxygen Consuption in Green Peas
Water Bath (35° C temp)

Initial		10 min		20 min		30 min	
G	NG	G	NG	G	NG	G	NG

Oxygen consumed in ml

REPORT

Optional Exercise

Repeat the above steps, using test tubes placed in an ice water bath and a 35° C warm water bath. Record your results.

Check Your Understanding

Q. In the experiment, what gas did the peas take up?

_____

_____

Q. What was the role of KOH in this experiment?

_____

_____

_____

Q. Compare and contrast the results obtained from the germinating and nongerminating peas.

_____

_____

_____

Q. Relate this experiment to the equation for aerobic cellular respiration.

_____

_____

_____

_____

Q. How did temperature influence the rate of oxygen consumption?

_____

_____

_____

 Student Activity—Carbon Dioxide in Living Systems

Measuring the Production of Carbon Dioxide in Living Systems

Plants and animals use oxygen to release energy from carbohydrates in cellular respiration. In the process, carbon dioxide is formed as a waste product. We have observed that, in the process of photosynthesis, the plants use carbon dioxide to produce oxygen and carbohydrates. The interrelationship between plants and animals is crucial to life.

In this activity, the chemical indicator Bromothymol blue (BTB) will be used to detect the production of carbon dioxide in closed systems. The color of the BTB will change from blue to greenish yellow in the presence of an acid. As carbon dioxide is released through respiration, the solution will become more acidic and will shift in color. To appreciate this concept fully, this activity will utilize eight closed systems.

- Two of the eight systems will serve as controls, containing only dechlorinated water and BTB.
- Two systems will contain the dechlorinated water, BTB, and a small water snail.
- Two systems will have the dechlorinated water, BTB, and a small leafy stem of *Elodea*.

- The last two systems will contain the dechlorinated water, BTB, a water snail, and a leafy stem of *Elodea*.

The role of light in the process also will be investigated by placing one of each group in a darkened environment (Fig. 11.7).

In strong artificial light

In a dark environment

Figure 11.7 Beakers showing light and dark environments.

Materials

- eight 250 ml glass beakers
- eight squares of aluminum foil (approximately 16 cm in size)
- dechlorinated water
- Bromothymol blue (BTB) solution
- four small water snails
- four leafy stems of *Elodea*
- glass dropper
- wax pencil
- artificial light source
- dark closet or box
- safety goggles

Procedure 11.5
Measuring Carbon Dioxide

1. Procure eight 250 ml glass beakers from your instructor.
2. Label the eight beakers as follows: 1c, 2c, 1s, 2s, 1e, 2e, 1se, 2se.
3. Add 150 ml of dechlorinated water to each beaker.
4. Using a glass dropper, add approximately eight drops of BTB to each beaker.
5. Cover the control beakers (1c and 2c) with aluminum foil.
6. Carefully place one small water snail in the beakers labeled 1s and 2s. Cover each beaker with aluminum foil (*Caution: Wash your hands after handling the snails*).
7. To beakers 1e and 2e add a leafy stem of *Elodea*, and cover each beaker with a piece of aluminum foil.
8. Carefully place one small water snail and one leafy stem of *Elodea* to the beakers labeled 1se and 2se. Cover each beaker with a piece of aluminum foil.
9. Place beakers 1c, 1s, 1e, and 1se under a strong artificial light.
10. Put beakers 2c, 2s, 2e, and 2se in a dark closet or a box.

11. After 24 hours, observe the closed systems and record your observations in the following table.

Yellow half opaque Light			Gray half opaque Dark		
System	Color	Organism	System	Color	Organism
1c			2c		
1s			2s		
1e			2e		
1se			2se		

12. Dispose of your materials as directed by the lab instructor.
13. Clean up your lab station.
14. Prepare a detailed laboratory report of this activity (see next page).

Optional Exercise

After 24 hours, record your observations, and place in the dark room or box the beakers that were in the dark under the strong artificial lights and the beakers that were under the artificial lights. After 24 hours, record your observations in the table.

Check Your Understanding

Q. After 24 hours, what color was the water in the control systems (beakers 1c and 2c)? Why?

_____

_____

_____

Q. What causes the color to change in some of the beakers?

_____

_____

_____

_____

Q. How did the beaker colors compare in light and dark conditions?

_____

_____

_____

_____

Q. Explain why there was a difference in the color of the water in the system's 1se and 2se.

_____

_____

_____

Q. Hypothesize what would happen if beaker 1se were placed in a dark room or box for 24 hours after having been under a light source for 24 hours. Explain your hypothesis.

_____

_____

_____

_____

REPORT

_____

_____

_____

_____

_____

_____

_____

_____

_____

_____

Name: _____ Date: _____ Section: _____

Review Questions

1. What is the overall equation for aerobic respiration?

_____

2. What are the origins of the reactants and the destiny of the products?

_____

_____

_____

3. Compare and contrast anaerobic and aerobic respiration. Include an account of the number of ATP molecules produced in each reaction.

_____

_____

_____

_____

4. What is the interrelationship between photosynthesis and cellular respiration?

_____

_____

_____

_____

5. Why do many scientists consider one of the most important events in the evolution of life to be the origin of aerobic respiration?

_____

_____

_____

6. Compare and contrast alcoholic fermentation and lactate fermentation.

_____

_____

_____

_____

7. What are three products of alcoholic fermentation and lactate fermentation?

_____

_____

_____

8. What is the significance of glycolysis?

_____

_____

_____

9. Outline the series of events that occur once pyruvate is committed to the aerobic pathway.

_____

_____

_____

10. How many ATP molecules are produced in the citric acid cycle?

_____

_____

_____

11. What is the significance of the electron transport chain?

_____

_____

12. What are the end products of the electron transport chain?

_____

_____

13. What is the role of oxygen in the electron transport chain?

_____

_____

14. Why are most life forms on the Earth intimately linked to a star that is 150 million kilometers away?

_____

_____

_____

Chapter 12
"Omnis Cellula e Cellula": Understanding Cell Reproduction

Student Outcome Objectives

At the completion of this exercise, the student will be able to:

1. Discuss the functions of cell division.
2. Describe the cell cycle.
3. Distinguish among the three stages of interphase.
4. Describe the stages of mitosis.
5. Draw and label the stages of interphase and mitosis.
6. Identify the stages of mitosis in an onion root preparation and a whitefish blastula, using a microscope.
7. Discuss control of the cell cycle.

Overview

How can a one-celled human zygote grow to eventually become an adult consisting of more than 100-trillion cells? How do those pesky weeds seem to pop up overnight in your yard? How does a one-celled organism such as an amoeba reproduce? What is cancer? These are a few of the many questions that can be explained with an understanding of the process of cell division.

In 1855, the German physician Rudolph Virchow restated his observations concerning the reproduction of cells as "Omnis cellula e cellula"—meaning that cells come from preexisting cells. Living organisms, as well as the cells that comprise tissues, are capable of reproduction and growth. **Cell division** is the mechanism by which new cells are produced, whether for growth, repair, replacement, or forming a new organism.

In 1875, the German botanist Eduard Strausberger first described cell division in plants, but the processes of cell division were not described in detail until 1876. In that year, the German zoologist Walhter Flemming, while describing the development of salamander eggs, coined the terms **chromatin** and **mitosis** (cell division) and also established the framework for understanding the stages of cell division. Today, mitosis is recognized as a major component of the process of cell division.

Where a cell arises, there a cell must have previously existed (omnis cellula e cellula), just as an animal can spring only from an animal, and a plant only from a plant.

—**Rudolph Virchow (1821–1902)**

Cell division is the biological process by which cellular and nuclear materials of **somatic cells** (non-sex cells) are divided between two **daughter cells** formed from an original **parent** cell. The resulting daughter cells are structurally and functionally similar to each other and to the parent cell. In **prokaryotic organisms** (bacteria and cyanobacteria) the distribution of exact replicas of genetic material is comparatively simple. In eukaryotic organisms such as animals, however, the process of cell division is much more complex. The complexity is the result of the presence of a larger cell, a nucleus, and more DNA (larger genome) on individual linear chromosomes. Thus, any study of cell division in eukaryotes will include a discussion of the **cell cycle** (Fig. 12.1) and its two components, **interphase** and **mitosis**.

THE CELL CYCLE

Cells that are dividing pass through a regular sequence of cell growth and division known as the **cell cycle**. Cell type, hormones, and growth factors, as well as external conditions, influence the time required to complete the cell cycle. The cell cycle is divided into two major stages: interphase, when the cell is not actively dividing, and mitosis, when the cell is actively dividing. At any given time, a cell exists in one of the stages of the cell cycle. The majority of cells in an organism are in interphase.

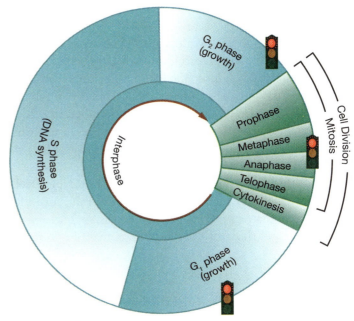

Figure 12.1 The cell cycle.

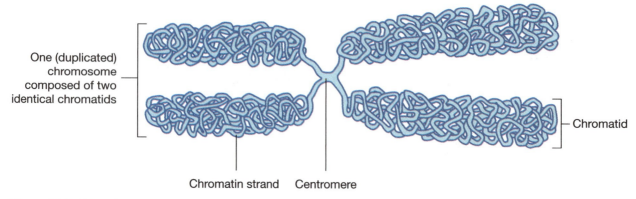

One (duplicated) chromosome composed of two identical chromatids

Chromatid

Chromatin strand   Centromere

Figure 12.2 Sister chromatids.

Interphase

Interphase typically accounts for 90% of the time that elapses during each cell cycle. Classically called the *resting stage*, interphase actually is a busy part of cell division. During interphase and before a cell begins the process of mitosis, it must undergo DNA replication, synthesize important proteins, produce enough organelles to supply both daughter cells, and assemble the structures used during cell division.

Interphase is divided into three phases: gap 1 (G_1), synthesis (S), and gap 2 (G_2). In recent years, a period known as G_0 has been described. G_0, or the **quiescent phase**, occurs after G_1 and represents the time that a cell is metabolically active but not proliferative. Actively dividing cells and cancer cells either skip G_0 or pass through the stage quickly. Some cells, such as nerve cells, may never exit G_0 once formed.

G_1 Phase

The G_1 **phase**, occurring after mitosis, serves as the cell's primary growth phase. During this time, the cell recovers from the previous division, increases in size, and synthesizes proteins, lipids, and carbohydrates. Also, the number of organelles and inclusions increases in number. In cells that contain **centrioles**, the two centrioles begin to form during G_1 (note that the cells of flowering plants, fungi, and nematodes do not contain centrioles). The G_1 phase occupies the major portion of the lifespan of a typical cell. Slow-growing cells, such as some liver cells, can remain in G_1 for more than a year. Fast-growing cells, such as epithelial cells and those of bone marrow, remain in G_1 for 16 to 24 hours.

S Phase

In the **S phase** of interphase, which follows the G_1 phase, each chromosome replicates to produce two daughter copies, termed **sister chromatids** (Fig. 12.2). The two copies remain attached at a point of constriction called the **centromere**. During this time, these structures are not visible under the light microscope. Among the numerous proteins manufactured during this phase are **histones** and other proteins that coordinate the various events taking place within the nucleus and cytoplasm. Duplication of the centrioles is completed, and they organize to migrate to the opposite poles of the cell. In addition, the microtubules that will become part of the spindle apparatus are synthesized.

G_2 Phase

The G_2 phase of interphase, which occurs after the S phase, involves further replication of membranes, microtubules, mitochondria, and other organelles. In addition, the newly replicated sister chromatids, which are diffusely distributed throughout the nucleus, begin to coil and become more compact. The start of chromosome condensation at the completion of G_2 signals the beginning of mitosis. By the end of G_2, the volume of the cell has nearly doubled.

Mitosis

Cell division includes both the division of the nucleus, called **karyokinesis,** and the division of the cytoplasm, **cytokinesis.** The overall action is to distribute the replicated chromosomes of the parent cell to the two newly forming identical daughter cells. Keep in mind that cell division is a dynamic and continuous process, but for ease in explanation and study, it has been divided into several stages—prophase, metaphase, anaphase, and telophase.

Prophase

Prophase is the longest and the first active stage of mitosis, characteristically accounting for 70% of the time that a cell spends in the mitotic process. During this stage, the chromosomes progressively become more visible as they shorten and thicken. Upon close examination in late prophase, the sister chromatids and their centromeres can be seen easily. Prophase also is characterized by migration of the centriole pairs toward opposite poles and disintegration of the nuclear envelope and nucleolus.

Short tubules known as **asters** appear and begin to radiate from the centrioles. The aster is believed to function to stiffen the point of microtubular attachment during the retraction of the spindle. Polar microtubules, between the centrioles, begin to form the **spindle fibers** that eventually will expand from one pole to another meeting at the equatorial plane.

The term **prometaphase** has become increasingly popular to describe events of late prophase. It usually is distinguished by the attachment of sister chromatids to the spindle fibers by means of a proteinaceous hook, or **kinetochore**.

Metaphase
Metaphase is the brief second stage of mitosis, when the centromere joining each pair of sister chromatids is attached to the spindle. Eventually the chromatid pairs appear to align along an imaginary line called the **equatorial plane**, or **metaphase plate**.

Anaphase
Anaphase is the third and most intriguing stage of mitosis. Although brief, this stage is characterized by the separation of the sister chromatids from their centromere, and the classic movement of the chromatids as they are pulled back along the spindle to the opposite poles. Errors during anaphase could result in an unequal distribution of genetic material.

Telophase
The final stage of mitosis, **telophase**, resembles a dumbbell with a set of chromosomes at each end. **Cytokinesis,** the division of cytoplasm upon completion of nuclear division, begins during this stage. Cytokinesis appears as a pinching-off of the cells along a cleavage furrow. During telophase, the spindle is disassembled, the nuclear envelope and nucleolus re-form, and cellular inclusions and organelles organize. Eventually the cleavage furrow is completed and two daughter cells are formed. Upon completion of telophase, the daughter cells enter the G_1 phase and the cell cycle starts anew. In plant cells, instead of undergoing cytokinesis, a **cell plate** is formed, dividing the mother cell into two daughter cells (Fig. 12.3 and Fig. 12.4).

Cell Regulation
The cell cycle is strictly regulated by a number of mechanisms that attempt to reduce the number of errors and ensure a smooth transition between the stages. Hormones such as the **growth hormone** control cell proliferation. Too much or too little growth hormone can influence an organism greatly. **Epidermal growth factors** (**EGFs**) are important in repairing wounds, and **fibroblast growth factors** are important in repairing and forming vessels.

Checkpoints in the cell cycle ensure that cell division is occurring properly. **Kinases** are enzymes, and **cyclins** are proteins that influence the activity of cell checkpoints. The first checkpoint, which occurs at the end of G_1, monitors the size of the cell and whether the DNA has been damaged. If it detects a problem, it can initiate repair, or **apoptosis** (cell suicide). The second checkpoint, which occurs at the end of G_2, essentially checks if DNA damage has occurred, that replication has occurred properly, and if the cell is ready to proceed to mitosis. The final checkpoint checks the spindle assembly and controls the onset of anaphase.

In humans, the *p53* gene is essential in the first checkpoint, in the apoptosis of damaged cells. If the *p53* does not function properly, damaged and mutated cells will be allowed to proliferate. Many forms of cancer in humans are related to a malfunctioning *p53* gene, and cancer can be thought of as cell division that has lost control.

Did You Know?
The Greek God of Medicine, Asclepius, was best known for prescribing rest and proper diet for his patients. Great temples of healing were built in his honor. In these temples, holy snakes and dogs were allowed to slither and walk among the sick and injured. Supposedly, the snakes would keep away evil spirits and the dogs would promote the healing by licking the wounds.

Today, it is known that dogs contain a great amount of **epidermal growth factors** (**EGFs**) in their saliva, which promote cell division and speed healing. The next time your dog detects a scratch on your arm, notice the concern! EGFs are an integral part of modern medicine, used in cornea transplants and burn treatment.

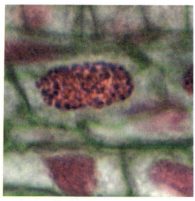

Early prophase
Chromatin begins to condense to
form chromosomes.

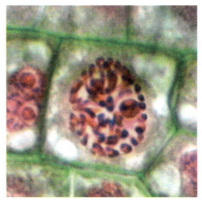

Late prophase
Nuclear envelope is intact,
and chromatin condenses into
chromosomes.

Early metaphase
Duplicated chromosomes are
each made up of two chromatids,
at equatorial plane.

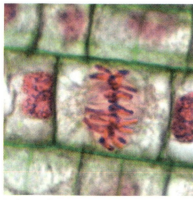

Late metaphase
Duplicated chromosomes are
each made up of two chromatids,
at equatorial plane.

Early anaphase
Sister chromatids are beginning
to separate into daughter
chromosomes.

Late anaphase
Daughter chromosomes are
nearing poles.

Telophase
Daughter chromosomes are at
poles, and cell plate is forming.
1. Cell plate

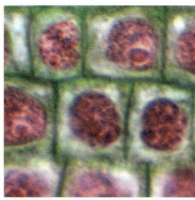

Interphase
Two daughter cells result from
cytokinesis.

Figure 12.3 The stages of mitosis in Hyacinth, *Hyacinthus*, root tip (All 430X).

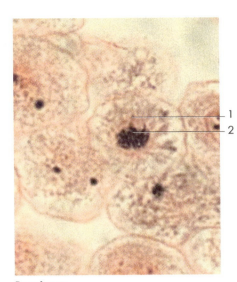

Prophase

Each chromosome consists of two chromatids joined by a centromere. Spindle fibers extend from each centriole.

1. Aster around centriole
2. Chromosomes

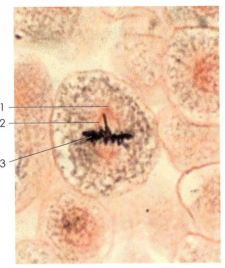

Metaphase

The chromosomes are positioned at the equator. The spindle fibers from each centriole attach to the centromeres.

1. Aster around centriole
2. Spindle fibers
3. Chromosomes at equator

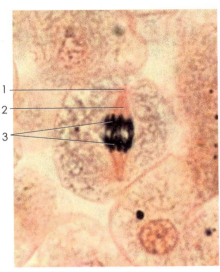

Anaphase

The centromeres split, and the sister chromatids separate as each is pulled to an opposite pole.

1. Aster around centriole
2. Spindle fibers
3. Separating chromosomes

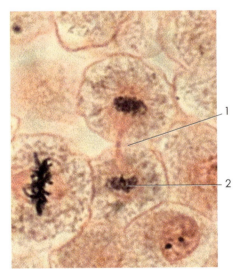

Telophase

The chromosomes lengthen and become less distinct. The cell membrane forms between the forming daughter cells.

1. New cell membrane
2. Newly forming nucleus

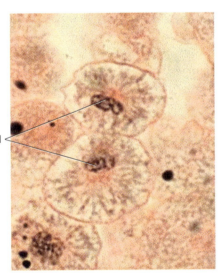

Daughter cells

The single chromosomes (former chromatids—see anaphase) continue to lengthen as the nuclear membrane reforms. Cell division is complete and the newly formed cells grow and mature.

1. Daughter nuclei

Figure 12.4 The stages of animal cell mitosis followed by cytokinesis. Whitefish blastula (All 500X).

Student Activity—Animal Cell Mitosis

Observing Mitosis in Animal Cells

In this exercise, you will observe the cell cycle in prepared slides of a whitefish blastula. The **blastula** occurs in the early stage in the embryonic development of an animal, seen as a ball of cells, each cell in one of the stages of interphase or mitosis.

Materials
- prepared slide of whitefish blastula
- compound light microscope
- colored pencils

Procedure 12.1
Observing Animal Mitosis

1. Procure a prepared slide of a whitefish blastula.
2. Obtain a compound light microscope, and follow the safety and handling procedures for a light microscope.
3. Observe the whitefish blastula first on low power. In this field you will be able to see many slices of depth of the blastula. Focusing on low power, you should see individual cells.
4. Switch to high power, and sketch and describe each of the stages of interphase and mitosis in the space below.
5. Draw each stage of cell division and describe what is happening within the cell in detail.

Interphase

Prophase

Metaphase

Anaphase

Telephase

Student Activity—Plant Cell Mitosis

Observing Mitosis in Plant Cells

In this exercise, you will observe the cell cycle in prepared slides of green onion (scallions) root tip mitosis. In plants, the root tips often are used to study the cell cycle. In plants, cell growth occurs in the **meristematic** regions, located at the tips of stems and roots.

Materials
- prepared slide of onion root tip mitosis
- compound light microscope
- colored pencils

Procedure 12.2
Observing Plant Mitosis

1. Procure a prepared slide of onion root tip mitosis.
2. Obtain a compound light microscope, and follow the safety and handling procedures for a light microscope.
3. Observe the onion root tip mitosis first on low power. In this field, you will be able to see many stages of interphase and mitosis. Focusing on low power, you should see individual cells.
4. Switch to high power, and sketch and describe each of the stages of interphase and mitosis, in the space provided on the following page.

Check Your Understanding

Q. When studying the cell cycle, why is it necessary to use an embryonic cell to observe the stages?

_____

_____

_____

_____

Q. Describe cytokinesis and the cell plate. Which stage of mitosis is associated with each?

_____

_____

_____

_____

_____

Interphase

Anaphase

Prophase

Telephase

Metaphase

Student Activity—Root Tip Squash Mount

Preparing a Slide of Plant Cell Mitosis

In this exercise, you will prepare a root tip squash of *Allium sativum* (onion root tips).

To observe the stages of the cell cycle, you will use a **squash mount**. To prepare a squash mount, tissue taken from the meristem region of the plant is removed and treated with a fixative to stop the cells from dividing. The onion root tissue then is placed on a microscope slide containing methyl green-pyronin Y stain. This stain contains two different stains to distinguish between the DNA and the RNA. The methyl green stains the DNA blue, and the pyronin Y stains the RNA pink. The onion root tip then is squashed, making a single layer of tissue. A coverslip is placed over the squashed tissue, and you then will view the squash mount under the microscope.

Materials
- safety goggles
- compound light microscope
- microscope slide
- coverslip
- forceps
- paper towels
- vial of pretreated *Allium* root tips
- methyl green-pyronin Y stain in a dropper bottle
- wooden macerating stick
- pencil with eraser
- watch or clock

Procedure 12.3
Squash Mount

1. Obtain materials from your instructor.
2. Put on your safety goggles.
3. Take a blank microscope slide and place it on top of several folded paper towels.
4. Carefully add two drops of the methyl green-pyronin Y stain to the center of the microscope slide. (*Caution: Be careful when handling the stain, as it will stain your clothing and your skin.*)

5. Using forceps, remove a prepared onion root tip from the vial and carefully transfer it to the stain on the microscope slide.
6. With the wooden macerating stick in your hand, gently but firmly smash the onion root tip into the stain, using a straight up-and-down motion.
7. Let the squashed root tip stain for 15 minutes. Add more stain if needed to keep the squashed root tip from drying out.
8. After the 15-minute period, carefully place a coverslip on top of the stained squashed onion root tip.
9. Place a folded paper towel over the coverslip and, using the eraser part of the pencil, press firmly straight down (do not twist) on the folded paper towel. (*Caution: Do not apply too much pressure, as the coverslip and slide could break. If you break the coverslip or the slide, notify your instructor.*)
10. Observe the slide, first using low power. Once you have the cells in focus, use caution when moving to a higher power. This is a thick mount.
11. Sketch and record your observations below.

12. After making your observations, return your materials to the proper place as directed by your lab instructor, and clean your lab station.

Check Your Understanding

Q. What are the most common stages of the cell cycle you observed?

_____

_____

_____

_____

_____

Q. What purpose does the methyl green-pyronin Y stain serve?

_____

_____

_____

Q. Compare the prepared slides of onion root tip mitosis to the slide that you prepared.

_____

_____

_____

NOTES

_____

_____

_____

_____

_____

_____

_____

_____

_____

_____

_____

_____

_____

_____

Name: _____ Date: _____ Section: _____

Review Questions

1. What is the significance of the S phase of interphase?

_____

_____

_____

2. Draw and describe the formation of sister chromatids.

_____

_____

_____

_____

_____

_____

3. Why did early scientists call interphase the "resting stage?"

_____

_____

_____

4. What are some differences between animal and plant mitosis?

_____

_____

_____

5. What is the relationship between cell division and cancer?

_____

_____

_____

6. How does the interphase set up the prophase?

_____

_____

_____

7. What is the difference between karyokinesis and cytokinesis?

_____

_____

_____

8. Draw and label the cell cycle.

9. What are the three checkpoints of the cell cycle, and what is the basic function of each?

_____

_____

_____

_____

10. What is the function of cell division?

_____

_____

Chapter 13
Ensuring Genetic Variation: Understanding Meiosis

Student Outcome Objectives

At the completion of this exercise, the student will be able to:

1. Define meiosis.
2. Identify which cells undergo the process of meiosis.
3. Explain the relevance of meiosis to sexual reproduction.
4. Describe how the chromosomes are reduced from diploid number (2n) to haploid number (1n) in meiosis I.
5. Define synapsis.
6. Describe the process of crossing over and its significance.
7. Explain why meiosis II often is referred to as a mitotic-like series of events.
8. Explain the events of spermatogenesis.
9. Describe the process of oogenesis.
10. Explain how meiosis differs from mitosis.
11. Compare and contrast aneuploidy and euploidy.

Overview

In addition to the evolution of aerobic respiration, one of the most significant events in the history of life was the "invention" of sexual reproduction. Sexual reproduction provides diversity to life. Offspring are no longer "chips off the old cell" but, rather, unique individuals with characteristics inherited from past generations along with new, hopefully beneficial, traits.

The underlying mechanism behind sexual reproduction is **meiosis.** Unlike mitosis, which involves a division of the somatic or body cells, meiosis involves **reduction division,** resulting in the formation of sex cells, or **gametes.** This occurs in two distinct nuclear divisions driven by mitosis and known as meiosis I and meiosis II. In addition, meiosis literally jumbles the genes, allowing for genetic variation.

With a few exceptions to the normal condition, species possess a genetically determined number of chromosomes. These chromosomes occur in pairs, one from each parent. The like chromosomes, one from each parent are the **homologous chromosome.** The **diploid number** (2n) represents the total number of chromosomes in the nucleus of a cell. In humans, the diploid number is 46. The number of pairs of chromosomes is the **haploid number** (n), which is one-half of the diploid number. In humans,

the haploid number is 23. Meiosis involves the reduction of the diploid state to the haploid state in gametes to ensure stability in the number of chromosomes passed from one generation to the next.

Table 13.1 Determine the diploid and haploid numbers of the following organisms		
Organism	**Diploid Number**	**Haploid Number**
Gold Fish		
Horse		
Bottlenosed Dolphin		
Cat		
Chimp		
Aspergillus		
Adders Tongue fern		
Ginkgo		
Strawberry		
Corn		

MEIOSIS I AND II

Sexual reproduction is based upon **fertilization**, the union of haploid male and female gametes to form a diploid **zygote**. The zygote possesses homologous chromosomes from each parent. The process of meiosis varies slightly in the Kingdoms Plantae, Fungi, and Animalia.

Meiosis consists of two distinct cell divisions designated **meiosis I** and **meiosis II**. In both meiosis I and meiosis II, the cell has progressed through interphase of the cell cycle.

Meiosis I

The first stage of meiosis I is **prophase I.** In this stage, homologous chromosomes pair up in a process called **synapsis** and form **bivalents** or **tetrads.** A **synaptonemal** complex

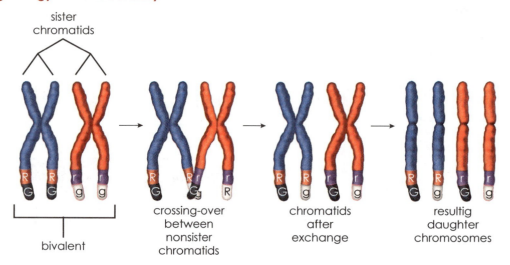

Figure 13.1 Crossing over.

is formed between homologous chromosomes to ensure proper pairing. The physical exchange of genetic material between homologous pairs, known as **crossing over**, results in the formation of new combinations of genetic material (Fig. 13.1). The arms of the sister chromatids are held together by **chiasmata**.

In **metaphase I** the bivalents line up independently along the metaphase plate. This independent alignment of sister chromatids allows for the genetic diversity seen in families. In contrast to metaphase during mitosis, the sister chromatids are lined up in a double row rather than a single row.

Anaphase I follows metaphase I. In this stage, homologous chromosomes separate and migrate toward opposite poles.

Telophase I, the final stage of meiosis I, is characterized by the sister chromatids reaching the opposite poles and eventually forming two new cells. Although the two cells produced at the end of meiosis I are considered haploid by convention, they possess pairs of homologous chromosomes (Fig. 13.2).

In some species, **interkinesis** follows telophase I. It is similar to interphase of the cell cycle, but DNA replication does not occur.

Meiosis II

Meiosis II is similar to the cell cycle and results in division of the sister chromatids from meiosis I. **Prophase II** is characterized by attachment of the sister chromatids to the spindle. During **metaphase II**, the sister chromatids line up along the metaphase plate. In **anaphase II**, the sister chromatids are separated and migrate toward the centrioles. **Telophase II** results in the formation of four unique daughter cells (Fig. 13.3).

Life Cycles in Plants

Plant life cycles are characterized by an alternation of generations with two distinct phases: the **sporophyte** generation (2n) and the **gametophyte** generation (n). The diploid sporophyte generation produces haploid spores that develop into a new organism. A pine tree is an example of a sporophyte. visible true mosses are an example of the gametophyte generation. The gametophyte generation produces haploid gametes that undergo fertilization to form a diploid zygote (Fig. 13.4).

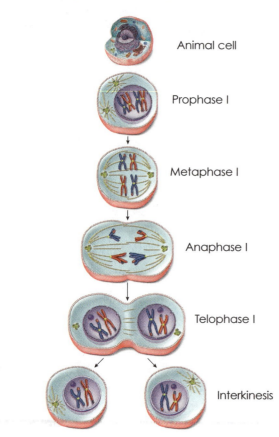

Figure 13.2 Overview of meiosis I in animal cells.

SEX DIFFERENTIATION IN ANIMALS

The majority of animals are **dioeciou**s—having separate sexes. Haploid cells produced in meiosis II become **gametes** (sperm or egg). The process of forming sex cells is termed **gametogenesis**. In males, **spermatogenesis** occurs in the seminiferous tubules in the **testes.** In females, **oogenesis**, the production of eggs, occurs in the **ovaries.**

Spermatogenesis

In males, diploid **primordial germ cells** divide by mitosis to form diploid **spermatogonia**. Some of the spermatogonia divide by mitosis, forming diploid **primary spermatocytes**. Others continue to divide, forming many spermatogonia.

   Spermatogenesis begins at puberty and continues throughout the male's lifetime. The primary spermatocyte will enter meiosis I. At the end of meiosis I, two haploid **secondary spermatocytes** are formed. The two secondary spermatocytes undergo meiosis II, eventually forming four **spermatids**. After undergoing metamorphosis or **spermiogenesis**, the spermatids mature into haploid sperm cells. The whole process of spermatogenesis, including spermiogenesis, takes about 2 months. From 200 to 600 million sperm are released during each ejaculation (Fig. 13.5).

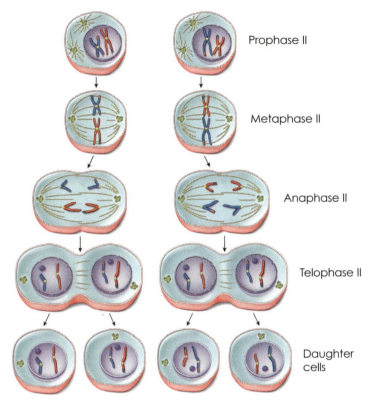

Prophase II

Metaphase II

Anaphase II

Telophase II

Daughter cells

Figure 13.3 Overview of meiosis II in animal cells.

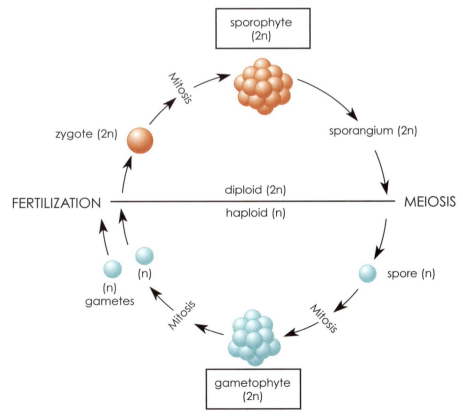

Figure 13.4 Alternation of generations in plants.

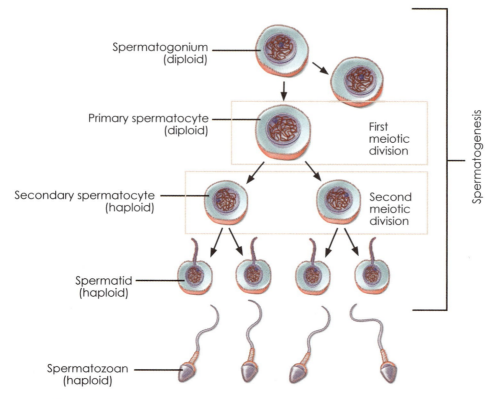

Figure 13.5 Overview of spermatogenesis.

Oogenesis

The ovaries are female endocrine glands. It is here that mature egg cells or ova are formed. **Oogenesis** is the process that results in egg formation. **Primordial germ cells** (diploid number) develop during the first month of embryonic development. These cells divide by mitosis and undergo differentiation to form diploid **oogonia**, an equivalent to the male spermatogonia. Oogonia undergo further differentiation to form diploid **primary oocytes**, again equivalent to the primary spermatocyte in males.

Unlike males, the primary oocytes form before birth. It is estimated that each ovary contains nearly a million primary oocytes at birth. Once formed, the primary oocyte begins meiosis I. This process occurs prenatally. Prophase I is an extended phase, not completed until the female reaches puberty. After reaching puberty, each month one primary ooctye that was held in suspension continues its meiotic division. A haploid **secondary oocyte** is formed along with a first **polar body.** This small haploid polar body is a result of unequal division of the cytoplasm. The first polar body from meiosis I also may divide to form two additional haploid polar bodies. The polar body or polar bodies eventually degenerate. Their biological importance is to provide the egg cell with massive amounts of cytoplasm and cellular organelles (Fig. 13.6).

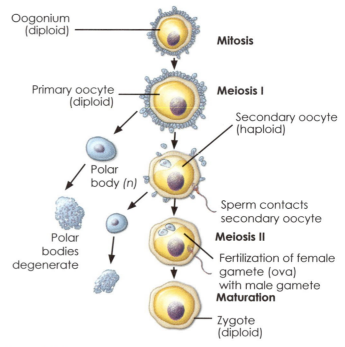

Figure 13.6 Overview of oogenesis.

Life thus forms a long, unbroken chain of generations, in which the child becomes the mother, and the effect becomes the cause.

—Rudolf Carl Virchow (1821–1902)

The secondary oocyte begins meiosis II. It is released from the Graafian follicle as a mature egg or ovum and begins its journey through the Fallopian tube, awaiting fertilization. If fertilization does not occur, the ovum is lost in the next menstrual cycle. If fertilization occurs, meiosis II is completed and a diploid zygote is formed and a haploid polar body is released. The zygote then can divide by mitosis and develop into an **embryo**.

Most humans exhibit **euploidy**, having the normal number of chromosomes. Unfortunately, one in 250 live births exhibit **aneuploidy**, resulting in an individual with the wrong number of chromosomes. Most aneuploid individuals abort spontaneously. Aneuploidy results from meiotic errors (**nondisjunction**), and the individual will have extra or too few chromosomes. Examples of aneuploid states are trisomy 21 (Down syndrome), Edward's syndrome, Patau syndrome, Klinefelter syndrome, and Turner syndrome.

If every one were cast in the same mould, there would be no such thing as beauty.

—**Charles Robert Darwin (1809–1882)**

 Check Your Understanding

Q. Describe several symptoms of each of the conditions listed above.

_____

_____

_____

Q. What is the relationship between the mother's age and Trisomy 21?

_____

_____

_____

Q. Describe three other aneuploid states in humans.

_____

_____

_____

 Student Activity—Stages of Meiosis in *Ascaris* Ovaries

Observing the Stages of Meiosis in *Ascaris* (Roundworm) Ovaries

In this activity, you will observe the stages of meiosis, using a series of prepared slides of *Ascaris*, a species of roundworm or nematode. *Ascaris* is a parasitic worm that can infect primarily the digestive system of animals, including humans. The female roundworm can exceed lengths of 30 cm. Female *Ascaris* have large, elongated ovaries that contain as many as 27 million ova. Within the nucleus of their cells, four chromosomes can be easily observed (Fig. 13.7).

Materials
• prepared slides of *Ascaris* ovaries
• compound light microscope
• colored pencils

 **Procedure 13.1
Meiosis in the *Ascaris***

1. Obtain a compound light microscope from your instructor.
2. Carefully procure the series of slides of meiosis in the *Ascaris* ovary.
3. Place your first slide on the stage of the microscope, and focus first on low power before moving the objective to a higher power.
4. Locate, sketch, and label the stages of meiosis that you observe in the space provided below.
5. Repeat steps 3 and 4 for the remaining slides.

Q. What are polar bodies? How are they formed?

_____

_____

_____

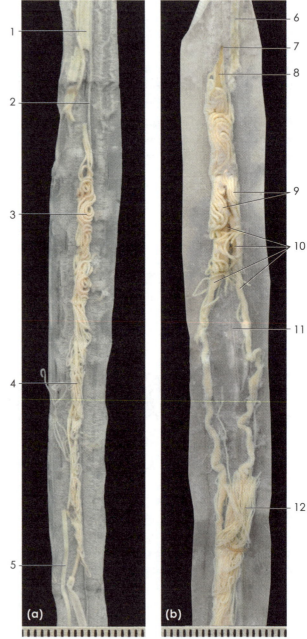

![Check Your Understanding icon] **Check Your Understanding**

Q. How many chromosomes does the nucleus of an _Ascaris_ ovum have?

_____

_____

_____

Q. How many pair of chromosomes will meiosis I have?

_____

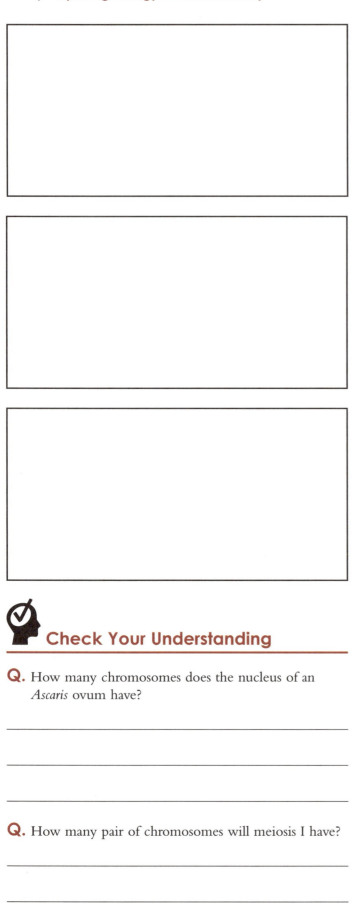

Figure 13.7 (a) The internal anatomy of a male _Ascaris_, (b) and the internal anatomy of a female _Ascaris_.

1. Intestine
2. Lateral line
3. Ductus deferens
4. Testes
5. Seminal vesicle
6. Intestine
7. Genital pore
8. Vagina
9. Oviducts
10. Uteri (Y-shaped)
11. Lateral line
12. Ovary

Student Activity—Sperm Cell Meiosis

Examining Meiosis in Sperm Cells

In this activity you will observe gametogenesis, using pre-pared slides from the testes of an animal. The formation of sperm (spermatogenesis) occurs in the seminiferous tubules of the testes.

Materials
- prepared slides of grasshopper, frog, bull, or other animal testis
- compound light microscope
- colored pencils

 **Procedure 13.2
Sperm Cell Meiosis**

1. Obtain a compound light microscope from your instructor.
2. Carefully procure the slide of a testis.
3. Place the slide on the stage of the microscope, and focus first on low power before moving the objective to a higher power.
4. Locate, sketch, and label the stages of meiosis that you observe, in the space below.

Check Your Understanding

Q. What is spermatogenesis?

_____

_____

_____

Q. Where in the testis does sperm formation take place?

_____

_____

_____

Q. Tell which of the following have the diploid number (2n) of chromosomes, and which possess the haploid number (n) of chromosomes: spermatogonia, primary spermatocytes, secondary spermatocytes, spermatids.

_____

_____

_____

_____

Q. During gametogenesis, a sperm cell undergoes struc-tural changes. Briefly describe these changes and how they relate to function.

_____

_____

_____

Student Activity—Plant Gametogenesis

Observing Plant Gametogenesis in a *Lilium* Anther and Ovary

Plants carry out sexual reproduction. Therefore, meiosis is integral in the formation of gametes. Their life cycle is different from animals in that plants undergo an **alternation of generations**, in which they alternate between the haploid and diploid forms. Meiosis occurs in the ovary of female plants. The male gametes (pollen grains) are produced in the anthers. In this lab activity, you will observe and describe stages of meiosis in prepared slides of the *Lilium* (lily) anther and the *Lilium* (lily) ovary (Fig. 13.8).

Figure 13.8 The anatomy of a lily.
1. Anther   4. Style
2. Filament   5. Petal
3. Stigma   6. Pedicel

Materials
- prepared slides of meiosis in the *Lilium* anther and *Lilium* ovary
- compound light microscope
- colored pencils

Procedure 13.3
Meiosis in the *Lilium*

1. Obtain a compound light microscope from your instructor.
2. Carefully procure the series of slides of meiosis in the *Lilium* anther and the *Lilium* ovary.
3. Place your first slide on the stage of the microscope, and focus first on low power before moving the objective to a higher power.
4. Locate, sketch, and label the stages of meiosis that you observe in the space provided below.

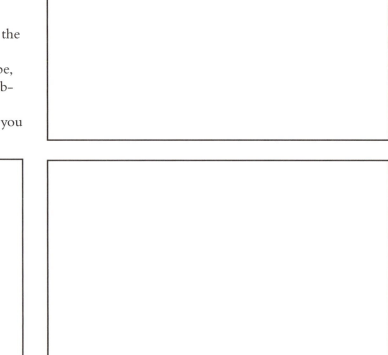

5. Repeat steps 3 and 4 for the remaining slides.
6. Sketch the slides in the space provided.

Check Your Understanding

Q. Why is sexual reproduction advantageous for the plant?

_____

_____

_____

Q. What is the function of the anther?

_____

_____

_____

Q. Name one difference between plant and animal meiosis.

_____

_____

_____

Student Activity—Stages of Meiosis in Flower Buds

Observing the Stages of Meiosis in Gathered Flower Buds

In this lab activity, students will prepare slides from the anthers of intact flower buds and observe different stages of meiosis in the cells. The student can collect very young buds from flowering plants or acquire them from a floral shop or plant nursery. The buds should be gathered in the spring before the anthers have produced pollen. Most flowering plants are good sources. Several young buds should be picked and stored in a stoppered flask containing a fixative for several days to bleach out the plant pigments. Once fixed, the buds can be used for up to a year. The students will follow the procedure outlined below for making approximately eight slides of the anthers and observe stages of meiosis in these slides. They can compare the slides they make to the prepared slides of the *Lilium* anther used in the subsequent activity. Refer to Figure 13.8.

Materials
- several young flower buds (beans, lilies, hibiscus, spiderworts, etc.)
- fixative solution (3:1 ethanol/acetic acid): Use approx. 300 ml of 100% ethanol or 100% methanol and 100 ml if glacial acetic acid. (*Caution: This is a flammable solution.*)

Materials (cont.)
- 250 ml flask with a stopper for the fixative and the young, intact flower buds
- 500 ml beaker
- Petri dishes
- eyedropper
- fine-tipped forceps
- scalpel
- acetocarmine stain (*Caution: This can stain skin and clothing.*)
- four glass slides and four coverslips
- dissecting microscope
- compound light microscope
- Bunsen burner of alcohol lamp or hot plate (*Caution: Follow safety rules for use of a Bunsen burner.*)
- clothespin
- paper towels

Procedure 13.4
Meiosis in Flower Buds

1. Procure very young flower buds (student or instructor).
2. Obtain the fixative solution from the instructor.
3. Carefully measure 200 ml of the fixative solution in the beaker, and pour the fixative into the flask.

4. Place the flower buds in the flask, and cover with a stopper.

5. Let the buds sit undisturbed in the fixative solution for 4 days.★

6. After the time allotted, remove the stopper from the flask.

7. Using the eyedropper, add a few drops of the fixative solution to a Petri dish to ensure that the bud does not dry out.

8. Using the forceps, carefully remove a bud from the fixative solution and place it in the Petri dish containing a small amount of fixative solution.

9. Using the forceps and scalpel, locate the anthers. (You may use a dissecting microscope to isolate the anthers.)

10. Remove the anthers, and carefully place them on the glass slide.

11. Cover the anthers with the acetocarmine stain.

12. Using a scalpel, mash the anthers in the stain.

13. While the anthers set in the stain, use the forceps to remove any cellular debris including the outer covering of the anthers.

14. Allow the anthers to set in the stain for 4 more minutes. Be sure to add more stain to the slide to keep the tissue from drying out.

15. Cover the stained anthers with a coverslip.

16. Attach a clothespin to one end of the glass slide containing the stained anther.

17. Using a Bunsen burner or an alcohol lamp or a hot plate (if using a hot plate, the temperature should be set at 80°C), move the slide in and out of the flame for approximately 4 minutes *Caution: The slide should be moved in and out of the flame quickly so that the stain does not boil.*

18. Continue to add acetocarmine stain dropwise to the edge of the coverslip so the cells do not dry out.

19. After the allotted heating time, let the slide cool for a few seconds.

20. Carefully place the slide between two layers of paper towels, and press firmly down to squash the cell's nuclei. *Caution: Press firmly straight down, but not so hard that you break the slide.*

21. The slide is now ready for viewing. Use a compound microscope to observe the meiotic cells. View first by using the low power objective and, once in focus, move the objective to a higher power.

22. Repeat steps 6–21 for the remaining slides.

23. Locate, sketch, and label the stages of meiosis that you observe, in the space provided.

★The instructor may have prepared the buds already.

Check Your Understanding

Q. Compare the prepared slides of the *Lilium* anther to the slides you have made.

_____

_____

_____

Q. What stage of meiosis was observed most readily in the slides you made?

_____

_____

Name: _____ Date: _____ Section: _____

1. What is meiosis?

_____

_____

_____

2. What cells undergo meiosis?

_____

_____

_____

3. Define and describe the terms diploid (2n), haploid (n), synapsis, and crossing over.

_____

_____

_____

4. Describe each stage of meiosis.

_____

_____

_____

5. Explain how the chromosome number is reduced.

_____

_____

_____

6. Describe the events of spermatogenesis.

_____

_____

_____

_____

7. Where does spermatogenesis occur?

_____

_____

_____

8. Explain the events of oogenesis.

_____

_____

_____

9. When does the process of oogenesis begin?

_____

_____

_____

10. What are polar bodies?

_____

_____

_____

11. Compare and contrast mitosis and meiosis.

_____

_____

_____

12. What is the evolutionary significance of meiosis?

_____

_____

_____

13. Describe several aneuploid conditions in humans.

_____

_____

_____

Chapter 14
The Developing Organism: Understanding Embryology

Student Outcome Objectives

At the completion of this exercise, the student will be able to:

1. Define embryology.
2. Discuss the role of cell division in embryology.
3. Define and describe cleavage.
4. Describe, draw, and label the basic stages of embryological development in starfish, including the unfertilized egg, zygote, two–cell stage, four-cell stage, eight-cell stage, morula, blastula, and gastrula.
5. Compare and contrast Protostomates and Deuterostomates.
6. Compare and contrast diploblastic and triploblastic.
7. Identify and describe the anatomy of a blastula and a gastrula.
8. Discuss the fate of the endoderm, ectoderm, and mesoderm.
9. Describe the anatomical features and functions of the parts of a chicken egg.
10. Describe the progression of development of a chicken embryo.
11. Using a microscope, identify the basic stages of embryological development in starfish and chickens.

Overview

One of the most fascinating and awe-inspiring events in nature is the development of a single-celled zygote into a multi-celled organism. **Embryology** is the study of the growth, early development, and differentiation of organisms. In animals, the basic patterns for early embryological development are remarkably similar. In fact, similarity in the development of embryos is significant in evolutionary biology.

In **fertilization**, the union of the sperm and the egg form a **zygote**. Following fertilization, a series of cell divisions (cleavage) occurs in the zygote, resulting in the formation of new cells. For several generations, the daughter cells are roughly half the size of the mother cell. These early cells are **biphasic**, alternating between the S phase of interphase and the mitotic cycle. These cells do not enter G1 and G2 of interphase.

The cleavage patterns vary in organisms, from **holoblastic cleavage** in some flatworms, most annelids, many molluscs, echinoderms, amphibians, and mammals where the whole zygote divides, to **meroblastic cleavage** in most arthropods, cephalopods, most fishes, reptiles, and birds where only part of the zygote divides. Specific types of these basic cleavage patterns also are recognized. The cleavage pattern is dependent upon the amount of yolk in the egg and the evolutionary history of the organism.

In many developing embryos, the cleavage divisions are synchronized, resulting in an embryo containing 2, 4, 8, 16, 32, etc. cells. Each new cell resulting from division is called a **blastomere**. At 16–32 cells, the growing embryo reminded early embryologists of a mulberry and hence was called a **morula**. Today it is more likely to be associated with a soccer ball with each pebble of the ball being a blastomere. At approximately 64 cells, the developing organism resembles a hollow sphere called the **blastula**. In some species, the blastula may not be hollow. The cells that make up the blastula are called blastomeres, and the central cavity is called the **blastocoel**. The blastula is like a basketball with each pebble of the ball being a blastomere.

Eventually, the migration and specialization of cells result in formation of the gastrula. The gastrulae are noted by the presence of the primitive gut, termed the **archenteron**, and formation of the **germ layers**. The germ layers consist of the **ectoderm**, **mesoderm**, and **endoderm**. Each germ layer gives rise to specific structures. After formation of the gastrula, development becomes more complex. For example, in chordates the **neurula** follows the gastrula and is characterized by development of the notochord, neural tube, and coelom. As embryological development continues, organs called **organogenesis** begin to form and the shape of the organism, **morphogenesis**, begins to take place.

In this laboratory, the student will examine early developmental stages in starfish and chicken embryos. The student must be able to identify these stages and describe the significance of each stage.

Did You Know?

In some organisms, mitosis occurs rapidly in early development. In some species of frog, an egg can divide into more than 37,000 cells in two days!

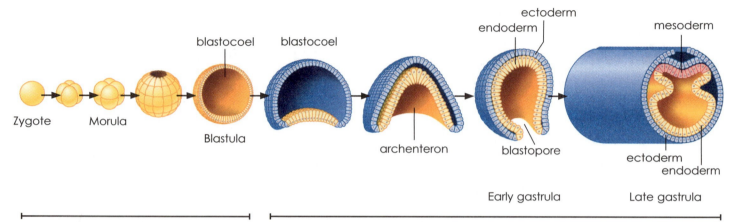

Figure 14.1 Stages of embryological development.

OBSERVING EMBRYOLOGICAL DEVELOPMENT

Unfertilized Egg

The unfertilized egg appears as a large spherical cell with a distinct nucleus and nucleolus (Fig. 14.2). The yolk in the unfertilized egg appears as evenly distributed particles throughout the cytoplasm. Starfish and placental mammals (including humans) possess this type of egg, known as an **isolecithal egg**.

Other Types of Egg

- *Mesolecithal egg*: contains a moderate amount of unevenly distributed yolk. This type of egg is found in cephalopod molluscs, amphibians, and some fish.
- *Telolecithal egg*: contains a large amount of yolk, usually concentrated at one end of the egg. This kind of egg can be found in many species of fish, reptiles, and birds.
- *Centrolecithal egg*: contains the yolk concentrated at the center of the egg. This kind of egg is found in most arthropods.

Fertilized Egg (Zygote)

The zygote is a fertilized egg and appears as a single-celled structure much like the unfertilized egg, except that the nucleus and the nucleolus are not visible (Fig. 14.3). Shortly after fertilization, a series of physical and physiological changes begin within the egg. Eventually the pronuclei of the sperm and egg unite to form the fertilization nucleus. In addition, fertilization triggers egg activation and initiates a variety of processes including cleavage.

Early Cleavage

Early cleavage follows predetermined patterns dependent upon the species. This results in the 2-, 4-, and 8-cell stage. The individual cells are known as blastomeres. In some organisms (the protostomes such as molluscs, and insects), the fate of the blastomere is determined from the onset of embryological development, known as **determinate cleavage**. In other organisms (deuterostomes such as echinoderms and chordates), the blastomeres have the ability to modify their fate under normal conditions, known as **indeterminate cleavage** (Fig. 14.4).

As the human fetus develops, its changing form seems to retrace the whole of human evolution from the time we were cosmic dust to the time we were single-celled organisms in the primordial sea to the time we were four-legged, land-dwelling reptiles and beyond, to our current status as large-brained, bipedal mammals. Thus, humans seem to be the sum total of experience since the beginning of the cosmos.

—**Jonas Salk (1914–1995)**

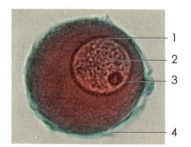

Figure 14.2 An unfertilized egg.
1. Nuclear membrane
2. Nucleus
3. Nucleolus
4. Cell membrane

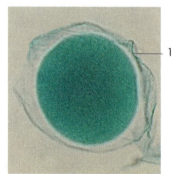

Figure 14.3 A fertilized egg.
1. Fertilization membrane

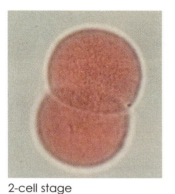
2-cell stage

4-cell stage

8-cell stage

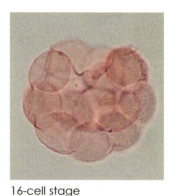
16-cell stage

Figure 14.4 2–16 cell stages of the sea star (All 100X).

Morula

The morula is a mulberry-like cluster of 16–32 blastomeres. Note that the morula is no larger than the fertilized egg (Fig. 14.5).

Blastula

As cleavage continues, the blastomeres begin to form around a fluid-filled inner cavity known as the **blastocoel**. Thus, the blastula is classically called a "hollow sphere of cell." (*Note:* Not all blastulas are hollow.) This event usually occurs around the 64-cell stage and continues until gastrulation. The blastula stage in mammals is termed the **blastocyst**. In starfish, the walls of the blastula usually are one cell layer thick (Fig. 14.6).

Gastrula

After formation of the blastula, a small depression begins to form at one end of the sphere of cells. This depression marks one of the most important developmental stages in the life history of an organism, **gastrulation**. During gastrulation, the embryo undergoes the reorganization that is critical to future development.

The invagination (depression) characteristic of the gastrula continues forming a cavity called the **archenteron**, or primitive gut. The opening of the gastrula is termed the **blastopore**. In **protostomates** (snails, segmented worms,

insects) the blastopore forms the mouth opening. In **deuterostomates** (echinoderms and mammals), the blastopore forms the anus.

In addition, during gastrulation, the germ layers form. **Diploblastic** (jellyfish) organisms possess two germ layers: the ectoderm and endoderm. **Triploblastic** (humans) organisms possess three germ layers: the ectoderm, endoderm, and mesoderm.

The germ layers eventually give rise to the specialized cells, tissues, and organs of the body. The **endoderm** gives rise to structures such as the linings of the urinary bladder, pharynx, pancreas, respiratory system, liver, intestine, and other structures. The **ectoderm** gives rise to the epidermis, sweat and oil glands, hair, inner ear, brain, spinal cord, and other structures. The **mesoderm** gives rise to the skeleton, muscles, dermis, kidneys, gonads, blood, heart, as well as other structures.

In starfish, the gastrula stage is reached in 2 days. Within a few more days, the free-swimming larval stage develops. A few larval stages may be found on the slides.

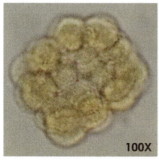
100X
Figure 14.5 Morula.

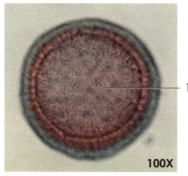
100X
Figure 14.6 A blastula.
1. Blastocoel

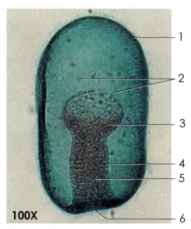
100X
Figure 14.7 A gastrula.
1. Ectoderm   5. Archenteron
2. Mesenchyme cells   (gastrocoel)
3. Coelomic sac   6. Blastopore
4. Endoderm

Student Activity—Embryological Development in Starfish

Observing Prepared Slides of Embryological Development in Starfish

In the following laboratory exercise, you will be viewing the early embryonic development of a starfish (*Asterias*). A number of stages of embryological development can be observed easily on each slide. The student has to be able to identify the unfertilized egg, fertilized egg, 2-cell stage, 4-cell stage, morula, blastula, and gastrula. In addition, the basic anatomical features of the blastula and gastrula should be labeled.

Materials
- prepared slides of starfish developmental stages
- compound light microscope
- colored pencils

Procedure 14.1 Starfish Development

1. Obtain a compound light microscope from your instructor.
2. Carefully procure the series of slides of starfish developmental stages.
3. Place a slide on the stage of the microscope, and focus first on low power before moving the objective to a higher power.
4. Locate, sketch and label the stages of development that you observe, on the following page.

Q. Are starfish protostomes or deuterostomes? Explain. List three protostomes and three deuterostomes.

_____

_____

_____

_____

_____

Q. Describe the three germ layers, and list several derivatives of each.

_____

_____

_____

_____

_____

_____

Q. Describe the function of the archenteron and the blastopore.

_____

_____

_____

_____

Q. Explain the significance of the statement made by Lewis Wolpert, 1983, "It is not birth, marriage, or death, but gastrulation, which is truly the most important event of your life."

_____

_____

_____

Speaking of Humans...

Two-celled stage: 3 hours after fertilization
Morula: 4 days after fertilization
Blastula: approximately 6 days after fertilization
Gastrula: approximately 15 days after fertilization

It is not birth, marriage, or death, but gastrulation, which is truly the most important event of your life.

—**Lewis Wolpert (1929–present)**

DRAWINGS

Unfertilized egg

Zygote

Early cleavage

Morula

Blastula

Gastrula

CHICK EMBRYOLOGY

The study of chick embryology is a great way to gain an understanding of embryological development in organisms. It takes approximately 21 days for a fertile egg to develop into a chick. The **egg shell**, composed of **calcium carbonate**, provides protection for the developing embryo. It is porous, allowing for the exchange of gases and moisture.

At the larger end of the inside of the egg is an **air space**, which acts as a cushion. The chick egg has inner and outer shell membranes. The membranes protect the inside from bacteria and other contaminates and from evaporation. The **outer shell membrane** is white and is attached to the shell. This will be carefully removed when a small hole is made in the surface of the egg. The **inner shell membrane** is below the air space. This membrane will remain intact (Fig. 14.8).

Inside the inner shell membrane is the **albumen**, or white of the egg. It provides nutrients to support the growing embryo. **Chalazae** are protein cords in the albumin and connect to the yolk, holding the yolk in place. The **yolk** is yellow in color. It contains fat, proteins, vitamins, and carbohydrates to support growth of the embryo. The yolk and the embryo are surrounded by an invisible **vitelline membrane**. A red spot, seen early in a fertilized egg, denotes the beginning of early embryological development. It is here that the embryo can be seen. The embryo eventually develops as a small white spot in an area called the **blastodisc**.

At 12 hours, one can observe formation of **Hensen's node**, the organizer for gastrulation in birds, which secretes cellular signals necessary for gastrulation, and the **primitive streak**, an elongated mass of cells found in avian, reptilian, and mammalian groups. The primitive streak is the first sign of gastrulation (Fig. 14.9).

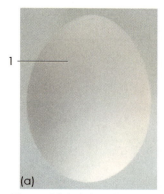 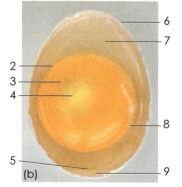

Figure 14.8 (a) An intact fertilized chicken egg and (b) a portion of the shell removed, exposing the internal structures.
1. Shell
2. Vitelline membrane
3. Yolk
4. Germinal disk
5. Shell membrane
6. Shell
7. Albumen (egg white)
8. Chalaza (dense albumen)
9. Air space

At 18 hours, the **notochord**, a flexible, rod-shaped structure found in the embryos of all chordates should be visible. In chickens, as in chordates, the notochord will be replaced by a vertebral column (Fig. 14.10).

By 28 hours, the nervous system is developing. A **head fold**, located at the anterior end of the embryo, and **neural fold**, located at the posterior end of the embryo, will be prominent, and the **neural tube**, which will close over time and become the spinal cord, is still in the process of forming. At this time, the embryo is developing primitive segments of muscle, **somites**, that have originated from masses of mesoderm. At 28 hours, the beginnings of a cardiovascular system can be seen with the formation of a **primordial heart** and blood vessels. A **foregut** is present, which will differentiate over time to become the structures of the mouth to the duodenum (Fig. 14.11).

By 48 hours, the embryo shows signs of **torsion**, or bending. The head is almost touching the heart. The heart has undergone further development. Also visible are the **ventricle**, or pumping chamber, an **atrium**, or receiving chamber, and **aortic arches**. The heart is actively contracting and pumping blood. **Vitelline arteries** and **veins** can be observed extending over the yolk. These vessels nourish the embryo, carrying nutrients from the yolk to the developing embryo (Fig. 14.12).

The brain has become more complex, having divided into the **forebrain**, **midbrain**, and **hindbrain**. Eye formation is prominent at 48 hours, with development of the **lens**, used for seeing. **Auditory pits**, signaling the formation of the inner ear, can be seen. By 48 hours, there are 24 pairs of somites.

At 56 hours, the cardiovascular system continues to show development, and **pharyngeal pouches** are found between the aortic arches. These pouches give rise to various structures including the thyroid and parathyroid glands and the Eustachian tube. **Limb buds,** which become the wings and hindlimbs, are appearing. Sensory structures continue to develop. The ears are more complex and noticeable. The eyes have a distinctive lens, and **olfactory pits** have formed, giving rise to nasal structures. The **allantois,** a sac-like structure that collects liquid wastes (excretion) from the embryo and exchanges gases (respiration), begins to develop at about this time. A **tail bud**, at the most posterior end of the embryo can be seen. At 56 hours, the embryo now has 36 pairs of somites (Fig. 14.13).

By 96 hours, torsion is complete. The neural tube is totally closed and now is called the spinal cord. The embryo has increased in size, and organ systems have become increasingly complex. At 96 hours, the embryo has approximately 38–41 somites (Fig. 14.14).

Over the next weeks, the chick embryo will show further development in size and complexity. By day 21, the chick is ready to emerge from the egg, using its **egg tooth** (a small protuberance or projecting structure on the beak) to peck its way out (Fig. 14.15).

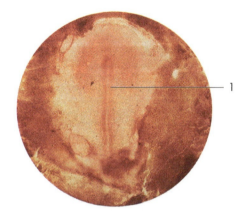

Figure 14.9 A 12-hour chick embryo.
1. Embryo main body formation

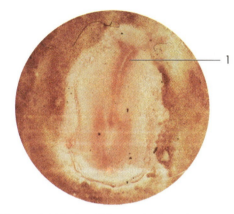

Figure 14.10 An 18-hour chick embryo.
1. Neurulation beginning

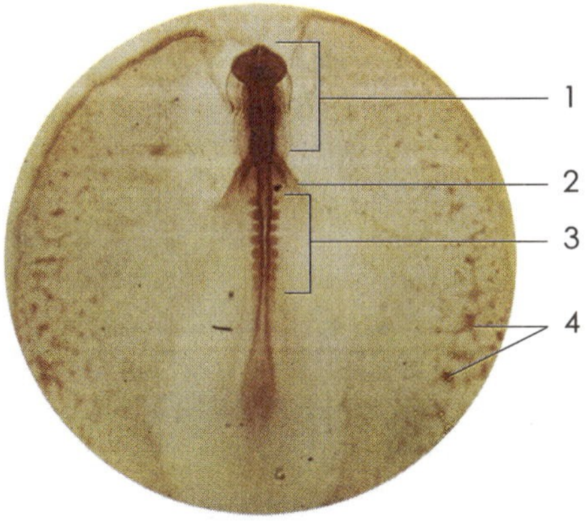

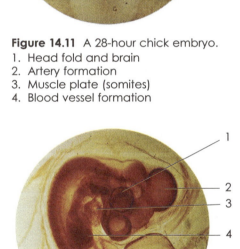

Figure 14.11 A 28-hour chick embryo.
1. Head fold and brain
2. Artery formation
3. Muscle plate (somites)
4. Blood vessel formation

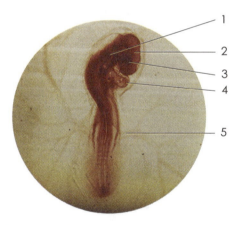

Figure 14.12 A 48-hour chick embryo.
1. Ear   4. Heart
2. Brain   5. Artery
3. Eye

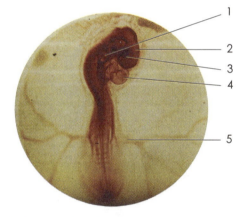

Figure 14.13 A 56-hour chick embryo.
1. Ear   4. Heart
2. Brain   5. Artery
3. Eye

Figure 14.14 A 96-hour chick embryo.
1. Eye   5. Fecal sac
2. Mesencephalon   (allantois)
3. Heart   6. Leg formation
4. Wing formation

Figure 14.15 21 days later, the final product.

Student Activity—Embryological Development in Chicks

Observing Prepared Slides of Embryological Development in Chicks

In this activity, the student will observe various stages in the development of a chick, using prepared slides. The student must locate and identify as many as possible of the anatomical structures described in the discussion of chick embryological development.

Materials
• prepared slides of chick developmental stages
• compound light microscope
• colored pencils

 ## Procedure 14.2
Chicken Development

1. Obtain a compound light microscope from your instructor.
2. Carefully procure the series of slides of chick developmental stages.
3. Place your first slide on the stage of the microscope, and focus first on low power before moving the objective to a higher power.
4. Locate, sketch, and label the stage of embryological development that you observe on the following page.
5. Repeat steps 3 and 4 for the remaining slides.
6. Sketch the specimens on the following page.

Q. What is the importance of the primitive streak?

_____

_____

_____

Q. In development, what does the notochord become?

_____

_____

_____

Q. Describe in general terms the development of the cardiovascular system.

_____

_____

_____

_____

_____

 Student Activity—Observing Embryological Development in Viable Eggs

Observing Embryological Development in Viable Chicken Eggs

In this activity, the student will gain an understanding of chick embryological development, using viable chicken eggs (*Gallus domesticus*). The fertilized eggs can be purchased from a scientific company or from a local hatchery year-round. An incubator will be necessary for this lab activity. The incubator can be purchased from a scientific company. A variety of incubators are available specifically for incubating chick eggs; however, a standard incubator may be used as well. The instructor will begin by incubating the viable eggs for 24 hours. In following the procedures outlined below, students will create a "window" in the upper surface of the egg and observe the chick as it goes through its stages of development. Students can compare what they observed in the activity "Observing Prepared Slides of Embryological Development in Chicks" to the stages of development in the viable chick egg.

DRAWINGS

12-hour chick embryo

18-hour chick embryo

28-hour chick embryo

48-hour chick embryo

56-hour chick embryo

96-hour chick embryo

Materials

- viable 24-hour-old incubated chicken eggs
- incubator
- autoclave or pressure cooker
- large (400 ml) beaker of water for incubator to provide humid environment
- 125 ml of 70% alcohol
- 12 Styrofoam™ coffee cups
- sterile cotton
- 12 sterile Petri dishes
- 12 sterile 400 ml beakers
- 12 sterile forceps
- 12 sterile dissecting probes
- 12 sterile razor blades
- safety goggles
- paper towels
- aluminum foil
- masking tape
- colored pencils

Pre-lab Preparation

The instructor will have to purchase the viable eggs a couple of days prior to the lab activity. The eggs must be incubated for 24 hours before the students proceed to create a window in the egg's surface.

1. Set the incubator temperature to 37°C or 99°F.
2. Place a 400 ml beaker containing tap water in the incubator to provide humidity. Refill with water as needed. The materials have to be sterile so the chick embryo does not become contaminated with bacteria, fungi, or other organisms. If an autoclave is not available, a pressure cooker can be used.

Figure 14.16 Apparatus for observing and incubating an egg.

3. One to 2 days before the lab activity, place in each 400 ml beaker a dissecting probe, a forceps, and a razor blade.
4. Cover the top of each beaker securely with aluminum foil, wrap each beaker with paper towels, and tape the outside well with masking tape.
5. Place the wrapped beakers in the pressure cooker and close the cooker. The sterilization process takes 15 minutes at 250°F.
6. Let equipment cool, then remove and store the materials in a clean, dry place.

Procedure 14.3
Observing a Chicken Egg

1. Procure an incubated chick egg from the instructor.
2. Locate the sterilized beaker containing the dissecting probe, razor blade, and forceps. Also obtain a sterile Petri dish, Styrofoam cup, sterile cotton, and 70% alcohol.
3. Put on safety goggles.
4. Using a razor blade (*Caution: This is a very sharp object*), cut one inch of the bottom off of the Styrofoam cup.
5. Take the cut bottom portion of the Styrofoam cup and place it upside down (bottom up) in the bottom of the sterile Petri dish.
6. Using the razor, cut a hole in the bottom of the Styrofoam cup large enough for the pointed end of the viable chick egg to sit.
7. Using the sterile cotton and the 70% alcohol, swab the cup bottom to sterilize it.
8. Carefully take the 24- to 48-hour incubated chick egg and wipe the egg with 70% alcohol.
9. Place the pointed end in the bottom of the Styrofoam cup (Fig. 14.16).
10. Using the blunt end of the sterile dissecting probe, tap the egg until it cracks.
11. Using the sterile forceps, carefully peel back the shell. You should see the air space under the shell. (*Caution: Be careful not to damage the membrane that covers the chick embryo.*) The opening should be the size of a quarter. The white membrane just underneath the shell can be removed.
12. Place the sterile 400 ml beaker over the egg and on the Petri dish.
13. Make your observations as quickly as possible, and place the Petri dish bottom containing the egg, styrofoam, and beaker back in the incubator.
14. Over the next few weeks, observe the egg as the chick develops.

15. Sketch what you see below. Be sure that you include the date of each observation.

16. At the end of the observation period, follow your lab instructor's directions for disposal of materials.

Q. Why does the egg shell have to be porous in nature?

_____

_____

Q. What are two functions of the inner membrane?

_____

_____

Q. Describe the food sources for the developing embryo.

_____

_____

_____

Q. What is the blastodisc?

_____

_____

Q. Why must the environment for the developing chick embryo be kept as sterile as possible?

_____

_____

Q. Describe which structures/systems developed first and which were observed to develop last.

_____

_____

_____

Note:
Biological companies make available frog eggs, salamander eggs, and toad eggs for the study of embryological development. These are seasonal and can be purchased only for the months of March and April. Scientific companies also have sea urchin and zebra fish embryology and development kits. The zebra fish is a good specimen for studying development because the eggs are large and transparent and it takes only one week to complete embryological development. Prepared slides depicting embryological development in the above-named organisms also can be purchased from biological companies.

NOTES

Name: _____ Date: _____ Section: _____

⍰ Review Questions

1. What are three derivatives of ectoderm, endoderm, and mesoderm?

_____

_____

_____

2. Draw and describe the three germ layers, archenteron, and blastopore found in a typical gastrula.

_____

_____

_____

_____

_____

_____

3. What factors determine the type of cleavage?

_____

_____

_____

4. Compare and contrast Protostomates and Deuterostomates.

_____

_____

_____

5. Trace the sequence of embryological development through the neurula in chordates.

_____

_____

_____

_____

6. What is the difference between determinate and indeterminate cleavage?

_____

_____

7. Using the microscope, how can one differentiate an unfertilized starfish egg and the zygote?

_____

_____

8. Why is embryology important in evolutionary biology?

_____

_____

_____

_____

_____

Chapter 15
It's All in the Genes: Understanding Basic Mendelian Genetics

Student Outcome Objectives

At the completion of this exercise, the student will be able to:

1. Discuss the role of genetics in biology and in society.
2. Define the following terms: chromosome, gene, allele, homozygous, heterozygous, genotype, phenotype, dominant trait, recessive trait.
3. Discuss expected results and observed results based on the inheritance of one trait.
4. Explain the inheritance of some common genetic traits in humans.
5. Discuss basic principles of population genetics.
6. Construct a simple pedigree.

Overview

"She has her father's eyes." "Do you think I'll be bald like Dad?" "Can two blue-eyed parents have a brown-eyed child?" "If my black-and-white cat mates with my neighbor's yellow cat, will we have any calico kittens?" "How can I increase the chances of my next litter of puppies being champions?" "What are the odds of two carrier parents having a child with cystic fibrosis?" The questions are endless when people begin discussing inheritance. People are naturally curious about how traits are inherited from one generation to the next. That's *genetics*!

As we approach the dawn of the age of genetics with its vast potential of understanding the human genome and engineering the genes themselves, it is hard to believe that the science of genetics had its humble origins in an obscure monastery garden in Austria in the mid-1800s. Our fundamental knowledge of genetics is a result primarily of the experiments conducted by an Austrian monk, Gregor Mendel (Fig. 15.1). His investigation into the inheritance patterns of certain characteristics of pea plants is the classical origin of modern genetics. Ironically, the works of Mendel were not appreciated until the early 1900s.

Each chromosome consists of thousands of structural and functional units called **genes** (units of heredity), which are long chains of DNA. Today, we are beginning to understand how genes work and where they are located on the chromosome. As examples, the gene for Alzheimer's disease was found on chromosome 21 and the gene for cystic fibrosis has been identified on

Figure 15.1 Gregor Mendel (1822–1884) has been called "The Father of Genetics."

chromosome 7. In a typical human (that's you, hopefully!) nearly 30,000 genes are responsible for producing the traits that make you human, as well as the characteristics that make you a unique member of the human race. A trait may be controlled by one pair of genes, or it may be controlled by more than one pair of genes.

Because genes are contained on homologous chromosomes, two members of a gene pair can be alike or different. The possible form that a gene may take is called an **allele**. If an individual possesses two identical alleles, they are said to be **homozygous**. If an individual possesses two different alleles, they are said to be **heterozygous**. An individual's genetic make-up, or **genotype**, in turn influences one's physical characteristics, the **phenotype**. In many cases, one allele may take over or prevent or mask the expression of another allele. This allele is called the **dominant allele**, and the allele that is not expressed is called the **recessive allele**. Other types of inheritance, such as incomplete dominance and codominance, as well as other variables that influence an individual's phenotype, will be discussed in lecture.

For simplicity, the following lab has been designed to consider traits that are controlled by only one pair of genes Keep in mind that the majority of human traits are controlled by more than one gene. In this laboratory experience, students' knowledge of the fundamental mechanisms of genetics will be reinforced by developing a simple model of inheritance and conducting a human genetic traits survey.

Student Activity—Understanding Heredity

Warming Up: Understanding Heredity

If the genotypes of both parents are known, the expected genotypes and phenotypes of their offspring can be calculated, either by mathematical methods or by using a **Punnett square**. This handy genetic device is named in honor of the British geneticist Reginald C. Punnett (1875–1967).

Expected results are specific figures and are not the result of chance. In nature, however, the **expected results** may not agree with the observed results. **Observed results** are those that appear in the offspring as a result of random combinations of the genes. This exercise has been developed to help students understand the concepts of expected results and observed results.

Students will work in teams of two. Assume that in pennies, heads are dominant to tails, assigning heads as (H) and tails as (h).

Materials
- paper
- pennies

Procedure 15.1 Understanding Heredity

1. Complete the following Punnett square based upon two heterozygous parents. Record the genotypes and phenotypes of the results below.

	H	h
H		
h		

2. Determine the expected genotypes and phenotypes for the Punnett square.

Genotypes: _____

Phenotypes: _____

3. Using the expected genotypes and phenotypes, predict the genotype and phenotype combinations for 100 tosses.

Genotypes: _____

Phenotypes: _____

4. Place two pennies in your hand, then toss them onto the tabletop. Tally the letter combinations below, and record your group's results, as well as class totals, on the charts provided.

GROUP DATA

Expected and Observed Genotypes

	Expected genotype 100 tosses	Observed genotype 100 tosses
HH		
Hh, hH		
hh		

Expected and Observed Phenotypes

	Expected phenotype 100 tosses	Observed phenotype 100 tosses
Heads		
Tails		

CLASS DATA

Expected and Observed Genotypes

	Expected genotype 100 tosses	Observed genotype 100 tosses
HH		
Hh, hH		
hh		

Expected and Observed Phenotypes

	Expected phenotype 100 tosses	Observed phenotype 100 tosses
Heads		
Tails		

Q. What did the pennies represent in the exercise? Were they an accurate representation? Why or why not?

_____

_____

Q. Why were two coins used?

_____

_____

Q. What is the difference between expected and observed results?

_____

_____

Q. Compare the observed genotypes and phenotypes, from your group and the class, to the expected genotypes and phenotypes.

_____

_____

_____

_____

Q. Did you notice any relationship between the number of tosses and the expected and observed results?

_____

_____

_____

_____

Q. In nature, what variables may affect the expected and observed results in a population?

_____

_____

_____

Q. What is the advantage of using larger populations in studies?

_____

_____

_____

_____

My scientific studies have afforded me great gratification; and I am convinced that it will not be long before the whole world acknowledges the results of my work.

—**Gregor Mendel (1822–1884)**

Student Activity—Population Genetics

Population Genetics

Evolution does not occur at the level of individuals but, rather, at the level of populations. It involves changes in the frequency (proportion) of a given allele in that population. For example, in a population of pea plants that originally has only alleles for yellow peas, a new allele might appear that causes peas to be green. Initially, the frequency of the green allele would be low, but over time the frequency might increase so eventually 10%, 50%, 75%, or even more of the alleles for pea color might be green. This change in allele frequency over time represents evolution in that population.

Allele frequencies are expressed as a fraction of 1.0. Thus, if 80% of the alleles are yellow and 20% of the alleles are green, the frequencies would be 0.8 and 0.2, respectively. When the frequency of alleles for a gene are constant over time (no change in allele frequency), that gene is said to be in **genetic equilibrium**. According to the **Hardy-Weinberg model**, genetic equilibrium will be maintained in a population if the following conditions are met:

1. There must be no mutation.
2. There must be no movement of individuals into or out of the population.
3. Mating between individuals must be completely random.
4. The population must be sufficiently large so the laws of probability apply.
5. There is no selection; no allele is favored over another allele.
   (Example: Green and yellow peas are equally likely to survive and reproduce.)

Part 1: Determining Allele Frequencies Over Several Generations

In this activity, we will be looking at allele frequencies over several generations in a population in which these conditions are generally met. Then we will study the effect of violating two of the above conditions on the allele frequencies in the population.

Materials (15.2–15.3)
- cup or beaker
- 50 red and 50 white plastic pieces
  [*Note:* Any contrasting material will suffice.]
- graph paper

Procedure 15.2
Population Genetics

1. Procure a cup or beaker, 50 red (allele R) and 50 white (allele r) plastic pieces from the instructor. This represents an allele frequency of 0.50 for R and 0.50 for r.
2. Shake the cup to mix the pieces, and without looking, pick out two pieces to represent the genotype of a single offspring from that population. Note and record the genotype (RR, Rr, rr) in Table 15.1.
3. Replace the pieces in the cup, shake to mix, and pull out two more pieces.
4. Continue this procedure until you have counted genotypes for a total of 50 individuals (this constitutes one generation).

Table 15.1 Measuring Hypothetical Population Genetics									
1.		11.		21.		31.		41.	
2.		12.		22.		32.		42.	
3.		13.		23.		33.		43.	
4.		14.		24.		34.		44.	
5.		15.		25.		35.		45.	
6.		16.		26.		36.		46.	
7.		17.		27.		37.		47.	
8.		18.		28.		38.		48.	
9.		19.		29.		39.		49.	
10.		20.		30.		40.		50.	

5. Calculate the allele frequencies for this next generation of the population. Each RR individual represents two copies of the R allele, each rr individual represents two copies of the r allele, and each Rr individual represents one copy of R and one copy of r.

Example:

RR 12 individuals = 24 copies of R

Rr 28 individuals = 28 copies of R + 28 copies of r

rr 10 individuals = 20 copies of r

Total 50 individuals: _____   Total 100 alleles: _____

The total number of R alleles is 24 + 28 = 52
    Allele frequency = 52 ÷ 100, or 0.52

The total number of r alleles is 20 + 28 = 48
    Allele frequency = 48 ÷ 100, or 0.48

In the example on the previous page, for the next generation to reflect this new allele frequency, you would add two red pieces to your cup and take out two white pieces to give a total of 52 red and 48 white.

Note: For your own experiment, you will be adding and subtracting whatever number of pieces are necessary to give you the appropriate allele frequencies—*not necessarily two!*

6. Repeat the entire experiment (50 individuals), starting with your new allele frequencies (0.52 and 0.48 for the example above) to find out the allele frequency of the next generation.

7. Continue this for a total of three generations, using your newly calculated allele frequencies to begin each subsequent generation.

8. Graph the frequency of the red allele (R) from the beginning (0.5) through the three generations on the graph provided below.

Part 2: Effects of Selection on Allele Frequencies

This experiment will be performed the same way as the first experiment, except that you will now assume that homozygous white (rr) is lethal. Individuals who have this lethal genotype will not survive and reproduce. In other words, r is being selected against, and R is being selected for. Beginning with allele frequencies of R = 0.5 and r = 0.5, select 50 individuals as before. This time, however, do not count the r alleles that occur in rr individuals.

Only time and money stand between us knowing the composition of every gene in the human genome.

—Francis Crick (1916–2004)

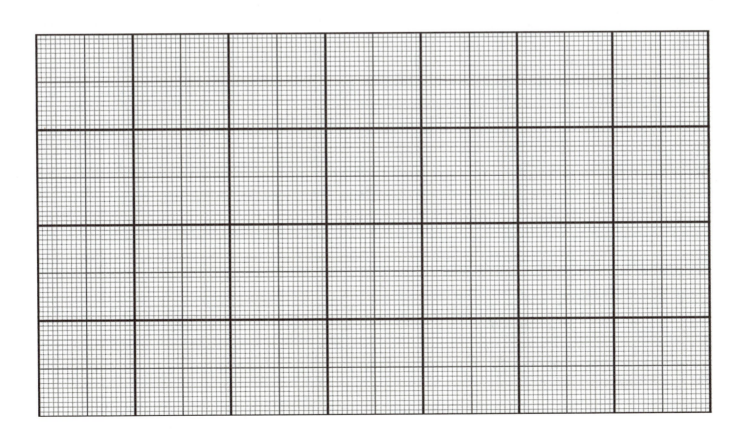

Example:

RR 12 individuals = 24 copies of R

Rr 28 individuals = 28 copies of R + 28 copies of r

rr 10 individuals = LETHAL

Total 50 individuals: _____   Total 80 alleles: _____

Total R alleles: 24 + 28 = 52
   Allele frequency of R= 52 ÷ 80 or 0.65

Total r alleles: 0 + 28 = 28
   Allele frequency of r = 28 ÷ 80 or 0.35

In this example, for the next generation, you would have to add 15 red pieces (for a total of 65) and take away 15 white pieces (for a total of 35) to give the proper allele frequencies. Again, continue the experiment for a total of three generations, and graph the frequency of the R allele over time on the same graph as the first activity.

Part 3: Effects of Small Population Size

This time you will be looking at allele frequencies in five individuals rather than 50. Simply pick out genotypes as before, but this time for only five individuals.

Procedure 15.3
Small Population Size

1. Begin the experiment with five red pieces and five white pieces. Calculate the allele frequencies.
2. Repeat this experiment six times (always beginning with five red and five white), and note the allele frequencies for each group.

Example:

RR 2 individuals = 4 copies of R

Rr 2 individuals = 2 copies of R and 2 copies of r

rr 1 individual = 2 copies of r

Total 5 individuals: _____   Total 10 alleles: _____

Frequency of R = 4 + 2 = 6   6 ÷ 10 = 0.6

Frequency of r = 2 + 2 = 4   4 ÷ 10 = 0.4

The next major explosion is going to be when genetics and computers come together. I'm talking about an organic computer—about biological substances that can function like a semiconductor.

—Alvin Toffler (1928–present)

Q. Discuss your findings below.

_____

_____

_____

_____

Q. Compare the change in the allele frequencies of R over three generations, both without selection and with selection. Which shows more change? Is evolution more or less likely to occur when selection is acting on a particular allele? Why?

_____

_____

_____

Q. Draw a chart that compares the allele frequencies of R and r from the large population without selection (Part 1) to the allele frequencies in the smaller population, also without selection (Part 3). Which shows more change in allele frequency over time? Is this the result you expected? Why?

_____

_____

_____

Q. In Part 2, will selection against the rr genotype ever lead to complete removal of the r allele from the population? Explain your answer.

_____

_____

_____

_____

Q. Discuss the overall results of these two activities, such as the effects of changing conditions on genetic equilibrium and on evolution.

_____

_____

_____

_____

 Student Activity—Corn Inheritance

The Corn Maize

The world is dependent upon corn as a source of food, industrial products, and fuel. Over the last 5,000 years, corn (*Zea mays*) has been developed through artificial selection and has changed tremendously from its ancestor teosinte. Among the many varieties of corn are sweet corn, popcorn, and Indian corn. One of the most colorful varieties of corn is Indian corn.

In this lab activity, students will investigate the inheritance patterns of color and sweetness in Indian corn. In Indian corn, kernel color (purple or yellow) is controlled by the alleles P and p (purple [P] being dominant), and carbohydrate content (starchy or sweet) is controlled by the alleles S and s (starchy [S] being dominant). Sweet kernels can be distinguished from starchy kernels because sweet kernels are wrinkled when they dry, and starchy kernels are smooth. The marbling observed in some of the kernels is a result of **transposons**, or jumping genes, first described by Barbara McClintock.

Materials
- ear of corn
- marking pen

 Procedure 15.4 Corn Inheritance

1. Obtain an ear of corn and a marking pen from your instructor.
2. Carefully count all the corn kernels (row by row), marking each kernel with your pen to ensure that the kernels are not counted twice.
3. As you count the kernels, tally their phenotype in the chart at the bottom of this page.
4. Once you have completed counting and tallying the kernels, determine the totals for your group and report your findings to the laboratory instructor.

Individual Tally				
Purple and Starchy	Purple and Sweet	Yellow and Starchy	Yellow and Sweet	Total kernels

5. The instructor will display each group's results for analysis (document them in the table below):

Genetics is about how information is stored and transmitted between generations.

—**John Moffat Smith (1835–1918)**

Group Tally					
Group number	Purple and Starchy	Purple and Sweet	Yellow and Starchy	Yellow and Sweet	Total kernels
1.					
2.					
3.					
4.					
5.					
6.					
7.					
8.					
9.					
10.					
Phenotypic Totals					
Ratios*					

*To determine ratio:
  a. Total each phenotypic column
  b. Calculate the total of all seeds counted
  c. Divide the total by 16
  d. Divide each phenotypic total by the value from 3 above

6. Discuss your findings.

_____

_____

_____

Q. What were the genotypes of the parental corn that was crossed in your ear of corn?

_____

_____

_____

Q. What are the possible genotypes for a purple/starchy kernel, purple/sweet kernel, yellow/starchy, and yellow/sweet kernel?

_____

_____

_____

Beyond the Lab—Discussing McKlintock

Discuss the significance of the work of Barbara McKlintock on Indian corn.

_____

_____

_____

_____

_____

_____

A-maize-ing Facts!

- An average ear of corn has approximately 800 kernels arranged in 16 rows.
- Corn has 10 chromosomes.
- The United States is the number-one consumer of corn per-capita in the world. The average American consumes 1,500 pounds of corn a year. Most of the corn goes to feeding livestock that is used by consumers. Other sources include high-fructose corn syrup used in the soft drink industry, the making of ethanol, and, of course, that tasty treat—sweet corn on the cob and popcorn. Other products that use corn include: adhesives, dyes, fireworks, tires, crayons, surgical dressing, lipstick, aspirin, shoe polish, metal platings, and pharmaceuticals.

 Student Activity—Human Genetics

How Do You Fit In?
(A Survey of Several Human Genetic Traits)

This activity asks students to participate in a survey of some common human characteristics (Fig. 15.2). In addition, the activity introduces students to the concept of genetic variability within a population.

 Procedure 15.5
Genetic Traits

1. Working with your lab partner, determine your phenotypes for the traits discussed in this activity. The traits in this activity are based upon two alleles and follow basic Mendelian dominance patterns. Record your phenotype and possible genotype in the chart on page 206.
2. The instructor will collect the class data and place it on a chart on the board. Record these data in the chart on page 206.
3. Answer the questions regarding the information you collected.

I got a hundred bucks says my baby beats Pete's baby. I just think genetics are in my favour.

—Andre Agassi (1970–present)

Characteristics to Be Observed

1. *Bent little finger*: Examine your little finger on each hand. If the last joint of your little finger bends toward your ring finger, you have the dominant gene "C" for bent little finger. A straight little finger gene "c" is recessive. The condition of bent little finger is called **clinodactyly**.

2. *Tongue rolling*: Extend your tongue, and attempt to roll it into a U-shape. The gene for tongue-rolling "R" is dominant to the gene for non-tongue-rolling "r".

3. *Widow's peak*: Think of Count Dracula or Eddie Munster. These individuals have a V-shaped hairline in the middle of the forehead. This characteristic comes from the dominant allele "W" for widow's peak. The recessive allele "w" results in a straight forehead hairline.

4. *Dimpled chin*: Now think of the New England Patriot's quarterback, Tom Brady or John Travolta. The presence of a dimpled chin "D" is dominant to the gene for a non-dimpled chin "d".

5. *Free earlobe*: If a portion of the ear lobe remains unattached below the point of attachment to the head, you have the dominant gene "E" for a free earlobe. The recessive gene "e" results in attached earlobes.

A Survey of Some Common Human Characteristics					
Trait	Your Phenotype	Possible Genotypes	Class N	Class Phenotypes	Class Percent Phenotypes
Bent little finger					
Tongue rolling					
Widow's peak					
Dimpled chin					
Free ear lobe					
Thumb crossing					
Hitchhiker's thumb					
Finger hair					
Dimpled cheeks					
Freckles					
Curly hair					
Eyebrow raising					
Ear wiggling					
Long toe					
PTC tasting					

Figure 15.2 Some common genetic characteristics.

6. *Thumb-crossing*: Swing your hands freely, and suddenly clasp your hands, interlocking your fingers. If your left thumb is uppermost, you possess the dominant gene "T". For the heck of it, force yourself to place your hands together the opposite way. Strange!

7. *Hitchhiker's thumb*: If you don't have the ability to bend your thumb back at a 60-degree angle, you are dominant for a straight thumb "S". If you can bend your thumb, you possess the recessive alleles "s" and have a hitchhiker's thumb.

8. *Finger hair*: If you possess hair on the middle segment of your fingers, you have the dominant allele "H" for mid-digital hair. If you do not have hair, you are recessive for the "h" allele.

9. *Dimpled cheeks*: The presence of dimples in one or both cheeks, "D", is dominant to dimple absence "d".

10. *Eyebrow raising*: Are you kin to Mr. Spock on Star Trek? Can you raise your eyebrows? If so, you have the dominant allele "Y". If not, you are recessive "y".

11. *Ear wiggling*: The gene for having the ability to wiggle your ears, "W", is dominant to non-wiggling, "w".

12. *Long toe*: Check out your second toe. If it is longer than your big toe, it is a result of a dominant allele "L". A shorter second toe is recessive "l".

13. *Curly hair*: Curly hair "A" is dominant over straight hair "a".

14. *Freckles*: The presence of freckles results from the dominant allele "Z". The allele "z" is recessive.

15. *PTC tasting*: Approximately 70% of the U.S. population has the ability to taste PTC (phenylthiocarbamide) paper. If you are chewing gum, remove it and wash out your mouth with water. Place a piece of PTC paper on your tongue. Record your results. PTC tasting comes from the dominant allele "P"; not tasting PTC indicates the recessive "p" allele.

Note: By the way, the chances of any two people in this class having the same phenotype for the above 15 characteristics is $(1/2)\ 15$.

Q. Look up or discuss five other characteristics that result from a single dominant gene.

_____

_____

_____

_____

_____

_____

Q. Describe three autosomal dominant conditions in humans.

_____

_____

_____

Q. Describe three autosomal recessive conditions in humans.

_____

_____

_____

Genetics explains why you look like your father and if you don't why you should.

—**Author unknown**

Student Activity—Blood Groups

Part 1: Determining Blood Type

All of us have one of the four blood types: A, B, AB, O. The **ABO blood types** are inherited proteins expressed on the surface of red blood cells and are called **antigens**. The antigens depend on multiple alleles: type A (I^AI^A, I^Ai), type B (I^BI^B, I^Bi), type AB (I^AI^B), and type O (ii) blood. The A, B, AB, O blood types provide a great example of **codominance**. The alleles I^A and I^B are codominant to each other and cause both antigens A and B to be expressed. They are both dominant over the alleles for type O blood (ii) (Fig. 15.3).

Blood typing is crucial in determining the safety of blood transfusions. In 1901, Karl Landsteiner developed the system for naming the blood groups. This is significant because a person with type A blood possesses anti-B proteins called **antibodies** in his or her plasma. Similarly, an individual with type B blood has anti-A antibodies.

Agglutination, or clumping of the red blood cells, can occur if anti-A antibodies from an individual with type B blood mix with red blood cells of a type A person, or if anti-B antibodies from a type A person mix with red blood cells from a person with type A blood. **Transfusion reactions** resulting from mixing these blood types can be deadly. Individuals with type AB blood do not possess antibodies and can receive blood from all blood groups. Individuals with type AB blood are called **universal recipients**. Those with type O blood have no cell surface antigens and are called **universal donors**. They have anti-A and anti-B antibodies in the plasma, but under emergency conditions, the red blood cells without plasma can be transfused safely.

Blood typing has become important in cases where paternity is in question. For example, if the mother has type AB blood (I^AI^B) and the father has type O blood (ii), the child will have either I^Ai or I^Bi because type A and type B are dominant over type O.

Q. Construct a Punnett square to determine the genotypes and phenotypes of the potential offspring of the above scenario.

Q. In a maternity ward, four babies become accidently mixed up. The ABO types of the four babies are known to be: O, A, B, and AB. If a mother is type O and the father is type B, what are the potential blood types of their offspring?

_____

_____

_____

Q. Recently Mary underwent extensive surgery and needs a transfusion. She has type A blood. Her friend Paul has type AB blood, and another friend, Thomas, has type O blood. Which of her friends' blood can she safely receive? Why?

_____

_____

_____

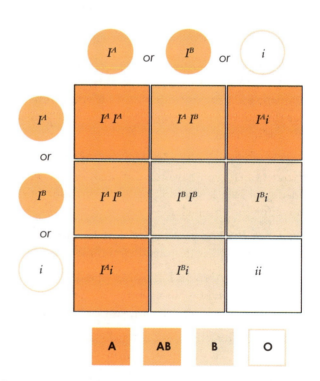

Figure 15.3 This figure shows different ABO blood types and the multiple alleles controlling the ABO blood types.

Part 2: Determining Rh Factor

Other antigens are found on the surface of red blood cells. In 1937, Karl Landsteiner and Alexander Wiener were studying blood from Rhesus monkeys and found other alleles on the surface of their red blood cells. He called these alleles Rhesus monkey factors (Rh factors). In humans, the Rh factor is the RhD antigen found on the surface of some the red blood cells of some individuals. Either an individual has the antigen (D positive) or does not have the antigen marker (D negative). Pregnant women are checked to see if they are Rh⁺ or Rh⁻. This is important, especially if the mother is Rh⁻ because a second pregnancy could cause **erythroblastosis fetalis**, a hemolytic disorder in the fetus that could be life-threatening.

In this lab activity, we will determine blood type and Rh factor. The instructor might provide the students with samples of synthetic blood if there are safety constraints, or students could safely gather a sample of their own blood to type through a finger stick. Kits are available through biological companies to conduct this lab activity safely.

Materials
- kit with samples of synthetic blood or kit for typing of student's blood
- kit containing anti-Rh serum
- safety glasses
- gloves
- paper towels

Procedure 15.6
Rh Factor

1. Procure the materials from the instructor.
2. Place on a paper towel the packet containing an alcohol swab, the sterile lancet, the card on which to place your blood drops, the card for determining Rh factor, three plastic color-coded stirs, the vials containing antiserum A and antiserum B, and the vial containing anti-Rh serum.
3. Taking the card, place a drop of antiserum A in the circle marked for type B blood and a drop of antiserum B in the circle on the card marked for type A blood.
4. On the card to test for Rh factor, place a drop of the anti-Rh serum in the circle.

Interesting Numbers

Blood types vary within populations. In the United States, approximately 46% of the people are type O, 41% are type A, 9% are type B, and 4% are type AB. Approximately 85% of Americans are Rh⁺ and 15% are Rh⁻.

5. If using your blood, open the packet containing the alcohol swab, and carefully clean the tip of your finger.
6. Open the packet containing the sterile lancet, hold the point against your fingertip, and press down quickly, piercing your finger.
7. Squeeze your finger gently, allowing a drop of blood to fall in the circle on the card marked A and a second drop of blood on the circle of the card marked B, and in the circle containing the anti-Rh serum.
8. Using a separate plastic stir for each well, stir the blood and antiserum in each well.
9. Observe any agglutination or clumping of blood cells in either well.
10. Follow instructor's rules for clean-up.

Q. What is your blood type (phenotype)? What is your potential genotype?

_____

_____

_____

_____

Q. What other blood types can you safely be transfused with? Why?

_____

_____

_____

_____

Q. If a father has type A blood and the mother has type B blood, what potential blood types would the offspring have?

_____

_____

_____

Q. How can blood type be used to establish the identity of an individual in a paternity suit?

_____

_____

_____

Q. What is your Rh factor?

_____

_____

_____

Q. Research and describe Rh incompatibility syndrome. Why is it important in family planning?

_____

_____

_____

 Student Activity—Developing Pedigrees

Developing Pedigrees

To many people, the term **pedigree** brings about images of championship dogs and thoroughbred horses. Essentially everything from fruit flies to watermelons has a pedigree. To geneticists, however, a pedigree is a valuable tool resembling a family tree that can be used to display family relationships and to track traits through a family. In medical genetics, pedigrees are helpful in understanding how disorders appear in families.

Pedigrees are particularly valuable in understanding the inheritance of **unifactorial** (single-gene) traits such as albinism, cystic fibrosis, and hemophilia. A pedigree can be used to visually represent Mendelian inheritance of a trait. A typical pedigree consists of universally accepted symbols connected by either horizontal or vertical lines. Generations are represented by Roman numerals, and Arabic numerals represent individuals. Filled shapes represent individuals who express a trait, and half shaded shapes represent carriers. A few common symbols appear below.

Symbols

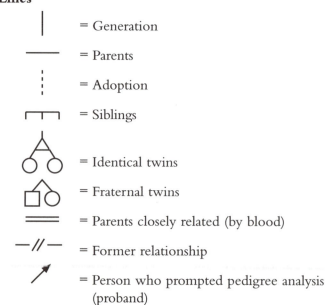

○ □ = Normal female, male

● ■ = Female, male who expresses trait

◐ ◧ = Female, male who carries an allele for the trait but does not express it (carrier)

∅ ⊘ = Dead female, male

◇ = Sex unspecified

∅ ⊘ (SB SB) = Stillbirth

○ □ ◇ = Pregnancy

△ = Spontaneous abortion (miscarriage)

△ = Terminated pregnancy (shade if abnormal)

Lines

| = Generation

— = Parents

⋮ = Adoption

⊓ = Siblings

= Identical twins

= Fraternal twins

= = Parents closely related (by blood)

—//— = Former relationship

↗ = Person who prompted pedigree analysis (proband)

Using a pedigree, autosomal recessive traits such as cystic fibrosis are easy to follow through the generations.

Q. With your knowledge of pedigrees, explain the inheritance of cystic fibrosis in two children of the third generation in the pedigree below.

_____

_____

_____

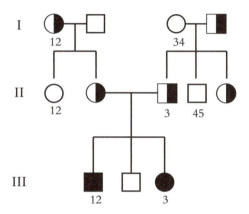

Explanation:
The inheritance of autosomal dominant traits also can be explored through pedigree analysis. **Polydactylism,** having extra digits, results from a dominant gene.

Q. Using the following pedigree, explain the appearance of polydactyly in children 1, 2, and 4 of generation 3.

_____

_____

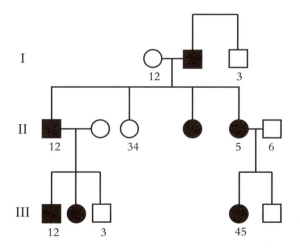

Explanation:
X-linked traits are carried exclusively on the X chromosome. Because a male possesses only one X chromosome, if he receives an X chromosome that carries an X-linked trait, he will express that trait. For a female to express an X-linked trait, she will have to inherit two copies of the gene, one on each X chromosome. Several conditions, such as hemophilia A and color-blindness, are X-linked recessive traits.

Knowing the following, construct a pedigree for color blindness in three generations of a family, on the next page, and answer the provided questions. In the first generation, neither parent was color-blind. The second generation had four children. In their birth order, one was a normal female, one was an affected male, another was an unaffected female, and one was a normal male. The first unaffected female had one unaffected daughter and one affected son. How did this happen, and what is the genotype of her husband? The affected male had one affected son and one affected daughter. How did this occur? The second unaffected female had three unaffected daughters and four unaffected sons.

Q. Explain the inheritance pattern. The normal male had three unaffected sons. How did this happen?

_____

_____

_____

Q. Construct your own pedigree on the next page. Using the traits discussed in Procedure 15.5, create a pedigree for your family, going as far back as possible. If you do not know the phenotype or genotype of your ancestors, construct a hypothetical tree for one of the traits.

_____

_____

_____

_____

I'm one of those people you hate because of genetics. It's the truth.

—**Brad Pitt (1963–present)**

Pedigree for color blindness in three generations of a family.

Pedigree of your family.

Name: _____ Date: _____ Section: _____

Review Questions

1. Why is genetics considered to be one of the most important disciplines of biology?

_____

_____

_____

2. What is the difference between genotype and phenotype?

_____

_____

_____

3. What is the Hardy–Weinberg equation, and when is it used?

_____

_____

_____

4. Why is inbreeding dangerous?

_____

_____

_____

5. What are the chances of two parents that carry the gene for albinism (an autosomal recessive disorder) having a child without albinism?

_____

_____

_____

6. Before considering starting a family, do you think it is reasonable to perform genetic testing on you and your spouse for common inherited disorders? Why or why not?

_____

_____

_____

7. In Manx cats, the alleles "TT" yield a cat with a normal long tail, the alleles "Tt" yield a cat with a short or absent tail, and the alleles "tt" are lethal to the embryo. Predict the genotypes and phenotypes of potential kittens resulting from the mating of two Manx cats.

_____

_____

8. American cat breeders are trying to establish a new breed of cat with unusual rounded, curled-back ears, to be known as the "curl cat." Suppose you found a curl cat and wanted to secretly start your own population. How would you determine whether the curl allele is dominant or recessive? How would you establish and maintain a true-breeding population based on whether the allele is dominant or recessive?

_____

_____

_____

9. Achondroplasia, the most common type of human dwarfism, affects 1 in 15,000 to 1 in 40,000 individuals. Many cases of achondroplasia are the result of a spontaneous mutation. Once the mutation occurs, however, it becomes an autosomal dominant trait. On a TV talk show, the audience was shocked to learn that two individuals with achondroplasia had a son who was of normal height and two other sons and a daughter with achondroplasia. Draw a pedigree of the family and explain how the parents can have a child with normal height.

10. The following pedigree illustrates the inheritance of Tay Sachs disease in four generations of a family. Interpret the pedigree and determine whether the trait is dominant or recessive. What happened in the third generation? What are the symptoms and incidence of Tay Sachs disease?

Chapter 16
Unraveling the Double Helix: Understanding DNA and the Genetic Code

Student Outcome Objectives

At the completion of this exercise, the student will be able to:

1. Name major contributors in the history of DNA.
2. Describe the composition of nucleotides.
3. Compare and contrast DNA and RNA.
4. Explain the molecular configuration of DNA.
5. Describe the make-up of chromosomes.
6. Trace the basic process of replication.
7. Describe the basic process of protein synthesis.
8. Relate the cause of and impact of mutations.
9. Describe mitochondrial and chloroplast DNA.
10. Explain the role of DNA in heredity, medicine, forensics, and evolution.
11. Trace the process of spooling DNA.

Overview

It is hard to believe that the humble beginnings of our knowledge of DNA can be traced back to an obscure physician studying the chemical composition of pus-soaked rags and the sperm of salmon. During his studies in the 1870s, the Swiss physician Johann Miescher (1844–1895) described a mysterious substance that he coined **nuclein**. Eventually, nuclein would be known as deoxyribonucleic acid, or **DNA**.

Scientists studying nuclein described this as a **nucleic acid**. Albert Kossel (1853–1927) labeled two distinct types of nucleic acids:

1. thymus nucleic acid (DNA), and
2. yeast nucleic acid (RNA).

Early scientists thought that DNA perhaps was involved in heredity and RNA as an energy source for cellular metabolism. Nucleic acids are large macromolecules composed of repeating units called **nucleotides** of a pentose sugar (5 carbon), a phosphate group, and a nitrogenous base (Fig. 16.1). The nucleotides in DNA contain deoxyribose sugar, and in RNA the nucleotides contain ribose sugar. In addition, the nitrogenous bases are different in DNA and RNA. Nucleotides are composed of five different bases. The bases are further divided into either double-ringed **purines—adenine** and **guanine**—or they are single-ringed **pyrimidines—thymine**, **cytosine**, and **uracil**. In RNA, uracil replaces thymine.

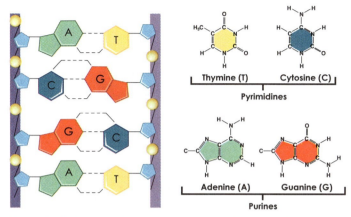

Figure 16.1 The structure of nucleotides found in DNA.

As a result of the contributions of Erwin Chargaff (1905–2002) and others, it was determined that in DNA, the purine adenine (A) is hydrogen-bonded to the pyrimidine thymine (T) and the purine guanine (G) is paired to the pyrimidine cytosine (C) by hydrogen bonds. In RNA, thymine is replaced by uracil (U).

By the 1950s, DNA was widely accepted as the heredity molecule, but its molecular configuration remained a mystery. As a result of the x-ray diffraction studies of Rosalind Franklin (1920–1958) and Maurice Wilkins (1916–2004), it was suspected that a molecule of DNA was shaped like a helix. Using the work of Chargaff, Franklin, Wilkins, and others, James Watson (1928–present) and Francis Crick (1916–2004) developed the double helix model of DNA (Fig. 16.2). Watson and Crick described DNA as a double-stranded helical structure with a backbone of deoxyribose sugar and phosphate and rungs of **complementary base pairs** (**ATCG**) held together by hydrogen bonds. As a result of bonding, the two strands of the DNA molecule run in opposite directions and accordingly are called **antiparallel**.

In eukaryotic cells, a **chromosome** consists of a continuous molecule of DNA and several types of associated proteins. Humans have 46 chromosomes. **Genes** are units of heredity located on specific chromosomes. For example, the cystic fibrosis transmembranal regulatory gene exists at a specific residence on chromosome 7. The collective

DNA is a double helix, the two intertwined strands resembling a "spiral" staircase. It is the sequence or ordering of the nucleotides along either of the constituent strands that is the language of life.

—Carl Sagan (1934–1996)

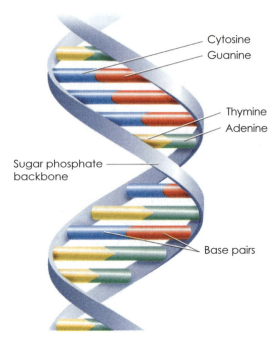

Figure 16.2 The DNA double helix.

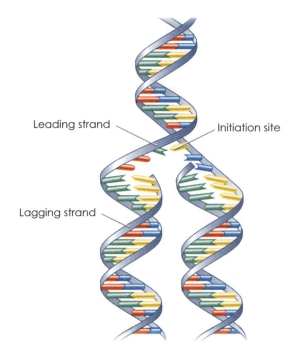

Figure 16.3 Overview of DNA replication.

genes that comprise an organism are called the **genome**. The human genome consists of fewer than 30,000 genes. In a chromosome, DNA is tightly coiled around proteins termed **histones**, which resemble a bead-like structure, forming a **nucleosome**. The nucleosome is composed of units of eight histone proteins capped by another histone known as a **linker**. The framework of DNA is maintained by specialized **scaffold proteins**. Collectively chromosomes make up chromatin, which consists of 60% DNA, 30% histones, 30% other proteins, and 10% RNA.

One of the most interesting characteristics of DNA is its ability to undergo **replication**, allowing DNA to make copies of itself. Usually, this action occurs with extreme fidelity and mistakes are minimal. DNA demonstrates **semiconservative replication**. One strand serves as a direct template for the new strand, and the other strand is pieced together. In 1958, **Matthew Meselson** and **Franklin Stahl**, while working with bacteria, proved that DNA was semiconservative.

An Amazing Fact!

If you were to start reciting the order of the 3 billion ATCGs in human DNA at a rate of 100 per minute, it would take more than 57 years to complete the task! And that is if you didn't take a break to sleep, eat, read this paragraph, or visit the bathroom!

STEPS IN REPLICATION

A human chromosome replicates at several hundred points along its length. In eukaryotes, from 500 to 5,000 base pairs are assembled per minute in up to 50,000 origins of replication.

1. Replication begins at a point called the **origin of replication**, or the **initiation site,** on the **parental strands** of DNA (Fig. 16.3). Here, an enzyme called **helicase** facilitates unwinding. Another enzyme, **gyrase**, prevents the strands of DNA from tangling. This unwinding results in a **replication fork**; gyrase forms a nick in the DNA that will repair later. The enzyme that repairs this nick is **ligase**.
2. One strand of the replication fork is called the **leading strand** (continuous) (3' to 5'), continuing in one direction. The other strand, the **lagging strand**

Crick and I were quick to grasp the intellectual significance of our discovery, but there is no way we could have foreseen the explosive impact of the double helix on science and society. Contained in the molecule's graceful curves was the key to molecular biology, a new science whose progress over the subsequent fifty years has been astounding. Not only has it yielded a stunning array of insights into fundamental biological processes, but it is now having an ever more profound impact on medicine, on agriculture, and on the law.

—**James Watson (1928–present)**

(discontinuous) (5' to 3'), continues in the opposite direction. Then, at the start of each DNA segment to be replicated, an enzyme called **RNA primase** builds a short piece of RNA called an **RNA primer**.

3. The RNA primer attracts an enzyme called **DNA polymerase**, which attracts the proper nucleotides to the **template**. The new strand grows as hydrogen bonds are formed. Copying occurs in two directions: In the 5' to 3' direction, **Okazaki fragments** are formed because of the nature of the molecule. The ends of these fragments are joined by ligase (Fig. 16.4).

4. Enzymes called **proofreading enzymes** are responsible for ensuring the fidelity of replication. DNA replication is accurate; only 1 in 10,000 base pairs is incorrect.

5. **Repair enzymes** also ensure fidelity. Excision and post-replication enzymes are common in the repair process. Unfortunately, repair systems can be damaged by prolonged exposure to sunlight and can bring about skin cancer. Examples of two genetic disorders that involve repair enzymes are xeroderma pigmentosum and ataxia telangiectasia.

RNA is best known as the chemical "ancestor" of DNA. Evolutionarily, RNA appeared before DNA. RNA plays a key role in building proteins. Five fundamental differences between DNA and RNA are the following.

1. DNA has deoxyribose sugar, and RNA has ribose sugar. Ribose has a hydroxyl instead of a hydrogen attached to the 2' carbon.

2. DNA has base pairs consisting of A-T, C-G, while RNA has A-Uracil, C-G.

3. DNA is a double-strand while RNA is a single strand.

4. DNA usually is longer than RNA.

5. DNA is more stable than RNA.

The three types of RNA are the following:

1. **Ribosomial RNA**: forms ribosomes. Ribosomial RNA is made inside of the nucleolus, whereas other RNA is made in the nucleus. Ribosomial RNA is 100–30,000 nucleotides long. Ribosomial RNA has two subunits: The **small subunit** binds the mRNA to the ribosome, and the **large subunit** attaches to the tRNA and helps bind it to the protein (Fig. 16.5).

2. **Messenger RNA**: a single strand of nucleotides whose bases are complementary to those of the template DNA to which the RNA was transcribed. Most mRNAs consists of 500–1000 bases. Working in groups of three, or **triplets**, the mRNA forms **codons** that specify specific amino acids in protein synthesis (Fig. 16.6).

3. **Transfer RNA:** connectors linking an mRNA codon to a specific amino acid. These consist of 75–80 nucleotide base pairs. A loop of tRNA has three bases called **anticodons**, which are complementary to the codon. The end opposite to the anticodon covalently bonds to a specific amino acid. tRNA always attaches to specific amino acids (Fig. 16.7).

Protein synthesis is the process of building proteins from amino acids. This process is directed and coordinated by DNA. Protein synthesis consists of two major steps: transcription and translation.

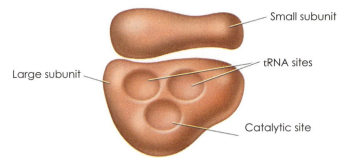

Figure 16.5 Ribosomial RNA.

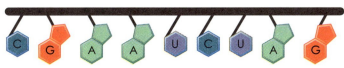

Figure 16.6 Messenger RNA.

DNA was the first three-dimensional Xerox machine.

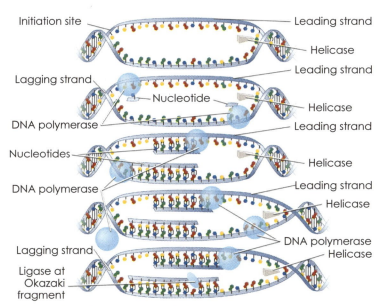

Figure 16.4 The process of DNA replication.

—**Kenneth E. Boulding (1910–1993)**

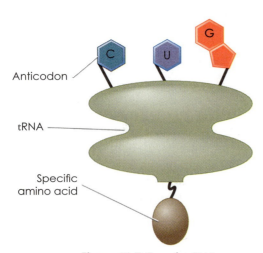

Figure 16.7 Transfer RNA.

Transcription

The process by which chemical information encoded in DNA is copied into RNA is called **transcription**. Generally, only one strand of the double helix of DNA is transcribed and is called the **sense strand**. The noncoding portion is called the **nonsense strand**. Most genes are composed of a **coding region** that is transcribed into RNA, and a **regulatory region** that oversees transcription in the coding portion. The **promoter** is a specific part of the regulatory region that serves as the starting point of transcription.

Building of the complementary strand of RNA is completed by large molecules of **RNA polymerase**. Beginning at the promoter, RNA polymerase unwinds the DNA, breaking the hydrogen bonds. Transcription ceases at a **transcription termination signal** on the DNA. Once RNA for a specific region is made, the DNA quickly re-forms and the RNA segment is displaced. This process can be prolific. This new RNA is called **messenger RNA**.

Messenger RNA now exits the nucleus and enters the cytoplasm via the nuclear pores. Once in the cytoplasm, the mRNA attaches to a ribosome. The messenger RNA works in units of three called **triplets**, which serve as code words called **codons**. Codons specify which one of 20 standard amino acids to pick up. For example, the codon GAG specifies glutamic acid. There are 64 different codons. One codon, **AUG**, encodes for methionine and serves as a **start signal** for building a protein. **UAA**, **UAG**, and **UGA** represents **stop codons**, ending the formation of a protein. These codons are responsible for the **genetic code** that, for the most part, is universal.

The genetic code is **degenerate**; that is, more than one codon can encode for a specific amino acid. These

DNA neither cares nor knows. DNA just is. And we dance to its music.

—**Richard Dawkins (1941–present)**

codons are **synonymous codons**. Glycine is coded by the codons GGU, GGC, GCA, and GGG. The first two nucleotides are the same in each codon. This phenomenon, described by Francis Crick, is known as the **wobble hypothesis**.

The next step, translation, cannot happen without **post-transcriptional modifications**. The beginning of the RNA sequence is called a **leader**, and the end part is called the **trailer**. The DNA sequence contains coding portions called **exons** and noncoding portions called **introns**. Before translation, introns are removed by protein complexes called spliceosomes. Introns range from 65 to 100,000 bases, and exons range from 100 to 300 bases. Many genes are riddled with introns. Many scientists consider the introns to be "genetic junk," such as old genes, which may be slices of viral material or could be the basis of future genes. Introns, it is thought, may even regulate other gene activity.

Translation

Translation is the actual process of expressing the genetic code and building a protein. The codons transcribe a complementary **anticodon** loop that can be found at the opposite end of protein attachment in the newly made **transfer RNA**. Transfer RNA will seek a specific amino acid in the cytoplasm as dictated by the codon. Translation is divided into three parts:

1. **Initiation,** the start of protein synthesis, begins when mRNA associates with a small ribosomial subunit. If all goes well, the AUG codon will pick up methionine to serve as the initiation point. The other codons then are read three at a time in the next step.
2. **Elongation** is the process by which all of the amino acids are joined by peptide bonds.
3. **Termination** is the point at which the stop sequence appears.

Protein folding and the final touches to the proteins occur after termination. The process of protein synthesis is accurate, but mistakes called **mutations** can happen. Protein synthesis is economical, as cells can produce large amounts of a protein from just one or two copies of a gene (Fig. 16.9). For example, a plasma cell in the human immune system can produce more than 2,000 identical antibodies.

First Base	Second Base				Third Base
	U	**C**	**A**	**G**	
U	UUU phenylalanine	UCU serine	UAU tryosine	UGU cysteine	U
	UUC phenylalanine	UCC serine	UAC tryosine	UGC cysteine	C
	UUA leucine	UCA serine	**UAA stop**	**UGA stop**	A
	UUU leucine	UCG serine	**UAG stop**	UGG tryptophan	G
C	CUU leucine	CCU proline	CAU histidine	CGU arginine	U
	CUC leucine	CCC proline	CAC histidine	CGC arginine	C
	CUA leucine	CCA proline	CAA glutamine	CGA arginine	A
	CUG leucine	CCG proline	CAG glutamine	CGG arginine	G
A	AUU isoleucine	ACU threonine	AAU asparagine	AGU serine	U
	AUC isoleucine	ACC threonine	AAC asparagine	AGC serine	C
	AUA isoleucine	ACA threonine	AAA lysine	AGA arginine	A
	AUG (start) methionine	ACG threonine	AAG lysine	AGG arginine	G
G	GUU valine	GCU alanine	GAU glutamine	GGU glycine	U
	GUC valine	GCC alanine	GAC glutamine	GGC glycine	C
	GUA valine	GCA alanine	GAA glutamine	GGA glycine	A
	GUG valine	GCG alanine	GAG glutamine	GGG glycine	G

Figure 16.8 The genetic code.

PROTEIN SYNTHESIS

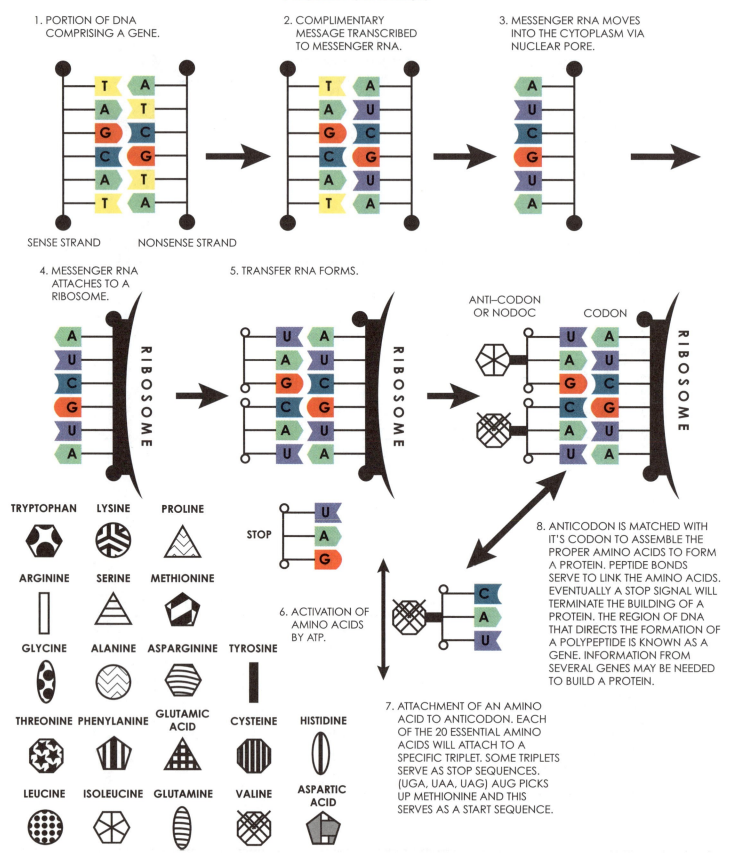

Figure 16.9 Overview of protein synthesis.

DNA, Mitochondria and Chloroplasts

Mitochondria are common organelles in the majority of eukaryotic cells (Fig. 16.10). Scientists have discovered that these organelles have their own unique DNA, **mtDNA**. This is illustrated by the **endosymbiont theory** proposed by Lynn Margulis in the 1970s. The theory holds that primordial cells established a symbiotic relationship with early purple bacteria. Eventually, the bacterial cell changed and became an important part of the cells machinery, providing the cell with energy.

The **mitochondrial genome** is significantly smaller than the nuclear genome. In many mammalian species including humans, the mitochondrial genome consists of 37 genes. With the exception of a few cases of the paternal transmission of mtDNA, the majority of mtDNA is transmitted maternally. Mitochondrial DNA is being used and studied in forensics, evolutionary studies, and medical genetics.

Chloroplasts in plant cells and some protists also have their own unique chloroplast DNA, **clDNA**. These organelles also were derived through the endosymbiont theory as the result of early cells engulfing cyanobacteria (Fig. 16.11). The clDNA genome is also very small. Recent studies have verified that chloroplasts are not inherited from pollen. In the future, this fact will have great impact on creating genetically modified plants.

More than a century after the work of Miescher and countless others, we are beginning to understand the role of DNA in heredity, medicine, forensics, and evolution. In private, Miescher speculated about the possible role of nuclein in heredity but was unable to validate his idea. Today, it is known that within the double helix of DNA lies the genetic code. Understanding the genetic code will allow scientists to understand the molecular basis of genetic disorders such as sickle cell anemia, cystic fibrosis, and Tay Sach's disease, and hopefully one day find a cure for these disorders. With an understanding of DNA, forensic science has grown tremendously. Knowledge of DNA will be used to capture criminals, solve crimes, and identify bodies. Recombinant DNA technologies have led to genetically engineered insulin and other medicines. The gene for biolumenescence in jellyfish has been introduced into nematodes, fishes, mice, and pigs causing them to glow in the dark showing that genes can be moved from one organism to another. Mutations of the genetic material serve as the mechanism of evolution. Knowledge of DNA has provided scientists with a means of understanding how evolution occurs and a greater appreciation of life itself.

With the description of the molecular configuration of DNA, Watson and Crick initiated a new era in the history of biology. From humble origins, the science of DNA has opened, and will continue to open, many new frontiers.

It is essential for genetic material to make exact copies of itself; otherwise growth would produce disorder, life could not originate, and favorable forms would not be perpetuated by natural selection.

—**Maurice Wilkins (1916–2004)**

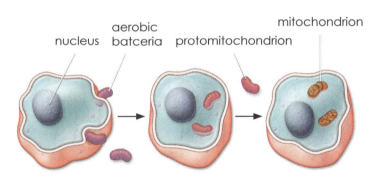

Figure 16.10 Mitochondria were derived from early purple bacteria that were engulfed by early cells.

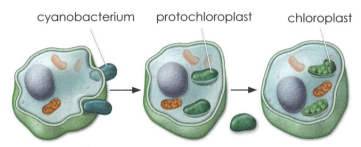

Figure 16.11 Chloroplasts were derived from early cyanobacteria that were engulfed by early cells.

Figure 16.12 A genetically engineered nematode that "glows in the dark."

Beyond the Lab—Visiting History

Describe the contributions of the following in establishing our understanding of DNA:

Frederick Griffith:

_____

_____

Archibald Garrod:

_____

_____

Oswald Avery, Colin Macleod, and Maclyn McCarty:

_____

_____

Alfred Hershey and Martha Chase:

_____

_____

George Beadle and Edward Tatum:

_____

_____

Matthew Meselson and Franklin Stahl:

_____

_____

_____

Linus Pauling and Vernon Ingram:

_____

_____

Severo Ochoa:

_____

_____

Kary Mullis:

_____

_____

Alec Jefferie:

_____

_____

Student Activity—Spooling DNA

Spooling DNA

In this activity you will isolate DNA from your cheek cells. This is a two-step process, requiring in the first step the removal of the cell and nuclear membranes, and in the second step isolating the DNA.

An Amazing Fact!

If the genome were a book, it would be the equivalent of 800 dictionaries. It would take a person typing 60 words per minute, 8 hours a day, around 50 years, to type the human genome. To hold all of this, you would need 3 gigabytes of storage space on a computer.

Materials
- 8 ounces of clear Gatorade™ or a .9% salt water solution (½ teaspoon salt added to 8 ounces water)
- 25% soap solution (5 ml dish liquid soap and 15 ml water)
- 15 ml of ice cold 95% ethanol (keep in the freezer or on ice until use)
- 6-ounce plastic cup
- 30–50 ml glass test tube
- 25 ml graduated cylinder
- glass stirring rod
- stop clock or timer
- test tube rack

Procedure 16.1
DNA

1. Procure materials from the lab instructor.
2. Using the graduated cylinder, pour 10 ml of Gatorade or .9% salt water solution into a plastic cup.
3. Swirl the contents of the cup in your mouth vigorously for 30 seconds. The vigorous swirling will allow you to slough off a large number of cheek cells.
4. Carefully spit the contents from your mouth into the cup.
5. Add 5 ml of the soap solution to a glass test tube.
6. Pour the contents from the cup into the test tube containing the soap solution.
7. Using a glass stirring rod, stir the contents for 3 minutes. Use a gentle motion to avoid forming bubbles. (In this step of the activity, the soap solution is used to break down the cell membranes.)
8. Remove the glass stirring rod, and carefully tilt the test tube at a 45-degree angle. Add 15 ml of the ice cold ethanol slowly down the side of the test tube. Do not shake or mix the ethanol with the contents of the test tube. The alcohol will form a layer on top of the solutions.
9. Place the test tube in the test tube rack and allow it to stand for 1 minute. You should observe a white fluffy or stringy mass of DNA precipitate out of the solution.
10. Using your glass rod, stir the DNA and spool the DNA (wind it onto the glass stirring rod, and observe) (Fig. 16.13).
11. Allow the DNA to air-dry for 10 minutes.
12. Follow the lab instructor's directions for disposal of all waste materials.

Q. Describe the action of the soap solution on the cellular and nuclear membranes.

_____

_____

_____

Figure 16.13 Spooling DNA.

Q. What does the addition of the ethanol to the contents in the test tube cause? Why?

_____

_____

_____

Q. A human cell contains 6 feet of DNA. Approximately how long was the strand of DNA that precipitated?

_____

_____

_____

We discovered the secret of life.

—**Francis Crick (1916–2004)**

SOLVING A WORLD WAR II MYSTERY: THE PLIGHT OF THE *LADY BE GOOD*

On the afternoon of April 4, 1943, the nine-member crew of the B-24 Liberator Bomber named the *Lady Be Good* began its first combat mission (Fig. 16.14). This plane was part of a 25-plane high-altitude raid from the 367th Bomb Group assigned to bomb the port at Naples, Italy. Because of strong winds and sandstorms blowing across the Sahara Desert, the bombers left in small groups from Soluch Airstrip in Libya. The *Lady Be Good* was one of the last three planes to leave. Shortly after take-off, the other two bombers were forced to return to base because of sand in their engines. Thus, the rookie crew members of the *Lady Be Good* were left on their own to catch up with the fragmented squadron.

The mission to Naples was to take 9 hours, and the bomber had 12 hours of fuel. To avoid detection from Nazi forces, the bombers maintained radio silence and constantly made course corrections. By the time the *Lady Be Good* reached Naples, it was night and the other Liberators had completed their mission and were returning to base. Around 10:00 p.m., the bomber aborted its mission and dropped its bombs harmlessly into the Mediterranean Ocean before heading back to Soluch.

Shortly after midnight on April 5, the *Lady Be Good* overflew the airbase at Soluch and headed southeast over the desert. At 2:00 a.m. the bomber was more than 400 miles from the airbase and running out of fuel. At this point, the crew bailed out of the ill-fated bomber. Eight members of the crew met in the darkness after reaching the ground. The bombardier was missing.

The plane crashed into the desert 16 miles away. Ironically, the plane suffered little damage from the crash and was well supplied with food, water, and a working radio. Unfortunately, the eight men had only half a canteen of water to share across the desert, where daytime temperatures could exceed 130° F. The eight men struggled against the desert for 5 days and walked more than 78 miles. At this point, five members of the crew, exhausted, stopped their trek and died in the desert sun. Three others struggled onward in an attempt to find help. Two of the three traveled nearly 30 more miles before succumbing to the desert heat. The fate of the third member, Staff Sergeant Vernon Moore is unknown.

Fifteen years after the incident, British petroleum explorers found the crash site of the *Lady Be Good* (Fig. 16.15). Despite the years, the plane was still in good condition, with food and water on board, as well as a working radio and machine gun. With the exception of Moore, the crew's remains were found between 1959 and 1960, either on the surface of a gravel plain or in the Calanscio Sand Sea. The navigator's broken body was found with a failed parachute, and it was apparent that he had died on impact.

The fate of Vernon Moore remains a mystery, but in 1953, on a gravel plain on the Calanscio Sand Sea, the mummified skeletal remains of a human body thought to be that of an Arab were found, photographed, and buried. Today, many people think this could be the remains of Moore. Although inconclusive, analysis of the photographs fueled the possibility that the remains were his. Today, researchers are searching again for the gravesite to identify the body.

In the 1950s and 1960s, the bodies of the crew members were identified by flight gear and dog tags. The person placed in the grave had neither clothes nor dog tags, so modern forensic techniques will be used to identify the body when it is re-discovered.

Figure 16.14 Crew of the *Lady Be Good;* Vernon Moore is the second to last airman.

Figure 16.15 The crash of the *Lady Be Good.*

Student Activity—Family Tree

A Hypothetical Family Tree of the Moore Family

In this activity, you will be working as a forensics scientist for the U.S. Army Central Identification Laboratory (www.cihi.army.mil). You have been given the task of identifying the mummified remains of a human body found buried in a shallow grave in the Calanscio Sand Sea in 1953. At the time, the skeletal remains were thought to be that of a young Arab man. Recently however, researchers believe it may be the body of *Lady Be Good* Staff Sergeant Vernon Moore.

 ### Procedure 16.2 Family Tree

1. Examine the following pedigree (Fig. 16.16) and, using your knowledge of the inheritance of mtDNA, identify the family members who should donate their mtDNA to solve this mystery.
2. Using a highlighter, draw lines to connect the family members to the great-grandmother.

Q. Describe the inheritance pattern of mtDNA.

_____

_____

_____

Q. If two brothers died in a crash, could you distinguish their remains based on mtDNA? Why or why not?

_____

_____

_____

_____

Q. Describe three other incidents in which mtDNA has been helpful to forensic scientists.

_____

_____

_____

_____

_____

_____

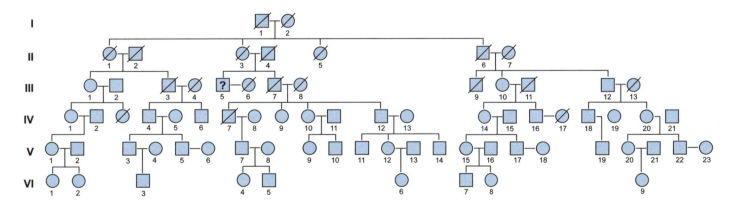

Figure 16.16 Hypothetical family tree of the Moore Family.

Beyond the Lab—Using the Internet

One of the most powerful sites on the Internet for studying human medical genetics is the On-line Mendelian Inheritance in Man (OMIM) Dictionary.

1. Using this site, describe the following:

Achoo syndrome:

_____

_____

_____

Jumping Frenchmen of Maine Disorder:

_____

_____

_____

Michelen Tire baby:

_____

_____

_____

Ichthyosis (type III):

_____

_____

_____

Hypertrichosis:

_____

_____

_____

2. Again, using OMIM, describe a genetic disorder in your family or a disorder in the family of an acquaintance.

_____

_____

_____

3. Describe the Human Genome Project site. How can you subscribe to the newsletter? How can you acquire a chromosome map poster?

_____

_____

_____

_____

4. Find three sites that have tutorials on DNA, replication, and protein synthesis.

_____

_____

_____

5. Name several sites that address the pros and cons of genetic engineering.

_____

_____

_____

_____

6. What is the genetic link between the Black Plague and HIV? Why are scientists excited about this link?

_____

_____

_____

_____

An Amazing Fact!

Less than 2% of the total DNA carries instructions to make proteins. The rest is misleadingly called "junk DNA," because it is a hodge-podge of sequences that do not seem to code for anything.

Name: _____ Date: _____ Section: _____

??? Review Questions

1. What is the chemical composition of DNA?

_____

_____

_____

2. What is replication, and what is its importance?

_____

_____

_____

3. What is the function of protein synthesis?

_____

_____

_____

4. What is a mutation? What is the significance of mutations in evolution?

_____

_____

_____

5. List and describe the function of three types of RNA.

_____

_____

_____

_____

6. What are some practical uses of our knowledge of DNA?

_____

_____

_____

7. What are the pros and cons of genetic engineering?

_____

_____

_____

8. How can our knowledge of DNA be important in the following disciplines:

a. Medicine

_____

_____

b. Forensics

_____

_____

c. Agriculture

_____

_____

d. Taxonomy

_____

_____

e. Evolution

_____

_____

f. Environmental science

_____

_____

9. Research and relate the nature and symptoms of xeroderma pigmentosum and ataxia telangiectasia.

_____

_____

_____

_____

Chapter 17
That Mystery of Mysteries: Understanding Evolution

Student Outcome Objectives

At the completion of this exercise, the student will be able to:

1. Define evolution.
2. Describe the contributions of Charles Darwin.
3. Compare and contrast artificial selection and natural selection.
4. Cite important biological and geological events from the major eras and periods of geologic time.
5. Define the terms *fossil* and *paleontology*.
6. Discuss the importance of studying fossils.
7. Cite factors that are favorable to fossilization and several means of fossilization.
8. Identify and discuss the natural history of several common fossils.
9. Explain how comparative anatomy serves as proof of evolution.
10. Explain how embryology can serve as proof of evolution.
11. Explain how biochemistry provides evidence for evolution.
12. Discuss how genetics is used in proving evolution.

Overview

Some questions have always intrigued enlightened minds: How and when did life begin? How did species form? Do we know our ancestors? The methods and tools of modern science are leading us to a better understanding of such questions. In 1973, a loyal defender of Darwinism, Theodosius Dobzhansky, stated that "nothing in biology makes sense except in the light of evolution." Today, with the continued controversy surrounding evolution, creationism, and intelligent design, this statement is a rallying point for modern biologists.

Unfortunately, many people do not understand or do not wish to understand what evolution is all about. Their concern seems to be only with the origin of humans, but we are one species among vast numbers of extinct and extant forms. Sometimes, in our anthropocentric view, we tend to forget the rest of life on Earth. Evolution is all around us. It explains why the HIV virus is so diabolical, how Methicillin-resistant *Staphylococcus aureus* (MRSA) is terrorizing our world, and even how roaches have developed resistance to our best pesticides.

Again—"Nothing in biology makes sense except in the light of evolution."

Classically, **evolution** has been defined as a change over time. Modern biologists have modified this definition in stating that evolution involves changes in gene frequency over time. These changes can be viewed on a small scale as changes in a single trait, or on a larger scale as the formation of a new species.

Among the many significant contributors in the history of evolutionary thought, **Charles Robert Darwin** (1809–1882), has been named "the Father of Evolution" (Fig. 17.1) Trained in theology and natural science, Darwin's idea of natural selection, that mystery of mysteries, has revolutionized modern science. His voyage on the *HMS Beagle* in 1831–1836 is a pivotal point in the history of evolution. As a young naturalist exploring the lands visited by the *Beagle*, Darwin made observations that eventually would shape his ideas on natural selection and descent with modification.

The domestication of dogs is thought to have begun approximately 15,000 years ago from wolf ancestors. Since then, humans have shaped a number of breeds of dog through **artificial selection**. This process involved selectively breeding for desirable traits. The American Kennel Club recognizes some 150 distinct breeds, from Boston

Figure 17.1 Charles Darwin 1809–1882.

terriers to malamutes. If artificial selection can be so powerful in 15,000 years, just think of the power of **natural selection** working for 3.8 billion years! This power is reflected in the diversity of life on Earth.

In 1859, the scientist T. H. Huxley, also known as "Darwin's Bulldog," stated, "How extremely stupid for me not to have thought of that!" after reading Darwin's *The Origin of Species*. He pointed out that Darwin's basic concepts outlined in the book were nothing more than common sense illustrated by brilliant and meaningful examples. In *The Origin of Species*, Darwin discussed how variation exists within a species, that organisms produce more young than can be expected to survive naturally, that there is a struggle for existence in nature, and how those with favorable characteristics have a better chance to survive.

After publication of *The Origin of Species*, the support for evolution has emerged from a variety of disciplines. A better understanding of the age of Earth has given scientists a framework for evolution, and the fossil record has provided many important fossils. Evidence from comparative anatomy and embryology has enlightened our knowledge of form and function. Biogeography has shed light on the distribution of species and their formation. Population genetics, classical genetics, neo-Mendelian genetics, and epigenetics have provided detailed information on inheritance and evolution, and molecular biology has produced evidence from the molecular basis of evolution.

Darwin and the Galapagos Islands

In his travels on the *HMS Beagle* to the Galapagos Islands, Darwin identified 13 different species of finches. Each species had its own distinctive beak and ecological niche on the Galapagos Islands. Molecular studies (comparing mitochondrial DNA) suggest that these finches probably evolved from an ancestral seed-eating warbler-type "dull-colored grassquit" that lived on the mainland several million years ago. **Adaptive radiation** refers to the process in which a species or group of related species evolves rapidly into many different species that occupy new habitats or geographic zones. Darwin's finches diverged in response to the availability of food in the different habitats.

There is grandeur in the view of life, with its several powers, having been originally breathed by the Creator into a few forms or into one; and that, whilst this planet has gone cycling on according to the fixed law of gravity, from so simple a beginning endless forms most beautiful and most wonderful have been, and are being evolved.

—**Charles Darwin (1809–1882)**

A Whale of a Tale!

In the first writing of *The Origin of Species*, Charles Darwin proposed that the ancestor of modern cetaceans was a bear-like creature (Fig. 17.2). Unable to substantiate his claim with fossils, Darwin dropped the statement from subsequent editions. In the late 1990s, however, Darwin was vindicated with a myriad of fossil discoveries of ancestral cetaceans from the shores of the ancient seas known as Tethys.

Today it is accepted that the ancestors of cetaceans are related to artiodactyls (pigs, for example). If you were to see the earliest land ancestor to a whale, you would declare—before you bolted for safety—that it resembled a wolf with hooves. The fossil record and molecular evidence continue to reinforce Darwin's brilliant vision.

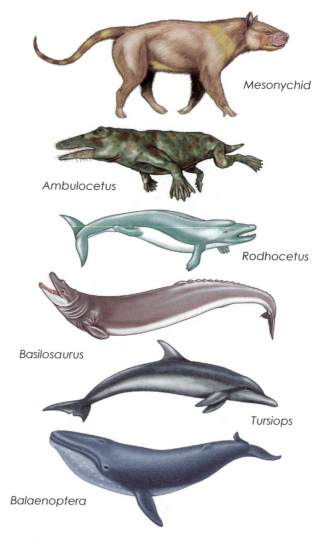

Mesonychid

Ambulocetus

Rodhocetus

Basilosaurus

Tursiops

Balaenoptera

Figure 17.2 Ancient and modern cetaceans.

When on board HMS Beagle *as naturalist, I was much struck with certain facts in the distribution of the organic beings inhabiting South America, and in the geological relations of the present to the past inhabitants of that continent. These facts…seemed to throw some light on the origin of species—that mystery of mysteries, as it has been called by one of our greatest philosophers.*

—Charles Darwin (1809–1882)

There is no mystery to Darwin's machine, it is no more than evolution plus time.

—Steve Jones (1944–present), *Darwin's Ghost*

Darwin's Finches

A particularly interesting example of contemporary evolution involves the 13 species of finches studied by Darwin on the Galapagos Islands, now known as Darwin's finches. A research group led by Peter and Rosemary Grant of Princeton University has shown that a single year of drought on the islands can drive evolutionary changes in the finches. Drought diminishes supplies of easily cracked nuts but permits the survival of plants that produce larger, tougher nuts. Drought thus favors birds with strong, wide beaks that can break these tougher seeds, which has produced populations of birds with these traits. The Grants (1991) estimated that if droughts occur about once every 10 years on the islands, a new species of finch might arise in only about 200 years.

Sources: "Natural Selection and Darwin's Finches," by Peter R. Grant, in *Scientific American*, 1991, pp. 82–87; *The Beak of the Finch*, by Jonathan Weiner (New York: Alfred A. Knopf, 1994).

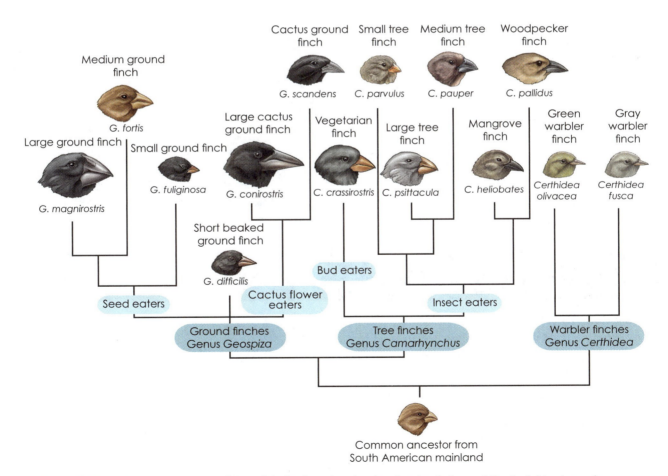

Figure 17.3 An Evolutionary tree of Darwin's finches showing beak adaptations of the individual species.

Beyond the Lab—Galapagos Islands

Investigating Galapagos Tortoises

Darwin examined many closely related species of tortoises on the islands in the Galapagos (Fig. 17.4). This activity asks students to research and report on Darwin's findings on tortoises on the Galapagos Islands.

Materials
* map or globe including Galapagos Islands

 **Procedure 17.1
Galapagos Tortoises**

1. After consulting online resources answer the following questions:
    a. Where are the Galapagos Islands located?

    _____

    b. How many islands make up the Galapagos Islands?

    _____

2. After consulting online resources on Darwin, answer the following:
    a. Where is the Darwin Research Station located, and what is its mission?

    _____

    _____

    _____

b. What are three Galapagos creatures other than finches and tortoises?

_____

_____

_____

c. What did the differences between these closely related species suggest to Darwin?

_____

_____

_____

_____

Figure 17.4 The major islands of the Galapagos.

 Student Activity—Geological Time

Studying Geologic Time

The geologic time scale is a valuable instrument used by geologists, biologists, and other scientists to relate various times in Earth's history. To understand evolution, we must understand the framework of geologic time. Eras represent vast amounts of time that are subdivided into periods and, eventually, epochs. In this activity, students are to refer to Table 17.1.

 **Procedure 17.2
Geological Time Scale**

1. Answer the following questions:
    a. If you were taking a time hike, in which each 2-foot step took you back 100 years, indicate how many steps and how long your hike would be to visit the following:

    _____ The Battle of Gettysburg (1863)

_____ Signing of the Declaration of Independence (1776)

_____ To watch Columbus launch for the Americas (1492)

_____ To meet Aristotle (circa 350 B.C.)

_____ To witness the end of the Pleistocene (12,000 years ago)

_____ To see Australopithicus (3.7 million years ago)

_____ To see the first whales (50 million years ago)

_____ To witness what really happened to the dinosaurs (65.5 million years ago)

_____ To see an Archaeopteryx (299 million years ago)

_____ To visit the end of the Permian (312 million years ago)

_____ To see giant insects of the Carboniferous (350 million years ago)

_____ To see the adaptive radiation of the amphibians (450 million years ago)

_____ To see eurypterids (465 million years ago)

_____ To see the first vertebrates (426 million years ago)

_____ To witness the Cambrian explosion (525 million years ago)

_____ To witness the origin of life (3.5 billion years ago)

b. Why is "Meet the Flintstones" misleading regarding man and dinosaurs?

_____

_____

_____

c. What is adaptive radiation? Give specific examples of adaptive radiation and its significance.

_____

_____

_____

d. What happened to the dinosaurs? Should we have renamed Jurassic Park "Cretaceous Park?" Why?

_____

_____

_____

e. Discuss the role of continental drift upon the history of life on Earth.

_____

_____

_____

f. Discuss the impact of mass extinctions upon life on Earth.

_____

_____

_____

g. What evidence of the various geologic periods can be found in this region?

_____

_____

_____

We are the product of 4.5 billion years of fortuitous, slow biological evolution. There is no reason to think that the evolutionary process has stopped. Man is a transitional animal. He is not the climax of creation.

—Carl Sagan (1934–1996)

Table 17.1 Geologic Time Scale

ERA	PERIOD		EPOCH	TIME	
ARCHEOZOIC	PRECAMBRIAN			4.5 to 2.5 billion years ago	
PROTEROZOIC				2.5 billion years ago to 542 million years ago	
PALEOZOIC Age of Amphibians	CAMBRIAN			542 to 499 million years ago	
	ORDOVICIAN			499 to 444 million years ago	
	SILURIAN			444 to 416 million years ago	
	DEVONIAN			416 to 359 million years ago	
	CARBONIFEROUS	MISSISSIPPIAN		359 to 315 million years ago	
		PENNSYLVANIAN		315 to 299 million years ago	
	PERMIAN			299 to 251 million years ago	
MESOZOIC Age of Reptiles	TRIASSIC			251 to 201 million years ago	
	JURASSIC			201 to 145 million years ago	
	CRETACEOUS			145 to 65.5 million years ago	
CENOZOIC Age of Mammals	TERTIARY		PALEOCENE	65.5 to 55.8 million years ago	
			EOCENE	55.8 to 39.9 million years ago	
			OLIGOCENE	39.9 to 29 million years ago	
			MIOCENE	29 to 5.3 million years ago	
			PLIOCENE	5.3 to 2.6 million years ago	
	QUARTERNARY		PLEISTOCENE	2.6 million to 12,000 years ago	
			RECENT	12,000 years ago to present	

ENVIRONMENTAL EVENTS	BIOLOGICAL EVENTS
Formation of the planet. Unstable and turbulent weather. Primitive atmosphere devoid of oxygen. Granite, sedimentary rocks, and red beds formed.	From absence of life through the chemical origin of life. Protobionts formed. Anaerobic and photosynthetic bacteria evolve. Heterotrophism established.
Vast volcanic activity. Mountain formations. Sedimentary rock formation. Glaciation with moist cool to dry cool climates. Oxygen released to atmosphere through photosynthesis. Ozone layer established.	Autotrophism established. Rise of eukaryotes. Algae diverse. Protozoans, worms and other softbodied animals present. Prokaryotic organisms still abundant.
Large oceans with a mild climate. Paleo-equator through west central North America.	All invertebrate phyla present. Many species of trilobites. Algae complex. **THE CAMBRIAN EXPLOSION**
Oceans were greatly enlarged. Much of the land submerged. Warm mild climates into higher latitudes.	Abundant trilobites, crinoids, brachiopods, cephalopods, and other marine invertebrates. First vertebrates appear. Primitive plants. **AGE OF INVERTEBRATES**
Relatively flat continents with continental seas. Mountain building in Europe with mild climates.	Invasion of the land by vascular plants and arthropods. Diversity of graptolites and marine invertebrates. Eurypterids were dominant. Fish were primitive.
Small inland seas with the formation of mountains. Variation of arid lands to lands with abundant rainfall.	First amphibians appear. Abundant sharks and fish. Presence of lungfish. Marine invertebrates abundant. Land plants increasing. **AGE OF FISHES**
Much mountain formation with inland seas. Warm to hot swampy environments.	Radiation of the amphibians. Many bony fish and sharks. First winged insects. Many crinoids and brachiopods. Great coal forests including club mosses, horsetails, and giant ferns.
Shallow inland seas and cool swamp forests. Glaciation in the Southern Hemisphere.	Origin of reptiles. Diversity of amphibians. Giant insects. Great coal forests including club mosses, horsetails, and giant ferns.
Wide spread mountains. Appalachians formed. Moist to cold dry climates. Pangaea intact. Ending with glaciation and an ice age.	Adaptive radiation of the reptiles displacing the amphibians. Major extinction of many species of plants and animals. Modern insects increase. Gymnosperms appear.
Two continental depositions separated by an intervening marine complex. Varied sandstone and limestone deposits. Climates arid with wide spread deserts.	Variety of reptiles. Thecodonts dominate. Presence of mammal-like reptiles and early mammals. First dinosaurs. Many marine invertebrates. Ferns, cycads, and conifers dominate.
Climates generally mild. Antarctica and Canada were temperate. Beginning of the Sundance Sea and the Gulf of Mexico. Continents with shallow seas. Sierra-Nevada mountains form.	Giant reptiles common. Diversity of dinosaurs and marine invertebrates. Early mammal populations increasing. Archaeopteryx and toothed birds. Ferns, cycads, and conifers dominate.
Spread of inland swamps and seas. Formation of the Andes, Rockies and Himalayas. Continued break up of Pangaea. Increased geologic instability. Climates cooling. Asteroid strikes near the Yucatan Peninsula.	Dinosaurs dominate before extinction. Extinction of the giant reptiles and many marine invertebrates. Elimination of one-fourth of all animal families. Reduction of cycads, ferns, and gymnosperms. Marsupials and placental mammal populations increase. Rise of flowering plants. **THE TIME OF DYING**
Mountain building. Temperate to subtropical climates. Separation of South America and Africa.	Archaic mammals dominate. Large flightless birds. Subtropical plants. Woody flowering plants present.
Heavy rainfall with mountain erosion. Land connection between Europe and North America.	Modern orders of animals. Adaptive radiation of placental mammals. First horses, giant birds, subtropical forests.
Climates mild, beginning to cool. Tethys Sea dominate. Nebraska was similar to modern Louisiana.	Whales and primates developing. Temperate plants. Giant foraminiferans. Archaic mammals extinct.
Development of plains. Climate moderate.	Vast grasslands with grazing mammals. Modern sub-families present. Radiation of the large carnivores. First hominids.
Continental elevation. Cool climates. Land bridge between North and South America. Asia and North America connected at Berring Strait.	The peak of mammals. Most genera of plants and animals present. Large carnivores.
Three ice ages in which glaciers covered most of North America, Europe, and Asia. The climates ranged from cold to mild.	Development of modern species of plants and animals. Decline of great mammals. Development of early humans.
End of the third ice age. Climates is warming. Geologic and geographical conditions similar to today. Remnants of the ice ages present in glaciers. Great Rift Valley in Africa forms.	Dominance of modern plants and animals. Rise of herbaceous plants. Extinction of giant mammals. Man has great impact on the planet.

PALEONTOLOGY

Stop! Don't throw that rock! You may be throwing away millions of years of geological and biological history. Your missile may contain a variety of fascinating fossils. A **fossil** is defined as the remains or traces of organisms that lived in the past, and the study of fossils is termed **paleontology**. Fossils may be complete organisms, parts of organisms, or even traces of organisms. Occasionally, a complete organism is found, such as fossils that have been preserved in ice. Most fossils, however, are parts of organisms such as shells, bones, and teeth, and traces of life in the past include tracks, burrows, and fossilized excrement.

If a fossil could talk, it would tell an incredible tale. Paleontologists listen to fossils and learn from their silent voices. Today, fossils are used to indicate evolutionary trends, morphology and anatomy of past organisms, injuries and disease of past organisms, behaviors of past organisms, characteristics of ancient environments, and ancient geography. In addition to being significant scientifically, many fossils are used for jewelry, for conversation pieces, in industry, and in the search for oil. To be classified as a fossil, a specimen has to be at least 10,000 years old.

Sometimes, looking for fossils is comparable to looking for an amoeba in a haystack. Although researching the fossil site thoroughly is important, finding that special fossil is often a matter of pure luck. Think about it: That old fossil in your driveway gravel was formed a long time before the dinosaurs roamed our planet, and you just bumbled upon it after hundreds of millions of years of geologic history. For a fossil to form requires a series of timely events.

In general, three major factors increase the chances of something fossilizing: possession of hard parts, escape from immediate destruction, and rapid burial.

1. Organisms that have hard parts such as bones, teeth, shells, and woody tissue are more likely to fossilize than organisms with soft parts. Even delicate organisms such as bacteria, sponges, worms, and jellyfish, however, may fossilize under the proper conditions.
2. The remains of most organisms are destroyed by mechanical forces such as weathering, crushing, and wave action, or by biological actions such as predators, scavengers, and mechanisms of decay. Paleontologists estimate that more than three-fourths of the life forms that ever lived did not form fossils.

Evolution is a dance to the music of time that invites all to join in.

—Steve Jones, *Darwin's Ghost* **(1944–present)**

3. Fossilization also can involve rapid burial in a medium that is capable of retarding decay. Organisms living in water or near water are likely to be buried in sediment that protects them from mechanical forces and decay. Tree sap, ice, volcanic ash, and even tar may preserve some organisms. Organisms living in arid regions may be preserved through dessication or mummification. A mummy of a dinosaur called a Trachodon (a species of duckbilled dinosaur) preserved in this manner indicated that these animals had leathery skin and webs between their toes on the front feet.

A tremendous fossil record has been unearthed through the years, and an immense number of fossils await future discovery. A look at fossils will reveal that fossilization can occur in many ways.

- Organisms that have been preserved in **ice** provide paleontologist with an ideal means of fossilization. Wooly mammoths that have been recovered from Alaska and Siberia look as if they were just taken out of the freezer!
- *Jurassic Park* introduced movie-goers to the concept of fossils in **amber**. In this case, tree sap may have trapped insects, feathers, or other biotic specimens and preserved them in a permanent encasement.
- The La Brea **tar pits** near Los Angeles have yielded an abundance of mammal fossils, including saber-toothed cats.
- **Carbonization** may reduce plants and animals to shiny films that actually are finer than tissue paper. Many fossil fern leaves are carbonized films.
- **Petrification** is a common form of fossilization that results from mineral matter soaking into every cavity and pore of a specimen and replacing the living material. The hard parts, however, are not totally replaced.
- Unlike petrification, **replacement** involves total replacement of the original structures. Many times, replacement fossils lack details.

In addition to the actual remains of organisms being preserved, certain traces of organisms are considered fossils. Tracks and trails of animals are one common type of **trace fossils**. Tracks from worms through giant dinosaurs have been collected. Fossilized excrement known as **coprolites** and gizzard stones, **gastrolyths**, are extremely valuable in studying the life history of animals. **Molds** and **casts** are two widespread types of trace fossils. Basically, molds are an impression of an organism or an organic structure in a hardened medium. Internal and external molds are common fossils. Casts result from a substance filling a mold. Many times, clam and snail shells dissolve, leaving a cast.

The age of fossils can be determined in two different ways: relative dating and absolute dating.

1. **Relative dating** involves age-dating fossils by association with known time periods, grouped into four eras: the Precambrian, the Paleozoic, the Mesozoic, and the Cenozoic. The eras represent distinct ages in the history of the Earth. Just as a movie can be age-dated by clothes, hair styles, and cars, fossils can be age-dated by strata and sediments. Relative dating does not result in numerical ages; it simply orders the appearance of organisms. Fossils can be dated through **stratigraphy**, a method of placing fossils in relative sequence to each other based on their location in the sedimentary strata. In beds of rock that have not been disturbed, the oldest rocks exist below the younger rock. The strata of one location can be correlated with another location by the presence of **index fossils**. These fossils are anatomically and morphologically distinct, common during a distinct period of time, widespread, and fossilize easily.

2. Although relative dating is important, **absolute dating** provides a more exact means of dating. Sometimes known as **radioactive dating**, absolute dating is based on the radioactive decay of certain isotopes. Every isotope has its own characteristic rate of decay. The time for one-half of an isotope to change to the more stable form is known as its **half-life**. The half-lives of isotopes vary from a few hours to millions of years. Because the half-life of an isotope does not vary, it is not influenced by variables such as temperature and pressure.

The age of a fossil is determined by comparing the percentage of the isotopes. For example, in carbon dating, the ratio of carbon 14 to carbon 12 is compared. Knowing that the half-life of carbon 14 is 5,730 years, the age of a fossil can be determined based upon percentages. Unfortunately, carbon dating is limited to specimens of less than 50,000 years. To date older specimens, other radioactive dating methods must be used.

Student Activity—Classifying Fossils

This activity asks students to identify, examine and classify fossils, working in groups of four. Students are to closely examine the fossils and answer the questions.

Materials
- model specimens (at least 8 for each collection)
- modeling clay

Procedure 17.3
Fossils

1. Procure the fossils or models provided by your instructor, and answer the following questions:

Q. Is the fossil a complete organism, a part of an organism, or a trace of an organism?

_____

_____

Q. In what kind of environment may this organism have lived?

_____

_____

Q. How did fossilization occur? Is it a cast or a mold?

_____

_____

_____

Q. What is your fossil?

_____

_____

Q. Where would you place the fossil on the geologic time scale?

_____

_____

Q. What are the closest living relatives to your fossil?

_____

_____

2. Sketch your selected fossil.

3. Choose a model specimen from your collection and, using modeling clay, make an impression of your fossil. Examine the model and answer the following:
   a. Identify your specimen.

   _____

   _____

   b. Did your impression reveal any surface features?

   _____

   _____

   c. Did you create an internal or an external mold?

   _____

   _____

   d. How can similar exercises aid paleontologists?

   _____

   _____

   _____

4. Many fossils are beautiful and quite expensive. Other fossils are valuable as scientific specimens. Describe several fossils common to your region.

   _____

   _____

   _____

5. If you have any fossils, bring some to class for discussion.

   a. _____

   b. _____

   c. _____

   d. _____

   e. _____

   f. _____

   g. _____

   h. _____

   i. _____

   j. _____

   k. _____

   l. _____

   m. _____

   n. _____

   o. _____

COMPARATIVE ANATOMY

In studying evolutionary relationships, scientists use a variety of methods to establish ancestry. Evolutionary biologists compare anatomical structures of organisms, embryological development, and the biochemical (macromolecular) make-up of organisms. Organisms then can be classified based on the extent of similarity or descent from a common ancestor.

Comparative anatomy yields convincing data to support the evolution of organisms and their structures. When structures share similar anatomical features and embryological development, they are said to be **homologous**. For example; the wing of a bat, the flipper of a whale, and the arm of a human are homologous structures—having evolved from a common ancestor but not always performing the same function (Fig. 17.6).

Although the type and number of fossils vary from region to region, the following images represent common fossils. Try to identify the fossil types in Figure 17.5 using the space provided on the previous page. Use the key below to check your answers.

Figure 17.5 Some common fossils are, (a) brachiopod, (b) bivalve, (c) colonial coral, (d) cephalopod, (e) fossiliferous limestone, (f) trilobite, (g) crinoid crowns, (h) crinoid, (i) horse tooth, (j) mastodon tooth, (k) megalodon tooth, (l) colonial coral, (m) bivalve, (n) ammonite, and (o) sea urchin.

It is possible for organisms to share a similar structure but to have evolved through different pathways. For example, the wing of a bird and the wing of an insect have similar structure and function but differ in their evolutionary history. The wings are said to be **analogous structures**.

Every individual alive today, the highest as well as the lowest, is derived in an unbroken line from the first and lowest forms.

—**August Weismann (1834–1914)**

 ## Student Activity—Homologous and Analogous Structures

In this activity, students will refer to Figure 17.6, then conduct some research on the vertebrate forelimbs of two species.

 ## Procedure 17.4
Like Structures

1. Examine the bones of the forelimbs pictured. For each organism, name the bones that resemble a common ancestor and the bones that differ.

    a. _____
    _____
    _____

    b. _____
    _____

    c. _____
    _____
    _____

    d. _____
    _____
    _____

    e. _____
    _____

2. Explain how modifications in the structure of some of the bones were necessary for the existence of the organism.

    _____
    _____
    _____

3. Sharks and dolphins share many analogous structures that allow them to meet the demands of their habitat. Why are these similar structures analogous rather than homologous?

    _____
    _____
    _____

STRUCTURES AS EVIDENCE OF EVOLUTION

Humans along with other organisms may have structures that have no apparent function and often are homologous to structures that function normally in other species. These structures, called **vestigial structures**, indicate common ancestry and, thus, provide evidence for establishing evolutionary pathways. In his book *The Descent of Man*, Darwin discussed a number of anatomical structures that he considered useless or nearly useless.

Vestigial structures in humans include the coccyx or tail bone, wisdom teeth, and third eyelid. Humans also have extrinsic ear muscles that serve no function but do have a function in monkeys (Fig. 17.7). Until recently, the appendix was thought to be a vestigial structure but now is thought to function as an "incubator" for good bacteria that colonize our colon. Vestigial structures can be found in other animals as well. Whales and dolphins contain remnants of hindleg bones, and pythons have pelvic spurs that are the remnants of legs.

When we are cold, goose bumps raise the body's hair to trap air to keep warm. This reflex is not considered vestigial, but when we get goose bumps that raise our hairs in response to stress, this is considered a vestigial reflex.

Check Your Understanding

Q. To what is this reflex linked in other organisms?

_____

_____

_____

Q. How many vestigial structures can be found in humans?

_____

_____

_____

Q. What are five specific vestigial structures in humans, and what is their importance in our ancestors?

_____

_____

_____

_____

_____

Q. What are some vestigial structures in other organisms (insects, whales, snakes, fish, and amphibians) that can be traced to earlier ancestors?

_____

_____

_____

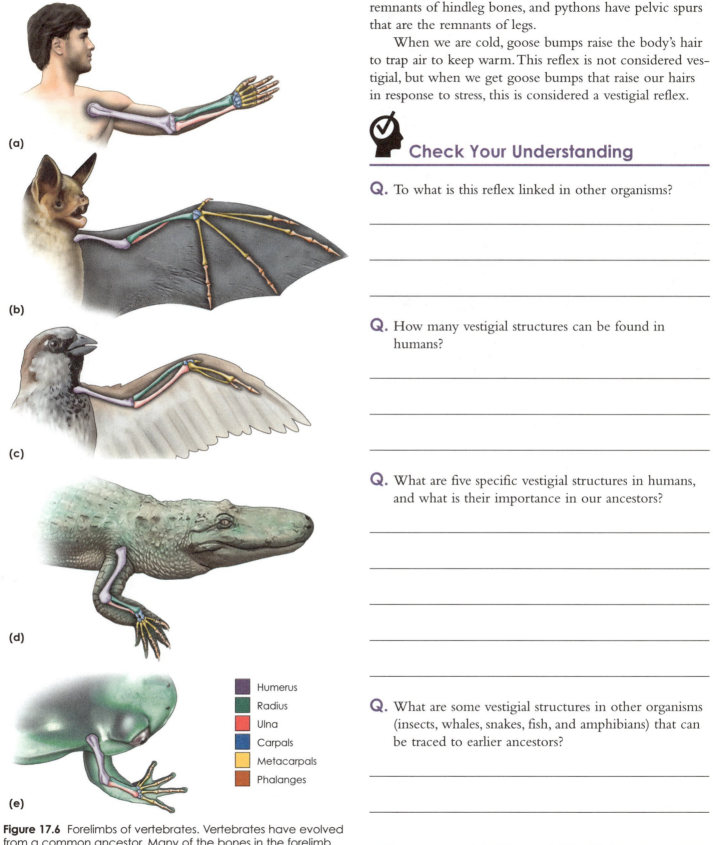

(a)

(b)

(c)

(d)

(e)

Humerus
Radius
Ulna
Carpals
Metacarpals
Phalanges

Figure 17.6 Forelimbs of vertebrates. Vertebrates have evolved from a common ancestor. Many of the bones in the forelimb are homologous structures.

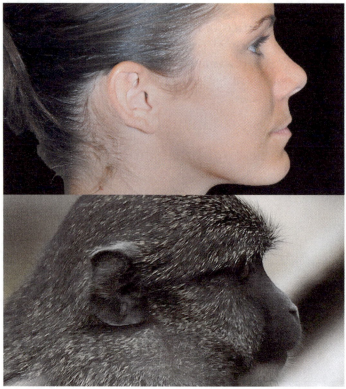

Figure 17.7 The muscles connected to the human ear do not develop enough to afford the same mobility as in monkeys.

Q. When your pet cat strangely opens its mouth when exploring some objects, what organ is it using? Do humans have this structure? If so, is it functional?

EMBRYOLOGY EVIDENCE OF EVOLUTION

Heinz Christian Pander, (1794–1865) was a Russian embryologist who studied the chick embryo. In 1817, through his research on chick embryos, he discovered the germ layers (the three distinct regions of the embryo—ectoderm, endoderm, and mesoderm—that give rise to the specific organ system).

Karl Ernst von Baer (1792–1876) was a German biologist who discovered the mammalian ovum. Through his observations of embryos of vertebrates (fishes, amphibians, reptiles, birds, and mammals), he concluded that all vertebrate embryos, especially in the early stages of development, are strikingly similar but that the adults are very different. Expanding on Pander's concept of germ layers in the chick embryo to include all vertebrates, von Baer laid

the foundation for comparative embryology. Today, von Baer is known as "the Father of Embryology." The laws of von Baer state that the general characters of a group to which an embryo belongs appear in development earlier than the special characters, and that the less general structural relations are formed after the more general ones, and so on, until the most specific appear.

Ernest Haeckel (1834–1919), a German zoologist, proposed the theory that ontogeny (changes in size and shape) recapitulates phylogeny (the evolutionary history of a species). What this meant was that advanced species repeated in their embryological development the stages that more primitive species ventured through. Haeckel supported this theory with drawings of embryological stages of various vertebrate species, claiming that they closely resembled each other (Fig. 17.8). Modern-day embryologists agree that some of Haeckel's drawings were misleading, that many of the features were overly exaggerated, and that the suggestion that ontogeny recapitulates phylogeny could not be taken literally. Still, they agree that all vertebrates do share certain characteristics. At some time during embryonic development, all have a post-anal tail, somites (body segments), and paired pharyngeal pouches (gill slits). The shared structures usually appear early in embryonic development and differ in latter stages of development.

Q. What happens to the pharyngeal pouches, post-anal tail, and somites in a shark, a turtle, a cow, and a human?

Q. What structure that is common among all vertebrates is one of the earliest structures laid out in all vertebrate embryos?

Q. What structure is the last to form in higher vertebrates?

_____

_____

_____

Q. Is it possible for a human to be born with gill slits, a tail, and webbed digits? Why or why not?

_____

_____

_____

Q. In human embryos, when will the heart, gonads, and eyes begin developing?

_____

_____

_____

Q. What is meant by the evo-devo revolution?

_____

_____

_____

EVIDENCE OF EVOLUTION: SIMILARITIES IN MACROMOLECULES

Although Darwin, Haeckel, von Baer, and others confirmed evolutionary relationships based on similarities in anatomical structures, they did not have the tools available today to observe organisms at the molecular level. Scientists now use a variety of tests to establish evolutionary relationships. Through genetic testing, scientists are able to verify common ancestry in species that share similarities in DNA. Results of some of the findings have led to changes in taxonomy for some organisms that previously were classified in one category but at the molecular level were found to share a common ancestor with an organism that may look quite different. For example, in the past decade, scientists have found that the elephant shrew, once thought to be more closely related to rodents, is genetically more closely related to aardvarks and elephants.

According to the **protein clock theory**, as DNA mutates, differences in proteins accumulate. Through examination of the amino acid sequences, researchers have concluded that the greater the similarity between the amino acid sequences of two species, the more closely related they are evolutionarily. Researchers have explored the amino acid sequences of specific proteins such as the hemoglobin molecule. Hemoglobin is a large protein found in red blood cells with the function of carrying oxygen and, to a lesser extent, CO_2 in the blood. Just one amino acid sequence in the hemoglobin separates humans from that of the gorilla's sequence.

Cytochrome c is a heme protein found in the mitochondria of many species of single-celled organisms, plants, and animals. Cytochrome c consists of a chain of approximately 100 amino acids and functions in the transfer of electrons. Thus, it is essential in the electron transfer chain. Scientists have studied the amino acid sequences of cytochrome c as a method of establishing common ancestry. Turkeys and chickens have been shown to share the same amino acid sequences as chimpanzees and humans.

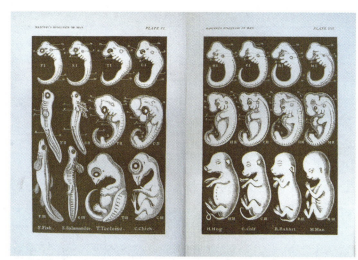

Figure 17.8 Haeckle's drawing of vertebrate embryological development, circa 1874.

The question now at issue, whether the living species are connected with the extinct by a common bond of descent, will best be cleared up by devoting ourselves to the study of the actual state of the living world, and to those monuments of the past in which the relics of the animate creation of former ages are best preserved and least mutilated by the hand of time.

—**Charles Lyell (1797–1875)**

Student Activity—Biochemistry

Relating Cytochrome c and Evolutionary Similarities

In this activity, students are asked to refer to Table 17.2, showing the amino acid sequence, then to follow the procedure below.

Procedure 17.5
Amino Acids

1. Study Table 17.2 and determine which vertebrate species show greater differences in cytochrome c sequences than that of humans. Relate this to the protein clock theory.

2. Relate the differences in the sequences of amino acids between the shark and the turtle.

3. Humans and chimpanzees share identical cytochrome c proteins in their amino acid sequences. How is it, then, that the two animals are different from each other?

Table 17.2 Cytochrome c Amino Acid Sequence

	Human	Shark	Horse	Turtle	Monkey	Chicken	Frog
42	Gln	Gln	Gln	Gln	Gln	Gln	Gln
43	Ala	Ala	Ala	Ala	Ala	Ala	Ala
44	Pro	Gln	Glu	Glu	Pro	Glu	Glu
46	Tyr	Phe	Phe	Phe	Tyr	Phe	Phe
47	Ser	Ser	Thr	Ser	Ser	Ser	Ser
49	Thr	Thr	Thr	Thr	Thr	Thr	Thr
50	Ala	Asp	Asp	Asp	Ala	Asp	Asp
53	Lys	Lys	Lys	Lys	Lys	Lys	Lys
54	Asn	Ser	Asn	Asn	Asn	Asn	Asn
55	Lys	Lys	Lys	Lys	Lys	Lys	Lys
56	Gly	Gly	Gly	Gly	Gly	Gly	Gly
57	Ile	Ile	Ile	Ile	Ile	Ile	Ile
58	Ile	Thr	Thr	Thr	Ile	Thr	Thr
60	Gly	Gln	Lys	Gly	Gly	Gly	Gly
61	Glu	Gln	Glu	Glu	Glu	Glu	Glu
62	Asp	Glu	Glu	Glu	Asp	Asp	Asp
63	Thr	Thr	Thr	Thr	Thr	Thr	Thr
64	Leu	Leu	Leu	Leu	Leu	Leu	Leu
65	Met	Arg	Met	Met	Met	Met	Met
66	Glu	Ile	Glu	Glu	Glu	Glu	Glu
100	Lys	Lys	Lys	Asp	Lys	Asp	Ser
101	Ala	Thr	Ala	Ala	Ala	Ala	Ala
102	Thr	Ala	Thr	Thr	Ala	Thr	Gly
103	Asn	Ala	Asn	Ser	Asn	Ser	Ser
104	Glu	Ser	Glu	Lys	Glu	Lys	Lys

Name: _____ Date: _____ Section: _____

❓ Review Questions

1. What are three types of fossilization?

   _____

   _____

   _____

2. Why are fossils important to scientists?

   _____

   _____

   _____

3. What are three proofs of evolution?

   _____

   _____

   _____

4. What is the role of natural selection in the process of evolution?

   _____

   _____

   _____

5. Using carbon dating, determine the age of a specimen that is 75% carbon 12 and 25% carbon 14. Determine the age of another specimen that is 85% carbon 12 and 15% carbon 14.

   _____

   _____

   _____

6. Explain the variations in the beaks of Darwin's finches and their importance.

   _____

   _____

   _____

7. What is meant by "survival of the fittest?" How does the environment influence survival of an organism?

_____

_____

_____

8. Define and contrast the terms *homologous* and *analogous* structures.

_____

_____

_____

_____

9. How does embryological development support evolution?

_____

_____

_____

10. Hypothesize what structure or structures we have today that one day might be considered vestigial. Support your hypothesis.

_____

_____

_____

_____

11. What is the protein clock theory, and why is it important in establishing ancestry?

_____

_____

_____

12. Compare and contrast artificial selection and natural selection. Using the Internet, research how breed was shaped for three breeds of dogs.

_____

_____

_____

_____

Chapter 18
Making Sense of Diversity: Understanding Classification

Student Outcome Objectives

At the completion of this exercise, the student will be able to:

1. Describe the role of taxonomy in modern biology.
2. Distinguish between taxonomy, systematics, and cladistics.
3. Describe the composition, derivation, and purpose of scientific names.
4. Discuss and outline the basic taxa used in taxonomy.
5. Classify humans from kingdom through species.
6. Identify organisms using a simple biological key.
7. Construct a simple biological key.
8. Construct and interpret a cladogram.

Overview

Look around—diversity abounds! Today, biologists face the awesome task of systematically identifying, studying, and developing the natural history of an estimated 5 million to 30 million different species of organisms. Some microbiologists estimate that possibly a billion species of bacteria exist. Zoologists already have classified more than 1.5 million species of animals, and approximately 13,000 new species are described yearly. Zoologists estimate that the number of species of animals named, so far, represents 10% to 20% of all the living animals and less than 1% of all of the animals that have ever lived. Botanist have classified more than 300,000 species of plants, and many new species are described yearly.

Despite exciting times in taxonomy, many organisms will remain unknown forever. Some organisms are so minute and elusive that only good luck or random chance will place them in the taxonomists' hands. Others live in regions of the world that have not been explored in depth. Still others are going extinct, especially in the tropical rainforests, faster than they can be classified.

Ancient people developed classification schemes based on the need to survive and on curiosity. Perhaps an ancient classification system was devised around the question, "Can I eat it, or will it eat me?" Early biologists realized that if they were ever going to make sense of the diversity of life, they must devise a logical system of classifying organisms.

God created, but Linnaeus classified.

—Carolus Linnaeus (1707–1778)

One of the early biologists, Carolus Linnaeus (1707–1778), known as "the Father of Taxonomy," developed a system of binomial nomenclature that is still used today. **Taxonomy** is defined as the study of the principles, procedures, and rules of scientific classification and the naming of organisms (Fig. 18.1). The Linnaean system of nomenclature will be discussed later in this chapter. It includes the application of distinctive names to each group of organisms.

The classification of living organisms is an arduous task. **Classification** is defined as the ordering of living organisms into groups, or **taxons**, on the basis of associations by contiguity, similarity, or both. Biological classification involves the two major steps of:

1. defining and describing organisms, and
2. arranging organisms into a logical classification scheme.

Classical taxonomy predates evolutionary biology, and many schemes of classical taxonomy are being challenged by newer theories. One of these is **systematics**, the scientific study of the kinds of organisms, their diversity, and their evolutionary relationships. Systematics examines organisms from many viewpoints, with the aim of understanding the evolutionary relationships between organisms and the construction of a phylogenetic tree that relates organisms.

Don't be surprised if you see variations in classification schemes as you thumb through different texts. One relatively new theme that you will discover in evolutionary taxonomy is termed **cladistics**, or **phylogenetic systematics**. This basically is a system of arranging taxa by analysis of primitive and derived characteristics so their arrangement will reflect phylogenetic relationships.

This lab experience has been designed to introduce students to the basic principles of taxonomy and cladistics. In addition, the student will gain practical experience using and developing biological keys.

BINOMIAL NOMENCLATURE

In the mid-19th century, the Swedish biologist Carolus Linnaeus (Fig. 18.3) devised a system of taxonomy based on two basic themes. First he assigned to each organism a two-part scientific name of Latin or Greek origin. This method was called **binomial nomenclature**. The first word of the name is the **genus** (plural, **genera**) to which

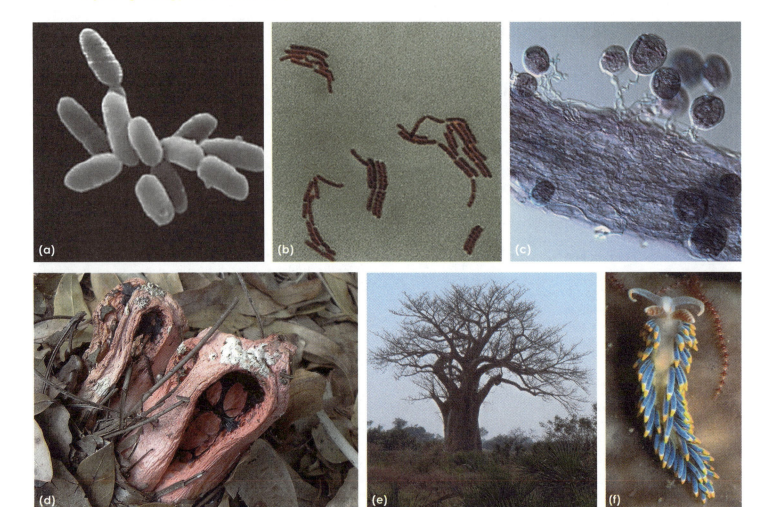

Figure 18.1 Scientific names: (a) *Halobacterium salinarum*, (b) *Bacillus megaterium*, (c) *Vorticella campanula*, (d) *Clathrus columnatus*, (e) *Adansonia digitata*, and (f) *Berghia coerulescens*.

a species belongs. The second word is specific to the organism and is called the **species epithet**. The genus and species epithet together make up the scientific name of an organism (Fig. 18.3).

In taxonomy, highly specific rules ensure reasonable uniformity and wide international acceptance of scientific names. Rigorous standards have been set for naming organisms. Whoever describes a genus or species has the honor of naming it. Some organisms are named after their founder or a namesake. These names are called **eponyms**. As examples, the rhea *Rhea darwinii* is named after Charles Darwin and the southern magnolia *Magnolia grandiflora* is named after Pierre Magnole.

Every biological species is a closed gene pool, an assemblage of organisms that do not exchange genes with other species. Thus insulated, it evolves diagnostic hereditary traits and comes to occupy a unique geographic range.

—Edward O. Wilson (1929–present)

Some scientific names relate the organism to a geographical region, such as the American alligator *Alligator mississippiensis*, the white-tailed deer *Odocoileus virginianus*, and the eastern hemlock *Tsuga canadensis*. Other scientific names are descriptive in nature, such as the mockingbird *Mimus polyglottis*, the flying squirrel *Glaucomys volans*, and the red maple *Acer rubrum*. Several scientific names are based in Greek mythology, such as the Louisiana flag iris *Iris versicolor*, named after Iris, the Goddess of the Rainbow. Some scientific names are based on word play, puns, and humor, such as a dung beetle named *Ytu brutus*, a rhinoceros beetle named *Enema pan*, a yaupon tree named *Ilex vomitoria*, and a spider named *Darthvaderum greensladeae*.

Many people ask why scientific names are necessary. Scientific names are universally accepted and serve to avoid confusion when discussing organisms. For example, bowfin, grindle, cypress trout, and choupique are colloquial names for the same fish, *Amia calva*.

Some Basics

In writing scientific names, the genus and species names are underlined when handwriting and are italicized when using word processing. The genus name always begins with an uppercase letter, and the species epithet always begins with a lowercase letter. For example, the scientific name for a great white shark is *Carcharodon carcharias*, and a ginkgo tree is *Ginkgo biloba*. The genus name serves as a noun, and the species name usually is an adjective that must agree with the genus. In the scientific name for the domestic cat, *Felis domestica*, *Felis* refers to a genus of cats, including wild and domestic species, and *domestica* refers to the common domesticated cat.

Names of genera apply to specific groups of organisms only, whereas the species epithet may be used with different genera. As examples, the yellow-billed cuckoo is *Coccyzus americanus*, the black bear is *Euarctos americanus*, and the American toad is *Bufo americanus*. They all share the same species epithet. The house mouse, *Mus musculus*, shares its species epithet with the blue whale, *Balaenoptera musculus*.

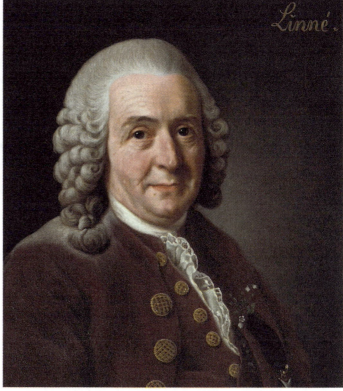

Figure 18.2 Carolous Linnaeus, "the Father of Taxonomy."

Check Your Understanding

Q. Name an animal and a plant with more than one colloquial name.

_____

_____

Q. If you were given 1,000 baseball cards, how would you develop a classification scheme?

_____

_____

_____

Q. Would everyone in the class develop the same scheme? What problems would you encounter?

_____

_____

_____

_____

Q. How might you classify living things?

_____

_____

_____

_____

The extent to which progress in ecology depends upon identification and upon the existence of sound systematic groundwork for all groups of animals cannot be too much impressed upon the beginner in ecology. This is the essential basis of the whole thing; without it, the ecologist is helpless, and the whole of his work may be rendered useless.

—**Charles Elton (1900–1991)**

Scientific names are relatively easy to find in biology books, on the Internet, in field guides, and in natural history books. Learning these names requires study, and sometimes making a game out of learning scientific names. For instance, if your friend reminds you of a squirrel, you might call him *Sciurus carolinensis* after the gray squirrel.

Figure 18.3 Using common and scientific names, (a) destroying angel mushroom, *Amanita virosa*, (b) pitcher plant, *Sarrancenia leucophylla*, and (c) bottlenose dolphin, *Tursiops truncatus*.

The Name Game

Answer the following:

Q. Find the scientific name of your state's tree, flower, and bird. Find five other organisms that are classically from your region. Choose any other state and do the same.

_____

_____

_____

_____

_____

_____

_____

_____

Naturalists try to arrange the species, genera, and families in each class, in what is called the Natural System. But what is meant by this system? Some authors look at it merely as a scheme for arranging together those living objects which are most alike, and for separating those which are most unlike.

—**Charles Darwin (1809–1882)**

HIERARCHICAL CLASSIFICATION SYSTEM

The second theme developed by Linnaeus is the **hierarchical classification system**. Basically, organisms are placed into taxa (major categories) and assigned a standard taxonomic rank. Today animals are placed in eight mandatory ranks. These ranks, starting with general characteristics and ending with the species, are:

domain
    kingdom
        phylum
            class
                order
                    family
                        genus
                            species

Taxonomists recognize more than 30 ranks in the animal kingdom. Note the basic classification of humans in Table 18.1.

Although the naming of organisms is changing constantly, three domains are recognized currently:

1. Archaea (ancient bacteria),
2. Bacteria (typical bacteria), and
3. Eukarya (protists, plants, fungi, and animals).

Today, at least six kingdoms are recognized:

1. Archaebacteria
2. Eubacteria
3. Protista★
4. Plantae
5. Fungi
6. Animalia

★ "Protista" is currently viewed as an artificial kingdom and will be used for convenience in this manual.

Interesting Scientific Names

Strigiphilus garylarsoni	biting louse
Abracadabrella birdsville	jumping spider
Agra vation	Peruvian beetle
Bullisichthys caribbaeus	pugnose bass
Rhodophthalmokytodermogammarus cinnamomeus	amphipod
Aha ha	Australian wasp
Montypythonoides riversleighensis	extinct python
Desmodus draculae	vampire bat
Cassiopea andromeda	upside–down jellyfish
Agkistrodon piscivorous	cottonmouth snake
Ba humbugi	snail

In the near future, the current six kingdoms will be further divided into as many as 15 new kingdoms.

List several of the proposed new kingdoms and give an example of each.

_____

_____

_____

Check Your Understanding

Q. Describe the following taxonomic ranks for humans:

   a. Superclass

_____

   b. Suborder

_____

   c. Infraorder

_____

   d. Subspecies

_____

It is the genus that gives the characters, and not the characters that make the genus.

—**Carl Linnaeus (1707–1778)**

Table 18.1 The Classification of Humans

Domain	**Eukarya**	Eukaryotic cells	protists, plants, fungi, and animals
Kingdom	**Animalia**	Multicellular heterotrophs	jellyfish, flatworms, roundworms, molluscs, insects, starfish, man, etc.
Phylum	**Chordata**	Presence of notochord, post-anal tail, dorsal hollow nerve cord, and gill slits	sea squirts, amphioxus, fish, amphibians, reptiles, birds, mammals, etc.
Subphylum	**Vertebrata**	Spinal cord enclosed in vertebrae	lampreys, sharks, perch, toads, sparrows, humans
Class	**Mammalia**	Young nourished with milk, presence of hair, warm-blooded	platypus, wombat, moles, bats, seals, dolphins, horses, lions, monkeys, humans
Order	**Primates**	Tree dwellers or their descendants fingers with flat nails, reduced olfaction	lemurs, tarsairs, baboons, monkeys, orangutans, gorillas, chimps, humans
Family	**Hominidae**	flat face, eyes focusing forward, color vision, bipedal	gorillas, chimps, ancient and modern humans
Genus	***Homo***	large brain, speech	*Homo erectus, Homo habilis, Homo ergaster, Homo sapiens*
Species	***sapiens***	Prominent chin, high forehead, sparse body hair	*Homo sapiens*

Q. How many other taxonomic ranks can you describe for humans?

_____

_____

_____

Q. How is Neanderthal man classified? Why is there a controversy?

_____

_____

_____

_____

Before Linnaeus, Latinized descriptions of organisms got out of hand. Try this name for a honeybee!

Apis pubescens, thorace subgriseo, addominate fusco, pedibus posticis glabris utrinque cliatus OR the fuzzy bee with the grayish thorax, dusky brown abdomen, hairless hind feet bordered with small hairs on both sides.

Q. Classify the Southern magnolia from domain through species.

_____

_____

_____

_____

_____

TAXONOMICAL KEYS

With the great diversity of life, few biologists have the expertise to identify more than a small group of organisms. Usually biologists become familiar with the common species in an area or those species that have been a part of their research. To identify organisms, biologists use a **taxonomical key** as an aid.

One simple means of keying out an organism is to compare the unknown organism to pictures and descriptions in a book. Various field guides and natural history books serve as an excellent means of identifying some unknown organisms. This method has been called "picture booking."

The majority of keys are formatted as a series of questions divided into two possible choices, called the **dichotomous key**, in which two contrasting choices are offered to the user and only one choice describes the specimen. At the end of the correct choice, the user will find a reference number to the next set of choices. The choices will lead to identification of the unknown organism.

 Student Activity—Taxonomy

A Mesozoic Menagerie

Ever since the discovery of the first dinosaur fossil, these giant reptiles have occupied a major part of our thoughts, imagination, and even movies. Dinosaurs were the dominant animal on our planet for more than 160 million years during the Mesozoic period. Although the giant dinosaurs get most of the attention, the average size of a dinosaur was about the size of a turkey. The fossil record has yielded great diversity of these fascinating creatures.

 ### Procedure 18.1
Fossils

The key in Table 18.2 is designed to demonstrate the mechanics of a dichotomous key, utilizing a variety of dinosaurs. The examples in Figure 18.4 represent a few of the more popular dinosaurs. Students will use the key in Table 18.2 to identify the dinosaurs in Figure 18.4. You probably have known the name of these animals since you were in grade school. Now see if you can key them out properly.

Table 18.2		Simple Dichotomous Key of the Dinosaurs in Figure 18.4	
1.	a.	Walks on two legs (bipedal)	go to 2
	b.	Walks on four legs (quadrapedal)	go to 3
2.	a.	Possesses short forelimbs	**Allosaurus**
	b.	Possesses large forelimbs and a large claw on the hindlimbs	**Velociraptor**
3.	a.	Possesses horns	go to 4
	b.	Does not possess horns	go to 5
4.	a.	Possesses one horn	**Styracosaurus**
	b.	Possesses three horns	**Triceratops**
5.	a.	Possesses plates or spines on dorsal side	go to 6
	b.	Does not possess plates or spines on dorsal surface	go to 7
6.	a.	Possesses two rows of plates on dorsal side	**Stegosaurus**
	b.	Possesses spines on dorsal side	**Ankylosaurus**
7.	a.	Forelimbs longer than hindlimbs	**Brachiosaurus**
	b.	Forelimbs and hindlimbs approximately of the same length	**Apatosaurus**

Taxonomy is often regarded as the dullest of subjects, fit only for mindless ordering and sometimes denigrated within science as mere "stamp collecting." If systems of classification were neutral hat racks for hanging the facts of the world, this disdain might be justified. But classifications both reflect and direct our thinking. The way we order represents the way we think. Historical changes in classification are the fossilized indicators of conceptual revolutions.

Figure 18.4 Follow the key in Table 18.2 and write the correct number and letter beside the dinosaur.

—**Stephen Jay Gould (1941–2002)**

Student Activity—Constructing a Key

In the chart below, construct a key to the insects illustrated in Figure 18.5.

Figure 18.5 Common insects. When the table is complete, write the correct number and letter beside the insect.

I believe that something more is included; and that propinquity of descent, the only known cause of the similarity of organic beings, is the bond, hidden as it is by various degrees of modification, which is partially revealed to us by our classifications.

—Charles Darwin (1809–1882)

CLADISTICS

Unlike Linnaean taxonomy, which classifies organisms in hierarchies (taxa) based on morphological similarities, **cladistics** (phylogenic systematics) groups organisms based on evolutionary ancestry. In 1950, William Henning, a German entomologist, pioneered the cladistics system of taxonomy. Cladists construct phylogenic trees called **cladograms** to discern evolutionary ties between species. **Cladograms** are tree-like diagrams that depict hypothetical evolutionary processes gained from studies at the molecular level of the organism (DNA and RNA) and morphological similarities.

Cladistics is sometimes called **phylogenetics** because the cladograms are generated from similarities in the molecular structure of organisms. Cladistics is based on the principle that groups of organisms are descended from a common ancestor and they do not mix ecology with phylogeny. The cladograms are branching diagrams showing evolutionary relationships among organisms. Each node (divergence of a population), or branching point, lists the derived characters of the descendants, **clades**, that follow the node. The derived characters are listed according to when it is hypothesized that they first appeared.

Cladists contend that evolution is the result of modifications in characteristics over time. Cladistics recognizes only **monophyletic groups**, which include all of the organisms that have descended from a common ancestor. For example, traditional biologists have separate taxonomical groups for reptiles and birds. Although traditional taxonomists certainly would agree that birds have inherited scales, amniotic eggs, and lungs from the reptiles, they place the birds into their own class Aves because of their differences. Birds are warm-blooded, and have specialized beaks and feet, which are adaptations they need to be able to survive in their various habitats.

Because cladists emphasize the importance of phylogeny, which places organisms in clades (monophyletic groups) based on their shared derived characters, birds are included in the cladogram with the reptiles because birds have descended from the reptiles. Figure 18.6 compares traditional taxonomy to that of the cladistics.

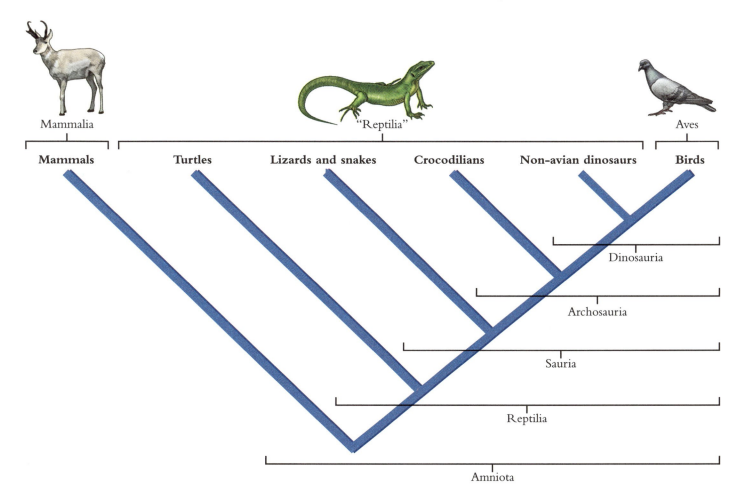

Figure 18.6 Traditional classification of tetrapods compared to that of the cladistics. Traditional classification places birds and reptiles in different classes; cladistic classification does not.

Q. Where would whales fit on this cladogram? Defend your answer.

_____

_____

_____

Q. The lamprey is considered to be an ancestor of fish having jaws. If a node were placed before the lamprey, what would be a possible derived character?

_____

_____

_____

Q. If a node were placed before the gorilla, what would be a possible derived character?

_____

_____

_____

Q. Besides liverworts, what is another example of an organism considered in the outgroup?

_____

_____

_____

Q. In which node would the following plants be found: Japanese magnolia, Spanish moss, Ginkgo, Hornworts, Cypress, Club mosses, Bleeding hearts, Spruce, and Daffodils. Defend your responses.

_____

_____

_____

_____

In all things of nature there is something of the marvelous.

—**Aristotle (384 BC–322 BC)**

Four billion years ago, the Earth was a molecular Garden of Eden. There were as yet no predators. Some molecules reproduced themselves inefficiently, competed for building blocks and left crude copies of themselves. With reproduction, mutation and the selective elimination of the least efficient varieties, evolution was well under way, even at the molecular level."

—**Carl Sagan (1934–1996)**

Q. Give three examples of ancestral organisms to the liverworts that have chloroplasts.

_____

_____

_____

NOTES

_____

_____

_____

_____

_____

_____

_____

_____

_____

_____

_____

_____

_____

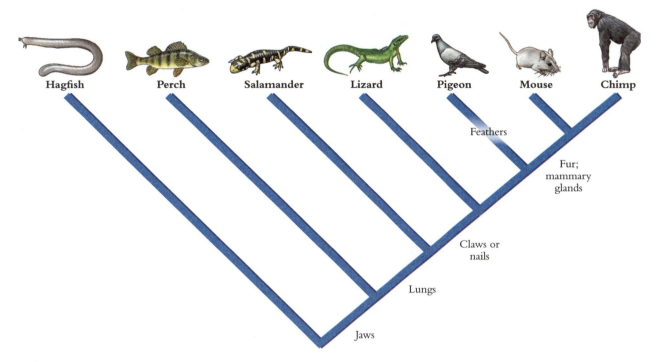

Figure 18.7 A cladogram depicting evolutionary relationships in vertebrates. Branches or nodes listing the derived characters are on one side of the cladogram, and the organisms (clades) sharing these derived characters follow each node.

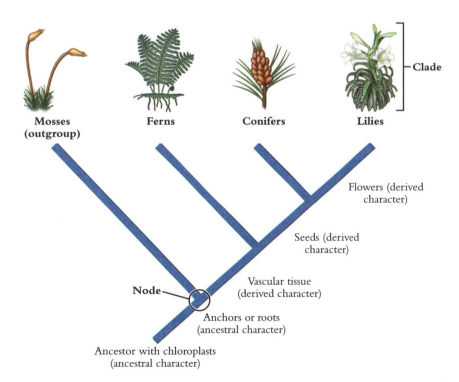

Figure 18.8 Cladogram of the evolutionary relationships of plants. Note the outgroup, in cladistics, the group that branched from the ancestor first and is less closely related to the other groups in the cladogram.

Student Activity—Making a Cladogram

Procedure 18.2
Cladogram

1. Draw a cladogram of the arthropods.
2. Include in your cladogram the nodes and clades, showing their relationships based on their shared derived characters. (See Figs. 18.7 and 18.8 for examples.)

What is a scientist after all? It is a curious man looking through a keyhole, the keyhole of nature, trying to know what's going on.

—**Jacques Yves Cousteau (1910–1997)**

Name: _____ Date: _____ Section: _____

Review Questions

1. What is systematics? How is it different from taxonomy?

_____

_____

_____

_____

2. What is the purpose of binomial nomenclature?

_____

_____

_____

_____

3. Give several examples of common names that might cause confusion if scientists were to use them.

_____

_____

_____

_____

4. What are the seven mandatory ranks?

_____

_____

_____

_____

_____

_____

5. Name the six kingdoms and give examples of each.

_____

_____

_____

_____

_____

6. What are the two major steps in biological classification?

_____

_____

_____

7. What are three reasons why scientific names are important?

_____

_____

_____

8. How are scientific names derived?

_____

_____

_____

9. If you want to identify the various species of birds in your backyard, what sources could you use to find their scientific names?

_____

_____

_____

10. How does the cladistics system of taxonomy differ from the hierarchical system of classification?

_____

_____

Chapter 19
On the Edge of Life: Understanding Viruses

Student Outcome Objectives

At the completion of this exercise, the student will be able to:

1. Define *virus* and *virion*.
2. Describe the anatomical features and shape of a typical virus.
3. Compare and contrast enveloped and naked viruses.
4. Compare and contrast lytic and lysogenic viruses.
5. Define the action of a typical bacteriophage attack.
6. Draw and label a typical bacteriophage.
7. Name significant families of viruses and the diseases that they cause.
8. Display competence in using the Centers for the Disease Control and the World Health Organization to gather pertinent information.
9. Describe prions and name several diseases caused by prions.
10. Describe viroids and name several diseases caused by viroids.
11. Describe how a disease can be transmitted in a population.

Overview

Feeling headachy, feverish, and just plain blah? Like nearly 50 million Americans each year, you may have contracted the flu. Most of those who suffer from the flu are the young, the very old, and those who are immunocompromised; however, everyone is susceptible. On the average, nearly 200,000 people are hospitalized yearly, and approximately 36,000 die from the flu annually. In 1918, the Spanish flu was responsible for nearly 25 million deaths worldwide. International health organizations constantly are monitoring and preparing for the possibility of devastating flu outbreaks.

The flu, like colds, yellow fever, rabies, smallpox, and herpes is caused by a **virus**. In Latin, the term *virus* means "poison." Today, many of the diseases that plague plants, animals, and humans are known to be caused by viruses. Our ancestors, without the tools of science, however, thought that these diseases were caused by poisons, evil spirits, and sin. For many years, the very nature of viruses was debated— whether they were considered living or nonliving entities.

In the early 1880s, the German scientist Adolf Mayer, while working with tobacco mosaic disease, discovered that the disease could be transmitted by rubbing sap from a plant infected with the disease onto a healthy plant. He concluded that tobacco mosaic disease was caused by tiny bacteria. At that time, it also was known that smokers could infect tomato plants with this disease. In fact, in the early ketchup industry, smokers were not allowed to handle tomatoes. The Russian scientist Dimitri Ivanovsky, after trying to filter infected sap, concluded that tobacco mosaic disease was not caused by bacteria but, instead, by something much smaller.

The Dutch botanist Martinus Beijerinck discovered that the minute disease-causing agent could not be cultivated like bacteria. He also concluded that these entities required the presence of a host cell to reproduce. Beijernck named the agent responsible for tobacco mosaic disease virus. In 1935, the American scientist Wendell Stanley crystallized the viral particles and viewed them for the first time with an electron microscope. Since this time, our knowledge of those once thought mysterious poisons has increased dramatically.

Viruses are simple. They are small, infectious particles consisting of a protein coat surrounding a core of DNA or RNA. One of the smallest viruses is only 20 nm in diameter, and the largest is barely visible under a light microscope. Millions of virus particles known as **virions** can fit in a typical human cell. Although the shape of viruses ranges from spherical to brick-shaped to helical, icosahedral, and polyhedral (Fig. 19.1), all viruses have an outer protein coat known as a **capsid**. Capsids are constructed from protein subunits termed **capsomers**.

Some viruses, called **envelope viruses**, are incased in an envelope derived from the host cell. Among envelope viruses are those responsible for HIV, herpes, smallpox, influenza and rabies. Those that do not contain an envelope are called **naked viruses**. Naked viruses are responsible for warts, polio, and Hepatitis A.

Even bacteria can fall victim to viruses known as **bacteriophages**, or simply **phages**. Their capsids are complex, and they possess an icosahedral head that encloses their genetic material (Fig. 19.2). These viruses resemble the shape of the lunar excursion module that was used to land on the moon. They have a tail sheath joined to the head in a region known as the collar. They also contain a base plate with tail fibers and pins that are used for attachment (Fig. 19.3).

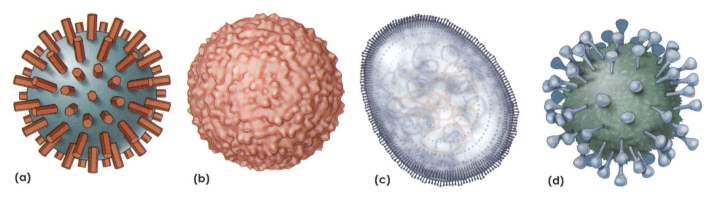

Figure 19.1 Common viruses (a) H1N1 virus (b) West Nile virus (c) Measles (d) SARS.

The majority of viruses have a limited **host range**. For example, the parvovirus, common in canines, will not infect felines or humans; however, some viruses, such as rabies, have a broader host range. The tobacco mosaic virus has been found to infect more than 150 species of plants.

Bacteriophage action has two major modes:

1. Viruses that have a **lytic** life cycle will cause the cell to burst or lyse, releasing the viral particles. Each particle in turn can then infect a healthy host cell. **Virulent** viruses undergo only the lytic cycle.
2. In contrast to lytic viruses, **lysogenic** viruses allow for the integration and replication of the viral genome in the host cell without destroying the host. Thus, the virus genome is passed along as the host cell reproduces.

Prophages are the result of integration of the viral genome into the bacterial genome. **Temperate** viruses undergo both lytic and lysogenic stages. Viruses work like a molecular pirate, using their genetic material to commandeer the molecular machinery of a host cell. Once commandeered, the host cell is at the mercy of the virus.

Classically, a typical bacteriophage life cycle consists of five phases (Fig. 19.4):

1. **Adsorption** or **attachment**. The virus recognizes the protein signature of a host cell and attaches to the host cell.
2. **Entry**, or penetration. During this stage, the viral genome is inserted into the host cell.
3. **Integration**. In this step, the viral genetic material combines with genetic material of the host cell, forming a prophage.
4. **Synthesis and assembly**. This phase involves replication of the viral genome and synthesis of the viral capsid. During assembly, the viral particles mature.
5. **Release**. This final stage usually results in lysis of the host cell. The number of viroids released during this stage is variable.

The term **latent** refers to a period of non-activity in a typical virus life cycle. For example, herpes simplex type I, responsible for cold sores, can remain latent in the spinal nerves until the first stage of activation.

Because viruses exist at the edge of life, their classification continues to undergo refinement, and several

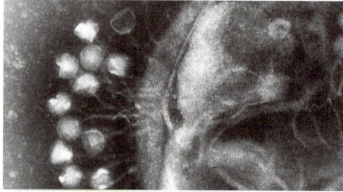

Figure 19.2 An SEM of bacteriophage.

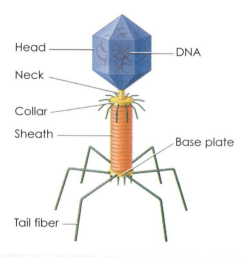

Head — DNA
Neck
Collar
Sheath — Base plate
Tail fiber

Figure 19.3 An illustration of bacteriophage.

classification schemes have been developed. One simple scheme divides the viruses into RNA and DNA viruses. David Baltimore, American biologist and Nobel Prize Laureate, developed a viral classification system that groups viruses into families depending on their type of genome and their method of replication. This system recognizes more than 80 families and accounts for more than 4,000 different viruses (Table 19.1).

Many scientists recognize entities beyond viruses. **Viroids** cause diseases in plants. Theodor Diener discovered and named these plant pathogens in 1971. Structurally, viroids consists of a circular strand of RNA only a few hundred bases in length. The viroids are transmitted by pollen grains or seeds by sucking insects such as aphids, and even from pruning. The smallest known viroid, sobemovirus, which causes rice yellow mottle disease, distorts plant growth; it is only 220 nucleobases in length. Viroids are devoid of a protein coat and are capable of various means of replication. Among viroid diseases are potato spindle tuber disease and coconut cadang.

In the early 1980s, Stanley Prusiner isolated an entity that is responsible for several diseases in animals including humans. He named this entity **prion**, short for *proteinaceous and infectious particles.* He named **PrP** (protease-resistant protein) to designate the specific protein of which the prion is composed. The PrP protein has different isoforms. **PrP^c** is the normal protein found in the membranes of cells. The c denotes the cellular or common form of PrP.

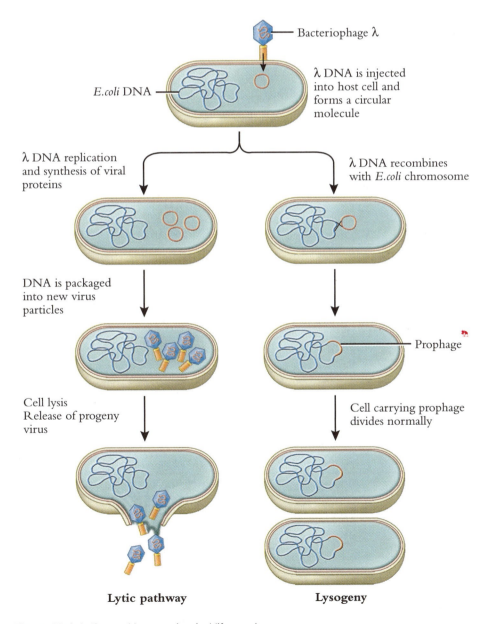

Figure 19.4 Lytic and lysogenic viral life cycles.

The **PrP**^{Sc} is the infectious form of the protein found in scrapie (Sc), a prion disease of sheep. It is hypothesized that prions infect and propagate by a mechanism whereby the diseased isoform interacts with the normal isoform and converts the normal form into the structurally abnormal form. For example, the PrP^{Sc} causes disease by converting the PrP^c into the infectious isoform of the disease by changing the conformation of the protein.

Prion diseases affect the nervous system, causing neurological damage and death. Plaques known as **amyloids**

Table 19.1 Survey of Noteworthy Viruses

Double-stranded DNA viruses (dsDNA)

Virus	Type	Diseases
Adenovirus	Naked	viral pneumonia, conjunctivitis
Papovirus	Naked	warts, human papilloma virus
Herpesvirus	Enveloped	herpes simplex type I – cold sores; herpes simplex type II—genital herpes, mononucleosis, Epstein Barr, shingles, Burkitt's lymphoma
Poxvirus	Enveloped	smallpox, monkeypox

Single-stranded DNA viruses (ssDNA)

Virus	Type	Diseases
Inovirus	Naked	M13 bacteriophage (*E. coli*)
Parvovirus	Naked	parvo in canines, feline panleukopenia

Double-stranded RNA viruses (dsRNA)

Virus	Type	Diseases
Cystovirus	Naked	Ph16 bacteriophage (*Pseudomonas* phage)
Reovirus	Naked	rotavirus, bluetongue in sheep, Colorado tick fever

Single-stranded RNA viruses (ssRNA)

Virus	Type	Diseases
Bunyavirus	Naked	hantavirus, Crimean Congo hemorrhagic fever
Calcivirus	Naked	Norwalk, feline herpes
Coronavirus	Enveloped	severe acute respiratory syndrome (SARS), canine coronavirus
Flavivirus	Enveloped	yellow fever, West Nile, hepatitis C
Filovirus	Enveloped	Ebola, Marburg, Reston
Orthomyxovirus	Enveloped	influenza virus, Thogotovirus
Paramyxovirus	Enveloped	mumps, measles, Newcastle's disease, canine distemper
Picornavirus	Naked	polio, hepatitis A, chronic fatigue syndrome, common cold (Rhinovirus)
Rhabdovirus	Enveloped	rabies, lettuce necrotic yellow virus
Retrovirus	Enveloped	human immunodeficiency virus (HIV), Simian immunodeficiency virus (SIV), feline immunodeficiency virus, mouse mammary tumor virus, chimpanzee foamy virus
Togavirus	Enveloped	rubella, Eastern equine encephalitis, O'nyongnyong virus

are formed within the diseased tissue. Examples of prion diseases include scrapie in sheep and goats, Creutzfeldt–Jakob disease (CJD), kuru in humans, and bovine spongiform encephalopathy (BSE), commonly known as mad cow disease, in cattle (Fig. 19.5).

Unlike living organisms and viruses, prions do have nucleic acids, thereby making it difficult to denature their infectious status. Traditional means of denaturation by proteases, radiation, routine sterilization, and formalin is not effective in the case of prions. The World Health Organization (WHO) has outlined specific procedures for the sterilization of surgical instruments contaminated with prions.

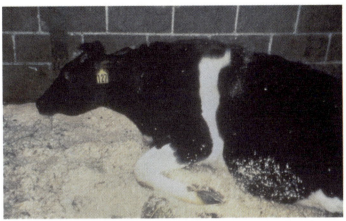

Figure 19.5 A cow with bovine spongiform encephalopathy (BSE).

 ## Student Activity—Communicable Disesase

Each year, viral diseases such as the cold virus and the flu virus cause disease in humans. These are communicable diseases spread from one person to another by contact through coughing and sneezing. Other viral diseases, such as hepatitis, AIDS, and genital herpes, are transmitted by sharing body fluids. This lab activity simulates the spread of disease in a population of people. Students will trace the transmission route and identify the original carrier of the disease. After three rounds of simulated exchange of body fluids, the students will calculate the number of individuals in the population who have become infected with the disease.

Materials

- test tubes
- test tube rack (one per lab station)
- dropper bottles with medicine droppers (one per student)
- markers and pencils for labeling test tubes and bottles
- distilled water
- 1 M NaOH (4g of sodium hydroxide in 100 mL of distilled H_2O)
- 2 dropper bottles with medicine droppers of a 1% phenolphthalein (dissolve 1 g of phenolphthalein with 50 mL of denatured alcohol, and then add to it 50 mL of distilled H_2O). This is the indicator solution.
- safety goggles
- lab coat

 ## Procedure 19.1
Transmission

1. Put on safety goggles and a lab coat.
2. Procure a clean test tube, a marker or pencil, and a numbered medicine bottle with dropper from the instructor.
3. Using the marker or pencil, write the number that is on the medicine bottle on your test tube.
4. Carefully add 3 full droppers of solution to your test tube. *Caution: Avoid getting any of the solution on your skin or clothing.*
5. Place the test tube in the test tube rack, and wait for instructions from the instructor.
6. When the instructor begins Round 1, choose a student in the lab to be your first contact person. Carefully pour the entire contents of your test tube into the test tube of the first contact person. Then pour half of the solution back into your test tube. In Table 19.2 write your test tube number under "Contact Person" and the number of the test tube of your first contact person.
7. Put your test tube back into the rack, and wait for your instructor to signal for Round 2. Choose a different contact person, and exchange contents of the test tubes as you did in Round 1. Record in Table 19.2 the test tube number of the individual you exchanged contents with.
8. Place your tube back into the test tube rack, and wait for the instructor to signal for Round 3. Then choose a different contact person again, and exchange the contents in the test tubes and record the contact's test tube number in Table 19.2.

9. Place your test tube in the test tube rack. The instructor will add an indicator solution to your test tube. If the solution changes to pink or red, you will be considered infected.

10. The instructor will ask for the test tube numbers of those infected, as well as the test tube numbers of the individuals that person contacted. Fill in Table 19.2 as the instructor completes the table on the board.

11. Carefully dispose of the contents of your test tube down the drain, and flush with water. Return materials to the instructor and clean up your work area.

12. To determine the number of individuals who would be infected at the end of each round, use the equation:

2^n = the number of person infected
where n = the number of the round

Our greatest concern must always rest with disadvantaged and vulnerable groups. These groups are often hidden, live in remote rural areas or shantytowns and have little political voice.

—**Dr. Margaret Chan (1947–present)**
WHO Director General

13. Using the equation above:
   a. How many people were infected at the end of Round 1?

   _____

   b. How many individuals were infected at the end of Round 2 and the end of Round 3?

   _____

   c. What percentage of the class was infected by the end of Round 3?

   _____

Q. Why is washing your hands with soap an effective means of disease prevention?

   _____

   _____

   _____

Table 19.2 Individuals Infected and their Contacts

Contact Person	Round 1	Round 2	Round 3

 Beyond the Lab—Getting to Know the CDC and WHO

The **Centers for Disease Control** (CDC) is a governmental agency that provides information on topics including diseases and health, workplace safety, emergency preparedness and response, travelers' health, environmental health, publications on emerging infectious diseases and morbidity and mortality, research opportunities, training, employment and data, and statistics on diseases and health.

 **Procedure 19.2
The CDC**

In this activity, students will refer to the Centers for Disease Control website, www.cdc.gov, to answer the following questions.

1. You are preparing for a trip to Kenya. What diseases might you be at risk for during your travel to this

country? What vaccinations and medications are required or recommended for individuals planning to travel to Kenya?

_____

_____

2. How far in advance of your departure date should you get your vaccinations and begin taking preventative medications? Why?

_____

_____

3. If you are traveling to a country such as Kenya, where you are at risk for contracting malaria, how far in advance of the departure date are you required to begin taking the preventive medication? How long must you continue to take the medication after returning home?

_____

_____

4. If you become ill with a fever or other symptoms after returning from a trip abroad (even up to a year after traveling abroad), why should you contact your physician and tell him/her about your travels?

_____

_____

5. Your family has decided to take a Caribbean cruise this summer. What diseases might you be at risk for during your vacation? What vaccinations and medications are required or recommended?

_____

_____

6. We are in great need of blood donors. Many individuals are restricted from donating blood because they have contracted certain diseases that are transmitted through the exchange of blood and other bodily fluids. West Nile disease is one of many diseases that can be transmitted through blood donation. Name five additional diseases that ban an individual from donating blood.

_____

_____

7. Individuals with tattoos or piercings must wait a period of time before donating blood. Why?

_____

_____

8. Define the terms _morbidity_ and _mortality_. On the Centers for Disease Control homepage, click on the _Morbidity and Mortality Weekly Report_ under "Publications." Summarize a current article in the _MMWR Weekly_.

_____

_____

_____

9. List five free publications that you can receive online from the CDC.

_____

_____

_____

The World Health Organization (WHO) is the directing and coordinating authority for health within the United Nations system. The WHO is responsible for providing leadership on global health matters, shaping the health research agenda, setting norms and standards, articulating evidence-based policy options, providing technical support to countries, and monitoring and assessing health trends.

To answer the questions below, the student is to log onto the WHO website at www.who.int/

Procedure 19.3
The WHO

1. What is the role of the WHO in public health?

_____

_____

Diseases are crises of purification, of toxic elimination.

—**Hippocrates (460 BC–370 BC)**

2. What are three key publications that the WHO provides to the public?

_____

_____

_____

3. What is the WHO Statistical Informational System (WHOSIS)?

_____

_____

4. Describe several categories of data generated by the WHO.

_____

_____

5. After viewing programs and projects on the WHO website, describe some career opportunities with the WHO.

_____

_____

_____

Student Activity—Getting to Know Prions and Viroids

For this activity, students can find information to answer the questions online at the CDC, WHO, and other medical websites.

Procedure 19.4
Prions and Viroids

1. Give examples of three viroid diseases, and describe their effects on plants.

_____

_____

2. Describe how viroids replicate.

_____

_____

3. Explain what is meant by the "protein-only hypothesis" in prions.

_____

_____

4. What are amyloids, and how do they disrupt the nerve tissue?

_____

5. Describe the prion diseases Kuru, Chronic Wasting Disease (CWD), and variant Creutzfeldt-Jakob Disease (vCJD).

_____

_____

_____

6. Why must hospitals take specific precautions when operating on a patient who is thought to have a prion disease?

_____

_____

7. Describe three procedures for the sterilization of surgical instruments contaminated with prions as recommended by the WHO.

_____

_____

_____

Name: _____ Date: _____ Section: _____

1. Define the term *virus*, and identify what organisms viruses can infect.

_____

_____

2. What are the anatomical features of a typical virus?

_____

_____

3. Describe the different shapes of viruses.

_____

_____

_____

4. Compare and contrast enveloped and naked viruses.

_____

_____

_____

_____

5. Diagram and explain the life cycles of lytic and lysogenic viruses.

_____

_____

6. Draw and label a typical bacteriophage.

7. Explain the action of a typical bacteriophage attack.

_____

_____

8. Discuss five significant families of viruses and the diseases they cause.

_____

_____

_____

_____

_____

9. Describe the structure of prions, and name several diseases caused by prions.

_____

_____

_____

10. Define viroids, and identify several diseases caused by viroids.

_____

_____

_____

11. Describe how a disease can be transmitted in a population.

_____

_____

Chapter 20
A Bacterial World: Understanding Bacteria

Student Outcome Objectives

At the completion of this exercise, the student will be able to:

1. Discuss the roles of bacteria in the environment and in medicine.
2. Describe the basic characteristics of bacteria.
3. Explain the difference between Archaea and Bacteria, and provide examples.
4. Identify several significant species of bacteria.
5. Describe the basic morphology of bacteria.
6. Discuss fundamental anatomical features of bacteria.
7. Demonstrate safe techniques in working with bacteria.
8. Demonstrate proficient techniques in observing bacteria.
9. Demonstrate the ability to Gram stain bacteria.
10. Describe the purpose and physiology of bioluminescence in bacteria.
11. Explain the role of antibiotics in bacteriology.
12. Construct an activity that addresses the effectiveness of antibiotics on bacterial growth.

Overview

To a visitor from another world conducting a survey of life on Earth, our planet would be considered the planet of the bacteria (Fig. 20.1). These prokaryotic organisms comprise an overwhelming percentage of the total biomass of life on Earth. Some soil samples contain more than 2 billion bacteria per gram, and a square centimeter of skin may contain nearly 100,000 bacteria. In fact, the total number of bacteria living on or in your body exceeds the total number of cells that make up your body. Bacteria are the most **cosmopolitan** of life forms, living in diverse environments including in the clouds, in deep basalt deposits more than 4,500 feet underground, at the frigid poles, and in extremely hot, deep, hydrothermal vents in the oceans.

Unfortunately, as the result of disease-causing, or **pathogenic**, bacteria, prokaryotes have acquired a bad reputation. The vast majority of bacteria are harmless, and many species are helpful. Bacteria are responsible for producing copious amounts of oxygen in the atmosphere and are vital members of the food chain. *Rhizobium* spp. is an

Figure 20.1 Earth is a planet of bacteria.

important bacterium that lives in a symbiotic relationship with plants known as **legumes** (clover, peanuts, etc.) and fixes atmospheric nitrogen to make it available for biological activities. In industry, bacteria are necessary in the production of many products including cheese, yogurt, wine, and several beneficial enzymes. Some bacteria are used to extend the shelf life of produce, and others to control mosquito larvae. Environmental applications include their use in cleaning up oil spills (bioremediation), in breaking down toxic wastes and herbicides, and in the production of biodegradable plastics.

Recently, one species *Thiobacillus ferroxidans* has been used in gold mining and initiated a new industry known as **biomining**. In medicine, some bacteria are used in producing antibiotics. The body harbors many harmless and beneficial bacteria. One milliliter of saliva may contain more than 40 million bacterial cells! Bacteria are essential in the proper functioning of the digestive system of humans, as well as other animals.

Bacteria cause a number of devastating diseases in plants and animals. Humans can fall victim to a number of bacterial infections. Some infections, such as the black plague and leprosy, have changed history. Unlike viruses, many bacterial diseases of animals can be transmitted to humans. These diseases, such as anthrax, are called **zoonotic** infections. Table 20.1 lists some dangerous bacterial infections in humans.

Love; before I heard the doctors tell the dangers of a kiss,
I had considered kissing you—the nearest thing to bliss.
But now I know biology and sit and sigh and moan; six
million mad bacteria and I thought we were alone.

—Aleister Crowley (1875–1947)

Table 20.1 A Review of Bacterial Pathogens Dangerous to Humans

Organism	Disease
Mycobactertium tuberculosis	Tuberculosis
Mycobacterium leparae	Hansen disease, or leprosy
Neisseria gonorrhoeae	Gonorrhea
Neisseria meningitidis	Meningitis
Pseudomonas aeruginosa	Lung and bladder infections
Staphylococcus aureus	Pimples, boils, toxic shock syndrome, MRSA
Streptococcus pyogenes	Strep throat
Corynebacterium diphteriae	Diphtheria
Bacillus anthracis	Anthrax
Salmonella typhii	Typhoid fever
Shigella spp.	Bacterial dysentery
Escherichia coli	Gastrointestinal problems
Legionella pneumoniae	Legionnaires disease
Vibrio cholerae	Cholera
Vibrio vulnificus	Flesh-eating, intestinal problems
Yersinia pestis	Bubonic or black plague
Haemophilus influenzae	Meningitis, pinkeye, otitis media
Chlamydia trachomatis	Chlamydia
Clostridium perfringens	Gangrene
Clostridium tetani	Tetanus
Clostridium botulinum	Botulism
Heliobacter pylori	Ulcers
Leptospira interrogans	Leptospirosis
Treponema pallidum	Syphilis
Borellia burgdorferi	Lyme disease
Mycoplasma pneumoniae	Atypical pneumonia
Rickettsia rickettsii	Rocky Mountain spotted fever
Rickettsia prowazekii	Epidemic typhus

Recall that bacteria are prokaryotic organisms (lacking a membrane-bound nucleus and organelles). The function of the organelles in eukaryotic organisms occurs primarily on the plasma membrane and in the cytoplasm of prokaryotes. Most prokaryotes are unicellular, although a few aggregate forms exist. In addition, compared to the cells of eukaryotes (protists, plants, fungi, animals) the cells of prokaryotes are small, ranging from 0.5–5 millimicrons in diameter. Prokaryotes display a number of nutritional modes, including **photoautotrophism**, **chemoautotrophism**, **photoheterotrophism**, and **chemoheterotrophism**.

Oxygen requirements vary in prokaryotes. Some bacteria are **obligative aerobes**, those requiring oxygen for survival; others, such as *Clostridium botulinum*, are **obligative anaerobes**, which do not live in an oxygen environment. Some bacteria, such as *Escherichia coli*, are **facultative anaerobes**, which can survive in both worlds. The vast majority of prokaryotes reproduce asexually through **binary fission**; however, three forms of **genetic recombination**, or horizontal gene transfer, have been described: transformation, transduction, and conjugation. This process is the basis of antibiotic resistance in pathogenic bacteria, as well as the evolution of prokaryotes.

The world of prokaryotes is diverse, and through the years, the classification of prokaryotes has been perplexing to scientists. Traditional classification is still important and is based upon observable characteristics such as shape, motility, and Gram staining. As the result of the contributions of molecular systematics, however, a new view of the prokaryotes is emerging. In 1977, based upon RNA sequencing technology, Carl Woese proposed dividing the traditional bacterial kingdom Monera into two distinct domains—Archaea and Bacteria. Today, the domains Archaea and Bacteria stand along with domain Eukarya as the foundation of the new systematics.

The Archaea were first described in the 1970s. Initially, they were classified along with the typical bacteria, but

Did You Know?

A giant prokaryote, *Thiomargarita namibiensis*, can be seen by the naked eye. It is a whopping 750 millimicrons in diameter. This giant is found in anaerobic, hydrogen sulfide-rich ocean floor sediments. The first forms were found off the coast of Nambia, but similar forms have been reported from the Gulf of Mexico. It is called the "sulfur pearl of Nambia."

further studies indicated that they were worthy of their own domain. Archaea differ from Bacteria in the composition of their cell wall and plasma membrane, as well as other features. The Archaea also possess characteristics in common with domain Eukarya, such as the presence of histone proteins in their DNA. No pathogenic Archaea have been described.

Archaea have been called **extremophiles** because the first studied specimens lived in harsh conditions. Examples of extremophiles are

1. **halophiles**, living in extremely salty conditions such as the Dead Sea;
2. **thermophiles**, living in extremely hot temperatures, 60°C–90°C, such as in the hot springs in Yellowstone Park;
3. **hyperthermophiles**, living in oceanic vents with temperatures exceeding 100°C;
4. **psychrophiles**, living in extremely cold conditions such as in the Antarctic;
5. **acidophiles**, living in acid conditions such as in pools of sulfuric acid; and
6. **alkaliphiles**, living in extreme base conditions such as slag dumps.

Some Archaea, however, can exist in soil, sediment, and other less extreme environments. These Archaea are called **mesophiles**.

One group of the Archaea, the **methanogens**, metabolize carbon dioxide to oxidize hydrogen and release methane gas as a waste product. Methanogens are decomposers in anaerobic, swampy soil (producing "marsh gas") and sewage treatment plants. Methanogens also can be found in the gut of several herbivorous animals including termites, and in cattle. In recent years, commercial **methane digesters** have been developed to utilize dung and compost, used to produce methane as an efficient source of fuel. But keep in mind that living next to a methane digester may be rather unpleasant!

The overwhelming majority of prokaryotes have been placed in domain Bacteria. This domain consists of many beneficial as well as pathogenic species. Bacteria are highly diverse and live in a variety of environments from the soil to your intestines. As systematics grows, so does the number of bacterial groups. Several well-known groups of bacteria are the cyanobacteria, proteobacteria, spirochaetes, chlamydias, and the Gram-positive bacteria.

1. The **cyanobacteria** were once described as blue-green algae, and can live as a single cell or in colonies. They live in aquatic environments and even on sidewalks, in trees, and on buildings. Cyanobacteria

Did You Know?

The methane generated from the manure of five pigs is enough to cook three meals a day for the average American family! A well-fed dairy cow produces nearly 100 pounds of manure a day! Domestic animals produce nearly 25 million tons of methane annually!

The polysaccharide cellulose is difficult to break down. Herbivores such as cattle have specific adaptations to help them digest these giant molecules. To help digest cellulose, not only do these animals possess unique digestive anatomy, but they also employ bacteria such as methanogens to aid the process. As the result of this bacterial activity, a typical cow produces 2 liters of gas a minute, which usually escapes as sweet-smelling belches. By the way—to keep the food moist, a cow also produces 60 liters of saliva daily!

saturated the early atmosphere with oxygen and still are important oxygen producers today. Common cyanobacteria include: *Oscillatoria*, *Anabaena*, and *Nostoc*.

2. The **proteobacteria** are a highly diverse group of bacteria that share many molecular traits. Perhaps the ancestor of mitochondria was a proteobacteria. The proteobacteria include the purple sulfur bacteria that utilize hydrogen sulfide, not water, as an electron donor in bacterial photosynthesis; the enteric bacteria such as *E. coli* and *Salmonella*, which are inhabitants of the digestive tract of many animals; and the nitrogen-fixing bacteria such as rhizobium. Other proteobacteria include: *Vibrio* spp., *Legionella* spp., *Yersinia pestis*, and *Neisseria* spp.
3. The **spirochaetes** have spiral-shaped bodies and move in a corkscrew motion. Examples are *Treponema* spp., *Leptospira* spp., and *Borrelia* spp.
4. **Chlamydias** are atypical bacteria that serve as intracellular parasites, such as *Chlamydia trachomatis*.
5. The **Gram-positive bacteria** comprise another large, diverse group that includes *Bacillus anthracis*, *Clostridium* spp., *Staphylococcus* spp., *Streptococcus* spp., and *Mycobacterium* spp. Other Gram-positive bacteria are the **actinomycetes**, found in the soil, and mycoplasms. Some actinomycetes are used in making antibiotics, and some mycoplasms are responsible for devastating lung infections.

The shape of the cell of bacteria and their cellular arrangement are used in identification. Bacteria that are rod-shaped are known as the **bacilli** (singular, bacillus).

Examples of bacilli are *E. coli, Bacillus anthracis,* and *Yersinia pestis.* The spherical bacteria are called the **cocci** (singular, coccus). The cocci can be highly variable in arrangement (Fig. 20.2).

- *Chlamydiae trachomatis* exists as a singular cocci.
- *Neisseria meningitides* exists in groups of two cells known as diplocci.
- *Lactobacillus* spp. and *Streptococcus pyogenes* occur in chains and are called streptocci.
- *Staphylococcus aureus* occurs in chains termed staphylocci.
- *Spirilli* (singular, spirillum) are spiral-shaped bacteria and include organisms such as *Leptospira interrogans* and *Treponema pallidum.*

Even though the anatomy of the prokaryotes is much simpler than that of an oak tree, a mushroom, or a dolphin, these unicellular organisms are capable of performing all of life's processes. The **cell wall** in prokaryotes is a means of protection, is involved in metabolic activity, and helps the cell maintain its shape. Archaea cell walls are different from those of Bacteria. In Archaea, the cell walls are composed of polysaccharides and proteins, compared to the cell walls of Bacteria, which are composed of peptidoglycans.

In 1884, the Danish physician Hans Christian Gram developed a simple laboratory technique to discriminate between various kinds of bacteria. Today, the **Gram stain** remains as a standard procedure in the bacteriology laboratory. This staining technique is based on characteristics of the cell wall. Many species of bacteria have a **capsule,** glycocalyx or slime layer, composed of polysaccharides encasing the cell wall. This layer helps the bacterium attach to a substrate to keep from drying out, or in some species protects the bacterium from being destroyed by predators or the host's immune system.

Some bacteria possess **flagella** for motility. These flagella are simpler than those of eukaryotes. Several species of bacteria possess **pili,** hair-like structures extending from the cell wall. Those that attach the bacterium to a substrate are called **fimbriae,** and **sex pili serve** in a primitive type of sexual reproduction called **bacterial conjugation.** A bilayered **plasma membrane** underlies the cell wall in bacteria. This membrane serves as a barrier in regulating substances that move in and out of the cell, and as a site of metabolic activities such as photosynthesis. Folds in the plasma membrane called **mesosomes** increase the surface area and are integral in metabolism. In some bacteria, structures called **magnetosomes** serve as a compass and are found in the plasma membrane.

Bacteria that are exposed to harsh conditions may produce an **endospore.** This thick-walled structure aids them in surviving intolerable conditions (disinfectant, radiation, acidic, dry) for long periods of time. Active endospores have been taken from the Egyptian pyramids and from a 7,500-year-old site in Minnesota. Examples of species that can form endospores are *Bacillus anthracis* and *Clostridium tetani.*

Because they are prokaryotic, bacteria do not have organelles, but various structures in the cytoplasm perform a variety of duties. Bacterial **ribosomes** serve as a site for building proteins. These ribosomes are different from eukaryotic ribosomes, as the former are smaller, and they differ in their protein structure. Antibiotics such as tetracycline and streptomycin are effective against certain bacteria because they inhibit the function of ribosomes.

From the bacteria's point of view, of course, we are a rather small part of them.... This is their planet, and we are only on it because they let us.

—**Bill Bryson (1951–present)**

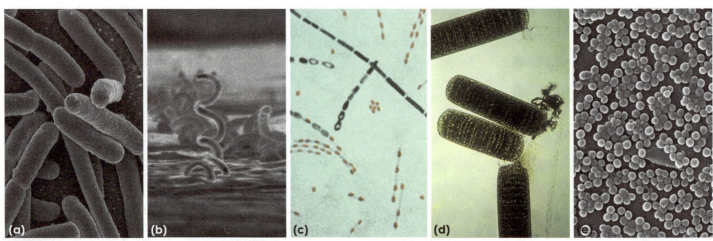

Figure 20.2 Micrographs of bacteria, (a) *Escherichia Coli,* (b) *Treponema pallidum,* (c) *Bacillus anthracis,* (d) *Oscillatoria,* and (e) *Staphylococcus aureus.*

In contrast to having many chromosomes, such as a eukaryote, bacterial genes usually are found in a single **bacterial chromosome**. This chromosome is housed in a non-membranous region called the **nucleoid**. Bacteria also may possess one or more small, ring-like structures containing DNA, called **plasmids**, which are integral in bacterial conjugation (Fig. 20.3).

For the first half of geological time our ancestors were bacteria. Most creatures still are bacteria, and each one of our trillions of cells is a colony of bacteria.

—**Richard Dawkins (1941–present)**

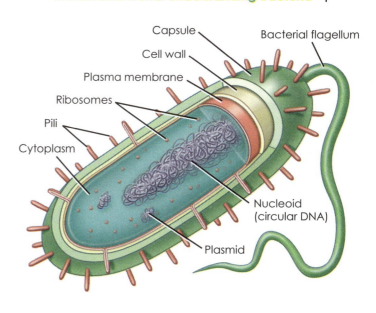

Figure 20.3 Basic bacterial anatomy.

Beyond the Lab—The World of Bacteria

As long as legitimate sources are used, a wealth of helpful material about bacteria is available on the Internet.

 ## Procedure 20.1
The World of Bacteria

Complete the following Internet challenge by answering the following:

Q. What are stromatolites, and what is their evolutionary significance?

_____

_____

_____

Q. What is the relationship between cyanobacteria and chloroplasts?

_____

_____

_____

Q. Why do many scientists think that domestic livestock such as cattle and their "emissions" are affecting our atmosphere negatively?

_____

_____

_____

Q. What is bacterial meningitis, and what are the symptoms of infection? Why should college students be aware of this disease and how can meningitis be prevented?

_____

_____

_____

_____

Q. What is MRSA? List several symptoms of a MRSA infection. Why is MRSA a good example of the evolutionary arms race?

_____

_____

Q. What was the impact of *Yersinia pestis* on humankind?

_____

_____

_____

Q. What is *Lactobacillus bulgaricus*, where is it found, and why is it considered a beneficial bacterium?

_____

_____

_____

Q. What is the difference between a bacterial endotoxin and a bacterial exotoxin? Provide an example of each.

_____

_____

_____

Q. A friend is going in for a botox injection and asks, "What is botox, and how does it work?" How would you answer this question?

_____

_____

_____

Q. Describe two bacterial species that are considered to be potential biological warfare agents.

_____

_____

_____

A Word on Safety!

Because bacteria are cosmopolitan and many species are potentially pathogenic, anyone working with bacteria must follow an aseptic laboratory technique and use common sense, to reduce the risk of contamination of yourself and the environment. Although the bacteria used in the activities in this chapter are nonpathogenic, they should be treated with proper aseptic technique. The following aseptic tips should be strictly followed in the laboratory.

- Don't hesitate to ask questions regarding safety matters.
- Report all potentially hazardous conditions, including spills and potential contamination, to your instructor.
- Don't bring any food or drink into the laboratory. Store your book bags, coats, and other personal items in the designated area.
- Wear a lab coat or lab apron over your clothes. Use eye protection if required.
- Thoroughly wipe your laboratory table with the provided disinfectant before and after use.
- Thoroughly wash your hands prior to following a procedure and at its completion. Any time you feel the urge to wash your hands—go for it!
- Always place disposable supplies in the proper container identified by your instructor. Never leave them lying around.
- Properly flame all qualified laboratory tools as indicated by your instructor.
- Report any injuries to your instructor.
- Don't remove material from the laboratory.

These microscopic organisms form an entire world composed of species, families, and varieties whose history, which has barely begun to be written, is already fertile in prospects and findings of the highest importance.

—Louis Pasteur (1822–1895)

Student Activity—Bacteria

Bacteria exist in a variety of shapes and arrangements. This observational activity acquaints students with several species of bacteria and techniques for observing bacteria. The instructor will provide prepared slides of the following bacteria: *Oscillatoria* spp., *Bacillus* spp., *Pseudomonas* spp., *Streptococcus* spp., *Staphylococcus* spp., *Neisseria* spp., *Treponema* spp., and *Anabaena* spp.

Materials
- microscope
- immersion oil
- prepared slides

 Procedure 20.2
Observing Bacteria

1. Before beginning, clean and disinfect your work area.
2. Procure a microscope, immersion oil, and the prepared slides.
3. Place the slide on the stage, and focus on the specimen using low power. Adjust the illumination if necessary. (Because bacteria are so small, only colored specks will be seen).
4. Rotate the high power objective into place, and focus with the fine adjustment only.
5. Rotate the oil immersion objective so it is halfway in position. Carefully place one drop of oil immersion oil on top of the specimen under the path of the light. Place the oil immersion objective directly into the oil.
6. Use only the fine adjustment to observe the specimen. Adjust the light if necessary.
7. In the space provided below and on the following page, sketch and label your specimen. In addition, briefly describe the basic morphology of the specimens.
8. Repeat the process for each specimen.
9. Upon completion of the activity, carefully and thoroughly clean the oil from the slide and the oil immersion objective.
10. Return the materials.

Oscillatoria spp.

Bacillus spp.

Actually bacteria are our symbiotic partners in both health and disease. They serve a useful role.

—**T.C. Fry (1926–1996)**

Pseudomonas spp.

Streptococcus spp.

Staphylococcus spp.

Neisseria spp.

Treponema spp.

Anabaena spp.

Student Activity—The Gram Stain

Without staining, bacteria are difficult to study. The microbiology laboratory uses a variety of classical as well as specialty stains to observe bacteria. Through the years, the Gram stain has been the stain most commonly used in studying bacteria.

In Gram staining, the specimen is treated with a primary stain called **crystal violet**, followed by **iodine** as a fixative or mordant. After iodine treatment, the specimen is flushed with alcohol to dehydrate peptidoglycans and trap the stain. Next, the specimen is treated with a second stain such as **safranin** or **fuchsin** (a pink stain).

Primarily, bacteria can be divided into Gram-positive and Gram-negative, based upon this simple procedure. The distinct variations are the result of differences in the cell wall. **Gram-positive bacteria** appear purple after staining. Their cell walls are simple and possess a relatively thick layer of the polymer peptidoglycan that traps the crystal violet/iodine complex. Examples of Gram-positive bacteria are *Staphylococcus aureus, Clostridium botulinum,* and *Bacillus anthracis.* Gram-negative bacteria appear pink from the safranin stain. Their cell walls, although thinner, are more complex and contain an outer layer of lipopolysaccharides. Examples of Gram-negative bacteria are *Neisseria meningitidis, E. coli,* and *Vibrio cholerae* (Fig. 20.4).

Materials
- 2 microscope slides
- inoculating loop
- Bunsen burner
- medicine dropper
- slide support
- crystal violet stain in a squirt bottle
- iodine solution in a squirt bottle
- alcohol solution (ethanol) in a squirt bottle
- safranin stain in a squirt bottle
- water in a squirt bottle
- Petri dish

 ## Procedure 20.3
Staining Bacteria

1. Before beginning, clean and disinfect your work area.
2. Procure the materials from your instructor. The instructor will supply the class with two cultures of nonpathogenic bacteria growing on a Petri dish.

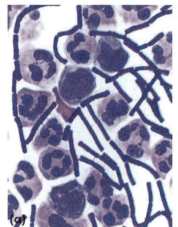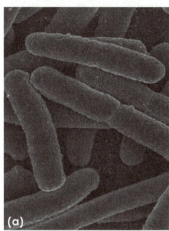

Figure 20.4 (a) Gram-positive bacteria, *Bacillus anthracis,* and (b) gram negative bacteria, *Escherichia Coli.*

3. Place one drop of water in the center of your slide with a clean medicine dropper.
4. Using the Bunsen burner, flame the inoculating loop by slowly passing the loop into the flame until it glows red.
5. Let the loop cool briefly, then gently touch the bacterial culture with the loop as indicated by the instructor.
6. Spread the bacteria in the water, gently forming an area about the size of a dime.
7. Flame the loop and put it away.
8. Allow the slide to dry for several minutes, then heat-fix the slide by slowly passing it over the flame several times with the smear side up.
9. Place your fixed slide across the holder in a sink. Squirt several drops of the crystal violet stain over the fixed region of the slide and allow it to sit for 90 seconds. *Caution: Be careful! Crystal violet will stain your skin. Gently wash away the excess stain with a light stream of water for 5 seconds.*
10. Squirt several drops of the iodine solution on the fixed region of the slide. Let this stand for 60 seconds.
11. Pour off the iodine solution and gently rinse the slide with water for 5 seconds. The specimen should be a violet-blue.
12. While holding the slide at an angle, let the alcohol solution (decolorizer) trickle down the slide until the violet-blue color disappears—perhaps 10 seconds. Too much alcohol could yield a false negative, and

too little alcohol could produce a false positive test. Gently rinse the slide with water for 5 seconds.

13. Squirt several drops of safranin stain over the fixed region of the slide, and allow it to sit for 90 seconds.

14. *Gently* wash away the excess stain with a light stream of water for 5 seconds.

15. Carefully blot the slide dry with bibulous paper (do not rub) or simply allow it to air-dry.

16. Repeat steps 3–15.

17. Observe the two slides under oil immersion, and record your observations below.

Specimen 1:

_____

_____

_____

_____

_____

Specimen 2:

_____

_____

_____

_____

_____

18. Return the materials to the designated place, properly dispose of the slides, rinse out the sink for several minutes, and thoroughly clean and disinfect your table.

A bacteriologist is a man whose conversation always starts with a germ of an idea.

—**Anonymous**

Check Your Understanding

1. Do all bacteria react to Gram staining? Why? If not, list several examples.

_____

_____

_____

2. Why is knowing whether a bacterium is Gram-positive or Gram-negative important in medicine?

_____

_____

_____

3. How do penicillin and cephalosporin kill bacteria?

_____

_____

_____

4. How does streptomycin and tetracycline kill bacteria?

_____

_____

_____

5. What are several sources of error in conducting Gram stains?

_____

_____

_____

_____

Student Activity—A Baterial World

Bacteria are inhabitants of nearly every environment. This activity asks students to detect whether bacteria, and perhaps fungi, can be found in the region around the laboratory and living in or on common objects. This activity is not designed to identify organisms but, instead, to examine the diversity of organisms that can be extracted from a specific environment.

Materials
- 5 sterile Petri dishes with growth medium
- sterile cotton swabs
- tape or Parafilm
- wax pencil
- colored pencils
- Bunsen burner

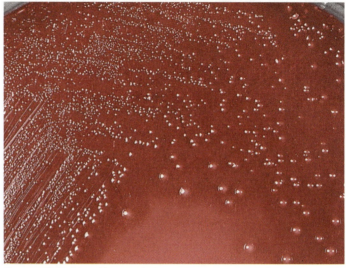
Figure 20.5 Streaking a plate

 Procedure 20.4
Where Can Bacteria Be Found

1. Discuss various areas that you would like to survey for bacterial growth. Some ideas: door knob, desktop, candy bar, lipstick, eyeshadow, top of an open soda can or bottle, hand soap, antibacterial soap, soil, a puddle of water. Perhaps leave an open Petri dish in a quiet corner of the room for 30 minutes. Choose five areas for testing. Students are to work in groups of 2 to 4.
2. Procure materials to perform the activity, and disinfect your work area. *Caution: Use aseptic techniques when collecting, culturing, and observing bacteria.*
3. Using the wax pencil, place a name on the bottom of each plate and number each plate. Record the number of the plate and the test area in Table 20.2 below.

4. Use a sterile cotton swab to rub over your study area (unless you choose to leave one dish exposed to air).
5. Carefully open your sterile Petri dish, and gently streak the cotton swab back and forth across the top of your dish. Place the used swab in the proper disposal container.
6. To spread the bacteria across the dish, flame your loop, let the loop cool, and place your loop gently on the end of the swabbed region, spreading the bacteria as shown in the diagram in Figure 20.5. The loop should be flamed between streaks.
7. Tape the Petri dish together as indicated by the instructor. Some instructors may prefer to seal the dish with Parafilm.
8. Repeat steps 6–7.

Table 20.2 Bacteria Observations	
Selected Environment	**Colony Description and Numbers**
1.	
2.	
3.	
4.	
5.	

9. Properly incubate the culture in the designated area for one week. Disinfect your work area, reflame the loop and return the material.

10. When you return to lab, observe the dishes. *Note:* Do not open the Petri dishes. Observe the number of colonies, their shape and color. Sketch the dishes and their colonies in the space on the following page, using a colored pencil. Place your results in Table 20.2 on the following page.

11. Visit the other lab stations and discuss your findings. Take notes on curious results.

12. Dispose of the Petri dishes as instructed, and disinfect your area. Wash your hands thoroughly.

Q. Which environment yielded the most colonies?

_____

_____

_____

Q. Which environment yielded the greatest diversity of colonies?

_____

_____

_____

Q. Based on your findings, discuss why we live on a bacterial planet.

_____

_____

_____

Q. Discuss several variables that may impact bacterial growth.

_____

_____

_____

Q. Why is hand-washing imperative?

_____

_____

_____

_____

Could you imagine a world without antibiotics? Our ancestors lived in such a world, and millions of deaths resulted from pathogenic bacteria. In some cases, the slightest scratch could mean death. Since the first commercial use of antibiotics at the end of World War II, millions of lives have been saved (perhaps yours!). Unfortunately, in our war against pathogens, many bacteria have mutated and have become resistant to the once effective antibiotic. This may have sobering results in the future.

Antibiotics comprise a group of biologicals, synthetics, or semi-synthetic drugs that inhibit the growth of or destroy bacteria. In 1928, Alexander Flemming brought the importance of antibiotics to the forefront with the discovery of penicillin extracted from the mold *Penecillium*. Antibiotics are classified as **bactericidal** if they kill bacteria directly and **bacteriostatic** if they inhibit bacterial growth. Antibiotics are administered only through a physician's prescription. Giving your antibiotics to someone else is irresponsible and illegal. When taking antibiotics, the prescribed regimen must be carried out. Ten days does not mean quit when you feel better 5 days later!

Without laboratories men of science are like soldiers without guns.

—Louis Pasteur (1822–1895)

Did You Know?

Nearly 3,000 years ago, the ancient Chinese used antibiotics and practiced methods of immunity. They utilized molds to prevent and treat disease. In addition, they were known to place a pus-soaked rag up the nose as a crude form of developing immunity. In the case of smallpox (a virus), the liquid beneath a crusty scab was excised and placed in a scratch on a healthy person.

BACTERIA OBSERVATIONS
(see Procedure 20.4)

Student Activity—Thanks for Antibiotics

This activity illustrates how certain antibiotics are able to affect the growth of selected bacteria. The bacteria used in this activity are nonpathogenic. Cultures recommended for this activity are nonpathogenic species of *Bacillus subtilis, Serratia marcescens, Eschercia coli,* and *Micrococcus spp.* Even so, aseptic techniques should be followed throughout the activity. An interesting extension of this activity involves placing a blank antibiotic disc soaked in garlic extract, along with the regular antibiotic disc, on each Petri dish.

Materials
- sterile Petri dishes with growth medium
- tape or Parafilm
- sterile cotton swab
- wax pencil
- metric ruler
- antibiotic dispenser
- antibiotic discs
- bacterial cultures
- garlic extract antibiotic disc

Procedure 20.5
Antibiotics and Bacterial Growth

1. Disinfect your work area, and wash your hands.
2. Collect your materials, and return them to your work area.
3. Using a wax pencil, label the bottom of your Petri dish with your name, date, and organism tested.
4. Your instructor will provide a pure culture of several species of nonpathogenic bacteria. Using a sterile cotton swab, take a sample from the provided culture.
5. Remove the lid from the Petri dish, and thoroughly swab the surface of the growing medium (agar), covering all of the edges. Rotate the dish and thoroughly swab again at a right angle to the original smear. The entire dish should be streaked uniformly.
6. Close the plate, and bring it to the antibiotic disk dispenser. Open the lid, and your instructor will inform you how to use the dispenser and what type of antibiotic is on each disc. Be sure to record the symbols on each disc.
7. Replace the lid on the Petri dish, and seal the lid with tape or Parafilm.
8. Repeat Steps 3–7 for your other cultures.
9. Incubate the dishes as instructed.

10. Return your materials, disinfect the work area, and wash your hands.
11. During the next class period, retrieve your cultures. Disinfect the work area and wash your hands. Observe the growth patterns on each dish.
12. A **zone of inhibition** or clear area of no growth should exist around some of the antibiotic discs. Measure the zone of inhibition in millimeters for each antibiotic on each of your cultures.
13. Discard the Petri dishes as instructed, disinfect the work area, and wash your hands.
14. Using the chart below, addressing the antibiotic used, organism, and whether it is Gm+ or Gm−. Discuss your findings for each organism.

Antibiotic Used	Organism	Gm+ or Gm−

_____

_____

_____

Q. How are similar activities important in medical laboratories?

_____

_____

_____

In other words, bacteria are bound to win the war against medicine. Bacteria show what natural selection needs and what it can do when it gets a chance.

—**Steve Jones (1944–present)**

Q. Name three major diseases and tell how they have been controlled by antibiotics.

_____

_____

_____

_____

_____

Q. List three common antibiotics used today and their target.

_____

_____

_____

_____

_____

Q. What is the difference between a narrow-spectrum and a broad-spectrum antibiotic? Give an example of each.

_____

_____

_____

_____

In the world of life, **bioluminescent** (light-producing) organisms evoke a sense of wonder. Common examples of bioluminescent organisms include fireflies on a summer night, "glow-in-the-dark algae," glow worms, jellyfish, and foxfire fungi. It is estimated that nearly 90 percent of deep-sea marine organisms are capable of bioluminescence.

Several species of bioluminescent bacteria exist as well. These bacteria can be found in the marine environment living in the water, as well as in a symbiotic relationship with several species of fishes and mollusks such as the cuttlefish. In this relationship, the fish or mollusk provides the bacteria with a substrate, nutrients, and oxygen and the fish uses the bioluminescent bacteria in attracting prey, illumination, defense, and communications.

Bacterial bioluminescence produces a pale blue-green glow because blue light penetrates farther in the depths of the ocean. Bioluminescence is a cold light emission and produces little thermal signature (Fig. 20.6). Recently, a gene for bioluminescence was identified. In bacteria, some fishes, and squid, the light-emitting chemical is a reduced riboflavin phosphate ($FMNH_2$) called **luciferin**.

The reduction reaction requires oxygen and is driven by the enzyme **luciferase**. As a result, the chemical **oxyluciferin** and light are produced. The process of bioluminescence requires a great amount of ATP; during the emission of light, perhaps as many as 50,000 molecules of ATP are used every second. The bioluminescent bacteria are Gram-negative bacilli. The most common organisms used in simple demonstrations and experiments, *Vibrio fischeri* and *Vibrio phosphoreum,* are nonpathogenic and provide stunning results.

Strange but True!

Cruise passengers occasionally witness a strange glow in their toilets. This is the result of seawater with bioluminescent organisms in the toilets.

(a)

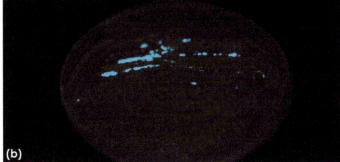
(b)

Figure 20.6 (a) Bioluminescent bacteria and (b) the bacteria glowing in the dark.

Student Activity—A Glowing Experience

Materials
- pure culture of *Vibrio fischeri* or *Vibrio phosphoreum*
- sterile cotton swab
- inoculating loop
- Bunsen burner
- tape or Parafilm
- wax pencil

Procedure 20.6
Bioluminescent Organisms

1. Clean and disinfect your workspace.
2. Bring the culture into a dark area and gently shake the culture tube. Record your observations.

_____

_____

3. Gently dip the end of a sterile cotton swab into the culture tube.
4. Use aseptic techniques (discussed in Activity 20.4) to prepare a smear of *Vibrio* spp.
5. Seal the Petri dish with tape or Parafilm. Using a wax pencil, place your name and the date on the Petri dish.
6. Place the Petri dish in the designated storage area, disinfect your work area, and wash your hands.
7. Return to the lab within 24 hours and observe the results in a dark place. If possible, view the colonies in a dark area with a dissecting microscope.
8. Record your observations below.

_____

_____

_____

A man who dares to waste an hour of time has not discovered the value of life.

—Charles Darwin (1809–1882)

Q. Describe bioluminescence in the following: flashlight fish, cookie cutter shark, dinoflagellates, krill, and vampire squid.

_____

_____

_____

Q. Recently, genetic engineers developed "glow-in-the-dark" mice and pigs. How was this accomplished?

_____

_____

_____

Q. What are several applications of genetically engineered organisms?

_____

_____

_____

Q. Why would the armed services be interested in cold light like that generated by bioluminescent organisms?

_____

_____

_____

Q. What are the main chemicals utilized in this activity, and their role?

_____

_____

_____

_____

Name: _____ Date: _____ Section: _____

Review Questions

1. Compare and contrast Archaea and Bacteria.

_____

_____

_____

_____

2. What are three diseases caused by bacteria?

_____

_____

_____

3. What are three helpful bacteria?

_____

_____

_____

4. Describe the major shapes of bacteria, and give an example of each.

_____

_____

_____

_____

_____

_____

5. What is the fundamental premise of Gram staining?

_____

_____

_____

6. What is the function of the following: cell wall, plasma membrane, plasmid, nucleoid, and pili.

_____

_____

_____

_____

_____

_____

_____

7. What are some safety factors to consider when working with bacteria?

_____

_____

_____

8. What is the difference between bioluminescence and luminescence? Name two luminescent organisms.

_____

_____

_____

9. Why are many scientists alarmed by the overuse of antibiotics? Provide an example of bacterial resistance, along with its dangerous impact on health and medicine.

_____

_____

_____

10. How are bacteria used in developing genetically engineered medicines?

_____

_____

Chapter 21
This Fine Mess: Understanding the Protists

Student Outcome Objectives

At the completion of this exercise, the student will be able to:

1. Describe the general characteristics of the protists.
2. Discuss why protists present a unique problem to taxonomists.
3. Compare and contrast traditional and modern protist classification.
4. Describe the traditional organization of the plant-like protists.
5. Provide common examples of the plant-like protists.
6. Describe the traditional organization of the fungus-like protists.
7. Provide common examples of the fungus-like protists.
8. Describe the traditional organization of the animal-like protists.
9. Provide common examples of the animal-like protists.
10. Draw and label the protist examples used in this chapter.

Overview

Oliver Hardy's statement to Stanley Laurel, "What a fine mess you have gotten us into" is appropriate for trying to make sense of the hodge-podge group of organisms known as **protists**. The term protist describes microscopic, unicellular organisms such as the amoeba, as well as gigantic multicellular organisms exceeding 600 feet in length, such as kelp. Protists are ancient eukaryotes that first appeared in the Precambrian Era nearly 2 billion years ago. A multitude of protists can live in a drop of water, in the terrestrial environment, or in the body of a host. In the environment, protists can function as **autotrophs**, **heterotrophs**, and even **decomposers**. Several protists are responsible for diabolical parasitic infections in humans.

The 100,000 named members that compose the protists are a **polyphyletic group** of organisms underlying the kingdoms Plantae, Fungi, and Animalia. This diversity is the main reason why the kingdom "Protista" is no longer considered viable, though the term protist is still used as a convenient, and widely understood, term to describe this diverse group (Fig. 21.1).

Classically, the protists have been categorized as plant-like protists, fungus-like protists, and animal-like protists, in accordance with their role in the environment. This scheme, although many consider it antiquated is still a useful tool in developing a basic understanding of the protists. Recently, studies using DNA sequencing and cytological analysis have painted a more complex picture of the protists. This new look at the protists suggests that the once singular kingdom should be divided into several supergroups that can be divided further into kingdoms (as many as 30, according to some authors). As time passes, a clearer picture of protist classification will emerge. Table 21.1 on page 295 provides an overview of the new view of the protists.

Although traditional classifications of protists tended to have little respect for phylogeny—neither in theory nor the data were available to make this possible—they did have one great advantage. They were designed to be user-friendly.

—Colin Tudge (1943–present)

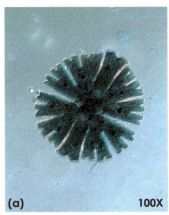

Figure 21.1 Example protists include (a) Desmid, *Micrasterias* sp., (b) kelp, *Laminaria*, sp., (c) slime mold, *Comatricha* sp., and (d) *Stentor*.

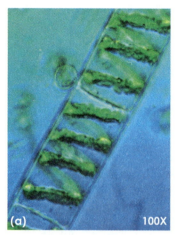
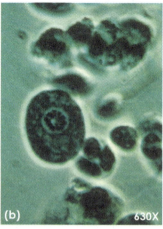
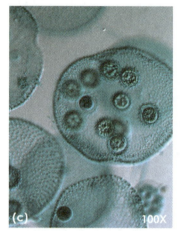

Figure 21.2 Examples from the phylum Chlorophyta include (a) *Spirogyra* sp., (b) *Chlamydomonas* sp., (c) *Volvox* sp., and (d) *Ulva* sp.

PLANT-LIKE PROTISTS

The plant-like protists commonly called the **algae** are extremely diverse, ranging from minute diatoms to giant multicellular kelp. These organisms exist as photoautotrophs possessing chlorophyll *a* and membrane-bound plastids. They are responsible for copious oxygen production, as well as the basis for the food chain. Traditionally, the names of the phyla were derived based upon accessory pigmentation.

Phylum **Chlorophyta**, known as the green algae, is composed of about 10,000 unicellular and multicellular species (Fig. 21.2). The majority of green algae are aquatic, but some species can be found growing on tree trunks, sidewalks, buildings, unwashed cars, and even snow. One species even lives on the fur of tree sloths, imparting a green sheen to them. The green algae are thought to be the ancestor of modern plants.

- *Spirogyra crassa* is a classic example of this phylum. This filamentous freshwater green algae can be found floating in mats on the surface of quiet ditches and ponds. Under the microscope, *Spirogyra crassa* appears ornamental with spiral chloroplasts occupying the inside of the filament.
- *Chlamydomonas* spp. is a freshwater green alga best known for its pair of flagella and light-sensitive eyespots.
- *Volvox* spp. is a colonial freshwater green algae that resembles a basketball under the microscope. It may be composed of as many as 5,000 to 50,000 cells.
- *Ulva* spp., or sea lettuce, is a multicellular marine green algae that may grow up to a meter in length.

Phylum **Rhodophyta** consists of what are known as the red algae. Approximately 5,000 species inhabit primarily marine environments, and fewer than 100 freshwater species have been described (Fig. 21.3). Although a few unicellular forms exist, the majority of red algae are filamentous. The tightly packed filaments form blade-like

structure, and when attached to a substrate with a holdfast, appear to be seaweeds. The largest species is about 3 feet long. The red color in red alga comes from accessory pigments known as **phycobilins**.

Some species of red algae are used to make **agar** (a growing medium for bacteria and fungi), and others are used to make **nori** (used to wrap sushi). Coraline red algae are noted for establishing and supporting coral reefs.

Figure 21.3 Examples from the phylum Rhodophyta include (a) *Rhodymenia* sp. and (b) Red algae, *Gelidium* sp.

Some species of red algae produce toxins called **terpenoids**, which may be used to control the growth of tumors and cancer.

Phylum **Phaeophyta** includes more than 1,500 species of mostly marine multicellular seaweeds known as brown algae. The color of the brown algae ranges from yellow to olive green to dark brown, depending on the concentration of **fucoxanthin.** The majority of species inhabit temperate-to-cool coastal marine waters (Fig. 21.4).

Giant kelp, a brown algae, can reach nearly 900 feet in length. One of the best-known giant kelp species, *Macrocystis pyrifera*, grows in tremendous "kelp forests" off the coast of California. *Laminaria digitata* is a kelp used in many Asian dishes. *Sargassum* spp. is a brown alga that grows in floating mats in a region of the Atlantic Ocean called the Sargasso Sea. Some fragments have washed up along the shore of the Northern Gulf of Mexico.

Many species of kelp have a plant-like body called a **thallus** (lacking roots, stems, and leaves). The thallus consists of a **holdfast** that attaches the kelp to a substrate, a **stipe** (stalk), and a flattened **blade. Bladders** filled with gas including carbon monoxide help to keep the blades afloat. A product of brown algae called **algin** has a variety of commercial uses including: a thickening agent in foods and medicines, an emulsifier in some salad dressings, a smoothing agent in some lotions, and in the filaments of tungsten light bulbs.

 Check Your Understanding

1. Why is algin used in beer?

2. Why is algin used in chocolate drinks and marshmallows?

3. How is algin used in paper?

4. Why is algin used in some surgical gut?

5. How is algin used in toothpaste?

In fact, the scaffolding of our body originated in a surprisingly ancient place: single-celled animals.

—**Neil Shubin (1960–present)**

Figure 21.4 Examples from the phylum Phaeophyta include (a) Brown algae, *Macrocystis* sp., (b) *Sargassum* sp., and (c) *Fucus* sp.

Phylum **Pyrrophyta** (Dinophyta) contains more than 3,000 species of unicellular algae. Some members of this phylum are photosynthetic, and others lack chlorophyll, leading to taxonomic confusion among biologists. Many pyrrophytes are considered to be dinoflagellates and members of the animal-like protists. Many species possess a series of cellulose plates resembling armor plating beneath the plasma membrane. These organisms also possess two distinct flagella, one serving as a rudder and the other as a locomotor structure. Photosynthetic species have two disc-shaped chloroplasta. Several species, including *Gonyaulax* and *Gymnodinium* are discussed with the animal-like protists.

Phylum **Bacillariophyta** consists of uniquely shaped algae called **diatoms,** which possess a **test** or shell made up of two halves. These organisms are the most numerous unicellular algae found in marine and freshwater environments. There are more than 10,000 species, and some biologists estimate a million species. Diatoms are exquisite organisms, appearing in a variety of geometrical shapes. A large component of the test is silica (SiO_2), a major component of glass incorporated into an organic mesh. Because their test is composed of silica, the fossil record of diatoms is well represented. They make up diatomaceous earth that is used in silverware polish, insulation, swimming pool filters, reflective paint, and even toothpaste. Some deposits of diatoms are more than 3,000 feet thick. A common diatom is *Cytotella stelligera*.

Certain organisms that we called fungi because they were slimy and repulsive and because we didn't know where else to put them—the slime molds, for example—have been exiled from the kingdom. That is not too hard to take, because few of us encounter slime molds with any regularity.

—**Gordon Grice** (*Discover* **Magazine, June 2008**)

Phylum **Euglenophyta** includes approximately 1,000 species of unicellular flagellated freshwater species. Nearly 40 genera of euglenoids have been described (Fig. 21.5). One-third of these genera possess chloroplasts, and the other two-thirds do not have chloroplasts. Nonphotosynthetic euglenoids gather food by particle feeding and absorption. Euglenoids do not store starch for energy. Instead, they utilize **paramylon bodies** to store energy. Some euglenoids are considered **mixotrophs** because they can undergo photosynthesis and can feed. Two common species are *Euglena deces* and *Phacus* spp.

FUNGUS-LIKE PROTISTS

Once considered fungi, the **slime molds** are now considered protists. Fungus-like protists are decomposers in forests and woodlands. Another member of this group, the water molds, serves as decomposers in aquatic environments. Slime molds are motile organisms and produce spores on **sporangia**, which in turn comprise **fruiting bodies**. The slime molds display complex life cycles in which they undergo remarkable morphological changes. Some slime molds measure a meter or more in diameter.

Phylum **Myxomycota** consists of about 500 species known as the **plasmodial slime molds**. These organisms exist as a multinucleate mass covered in a sheath of slime. Many are colored a bright yellow or orange and resemble a giant amoeba sliming ever so slowly across the forest floor. *Physarum* spp. is a common slime mold found in the woodlands of North America (Fig. 21.6).

The **cellular slime molds** have been placed in Phylum **Acrasiomycota**. These organisms exist as solitary amoeboid cells in the soil. During adverse conditions, they aggregate, forming a mass called a **pseudoplasmodium**. This temporary stage gives rise to fruiting bodies that produce spores. *Dictyostelium discordeum* is a common cellular slime mold.

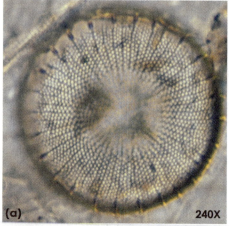
(a) 240X

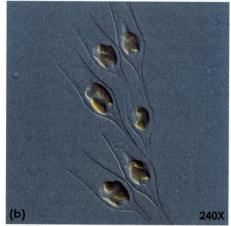
(b) 240X

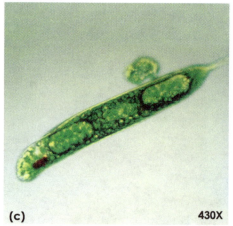
(c) 430X

Figure 21.5 Examples from the phylum Euglenophyta include (a) Diatom, *Stephanodiscus* sp., (b) Dinobryon, *divergens* sp., and (c) *Euglena* sp.

Figure 21.6 (a) Slime mold, *Physarum* sp., growing on grass. (b) Water mold, *Saprolegnia* sp., on a dead fish.

Water molds are placed in Phylum **Oomycota.** They can be easily observed as a cotton-like filamentous mass growing on dying or dead fish. *Saprolegnia* spp. is a common water mold that helps decompose aquatic animals. The water molds have a cell wall composed of cellulose as opposed to fungi, which have a cell wall composed of chitin. One species, *Phytophthora infestans*, was responsible for the Irish potato famine in the 1840s. This organism had a direct impact on U. S. history as a large number of Irish people emigrated to America.

ANIMAL-LIKE PROTISTS

Remember the first time you examined ditch water with a microscope? Wow! It was like viewing a miniature version of *Star Wars* with all of the tiny organisms gliding around in a thin film of water. Antony van Leeuwenhoek shared

> *A drop of pond water, viewed through a microscope, swarms with tiny organisms, some spinning, some crawling, some whizzing across the field of vision like rockets.*
>
> **—David Attenborough (1926–present)**

your fascination centuries before, terming these organisms "animacules and cavorting wee beasties." Today we know that many of Leeuwenhoek's "wee beasties" actually were animal-like protists sometimes called protozoans (Fig. 21.7). Classically, the animal-like protists are divided into four major groups:

1. those that move by false-feet,
2. those that move by flagella,
3. those that move by cilia, and
4. nonmotile forms.

Some animal-like protists possess cytoplasmic extensions called **pseudopodia.** These specialized structures are used for locomotion and for capturing food. Members of this group are found primarily in freshwater and marine environments, although several parasitic species exists. Classically, amoeboid protists such as the common freshwater amoeba (*Amoeba proteus*) have been placed in Phylum **Sarcodina** (**Rhizopoda**). The sarcodines do not have a wall or pellicle surrounding their plasma membrane. The majority of sarcodines reproduce asexually through fission. One species, *Entamoeba histolytica*, is an

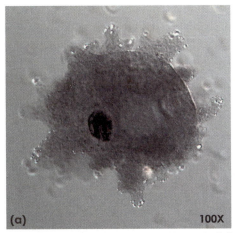
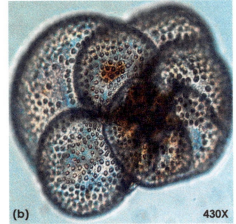
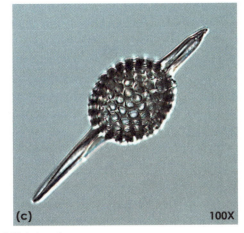

Figure 21.7 Animal-like protists include (a) *Amoeba proteus*, (b) *Globigerina* sp., and (c) *Stylatractus* sp.

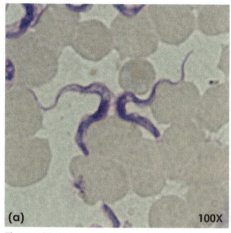
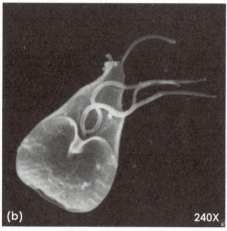

Figure 21.8 Flagellated protists include (a) *Trypanosoma brucei*, (b) *Giardia intestinalis* (*lamblia*), and (c) *Gymnodinium* sp.

intestinal parasite responsible for amoebic dysentery, spread by contaminated food.

The foraminiferans placed in **Phylum Foraminifera** are an intriguing group of aquatic and marine amoeboid protists with an elaborate, colorful shell composed of calcium carbonate. These organisms have thread-like branched pseudopods that extrude from pores in the shell. Foraminiferans are found in tremendous numbers in the oceans. It is estimated that about one-third of the sea floor consists of the shells or cast from dead foraminiferans such as *Globigerina* sp. Fossil foraminiferans, or forams, are responsible for forming limestone. The White Cliffs of Dover are composed primarily of the remains of foraminiferans.

The radiolarians (**Phylum Actinopoda**) possess an endoskeleton composed of silicon dioxide. Radiolarians, with their thin, relatively stiff pseudopods, are found in the marine environment. Many consider the radiolarians, with their geometric endoskeletons, to be the most beautiful organisms on Earth. Some radiolarians, such as *Lychnaspis miranda*, resemble snowflakes. In some parts of the ocean, radioloarians ooze on the floor can be over 4,000 meters thick. Fossil foraminiferans and radiolarians such as *Stylatractus* sp. are helpful in correlating the age of various geologic strata.

Perhaps the flagellated protists are the most confusing to taxonomists (Fig. 21.8). Classically, these are subdivided into the **phytomastigophorans** (photosynthetic flagellates) and the **zoomastigophorans** (animal-like flagellates). Recently, they were placed into several supergroups, as indicated in Table 21.1 on page 295. For convenience, the well-known **Phylum Zoomastigophora** (**Sarcomastigophora**) is used here to describe the animal-like protists that move by flagella.

The vast majority of zoomastigophorans are free-living, residing in freshwater, the marine environment, or in the soil. The species described here are well-known

parasites. *Trypanosoma brucei* is responsible for causing African sleeping sickness. The vector for this disease is the tsetse fly. *Trypanosoma cruzi* is the infective agent that causes Chaga's disease in South and Central America. The vector is the triatomine, or kissing bug. Chaga's disease kills approximately 50,000 people annually.

Leishmaniasis is a serious infection transmitted by sand flies. *Leishmania donovani* causes "kala-azar," a lethal disease in India, China, Bangladesh, Ethiopia, and the Sudan. *Leishmania braziliensis*, a mucocutaneous form, causes grotesque facial deformities in people in South America. *Leishmania Mexicana*, responsible for the chiclero ulcer or sores, is commonly contracted by unsuspecting tourists visiting Mexico, South America, and Central America.

Giardia intestinalis (*lamblia*) causes **giardiasis**, an intestinal infection that causes intense diarrhea and dehydration. Giardiasis is a common waterborne infection worldwide. Many *Giardia* infections result from swimming in or drinking from contaminated waterways or even swimming pools. *Giardia* also is interesting in that it has modified mitochondria called **mitosomes**, which do not generate ATP directly. In modern schemes, *Giardia* occupies its own kingdom known as **Diplomonadida**.

Another unusual flagellate is *Trichomonas vaginalis*, a sexually transmitted parasite that causes urogenital infections. *Trichomonas* possesses modified mitochondria, termed **hydrogenosomes**, and is placed in its own kingdom called **Parabasala.** One species, *Trichonympha* spp., a large protozoan that lives in a symbiotic relationship in the gut of termites, helps the termite break down cellulose.

Many biologists also consider the **dinoflagellates** to be zoomastigophorans, but several schemes place them in the plant-like protist phylum **Pyrrophyta**. The dinoflagellates are primarily marine, but a few freshwater species exist. These protists possess a faceted hard cellulose case

Table 21.1 A New View of the Protists		
Super Group	**Example**	**Representative Organism**
Excavata	Diplomonadida Parabasala	*Giardia intestinalis* *Trichomonas vaginalis*
Euglenozoa (Discicristates)	Euglenida Kinetoplastea	*Euglena deces* *Trypanosoma brucei*
Alveolata	Ciliophora Apicomplexa Dinoflagellata	*Paramecium caudatum* *Plasmodium vivax* *Gonyaulax catanella*
Amoebozoa	Amoebas Cellular slime molds Plasmodial slime molds	*Amoeba proteus* *Dictyostelium discordeum* *Physarum polycephalum*
Stramenopila (heterokonts)	Bacillariophyta Oomycota Chrysophyta Phaeophyta	*Cytotella stelligera* *Phytophythora infestans* *Dinobryon sociale* *Macrocystis pyrifera*
Rhizaria (cercozoa)	Radiolaria Foraminifera Chloroarachniophyta	*Lychnaspis miranda* *Globigerina falconensis* *Chlorarachnion reptans*
Archaeplastida	Rhodophyta Chlorophyta Glaucophyta Vascular plants*	*Gelidium amansii* *Spirogyra crassa* *Cyanophora paradoxa* *Acer rubrum*
Opisthokonta	Choanomonada Fungi* Animalia*	*Monosiga brevicollis* *Amanita verna* *Homo sapiens*

* A separate Kingdom

and a single long flagellum. Although some are capable of photosynthesis, the majority are heterotrophs. Some, such as *Gymnodinium breve* and *Gonyaulax catenella*, are responsible for the devastating "red tide" in warm marine waters. *Gonyaulax* also can cause paralytic shellfish poisoning, which can be lethal to humans. *Pfiesteria piscida*, living in the brackish water along the Atlantic coast, causes deadly infections in fishes, as well as humans, Discovered in 1988, *Pfiesteria* has a complex life cycle, with more than 20 body types. Symbiotic species (zooxanthellae) live in sponges, jellyfish, anemones, and coral, which provide a supply of energy to their hosts. *Noctiluca scintillins* is a bioluminescent dinoflagellate that amazingly lights up marine waters at night.

Phylum **Ciliophora** consists of approximately 8,000 species of freshwater and marine organisms called **ciliates**. Their name is derived from the hair like-projections called **cilia** that cover their body surface or oral region. These structures are used in locomotion and feeding. The majority of ciliates are free-living, but there are several species of **sessile** (attached to a substrate), colonial, and parasitic forms.

The ciliates are considered the most structurally complex animal-like protists. The outer covering, or **pellicle**, of ciliates is usually extremely tough. Most ciliates obtain their food using a **cytostome** (mouth) and an **oral groove**. The oral groove opens into a short canal, the **cytopharynx**, and ultimately into a region where **food vacuoles** are formed. The **cytoproct**, posterior to the oral groove, is responsible for elimination of fecal material. **Contractile vacuoles** are involved in **osmoregulation**. **Radiating canals** collect fluids and empty into the contractile vacuole.

Ciliates are multinucleate—possessing at least one **macronucleus** and one **micronucleus**. Macronuclei perform the basic nuclear duties, and the micronuclei are exchanged during **conjugation**, a form of sexual reproduction. Some ciliates possess trichocysts and toxicysts. **Trichocysts** lie beneath the pellicle and can be discharged like a harpoon for defense. **Toxicysts** are

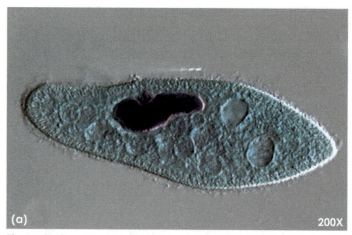

Figure 21.9 Examples from the phylum Ciliophora include (a) *Paramecium caudatum*, and (b) *Ichthyophthirius multifilis*, known as "ich" seen here as small white specs on a freshwater fish.

different from trichocysts in that they release a poison that can immobilize a prey. Just think—if a paramecium were as big as a dolphin, it would be formidable!

The classic example of a free-living ciliate is *Paramecium caudatum*, a common inhabitant of freshwater environments. *Stentor* spp. is a sessile, vase-shaped ciliate found in the freshwater environment. Freshwater fishes may suffer from "ich," caused by the ciliate *Ichthyophthirius multifilis* (Fig. 21.9).

Phylum **Apicomplexa** consists of endoparasitic protozoans that live in or between the cells of their host. These organisms lack locomotor structures in the adult

form. This phylum is so named because its members possess microtubules and organelles at the apical end of the organism (this portion is visible only with the electron microscope). The complex life cycle usually has a sexual phase and an asexual phase. In many species, the life cycle involves multiple hosts.

The best-known apicomplexan is responsible for causing **malaria** ("bad air"). The most common symptoms of malaria are malaise, chills, fever, headache, and flu-like symptoms. Between 350 and 500 million clinical episodes of malaria occur yearly, resulting in at least

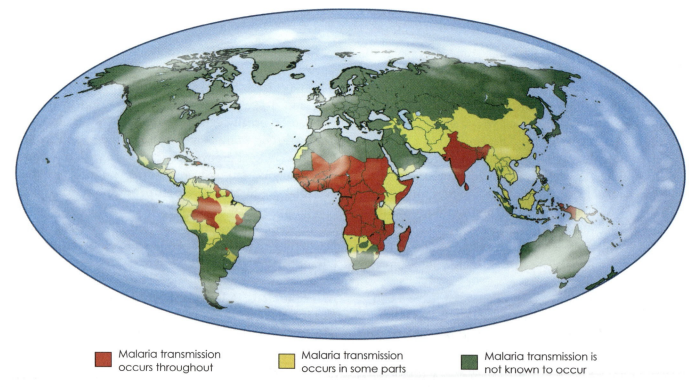

■ Malaria transmission occurs throughout   ■ Malaria transmission occurs in some parts   ■ Malaria transmission is not known to occur

Figure 21.10 World map showing the incidence of malaria, *Plasmodium falciparum*.

a million deaths. Malaria is the leading cause of death in children under 5 years of age. The majority of cases occur in Africa south of the Sahara Desert (Fig. 21.10). The vector for malaria is the female *Anopheles* spp. mosquito.

Four species of apicomplexans are responsible for malaria:

1. *Plasmodium vivax*,
2. *Plasmodium, malariae*,
3. *Plasmodium ovale*, and the most deadly form,
4. *Plasmodium falciparum*.

Alexander the Great is thought to have died from *Plasmodium falciparum*. Malaria has been a formidable disease throughout human history, and presently seems prime for a comeback, considering the ease of world travel.

Another ampicomplexan of note is *Toxoplasma gondii*, responsible for causing toxoplasmosis in humans. The Centers for Disease Control (CDC) estimates that perhaps one-third of the world's human population carries this parasite. Although it infects a variety of mammals, cats are the primary host. Healthy adults rarely contract the disorder; individuals with a weakened immune system and fetuses are at the greatest risk. Toxoplasmosis can cause severe nerve damage, hydrocephaly, and death in newborns. Therefore, pregnant mothers should bring their cats to the vet for a check-up and avoid changing litter boxes and gardening in areas where cat feces are common.

The **choanoflagellates** are protists that perhaps are the direct relatives of multicellular animals. Choanoflagellates can be single-celled or reside in colonies in freshwater and marine environments. Their basic anatomy is similar to the **collar cells** found in multicellular sponges. The choanoflagellates possess a single flagellum surrounded by a collar made of a number of cytoplasmic extensions called **tentacles**. They are placed in the supergroup **Opisthokonta**, which also includes fungi and animals (Fig. 21.11).

Figure 21.11 Molecular evidence suggests that the choanoflagellates, *Sphaeroeca* sp. colony, perhaps are the ancient ancestors of both Kingdom Fungi and Kingdom Animalia.

 Student Activity—Sharing Leeuwenhoek's Enthusiasm

Antony van Leeuwehoek turned his simple microscope on to anything that he could find. He looked for spears in hot pepper (that supposedly burned the tongue), fleas, ditch water, and even his own diarrhea, where he found a multitude of *Giardia intestinalis*. In this activity, students will share Leeuwenhoek's wonder by observing life in a drop of water—but we won't try to search for *Giardia*!

Materials
- pond water
- microscope slides and coverslips
- eyedropper or pipette
- forceps
- microscope
- Protoslo™, soap, or methyl cellulose
- colored pencils

 **Procedure 21.1
Pond Water**

1. Procure the needed materials and equipment and bring them to your laboratory table.
2. Using the techniques developed in Chapter 4, make a wet mount of pond water.
3. Observe several slides of the pond water, and make your observations and sketches in the space provided below. You may have to use Protoslo™, soap, or methyl cellulose to slow down the protists.
4 Using the figures on page 43, identify the organisms that you observe.
5. Return the materials as directed by the instructor, clean your laboratory table, and thoroughly wash your hands.

In the year 1657, I discovered very small living creatures in rainwater.

—**Anton von Leeuwenhoek (1632–1723)**

6. Record your observations and place your sketches below.

Check Your Understanding

Q. Did you observe protists in the water? If so, were they plentiful?

_____

_____

_____

Q. Describe the movements of the protists.

_____

_____

_____

Q. Which phyla were represented in your water?

_____

_____

_____

The role of biology today, like the role of every other science, is simply to describe, and when it explains it does not mean that it arrives at finality; it only means that some descriptions are so charged with significance that they expose the relationship of cause and effect.

—Donald Peattie (1898–1964)

Student Activity—The Hay Infusion

Growing "Wee-beasties"

Depending on a number of variables, when searching for protists in a drop of pond water or ditch water, the observer many times is disappointed in the number or density of "wee-beasties" to be found. A tried-and-true method to increase the density of protists is to make a classical **hay infusion**. If the hay infusion activity is carried out for several weeks, an interesting succession of organisms will appear.

Materials
- ditch water, pond water, or stream water
- large-mouth mayonnaise or disposable jar
- probe
- hay or grass clippings
- yeast (optional)
- eyedropper or pipette
- microscope slides and coverslips
- microscope
- light source (optional)

 ## Procedure 21.2
Wee-beasties

1. Carefully collect water from a local ditch, pond, or stream to fill about half a mayonnaise jar. After collecting the water, rinse the outside of the jar and wash your hands. (Or your instructor may provide a container of water for the class.)
2. Making sure that the hay or grass clippings to be used in this activity are free of herbicides or pesticides, place a small handful of dried hay or grass into the jar, gently submerging it with a probe into the water.
3. Your instructor may suggest adding a small amount of yeast to your infusion. Yeast encourages bacterial growth that will feed the protists.
4. Let the jar incubate at room temperature for the next week without a lid. It is highly recommended that air be bubbled into the solution once a day by blowing into a pipette. If time permits, check the solution every 2 days, and record your observations.

5. As the week comes to an end, the solution will begin to smell quite foul. To check your infusion, using an eyedropper or a pipette, take samples from the surface of the water, from the bottom of the jar, and near debris.
6. Using the aesptic techniques acquired previously, properly prepare a microscope slide, then record your observations below. Always remember to clean your laboratory area and wash your hands thoroughly.

7. Discuss and illustrate your observations below. Referring to a key or the figures on page 43 of this manual, identify the organisms living in the infusion.

Optional Exercise

1. Start another infusion, and expose it to a light source. This will encourage photosynthetic protists to grow. Add a small amount of the original water to compensate for evaporation.
2. Record your observations below.

3. Maintain your infusion for several weeks or longer. Keep records regarding populations of protists. Remember to add a small amount of the original water to compensate for evaporation.

4. Record your observations below.

_____

_____

_____

Q. In the infusion, is there a change in the number and species of organisms?

_____

_____

_____

5. Your instructor will provide information about the proper cleanup and disposal of the material used in this activity.

On the Kingdom Protista:
The kingdom of primitive forms. Boundary kingdom intermediate between the animal and vegetable kingdoms. Containing organisms, neither animals nor plants.

—Ernst Haeckel (1834–1919)

Check Your Understanding

Q. What is the function of a hay infusion?

_____

_____

_____

Q. Describe the density of protists living in the hay infusion.

_____

_____

_____

Q. Describe several specimens living in the hay infusion.

_____

_____

_____

_____

_____

Student Activity—The Plant-like Protists

The plant-like protists are a diverse group of organisms commonly called the algae. The following organisms are typical examples of the plant-like protists.

Observations of *Spirogyra* and *Spirogyra* Conjugating
Spirogyra, sometimes called watersilk, is a filamentous freshwater species that possesses ornate spiral chloroplasts (Fig. 21.12). *Spirogyra* is capable of a type of sexual reproduction known as **conjugation**. During this process, a conjugation tube forms between adjacent filaments. Protoplasts from one tube migrate via the tube to another filament to fuse with a waiting protoplast. The mobile protoplast is considered male, and the stationary protoplast is considered female. A thick-walled zygote results. Eventually the zygote undergoes meiosis, forming four haploid cells. Three disintegrate, and the remaining cell forms a new *Spirogyra* filament.

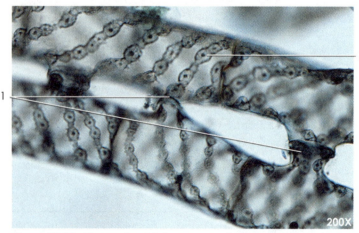

Figure 21.12 Filaments of *Spirogyra* sp. showing initial contact of conjugation tubes.
1. Conjugation tube   2. Pyernoid in chloroplast

Materials
- compound microscope
- prepared slides of *Spirogyra* and *Spirogyra* conjugating
- herbarium sheets
- wet mounts
- microscope slides and coverslips
- eyedropper
- colored pencils
- culture of *Spirogyra*

Procedure 21.3
Spirogyra and *Spirogyra* Conjugating

1. Procure the equipment and specimens.
2. Using the compound microscope and a prepared slide of *Spirogyra*, describe and sketch *Spirogyra* to the right.
3. Using the compound microscope and a prepared slide of *Spirogyra* undergoing conjugation, describe and sketch *Spirogyra* conjugation to the right.
4. Using the compound microscope prepare a wet mount of *Spirogyra* from the provided culture. Note your observations and sketches below.

Student Activity—*Volvox*

Observation of *Volvox*

Volvox is a common colonial freshwater algae found in ponds, ditches, and puddles (Fig. 21.13). The colony is composed of as many as 50,000 individual flagellated cells that form a hollow sphere. **Protoplasmates**, strands of cytoplasm, connect adjacent cells. Each cell has an eyespot, which helps to locate the light necessary for photosynthesis.

Some colonies are asexual, composed of non-reproducing vegetative cells and gonidia that produce new **daughter colonies**. *Volvox* is capable of sexual reproduction, with male colonies producing and releasing **microgametes** and female colonies producing female sex cells, **macrogametes**.

Materials
- compound microscope
- prepared slides of *Volvox*
- microscope slides and coverslips
- eyedropper
- colored pencils
- culture of *Volvox*

Procedure 21.4
Volvox

1. Procure the equipment and specimens.
2. Using the compound microscope and a prepared slide of *Volvox*, describe and sketch *Volvox* below.

3. Using the compound microscope, prepare a wet mount of *Volvox* from the provided culture. Note your observations and sketches below.

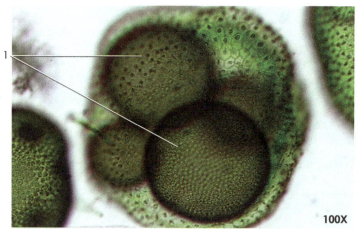

Figure 21.13 *Volvox* sp. is a single organism with several large daughter colonies.
1. Daughter colonies

Student Activity—Diatoms

Observation of Diatoms

Diatoms are unicellular algae that possess a unique silica cast composed of silicon dioxide and an organic matrix. The majority of diatoms live in aquatic and marine environments. Millions of diatoms may exist in a liter of water. Diatoms are known for their beauty (Fig. 21.14).

Materials
- compound microscope
- prepared slides of diatoms
- microscope slides and coverslips
- eyedropper
- colored pencils
- culture of diatoms

Biddulphia sp., a colony forming colonies. These cells are beginning cell division.

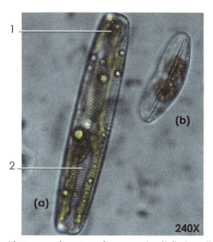

Live specimens of pennate (bilaterally symmetrical) diatoms. (a) *Navicula* sp., and (b) *Cymbella* sp.

1. Chloroplast   2. Striae

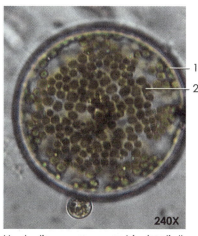

Hyalodiscus sp., a centric (radially symmetrical) diatom, from a freshwater spring in Nevada.

1. Silica cell wall   2. Chloroplasts

Epithemia sp., a distinctive pennate freshwater diatom.

Stephanodiscus sp., a centric diatom.

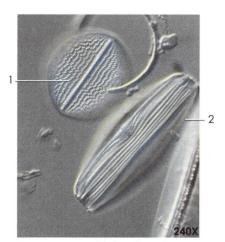

Two common freshwater diatoms.
1. *Cocconeis* sp.   2. *Amphora* sp.

Figure 21.14 Several examples of diatoms.

Procedure 21.5
Diatoms

1. Procure the equipment and specimens.
2. Using the compound microscope and a prepared slide of diatoms, prepare a wet mount of diatoms from the culture provided. Note your observations and sketches below.

Student Activity—*Polysiphonia*

Observation of *Polysiphonia*

Polysiphonia, or "mermaid's hair," is a common filamentous red alga found throughout the world. It is capable of growing in deep waters. The body of *Polysiphonia* appears feathery, with many "pipes," or branch-like structures, and long hair-like **trichoblasts** growing from the branches (Fig. 21.15). Sexual reproduction in *Polysiphonia* is complex and has many stages. *Polysiphonia* is commonly harvested for food, usually eaten raw or dried and also used in soups, salads, and sushi.

Materials
- compound microscope
- prepared slide of *Polysiphonia*
- colored pencils

Procedure 21.6
Polysiphonia

1. Procure the equipment and specimens.

2. Using the compound microscope and a prepared slide of *Polysiphomia*, describe and sketch *Polysiphonia* below.

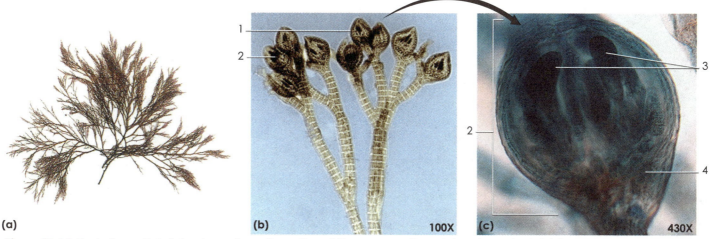

Figure 21.15 Red algae, *Polysiphonia* sp., has alternation of three generations: (a) mature plant, (b) female gametophyte with attached carposporophyte generation, and (c) a closeup of the cystocarp plant.

1. Pericarp   2. Cystocarp   3. Carpospores   4. Carposporophyte

---

 Student Activity—*Euglena*

Observation of *Euglena*

Members of the genus *Euglena* are unicellular algae found primarily in freshwater environments. They are capable of photosynthesis and can capture their own food—which presents taxonomists with a paradox. Chloroplasts serve as the site of photosynthesis. They are characterized by having a **pellicle** and lacking a rigid cell wall, possessing two **flagella** (a short flagella and a long flagella, used for locomotion), having a **paramylon body** for carbohydrate storage, possessing a gullet through which food can be ingested, and having a red **stigma**, or eyespot. The eyespot aids in light detection.

Materials
- compound microscope
- prepared slide of *Euglena*
- microscope slides and coverslips
- eyedropper
- colored pencils
- culture of *Euglena*

 Procedure 21.7
Euglena

1. Procure the equipment and specimens.
2. Using the compound microscope and a prepared slide of *Euglena*, describe and sketch *Euglena* below.

3. Using the compound microscope, prepare a wet mount of *Euglenax* from the culture provided. Place your observations and sketches below and on the next page.

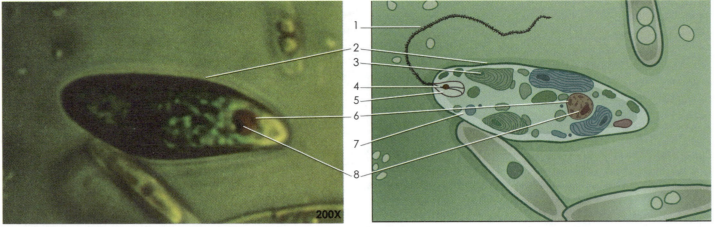

Figure 21.16 Species of *Euglena* from a brackish lake in New Mexico.
1. Flagellum   3. Chloroplast   5. Reservoir   7. Contractile vacuole
2. Pellicle   4. Photoreceptor   6. Nucleus   8. Nucleolus

Q. Describe the movement of *Volvox* and the *Euglena*.

_____

_____

_____

_____

Q. What are the advantages of living in a colony?

_____

_____

_____

_____

Check Your Understanding

Q. Compare and contrast the basic biology of the organisms observed in this activity.

_____

_____

_____

Q. Why do some scientists regard *Euglena* as a mixotroph?

_____

_____

_____

_____

Student Activity—Slime Mold

Growing *Physarum polycephalum*

Students will grow the plasmodium slime mold *Physarum polycephalum* on a Petri dish with agar (growth media) and oatmeal flakes. Over a period of a week, they will be able to observe the active **plasmodial** stage of the slime mold, including protoplasmic streaming.

When nutrients (growth media/oatmeal flakes) are no longer available, the slime mold will enter into the **sclerotia** stage (Fig. 21.17). This is the inactive or dormant stage, and the mold can remain in this state for years. If nutrients and proper environmental conditions recur, fruiting bodies form within 12 hours and **sporulation** will occur.

Materials

- sterile Petri dish containing nutrient agar
- sterile oatmeal flakes
- culture of *Physarum polycephalum*
- sterile forceps
- sterile scalpel
- marker for labeling Petri dish
- dissecting microscope
- colored pencils
- wax pencil

Procedure 21.8
Physarum polycephalum

1. Procure the materials from your instructor.
2. Using a sterile scalpel, carefully cut a tiny cube of the *Physarum* culture from the stock dish and place it upside down on your sterile Petri dish containing agar.
3. Using the sterile forceps, place two oatmeal flakes one inch away from the *Physarum* culture.
4. Place the lid back onto the dish and, with a wax pencil, label your name/class section on a corner of the lid.
5. Each day, add two or three more oatmeal flakes, following the same procedure as in step 3.
6. Observe and record the growth and streaming of the plasmodium.

_____

_____

_____

7. Describe the growth of the slime mold over a period of a week.

_____

_____

_____

8. What color is the plasmodium? Describe the morphology of the slime mold.

_____

_____

_____

Slime molds are so close to being like plant and animal it's like they can't make up their minds. And they're thinking now that maybe it's this who's been running the earth all this time.

—**From the movie *Spinal Tap* (1984)**

Did You Say "Dog Vomit?"

Utilize online resources to research the plasmodium slime mold *Fuligo septic*, or "Dog vomit slime mold," and answer the following questions:

- In what order is *Fuligo* found?

_____

- Where is this slime mold usually found?

_____

- Describe the life cycle of *Fuligo septica*.

_____

_____

- Why did it get the uncomplimentary name "dog vomit slime mold?"

_____

_____

9. Using a dissecting scope, describe the macroanatomy of your specimen.

_____

_____

_____

10. Record your observations in the space next to each specimen.

_____

_____

_____

10. Clean up and dispose of your materials according to the instructor.

Figure 21.17 _Physarum_ sp., a slime mold.

 Student Activity—*Saprolegnia*

Observing *Saprolegnia*

Saprolegnia is a common fungus-like protist that grows on dying and dead fishes, insects, and other organisms. It also will grow on bits of dog food and other morsels placed in an aquarium (Fig. 21.18).

Materials
- specimen with *Saprolegnia*
- dissecting scope
- Petri dish
- probe

 ## Procedure 21.9
Saprolegnia

1. The laboratory instructor will provide a specimen that is growing *Saprolegnia*.
2. Place the specimen in a Petri dish and observe it under the dissecting scope.
3. Draw your specimen below.

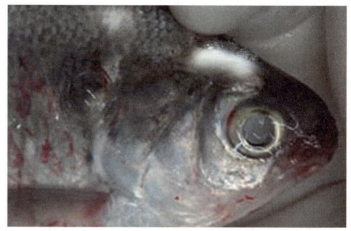

Figure 21.18 *Saprolegnia*, a water mold.

4. Clean your laboratory area, discard the specimens, and wash your hands thoroughly.

 ## Student Activity—The Animal-like Protists

Part 1: The Amoeboids

The amoeba was first described by the German naturalist August Johann Rösel von Rosenhof in 1757. Early naturalists called the amoeba *Proteus animalcule* after the Greek god Proteus, who could change his shape. *Amoeba proteus* is a relatively common animal-like protist that inhabits aquatic environments. Amoebae are the ultimate shape-shifters of the protist world, because they move by false feet, or **pseudopodia.** The pseudopods in the amoeba are lobe shaped and tipped by a **hyaline cap.**

In addition to locomotion, the pseudopodia are important in aiding the amoeba surround its food through **phagocytosis.** The amoeba is surrounded by a plasma membrane that possesses two distinct parts. The **ectoplasm** is a thin, clear, non-granular region of cytoplasm directly beneath the plasma membrane and the endoplasm is a granular region of cytoplasm that makes up the majority of the amoeba. **Food vacuoles** (phagosomes) are also apparent in the amoeba. They serve to hold food particles and fuse with a **lysosome** that aids in digesting

the food. Waste is usually eliminated by excocytosis. The **nucleus** or in some species several nuclei are easily seen in amoebae. Many times the smaller **nucleolus** is also apparent within the nucleus. Occasionally, a **contractile vacuole** can be seen within an amoeba. It is important in osmoregulation (Fig. 21.19).

This activity asks students to observe and record the basic anatomy of several animal-like protists from prepared slides and live-mounts. The students are to record their observations in the spaces provided below each specimen, and answer the questions in each section.

Materials
- compound microscope
- prepared slide of *Amoeba* sp.
- microscope slides and coverslips
- eyedropper
- colored pencils
- culture of *Amoeba* sp.

Procedure 21.10
The Amoeboids

1. Procure a slide of the provided species of amoeba and record which species is used in the space below.

_____

_____

2. Observe the specimen using low and high power. Locate the pseudopods, plasma membrane, hyaline cap, ectoplasm, endoplasm, nucleus, nucleolus, food vacuoles, and contractile vacuole.

3. Describe, sketch, and label your prepared specimen below.

_____

_____

_____

4. Make a wet mount from the culture containing amoebae. Identify, describe, and sketch your specimen below.

_____

_____

_____

Figure 21.19 (a) *Amoeba proteus,* and (b) an illustration.

1. Cell membrane
2. Ectoplasm
3. Endoplasm
4. Nucleus
5. Food vacuole
6. Pseudopodia

430X

(a)

(b)

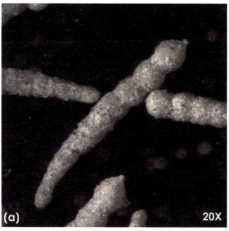 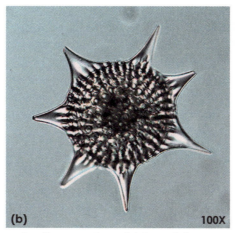 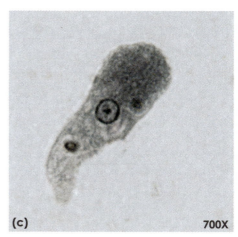

Figure 21.20 Examples of some common amoeba-like organisms, (a) foraminiferan, *Arenaceous uniserial*, (b) radiolarian, and (c) *Entamoeba histolytica*.

Part 2: Other Amoeba-like Organisms of Interest

While referring to the descriptions of foraminiferans, radiolarians, and *Entamoeba histolytica* that appeared earlier in this chapter, observe each, and describe the natural history of these organisms in Procedure 21.12 below (Fig. 21.20).

Materials

- compound microscope
- prepared slide of foraminiferans
- prepared slide of radiolarians
- prepared slide of *Entamoeba hystolitica*
- colored pencils

 ## Procedure 21.11
Other Amoeba-like Organisms

1. Procure the following prepared slides: foraminiferans, radiolarians, and *Entamoeba histolytica*.
2. Observe the prepared slide of foraminiferans on both low and high power. Place your observations and sketches below.

3. Observe the prepared slide of radiolarians on both low and high power. Place your observations and sketches below.

4. Observe the prepared slide of *Entamoeba hystolytica* on both low and high power. Place your observations and sketches below.

5. Describe the natural history of the organisms you observed.

_____

_____

_____

_____

_____

_____

_____

_____

_____

_____

Straight from Science Fiction!

Some science fiction movies feature a critter that eats the brain of its host. In the biological world, we don't have to look far to find a "brain-eating amoeba." *Naegleria fowleri* is an amoeba that inhabits the warm freshwaters of many southern states. It is responsible for the rare but fatal brain disorder Amebic Meningoencephalitis (PAM). It infects humans by entering the body through the nose while engaged in activities such as swimming or diving.

Naegleria travels up the nose to the brain and spinal cord, where it destroys nervous tissue. Early symptoms of PAM begin 1 to 14 days after *Naegleria* enters the brain and may include headache, fever, nausea, vomiting, and a stiff neck. As the disease progresses, symptoms include confusion, loss of balance, seizures, and hallucinations. After the start of symptoms, the disease progresses rapidly and usually causes death within 3 to 7 days.

• What is the treatment for PAM?

_____

_____

• What can be done to reduce *Naegleria* infections?

_____

_____

 ## Student Activity—The Flagellates

The Trypanosomes are protists that are responsible for a number of diseases in humans and other vertebrates. In humans, they cause African sleeping sickness and Chaga's disease. Trypanosomes are endoparasites that are transmitted by arthropod vectors and live in the blood plasma of their host **hemoflagellates**. Trypanosomes are characterized anatomically as having an elongated shape with a flagellum supported by microtubules originating in the posterior region. A **kinetosome** is a self-duplicating structure found at the base of the flagellum.

Trypanosomes possess a **kinetoplast** near the kinetosome. It is derived from a large mitochondrion and forms the basis for the potentially new kingdom Kinetoplastea. In addition, an **undulating membrane** is prominent in Trypanosomes (Fig. 21.22).

In this activity, students will observe a representative Trypanosome from a prepared blood smear and record their observations.

Materials
• compound microscope
• prepared slide of *Trypanosoma* sp.
• colored pencils

Procedure 21.12
The Flagellates

1. Procure a slide of the provided species of *Trypanosome* sp. and record which species is used in the space to the right. What disease is this specimen associated with?

_____

_____

_____

2. Observe the specimen using low and high power. Locate the flagellum, undulating membrane, and nucleus.

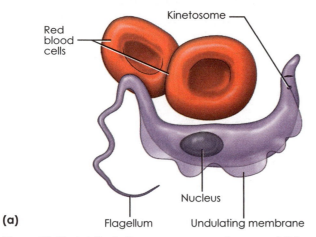

(a)

3. Draw and label your specimen below.

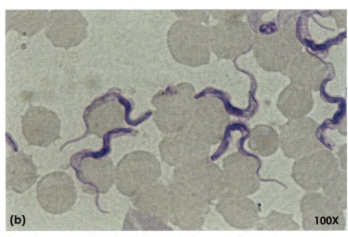

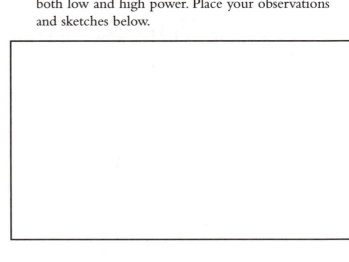

(b)   100X

Figure 21.21 (a) Basic Trypanosome anatomy and (b) a micrograph of Trypanosomes in the blood plasma of a host.

Part 1: *Gonyaulax cantenella* and *Giardia lamblia*
Gonyaulax catenella, Giardia lamblia (intestinalis), and *Trichonympha* spp. are interesting flagellates (Fig. 21.22). (See the description of these organisms in the introduction to Phylum Zoomastigophora, page 294.)

Materials
- compound microscope
- prepared slide of *Gonyaulax catenella*
- prepared slide of *Giardia lamblia (intestinalis)*
- colored pencils

Procedure 21.13
Other Flagellates

1. Procure prepared slides of *Gonyaulax catenella*, and *Giardia lamblia (intestinalis)*.

2. Observe the prepared slide of *Gonyaulax catenella* on both low and high power. Place your observations and sketches below.

_____

_____

3. Observe the prepared slide of *Giardia lamblia (intestinalis)* on both low and high power. Place your observations and sketches below.

Q. Write a brief paragraph describing the natural history of each organism that you observed.

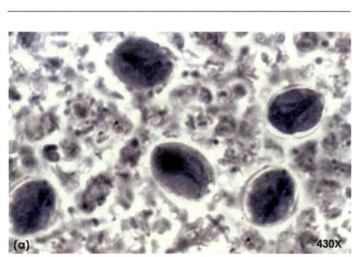
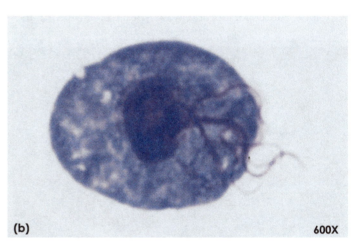

Figure 21.22 Two other types of flagellates are (a) *Giardia lamblia intestinalis* and (b) *Trichomonas vaginalis uniserial*.

Part 2: *Trichonympha*

Trichonympha spp. is essential to termites. This flagellate, living in a symbiotic relationship with the termite, helps them digest cellulose. If termites are available, trichonymphs can be easily observed in their gut (Fig. 21.23).

Materials
- compound microscope
- microscope slides and coverslips
- termites
- colored pencils

Procedure 21.14
Trichonympha

1. Procure the equipment and a termite from the instructor.
2. Place the termite on the center of a slide, and quickly place a coverslip over the termite. Press down on the coverslip, squashing the termite.
3. Search through the remains of the termite. In the former gut region, you should see these unusual protists.

4. Sketch *Trichonympha* spp. and place your observations below.

_____

_____

_____

5. Clean your laboratory area, properly dispose of the slide and coverslip, and wash your hands thoroughly.

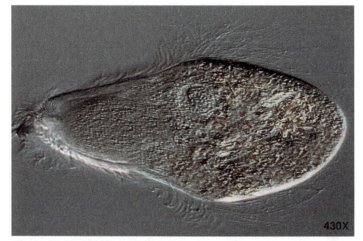

Figure 21.23 The flagellate *Trichonympha*.

Check Your Understanding

Q. Describe three parasitic flagellates and their vector.

_____

_____

_____

_____

_____

_____

Q. Why do some biologists regard Euglena as a mixotroph?

_____

_____

_____

_____

Q. What is unique about *Giardia*? How is it changing taxonomy?

_____

_____

_____

_____

Student Activity—The Ciliates

In this activity students will examine a common ciliate protozoan known as *Paramecium caudatum* (Fig. 21.24). These organisms are common in freshwater environments such as ditches and ponds. Paramecia are more abundant in environments containing aquatic plants and decaying organic matter. The paramecium is slipper-shaped, transparent, and colorless. Paramecia appear to be highly active under the microscope, reaching speeds of 1 to 3 mm a second. These organisms have a pointed posterior end and a rather blunt anterior end. The **oral groove** is distinct, running backward to the ventral side. Paramecia can be observed using their flagella to channel food into the oral groove. The living **pellicle** is distinct and is covered with **cilia**.

In paramecia, the cilia near the oral groove create currents that bring water, and potentially food, into the oral groove. **Trichocysts** are used to gather food and for defense in paramecia. Contractile vacuoles in the paramecium act as an **osmoregulatory** apparatus, or bilge pump. Living in a freshwater (hypotonic) environment, the organism has to remove excess water as it diffuses through the cell membrane.

Materials

- prepared microscope slides of paramecia and paramecia undergoing conjugation
- compound microscope
- eyedropper
- culture of *Paramecium* spp.
- blank slides and coverslips
- pipettes
- vinegar
- India ink
- Protoslo™
- probe or chemicals
- yeast solution stained with methylene blue
- colored pencils
- prepared slide of *Paramecium* conjugation

Procedure 21.15
The Ciliates

1. Procure the equipment and material listed above.
2. Examine the prepared slide of *Paramecium caudatum* under both low and high power. Be able to identify the following structures: pellicle, cilia, trichocyst, oral groove, macronucleus, micronucleus, and contractile vacuole.
3. Draw and label your specimen in the space provided below.

4 Prepare a wet mount of paramecia and describe their movement, the action of the contractile vacuole, and the organism in general. Paramecia are "speed demons," so Protoslo™ or methyl-cellulose can be used to slow them down significantly.

_____

_____

_____

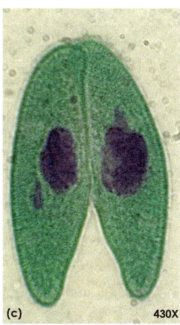

Figure 21.24 Examples of ciliates include a) *Paramecium caudatum*, (b) illustration of *Paramecium caudatum*, and (c) *Paramecium* in conjugation.

1. Cilia	3. Food vacuole	5. Macronucleus	7. Tricocyst
2. Oral groove	4. Contractile vacuole	6. Pellicle	

5. Place a tiny drop of India ink near the oral groove of a paramecium slowed by Protoslo™. Describe how the paramecium "tests' the water by drawing the ink into the oral groove.

_____

_____

_____

6. Your instructor will provide you with a solution of yeast that has been stained with methylene blue.
7. Capture and place a single paramecium on a slide, and place a coverslip over the liquid. Next, squeeze a small drop of the yeast solution along the edge of the coverslip. Describe how the paramecium feeds.

_____

_____

_____

8. Paramecia are sensitive to chemicals and objects in their environment. If a negative stimulus appears, the paramecium always turns to the left (the side away from the oral groove). Attempt to create a negative stimulus, using a probe or chemicals, and describe the avoidance behavior in the paramecium, in the space below.

_____

_____

_____

9. Capture and place a single paramecium on a slide, and place a coverslip over the liquid. Then squeeze a small drop of vinegar along the edge of the coverslip. Below, describe how the trichocysts react.

_____

_____

_____

10. Count the number of times the contractile vacuole contracts in a slowed paramecium in its normal environment. After several trials, capture the paramecium with a pipette, and place it on a slide with distilled water. Count the number of times the contractile

vacuole contracts in the distilled water. Capture another paramecium and repeat the process placing the paramecium in a saltwater solution. Record your observations below, and explain why there is a difference in the number of contractions.

_____

_____

_____

11. Paramecia are capable of conjugation, a crude means of exchange of genetic material between two individuals. Observe a prepared slide labeled _Paramecium_ conjugation, and draw your specimen below.

[drawing box]

12. Clean up your lab area, return the supplies, dispose of the material as instructed, and wash your hands.

Other Ciliates of Interest
Stentor spp. and _Vorticella_ spp. are two relatively common ciliates (Fig. 21.25). They can be found attached to a substrate in the freshwater environment. Many times _Vorticella_ can be found living on the fins of fish.

Check Your Understanding

Q. Why do some people describe paramecia as miniature battleships?

_____

_____

_____

Q. Describe the action of the contractile vacuole in a hypotonic solution.

_____

_____

_____

Q. Briefly describe the process of conjugation in paramecia.

_____

_____

_____

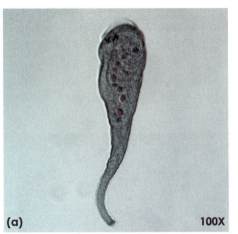 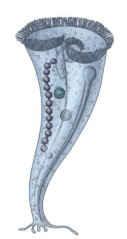 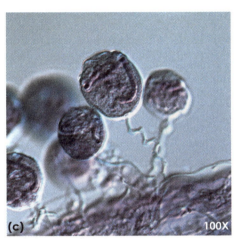

Figure 21.25 Common ciliates are (a) *Stentor* sp., (b) an illustration of *Stentor*, and (c) *Vorticella* sp.

 ## Student Activity—The Apicomplexans

In this activity, a member of the Phylum Apicomplexa will be examined. The representative organism will be *Plasmodium* spp., the endoparasite responsible for malaria. Four species of Plasmodium infect humans. The vector for malaria is the female *Anopholes* mosquito. The life cycle of *Plasmodium* is complex and consists of an asexual and a sexual stage (Fig. 21.26).

General Life Cycle of *Plasmodium* sp. Asexual Stage

1. A female *Anopheles* spp. mosquito feeds on the blood of a host. During feeding, the mosquito injects the infective stage of malaria known as the **sporozoite** into the bloodstream of the host.

2. The sporozoites that survive the immune system enter the cells of the liver. While in the liver, the sporozoites develop into large **multinucleate schizonts**. Through the asexual process of **schizogony** (multiple fission), 5,000–10,000 new cells are formed. When the schizonts rupture, they release thousands of cells called **merozoites** into the bloodstream. Two species,

Plasmodium vivax and *Plasmodium ovale*, produce **hypnozoites**, which remain in the liver and can cause relapses.

3. When a merozoite enters a blood cell, it becomes an amoeboid **trophozoite**. The trophozoite, in turn, feeds on the hemoglobin in the blood, producing the waste product **hemozoin**. Destruction of the erythrocytes by the trophozoites causes anemia in the host. Each trophozoite grows and undergoes **schizogony** producing 6–36 new merozoites depending upon the species of Plasmodium. The blood cell bursts, releasing merozoites to infect other red blood cells. The release of the merozoites and the metabolic wastes triggers the chills and fever associated with malaria. Usually, the first release is spread over time, producing mild fever, but later releases are synchronized with the daily rhythm of the victim's body, causing an onset of the symptoms every 48–72 hours, depending upon the species.

4. After a few cycles of schizogony in the red blood cells, trophozoites develop inside some blood cells

into male **microgametocytes** and female **macroga-metocytes** rather than merozoites.

5. The microgametocytes and macrogametocytes are released into the blood and could be picked up by another female *Anopheles* spp. mosquito. Changes in pH cause the gametocytes to develop into mature gametes. The gametes, in turn, develop flagella through exoflagellation and proceed to the gut of the mosquito.

Sexual Stage

6. In the gut of the mosquito, the male gamete fertilizes the female gamete, forming a **zygote**. The zygote burrows into the gut lining of the mosquito, forming an **oocyst**. Within each oocyst, thousands of sporozoites develop through **sporogony** from the zygote. (Keep in mind that sporozoites are not spores!)

7. The sporozoites migrate to the salivary glands to await another victim.

Materials
- prepared slides of *Plasmodium* spp.
- compound microscope
- immersion oil (optional)

Procedure 21.16
The Apicomplexans

1. Procure the necessary equipment.
2. Select several slides labeled *Plasmodium* spp. Examine the slides under high power or oil immersion.

Did You Know?

Female mosquitoes are the blood feeders, while the male mosquitoes eat pollen. The male mosquito can be distinguished by a more elaborate antenna system.

One can distinguish between the *Anopholes* spp. mosquito and other species of mosquitoes by the way they land. The *Anopholes* mosquito holds its abdomen at a 45-degree angle to the host, whereas other species keep their abdomen parallel to the host. Knowing this, don't give the "guys" a break, because it takes two to make more mosquitoes. Also, don't give any mosquito a break by the way it lands.

Other species carry other diseases including the West Nile virus.

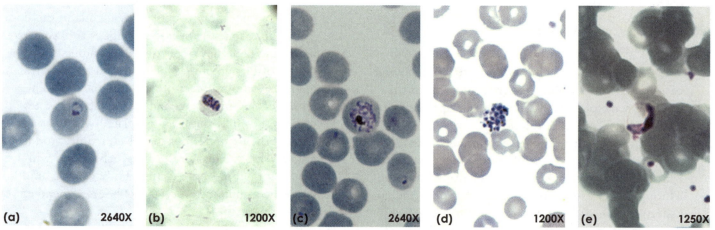

Figure 21.26 These images represent several stages in the life cycle of *Plasmodium* spp. (a) Ring stage (a young trophozoite) of *Plasmodium falciparum*, (b) a band trophozoite of *Plasmodium malariae*, (c) developing schizont of *Plasmodium falciparum*, (d) mature schizont of *Plasmodium vivax*, and (e) a gametocyte of *Plasmodium falciparum* in an erythrocyte.

(a) 2640X (b) 1200X (c) 2640X (d) 1200X (e) 1250X

3. Draw the various stages of the life cycle of *Plasmodium* in the space provided.

4. Properly clean the slide from immersion oil, and return the supplies.

NOTES

_____

_____

_____

_____

_____

_____

_____

_____

Beyond the Lab—Protist Stars on YouTube

Many high-quality, as well as poor, videos of protists can be found on YouTube.

Procedure 21.17
YouTube

1. Enter the following terms into the "search field" on the website youtube.com and observe the resulting video clips. Record your observations about the clips in the space provided.

Stentor Feeding on *Euglena*

_____

_____

_____

The Coolest *Stentor* Ever

_____

_____

_____

Vorticella Feeding

_____

_____

_____

Paramecium Dividing

_____

_____

_____

Paramecium Conjugating

_____

_____

_____

Paramecium Feeding

_____

_____

_____

Volvox Dances

_____

_____

_____

2. Describe five other YouTube sites that feature protists, such as diatoms, etc.

_____

_____

_____

_____

_____

_____

Name: _____ Date: _____ Section: _____

Review Questions

1. Why is Protista considered a "hodgepodge" kingdom? What is the future of kingdom Protista?

_____

_____

_____

_____

2. How have the protists been traditionally classified?

_____

_____

_____

3. Draw and label the anatomy of *Euglena* spp.

4. Why are geologists interested in foraminiferans and radiolarians?

_____

_____

_____

5. Why is malaria making a comeback?

_____

_____

_____

6. Why are phylum Pyrrophyta and Dinoflagellata a paradox?

_____

_____

_____

7. Name five uses of the algae.

_____

_____

_____

_____

_____

8. Describe three parasitic protists.

_____

_____

_____

9. What is _Phytophthora infestans,_ and how did it change U.S. history?

_____

_____

_____

10. Draw and label a typical paramecium.

Chapter 22
The Green Machine: Part I
Understanding the Nonvascular Plants

Student Outcome Objectives

At the completion of this exercise, the student will be able to:

1. Discuss the importance of plants.
2. Describe the origin of plants.
3. Compare and contrast nonvascular and vascular plants.
4. Describe the process of alternation of generations.
5. State the characteristics of a nonvascular plant.
6. Describe the characteristics of and give examples of *Hepatophytes*.
7. Compare and contrast thallus and leafy liverworts.
8. Discuss the biology of and label the anatomy of *Marchantia*.
9. Describe the characteristics of and give examples of Antocerophytes.
10. Discuss the biology of and label the anatomy of *Antoceros*.
11. Describe the characteristics of and give examples of Bryophytes.
12. Discuss the biology of and label the anatomy of *Polytrichum*, *Mnium* and *Sphagnum*.
13. Compare and contrast water absorption in *Sphagnum* and various other items.

Overview

A wise man once declared that "we need the plants much more than they need us." This statement is not only accurate but also meaningful. Have you ever stopped to thank a plant? Think about it! Plants provide oxygen, food, shelter, shade, erosion control, and commercial products for human uses such as timber, medicine, and even the paper that you are looking at right now. All of this—plus many species of plants are aesthetically pleasing. As Ralph Waldo Emerson said, "The Earth laughs with flowers." Remember—the next time that you are sitting under a majestic tree—say thank you!

Plants are a diverse group of eukaryotic, multicellular, photosynthetic autotrophs that inhabit a myriad of environments from lush tropical rainforests to scorching deserts. Biologists have identified approximately 300,000 species of plants and estimate that perhaps 400,000 species may exist. Plants vary in size from the smallest flowering plant *Wolffia angusta* (a duckweed), measuring less than 1 mm in diameter, to the giant *Sequoia sempervirens*, which measures nearly 120 m tall. The oldest plant in the world is thought to be the King Holly (*Omatia tasmanica*), a shrub that lives in Tasmania. This remarkable plant is estimated to be more than 43,000 years old (Fig. 22.1)!

Paleobotanists theorize that plants evolved from freshwater green algae known as **charophytes** during the Paleozoic Era, approximately 450 million years ago. To make the transition from the aquatic environment, ancestral plants had to evolve mechanisms that prevent desiccation, anchor the plant body, transport water and nutrients, and ensure propagation of the species.

The absence or presence of specialized conducting tissue known as **vascular tissue** is a common way to distinguish plants. **Nonvascular plants** lack specialized conducting tissues to transport water and nutrients throughout the plant's body. In addition, these plants lack true roots, stems, and leaves. Examples of nonvascular plants are liverworts, stoneworts, and true mosses. Presently, about 25,000 species of nonvascular plants have been identified. The vast majority of living plants are considered to be vascular plants because they possess an extensive conducting system composed of specialized tissues.

Figure 22.1 Bristlecone pines (*Pinus longaeva*) may live for more than 4,500 years.

Vascular plants can be divided into the seedless vascular plants and the seed plants. The **seedless vascular plants** include the club mosses and the ferns. The **seed plants** comprise the largest group of vascular plants and include plants such as, ginkgos, cycads, conifers, and zinnias.

The life cycle of plants is characterized by an **alternation of generations** (see Figure 22.2). In this process, two distinct generations give rise to each other. The haploid (n) **gametophyte** generation is characterized by the production of male and female gametes through cell division. The male and female gametes fuse during fertilization, forming a diploid **sporophyte.** The sporophyte generation is diploid (2n). They produce haploid spores that undergo cell division to form a gametophyte. In nonvascular plants the gametophyte generation is dominant,

In a very real sense, we're all made out of sunlight. Sunlight radiating heat, visible light, and ultraviolet light is the source of almost all life on Earth. Everything you see alive around you is there because a plant somewhere was able to capture sunlight and store it. All animals live from these plants, whether directly (as with herbivores) or indirectly (as with carnivores). This is true of mammals, insects, birds, amphibians, reptiles, and bacteria … everything living. Every life-form on the surface of this planet is here because a plant was able to gather sunlight and store it, and something else was able to eat that plant and take that sunlight energy in to power its body.

—**Thom Hartmann (1951–present)**

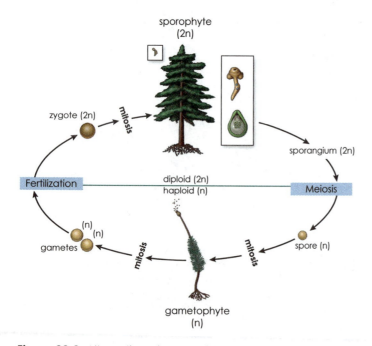

Figure 22.2 Alternation of generations.

but in seedless vascular plants and seed plants, the sporophyte generation dominates. A fern, a pine tree, and a tulip are all examples of sporophytes.

THE NONVASCULAR PLANTS

In the history of plants on Earth, the nonvascular plants played an important role in establishing the transition from water to land and the gametophyte lifestyle. The nonvascular plants are generally small and **herbaceous** (non-woody). Although the majority of these plants are found in moist environments, some species can survive in arid environments. Three distinct phyla of nonvascular plants have been established:

1. Phylum **Hepatophyta**, the liverworts,
2. Phylum **Anthocerophyta**, the hornworts, and
3. Phylum **Bryophyta**, the true mosses.

In recent years, the term **bryophyte** has been used to describe all of the nonvascular plants, causing some confusion.

PHYLUM HEPATOPHYTA

Phylum Hepatophyta includes approximately 8,000 species of small nonvascular plants commonly known as liverworts. The term **wort** is derived from an Old English term referring to herb. Medieval physicians thought that because the flattened, liver-shaped body of these plants resembled the liver, they could be used in treating liver ailments. The vast majority of liverworts are terrestrial, living in moist environments, but several aquatic species have been described.

Two distinct types of liverworts have been described:

1. the **thallus liverworts,** which possess leaf-like lobed bodies (thalli) and usually are found living along creek banks or on moist soil.
2. the majority of liverworts, known as **leafy liverworts**, which at first glance resemble mosses; the leafy liverworts are more commonly found in tropical environments.

Marchantia is an example of a thallus liverwort. The body, or **thallus**, is approximately 30 cells thick in the center, and the edges may be as thin as 10 cells. *Marchantia* thalli branch out from the center in a flattened pattern. The top surface of the small plant is characterized by a diamond-shaped pattern of segments that correspond to chambers below the surface. Each segment of *Marchantia* features a small pore leading to the interior of the plant. On the lower surface, hair-like structures called **rhizoids** can be found. The rhizoids extend into the soil and anchor the thallus (Fig. 22.3).

Figure 22.3 The liverwort (a) *Marchantia* sp., (b) gemmae cupules, and (c) transverse section through a gemma cupule.

1. Antheridial receptacles
2. Gametophyte thallus
3. Rhizoids
4. Gemmae cupules
5. Gemmae cupule
6. Gemmae

Marchantia is capable of both asexual and sexual re-production. Asexually, small cup-shaped structures called **gemma cups** (or splash cups) appear on the upper surface of the thallus and contain tiny pieces of tissue, **gemmae**. In turn, the gemmae detach (perhaps through the action of raindrops), establish a new location, and grow into a new plant. Sexually, like all plants, the life cycle of *Marchantia* has a sporophyte and a gametophyte generation. In the nonvascular plants, the gametophyte generation is larger (Fig. 22.4 and 22.5).

During the gametophyte generation, male and female sex cells are produced on separate umbrella-shaped structures that rise above the plant, called **gametophores**. The top portion of the male gametophores, the **antheridophore**, is composed of **antheridia** containing numerous sperm cells. The female **archegoniophore** possesses **archegonia** containing a single egg. Raindrops or splashing water carry sperm to the awaiting egg (Fig. 22.6).

After fertilization, the zygote eventually develops into a sporophyte. The maturing sporophyte is anchored to the archegoniophore by a knob-like **foot**. A thin **seta**, or stalk, connects the foot to the main body, called the **capsule**. **Sporocytes** within the capsule undergo meiotic division and produce haploid **spores**, or **meiospores**. Initially, the immature sporophyte is protected by a structure called the **calyptra**. The spores possess structures called **elaters**, which aid their dispersal into new environments. The spores germinate, forming a new gametophyte (Fig. 22.7–22.9).

Approximately 80% of liverworts are considered leafy. These small plants are more common in tropical regions of the world. Many people confuse them with mosses, but their overlapping leaf-like structures contain prominent oil bodies and many lobes and folds. The sex cells are produced in **archegonia** or **antheridia** in the axial region of the leaf-like structures (Fig. 22.10). After a spore germinates, a thin photosynthetic filament called a **protonema** develops and soon gives rise to a mature gametophyte.

Figure 22.4 (a) *Marchantia* sp. with prominent male antheridial receptacles, and (b) showing archegonial receptacles.

1. Antheridial receptacles
2. Gametophyte thallus
3. Archegonial receptacles

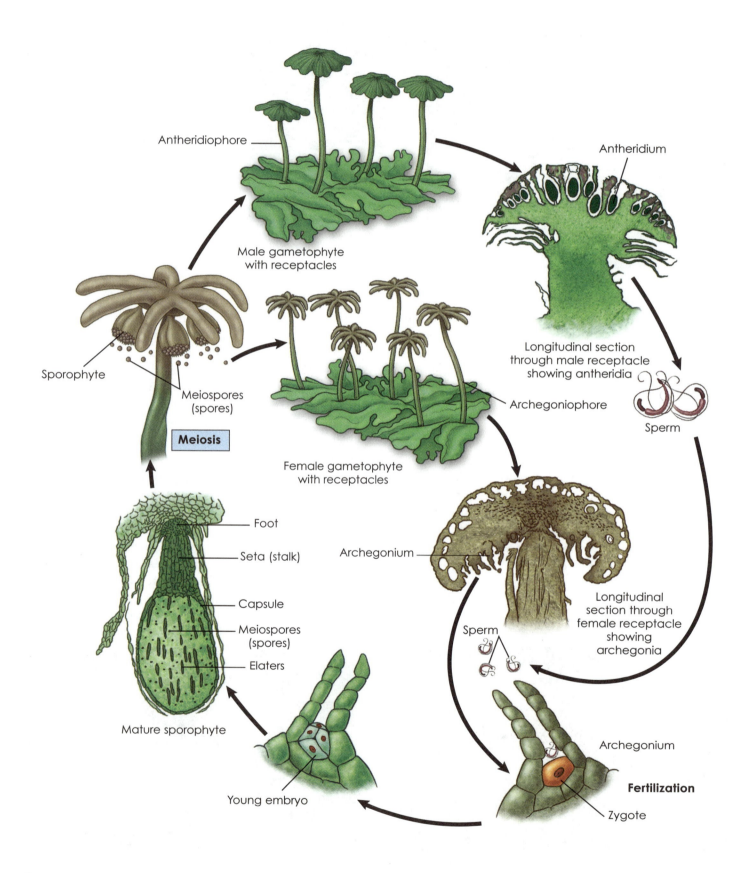

Figure 22.5 Life cycle of the thalloid liverwort, *Marchantia* sp.

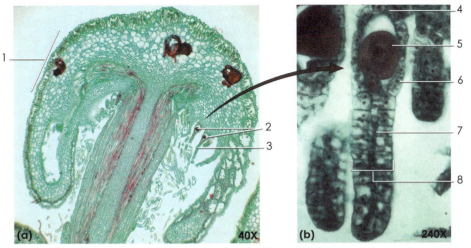

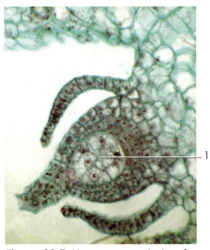

Figure 22.6 (a) Archegonial receptacle of a liverwort, *Marchantia* sp., in a longitudinal section. (b) Archegonium with egg.

1. Archegonial receptacle
2. Eggs
3. Neck of archegonium
4. Base of archegonium
5. Egg
6. Venter of archegonium
7. Neck canal
8. Neck of archegonium

Figure 22.7 Young sporophyte of *Marchantia* sp.
1. Young embryo

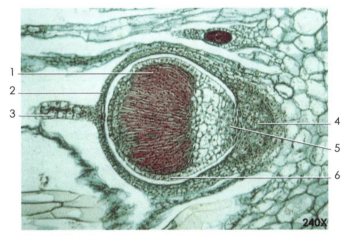

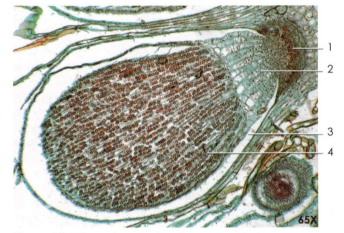

Figure 22.8 Young sporophyte of *Marchantia* sp., in longitudinal section.
1. Sporogenous tissue (2*n*)
2. Enlarged archegonium (calyptra)
3. Neck of archegonium
4. Foot
5. Seta (stalk)
6. Capsule

Figure 22.9 Immature and mature sporophytes.
1. Foot
2. Seta (stalk)
3. Sporangium (capsule)
4. Spores (*n*) and elaters (2*n*)

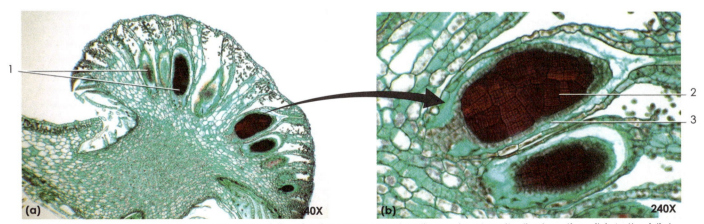

Figure 22.10 (a) Male receptacle with antheridia of a liverwort, *Marchantia* sp., in a longitudinal section. (b) Antheridial head showing a developing antheridium.
1. Antheridia
2. Spermatogenous tissue
3. Antheridium

Student Activity—Observations of *Marchantia*

First, students will observe and describe the macroanatomy and microanatomy of *Marchantia*.

Materials
- dissecting microscope
- compound microscope
- living specimen of *Marchantia*
- prepared slides of *Marchantia*
- colored pencils

 Procedure 22.1
Marchantia

1. Procure the necessary materials and supplies.
2. Using a dissecting microscope, draw and label the lower and upper surface of the thallus. Pay particular attention to the diamond-shaped sections, pores, and rhizoids. If your specimen has gemma cups, atheridiophores, or archegoniophores, draw and label these structures.
3. Place your labeled sketches and observations below.

4. Using a microscope on both low and high power, observe the archegonial receptacle, young sporophyte, immature sporophyte, and atheridial structures of *Marchantia*.
5. Draw and label your observations below. Compare your observations to Figure 22.6.

If you think in terms of a year, plant a seed; if in terms of ten years, plant trees; if in terms of 100 years, teach the people.

—Confucius (551 BC–479 BC)

Student Activity—Observations of Leafy Liverworts

In this activity, students will make specimens of leafy liverworts using living or prepared specimens.

Materials
- dissecting microscope
- compound microscope
- living specimen of a leafy liverwort (*Lepidozia, Frullamia, Plagiochilia,* or *Blepharostoma*)
- prepared slides of a leafy liverwort (*Lepidozia, Frullamia, Plagiochilia,* or *Blepharostoma*)
- colored pencils

 Procedure 22.2
Leafy Liverworts

1. Procure the needed equipment and supplies.
2. Using a dissecting microscope, observe a specimen of *Lepidozia, Frullamia, Plagiochilia,* or *Blepharostoma*. Record your observations below.

 Check Your Understanding

Q. What is the function of a rhizoid?

Q. Describe asexual and sexual reproduction in *Marchantia*.

Q. Compare and contrast thallus and leafy liverworts.

Calopegia sp.

Conocephalum sp.

Bazzania sp.

Porella sp.

Riccia sp.

Scapania sp.

Figure 22.11 Examples of liverworts.

PHYLUM ANTHOCEROPHYTA

The hornworts are a relatively obscure group of non-vascular plants belonging to Phylum Anthocerophyta. Approximately 100 species have been described (Fig. 22.12). Hornworts usually live on moist ground in the shade. They are small, only 2 centimeters in diameter and up to 5 cm high, and easy to overlook. The thallus body has pores and cavities filled with mucilage. In many hornworts, nitrogen-fixing **cyanobacteria** live in the rich mucus. As in liverworts, rhizoids help anchor the minute plant.

Asexually, many species reproduce through fragmentation. In hornworts, the sporophytes are distinct, looking rather like a green tentacle arising from a thin thallus gametophyte. The archegonia and anthergonia are located in rows beneath the upper surface of the thallus gametophyte. The most studied hornwort is *Anthoceros* sp. (Fig. 22.13–22.15).

Figure 22.12 *Anthoceros*, a common hornwort.

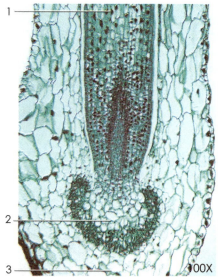

Figure 22.13 Longitudinal section of a portion of the sporophyte of the hornwort, *Anthoceros* sp.
1. Meristematic region of sporophyte
2. Foot
3. Gametophyte

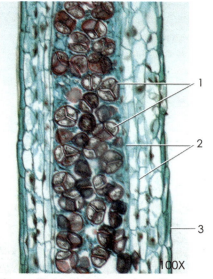

Figure 22.14 Longitudinal section of the sporangium of a sporophyte from the hornwort, *Anthoceros* sp.
1. Spores
2. Elater-like structures (pseudoelaters)
3. Capsule

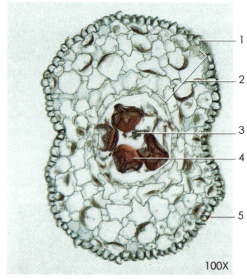

Figure 22.15 Transverse section through the capsule of a sporophyte of the hornwort, *Anthoceros* sp.
1. Epidermis
2. Photosynthetic tissue
3. Columella
4. Tetrad of spores
5. Pore (stomate)

Student Activity—Observations of *Anthoceros*

Materials
- dissecting microscope
- compound microscope
- living specimen of *Anthoceros*
- prepared slides of *Anthoceros*
- colored pencils

Procedure 22.3
Anthoceros

1. Procure the necessary equipment and specimens.
2. Using a dissecting microscope, draw and label the lower and upper surfaces of the hornwort. Pay particular attention to the surface, rhizoids, and reproductive structures.
3. Place your labeled sketches and observations below.

4. Using a microscope on both low and high power, observe a prepared longitudinal section of the sporophyte and a transverse section through the capsule of *Anthoceros*. Draw and label your observations below, comparing them to Figures 22.13–22.15.

If my IQ had been two points lower, I would have been a plant somewhere.

—Lee Trevino (1939–present)

PHYLUM BRYOPHYTA

True mosses have been placed in Phylum Bryophyta. Presently, nearly 15,000 species of mosses have been identified, but many small plants that are called mosses do not fall into this group. For instance, Spanish moss is actually a member of the pineapple family, reindeer moss is actually a lichen, Irish moss is a red algae, and club mosses are lower vascular plants.

The gametophyte stage of mosses consists of small, spirally arranged **leaf-like** structures surrounding a central axis. The **blades** of the leaf-like structure are one cell layer thick, lack vascular tissue and stoma, and surround a thickened **midrib**. Rhizoids anchor mosses to their substrate.

Mosses are capable of asexual reproduction through **fragmentation**, but they undergo an alternation of generation with gametocyte and sporocyte stages. The "leafy" gametocytes are either male-bearing antheridia or female-bearing archegonia. Flagellated sperm cells exit the antheridia and travel, with the aid of water, to the archegonia, where a single egg is fertilized. The zygote undergoes cell division, forming spores that are housed in the sporophyte (Fig. 22.16). The sporophyte appears as a tall stalk topped by a **sporangium**, or **capsule** (Fig. 22.17). The **calyptra** protects the capsule. A **foot** connects the **seta**, or stalk, of the sporophyte to the leafy gametophyte. In some mosses, a single capsule may contain 50 million spores.

The tip of the capsule consists of a lid-like structure called the **operculum**. Teeth-like structures comprise the **peristome** and lock the operculum to the capsule (Fig. 22.18). During dry conditions, they unlock and allow spores to be carried by the wind. With a hand lens, the peristome appears ornate and usually is orange or red (Fig. 22.19). Immature spores land on a substrate and under suitable conditions develop into a filamentous **protonema**. The protonema eventually develops into the "leafy" gametocyte, and the cycle begins again (Fig. 22.20).

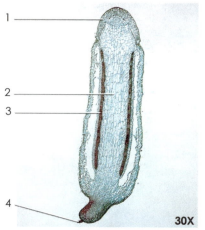

Figure 22.17 Capsule of the moss *Mnium* sp.
1. Operculum
2. Columella
3. Spores
4. Seta

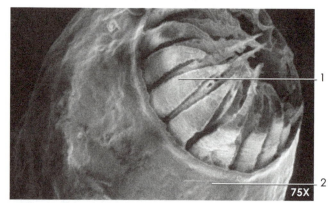

Figure 22.18 Scanning electron micrograph of the peristome of the moss *Mnium* sp. The operculum is absent in the specimen.
1. Peristome
2. Capsule

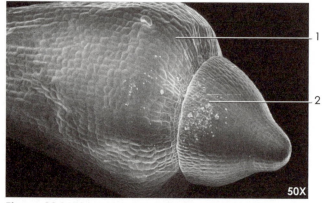

Figure 22.16 Scanning electron micrograph of the sporophyte capsule of the moss *Mnium* sp.
1. Capsule
2. Operculum

Figure 22.19 Scanning electron micrograph of the peristome of the moss *Mnium* sp.
1. Outer teeth of peristome
2. Capsule
3. Inner teeth of peristome

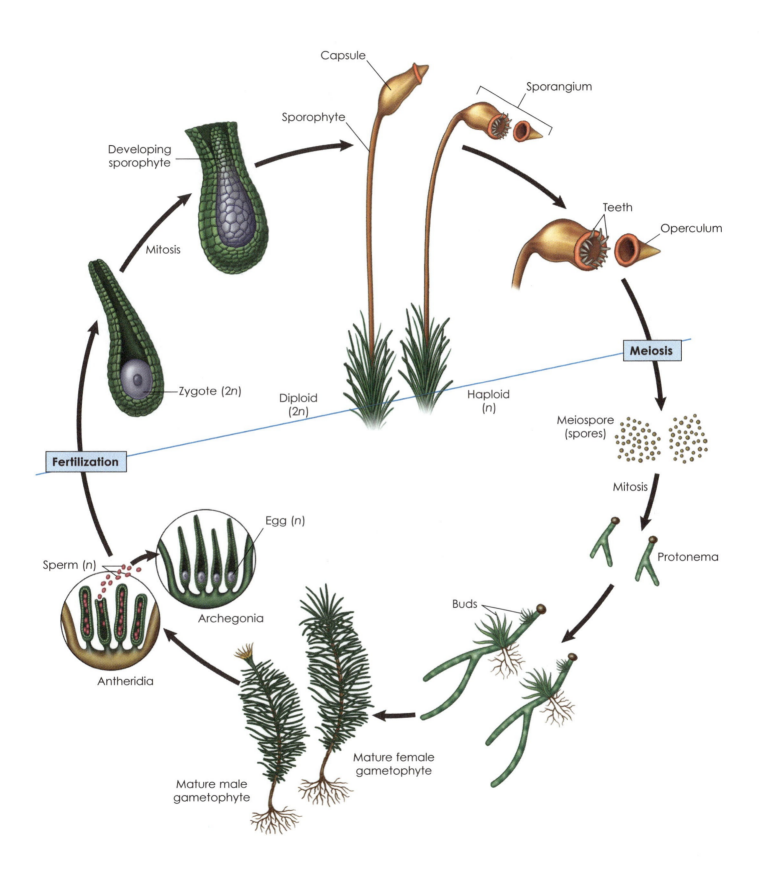

Figure 22.20 Life cycle of *Polytrichum,* a typical moss.

The majority of mosses live in moist environments of the temperate zone, although several species have been found living in the Arctic, Antarctic, and even deserts. Frequently, mosses can be seen growing on the trunks of trees and on the sides of buildings. After a fire or volcano, some mosses serve as pioneer species and help to form soil.

Commercially, one of the most valuable species of moss is *Sphagnum*, or peat moss (Fig. 22.21). In many regions, deposits of peat are mined for their use as packing material and fuel. Peat can absorb great amounts of water and is used in the gardening industry to enhance the water-holding capacity of soil and potted plants.

The use of peat moss, or *Sphagnum*, has been mentioned in folklore as far back as the 11th century. Native Americans used peat moss in diapers, and some ancient societies used it in menstrual pads. Medical texts in the 1800s mentioned that *Sphagnum* could be used as bandages or to pack abscesses. In 1904–1905 in the Russo-Japanese War, *Sphagnum* was commonly used as bandage material. In France, during World War I in a time of intense fighting, hospital staffs resorted to packing wounds and making bandages with *Sphagnum*. To their surprise, the peat moss was not only more absorbent than cotton, but the infection rate among the wounded also was reduced significantly (Fig. 22.22). The remarkable peat moss was used in World War II in a similar fashion.

Figure 22.21 The moss *Sphagnum* growing in the Pacific Northwest.

Figure 22.22 The moss *Sphagnum* was used as bandages during World War I.

 Beyond the Lab—Moss and Ancient Man

1. A surprising relationship between the 5,300 ice mummy found in the Alps and moss existed. How do archaeologists think that Iceman (Otzi) used moss?

_____

_____

_____

_____

_____

_____

_____

2. Who is Tollund Man, and how did he form?

_____

_____

_____

_____

_____

_____

_____

Student Activity—Observations of *Polytrichum*

Polytrichum sp., haircap moss, is a common moss that can be found living in bogs. It is a relatively large moss, perhaps reaching 10 cm in length in certain environments. It is distinguished by having tall sporangium with a golden calyptra sitting atop the seta and protecting the capsule (Fig. 22.23).

Polytrichum tea has been taken to dissolve gallbladder and kidney stones in some cultures. Also, *Polytrichum* has been used as a shampoo and to make brooms, mats, rugs, and baskets.

Materials
- dissecting microscope
- blank slides and coverslips
- medicine dropper
- living specimen of *Polytrichum*
- colored pencils

Procedure 22.4
Polytrichum

1. Procure the necessary supplies and equipment.
2. Using a dissecting microscope, draw and label the specimen of *Polytrichum*. Pay particular attention to the surface, rhizoids, and reproductive structures. Place your labeled sketches and observations to the right.

3. Prepare a wet mount of the "leaflet" of your specimen. In the space below, draw the specimen under low and high power. Record your observations of the margins of the blade and describe the thickness of the blade. Can you find any stoma? Describe the midrib.

Figure 22.23 (a) *Polytrichum* sp., a common moss often used in course work. (b) Gametophyte plants with sporophyte plant attached. (c) Sporophyte plant and capsule.

1. Calyptera
2. Capsule of sporophyte (covered by calyptera)
3. Stalk (seta)
4. Gametophyte
5. Operculum
6. Capsule of sporophyte (with calyptera absent)
7. Stalk (seta)

Student Activity—*Sphagnum* and the Water Absorption Tests

Many gardeners use *Sphagnum* in their yards and pots because of its ability to absorb water. In this activity, students will compare the absorption rate of *Sphagnum* (peat moss) to equivalent masses of a true sponge, an artificial sponge, and a paper towel. When measuring the amount of water absorbed, 1 milliliter of water weighs one gram.

Materials
- 5 grams of *Sphagnum*
- 5 grams of a true sponge
- 5 grams of an artificial sponge
- 5 grams of a paper towel
- scale
- Four 250 ml beakers
- 100 ml graduated cylinder
- wax pencil
- stopwatch

Procedure 22.5
Sphagnum and Water

1. Weigh out 5-gram portions of: *Sphagnum*, a true sponge, an artificial sponge, and a paper towel.
2. Add 250 ml of water to four separate beakers. Label the beakers *Sphagnum*, true sponge, artificial sponge, and paper towel.
3. Add to the appropriate beakers the 5-gram portions of *Sphagnum*, a true sponge, an artificial sponge, and a paper towel.
4. Allow the material to sit in the beaker for 2 minutes, then remove the materials from each beaker.
5. Pour the water left over from each beaker into the 100 ml graduated cylinder, and record your measurement for each material.

Give me the man who will surrender the whole world for a moss or a caterpillar, and impracticable visions for a simple human delight.

—Bruce Frederick Cummings (1889–1919)

6. Record your observations in the chart below.

Sphagnum	ml
True sponge	ml
Artificial sponge	ml
Paper towel	ml

7. In the space provided below, discuss your findings.

_____

_____

_____

_____

_____

_____

_____

_____

_____

_____

Name: _____ Date: _____ Section: _____

Review Questions

1. What are 10 reasons that you should "thank a plant?"

_____

_____

_____

_____

_____

_____

_____

_____

_____

_____

_____

_____

2. To conquer the land, what sort of adaptations did early land plants have to develop?

_____

_____

3. Compare and contrast nonvascular plants with vascular plants.

_____

_____

_____

4. Describe the process of alternation of generation.

_____

_____

_____

5. Name several characteristics and give several examples of Phylum Hepatophyta.

_____

_____

_____

6. Name several characteristics and give several examples of Phylum Anthocerophtya.

_____

_____

_____

7. Name several characteristics and give several examples of Phylum Bryophyta.

_____

_____

_____

8. Name several uses of mosses.

_____

_____

_____

9. Sketch and label *Polytrichum*.

Chapter 23
The Green Machine: Part II
Understanding the Seedless Vascular Plants

Student Outcome Objectives

At the completion of this exercise, the student will be able to:

1. Briefly describe the flora and fauna of the Silurian, Devonian, and Carboniferous Periods.
2. Describe how the vascular plants adapted to life on the land.
3. Discuss the function of xylem and phloem.
4. Discuss the functions of roots, stems, and leaves.
5. Explain reproduction in seedless vascular plants.
6. Name the characteristics of and give examples of Phylum Lycophyta.
7. Describe the natural history of and basic biology of *Lycopodium*.
8. Discuss medical and commercial uses of *Lycopodium*.
9. Describe the natural history of and basic biology of *Selaginella*.
10. Discuss the biology of *Lepidodendron*.
11. Identify macroscopic and microscopic anatomical features of *Lycopodium*.
12. Identify macroscopic and microscopic anatomical features of *Selaginella*.
13. Identify and discuss the biology of fossil *Lepidodendron*.
14. Describe the characteristics of and give examples of Phylum Psilotophyta.
15. Describe the natural history of and basic biology of *Psilotum*.
16. Identify macroscopic and microscopic anatomical features of *Psilotum*.
17. Describe the characteristics of and give examples of Phylum Sphenophyta.
18. Describe the natural history of and basic biology of *Equisetum*.
19. Discuss medical and commercial uses of *Equisetum*.
20. Identify macroscopic and microscopic anatomical features of *Equisetum*.
21. Identify and discuss the biology of fossil *Calamites*.
22. Describe the characteristics of and give examples of Phylum Pterophyta.
23. Describe the life cycle of a typical fern.
24. Describe the natural history of and basic biology of a typical fern.
25. Discuss medical and commercial uses of ferns.

Overview

Imagine the world of the Silurian Period 443 to 416 million years ago. The ancient oceans were teeming with life. Coral reefs were beginning to form, eurypterids (sea scorpions) cruised the seas, trilobites were abundant, and the ancestors of spiders and centipedes invaded the land. During this time, jawless fishes were common in the oceans, and fishes with jaws made their first appearance. Early in this period, the bryophytes lived near the waters edge, helping turn rock into soil. During the Silurian, plants took a giant leap toward conquering the land.

Paleontologists have discovered that members of the extinct plant Phylum Rhyniophyta such as *Cooksonia* had their humble origins during this time (Fig. 23.1). *Cooksonia* was a small plant only a few centimeters tall and possessed simple vascular tissues for conducting water and nutrients. It appeared as a branching stem with no roots or leaves. Spore-producing bodies known as **sporangia** appeared at the tips of the branches. The sporangia, like those of the bryophytes, produced only one type of spore, and thus were known as **homosporous**. *Cooksonia* and other primitive plants such as *Rhynia* established the foundations of the vascular plants.

Vascular plants possess specialized tissues for conducting water and nutrients throughout the plant. In

Figure 23.1 *Cooksonia* is one of the oldest vascular land plants.

modern vascular plants, **xylem** conducts water and dissolved minerals and **phloem** conducts nutrients such as sucrose, hormones, and other molecules. The cells responsible for conducting water are often strengthened by the polymer **lignin**, which allows the plant to grow tall. In vascular plants, the **sporophyte** generation is dominant and possesses the vascular tissues. Vascular plants have a waxy **cuticle** for protection against desiccation, as well as small openings called **stomata** on photosynthetic structures to allow for gas exchange. Vascular plants also possess true roots, stems, and leaves.

1. **Roots** are plant organs that absorb water and nutrients from the soil and, in addition, anchor a plant.
2. **Stems** are vascular plant organs that support leaves and reproductive structures.
3. **Leaves** are the primary photosynthetic organs of plants.

The two major types of vascular plants are the seedless vascular plants and the seed plants. The **seedless vascular plants**, as the name indicates, do not produce seeds but, instead, reproduce through the production of spores as their ancestors did. These plants feature a dominant sporophyte generation and a reduced gametophyte generation. The seedless vascular plants dominated the landscape during the Devonian and Carboniferous periods. Unlike the seedless vascular plants of today, which are relatively small, some species of seedless vascular plants such as

Lepidodendron (scale tree) reached heights exceeding 30 meters. The vast coal deposits of the Carboniferous Period are the result of carbonization of seedless vascular plants that resided in giant swamp forests (Fig. 23.2). Today, the seedless vascular plants are represented by the following phyla: Lycophyta (club mosses), Sphenophyta (horsetails), Psilotophyta (whisk ferns), and Pterophyta (ferns).

PHYLUM LYCOPHYTA

Approximately 1,150 species of small plants known as ground pines, club mosses, quillworts, and spike mosses comprise Phylum Lycophyta. Keep in mind that the term "moss" in this instance is a misnomer. This small group of plants represents the vestiges of a much larger phyla that dominated the Carboniferous Period. They are considered to be the most ancient group of seedless vascular plants. Unlike the large tree lycophytes, the smaller lycophytes survived the changes of time and are found mostly in the tropics and moist, temperate regions of Earth. The two best known genera of lycophytes are *Lycopodium* and *Selaginella*.

Members of genus *Lycopodium* (ground pines) are relatively common inhabitants of moist forest floors in temperate regions. Although the majority of ground pines are only 30 cm in length, they are called ground pines because they resemble small pine trees or Christmas trees (Fig. 23.6 and 23.7). In ground pines, an underground stem, the **rhizoid**, branches horizontally across the surface, producing **aerial stems** and **underground roots**. The surface of the aerial stem is covered by small closely and spirally packed scale-like leaves. Ground pines are homosporous. The reproductive sporangia, called **sporophylls**, are located on the surface of leaves. A cone-shaped structure, a **strobilus**, sits at the tip and contains spores (Fig. 23.6–Fig. 23.11).

Figure 23.2 Representative plants of a Carboniferous swamp forest.

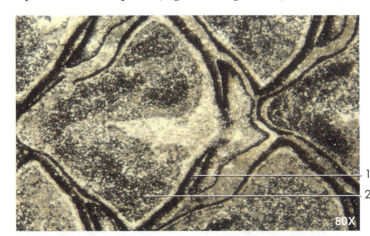

Figure 23.3 A longitudinal section through a fossil strobilus of the lycophyte *Lepidostrobus* sp., from approximately 300 million years ago.

1. Sporangium   2. Sporogenous tissue

Through the years, the body of *Lycopodium*, as well as its spores, have had a number of medical and commercial purposes. Medically, it has been used in homeopathic remedies including as an emetic, as a baby powder, and as a worming agent. At Christmas, ground pines are used for decorations. *Lycopodium* spores have been used in controlled explosions. *Lycopodium* has been used as a stabilizer in ice cream. The spores have also been used as a fingerprint powder and in fireworks.

I never saw a discontented tree. They grip the ground as though they liked it, and though fast rooted they travel about as far as we do. They go wandering forth in all directions with every wind, going and coming like ourselves, traveling with us around the sun two million miles a day, and through space heaven knows how fast and far!

—**John Muir (1838–1914)**

So That's How It Was Done!

Have you ever seen an old movie in which the photographer had a flash bar attached to an old camera? Prior to flash, cameras used slow shutter speeds, and portraits required time and patience. Thus, many times a stuffed bird (Watch the birdie!) was mounted near the lens so that the person being photographed would concentrate and stay still. In addition, a neck brace was used to hold the subject's head still. *Lycopodium* spores constituted the flash powder in early flash bars. When tightly packed and exposed to a spark or electric current, the packed spores would produce a bright inflammable explosion. Most early flashes were set off by a flash igniter. The photographer would wind it up like a wind-up toy, and when a trigger was pulled, a wheel would spin against a flint, generating sparks that ignited a small pile of flash powder (Fig. 23.4).

Figure 23.4 Early flash bars used *Lycopodium* spores.

Figure 23.5 A specimen of a lycopod, *Lycopodium* sp, growing in a greenhouse.

NOTES

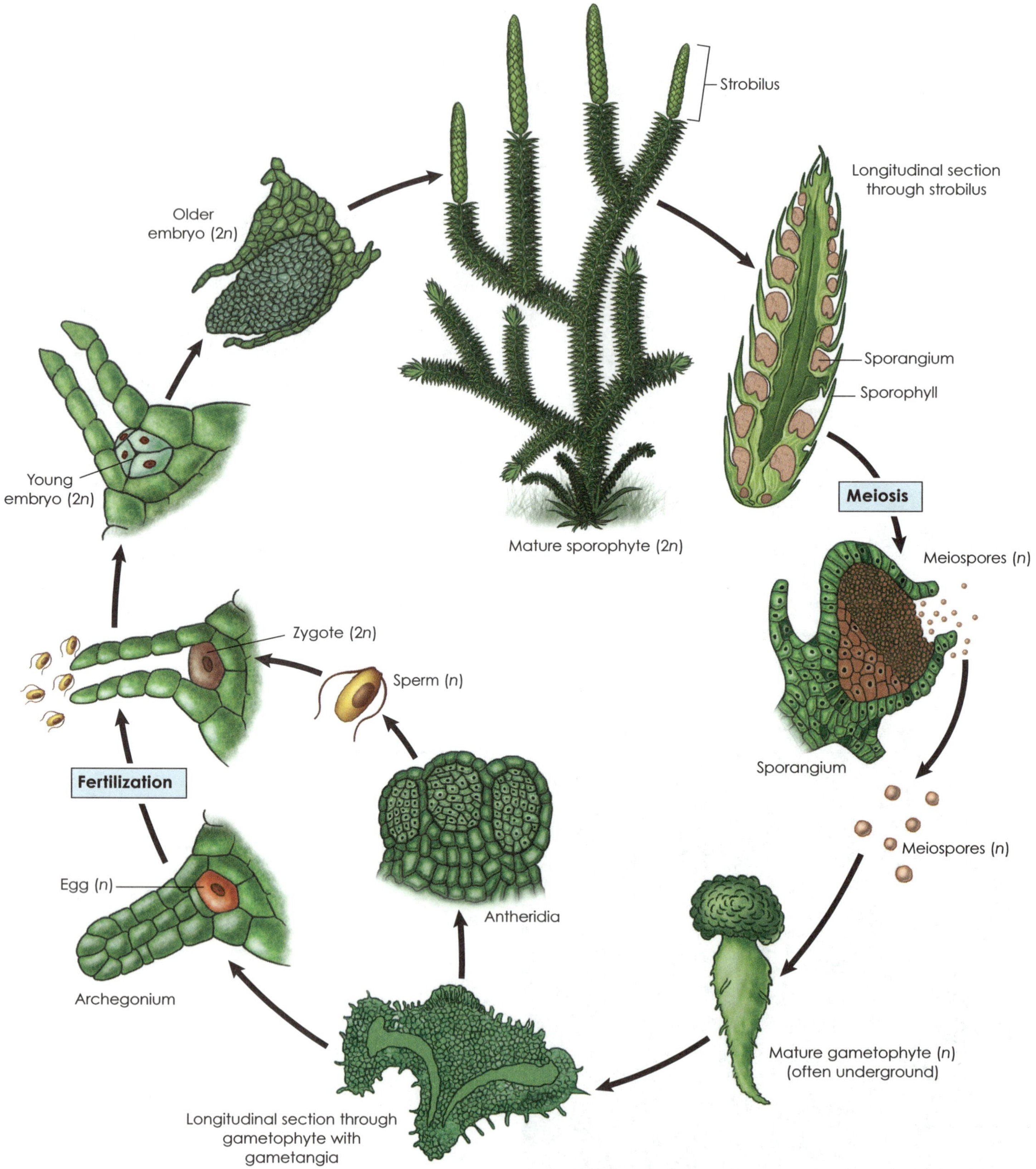

Figure 23.6 The life cycle of the homosporous clubmoss, *Lycopodium* sp.

Figure 23.7 A specimen of a lycopod, *Lycopodium clavatum*, (a) plant and (b) strobilus. *Lycopodium* occurs from the arctic to the tropics (scale in mm).
1. Strobilus   2. Stem

Figure 23.8 An enlargement of a specimen of *Lycopodium* sp., showing branch tip with sporangia on the upper surface of sporophylls (scale in mm).
1. Sporangia
2. Sporophylls (leaves with attached sporangia)

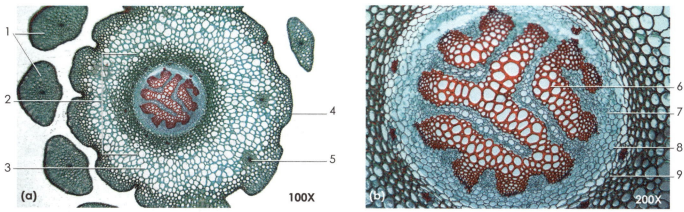

Figure 23.9 (a) A transverse view of an aerial stem of the clubmoss, *Lycopodium* sp. (b) A magnified view of the stele.

1. Leaves (microphylls)	4. Epidermis	5. Leaf trace	7. Phloem	9. Endodermis
2. Stele	3. Cortex	6. Xylem	8. Pericycle	

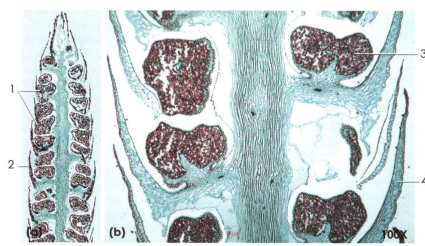
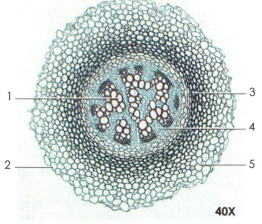

Figure 23.10 (a) A longitudinal section of the strobilus (cone) of the clubmoss *Lycopodium* sp., and (b) a magnified view of the strobilus showing sporangia.
1. Sporangia   3. Sporangium
2. Sporophyll   4. Sporophyll

Figure 23.11 A transverse section of a rhizome of *Lycopodium* sp. The rhizome of Lycopodium is similar to an aerial stem, but it lacks the microphylls.
1. Xylem   3. Endodermis   5. Cortex
2. Epidermis   4. Phloem

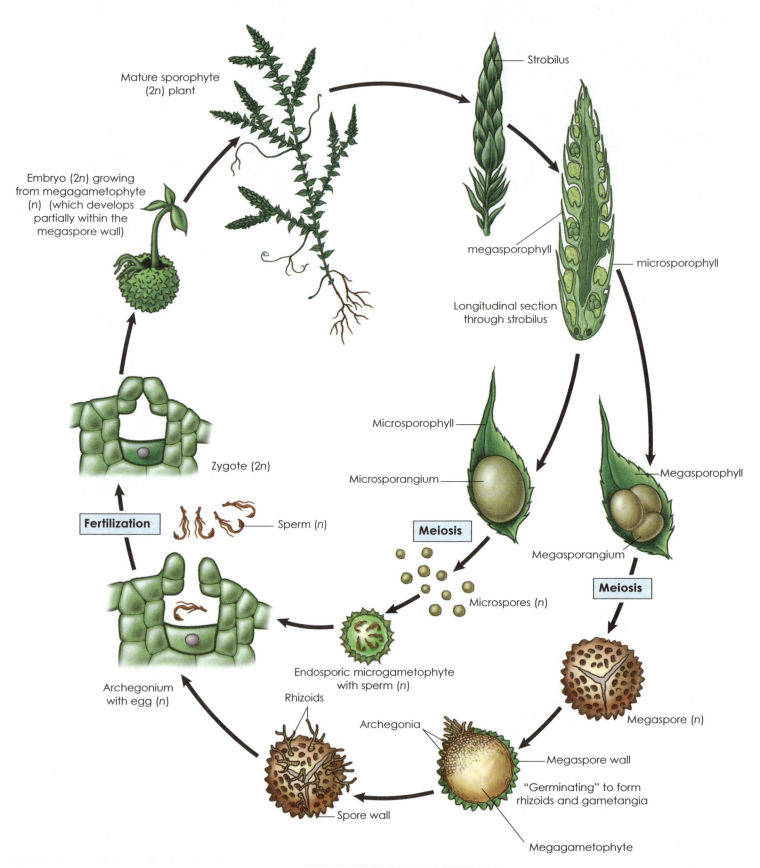

Figure 23.12 The life cycle of *Selaginella* sp., which is heterosporous.

Selaginella is a member of Phylum Lycophyta, commonly called spike moss. Approximately 700 species live in moist regions worldwide. *Selaginella* appears to creep along the ground with simple, scale-like leaves on branching stems from which roots also arise. The leaves of *Selaginella* have a distinct **ligule**, or tongue, on their upper surface and generally appear larger than *Lycopodium*. The plants are heterosporous, producing female megaspores and male microspores. Megasporophylls bear megasporocytes containing megaspores, and microsporophylls bear microsporangia housing microspores (Fig. 23.12 and 23.17).

Figure 23.13 The spikemoss, *Selaginella kraussiana*

Figure 23.14 The spikemoss, *Selaginella kraussiana* showing strobili (cones).
1. Strobili (cones)
2. Sporaphyll with sporangium

Figure 23.15 The spikemoss, *Selaginella pulcherrima*.

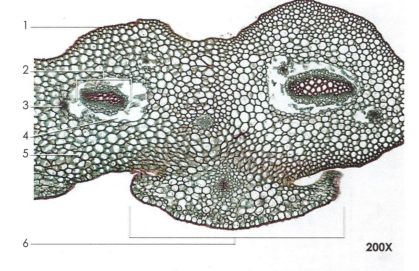

200X

Figure 23.16 A transverse section through stem of *Selaginella* sp. immediately above dichotomous branching.
1. Epidermis
2. Protostele (surrounded by endodermis)
3. Root trace
4. Air cavity
5. Cortex
6. Leaf base

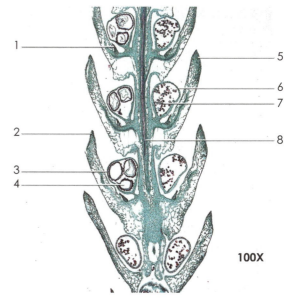

100X

Figure 23.17 A longitudinal section through the strobilus of *Selaginella* sp.
1. Ligule
2. Megasporophyll
3. Megasporangium
4. Megaspore
5. Microsporophyll
6. Microsporangium
7. Microspore
8. Cone axis

Student Activity—Observations of *Lycopodium*

Materials
- dissecting microscope
- compound microscope
- living specimen of *Lycopodium*
- prepared slides of *Lycopodium*
- colored pencils

Procedure 23.1
Lycopodium

Consider photographing this activity.

1. Procure the needed equipment and supplies.
2. Using a dissecting microscope, draw and label a living specimen of *Lycopodium*.
3. Place your labeled sketch and observations below.

4. Using a microscope on both low and high power, observe a transverse section of an aerial stem of *Lycopodium*, a longitudinal section of a strobili of *Lycopodium*, and a transverse section through the rhizome of *Lycopodium*. Draw and label your slides below, and compare your observations to Figure 23.9 through 23.11 on page 345.

Transverse section of an aerial stem of *Lycopodium*

Longitudinal section of a strobili of *Lycopodium*

Transverse section through the rhizome of *Lycopodium*

Student Activity—Observations of *Selaginella*

Materials
- dissecting microscope
- compound microscope
- living specimen of *Selaginella*
- prepared slides of *Selaginella*
- colored pencils

Procedure 23.2
Selaginella

Consider photographing this activity.

1. Procure the needed equipment and supplies.
2. Using a dissecting microscope, draw and label a living specimen of *Selaginella*. Pay particular attention to the ligules. Place your labeled sketch and observations below.

Living specimen of *Selaginella*

3. Using a microscope on both low and high power, observe a transverse section of a stem of *Selaginella* and a longitudinal section of a strobili of *Selaginella*. Draw and label your slides below, and compare your observations to Figure 23.16 and 23.17 on page 347.

Transverse section of a stem of *Selaginella*

Longitudinal section of a strobili of *Selaginella*

Science is the father of knowledge, but opinion breeds ignorance.

—Aristotle (460 BC–377 BC)

PHYLUM PSILOTOPHYTA

Phylum Psilotophyta is a rather obscure phylum of plants known as the whisk ferns. Only two genera survive today—*Psilotum* and *Tmesipteris*. *Psilotum* is commonly found in the southern United States, and *Tmesipteris* is confined to the islands of the South Pacific including New Zealand and Australia. The *Psilotum* sporophyte appears to be a vestige from the Devonian Period. The small plant has no leaves or roots, and a dichotomously branching green stem with small scales that bears bright yellow **synangia** on lateral branches. A synagia is formed from three fused sporangia on short lateral branches. *Psilotum* is homosporous. A horizontal rhizome gives rise to an aerial stem (Fig. 23.18–Fig. 23.28).

Figure 23.18 A *Tmesipteris* sp., growing as an epiphyte on a tree fern in Australia.

Figure 23.19 A whisk fern, *Psilotum nudum,* is a simple vascular plant lacking true leaves and roots.

Figure 23.20 The branches (axes) of *Psilotum nudum* (scale in mm).
1. Aerial axis   2. Rhizome

Figure 23.21 A sporophyte of the whisk fern, *Psilotum nudum.* The axes of the sporophyte support sporangia (synangia), which produce spores (scale in mm).
1. Branch (axis)   2. Sporangia (synangia)

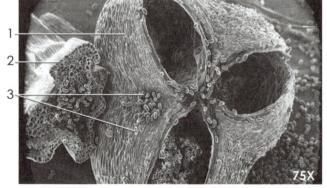

Figure 23.22 A scanning electron micrograph of a ruptured synangium (3 fused sporangia) of *Psilotum* sp., which is spilling spores.
1. Sporangium (often   2. Axis   3. Spores
   called synangia)

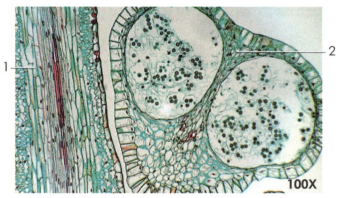

Figure 23.23 A longitudinal section through a stem and sporangium (synangium) of *Psilotum* sp.
1. Axis   2. Sporangia (synangium)

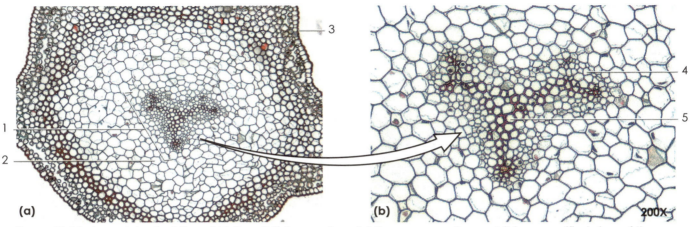

Figure 23.24 An aerial axis of the whisk fern, *Psilotum nudum*. (a) Transverse section and (b) a magnified view of the vascular cylinder (stele).

1. Stele	2. Cortex	3. Epidermis	4. Phloem	5. Xylem

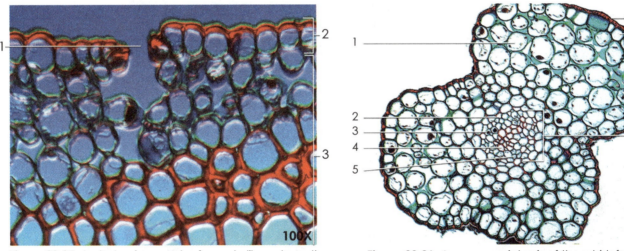

Figure 23.25 A photomicrograph of a scale-like outgrowth from the axis of the whisk fern *Psilotum nudum*.

1. Stoma	2. Epidermis	3. Ground tissue

Figure 23.26 A young aerial axis of the whisk fern, *Tmesipteris* sp.

1. Cortex	4. Xylem	7. Epidermis
2. Endodermis	5. Phloem	8. Protostele
3. Pericycle	6. Cuticle	

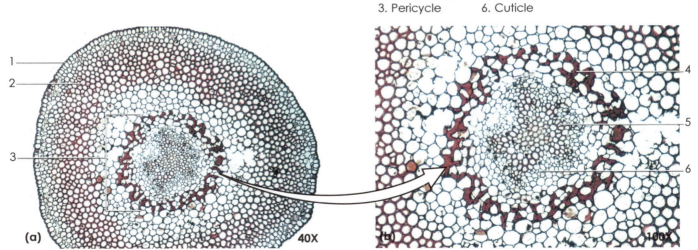

Figure 23.27 An older aerial axis of the whisk fern, *Tmesipteris* sp. The genus *Tmesipteris* is restricted to distribution in Australia, New Zealand, New Caledonia, and other South Pacific islands. (a) Axis arising from the aerial axis and (b) a magnified view of the stele.

1. Epidermis	2. Cortex	3. Stele	4. Endodermis	5. Xylem	6. Phloem

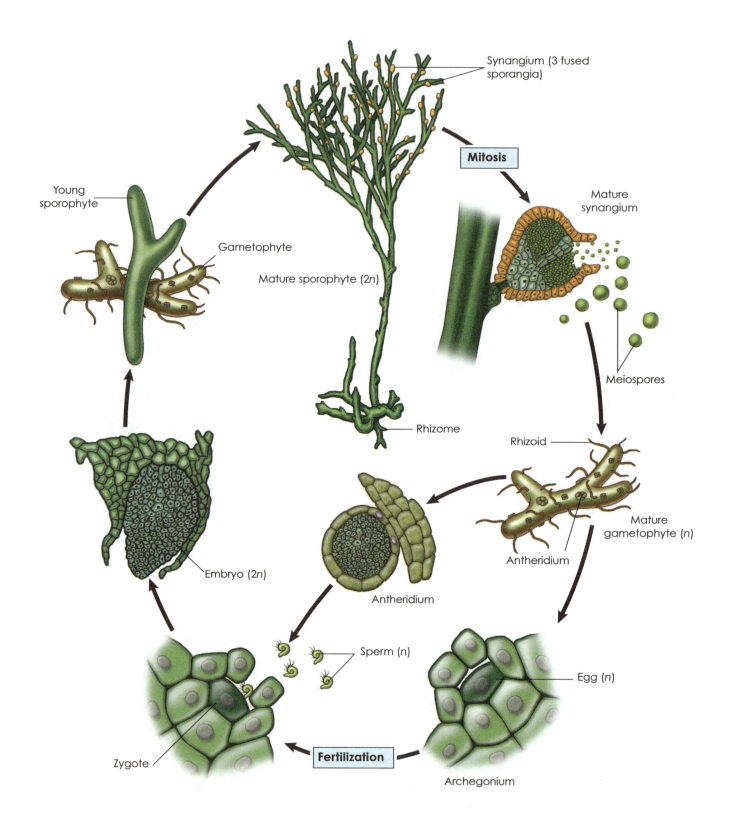

Figure 23.28 The life cycle of the whisk fern, *Psilotum* sp.

Student Activity—Observations of *Psilotum*

Materials
- dissecting microscope or magnifying glass
- compound microscope
- living specimens of *Psilotum*
- prepared slides of *Psilotum*
- colored pencils

 Procedure 23.3
Psilotum

Consider photographing this activity.

1. Procure the necessary equipment and supplies.
2. Draw an overview of *Psilotum*. Using a dissecting microscope or magnifying glass, draw and label your specimen. Place your labeled sketches and observations below.

3. Using a microscope on both low and high power, observe a longitudinal section through a stem and sporangium of *Psilotum*, and a cross-section and transverse section of the stem of *Psilotum*. Draw and label your slides below and on the next page and compare your observations to Figure 23.24 through 23.27 on 351.

PHYLUM SPHENOPHYTA

Members of Phylum Sphenophyta are known as horsetails or scouring rushes. Only one genus, *Equisetum* (Latin, *equus* "horse" + *seta* "bristle"), represented by 25 species, remains today. Members of Phylum Sphenophyta commonly live in wet environments throughout the world. Although modern species of *Equisetum* are herbaceous and grow to about 1.5 m in height, fossil relatives such as *Calamites* grew to

Figure 23.29 (a) Illustration of *Calamites*; (b) fossil *Calamites*.

heights of more than 20 m (Fig. 23.29). Some species of horsetail have whorled branches at each **node**, and other species are unbranched.

Both types have small, scale-like leaves, **microphylls**, arranged in a whorl at the nodes. The tiny leaves are fused, forming a collar that turns brown as the plant ages. The aerial stems of horsetails are deeply ribbed, and stomata occur in the grooves between the ribs. Nodes and internodes are obvious in horsetails. The center of the stem, the **pith**, is hollow. Rhizomes run horizontally across the ground and give rise to an aerial stem and roots.

Horsetails can reproduce asexually through fragmentation. Sexually, sporophyte horsetails produce strobili at the tip of the stem, composed of scale-like **sporangiophores**. Beneath the sporangiophores, **sporangia** produce **spores**. The spores are distinctive, green in color and possess wing-like structures called **elaters**. The spores are carried by wind, and if they land on a suitable substrate, germinate within a week, forming a small gametocyte. The gametocytes can have antheridia, producing sperm cells, or archegonia, producing eggs. The sperm and egg undergo fertilization, producing a zygote. The zygote eventually forms the sporophyte generation.

Through the centuries, horsetails have been used as food and medicine. Today, however, the consumption of horsetail is discouraged. Medically, ancient cultures used horsetail as diuretics, as astringents, as an agent to discourage lice, fleas, and mites, and as a cure for diarrhea. Pioneers used horsetail to scrub pots and sharpen knives (Fig. 23.30–Fig. 23.43).

Figure 23.30 An *Equisetum telmateia* showing lateral branching.

Figure 23.31 A close-up of *Equisetum telmateia* showing lateral branches growing through leaf sheath.

Figure 23.32 The stems of *Equisetum* sp. without lateral branching and showing a prominent leaf sheath at the node.
1. Stem   2. Leaf sheath

(a)   (b)   (c)   (d)

Figure 23.33 The horsetail, *Equisetum* sp. Numerous species of Equisetophyta were abundant throughout tropical regions during the Paleozoic Era, some 300 million years ago. Currently, Equisetophyta are represented by this single genus. The meadow horsetail, *Equisetum* sp., showing (a) an immature strobilus, (b) mature strobilus, shedding spores, (c) an open strobilus, and (d) a sporangiophore with its spores released.
1. Sporangiophores
2. Separated sporangiophores revealing sporangia
3. Sporangiophores after spores are shed
4. Open sporangia with spores shed

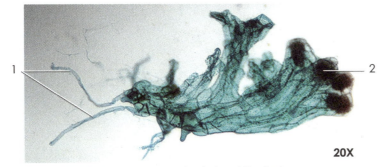

20X

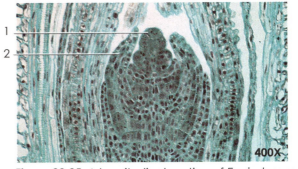

400X

Figure 23.34 A young gametophyte of *Equisetum* sp.
1. Rhizoids   2. Antheridium

Figure 23.35 A longitudinal section of *Equisetum* sp. shoot apex.
1. Apical cell   2. Leaf primordium

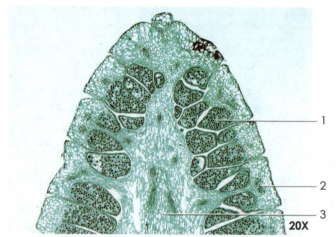

Figure 23.36 A longitudinal section of *Equisetum* sp. strobilus.
1. Sporangium   2. Sporangiophore   3. Strobilus axis

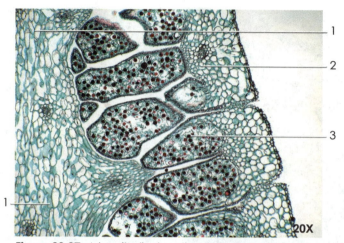

Figure 23.37 A longitudinal section through *Equisetum* sp. strobilus.
1. Axis of the strobilus   2. Sporangiophore   3. Sporangium

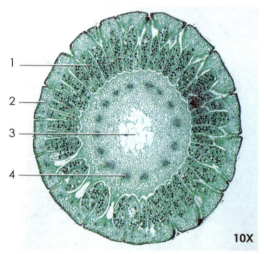

Figure 23.38 A transverse section of the strobilus of *Equisetum* sp.
1. Sporangium   3. Strobilus axis
2. Sporangiophore   4. Vascular bundle

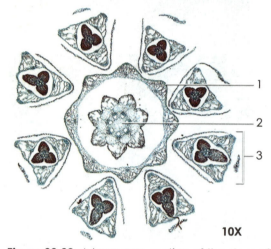

Figure 23.39 A transverse section of the stem of *Equisetum* sp. just above a node.
1. Leaf sheath   2. Main stem   3. Branch

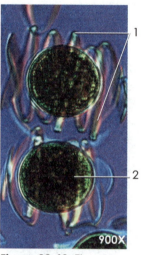

Figure 23.40 The spores of *Equisetum* sp.
1. Perispore (elater)
2. Spore

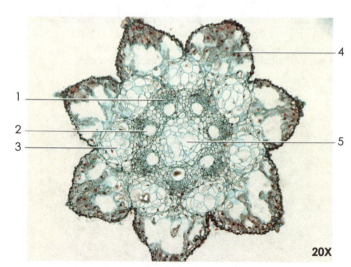

Figure 23.41 A transverse section of *Equisetum* sp. young stem.
1. Vascular tissue   3. Future air canal   5. Pith
2. Air canal   4. Cortex

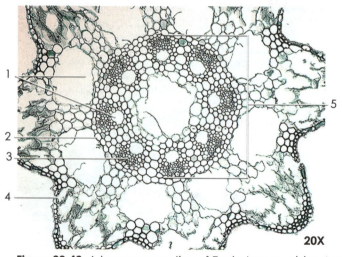

Figure 23.42 A transverse section of *Equisetum* sp. older stem.
1. Air canals   3. Vascular tissue   5. Fustele
2. Endodermis   4. Stomate

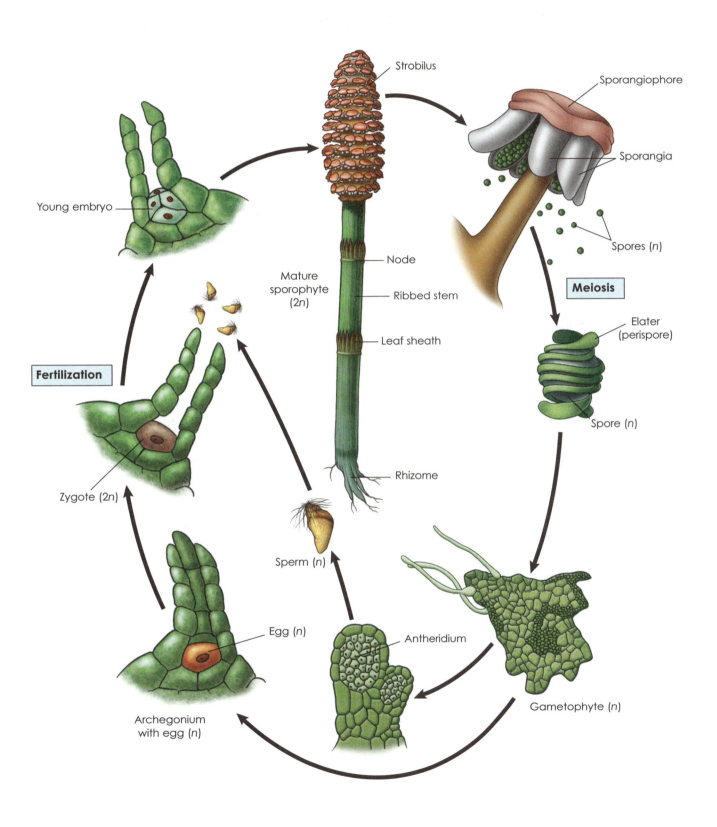

Figure 23.43 The life cycle of the horsetail, *Equisetum* sp.

Student Activity—Observations of *Equisetium*

Materials
• dissecting microscope
• compound microscope
• living specimens of *Equisetium*, one branched and one unbranched
• prepared slides of *Equisetium*
• colored pencils

Procedure 23.4
Equisetium

Consider photographing this activity.

1. Procure equipment and supplies.
2. Using a dissecting microscope, draw and label living specimens of branched and unbranched *Equisetium*. Pay particular attention to the strobili, sporangiophores, and spores. Place your labeled sketches and observations below.

3. Compare and contrast the external anatomy of the branched and unbranched forms of *Equisetium*.

4. Describe the texture of the stem.

5. Using a microscope on both low and high power, observe and sketch a young gametophyte of Equisetium, a longitudinal section of the shoot apex and strobilus of Equisetium, a transverse section of the strobilus of Equisetium, a transverse section of the stem of Equisetium just above a node, and transverse sections of young and older Equisetium stems. If available, observe a prepared slide of Equisetium spores. Draw and label your slides on the next page, and compare your observations to Figures 23.41 and 23.42 on page 356.

Young gametophyte of *Equisetium*

Longitudinal section of the shoot apex
and strobilus of *Equisetium*

Transverse section of the strobilus of *Equisetium*

Transverse section of the stem of *Equisetium* just above a node

Transverse sections of young *Equisetium* stems

Transverse sections of older *Equisetium* stems

Student Activity—Observations of *Calamites*

Materials
- dissecting microscope or magnifying glass
- fossil specimen of *Calamites*
- colored pencils

Procedure 23.5
Calamites

Consider photographing this activity.

1. Procure the necessary equipment and supplies.
2. Draw an overview of *Calamites*.

3. Using a dissecting microscope or magnifying glass, draw and label your specimen. Place your labeled sketch and observations below.

PHYLUM PTEROPHYTA

Ferns placed in **Phylum Pterophyta** (Polypodiophyta) are the most abundant group of seedless vascular plants. The majority of fern species live in moist tropical regions of the Earth, although some species reside in temperate regions, as well as the Arctic Circle. Some ferns can even live in aquatic environments and dry areas. Ferns range in size from giant tropical tree ferns that can exceed 28 meters

Figure 23.44 (a) The tree fern, *Cyathea* sp., and (b) the aquatic fern, *Azolla* sp.

in height to *Azolla*, a diminutive aquatic fern that measures less than a centimeter in diameter (Fig. 23.44). It is thought that ancestral ferns first appeared in the Devonian Period approximately 375 million years ago. Today, plant taxonomists have identified approximately 11,000 species of ferns.

The sporophyte stage is dominant in ferns. The leaves of ferns, known as **fronds**, arise from rhizomes. Immature fronds develop from the tip of a rhizome and appear as a tightly coiled and rolled-up structure called a **fiddlehead**. **Compound frond ferns** possess ornate leaflets or **pinnae**. **Simple frond ferns** have leathery, broad, unbranched, strap-like fronds. The pinnae are attached to a **midrib**, sometimes called a **rachis**. A **petiole**, or stalk, attaches the pinnae to the rhizome. Other ferns have leathery, broad, unbranched, strap-like fronds. Branching roots also arise from the rhizomes.

The majority of fern species are **homosporous.** The spores are produced in sporangia, appearing as distinct brown spots on the underside of the frond, called **sori**. To an untrained eye, the sori may appear as a fungus or as insect eggs. In some species, the sori are protected by a colorless flap called an **indusium**. The **annulus**, a fuzzy region of the sori, catapults the mature spores. Adder's tongue fern may produce more than 15,000 spores per sori. The collective production of spores in some ferns exceeds 50 million. The water fern *Marsilea* and several other species are heterosporous, producing spores in a **sporocarp**. Spores that land in a favorable environment germinate, producing heart-shaped gametocytes,

God made ferns to show what He could do with leaves.

—**Henry David Thoreau (1817–1862)**

or **prothalli.** The gametocytes possess rhizoids, which anchor them to their substrate. Flagellated sperm cells are produced in the antheridia of the gametocyte. The sperm are released and swim to the archegonia, where they fertilize the awaiting egg, forming a zygote. The zygote develops into the sporophyte generation, completing the life cycle of a fern (Fig. 23.47).

Ferns are cherished for their ornamental value. They are used for indoor as well as outdoor decorations, and by florists to construct bouquets. The Environmental Protection Agency suggests that ferns are valuable in filtering formaldehyde and other toxins from the air. Fern rhizoids and fronds are foods in many cultures. Bracken fern fronds were used in the past to thatch roofs. Medicinally, ferns and their products have been used in the treatment of leprosy, parasitic worms, labor pains, sore throat, diabetes, dandruff, and many other maladies.

Back to Life

Polypodium polypoidioides is a common fern that grows on the trunks and branches of trees, particularly in the southeastern United States. Being an air fern, it receives its water and nutrients from the surface of the bark on the host plant. During periods of a long drought, the fronds turn brown and curl up, appearing dead. When they are exposed to water, they seem to return to life miraculously. Thus, this remarkable fern is called the *resurrection fern* (Fig. 23.46).

Figure 23.45 Fossil of an ancient fern.

Figure 23.46 Dry and green *Polypodium*.

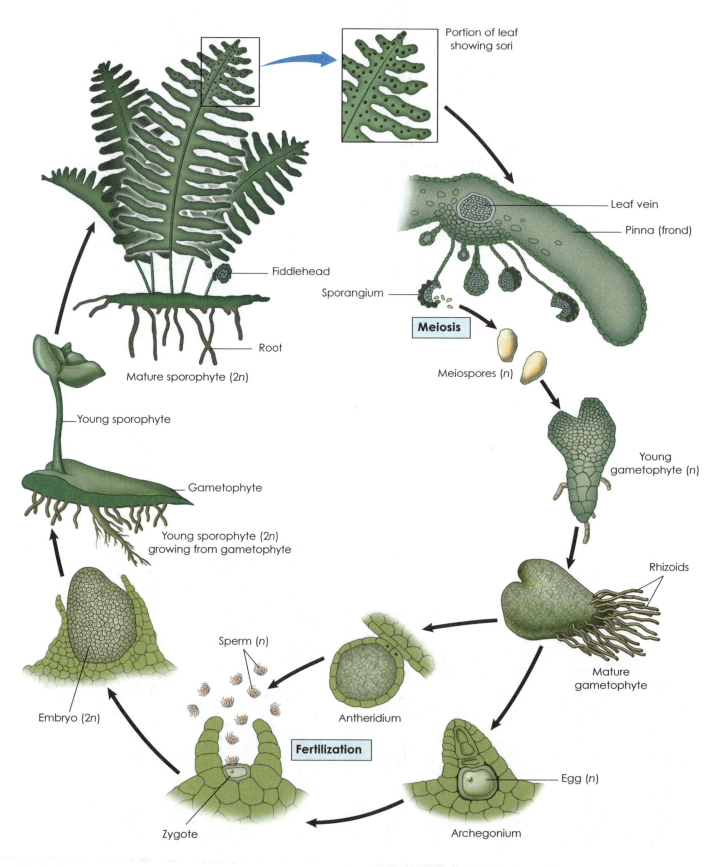

Portion of leaf showing sori

Leaf vein

Pinna (frond)

Meiosis

Sporangium

Meiospores (*n*)

Young gametophyte (*n*)

Fiddlehead

Root

Mature sporophyte (2*n*)

Young sporophyte

Gametophyte

Rhizoids

Young sporophyte (2*n*) growing from gametophyte

Mature gametophyte

Sperm (*n*)

Antheridium

Embryo (2*n*)

Fertilization

Egg (*n*)

Zygote

Archegonium

Figure 23.47 The life cycle of a fern.

Figure 23.48 A view of a new (a) compound and (b) simple fern leaf showing circinate vernation forming a fiddlehead.

Figure 23.49 The fronds of the staghorn fern, *Platycerium alcicorne*.

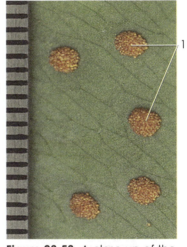

Figure 23.50 A pinnate leaf showing pinnate venation in the leaflets of a fern.
1. Venation   3. Leaf
2. Pinnae

Figure 23.51 The leaf of the fern *Polypodium virginianum*.

Figure 23.52 The leaf of the fern *Polypodium virginianum*, showing sori (groups of sporangia).
1. Pinna   2. Sori

Figure 23.53 A close-up of the fern pinna of *Polypodium virginianum* (scale in mm).
1. Sorus

Figure 23.54 The fern *Polypodium* sp. (a) Sori on the undersurface of the pinnae, and (b) a scanning electron micrograph of a sorus.
1. Pinna   2. Sori   3. Annulus   4. Sporangium

Figure 23.55 A magnified view of the fern pinna of *Pteridium* sp. showing numerous scattered sporangia.
1. Sporangia

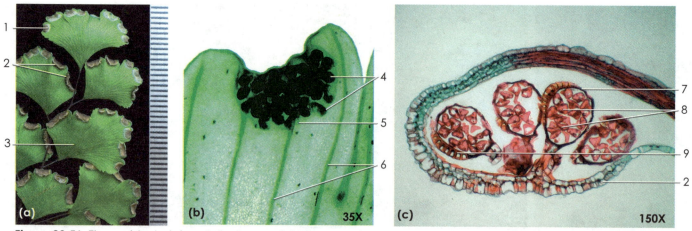

Figure 23.56 The maidenhair fern *Adiantum* sp. (a) Pinnae and sori. (b) Magnified view of the tip of a pinna folded under to form a false indusium that encloses the sorus. (c) Sorus with sporangia containing spores (scale in mm).
1. False indusium
2. Sori
3. Pinna
4. Sporangia with spores
5. False indusium enclosing a sorus
6. Vascular tissue (veins)of the pinna
7. Sporangium
8. Spores
9. Annulus

Figure 23.57 A young fern gametophyte.
1. Gametophyte
2. Spore cell wall
3. Rhizoid

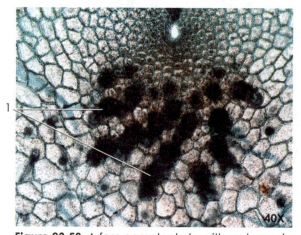

Figure 23.58 A fern gametophyte with archegonia.
1. Archegonia

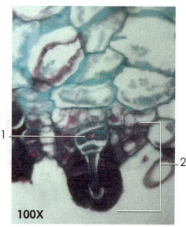

Figure 23.59 A fern gametophyte showing archegonium.
1. Egg
2. Archegonium

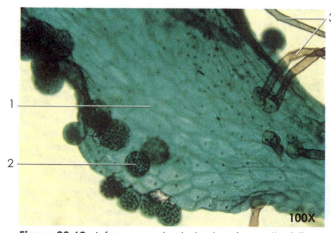

Figure 23.60 A fern gametophyte showing antheridia.
1. Gametophyte (prothallus)
2. Antheridium with sperm
3. Rhizoids

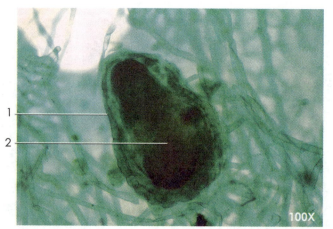

Figure 23.61 A fern gametophyte with a young sporophyte attached.
1. Expanded archegonium
2. Young sporophyte

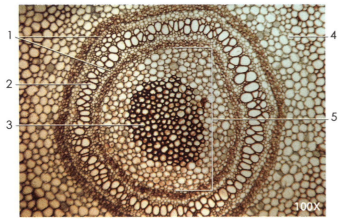

Figure 23.62 A transverse section through the stem of a fern, *Dicksonia* sp. showing a siphonostele.
1. Phloem   3. Sclerified pith   5. Pith
2. Xylem   4. Cortex

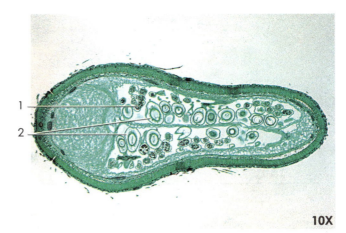

Figure 23.63 A transverse section of a sporocarp of the water fern, *Marsilea* sp., which is one of the two living orders of heterosporous ferns.
1. Microsporangium with microspores
2. Megasporangia with megaspores

 Student Activity—Observations of Fern External Anatomy

Materials
- dissecting microscope
- living specimens of several species of ferns
- scalpel
- Petri dishes
- water
- dropper
- microscope
- microscope slides and coverslips
- acetone
- rubbing alcohol
- salt water
- distilled water
- gloves
- colored pencils

 **Procedure 23.6
Fern Anatomy**

Consider photographing this activity.

1. Procure the necessary supplies and equipment. Wear gloves in this activity.
2. Using a dissecting microscope, draw and label living fern specimens. Pay particular attention to the rhizome, roots, pinnae, midrib, petiole, sori, indusium, and annulus. Place your labeled sketches and observations below.

5. Locate the sori on the underside of the frond of one of your ferns. Using the scalpel, scrape a single sorus, dropping the spores into a drop of water on a blank slide. Cover your specimen with a coverslip. Record your observations and sketches below.

3. With the scalpel, remove a sorus from the underside of a frond. Place the sorus in a small amount of water in a Petri dish.

4. Using a dissecting microscope, observe the anatomy of the sorus. Locate the sporangium, annulus, indusium, and annulus. Record your observations and labeled sketches in the space below.

6. Remove another sorus from the pinnae with a scalpel, and place it in a Petri dish. Place one drop of acetone on the sorus, and observe with a dissecting microscope. What happened?

7. Repeat Steps 3–6, using rubbing alcohol, salt water, and distilled water.

8. Observe the fiddleheads of a fern, and place your observations and sketches below.

Student Activity—Observations of Fern Internal Anatomy

Materials

- microscope
- prepared microscope slides: pinnae and sori of a maidenhair fern, a fern gametophyte, archegonia and antheridia of a gametophyte, a transverse section through a stem of a fern, and a transverse section through a sporocarp of *Marsilea*

Procedure 23.7
Internal Fern Anatomy

1. Procure the needed equipment and supplies.
2. Make observations and sketches of the following prepared slides (Fig. 23.56 through 23.63):

fern gametophyte

pinnae and sori of a maidenhair fern

archegonia and antheridia of a gametophyte

transverse section through a stem of a fern

transverse section through a sporocarp of *Marsilea*

Student Activity—Bringing *Polypodium* "To Life"

Materials
- dry *Polypodium*
- water
- paper towels
- plastic container

Procedure 23.8
Polypodium Back to Life

Consider photographing this activity.

1. Procure some living dry *Polypodium*.
2. Place the *Polypodium* on a dry paper towel placed in a plastic container.
3. Sprinkle the *Polypodium* heavily with water.
4. Place the container is a safe place.
5. Return the next day, and make your observations.

6. If no results are noticed, repeat Step 3.
7. Repeat up to three times if necessary.

Check Your Understanding

Q. What is the function of sori?

Q. How does dried *Polypodium* react to water?

Q. Sketch and label a fern frond.

Review Questions

1. What are the general characteristics of seedless vascular plants?

_____

_____

_____

2. Describe the characteristics of and provide examples of the following phyla: Lycophyta, Sphenophyta, Psilotophyta, and Pterophyta.

_____

_____

_____

3. Sketch and label *Equisetium*.

4. Distinguish between the sporophyte and the gametophyte generation in seedless vascular plants.

_____

_____

_____

5. What are some commercial and medical uses of seedless vascular plants?

_____

_____

_____

6. Compare and contrast *Lycopodium* and *Selaginella*.

_____

_____

_____

7. How did the pioneers use horsetail?

_____

_____

_____

8. Trace the life cycle of a typical fern.

_____

9. What is the significance of the vast fern forests of the Carboniferous?

_____

_____

_____

10. What is the function of a sporangia?

_____

_____

Chapter 24
The Green Machine: Part III
Understanding the Seed Plants (Gymnosperms)

Student Outcome Objectives

At the completion of this exercise, the student will be able to:

1. Briefly describe the taxonomical organization of the seed plants.
2. Describe landmarks in the evolution of the seed plants.
3. Compare and contrast gymnosperms and angiosperms.
4. State the impact of seed plants on human civilization.
5. Describe the fundamental characteristics of gymnosperms.
6. Explain the basic biology of Phylum Cycadophyta.
7. Trace the natural history of and identify anatomical features of a cycad.
8. Describe and identify selected histological features of a cycad.
9. Describe the basic biology of Phylum *Ginkgophyta.*
10. Trace the natural history of and identify anatomical features of *Ginkgo biloba.*
11. Describe and identify selected histological features of *Ginkgo biloba.*
12. Describe the basic biology of Phylum *Gnetophyta.*
13. Trace the natural history and basic biology of *Gnetum gnemon.*
14. Describe the natural history of and basic biology of *Welwitschia mirabilis.*
15. Trace the natural history of and identify anatomical features of *Ephedra* sp.
16. Describe and identify selected histological features of *Ephedra* sp.
17. Describe the basic biology of Phylum Coniferophyta.
18. Describe and identify selected specimens of Phylum Coniferophyta,
19. Describe the macroanatomy and microanatomy of conifer cones.
20. Describe the macroanatomy and microanatomy of conifer needles and leaves.
21. Determine the age of a tree by analyzing annual rings.

Love not the flower they pluck and know it not, and all their botany is Latin names.

—Ralph Waldo Emerson (1803–1882)

Overview

Just look around: The majority of plants that comprise your world are the seed plants. From picnicking on soft grass, to taking a stroll in the park admiring the ornamental azaleas and smelling the roses, to walking through the majestic forest appreciating the splendor of the giant oaks and pines, to shopping for fruits and vegetables at the local market—seed plants are all around you. Today, the seed plants are the dominant group of plants on Earth, with an estimated 300,000 species.

The seed plants, **spermatophytes**, of today are divided into two major groups, the **gymnosperms** and the **angiosperms** (Fig. 24.1). The gymnosperms consists of the following four living phyla:

1. **Cycadophyta**, the cycads and sago palms
2. **Ginkgophyta**, only one living species, *Ginkgo biloba*
3. **Gnetophyta**, three genera of unusual plants
4. **Coniferophyta** the largest phylum, consisting of plants including pine, spruce, sequoia, juniper, cedar, and cypress.

The angiosperms are flowering plants. The majority of living plants on Earth are angiosperms, placed in Phylum **Magnoliophyta**. Examples of angiosperms are cacti, oaks, grasses, tulips, sycamores, and magnolias.

Seed plants first appear in the fossil record in the late Devonian Period, approximately 360 million years ago. The oldest known fossil is a "seed fern," *Elkinsia polymorpha*

Figure 24.1 (a) The pine tree is an example of a gymnosperm, and (b) the oak tree is an example of an angiosperm.

(a)

Figure 24.2 Illustration of *Elkinsia polymorpha*.

(b)

Figure 24.3 A fossil leaf of an Eocene gymnosperm.

WOW!

A 2,000-year-old seed of an extinct Judean date palm tree was germinated successfully in Israel in 2005. It was found in King Herod's palace on Mount Masada, near the Dead Sea. The age of the seed was determined by carbon dating. Scientists hope that the unique seedling, named "Methuselah," will one day yield vital information about the medicinal properties of the fruit of the date tree.

(Fig. 24.2). This ancient plant featured **ovules** (structures of seed plants containing the female sex cells with the potential to develop into seeds) at the tips of their slender branches. The tips formed **cupules** for the development of seeds. Another ancient seed plant was *Archaeosperma arnoldii*, which possessed obvious cupules containing two ovules surrounded by prominent claw-like appendages. These early seed plants did not have cones or flowers. Seed plants continued to develop during the Carboniferous Period but were overshadowed by the giant seedless vascular plants.

Paleobotanists agree that the closest relatives to the gymnosperms developed during the Permian Period. By the time of the Triassic Period, all of the phyla of the seed plants were flourishing, with the exception of the flowering plants, the angiosperms. The flowering plants made their appearance about 140 million years ago, during the Cretaceous Period, and became the dominant plants on Earth during the Paleocene Epoch of the Cenozoic Era, approximately 60 million years ago.

The seed plants feature a life cycle dominated by the sporophyte generation. Examples of this generation are giant redwood and the tiny duckweed. The sporophyte produces two distinct types of gametophytes and is **heterosporous**. Multicellular male gametocytes (microspores) are called **pollen grains**. In nature, **pollination** occurs when pollen is carried to a waiting female gametocyte (megaspore) in a number of ways including wind, insects, and birds. Plants also can be pollinated artificially by humans. *So that's how Mendel did it!*

In seed plants, a pollen tube forms, allowing the sperm in the pollen grain to unite with the female gametocyte in the ovule. The ovule is a sporangium enclosed by modified leaves called the **integument**. The fertilized female gametocyte becomes the embryo, and the ovule's integument forms a protective seed coat. The **seed** provides the embryonic plant with essential nourishment and protection. Thus, the seed can withstand harsh conditions and stay dormant for many years.

Gymnosperm and angiosperm seeds are distinctly different (Fig. 24.3). The term "gymnosperm" literally means "naked seed leaf." In these plants, the seeds are not enclosed in an ovule, and they mature on the surface of a cone scale such as a pine cone. The nutritive material in gymnosperms accumulates prior to fertilization. In angiosperms, the nutritive material is stored only after fertilization. In angiosperms, double fertilization occurs, producing an embryo and a nutritive **endosperm**. The seeds of angiosperms are encased in a **fruit**. In both cases, the parental sporophyte generation provides nutrition to potential offspring, giving them a distinct advantage over seedless plants.

Evolution of the seed has changed the destiny of plants as well as humans. Seeds have allowed seed plants to become the dominant plants on Earth by allowing them to literally "get a head start" on life. Seed plants provide food for animals and humans. Neolithic human societies approximately 12,000 years ago utilized seed plants such as wheat, figs, corn, and squash, and shaped their destinies through artificial selection.

THE GYMNOSPERMS

The gymnosperms first appear in the fossil record approximately 305 million years ago during the Carboniferous Period. During the Mesozoic Era, the gymnosperms dominated plant life. Although angiosperms dominate the Earth today, the gymnosperms are still important plants in many ecosystems. The gymnosperms are generally characterized by lacking flowers and fruits; the seed develops in association with a scale that exists on a cone.

A good gardener always plants 3 seeds—one for the bugs, one for the weather and one for himself.

—**Leo Aikman (1908–1978)**

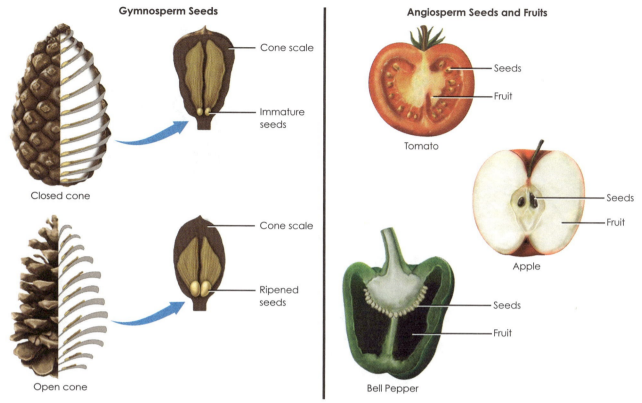

Figure 24.4 Comparison of gymnosperm and angiosperm seeds.

PHYLUM CYCADOPHYTA

Approximately 300 members of **Phylum Cycadophyta** survive today. During the Mesozoic, however, they were a dominant plant (Fig. 24.5). The cycads often are mistaken as ferns or palms because of their distinctive, large, palm-like leaves and unbranched trunks. Cycads are slow-growing plants, usually stout in appearance. Some species, however, can reach heights of more than 15 meters.

Cycads are **dioecious** (having separate sexes) and produce distinct male pollen cones and female seed cones. The male pollen cones usually are elongated, and the female seed cones are rounded and may contain dozens of large seeds. The sperm of cycads is the largest living in the world and may possess more than 10,000 flagella. The scales surrounding the sometimes colorful seeds may be covered by felt-like hairs that can be highly irritating to the skin. Thrips and beetles serve as insect pollinators in cycads.

Cycads are found primarily in tropical and subtropical forests. Only one species, *Zamia pumila,* originally found in Florida, is native to North America. Today, cycads are used mostly as ornamental plants. Some species, such as the sago palm *Cycas revolute,* is common in southern landscapes. Although they are a rich source of starch, cycads should not be consumed (Fig. 24.6–24.21).

Figure 24.5 Cycads were abundant during the Mesozoic Era. Currently, there are 10 living genera, with about 100 species, that are found mainly in tropical and subtropical areas. (a) *Cycas revoluta.* (b) *Encephalartos princeps,* and (c) *Zamia furfuracea.*

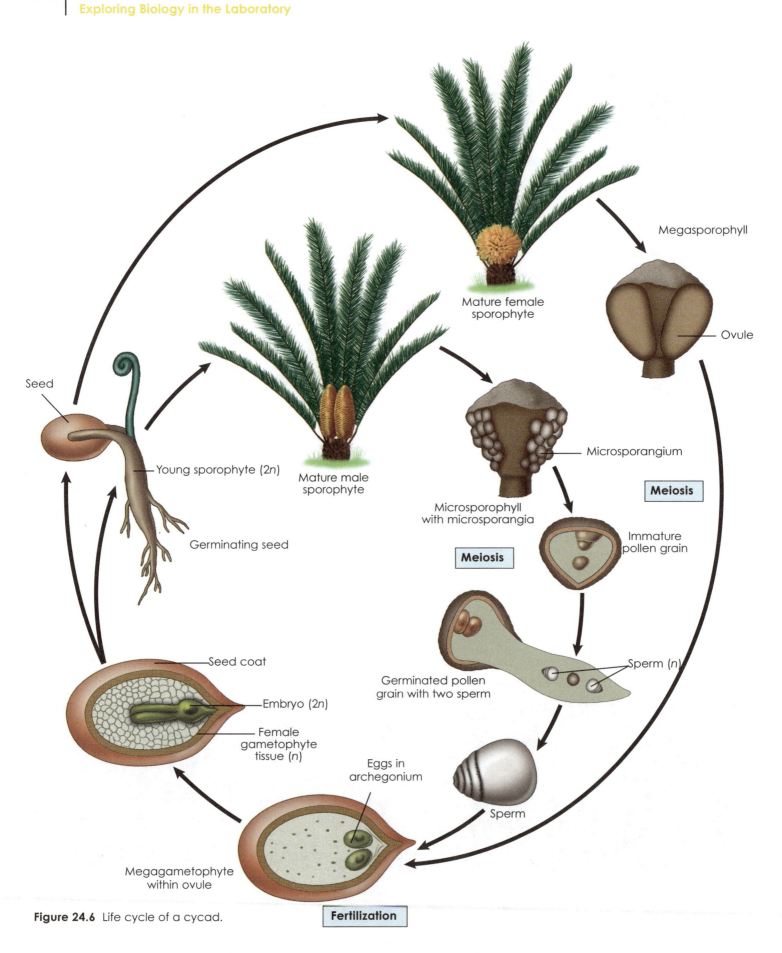

Figure 24.6 Life cycle of a cycad.

Megasporophyll

Ovule

Mature female
sporophyte

Microsporangium

Meiosis

Immature
pollen grain

Microsporophyll
with microsporangia

Meiosis

Mature male
sporophyte

Young sporophyte (2*n*)

Seed

Germinating seed

Sperm (*n*)

Germinated pollen
grain with two sperm

Sperm

Seed coat

Embryo (2*n*)

Female
gametophyte
tissue (*n*)

Eggs in
archegonium

Megagametophyte
within ovule

Fertilization

Figure 24.7 *Cycas revolute* showing (a) the leaf, (b) the pollen (microsporangiate) cone, and (c) the seed (megasporangiate) cone.

Figure 24.8 A male cone of *Cycas revoluta*. (a) Before the release of pollen, and (b) after the release of pollen.
1. Cones

Figure 24.9 (a) *Cycas revoluta* showing a close-up view of a female cone with developing seeds. (b) *Cycad* sp. showing a close-up view of a female cone during seed dispersal.
1. Seeds   2. Megasporophyll

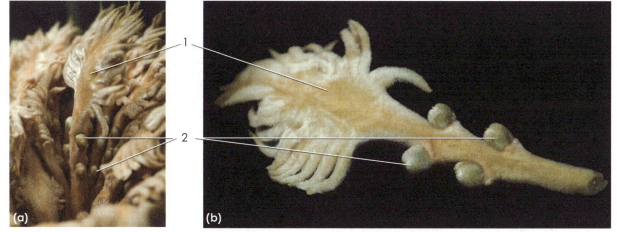

Figure 24.10 The megasporophyll and ovules of *Cycas revoluta*. (a) attached to the megasporangiate cone and (b) removed.
1. Megasporophyll   2. Ovules

Figure 24.11 (a) The leaf of the cycad *Zamia* sp., and (b) a microsporangiate cone. (c) A longitudinally sectioned microsporangiate cone of the cycad, *Zamia* sp., and (d) A microsporangiate cone of a cycad showing megasporangia on microsporophylls.

1. Microsporangia   2. Microsporophyll

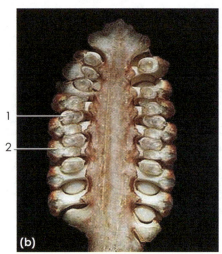

Figure 24.12 (a) Megasporangiate cones of the cycad *Zamia* sp., and (b) a longitudinally sectioned cone showing the showing the ovules and megasporophyll.

1. Ovule   2. Megasporophyll

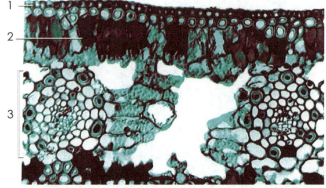

Figure 24.13 Transverse section of the leaf of the cycad *Zamia* sp.

1. Upper epidermis   3. Vascular bundle (vein)
2. Palisade mesophyll

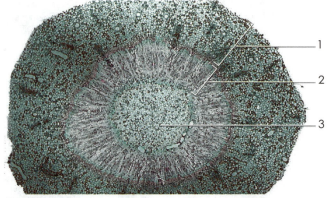

Figure 24.14 Transverse section of the stem of the cycad *Zamia* sp.

1. Cortex   3. Pith
2. Vascular tissue

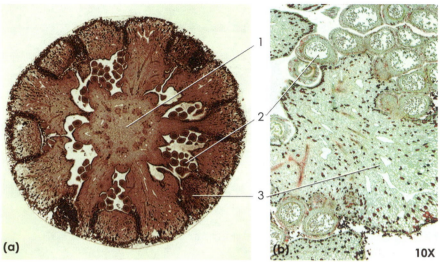

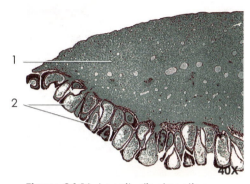

Figure 24.16 Longitudinal section of a microsporophyll of the cycad *Cycas* sp. Note the microsporangia develop on the undersurface of the microsporophyll.
1. Microsporophyll   2. Microsporangia

Figure 24.15 Transverse sections of a microsporangiate cone of the cycad *Zamia* sp. (a) A low magnification, and (b) a magnified view.
1. Cone axis   2. Microsporangia   3. Microsporophyll

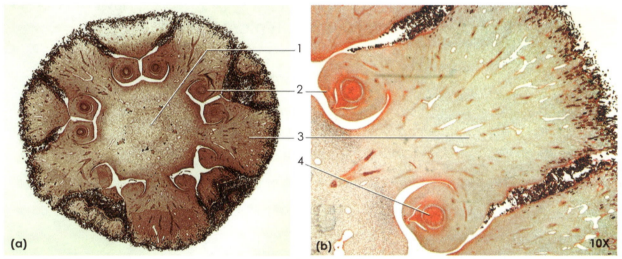

Figure 24.17 Transverse sections of a megasporangiate cone of the cycad *Zamia* sp. (a) A low magnification, and (b) a magnified view.
1. Cone axis   2. Ovule   3. Megasporophyll   4. Megasporocyte

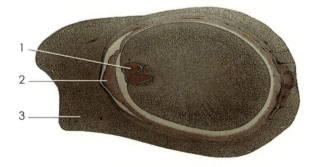

Figure 24.18 An ovule of the cycad *Zamia* sp. The ovule has two archegonia and is ready to be fertilized.
1. Archegonium
2. Megasporangium (nucellus)
3. Integument (will become seed coat)

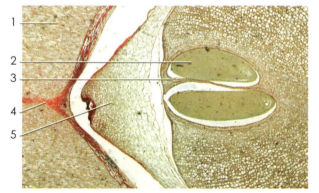

Figure 24.19 A magnified view of the ovule of the cycad *Zamia* sp. showing eggs in archegonia.
1. Integument   4. Micropyle area
2. Egg   5. Megasporangium
3. Archegonium

Figure 24.20 An ovule of the cycad *Zamia* sp. The ovule has been fertilized and contains an embryo. The seed coat has been removed from this specimen.
1. Female gametophyte
2. Embryo

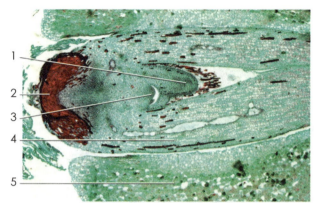

Figure 24.21 A magnified view of the ovule of the cycad *Zamia* sp. showing the embryo.
1. Leaf primordium   4. Cotyledon
2. Root apex   5. Female gametophyte
3. Shoot apex

 Student Activity—Observations of a Typical Cycad

Materials
- dissecting microscope
- compound microscope
- colored pencils
- a transverse section of a microsporangiate cone, a transverse section of a megasporangiate cone, an unfertilized ovule, and a fertilized ovule from a living specimen of a cycad
- leaf, pollen cone, microsporophyll, megasporophyll, and a seed from a living specimen of a cycad
- prepared slides of a typical cycad species such as *Cycas revoluta* or *Zamia pumila*

 Procedure 24.1
Cycas revoluta

1. Procure the needed equipment and supplies.
2. Observe a leaf, pollen cone, microsporophyll, megasporophyll, and seed from a living specimen of a cycad. Using a dissecting microscope, observe and draw the anatomical features of your specimen. Record your labeled sketches and observations.

A leaf of a cycad

A pollen cone of a cycad

_____

_____

A microsporophyll of a cycad

_____

_____

A megasporophyll of a cycad

_____

_____

A seed of a cycad

3. Using a microscope on both low and high power, observe a transverse section of a microsporangiate cone, a transverse section of a megasporangiate cone, an unfertilized ovule, and a fertilized ovule of a cycad. Draw and label your slides, and compare your observations to Figure 24.15 to 24.21 on pages 377–378.

Transverse section of a
microsporangiate cone of a cycad

_____

_____

A transverse section of a
megasporangiate cone of a cycad

_____

_____

An unfertilized ovule of a cycad

A fertilized ovule of a cycad

PHYLUM GINKGOPHYTA

How did you do on your last quiz? Oh no! Some people claim that you should have taken your *Ginkgo biloba* to improve your memory. Others claim that it is a gimmick. *Ginkgo biloba* is the last living member of **Phylum**

For in the true nature of things, if we rightly consider, every green tree is far more glorious than if it were made of gold and silver.

—**Martin Luther (1483–1546)**

Ginkgophyta. The first member of the genus Ginkgo appeared in the Jurassic Period, but by the Pliocene Epoch of the Cenozoic Era, with the exception of a small population in central China, all ginkgos had become extinct (Fig. 24.22).

Charles Darwin termed *Ginkgo biloba* a living fossil. Modern Ginkgo *biloba* trees are thought to have descended from seeds collected in a Japanese temple garden. Today, *Ginkgo biloba*, known as the maidenhair tree because its distinct, notched, fan-shaped leaves resemble the pinnae of maidenhair ferns. (Fig. 24.23) In Chinese, the term *Ginkgo* literally means "silver apricot."

Ginkgo biloba is a unique plant in many ways. The leaves do not possess a **midrib** (central vein) and have **dichotomous** (forked) venation. The tree is deciduous (shedding leaves yearly), and the leaves turn bright yellow before **abscission** (shedding) in the fall. *Ginkgo* trees have two types of shoots (Fig. 24.28–24.30): **Short shoots** or spurs appear knobby and feature clusters of leaves and immature ovules. The leaves of slow-growing short shoots usually are unlobed or slightly bilobed. The leaves of fast-growing long shoots usually are deeply bilobed. *Ginkgo* trees are **dioecious**, having separate sexes. Male trees produce pollen in their strobilli (Fig 24.24 and Fig. 24.29). The pollen is carried by wind to a waiting ovule. Pollen tubes form and travel through the ovule. When the pollen tube bursts, flagellated sperm make their way to the egg.

After fertilization, embryos form and the integument develops into an extremely bad-smelling, fleshy seed coat (Fig. 24.31 and 24.32). The seed smells so bad that female ginkgos have been called stink-bomb trees! The majority of ginkgo trees planted in populated areas are male because of the females' nauseating odor. The nut within the seed, though, is tasty and is prized in the Orient.

Ginkgo trees can reach a height of 30 meters or more. The trunk can exceed 3.5 meters in diameter. The trunks of *Ginkgo* trees are straight, columnar, and branched sparingly. *Ginkgo* trees are popular in cities because they are beautiful, are hardy, have an appealing growth pattern, and are thought to improve air quality. *Ginkgo biloba* may live for more than a thousand years. The oldest *Ginkgo* tree, in China, is more than 3,500 years old (Fig. 24.24–Fig. 24.32).

Figure 24.22 Fossil *Ginkgo biloba* leaf impression from Paleocene sediment. This specimen was found in Morton County, North Dakota.

Figure 24.23 Leaf from the *Ginkgo biloba* tree. The fan-shaped leaf is characteristic of this species.

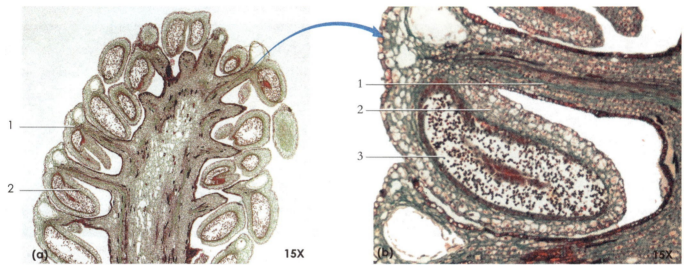

Figure 24.24 Microsporangiate strobilus of *Ginkgo biloba*. (a) A longitudinal section and (b) a magnified view showing a microsporangium.

1. Sporophyll   2. Microsporangium   3. Pollen

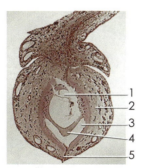

Figure 23.25 Longitudinal section of an ovule of *Ginkgo biloba* prior to fertilization.
1. Megagametophyte
2. Integument
3. Pollen chamber
4. Nucellus
5. Micropyle

Figure 23.26 Longitudinal section of a seed of *Ginkgo biloba* with the seed coat removed.
1. Megagametophyte
2. Developing embryo

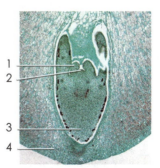

Figure 23.27 Magnified view of the ovule of *Ginkgo biloba* showing the embryo.
1. Leaf primordium
2. Shoot apex
3. Root apex
4. Megagametophyte

Figure 24.28 Leaves and immature ovules on a short shoot of the ginkgo tree, *Ginkgo biloba*.
1. Leaf   3. Short shoot
2. Immature ovules   4. Long shoot

Figure 24.29 The pollen strobili of the ginkgo tree, *Ginkgo biloba*.
1. Leaf   3. Long shoot
2. Short shoot   4. Pollen strobilus

Figure 24.31 Branch of a *Ginkgo biloba* tree supporting a mature seed.
1. Short shoot (spur)   3. Mature seeds
2. Long shoot

Figure 24.30 Transverse and longitudinal sections through a living immature seed of *Ginkgo biloba* showing the green megagametophyte.
1. Fleshy layer of integument   3. Stoney layer of integument
2. Megagametophyte

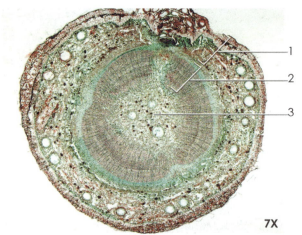

7X

Figure 24.32 Transverse section of a short branch from *Ginkgo biloba*.
1. Cortex   3. Pith
2. Vascular tissue

Student Activity—Observations of *Ginkgo biloba*

Materials

- dissecting microscope
- compound microscope
- colored pencils
- stems and leaves of *Ginkgo biloba*
- prepared slides of *Ginkgo biloba*: a transverse section of a short branch, a microsporangiate strobilus, a longitudinal section of an ovule, a longitudinal section of a seed with the seed coat removed, a section of the ovule

Procedure 24.2
Ginkgo biloba

1. Procure living specimens of *Ginkgo biloba*. Using a dissecting microscope, observe and draw the anatomical features of your specimens. Describe and sketch the short shoots, long shoots, leaves, and, if available, the sexual structures. Place your labeled sketch and observations below.

2. Using a compound microscope on both low and high power, observe the following slides of *Ginkgo biloba*: a transverse section of a short branch, a microsporangiate strobilus, a longitudinal section of an ovule, a longitudinal section of a seed with the seed coat removed, and a section of the ovule. Record your observations and sketches below.

A transverse section of a short branch of *Ginkgo biloba*

A microsporangiate strobilus of *Ginkgo biloba*

A longitudinal section of an ovule of *Ginkgo biloba*

A longitudinal section of a seed
with the seed coat removed of *Ginkgo biloba*

A a section of the ovule of *Ginkgo biloba*

Figure 24.33 Ginkgo, or maidenhair tree, *Ginkgo biloba* is the sole member of the phylum Ginkgophyta. *Ginkgo biloba* may have the longest genetic lineage among seed plants.

Check Your Understanding

Q. How are cycads pollinated?

Q. Why are female Ginkgo trees considered undesirable in urban areas?

Q. Sketch and describe a ginkgo leaf.

Ginkgo biloba:
The leaf of this Eastern tree
Which has been entrusted to my garden
Offers a feast of secret significance,
For the edification of the initiate.
Is it one living thing
That has become divided within itself?
Are these two who have chosen each other,
So that we know them as one?
I think I have found the right answer
To these questions;
Do my songs not make you feel
That I am both one and twain?

—**Johann Wolfgang von Goerthe (1749–1832)**

PHYLUM GNETOPHYTA

Phylum Gnetophyta is composed of three genera and 71 species of relatively obscure gymnosperms. The three genera are *Gnetum, Welwitschia,* and *Ephedra* (Fig. 24.34 and Fig. 24.35). The gnetophytes are unique gymnosperms because their wood contains conducting cells, known as **vessel elements,** similar to the angiosperms or flowering plants. In addition, gnetophytes such as angiosperms undergo **double fertilization**. It is thought that the gnetophytes perhaps are ancestral to the angiosperms.

Xylem forms the wood in woody plants and is responsible for conducting dissolved minerals and water throughout the plant. Xylem consists of two basic types of cells: **tracheids** and **vessel elements.** Tracheids are the only water-conducting cells of all gymnosperms with the exception of the gnetophytes. Tracheids appear as long, slender cells with tapered overlapping ends. **Bordered pits** in the cell wall allow for the passage of water. **Vessel elements** are more advanced than tracheids and, with the exception of gnetophytes, are exclusive structures in angiosperms. Vessel elements are shorter and wider than tracheids and are stacked end-to-end to form vessels. Vessels are more efficient than tracheids in conducting water throughout the plant. **Phloem,** responsible for transporting nutrients in plants, is composed of two types of cells, sieve tube elements, and companion cells. Sieve tube elements are narrow tubes existing end-to-end and conduct nutrients. **Porous sieve plates** are found between adjacent sieve tubes. Narrow **companion cells,** adjacent to sieve tubes, help to control their function.

Most gnetophytes are dioecious; the flowers possess both sexes. In some species, the nectar attracts pollinating insects. In gnetophytes, the sperm cells are non-motile. Like angiosperms, gnetophytes undergo double fertilization. One sperm cell fertilizes the waiting egg in the female gametophyte, and the other cell fuses with another cell in the female gametophyte. The second structure disintegrates in gnetophytes instead of forming supportive endosperm as in angiosperms.

Members of the genus *Gnetum* are mostly vine-like plants, with the exception of *Gnetum gnemon,* a tree that grows up to 10 meters in height. These plants are found in the tropical forests of Southeast Asia, South America, and Africa. The genus *Welwitschia* is represented by one member, *Welwitschia mirabilis,* which is native to the extremely dry Namib and Mossamedes deserts of southwestern Africa. It is rather strange in appearance, possessing a long taproot and a short stem that usually supports two permanent strap-like leaves. *Welwitschia mirabilis* may live for more than 1,000 years.

Ephedra consists of 35 species of short, stubby plants residing on every continent except Australia (Fig. 24.36– Fig. 24.42). At first glance, the leafless appearance and jointed stems resemble the horsetails. Thus, *Ephedra* commonly is called a joint fir. *Ephydra* is not leafless; the mature leaves occur in groups of two or three at the nodes and are small, brown, and non-photosynthetic. Photosynthesis takes place in the green, rounded stems.

In China, one species of *Ephedra sinica,* "ma huang," has been a medicine for more than 5,000 years, as a stimulant, as a cure for respiratory problems, and as a diuretic. Native Americans prepared the plant as a medicine for intestinal disorders, colds, fever, and headache. In the southwestern United States, several species of *Ephydra* (primarily *Ephedra nevadensis*) are known as Mormon tea or Brigham tea. Early Mormon settlers, who abstained from drinking contemporary tea and coffee, drank tea made from this plant. Early settlers also used this tea as a decongestant and to address urinary tract problems.

Figure 24.34 A specimen of *Welwitschia mirabilis.*

20X

Figure 24.35 Transverse section through a young stem of *Welwitschia mirabilis.* Cone-bearing branches arise from meristematic tissue on the margin of the disk.

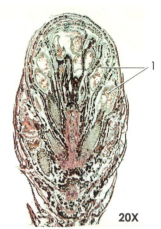

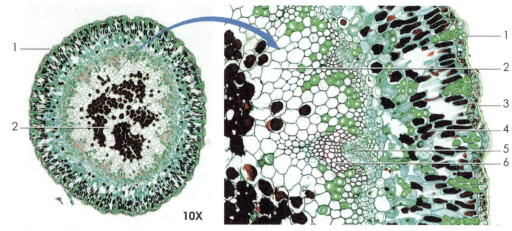

Figure 24.36 Longitudinal section through a microsporangiate cone of *Ephedra* sp.
1. Microsporangia

Figure 24.37 Transverse section through a stem of *Ephedra* sp. Note that unlike most other gymnosperms, *Ephedra* has vessel elements in the xylem similar to those found in angiosperms.

1. Epidermis	3. Cuticle	5. Phloem
2. Pith	4. Cortex	6. Xylem

Figure 24.38 Microsporangiate cones of *Ephedra* sp.

1. Microsporangiate cones

Figure 24.39 Stem of *Ephedra* sp. with several microsporangiate cones attached.
1. Stem   2. Microsporangiate cone

Figure 24.40 (a) *Ephedra* sp. with attached megasporangiate (ovulate) cones about the time of pollination, and (b) microsporangiate cones.
1. Cones   2. Stem

Figure 24.41 Stem of *Ephedra* sp. with several attached megasporangiate cones.
1. Cones   2. Stem

Did You Know?
Ephedra and Medicine

Ephedra sinica contains the alkaloids ephedrine and pseudoephedrine, which stimulate the central nervous system and cause bronchodilation and vasoconstriction. In recent years, *Ephedra*-supplemented dietary products have been removed from the market by the Food and Drug Administration because it has been associated with central nervous system excitation, dehydration, hypertension, tachycardia, arrhythmia, heart attack, stroke, and death. In the Old West, *Ephedra* also was called Whorehouse tea because it was thought to be a cure for gonorrhea, syphilis, and other venereal diseases.

Figure 24.42 *Ephedra* sp. is one of three genera of shrubs within the phylum Gnetophyta. Although found throughout most arid or semiarid regions of the world, *Ephedra* sp. is the only one of the three genera of gnetophytes found in the United States. It is a highly branched shrub with very small leaves.

PHYLUM CONIFEROPHYTA

Pines, cypresses, spruces, redwoods, cedars, hemlocks, junipers, and yews are common gymnosperms that have been placed in **Phylum Coniferophyta** (Pinophyta) (Fig. 24.43). This phylum is composed of approximately 600 species of woody, mostly evergreen, cone-bearing plants that most often are found in cold and temperate climates. A number of conifers are record-setters. *Pinus longaeva*, the bristlecone pine, is the oldest known non-clone living organism on Earth. One specimen, "Methuselah," is nearly 5,000 years old. In California, a coastal redwood, *Sequoia sempervirens*, is the tallest tree in the world, measuring above 110 meters in height.

The General Sherman, *Sequoiadendron giganteum,* a giant sequoia in California, measures more than 83 meters in height and has a circumference at the ground exceeding 31 meters. It is the largest tree on Earth by volume.

Some conifer species are sources of lumber, paper, wood alcohol, turpentine, and resin. Several species, such as juniper and yew, are ornamentals. In this regard, bonsai conifer plants are popular, along with spruce and fir Christmas trees. Oils from conifers are used in soaps and air fresheners. Humans eat some seeds, such as pine nuts. A number of conifer products, such as taxol (a cancer treatment), are used in medicine.

Figure 24.43 Various conifers: (a) A bristlecone pine, *Pinus longaeva*, (b) A bald cypress, *Taxodium distichum*, and (c) A Colorado blue spruce, *Picea pungens*.

Student Activity—Macroscopic Anatomy of Selected Conifers

Materials
- dissecting microscope or hand lens
- colored pencils
- specimens of select conifers such as: pine, bald cypress, cedar, spruce, juniper, arborvitae, fir, and other specimens provided by the instructor.

Procedure 24.3
Macroanatomy of Conifers

1. Procure needed equipment and supplies.
2. Observe the overall specimens, leaf arrangement, bark, cones, and distinguishing characteristics. Compare your observations to the images of conifers in this chapter. Place your observations and sketches in the space provided.

The majority of conifers produce two distinct types of cones as a sporophyte: the **microsporangiate pollen cone** (male) and the **ovulate seed cone** (female). Pollen cones are soft, scale-like structures usually found on the tips of branches. Pollen develops within the microsporangia of the pollen cone. In pines, the pollen grain has a pair of **air bladders,** or wings, to aid in dispersal. A single group of pollen cones at the tip of a branch may be capable of producing more than a million pollen grains. Seed cones are more distinctive and variable in appearance.

In the seed cone, the **megasporangia** produce eggs in the **archegonia**, forming the ovule. Initial seed cones are small, scaly, and slightly opened to allow pollen to enter the **ovule**-bearing scales. Pollen enters the mature seed cone and forms sperm cells that ultimately are delivered to the waiting egg by means of a pollen tube. The egg is fertilized and forms a zygote. The seed cone closes and increases in size as the seeds develop.

Many seed cones, such as in pine, are woody. Others, such as in junipers, are fleshy. The seeds of conifers released from the seed cone are winged and require air dispersal. The dry, scaly, woody cones beneath a pine tree represent the spent seed cones. After a seed lands on a suitable substrate, it germinates and develops into a new sporophyte (Fig. 24.44–Fig. 24.62).

Astonishing Numbers!

- In 2007, approximately 31.3 million Christmas trees were purchased in the United States, costing more than 1.3 billion dollars!
- An average American uses about 750 pounds of paper yearly.

Check out http://www.50states.com/tree/ and find your state tree.

Pinus sp.

Aibes sp.

Picea sp.

Taxodium sp.

Taxus sp.

Thuja sp.

Figure 24.44 Images of the seed cones of conifers.

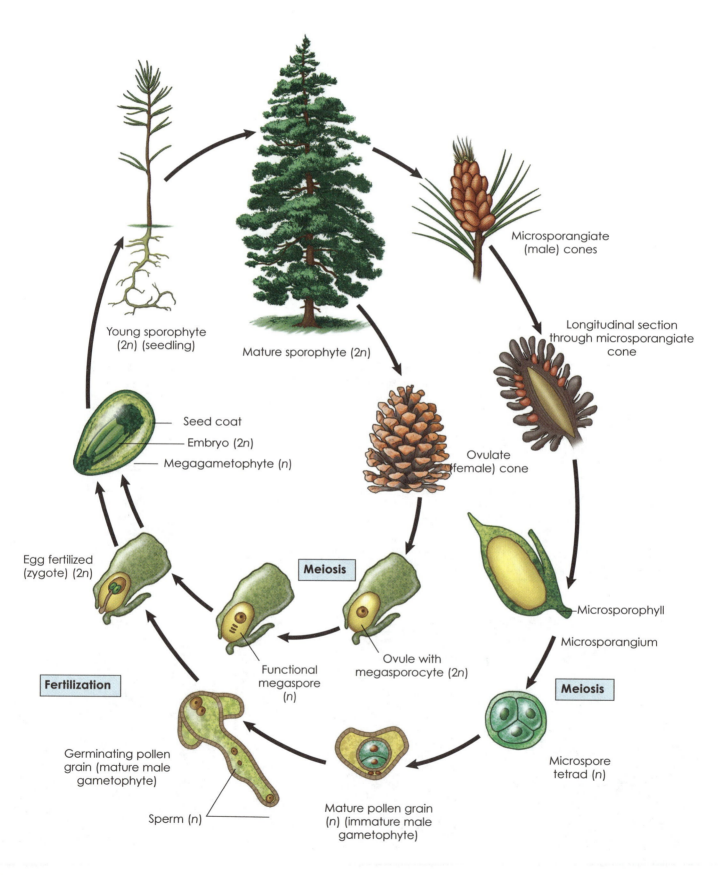

Young sporophyte (2n) (seedling)

Mature sporophyte (2n)

Microsporangiate (male) cones

Longitudinal section through microsporangiate cone

Seed coat

Embryo (2n)

Megagametophyte (n)

Ovulate (female) cone

Egg fertilized (zygote) (2n)

Meiosis

Microsporophyll

Microsporangium

Fertilization

Functional megaspore (n)

Ovule with megasporocyte (2n)

Meiosis

Germinating pollen grain (mature male gametophyte)

Sperm (n)

Mature pollen grain (n) (immature male gametophyte)

Microspore tetrad (n)

Figure 24.45 The life cycle of the pine, *Pinus* sp.

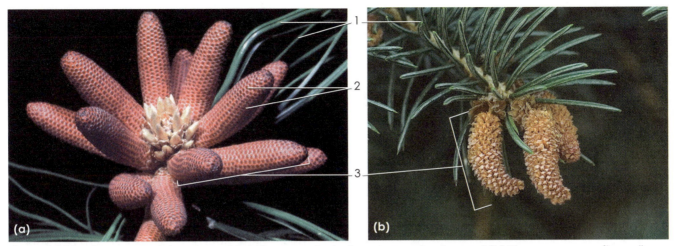

Figure 24.46 Microsporangiate cones of (a) *Pinus* sp. prior to the release of pollen and (b) *Picea pungens* after pollen has been released. The pollen cones are at the end of a branch.

1. Needle-like leaves   2. Microsporophylls   3. Pollen cone

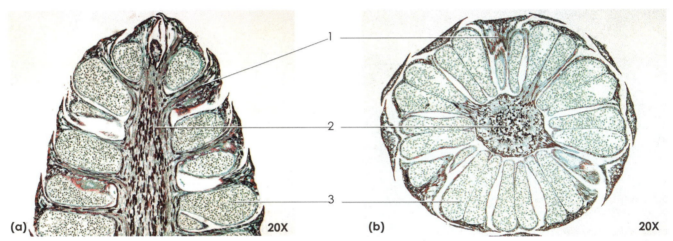

Figure 24.47 (a) Longitudinal section through the tip of a microsporangiate cone of *Pinus* sp. and (b) a transverse section.
1. Sporophyll   2. Cone axis   3. Microsporangium

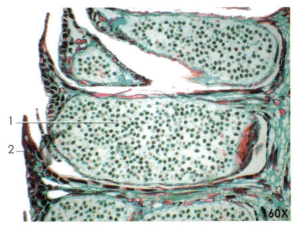

Figure 24.48 Close up of a microsporangiate cone scale andmicrosporangium of *Pinus* sp.

1. Microsporangium   2. Microsporophyll
   with pollen grains

Figure 24.49 Micrograph of stained pollen grains of *Pinus* sp. showing wings.

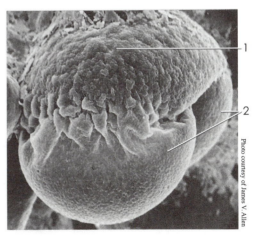

Figure 24.50 Scanning electron micrograph of a Pinus sp. pollen grain with inflated bladder-like wings.
1. Pollen body   2. Wings

Photo courtesy of James V. Allen

Figure 24.51 First-year ovulate cone in *Pinus* sp.

1. Pollen cones   2. First-year ovulate cone

Figure 24.52 Transverse section through a first-year ovulate cone in *Pseudotsuga* sp.

1. Cone scale bracts   2. Immature ovule

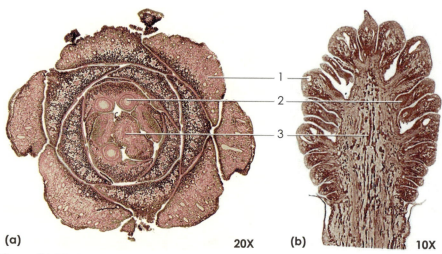

(a)   20X   **(b)**   10X

Figure 24.53 Ovulate cones of a *Pinus* sp. (a) Transverse section, and (b) longitudinal section.

1. Ovuliferous scale   2. Ovule   3. Cone axis

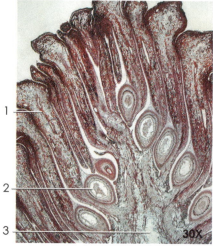

30X

Figure 24.54 Magnified view of a *Pinus* sp. ovulate cone (longitudinal view).

1. Ovuliferous scale   3. Cone axis
2. Ovule

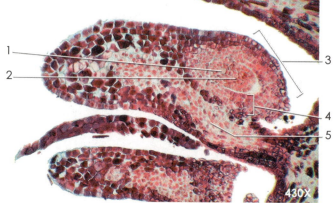

430X

Figure 24.55 Magnified view of a *Pinus* sp. ovule (immature).

1. Megaspore mother cell   4. Integument
2. Nucellus   5. Cone scale
3. Ovule

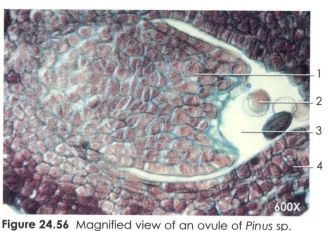

600X

Figure 24.56 Magnified view of an ovule of *Pinus* sp. with pollen grains in the pollen chamber.

1. Nucellus   3. Pollen chamber
2. Pollen grain   4. Integument

Figure 24.57 Close up of an ovulate cone scale in *Pinus* sp.
1. Mature seeds (wings)
2. Ovulate cone scale
3. Seed (containing embryo within seed coat)

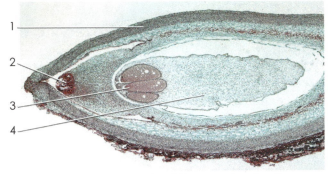

Figure 24.58 Young ovule of *Pinus* sp. showing the megaagmetophyte.
1. Ovule
2. Micropyle
3. Archegonium
4. Megagametophyte

Figure 24.59 Young ovule of *Pinus* sp. showing the egg in archegonium.
1. Egg
2. Nucleus

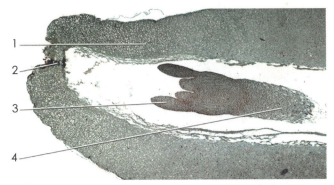

Figure 24.60 Magnified view of the ovule of *Pinus* sp. showing the embryo.
1. Integument
2. Micropyle
3. Leaf primordium
4. Root primordium

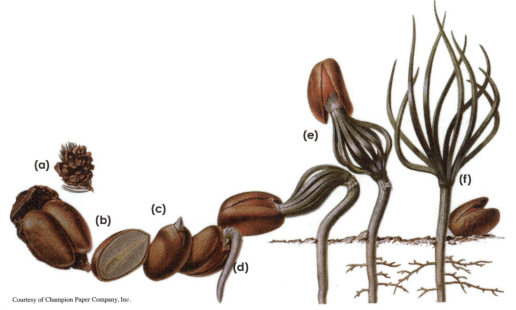

Courtesy of Champion Paper Company, Inc.

Figure 24.61 A diagram of pinyon pine seed germination producing a young sporophyte. (a) The seeds are protected inside the cone, two seeds formed on each scale. (b) A sectioned seed shows an embryo embedded in the female gametophyte tissue. (c) The growing embryo splits the shell of the seed, enabling the root to grow toward the soil. (d) As soon as the tiny root tip penetrates and anchors into the soil, water and nutrients are absorbed. (e) The cotyledons emerge from the seed coat and create a supply of chlorophyll. Now the sporophyte can manufacture its own food from water and nutrients in the soil and carbon dioxide in the air. (f) Growth occurs at the terminal buds at the base of the leaves.

Figure 24.62 A young sporophyte (seedling) of a pine, *Pinus* sp. (scale in mm).
1. Seedling leaves (needles)
2. Young stem
3. Young roots

Student Activity—Observing the Cones of Selected Conifers

Materials

- dissecting microscope
- compound microscope
- colored pencils
- white paper
- sterile microscope slides
- dropper
- water
- cover slip
- pollen cones from select conifers
- pollen from select conifer
- seed cones from select conifers
- seeds from select conifers
- microscope slides of the following: longitudinal section through the tip of a microsporangiae cone of *Pinus* sp., pollen from *Pinus* sp., transverse and longitudinal sections of the seed cones of *Pinus* sp., ovule of *Pinus* sp.

Procedure 24.4
Conifer Cones

1. Procure the needed equipment and supplies.
2. Examine pollen cones provided by the instructor, and record your observations below.

_____

_____

3. If permitted, shake the cone, and gently bump it over a piece of white paper. Did you collect any pollen? If so, place the pollen on a clean, blank microscope slide, carefully place a drop of water over the pollen, and place a coverslip over the drop. Record and sketch your observation.

4. Observe a prepared slide of *Pinus* sp. pollen, and record your observation below. Compare your slide with Figure 24.48 on page 391. Label the pollen grain and bladders.

5. Observe, compare, sketch, and label select seed cones from conifers provided by the instructor. Record your sketches and observations below.

_____

_____

_____

_____

6. If possible, observe and collect seeds from a seed cone. Or the instructor will provide seeds for observation. Describe and sketch the seeds and note how they are distributed.

_____

_____

7. Drop a seed from *Pinus.* sp. from above your head. Describe the action as it falls.

_____

_____

8. Examine a prepared slide of a longitudinal section through the tip of a microsporangiae cone of *Pinus* sp. Compare your slide with Figure 24.47 on page 391. Sketch and record your observations.

_____

_____

9. Examine a prepared slide of transverse and longitudinal sections of the seed cones of *Pinus* sp. Compare your slide with Figures 24.53 through 24.56 on page 392. Sketch and record your observations below.

_____

10. Using a compound microscope, examine a prepared slide of an ovule of *Pinus* sp. on both low and high power. Compare your slide with Figures 24.58 through 24.60 on page 393. Sketch and record your observations below.

_____

_____

Student Activity—Germination and Seedlings

Materials
- pine seeds
- pine seedlings
- sand
- sectioned potting container with drains
- potting soil
- small pot
- pencil with eraser

 Procedure 24.5
Germination

1. Place sand in 10 sections of a well-drained potting container.
2. Using a pencil eraser, make a shallow hole in the center of each section.
3. Place a pine seed in each section.
4. Sprinkle water over each section.
5. Place the container in a designated place.
6. Every 2 days for the next 3 weeks, make detailed observations of the container. Record your observations.

Extension Activity
1. Procure five pine seedlings from the instructor.
2. Sketch and describe the seedlings.

3. Plant the seedlings in a small container filled with potting soil.
4. Once every 5 days, carefully remove a seedling and note any changes. Treat the roots with care.
5. Carefully replant your seedling.
6. Record changes in the seedling.

Figure 24.63 Monkey puzzle pine, *Araucaria auracana*, is a primitive conifer characterized by sharp, thick spine-like leaves. (a) Tree, (b) stems, and (c) trunk.

Figure 24.64 Leaves of most species of conifers are needle-shaped such as those of the (a) blue spruce, *Picea pungens*, and (b) *Podocarpus* sp. has strap-shaped leaves. (c) *Araucaria heterophyla*, Norfolk Island pine, however, has awl-shaped leaves,.

Figure 24.65 (a) A cluster of pine needles, (b) a flat arborvitae leaf, and (c) a feathery bald cypress leaf.
1. Needle   2. Fascicle

The leaves of conifers are distinct. In pines, the leaf is called a **needle**. Pine needles reside in bundles called fascicles. In the majority of pine species, the **fascicles** contain between two and five needles. A fascicle is a short shoot with brown nonphotosynthetic leaves at its base. Some conifers, such as firs, spruces, and redwoods, possess long, narrow leaves and do not have fascicles. In contrast, bald cypress (*Taxodium distichum*) leaves are feathery, and arborvitae (*Thuja*) leaves are flattened. The majority of conifers are evergreen, slowly shedding their needles. Bald cypress and the dawn redwood are **deciduous** species, shedding their leaves yearly (Fig. 24.63–Fig. 24.65).

Figure 24.66 Transverse section of a leaf (needle) of *Pinus* sp.
1. Stoma
2. Endodermis
3. Resin duct
4. Photosynthetic mesophyll
5. Epidermis
6. Phloem
7. Xylem
8. Transfusion tissue

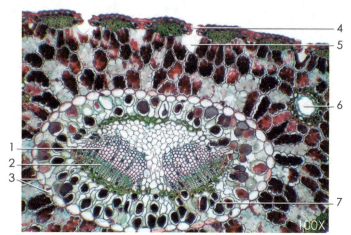

Figure 24.67 Transverse section through the leaf (needle) of *Pinus* sp.
1. Xylem
2. Phloem
3. Endodermis
4. Sunken stoma
5. Sub-stomatal chamber
6. Resin duct
7. Transfusion tissue (surrounding vascular tissue)

Student Activity—Observing the Needles and Leaves of Selected Conifers

The internal anatomy of a typical pine needle is complex. The outer portion of the needle is covered by an **epidermis** with numerous sunken **stoma**. The **mesophyll** in the needle is primarily responsible for photosynthesis. Pine needles have an **endodermis**, which separates the **transfusion** tissue from the mesophyll. The vascular tissue (xylem and phloem) is located within the transfusion tissue. **Resin ducts** are conspicuous structures in pine needles. Resin is a liquid containing terpenes, resin acids, and other compounds. It serves as a defense against insects and other animals that may want to eat the needle (Fig. 24.66–24.67).

Procedure 24.6
Needles and Leaves of Conifers

1. Procure the needed equipment and supplies.
2. Examine selected specimens of conifer needles and leaves provided by the instructor. Note if fascicles are present. If so, how many needles are associated with the fascicle? Record your sketches and observations.

Materials
- dissecting microscope
- compound microscope
- colored pencils
- selected specimens of conifer needles and leaves
- prepared microscope slide of the transverse section of a needle from *Pinus* sp.

[blank box]

_____

_____

3. If your conifer needle is fresh, smell your specimen. Describe the smell. What is responsible for the distinct smell?

_____

_____

4. Using a prepared microscope slide of a transverse section of a needle from *Pinus* sp., compare your slide with Figures 24.66 and 24.67. Place your observation and sketch below.

[blank box]

_____

_____

The wood of conifers is considered **softwood**, and the wood of angiosperms is considered **hardwood**. Softwood is composed primarily of tracheids and rays (lateral conduction structures), whereas hardwoods have tracheids, rays, and vessel elements. As a result, softwoods are relatively light and usually less dense than hardwoods, with the exceptions of balsa and basswood, among several other types. Softwoods contain vertical resin canals that occur either naturally or as a result of injury. Generally, softwoods are easier to work with in the building and furniture industry. Softwoods also are used in the production of paper, medium density fiberboard (MDF), and the majority of plywood. Hardwood is used in fine furniture and flooring and typically is more expensive and durable than softwood.

The trunks (stems) of conifers and angiosperms share the same basic anatomy. The **bark** consists of the cork, cork cambium, and phloem. The visible outer bark, **periderm**, protects the tree against water loss, extreme temperature, and infestations of insects and fungi. It is composed primarily of cork-producing cells of the **cork cambium** and non-living **cork** cells. The inner bark consists of **phloem**, which transports nutrients throughout the plant. The **vascular cambium**, located inner to the phloem, produces new phloem and xylem. The vascular cambium produces the visible annual rings. The **secondary xylem**, beneath the vascular cambium, transports water and provides support (Fig. 24.68).

Annual rings are composed of a visible band of spring wood and summer wood. **Spring wood** is lighter in color, with larger cells. **Summer wood** is darker in color, with smaller cells. In many species, the age of a tree can be determined by counting the summer wood bands. In addition to aging a tree, the annual rings can tell the life story of a tree (Fig. 24.69). In many trunk cross-sections, two distinct regions of xylem are present: Sapwood is the lighter, outer xylem tissue that actively transports water; **heartwood** is the darker inner xylem tissue that serves primarily as a reservoir for gum, resin, and tannin. The heartwood clogs up over time (Fig. 24.71–Fig. 24.77).

Courtesy of Champion Paper Company, Inc.

Figure 24.68 Diagram of the tissues in the stem (trunk) of a conifer. The periderm and dead secondary phloem (outer bark) protects the tree against water lost and the infestation of insects and fungi. The cells of the phloem (inner bark) compress and become nonfunctional after a relatively short period. The vascular cambium annually produces new phloem and xylem and accounts for the growth rings in the wood. The secondary xylem is a water transporting layer of the stem and provides structural support to the tree.

1. Outer bark
2. Phloem
3. Vascular cambium
4. Secondary xylem

Courtesy of Champion Paper Company, Inc.

Figure 24.69 Stem (trunk) of a pine tree that was harvested in the year 2000 when the tree was 62 years old. The growth rings of a tree indicate environmental conditions that occurred during the tree's life.

1. **1939**—A pine seedling.
2. **1944**—Healthy, undisturbed growth indicated by broad and evenly spaced rings.
3. **1949**—Growth disparity probably due to the falling of a dead tree onto the young healthy six-year old tree. The wider "reaction rings" on the lower side help support the tree.
4. **1959**—The tree is growing straight again, but the narrow rings indicate competition for sunlight and moisture from neighboring trees.
5. **1962**—The surrounding trees are harvested, thus permitting rapid growth once again.
6. **1965**—A burn scar from a fire that quickly scorched the forest.
7. **1977**—Narrow growth rings resulting from a prolonged drought.
8. **1992**—Narrow growth rings, resulting from a sawfly insect infestation, whose larvae eat the needles and buds of many kinds of conifers.

Figure 24.70 (a) A knot in a section of redwood, *Sequoia sempervirens*, and (b) knees of the bald cypress, *Taxodium distichum.*

Of Knots and Knees

Look at a piece of flooring or furniture and observe the knots. The pattern of knots seems to give wood its character. A knot is where the base of a branch has been overtaken by the lateral growth of the trunk. The bald cypress *(Taxodium distichum)* is characterized by the presence of knees, or the pneumatophores. Although the exact function of cypress knees is unknown, they are thought to provide stability in wet soils and aerate the roots, providing oxygen.

Each generation takes the earth as trustees. We ought to bequeath to posterity as many forests and orchards as we have exhausted and consumed.

—**J. Sterling Morton (1832–1902)**

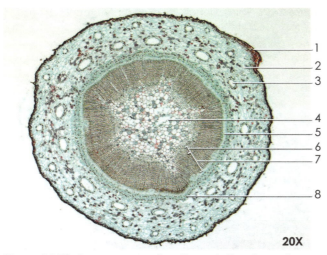

20X

Figure 24.71 Transverse section through the stem of a young conifer showing the arrangement of the tissue layers.
1. Epidermis
2. Cortex
3. Resin duct
4. Pith
5. Cambium
6. Primary xylem
7. Spring wood of secondary xylem
8. Primary phloem

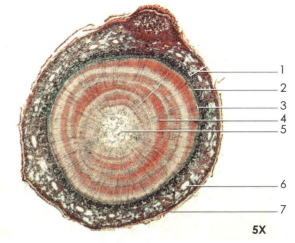

5X

Figure 24.72 Transverse section through the stem of *Pinus* sp., showing secondary stem growth.
1. Bark (cortex and periderm)
2. Secondary phloem
3. Vascular cambium
4. Secondary xylem
5. Pith
6. Resin duct
7. Epidermis

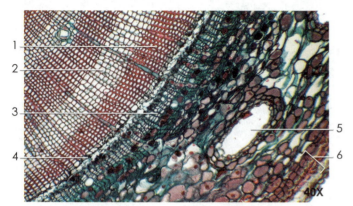

Figure 24.73 Enlarged view of the stem of *Pinus* sp. showing tissues following secondary growth.

1. Secondary summer wood
2. Secondary spring wood
3. Secondary phloem
4. Vascular cambium
5. Resin duct
6. Periderm

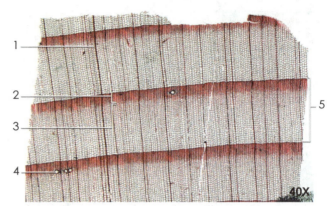

Figure 24.74 Transverse section of mature wood of *Pinus* sp. showing three growth rings.

1. Ray
2. Secondary summer wood
3. Secondary spring wood
4. Resin duct
5. One ring

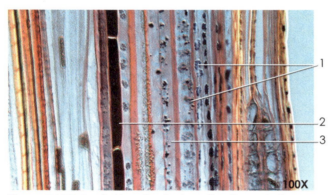

Figure 24.75 Radial longitudinal section through the phloem of *Pinus* sp.

1. Sieve areas on a sieve cell
2. Storage parenchyma
3. Sieve cell

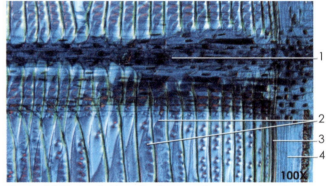

Figure 24.76 Radial longitudinal section through a stem of *Pinus* sp., cut through the xylem tissue.

1. Ray parenchyma
2. Tracheids
3. Vascular cambium
4. Sieve cells

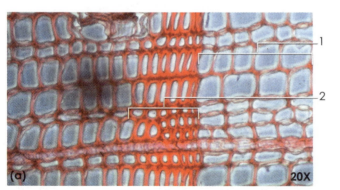

Figure 24.77 Growth rings in *Pinus* sp. (a) Transverse section through a stem; and (b) radial longitudinal section through a stem.

1. Spring wood
2. Summer wood

Student Activity—Observing the Microscopic Anatomy of Conifer Stems

Materials
- compound microscope
- colored pencils
- prepared microscope slide of the following: transverse section through the stem of a young conifer, transverse section through the stem of *Pinus* sp., radial longitudinal section through the phloem of *Pinus* sp., and radial longitudinal section through the xylem of *Pinus* sp.

Procedure 24.7
Microanatomy of Conifer Stems

1. Procure the needed equipment and supplies.
2. Examine a prepared microscope of a transverse section through the stem of a young conifer. Compare your slide to Figure 24.71 and 24.72 on page 401. Sketch and label your specimen below.

3. Examine a prepared microscope slide of the transverse section through the stem of *Pinus* sp. Compare your slide to Figure 24.73. Sketch and label your specimen below.

4. Examine a prepared microscope slide of a radial longitudinal section through the phloem of *Pinus* sp., Compare your slide to Figure 24.75. Sketch and label your specimen below.

5. Examine a prepared microscope of a radial longitudinal section through the xylem of *Pinus* sp. Compare your slide to Figure 24.76. Sketch and label your specimen below.

Student Activity—Observing and Aging a Conifer Stem

Materials
• Cross-section of a tree trunk

Procedure 24.8
Conifer Stem Aging

1. Procure a cross-section of a trunk.
2. Identify the bark, rays, heartwood, sapwood, summer wood, and spring wood.

3. Determine the age of your specimen.

_____

_____

4. Examine the annual rings. Was the tree exposed to any interesting conditions? How do you know, and when in the life of the tree did it occur?

_____

_____

_____

_____

_____

_____

NOTES

Name: _____ Date: _____ Section: _____

Review Questions

1. Compare and contrast the seeds of gymnosperms and angiosperms.

_____

_____

2. How have seed plants influenced the development of humans?

_____

_____

3. What are the general characteristics of gymnosperms?

_____

_____

4. List and give examples of each phylum of gymnosperms.

_____

_____

_____

5. What are some unique characteristics and facts about *Ginkgo biloba*?

_____

_____

6. What is unique about *Welwitschia*?

_____

_____

7. What characteristic do members of Phylum Gnetophyta have in common with angiosperms?

_____

_____

8. Describe reproduction in cycads.

_____

_____

9. Sketch and label a section of a tree trunk.

10. Describe the life cycle of a pine.

_____

_____

_____

12. Discuss the economic importance of conifers.

_____

_____

13. Trace the life cycle of a pine tree.

14. Why is bald cypress different from other conifers?

_____

Chapter 25
The Green Machine: Part IV
Understanding the Seed Plants (Angiosperms)

Student Outcome Objectives

At the completion of this exercise, the student will be able to:

1. Describe the fundamental characteristics of angiosperms.
2. Explain the distribution of angiosperms.
3. Discuss the lifestyles of angiosperms.
4. State the importance of angiosperms to human civilization.
5. Compare and contrast herbaceous and woody plants.
6. Compare and contrast annuals, biennials, and perennials.
7. Describe the characteristics of Phylum Magnoliophyta.
8. Discuss difficulties in the classification of angiosperm.
9. Describe and provide examples of basal dicots.
10. Describe and provide examples of eudicots.
11. Describe and provide examples of monocots.
12. Compare and contrast fundamental characteristics of classic dicots and monocots
13. Identify representative basal dicots.
14. Identify representative eudicots.
15. Identify representative monocots.
16. Photograph given plant specimens.

Overview

Charles Darwin called the sudden appearance of modern flowers in the fossil record "an abominable mystery" because their abrupt emergence was difficult to explain. Since the time of Darwin, however, the fossil record has increased tremendously. The discovery in 1998 of an angiosperm fossil collected in Liaoning Province, China—*Archaefructus sinensis* from the Cretaceous Period approximately 125 million years ago—and other discoveries have provided scientists with clues as to how angiosperm characteristics may have been acquired in a series of steps as Darwin predicted. These steps began in the Jurassic, and today the flowering plants are the dominant plants on Earth, representing more than 90% of all living plant species and 18% of all species. Development of the angiosperms transformed the face of the planet. Today the landscape is painted with colorful flowers as plants advertise themselves to pollinators.

The **angiosperms** (Greek − *vessel, seed*) are known as the flowering plants. They comprise the most diverse and numerous groups of plants on Earth (Fig. 25.1). Botanists have identified more than 280,000 species of angiosperms thus far. Like the gymnosperms, the sporophyte generation is the dominant portion of the life cycle of angiosperms. Angiosperms can be found in a number of environments, including the desert, rainforests, and salt-marshes. They range in size from a duckweed *Wolffia angusta*, smaller than

Figure 25.1 Angiosperm diversity: (a) an orchid, *Phalaenopsis* sp., (b) a sundew, *Drosera binata*, (c) a cactus, *Carnegiea gigantea*, (d) a vine, *Parthenocissus quinquefolia*, and (e) a baobab tree, *Dansonia digitata*.

1 millimeter in diameter, to Australia's mountain ash tree *Eucalyptus regnans*, which that stands nearly 100 meters in height and rivals the great height attained by some giant sequoias.

Angiosperms also vary in form, including delicate orchids, strange insectivorous sundews, succulent cacti, wire-like vines, and the majestic baobab tree. The vast majority of angiosperms are autotrophic. Several species of angiosperms are parasitic. Mistletoe (*Phoradendron* sp.) is a **hemiparasite** undergoing photosynthesis and parasitizing its host plant, and dodder (*Cuscuta* sp.) is a true parasitic plant. Indian-pipe (*Monotropa uniflora)* and the snowplant (*Sarcodes sanguinea*) are two of several **saprophytic** species of angiosperms. Spanish moss (*Tillandsia usneoides*) and some orchids, cacti, and ferns are **epiphytes**, or "air plants," which attach to a substrate such as another plant or the side of a building. The angiosperms are important to humans as sources of food, medicine, aesthetic beauty, cotton, lumber, and other commercial products.

The classification of angiosperms has been affected radically by molecular studies. The unprecedented increase in knowledge has led to many changes in the classification of flowering plants. This became a source of confusion and to scientists and laypeople alike. In general, the angiosperms have several distinguishing characteristics. The angiosperms are noted for their complex reproductive structures called **flowers**. The **ovules** of angiosperms are encased within diploid tissue (integuments) supplied by the parent plant. Angiosperms and their oldest living relatives, the gnetophytes, undergo double fertilization, in which a fertilized egg and nutritive endosperm form. In addition to tracheids, the xylem of angiosperms possesses vessels that transport water efficiently throughout the plant.

The majority of angiosperms are deciduous, losing their leaves during the winter or perhaps during a drought. Angiosperms such as a zinnia are **herbaceous**, possessing little or no woody tissue, and others, such as an apple tree, are **woody**. Flowering plants can cycle from germination to mature plant in less than a month or as long a 150 years or more. In **annuals** such as geraniums, the life cycle of a plant is completed in one season. **Biennials** such as foxglove complete their life cycle in two growing seasons. **Perennials** such the Black-eyed Susan may take more than two seasons to complete their life cycle.

The angiosperms, or flowering plants, have been placed in either Phylum **Magnoliophyta** or Phylum **Anthophyta**. Traditionally, Phylum **Magnoliophyta** has been divided into two major ranks, or classes: the

How About a Kiss Under the Mistletoe?

In Anglo-Saxon, mistletoe literally means "dung-on-a-twig." It was thought to appear through spontaneous generation from bird feces. The etymology does not exactly reflect the romantic reputation of the mistletoe plant! Mistletoe has been considered one of the most magical plants in European folklore. It was thought to bestow fertility and life to people, protect against poisons, and even serve as an aphrodisiac.

The first recorded tradition of kissing under the mistletoe is associated with the Greek festival of Saturnalia and early marriage rites. Mistletoe supposedly brought about fertility and long life. In Norse mythology, two enemies were said to find peace by kissing under the mistletoe. In Victorian England at Christmas, a young lady would stand under a kissing ball of mistletoe and await a kiss. If she was kissed, it could mean romance or close friendship. If she was not kissed, it meant that she would not marry during the next year. Today, the tradition of kissing under the mistletoe appears throughout the holiday season and signifies love or lasting friendship. Ironically, mistletoe berries are poisonous!

Dicotyledones (Magnoliopsida) and the **Monocotyledones** (Liliopsida). Although this scheme is "user-friendly" and is still used extensively, new molecular evidence and advanced observations are paving the way for a new classification of angiosperms. Classically, Dicotyledones, or **dicots**, are flowering plants whose seed contains two embryonic leaves, or **cotyledons**, and Monocotyledones, or **monocots**, are flowering plants with a single cotyledon. A cotyledon is a seed leaf containing nutrients that nourish the developing embryonic plant. This is the first leaf evident upon germination.

A number of newer schemes are attempting to make sense of the diversity of angiosperms. A commonly used scheme divides the angiosperms into the basal dicots, the eudicots, and the monocots (Fig. 25.2). The **basal dicots**, or paleodicots, such as water lilies, avocado, black pepper,

Figure 25.2 (a) The Magnolia tree, *Magnolia* sp., is an example of a basal dicot; (b) the American sycamore, *Platanus occidentalis*, is an example of a eudicot, and (c) the coconut palm, *Cocos nucifera*, is an example of a monocot.

Beyond the Lab—Angiosperms

Look up the following:

1. Five angiosperms that have a strong role in medicine.

   foxglove → digitalis
   digitalis purpurea

2. Five angiosperms that are ornamental plants.

3. Five wild edible angiosperms.

4. Five poisonous angiosperms.

5. Five angiosperms used as food or seasoning.

   rice maeze wheat
   apple
   ginger
   brasseca

Did You Know?

The magnolia tree is named after the French botanist Pierre Magnol (1638–1715).

and magnolias, possess **monosulcate pollen**, as do the monocots (Fig. 25.3). Monosulcate pollen has a linear, thin, furrow-like groove, or sulcus, on the distal surface of the grain and one pore. The true dicots, **Eucotyledones** or **eudicots**, sometimes called **tricolpates**, are the largest group of angiosperms and possess **tricolpate pollen**. This pollen has three long, grooved apertures or pores on the surface. Example eudicots are roses, sycamore trees, citrus trees, sunflowers, peanuts, and cabbage. The monocots are thought to have evolved from the dicots. Example monocots are cattail, corn, lilies, orchids, palms, and bananas.

Despite major problems and much confusion in organizing angiosperm taxa, the classical distinctions between dicots and monocots are helpful. Keep in mind that Table 25.1 is based upon generalizations with many exceptions.

THE BASAL DICOTS

In recent years the dicots have been split into the basal dicots and the eudocts. The basal dicots or paleodicots are divided into several groups:

1. The **Nymphaeales** are aquatic plants, most of which have floating leaves, such as water lilies and lotus plants.
2. The **Piperales** are a group of herbs, shrubs, and small trees, such as black pepper, lizard's tail, vine pepper, and wild ginger.
3. The **Laurales**, such as sassafras, cinnamon, and avocado, at one time were placed with magnolias.
4. The **Magnoliales**, the best-known group, consist of magnolia, sweet bay, nutmeg, and tulip tree.

Several other basal eudicot groups have been identified in recent years.

Figure 25.3 Examples of basal dicots, (a) water lily *Nymphaea* sp., (b) sassafras *Sassafras* sp., (c) lizard's tail, *Saururus* sp., and (d) tulip tree *Liriodendron tulipifera*.

Table 25.1 Comparison of Dicots and Monocots

Dicots	Monocots
Embryo with two cotyledons	Embryo with one cotyledon
Pollen with three apertures (except basal dicots)	Pollen with sulcus and one pore
Flower parts in multiples of four or five	Flower parts in multiples of three
Netted venation of leaf veins	Parallel venation of leaf veins
Vascular bundles arranged in a ring	Vascular bundles scattered
Secondary growth present	Secondary growth absent

Student Activity—Observation of Basal Dicots

Materials
- dissecting microscope
- compound microscope
- hand lens
- colored pencils
- wet mount
- prepared slides of basal dicots: stem, leaves, flowers, cones, and pollen from several basal dicots such as a water lily, lizard tail, sassafras, avocado, magnolia, tulip tree, and others
- prepared slide of pollen from a eudicot
- microscope slides and coverslips

 Procedure 25.1
Basal Dicots

1. Procure select specimens of the basal dicots and equipment provided by the instructor. Using a dissecting microscope or hand lens, observe and draw the anatomical features of your specimens. Place your labeled sketches and observations below. Remember to label your specimen.

2. Prepare a wet mount of pollen from one of the flowers, or use a prepared slide of pollen from a basal dicot. Using a compound microscope, observe, describe, and sketch a sample of magnolia pollen in the space below.

The rapid development as far as we can judge of all the higher plants within recent geological times is an abominable mystery.

—Charles Darwin (1809–1882)

Check Your Understanding

Q. What are the fundamental characteristics of basal dicots?

_____

_____

Q. List five plants that are considered basal dicots.

_____

_____

_____

_____

_____

The cutting of the primeval forest and other disasters fueled by the demands of growing human populations, are the overriding threat to biological diversity everywhere.

—Edward O. Wilson (1929–present)

Q. Describe four groups of basal dicots and provide two examples of each.

_____

_____

_____

_____

_____

_____

_____

_____

THE EUDICOTS

The majority of plants traditionally recognized as "dicots" are now considered eudicots. The eudicots consist of an extremely large, diverse group of angiosperms that have an incredible range of geographic distribution, variation within habitats, anatomy, morphology, and biochemistry (Fig. 25.4). Table 25.2 lists common families of eudicots with examples of each.

Student Activity—Observations of Eudicots

Materials
- dissecting microscope
- compound microscope
- hand lens
- colored pencils
- stem, leaves, flowers, fruits, and pollen from several eudicots such as a rose, maple, oak, squash, sunflower, willow, and others
- prepared slide of pollen from a eudicot
- microscope slides and coverslips

Procedure 25.2
Eudicots

Consider photographing this activity.

1. Procure select specimens of the eudicots and equipment provided by the instructor. Using a dissecting

microscope or hand lens, observe and draw the anatomical features of your specimens. Place your labeled sketches and observations below and to the right.

Table 25.2 Representative Eudicots

Family	Examples	Family	Examples
Aceraceae	sugar maple, red maple	Linaceae	blue flax, yellow flax
Altingiaceae	sweetgum, alligator-wood	Meliaceae	mahogany, chinaberry
Anacardiaceae	poison ivy, cashews, pistachios	Malvaceae	hibiscus, mallow, cotton
Aquifoliaceae	American holly, English holly	Moraceae	mulberry, fig
Asclepiadaceae	milkweed, twinevine	Myrtaceae	myrtle, bottlebrush, eucalyptus
Asteraceae	zinnia, aster, sunflower, daisy, thistle, lettuce	Oleaceae	olives, ashes, lilacs
Betulaceae	alder, birch	Oxalidaceae	oxalis, wood sorrel
Bignoniaceae	bigonia, desert willow	Papaverageae	poppy, bloodroot
Brassicaceae	radish, broccoli, cabbage, turnip	Passifloraceae	passion flower, love-in-a-mist
Cactaceae	saguaro cactus, prickly pear cactus	Platanaceae	sycamore, plane tree
Caprifoliaceae	honeysuckle, elderberry	Primulaceae	primrose, shootingstar
Convolvulaceae	morning glory, sweet potato	Ranunculaceae	buttercup, columbine, goldenseal
Cornaceae	dogwood, tupelo	Rosaceae	roses, apples, strawberries, almonds
Cucurbitaceae	pumpkin, squash, gourd	Rubiaceae	coffee, chinchona, madder
Cuscutaceae	dodder	Rutaceae	orange, lime, lemon, grapefruit
Droseraceae	sundew	Salicaceae	willow, poplar, cottonwood
Ericaceae	azaleas, heath, huckleberry, mountain laurel	Sarraceniaceae	white-top pitcher plant, crimson pitcher plant
Euphorbiaceae	spurge, cassava, rubber plant, poinsettia	Smilaceae	green briar, kudzu, sarsaparilla
Fabaceae	beans, peas, peanuts, clover, mimosa, licorice	Solanaceae	potato, tomato, tobacco
Fagaceae	live oak, water oak, red oak, beech	Theaceae	camellia, stewartia
Geraniaceae	geranium, pelargonium	Ulmaceae	elm, hackberry
Hydrangeaceae	hydrangia, whipplevine	Umbelliferae	carrot, parsley
Juglandaceae	walnut, hickory, pecan	Vitaceae	grape, muscadine
Lamiaceae	mint, catnip, lavender, sage, oregano		

(a)

(b)

(c)

Figure 25.4 Examples of eudicots: (a) passion flower, *Passiflora* sp, (b) Russian olive, *Elaeagnus Angustifolia,* (c) and rose, *Rosa* sp.

sketch a sample of eudicot pollen. Record and sketch your observations below.

2. Prepare a wet mount of pollen from one of the flowers, or use a prepared slide of pollen from a eudicot. Using a compound microscope, observe, describe, and

Figure 25.5 Examples of monocots: (a) papyrus, *Cyperus papyrus*, (b) water hyacinth, *Eichhornia* sp., (c) and duckweed, *Lemna* sp.

THE MONOCOTS

Monocots comprise about one-fourth of all living angiosperms. The monocots diverged from the basal dicots approximately 90 million years ago. The majority of plants used in the agricultural industry are monocots. Monocots play a significant role in the floral and horticulture industry (Fig. 25.5). Table 25.3 lists common families of monocots and gives examples of each.

 Student Activity—Observations of Monocots

Materials
- dissecting microscope
- compound microscope
- hand lens
- wet mount
- colored pencils
- stem, leaves, flowers, and pollen from several monocots such as corn, orchids, azaleas, lilies, bananas, tulips, and others
- prepared slide of pollen from a monocot
- microscope slides and coverslips

 **Procedure 25.3
Monocots**

Consider photographing this activity.

1. Procure select specimens of monocots and equipment provided by the instructor. Using a dissecting

Table 25.3 Representative Monocots	
Family	**Examples**
Agavaceae	century plant, agave
Alismataceae	arrowhead, sagittaria
Aloeaceae	aloe
Amaryllidaceae	spider lily, amarylis
Araceae	caladium, philodendron, skunk cabbage, peace lily, Jack-in-the-Pulpit
Arecaceae	coconut palm, palmetto
Bromeliaceae	pineapple, Spanish moss
Commelinaceae	tradescantia, spiderwort
Cyperaceae	water chestnut, papyrus, spikerush, sawgrass
Hydrocharitaceae	elodea
Iridaceae	iris, gladiolus, crocus
Juncaceae	brown bog rush, blue rush
Lemnaceae	duckweed, wolffia
Liliaceae	lily, agave, tulip, asparagus, onion, hyacinth
Musaceae	banana, cannaplant, bird-of-paradise
Poaceae	corn, sugar cane, rice, bluegrass, wheat, oats, bamboo, crab grass, broom sedge, foxtail
Pontederiaceae	pickerel weed, water hyacinth
Orchidaceae	orchids
Typhaceae	cattail
Trilliaceae	trillium

microscope or hand lens, observe and draw the ana–tomical features of your specimen. Place your labeled sketches and observations below.

2. Prepare a wet mount of pollen from one of the flow-ers, or use a prepared slide of pollen from a monocot. Using a compound microscope, observe, describe, and

sketch a sample of monocot pollen. Record your observations and sketches below.

Photography has always been a fascinating hobby and profession. Taking photographs of family, friends, pets, and vacation is a fun activity. Why not try a little biophotography with your digital camera? You do not have to use an expensive digital single-lens reflex (DSLR) camera to complete this assignment. Think of the following activity as a digital field trip or a scavenger hunt. The activity is green, too—no plants will be harmed in this activity. Find the plants in question, take your time to compose a photograph of the plant, use good photographic technique, check out the quality of the image on the camera LCD, transfer the image to a computer, and develop a labeled presentation (Fig. 25.6).

Suburbia is where the developer bulldozes out the trees,
then names the streets after them.

—**Bill Vaughan (1915–1977)**

 ## A Digital Field Trip—Pictures of Angiosperms

Materials
- digital camera
- notebook to record data
- computer
- presentation software
- storage device such as a jump drive or CD
- projector and screen

 ## Procedure 25.4
Digital Field Trip

1. Referring to the lists of basal dicots, eudicots, and monocot families in this manual, locate a representative member of two different basal dicots, eight different eudicots, and five different monocot families.
2. In addition, for a wild card, photograph five other families of your choice—three gymnosperms plus one nonvascular plant and one seedless vascular plant. If you have not used your state flower and tree in this endeavor, include them in your slide show as well. You may find these plants on campus, in your yard, in the woods, in a greenhouse, at an arboretum, or in a botanical garden.
3. Take photographs of a plant. The photographs should include the entire plant, as well as its distinguishing features such as flowers, bark, and growth pattern.

4. Using computer and presentation software, develop a digital slide show of your plants. Label your slides clearly. Include whether the plant is a basal dicot, an eudicot, or a monocot, as well as the family and common and scientific names. Be sure to label your wild cards! And how about adding some music to your presentation?
5. Place your slide show on a storage device or CD. The instructor may allow you to present your slide show to the class.

Figure 25.6 A student out taking pictures.

Name: _____ Date: _____ Section: _____

Review Questions

1. How has the evolution of angiosperms transformed the face of the planet?

_____

_____

_____

2. Differentiate hemiparasitic and true parasitic plants, epiphytes, and saprophytic plants.

_____

_____

_____

3. Describe annuals, biennials, and perennials.

_____

_____

_____

4. What is a basal dicot? Provide several examples.

_____

_____

_____

_____

5. Describe the fundamental characteristics of a eudicot. Provide 10 examples of eudicots found in your region.

_____

_____

_____

_____

_____

_____

6. What are the fundamental characteristics of a monocot? Provide 10 examples of monocots found in your region.

_____

_____

_____

_____

_____

7. What is the basis for reorganization of the angiosperms?

_____

_____

_____

8. Why was Darwin perplexed with the evolution of angiosperms?

_____

_____

9. Name 10 important angiosperms and their uses.

_____

_____

_____

_____

_____

_____

_____

_____

_____

10. How can biophotography enhance your laboratory experience?

_____

_____

Chapter 26
From the Ground Up:
Understanding Roots, Stems, and Leaves

Student Outcome Objectives

At the completion of this exercise, the student will be able to:

1. Discuss the vegetative organs of a typical plant.
2. Describe the function of roots.
3. Compare and contrast the root systems of eudicots and monocots.
4. Describe the function of root hairs.
5. Identify and describe the macroscopic anatomy of roots.
6. Identify and define the function of microscopic anatomical features of roots.
7. Identify and define the function of specialized roots.
8. Discuss the relevance of roots to humans.
9. Perform an activity that illustrates the growth of roots.
10. Explain the function of stems.
11. Compare and contrast the stems of eudicots and monocots.
12. Identify and describe anatomical features of a typical woody stem.
13. Identify and define the function of microscopic anatomical features of various stems.
14. Identify and define the function of specialized stems.
15. Describe the movement of water up a stem.
16. Discuss the relevance of stems to humans.
17. Determine the age of a stem.
18. Describe the function of leaves.
19. Compare and contrast the leaves of eudicots and monocots.
20. Identify and describe anatomical features of a typical leaf.
21. Locate and describe the function of stomata.
22. Classify leaves based upon phyllotaxy.
23. Classify leaves based upon shape, type of margin, type of apex, and type of base.
24. Identify and define the function of microscopic anatomical features of various leaves.
25. Identify and define the function of specialized leaves.
26. Discuss the relevance of leaves to humans.

Overview

Although the flowering plants vary tremendously in habitat, size, and morphology, they share the same fundamental vegetative organs: roots, stems, and leaves (Fig. 26.1). Working together, these three organs build the plant from the ground up. The basic tissues that comprise the roots, stems, and leaves of the eudicots and monocots are arranged differently and provide a simple means of differentiating these two distinct groups of angiosperms. The roots and their ancillary structures comprise the **root system** of a plant, and the stem and leaves and their ancillary structures comprise the **shoot system** of a plant. In addition to vegetative structures, plants possess **reproductive** or **germinative structures** such as flowers, fruits, and seeds, will be discussed in Chapter 27.

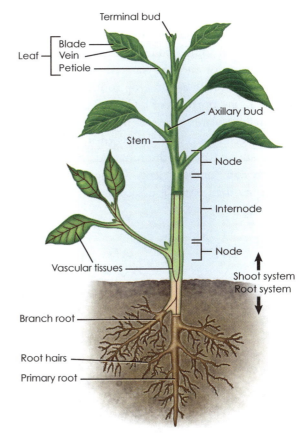

Figure 26.1 Basic organization of the vegetative organs of a generalized plant.

ROOTS

Roots are plant organs that evolved to anchor and support a plant, absorb water and necessary minerals, store food, and produce growth-stimulating hormones. The majority of the root system is found underground, although some plants, such as the tropical fig, possess extensive aerial roots. Generally, the extent of the root system is equal to or exceeds that of the shoot system. In some plants, including many grasses, the root system is just a few millimeters beneath the soil, and in some dry-climate plants such as *Juniperus monosperma*, the root system exceeds 20 meters in depth. The number of roots produced by a plant can be staggering. A single ryegrass plant, for example, may have up to 15 million roots and a surface area larger than a volleyball court.

In plant development, when a seed germinates, a small root-like structure, the **radicle**, emerges from the embryo and forms the first root. The radicle usually gives rise to a single, tapered **taproot** with many small lateral branches. Taproot systems are common in conifers and many species of eudicots. Taproots anchor the plant and seek deep water supplies. The fleshy portion of a carrot is an example of a taproot that stores food in the form of carbohydrates.

Monocots and some eudicots possess a **fibrous root system**. The stem or another plant part produces **adventitious roots**, such as the prop roots of corn. Sweet potatoes are fleshy portions of a fibrous root system. Many species of mature plants have a combination of a taproot and fibrous roots. Roots also possess **root hairs**, which increase the surface area of the root and, therefore, its ability to absorb water and nutrients. A single ryegrass plant may have more than 15 billion root hairs. In removing

plants, the root system should not be injured, as the delicate roots and root hairs are vital features (Fig. 26.2).

The **root cap** is a group of specialized cells at the tip of the root. Its major function is to protect the delicate inner root from abrasive soil. The cells also produce a muscilaginous lubricant that helps the root pass through the soil and aids the growth of nitrogen-fixing bacteria in some plants, such as clover, peas, and peanuts. Cells in the root cap also are thought to orient the root growth toward gravity, called **gravitropism**.

A typical root has four distinct regions, or zones:

1. The **zone of cell division**, the apical meristem or meristematic region, is found behind the root cap. Cells in this region are prolifically undergoing cell division. Three distinct types of tissue are produced in the apical meristem:

   a. the **protoderm,** which eventually gives rise to the epidermis;

   b. the **ground meristem,** which produces the parenchyma cells of the cortex; and

   c. the **procambium,** which gives rise to primary xylem and phloem.

2. The **zone of elongation** is found above the zone of cell division. In this region, the cells greatly increase in length, and much less in width, as small vacuoles merge, filling up more than 90% of the cell. The elongation in these cells helps to push the root cap through the soil. Some roots can push through the soil at a rate of 4 cm a day.

3. In the **zone of maturation**, the cells produced in the zone of elongation become differentiated and mature, forming the epidermis, cortex, endodermis, and vascular cylinder. The outermost cells differentiate into the single-layered **epidermis**. Interior to the epidermis is the **cortex**, composed of parenchyma cells. These cells possess starch granules and are used primarily in food storage.

4. The endodermis consists of a single layer of cells that serves as the boundary between the cortex and the fourth region, the **vascular cylinder**. The primary cell walls of the endodermis contain lignin and suberin, which form an impermeable layer called the **Casparian strip**. The strip blocks the passage of water and minerals between adjacent cells and regulates the movement of water and minerals into the **vascular cylinder**. The vascular cylinder contains the xylem and phloem, which, in eudicots, sometimes is referred to as the **stele**. The **pericycle** is the first layer of cells in the vascular cylinder.

Taproot
(shrubs)

Fibrous root system
(grasses)

Figure 26.2 Taproots and fibrous root systems anchor plants and absorb water and essential minerals.

The cells of the pericycle can divide and initiate the development of lateral roots. In eudicots, the xylem is star-shaped with several radiating arms. The phloem is located between the radiating arms. In monocots, ground tissue forms the **pith**, which is centrally located. Vascular tissue is located in bundles in a ring surrounding the pith, with the xylem oriented exteriorly and the phloem oriented interiorly. Roots can undergo **primary growth** and lengthen, and **secondary growth** can increase the diameter. Over time, eudicot roots may develop concentric rings of xylem, and in woody eudicot roots, everything outside the stele is replaced by bark (Fig. 26.3–Fig. 26.17).

Check Your Understanding

Q. Name several functions of roots.

_____

_____

Q. List and describe the major zones of a typical root.

_____

_____

Q. What is the difference between primary and secondary growth?

_____

_____

_____

_____

Q. Describe several specialty roots.

_____

_____

_____

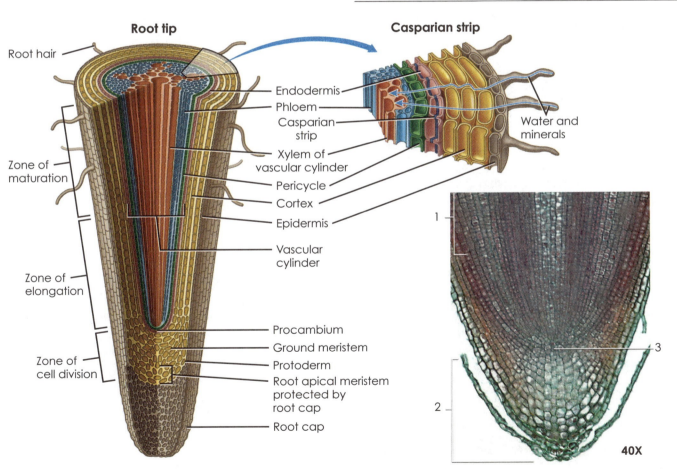

Figure 26.3 Diagram of a root tip.

Figure 26.4 Photomicrograph of the root tip of a pear, *Pyrus* sp., seen in longitudinal section.
1. Elongation region   3. Apical meristem
2. Root cap

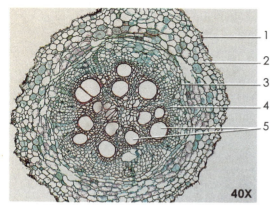

Figure 26.5 Transverse section of a sweet potato root, *Ipomaea* sp.
1. Remnants of epidermis   4. Phloem
2. Cortex   5. Xylem
3. Endodermis

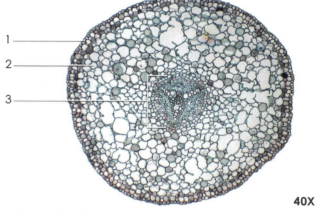

40X

Figure 26.6 Transverse section of a young root of *Salix* sp.
1. Epidermis   3. Stele
2. Cortex

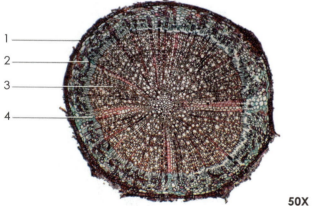

Figure 26.7 Transverse section of an older root of *Salix* sp., showing early secondary growth.
1. Epidermis   3. Vascular tissue
2. Cortex

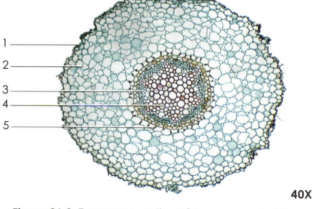

40X

Figure 26.8 Transverse section of a young root of *Pyrus* sp.
1. Epidermis   4. Primary xylem
2. Cortex   5. Endodermis
3. Primary phloem

50X

Figure 26.9 Transverse section of the root, *Pyrus* sp., showing secondary growth.
1. Periderm   3. Secondary xylem
2. Secondary phloem   4. Vascular cambium

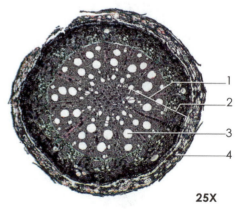

25X

Figure 26.10 Transverse section of the root of basswood, *Tilia* sp., showing secondary growth.
1. Secondary xylem   3. Vessel element
2. Secondary phloem   4. Periderm

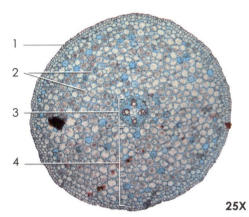

Figure 26.11 Low magnification of the root of a buttercup, *Ranunculus* sp.
1. Epidermis
2. Parenchyma cells of cortex
3. Stele
4. Cortex

25X

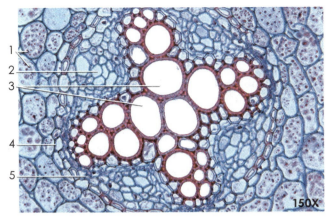

Figure 26.12 High magnification showing a transverse section of the stele.
1. Starch grains within parenchyma cells
2. Primary phloem
3. Primary xylem
4. Endodermis
5. Pericycle

150X

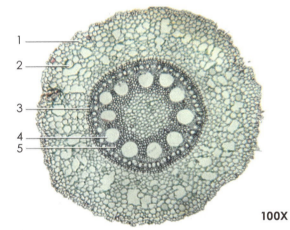

Figure 26.13 Transverse section of the root of the monocot *Smilax* sp., low magnification.
1. Epidermis
2. Cortex
3. Endodermis
4. Xylem
5. Phloem

100X

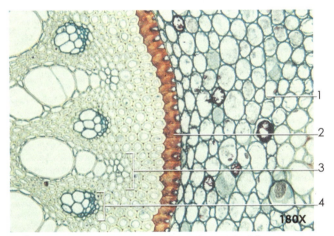

Figure 26.14 High magnification of a root of the monocot *Smilax* sp.
1. Cortex
2. Endodermis
3. Xylem
4. Phloem

180X

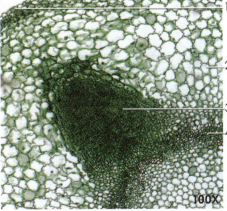

Figure 26.15 Longitudinal section of a willow species showing lateral root formation.
1. Lateral root
2. Epidermis
3. Cortex
4. Vascular tissue

110X

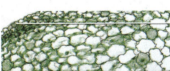

Figure 26.16 Transverse section branch root formation of *Phaseolus* sp.
1. Epidermis
2. Cortex
3. Branch root
4. Vascular tissue (stele)

100X

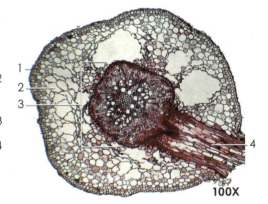

Figure 26.17 Transverse section of the root of *Salix* sp. showing branch root development.
1. Epidermis
2. Cortex
3. Stele
4. Branch root

100X

Throughout the plant kingdom, roots vary in form and function. Corn (*Zea mays*) produces a type of adventitious root known as a **prop root** from its lower stem. Prop roots help to anchor and brace the plant against wind. Dodder (*Cuscuta* sp.) possesses parasitic roots, called **haustoria**, which parasitize a host plant. Some members of the pumpkin family (*Curcurbitaceae*) that live in dry climates produce large **water storage roots**. Certain species of figs and swamp trees, such as the tupelo and bald cypress, have expanded **buttress roots** for stability in wet environments.

Some **food storage roots**, such as horseradish, dandelion, beet, radish, turnip, and carrot, have expanded food storage capability. Lily bulbs have **contractile roots** that help to pull the bulb deeper into the soil. **Aerial roots** are diverse; examples are English ivy, Virginia creeper, banyan trees, and tropical fig trees. **Pneumatophores** are spongy roots that extend out of the water in swamp plants such as the black mangrove (*Avicennia nitida*). At one time, the **knees** of bald cypress (*Taxodium distichum*) were thought to be pneumatophores. Today, their complete function is unknown, but they are thought to be primarily for support (Fig. 26.18).

Some roots display complex relationships with other organisms. Members of the legume family Fabaceae, such as peanuts, clover, beans, and peas, possess small **root nodules** containing nitrogen-fixing bacteria. The bacteria produce enzymes that convert atmospheric nitrogen into nitrates and nitrogenous substances that can be absorbed by the roots. **Root knots** are swellings found in tomatoes and several other plants that may house parasitic roundworms, or nematodes. The roots of many plants have a mutualistic relationship with fungi, known as **mycorrhizae**. The plant supplies the mutualistic relationship with sugars and amino acids, and the fungus helps the plant metabolize phosphorus.

In the environment, roots hold soil together. This is particularly important in coastal regions. Roots are used for food by animals and humans. Edible roots include, among others, carrot, sweet potato, cassava, beet, horseradish, turnip, radish, and rutabaga. Several roots are used for spices and include licorice, ginger, and sassafras. Sugar beets are a primary source of sugar, and the roots of yams have been a source of estrogen compounds in making birth control pills. The roots of many plants, including ipecac, gentian, ginsing, and reserpine, are used in medicines. The insecticide rotenone is derived from the roots of *Lonchocarpus* sp. Other roots are used to weave baskets and in the souvenir industry.

I never saw a discontented tree. They grip the ground as though they liked it, and though fast rooted they travel about as far as we do. They go wandering forth in all directions with every wind, going and coming like ourselves, traveling with us around the sun two million miles a day, and through space heaven knows how fast and far!

—John Muir (1838–1914)

Figure 26.18 Specialized roots, (a) prop root of corn, (b) haustoria of dodder, and (c) bald cypress knees.

Student Activity—Macroscopic Anatomy of Roots

Materials

- dissecting microscope
- hand lens
- compound microscope
- slides and coverslips
- scalpel
- colored pencils
- Instructor's choice: root specimens such as a plant with a taproot, a plant with a fibrous root system, specimen with numerous lateral roots and root hairs, and various specialty roots
- roots with prominent root nodules, root knots, and mycorrhizae
- a germinating seed with a prominent radical
- camera or camera phone (optional)

Procedure 26.1
Macroanatomy of Roots

1. Procure the specimens and equipment provided by the instructor. Describe the specimens in detail and, using a dissecting microscope and a hand lens, observe, draw, and label the anatomical features of your specimens discussed above in the space provided.

2. Observe root nodules, root knots, and mycorrhizae. Record and sketch your observations below.

To find water, a plant has to position its roots with just as much precision as it arranges its leaves.

—David Attenborough (1926–present)

3. Carefully crush or cut the nodule, knot, or mycor-rhizae. Make a wet mount, and observe the specimen with a compound microscope. Record and sketch your observations below.

4. Procure a germinating seed with a prominent radicle. Record and sketch your observations below.

Student Activity—Microscopic Anatomy of Roots

Materials
- compound microscope
- prepared slides of eudicot and monocot roots
- prepared slides of lateral root growth
- colored pencils

 ## Procedure 26.2
Microanatomy of Roots

1. Procure equipment and prepared slides of various eudicot and monocot roots provided by the instructor.
2. Observe, sketch, and label your eudicot root slides below. Compare your observations to Figures 26.5 through 26.10 on page 422.

3. Observe, sketch, and label your monocot root slides below. Compare your observations to the Figures 26.13 and 26.14 on page 423.

4. Observe, sketch, and label your lateral root slides below. Compare your observations to Figures 26.15 through 26.17 on page 423.

You've got to go out on a limb sometimes because that's where the fruit is.

—**Will Rogers (1879–1935)**

Figure 26.19 External anatomy of a woody twig.

1. Terminal bud
2. Internode
3. Terminal budscale scars
4. Lenticel
5. Stem
6. Axillary bud
7. Node
8. One years growth
9. Bud Scale
10. Leaf scar

STEMS

The trunk of a giant tree and all of its numerous branches and twigs are **stems**, as is the delicate body of a dandelion. Stems produce and support flowers and leaves, provide for the plant's growth, carry water and minerals up from the roots to the leaves to be used in photosynthesis, and carry food back down the plant, to be stored and distributed as needed. **Herbaceous stems** are usually green, soft, and succulent, compared to the harder, lignified **woody stems**. The majority of monocots, as well as several species of eudicots, are herbaceous. Many species of eudicots possess woody stems.

A simple twig can yield the basic external anatomy of a woody stem. The **terminal bud**, located at the tip of the twig, contains the tip of the shoot. The terminal bud is protected by modified leaves called **bud scales**. The **apical meristem** within the terminal bud is enveloped by immature leaves called **leaf primordia**. The bud scales associated with the terminal bud leave a distinct scar, the **terminal bud scale scar**. In many twigs, counting the number of terminal bud scale scars can denote the age of the twig.

Nodes mark the region of the stem where a leaf or leaves were attached by a stalk called a **petiole**. Close examination of the nodes may yield **bundle scars** from past vascular tissue. The region between the nodes is the **internode**, which increases in length as a stem grows. **Axillary buds** that can give rise to new branches or flowers are located between the petiole and the stem. Close examination of a woody twig also may yield small, slightly raised structures on the twig, known as **lenticels**. These structures allow for gas exchange from the interior of the plant to the external environment.

When a twig breaks dormancy and begins to grow, cell division occurs within the apical meristem, giving rise to three types of **primary meristem**: protoderm, procambium, and ground meristem. The primary meristem adds to the length of a stem.

1. The **protoderm**, the outermost portion of the primary meristem, forms the epidermis. A waxy **cuticle** covers the epidermis in the herbaceous plants.

2. The **procambium**, beneath the protoderm, forms primary xylem and phloem. A distinct band of cells between the primary xylem and phloem becomes the **vascular cambium**, which helps to form the girth of the twig.

3. The **ground meristem** forms the center of the stem, called the **pith**. In some plants the pith breaks down, forming a hollow cylinder, or in some woody

plants is crushed as new tissues add girth to the stem. The **cortex**, also formed by the ground meristem, is used primarily for storage.

4. **Cork cambium** arises within the cortex or from the phloem or epidermis. It produces cork cells, which, in turn, comprise the **bark**.

One of the characteristics that distinguishes eudicots from monocots is the arrangement of the vascular bundles. The majority of conifers and flowering plants have **eusteles**, in which the primary xylem and phloem exist in distinct vascular bundles. Ultimately, the arrangement of the vascular bundles depends upon the stem's ability to undergo secondary growth (girth). In eudicots, the vascular bundles form a ring around the outside of the stem. In most monocots (except some grasses with a ground tissue cavity, such as wheat), the vascular bundles are spread throughout the ground tissue. In both the eudicots and the monocots, the phloem faces outward and the xylem faces inward. As discussed in Chapter 24, the annual patterns of vascular cambium form the growth or annual rings of a tree (Fig. 26.20–Fig. 26.27).

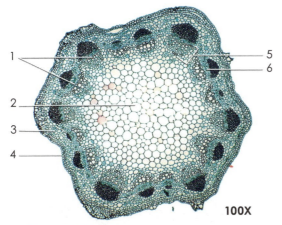

100X

Figure 26.20 Transverse section through the stem of a young sunflower, *Helianthus* sp.
1. Vascular bundles
2. Pith
3. Cortex
4. Epidermis
5. Xylem
6. Phloem

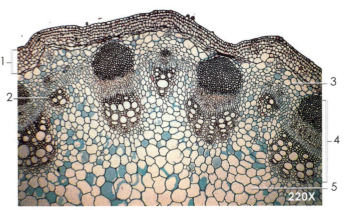

220X

Figure 26.21 Tranverse section through an older stem of a sunflower, *Helianthus* sp. at high magnification.
1. Collenchyma
2. "New" vascular bundle
3. Cortex
4. "Original" vascular bundle with secondary growth
5. Pith

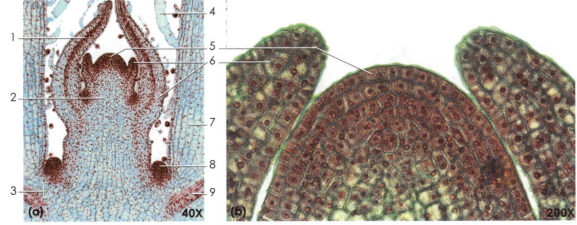

(a) **40X** (b) **200X**

Figure 26.22 Longitudinal section of the stem tip of the common houseplant *Coleus* sp.
1. Procambium
2. Ground meristem
3. Leaf gap
4. Trichome
5. Apical meristem
6. Developing leaf primordia
7. Leaf primordium
8. Axillary bud
9. Developing vascular tissue

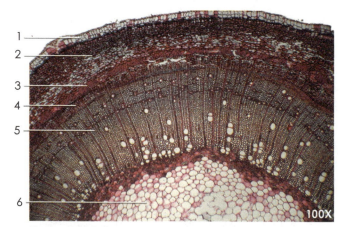

Figure 26.23 Transverse section through one-year-old *Fraxinus* sp. stem showing secondary growth.
1. Periderm
2. Cortex
3. Phloem fibers
4. Secondary phloem
5. Secondary xylem
6. Pith

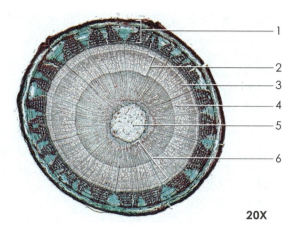

Figure 26.24 Transverse section through a three-year-old *Tilia* sp. stem showing secondary growth.
1. Bark
2. Annual ring
3. Summer wood
4. Spring wood
5. Pith
6. Wood

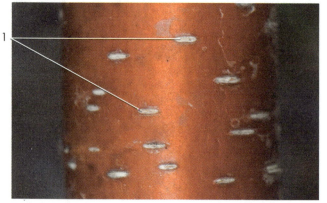

Figure 26.25 Bark of a birch tree, *Betula occidentalis*, showing lenticels. Lenticels are spongy areas in the cork surfaces that permit gas exchange between the internal tissues and the atmosphere.
1. Lenticels

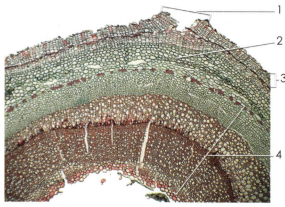

Figure 26.26 Transverse section of a dicot stem showing a lenticel and stem tissues.
1. Lenticel
2. Cortex
3. Periderm
4. Vascular tissue

Figure 26.27 Sample bark patterns of representative conifers and angiosperms.
(a) **Redwood**—The tough, fibrous bark of a redwood tree may be 30 cm thick. It is highly resistant to fire and insect infestation.
(b) **Ponderosa pine**—The mosaic-like pattern of the bark of mature ponderosa pine is resistant to fire.
(c) **White birch**—The surface texture of bark on the white birch is like white paper. The bark of the white birch was used by Indians in Eastern United States for making canoes.
(d) **Sycamore**—The mottled color of the sycamore bark is due to a tendency for large, thin, brittle plates to peel off, revealing lighter areas beneath. These areas grow darker with exposure, until they, too, peel off.
(e) **Mangrove**—The leathery bark of a mangrove tree is adaptive to brackish water In tropical or semi-tropical regions.
(f) **Shagbark hickory**—The strips of bark in a mature shagbark hickory tree gives this tree its common name.

Many angiosperms possess modified stem systems that perform specialized tasks. Although the appearance of these stems may vary, they all have nodes, internodes, and axillary buds, distinguishing them from roots (Fig. 26.28).

1. **Rhizomes** are a type of modified stem in which the stem grows horizontally below the ground, resembling a root. The rhizome may be slender, as in some grasses and ferns, or a thick structure, as in some irises. Some plants possess runners and stolons.

2. **Runners** are similar to rhizomes except that they are above ground. Strawberry plants produce runners after they flower. In philodendrons, runners can be seen giving rise to new plants weeping down from a basket.

3. **Stolons** are runner-like, growing beneath the surface and in different directions. Some botanists do not distinguish between runners and stolons.

4. **Tubers** are swollen extensions of stolons, modified to store carbohydrates. The Irish potato is an example of a tuber. The eyes of the potato are actually axillary buds.

5. **Bulbs**, as in onions, tulips, and lilies, possess small underground stems with large buds. Adventitious roots grow from the bottom of the stem, and fleshy leaves make up most of the bulb.

6. **Corms** look like bulbs, but they do not have fleshy leaves. The only leaves are thin, papery, brown structures on the outside of the corm. Examples of corms are gladiolus and crocus.

7. **Cladophylls** are flattened photosynthetic stems. In cacti, the cladophylls are the broadened green structure and the spines are modified leaves.

8. **Tendrils** are common in climbing plants such as grapes and green briar. Some tendrils such as those of pumpkins and peas are modified leaves.

9. The **thorns** of honey locus are a modified stem.

Stems are integral in human civilization. Some stems are a food source for animals as well as humans. Stems are used in wood products from lumber to toothpicks. Wood is used in pulp for making paper, fibers, and even filler for ice cream and bread. Sugar cane is a major source of sugar in syrups, soft drinks, and other foods. Cinnamon spice is derived from the stem of *Cinnamomum* spp., and the anti-malarial drug quinine is derived from the bark of *Cinchona* spp.

In roots, stems, and leaves, xylem transports water and minerals throughout the plant. In eudicots and monocots, water is transported by tracheids and vessel elements. Water moves up the vascular tissue by its cohesive and adhesive properties. This illustrates the movement of water up a stem.

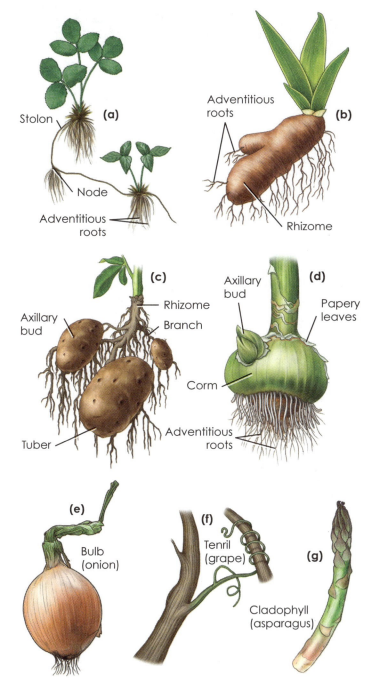

Figure 26.28 Examples of the variety and specialization of angiosperm stems, (a) runners, (b) rhizomes, (c) tubers, (d) corms, (e) bulbs, (f) tendrils, and (g) cladophyll. The stem of an angiosperm is often the ascending portion of the plant specialized to produce and support leaves and flowers, transport and store water and nutrients, and provide growth through cell division. Stems of plants are utilized extensively by humans in products including paper, building materials, furniture, and fuel. In addition, the stems of potatoes, onions, cabbage, and other plants are important food crops.

Student Activity—Macroscopic Anatomy of Stems

Materials
- dissecting microscope
- hand lens
- colored pencils
- Instructors choice: stem specimens such as a woody twig, a herbaceous eudicot, a monocot, and cross-section of a tree trunk
- camera or camera phone (optional)

Procedure 26.3
Macroanatomy of Stems

Consider photographing this activity.

1. Procure specimens and equipment provided by the instructor.
2. Describe the specimens in detail and, using a dissecting microscope and a hand lens, observe, draw, and label the anatomical features of your specimens in the spaces provided below.

Woody twig

Herbaceous eudicot stem

Monocot stem

Cross-section of a tree trunk

Student Activity—Modified Stems

Materials
- dissecting microscope
- hand lens
- colored pencils
- Instructor's choice of modified stems such as: rhizome, runner, stolon, tuber, bulb, corm, cladophyll, tendril, thorn
- camera or camera phone (optional)

Procedure 26.4
Stem Modifications

Consider photographing this activity.

1. Procure the specimens and equipment provided by the instructor.
2. Identify your specimen and the type of modified stem that it has.

_____

_____

_____

3. Sketch each specimen, and discuss the function of each in the space below.

Student Activity—Microscopic Anatomy of Stems

Materials
- compound microscope
- prepared slides of eudicot and monocot stems
- colored pencils

Procedure 26.5
Microanatomy of Stems

1. Procure equipment and prepared slides of various eudicot and monocot stems provided by the instructor.
2. Observe, sketch, and label the eudicot stem slides below. Compare your observations to the eudicot stem images in Figure 26.21 on page 429.

3. Observe, sketch, and label the monocot stem slides below. Compare your observations to the monocot stem images in Figure 26.23 on page 430.

There is nothing in the world more beautiful than the forest clothed to its very hollows in snow. It is the ecstasy of nature, wherein every spray, every blade of grass, every spire of reed, every intricacy of twig, is clad with radiance.

—William Sharpe (1855–1905)

Check Your Understanding

Q. Describe several functions of stems.

_____

_____

_____

Q. Describe several specialty stems.

_____

_____

_____

LEAVES

Albert Camus (1913–1960) once exclaimed, "Autumn is a second spring where every leaf is a flower." **Leaves** are the most conspicuous structures of a plant. Their shape and color intrigue us and indicate the different seasons. The majority of leaves are green as a result of chlorophyll; however, the brilliant colors of autumn leaves emerge when chlorophyll is broken down and pigments such as carotene (orange), xanthrophyll (yellow), and anthocyanin (red) show through (Fig. 26.29). The leaves of oaks and some trees turn brown or tan as the result of a reaction between the tannin stored in vacuoles and the proteins in the leaf. At the beginning of fall, many trees lose their leaves. The process by which plants seasonally lose their leaves or lose their leaves from injury or drought is called **abscission**. This process begins in a specific region of the petiole called the **abscission zone**, and is controlled by auxin and ethylene.

Leaves are the primary food factories in a typical plant. Powered by the sun's energy, leaves take in water and carbon dioxide and give off oxygen and glucose through photosynthesis. Because leaves require daylight to function, the leaves have to be efficient at gathering and utilizing light energy. To capture sunlight, leaves have evolved many novel methods. Some leaves have attained tremendous size; examples are the aroid (giant elephant ear) plant of Borneo (its heart-shaped leaves are 10 m across) and the raffia palm *Raphia regalis* of tropical Africa (with leaves up to 24 m in length). Others, such as the sunflower, follow the sun during the course of the day. Some leaves have purple or red undersides to reflect the light back through the thickness of the leaf. In certain begonias, some surface cells are transparent, acting as tiny lenses to focus light into the plant.

Raking the lawn reinforces trees' capability to produce many leaves. A mature oak tree may produce more than 500,000 leaves in a year, and a mature elm, several million. Each year, leaves produce an estimated 200 billion tons or more of sugar worldwide.

Despite great diversity in size and shape, all leaves originate in the **leaf primordia** of a bud. Over time, the immature leaf takes the characteristic shape of its species. The majority of leaves are attached to a stem by a stalk-like **petiole** and possess a flattened **blade**, the **lamina**. The blade can vary in shape and size. Even within a tree leaf, size can vary; **shade leaves** are larger and thinner, and **sun leaves** are smaller and thicker. In some leaves, a pair of thorn-like structures called **stipules** are present at the base of the petiole (Fig. 26.30).

The changing leaves of autumn colored brightly red and gold,
Also bring a warning of the coming winter cold.
As the wind is changing from the milder summer breeze,
You can feel the difference as it echoes through the trees.

—**James A. Kisner (1947–2008)**

Figure 26.29 The brilliant colors of autumn are caused by pigments such as carotene, xanthrophyll, and anthocyanin.

A leaf may be classified as a **simple leaf** if it has a single blade, such as azalea and birch leaves, or a **compound leaf**, such as ash trees and pecans, if it is divided into smaller **leaflets**. In **pinnately compound** leaves such as black walnut and locust, leaflets occur in pairs along the **rachis** (extension) of the petiole, and in **bipinnately compound** leaves such as the silk tree and mimosa, the leaflets are subdivided into smaller leaflets. In **palmately compound** leaves, such as buckeye and Virginia creeper, all of the leaflets are attached at the same origin (Fig. 26.31).

Leaves are attached to the stem at the **node**. The arrangement of leaves on a stem is the **phyllotaxy**. In plants with **opposite leaf attachment**, such as dogwood and maple, two leaves are attached at each node. In **alternate leaf arrangement**, such as in poplar and aspen, a single leaf appears at each node. The majority of plants have the alternate leaf arrangement. In **whored leaf arrangement**, as in bedstraw and oleander, three or more leaves are attached at a single node (Fig. 26.32).

The veins of a leaf are composed of vascular tissue. In basal dicots and eudicots, the veins are arranged in a **netted** or reticulate pattern. **Pinnately veined** leaves, such as in apple and cabbage, have one prominent primary vein, or **midrib**, and secondary veins branch off the midrib. In **palmately veined** leaves, such as in sycamore and sweetgum, several primary veins branch out from a single point. Monocots, such as corn and sugarcane, have **parallel venation**, in which the veins are arranged nearly parallel to each other. Interestingly, *Ginkgo biloba* has no midrib and the veins fork out from the base of the blade. *Ginkgo* exhibits **dichotomous venation** (Fig. 26.33–Fig. 26.37).

Figure 26.30 Basic leaf anatomy of a cottonwood, *Hibiscus tiliaceus*, leaf.
1. Lamina (blade) 4. Veins
2. Serrate margin 5. Petiole
3. Midrib

Figure 26.31 Leaf arrangements on stems: (a) opposite, (b) alternate, and (c) whorled.

Figure 26.32 Leaf complexity: (a) palmately compound, (b) simple, (c) pinnately compound, and (d) bipinnately compound.

Figure 26.33 Leaf venation: (a) pinnate, (b) parallel, (c) palmate and (d) dichotomous.

Broadly	Narrowly	Abruptly		Broadly	Narrowly		Apiculate		Aristate
	Acuminate				**Acute**				

Caudate	Cuspidate	Emarginate	Mucronate	Obtuse	Retuse

Figure 26.34 Examples of leaf apicies.

Never say there is nothing beautiful in the world anymore.
There is always something to make you wonder in the
shape of a tree, the trembling of a leaf.

—Albert Schweitzer (1875–1965)

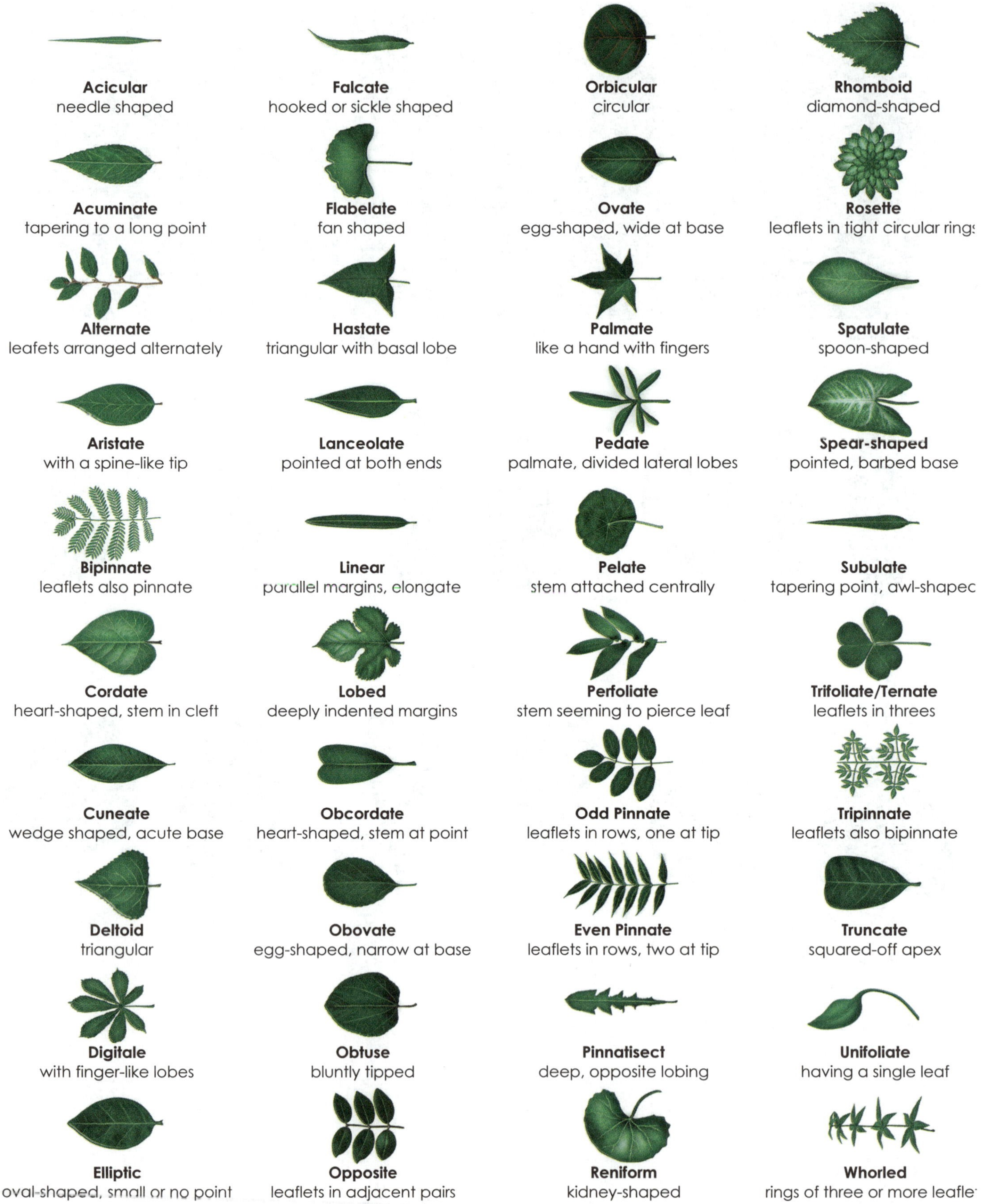

Acicular
needle shaped

Falcate
hooked or sickle shaped

Orbicular
circular

Rhomboid
diamond-shaped

Acuminate
tapering to a long point

Flabelate
fan shaped

Ovate
egg-shaped, wide at base

Rosette
leaflets in tight circular rings

Alternate
leafets arranged alternately

Hastate
triangular with basal lobe

Palmate
like a hand with fingers

Spatulate
spoon-shaped

Aristate
with a spine-like tip

Lanceolate
pointed at both ends

Pedate
palmate, divided lateral lobes

Spear-shaped
pointed, barbed base

Bipinnate
leaflets also pinnate

Linear
parallel margins, elongate

Pelate
stem attached centrally

Subulate
tapering point, awl-shaped

Cordate
heart-shaped, stem in cleft

Lobed
deeply indented margins

Perfoliate
stem seeming to pierce leaf

Trifoliate/Ternate
leaflets in threes

Cuneate
wedge shaped, acute base

Obcordate
heart-shaped, stem at point

Odd Pinnate
leaflets in rows, one at tip

Tripinnate
leaflets also bipinnate

Deltoid
triangular

Obovate
egg-shaped, narrow at base

Even Pinnate
leaflets in rows, two at tip

Truncate
squared-off apex

Digitale
with finger-like lobes

Obtuse
bluntly tipped

Pinnatisect
deep, opposite lobing

Unifoliate
having a single leaf

Elliptic
oval-shaped, small or no point

Opposite
leaflets in adjacent pairs

Reniform
kidney-shaped

Whorled
rings of three or more leaflet

Figure 26.35 Examples of leaf shape.

Ciliate
with fine hairs

Crenate
with rounded teeth

Dentate
with symmetrical teeth

Denticulate
with fine dentition

Doubly Serrate
serrate with sub-teeth

Entire
even, smooth throughout

Lobate
indented, but not to midline

Serrate
teeth forward-pointing

Serrulate
with fine serration

Sinuate
with wave-like indentations

Spiny
with sharp stiff points

Undulate
widely wavy

Figure 26.36 Examples of leaf margins.

Attenuate Auriculate Clasping Cordate Cuneate Hastate

Oblique Peltate Perfoliate Rounded Sagittate Truncate

Figure 26.37 Examples of leaf bases.

The internal anatomy of a typical eudicot leaf is complex, consisting of three basic types of tissues: epidermal, ground, and vascular tissue. The **epidermis** of a typical leaf is present on both sides of the blade. It is transparent and does not undergo photosynthesis. In a horizontal leaf, a number of openings, or **stomata**, can be found mostly on the underside. An oak leaf may have nearly 60,000 stoma, and a lettuce leaf nearly 11 million stoma. The frequency and arrangement of stoma are determined genetically.

Stoma are essential in gas exchange. The opening of a stoma is regulated by a pair of adjacent **guard cells**. A waxy cuticle can be observed above the epidermis on many leaves. The wax, **cutin**, is important in water relationships and in discouraging insect predation. **Leaf hairs**, a type of **trichome**, are extensions of the epidermis that increase the surface area of the leaf. Other trichomes may be glandular, producing oils.

The **mesophyll**, or ground tissue, in a typical leaf is composed of **palisade mesophyll** (**parenchyma**) and **spongy mesophyll** (**parenchyma**). The palisade mesophyll is made of tightly packed parenchyma cells, and the spongy mesophyll of loosely packed cells. The majority of photosynthesis occurs in the mesophyll. **Veins** held within **bundle sheaths** house the vascular tissue of a plant. The veins of a leaf are composed of **xylem** on the topside of a vein and **phloem** on the bottomside of a vein (Fig. 26.38–Fig. 26.50).

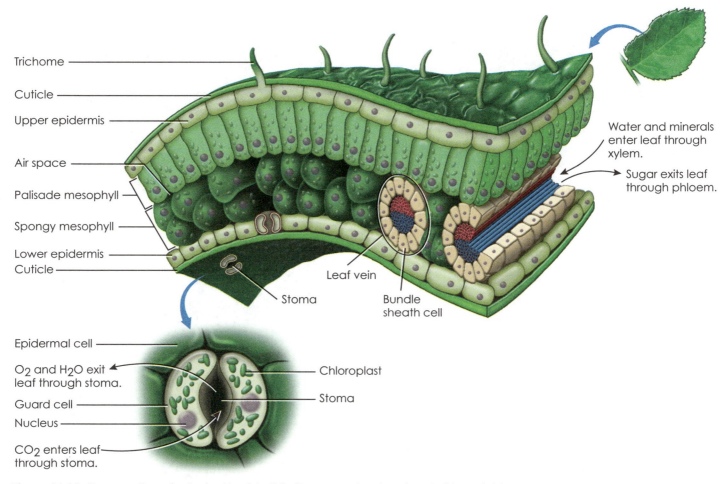

Trichome

Cuticle

Upper epidermis

Air space

Palisade mesophyll

Spongy mesophyll

Lower epidermis

Cuticle

Water and minerals enter leaf through xylem.

Sugar exits leaf through phloem.

Leaf vein

Stoma

Bundle sheath cell

Epidermal cell

O_2 and H_2O exit leaf through stoma.

Guard cell

Nucleus

CO_2 enters leaf through stoma.

Chloroplast

Stoma

Figure 26.38 Cross-section of a typical leaf. In this diagram, xylem is red and phloem is blue.

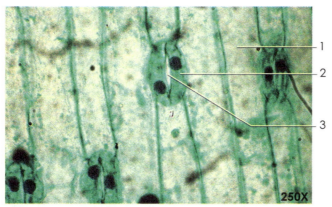

Figure 26.39 Face view of the epidermis of onion, *Allium* sp. Note the twin guard cells with the stoma opened.
1. Lower epidermis 3. Stomate
2. Guard cell

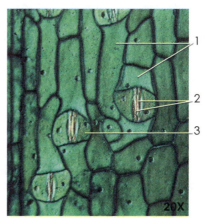

Figure 26.40 Surface view of the leaf epidermis of *Tradescantia* sp.
1. Epidermal cells
2. Guard cells surrounding stomata
3. Subsidiary cells

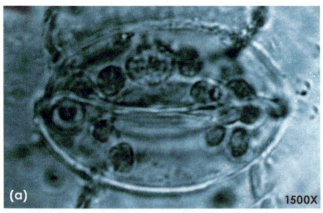

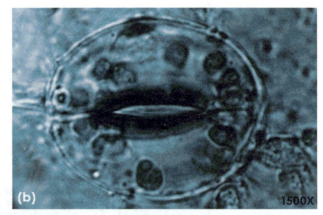

Figure 26.41 Guard cells in many plants regulate the opening of the stomata according to the environmental factors, as indicated in this diagram. (a) Face view of a closed stoma of a geranium, and (b) an open stoma.

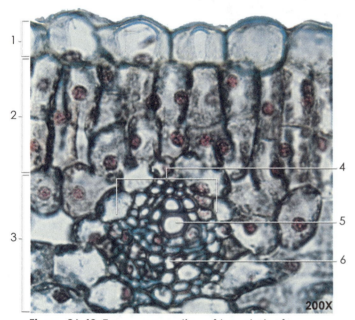

Figure 26.42 Transverse section of tomato leaf, *Lycopersicon* sp.

1. Upper epidermis
2. Palisade mesophyll
3. Spongy mesophyll
4. Leaf vein (vascular bundle)
5. Xylem
6. Phloem

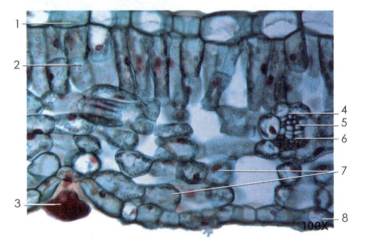

Figure 26.43 Transverse section through the leaf of the common hedge privet Ligustrum sp. The typical tissue arrangement of a leaf includes an upper epidermis, a lower epidermis, and the centrally located mesophyll. Containing chloroplasts, the cells of the mesophyll are often divided into palisade mesophyll and spongy mesophyll. Veins within the mesophyll conduct material through the leaf.

1. Upper epidermis
2. Palisade mesophyll
3. Gland
4. Bundle sheath
5. Xylem
6. Phloem
7. Spongy mesophyll
8. Lower epidermis

Figure 26.44 Transverse section through the leaf of barberry, *Berberis* sp.

1. Upper epidermis
2. Palisade mesophyll
3. Spongy mesophyll
4. Lower epidermis
5. Leaf vein (midrib)

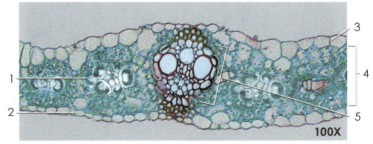

Figure 26.45 Transverse section through a corn leaf.

1. Bundle sheath
2. Lower epidermis
3. Upper epidermis
4. Mesophyll
5. Leaf vein (midrib)
6. Xylem

Figure 26.46 Transverse section of common garden flower, *Dianthus* sp., through leaf.
1. Upper epidermis
2. Upper palisade mesophyll
3. Spongy mesophyll
4. Lower palisade mesophyll
5. Lower epidermis
6. Leaf vein

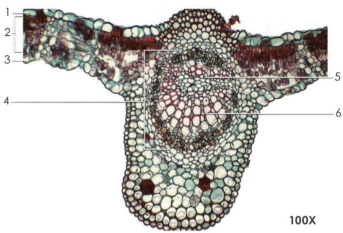

Figure 26.47 Transverse section through the leaf of basswood, *Tilia* sp.
1. Upper epidermis
2. Mesophyll
3. Lower epidermis
4. Leaf vein (midrib)
5. Phloem
6. Xylem

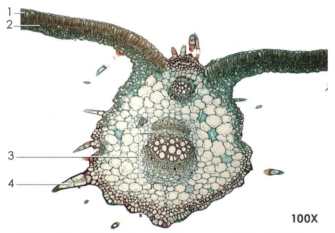

Figure 26.48 Transverse section through the leaf of cucumber, *Cucurbita* sp.
1. Palisade mesophyll
2. Spongy mesophyll
3. Leaf vein (midrib)
4. Trichome

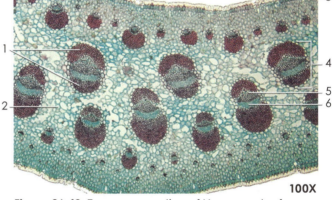

Figure 26.49 Transverse section of *Yucca* sp. leaf.
1. Bundle caps (fibers)
2. Ground tissue
3. Epidermis
4. Vascular bundle
5. Xylem
6. Phloem

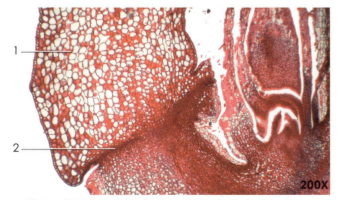

Figure 26.50 Longitudinal section of privet, *Ligustrum* sp., through the stem and petiole at the abscission layer.
1. Leaf petiole
2. Abscission layer

Kitty Craze!

Have you ever given your cat some catnip? When cats smell catnip, they display a variety of strange behaviors including head shaking, drooling, turning over, rubbing, and acting bizarre. An estimated 70% of cats have catnip receptors. Catnip is a weed-like perennial herb, *Nepeta cataria*, belonging to the mint family *Labiatae*. Nepetalactone, the active ingredient in catnip, is thought to mimic the effects of a pheromone that causes a variety of psychosexual behaviors. Catnip is commonly called "kitty cocaine." Catnip also has been used in human medicine for bronchitis, insomnia, diarrhea, and colic.

Student Activity—Macroscopic Anatomy of Leaves

Materials
- dissecting microscope
- hand lens
- colored pencils
- Instructor's choice: leaf specimens representing a variety of types of leaves discussed above
- camera or camera phone (optional)

 Procedure 26.6
Macroanatomy of Leaves

1. Procure specimens and equipment provided by the instructor.
2. Describe the specimens in detail and, using a dissecting microscope and a hand lens, observe, draw, and label the anatomical features, in the space provided below. Remember to include both the scientific name and the common name of your specimen.

3. Using the dissecting microscope, make observations of the stoma on the leaves and sketch your observations below. Were the stomata open on your specimens?

Student Activity—Classification of Leaves

Materials

- hand lens
- colored pencils
- Instructor's choice: leaf specimens representing a variety of leaf shapes, leaf margins, leaf apicies, and leaf bases
- camera or camera phone (optional)

 Procedure 26.7
Leaf Classification

1. Procure the leaf specimens and equipment from the instructor.
2. Using Figures 26.31–26.37 on pages 436–440, classify the shape, margin, apices, and base of the leaves provided by the instructor. Be sure to include both the scientific name and the common name of your specimen.
3. Record your observations below. Remember to include both the scientific name and the common name of your specimen.

_____

_____

_____

_____

_____

 Student Activity—Microscopic Anatomy of Leaves

Materials

- compound microscope
- colored pencils
- prepared slides of various monocot and dicot leaves such as transverse sections through various dicot and monocot leaves

 Procedure 26.8
Microanatomy of Leaves

1. Procure slides and equipment provided by the instructor.
2. Describe the slide in detail, using a compound microscope on both low and high power.

_____

_____

_____

3. Compare the microscopic anatomy of your slides with the micrographs in the leaf section in Figures 26.41–26.49 on pages 442–443. Observe, draw, and label the anatomical features in the space provided.

A myriad of specialized leaves are found in the plant world (Fig. 26.51), performing a number of duties. Thick, water-retaining **succulent leaves**, such as aloe, jade tree, and yuccas, can be found in plants from arid regions to salty seashores. The **spines** of cacti, which are effective defenses, are actually modified leaves that evolved to conserve water. Many plants living in aquatic regions, such as the water hyacinth, have large air spaces of **aerenchyma tissue** that allow them to float. The sensitive plant, *Mimosa pudica*, droops when touched. The drooping, or **nastic movement**, is the result of changes in turgor pressure. In some plants, such as the garden pea, leaves are modified, forming **tendrils** to enable climbing.

Bracts are specialized leaves at the base of a flower. The brightly colored bracts of poinsettia are popular at Christmas. The obvious white or pink bracts of dogwood resemble flower petals. Approximately 200 species of plants have modified **insect-trapping leaves**; these include pitcher plants, sundews, and Venus flytraps.

Besides the fact that leaves are an essential source of oxygen and food for animals, humans use leaves and their products in a number of ways. Many medicinal products are derivatives of leaves. These include aloe for burns; digitalis, which is a heart stimulant, from foxglove; and atropine, which lowers parasympathetic activities from belladonna, jimsonweed, and mandrake. Tobacco, marijuana, and cocaine also are leaf products. In the past, plants such as prickly pear, bearberry, henna, sassafras, and indigo were used for dyes. The leaves from some plants are used in beverages, including agave in tequila and camellia leaves in tea. Spices are derived from a number of plants including peppermint, oregano, spearmint, wintergreen, bay, and basil, among others.

Succulent leaf of an aloe.

Spines of a cactus.

A sensitive plant.

Floating leaves, (a) giant water lily and (b) water hyacinth.

Tendrils of a garden pea.

Bracts of a poinsettia.

Insect trapping leaves of (a) a pitcher plant, (b) sundew, and (c) venus fly trap.

Figure 26.51 Examples of modified leaves.

Student Activity—Specialized Leaves

Materials
- hand lens
- colored pencils
- Instructor's choice of select specialized leaves

 Procedure 26.9
Leaf Specialization

1. Procure specimens and equipment provided by the instructor.
2. Using a hand lens, describe the specimens in detail. Observe, draw, and label the specialized leaves in the space provided below. Remember to include both the scientific name and the common name of your specimen and the function of the specialization.

Check Your Understanding

Q. Describe several specialized leaves.

Q. Sketch and label a typical transverse section of a dicot leaf.

A Digital Field Trip—Pictures of Plants

Think of the following activity as a digital field trip or a scavenger hunt. You may find these plants on campus, in your yard, in the woods, in a greenhouse, at an arboretum, or at a botanical garden. Locate the specimens in question, take your time to compose a photograph of the specimen, use good photographic technique, check out the quality of the image on the camera LCD, transfer the image to a computer, and develop a labeled presentation.

Materials
- digital camera
- notebook to record data
- computer
- presentation software
- storage device such as a jump drive or CD
- projector and screen
- objects for photographing

 Procedure 26.10
Pictures of Plants

1. Locate and photograph the following:
   - ☐ two herbaceous eudicots
   - ☐ two woody eudicots
   - ☐ two monocots
   - ☐ a taproot of a seedling
   - ☐ a fibrous root system
   - ☐ two kinds of modified roots
   - ☐ a twig
   - ☐ a close-up of a terminal bud
   - ☐ a close-up of a lenticel
   - ☐ three types of modified stems
   - ☐ an image illustrating the anatomy of a leaf
   - ☐ a simple leaf
   - ☐ a pinnately compound leaf
   - ☐ a palmately compound leaf
   - ☐ a twig with opposite leaf arrangement
   - ☐ a twig with alternate leaf arrangement
   - ☐ a twig with whorled leaf arrangement
   - ☐ a leaf with parallel venation
   - ☐ a leaf with pinnate venation
   - ☐ a leaf with palmate venation
   - ☐ three examples of leaf shape
   - ☐ two examples of leaf margins
   - ☐ two examples of leaf apices
   - ☐ two examples of leaf bases
   - ☐ three specialized leaves

2. Photograph the specimens. The photographs should include the entire specimen, as well as distinguishing features.

3. Using a computer and presentation software, develop a digital slide show of your projects. Be sure to clearly label your slides. Include the title of your specimen, label important parts, and use both the common and the scientific names of the plants. And how about adding some music to your presentation?

4. Place your slide show on a storage device or CD. The instructor may allow you to present your slide show to the class.

NOTES

_____

_____

_____

_____

_____

_____

_____

_____

_____

Name: _____ Date: _____ Section: _____

Review Questions

1. Compare and contrast the roots, stems, and leaves of eudicots and monocots.

_____
_____
_____
_____
_____

2. Label the following root micrographs.

a. _____

b. _____

c. _____

d. _____

e. _____

a. _____

b. _____

c. _____

d. _____

e. _____

3. What are several functions of roots?

_____
_____
_____
_____

4. Name and describe three types of specialized roots.

_____
_____
_____
_____

5. Sketch a woody twig, and label the following structures: terminal bud, lenticels, node, internode, terminal bud scale scars, lateral bud, leaf scar.

6. List the three types of primary meristems and derivatives of each.

_____

_____

_____

_____

_____

7. Describe the arrangement of vascular bundles in eudicots and monocots.

_____

_____

_____

8. Name and describe three types of specialized stems.

_____

_____

_____

Name: _____ Date: _____ Section: _____

9. Label the following stem micrographs.

a. _____

b. _____

c. _____

d. _____

e. _____

f. _____

a. _____

b. _____

c. _____

d. _____

e. _____

f. _____

10. How do stoma work?

_____

_____

_____

11. Describe and sketch a basic leaf phyllotaxy.

_____

_____

_____

12. Describe and sketch several types of leaf venation.

_____

_____

_____

_____

13. Name and describe three types of specialized leaves.

_____

_____

_____

14. Describe the orientation of xylem and phloem in roots, stems, and leaves.

_____

_____

_____

15. Label the following micrographs of leafs.

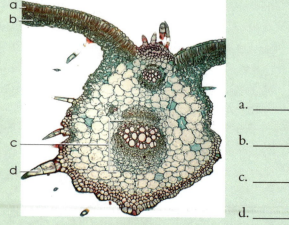

a. _____

b. _____

c. _____

d. _____

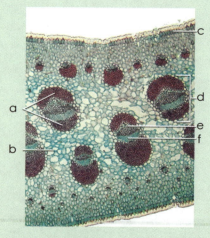

a. _____

b. _____

c. _____

d. _____

e. _____

f. _____

Chapter 27
Flowers, Fruits, and Seeds: Understanding Flowers, Fruits, and Seeds

Student Outcome Objectives

At the completion of this exercise, the student will be able to:

1. Define a flower, a fruit, and a seed.
2. Compare and contrast fruits and vegetables, and give examples of each.
3. Identify and describe the macroscopic anatomy of a flower.
4. Compare, contrast, and identify hypogynous, epigynous, and perigynous ovary configurations.
5. Compare, contrast, and identify complete and incomplete flowers.
6. Compare, contrast, and identify perfect and imperfect flowers.
7. Compare, contrast, and identify actinomorphic and zygomorphic flowers.
8. Compare, contrast, and identify solitary flowers, inflorescence, and catkins.
9. Describe the macroanatomy of a grass flower.
10. Identify and describe the basic external and internal structures of various fruits.
11. Classify fruits as to group and type.
12. Identify and describe the basic external anatomy and internal anatomy of typical seeds.
13. Compare and contrast eudicot and monocot germination.
14. Describe the role of the cotyledon in germination.
15. Discuss the effect of pollution upon seed germination.
16. Describe gravitropism in root development during germination.

Overview

Ralph Waldo Emerson once observed that "the Earth laughs in flowers." **Flowers** are the exclusive reproductive organs of angiosperms. While we cherish their beauty, they are elaborate reproductive structures solely programmed to propagate the species. Their color, texture, and nectar are designed to ensure pollination. Evolutionarily speaking, the appearance of flowers during the Cretaceous Period changed the face of the planet and opened the door for the co-evolution of many pollinators and florists alike (Fig. 27.1).

A walk through the market yields a variety of fruits and vegetables (Fig. 27.2). This raises a good question: What exactly is a **fruit**, and what is a vegetable? A fruit is a structure derived from the ovary of a plant and its accessory tissues. Fruits are the products of flowers and are exclusive to angiosperms. Fruits house, protect, nourish, and aid in the dissemination of seeds. Examples of fruits include the obvious: strawberries, apples, oranges, grapes, and peaches, as well as the not so obvious: tomatoes, cucumbers, pecans, rose hips, and beans. A **vegetable** is an edible part of a plant derived from petioles, leaves, specialized leaves, roots, stems, or flowers. Examples of vegetables are sweet potatoes, carrots, broccoli, turnips, onions, cauliflower, and parsley.

Darwin noted that organisms produce more young than naturally can be expected to survive. The vast numbers of seeds produced by a mustard plant, watermelon, dandelion, pine tree, or oak tree are examples of this principle. For example, of the thousands of acorns dropped by a typical mature oak tree, only one in 10,000 will become a tree. A **seed** is a ripened ovule of a plant that contains an embryo housed in a protective coat and nourished by stored food. Seeds have ensured the success of gymnosperms and angiosperms in populating the planet (Fig. 27.3).

Figure 27.1 Flowers changed the face of the Earth.

Figure 27.2 A trip to a nearby market yields a myriad of fruits and vegetables.

FLOWERS

In 1878, the Italian botanist Odoardo Beccari recorded his finding a corpse flower, or titan arum *Amorphophallus titanium,* in the rainforest of Sumatra, with a circumference of more than 1.5 m and a height exceeding 3 m. The fragrance of the flower is not as delightful as a rose. Actually, it smells more like a rotting corpse—hence its common name. The corpse flower is the largest unbranched **inflorescence** (cluster of flowers) in the world. The largest *branched* inflorescence is produced by the tailpot palm *Coryphyta umbraculifera,* native to Southeast Asia. Its inflorescence can attain a length of 8 m and have several million individual flowers. The largest single flower is produced by *Raffiesia arnoldii,* also known as the corpse flower. It resembles a giant reddish-brown mushroom and can attain a diameter of more than a meter and weigh 11 kg. It is a native of the rainforest of Southeast Asia.

The smallest flower in the world is produced by *Wolffia globosa,* a type of duckweed. The mature plant weighs about the same as two grains of salt. A bouquet of a dozen of these tiny flowers would be about the size of the head of a pin. Despite the tremendous range in size, color, and shape, all flowers are made primarily of the same components (Fig. 27.4).

Figure 27.3 In North America, the century plant *Agavi americana* can produce a spike nearly 13 meters high. The term "century plant" is a misnomer, as the plant takes about 15 years to build up enough resources to produce a flower.

Columbine   Bird of Paradise   Hibiscus   Spanish Moss

Pitcher plant   Periwinkle   Dutch clover   Tulip Poplar (Yellow Poplar)

Grass flower   Cattail   Passion flower   Rhododendron

Figure 27.4 Angiosperm flowers are extremely diverse.

Flowers begin to develop from a specialized stalk, the **peduncle**, or from several smaller stalks, **pedicels**. The peduncle or pedicels form the **receptacle**, a swollen region that contains the other floral parts, arranged in **whorls**. Keep in mind that in eudicots, the flower parts are arranged in multiples of four or five, and in monocots in multiples of three. The outermost whorl is composed of leaf-like, green **sepals**. The sepals, in turn, form the **calyx**. In many species the calyx serves to protect the flower while it develops within the bud. The **petals**, the most conspicuous part of a flower, range in color, shape, size, and fragrance. The petals collectively form the **corolla**. The color of the petals and the shape of the corolla are significant in pollination. The calyx and corolla comprise the **perianth**.

The "male" portion of the flower, the **stamen**, consists of a slender stalk, the **filament**, and the sac-like **anther**, where **pollen** is produced. Collectively, the stamens comprise the **androecium**. Usually, anthers release pollen by splitting open (such as in a daisy) but in some species (such as azaleas), the pollen is released from pores at the tip of the anther.

The most obvious female portion of the plant is the centrally located **pistil**. (Some texts refer to the pistil as a carpel). It is composed of a sticky knob that receives pollen and is known as the **stigma**, which sits atop a slender tube called the **stalk**, which leads to the **ovary** (Fig. 27.5). The position of the ovary is significant in plant classification. If the calyx and corolla are attached to the receptacle at the base of the ovary, the ovary is classified as **superior** or **hypogynous**, as in grape, honeysuckle, and quince. In an **inferior** or **epigynous** ovary, the calyx and corolla appear to be attached at the top of the ovary, such as in blueberry, watermelon, and pear. In a **semi-inferior** or **perigynous** ovary, the calyx and corolla are found on a cup-shaped structure surrounding the receptacle, such as in crape myrtle, cherry, and *Pyracantha* (Fig. 27.6). An ovary eventually forms the fruit. One or more **carpels** serve as the main portion of the ovary. The carpel or carpels are known collectively as the **gynoecium**. **Ovules**, produced on the carpels, contain the female gametocyte. Generally, the number of carpels is related to the number of divisions of the stigma. For example, each section of a tomato or a grapefruit represents a carpel (Fig. 27.7–Fig. 27.12).

Tantus amor florum
(So great is the love of flowers)

—**Gilbert White (1720–1783)**

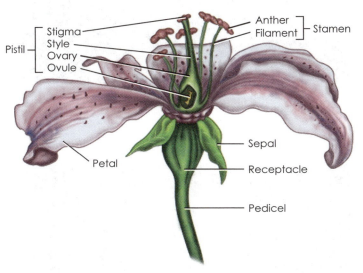

Figure 27.5 Basic flower anatomy.

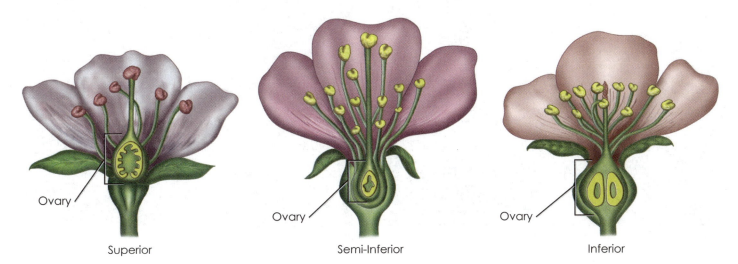

Superior

Semi-Inferior

Inferior

Figure 27.6 Position of the ovary in angiosperms.

Figure 27.7 Floral structure of a tulip, *Tulipa* sp.
1. Petal
2. Anther
3. Stigma
4. Filament
5. Style

Figure 27.8 Structure of a dissected cherry, *Prunus* sp., showing a perigynous flower.
1. Petal
2. Filaments
3. Sepal
4. Anther
5. Stigma
6. Style
7. Floral tube

Figure 27.9 Structure of a dissected pear, *Pyrus* sp., showing an epigynous flower.
1. Petal
2. Anther
3. Filament
4. Style
5. Sepal
6. Ovary

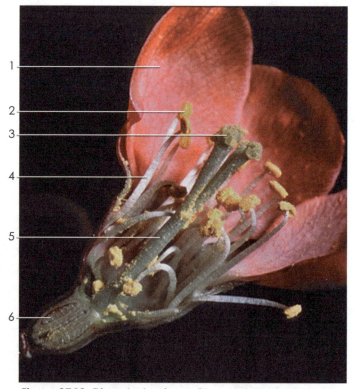

Figure 27.10 Dissected quince, *Chaenomeles japonica*, showing an hypogynous flower.
1. Petal
2. Anther
3. Stigma
4. Filament
5. Style
6. Ovules

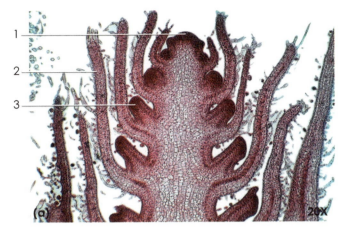

Floral bud of Coleus, *Coleus* sp.
1. Apical meristem   3. Floral bud
2. Bract

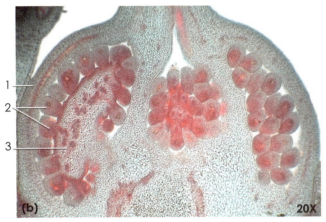

Ovary of tomato, *Lycopersicon* sp., with developing ovules.
1. Ovary wall   3. Placenta
2. Ovules

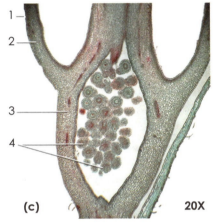

Floral bud of a currant, *Ribes sp.,* showing an inferior ovary with developing ovules.
1. Style   3. Ovary
2. Petal   4. Ovules

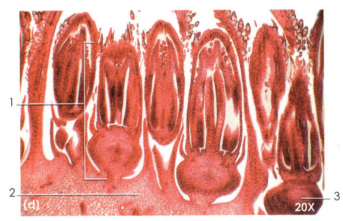

Floral bud of sunflower, *Helianthus* sp., with several immature flowers.
1. Individual flower   3. Ovary of individual flower
2. Receptacle

Figure 27.11 Floral bud arrangement in (a) *Coleus*; (b) *Lycopersicon* (tomato), *Rilies* (currant), and *Helianthus* sp. (sunflower).

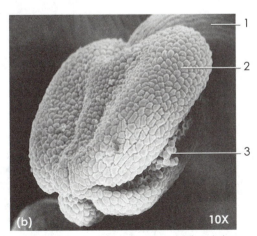

Figure 27.12 The stigma is the location where pollen grains adhere and germinate to produce a pollen tube. (a) Scanning electron micrograph of the stigma of an angiosperm pistil, (b) Scanning electron micrograph of the anther of candy tuft, *Lobularia* sp. The anther has ruptured, resulting in the release of pollen grains.
1. Filament   2. Anther   3. Pollen grains

I perhaps owe having become a painter to flowers.

—Claude Monet (1840–1926)

In joy or sadness, flowers are our constant friends.

—Kozuko Okakura (1862–1919)

Flower terminology and arrangement are complex. **Complete flowers** such as magnolia, tulips, apples, azaleas, and lilies possess sepals, petals, stamen, and pistils. **Incomplete flowers** lack one or more sepals, petals, stamen, or pistils; examples are squash, begonia, oak, and walnuts. A **perfect flower** such as a dandelion, lily, banana, or pea possesses both stamens and pistils. An **imperfect flower** possesses only one sex because it lacks either the stamens or the pistil. **Staminate flowers** have only stamens, and **pistillate flowers** have only pistils. Imperfect flowers may appear on separate plants, such as holly, mulberry, and persimmon, or on the same plant, such as cattail, oak, and corn (Fig. 27.13).

Floral symmetry refers to the arrangement of flowers along a plane. **Actinomorphic** (radially symmetrical) flowers can be divided into symmetrical halves by more than one longitudinal plane passing through the axis. In these flowers the petals are similar in shape and size. Examples of actinomorphic flowers are azaleas, buttercups, and roses. **Zygomorphic** (bilaterally symmetrical) flowers can be divided by a single plane into two mirror-image halves. Zygomorphic flowers generally have petals of two or more different shapes and sizes. Examples of zygomorphic flowers are orchids, foxglove, and snapdragon.

Flowers may be **solitary,** such as a petunia and a camellia, or appear in clusters known as an **inflorescence,** such as in oak, sunflower, and willow. **Catkins** (Dutch = kitten) are a drooping, slim inflorescence, lacking petals or having inconspicuous petals that resemble a kitten's tail. They contain many, usually unisex, flowers, arranged closely along a central stem. In some plants, such as willow, mulberry, and oak, only the male flowers form catkins and the female flowers are solitary. In other plants, such as poplar, both male and female flowers are borne in catkins (Fig. 27.14–Fig. 27.16).

Many people do not realize that grasses produce flowers. In the summer, just let your yard get out of control and notice the flowers and seed heads (for example, my yard while I am writing this manual!). The flowers of grasses are inflorescence. In bluegrass, wheat, rice, and other herbaceous grasses, each leaf consists of a **basal sheath** that encompasses the **culm** (grass stem) down to its point of origin, the **node**. The **internodes** of grasses typically are hollow, such as in bamboo. The leaf blade usually grows away from the culm. A membranous scale called the **ligule** can be found at the junction of the basal sheath and leaf blade. The tiny projections near the base of the leaf blade are known as **auricles**.

Figure 27.13 Example of (a) a complete flower, lily and (b) an incomplete flower, orchid.

Figure 27.14 Example of (a) a perfect flower and (b) an imperfect flower.

Figure 27.15 Example of (a) Actinomorphic symmetry and (b) zygomorphic symmetry in flowers.

Figure 27.16 (a) A solitary flower, dahlia, and inflorescent flowers, (b) a sunflower, and (c) a walnut catkin.

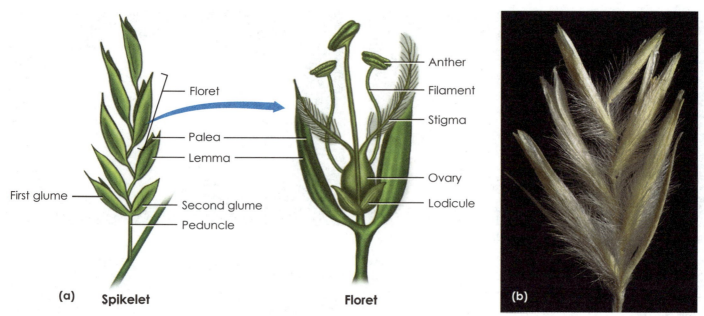

Figure 27.17 Floral structure of grasses (a) and (b) *Elymus flavescens*, showing spikelets with six florets.

A spikelet is a conspicuous extension of the peduncle, consisting of many small florets. A **glume**, designated as the first and second glume or protective husk, appears externally in the **floret**. A floret consists of two bracts—the **lemma** and the palea. Within the floret are one pistil, three stamens, the ovary, and the scale-like lodicule. After fertilization, the ovary develops into a one-seeded fruit, a grain or **caryopsis**. In weeding the yard, if the peduncle is cut, the seed head will not repair (Fig. 27.17).

The **sporophyte** is the dominant generation in angiosperms. In the male portion of the plant, anthers have four **pollen sacs** containing numerous microsporocytes. Each microsporocyte produces four haploid microspores. After cell division, the haploid nuclei of the microspores produce a **pollen grain**. In the female portion of the plant, one or several ovules can be found in the ovary.

Within an ovule, four megaspores are formed. One of the megaspores becomes an **embryo sac,** or female gametocyte.

During **pollination**, a two-celled pollen grain lands on the stigma of the same species of plant; one cell forms a tube cell and the other a generative cell. The tube cell will form the pollen tube, and the generative cell will produce two sperm cells. The pollen tube moves down the style to the ovary to an awaiting ovule. Of the two sperm cells, one fertilizes the egg and the other fuses with two polar nuclei to form a 3n endosperm. The endosperm nourishes the developing embryo. The ovule develops into a seed containing the embryonic plant (sporophyte) and the endosperm. Flowering plants undergo **double fertilization**, forming an embryo and the endosperm.

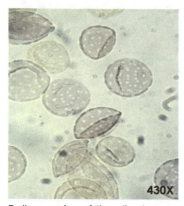

Pollen grains of the dicot pigweed, *Amaranthus* sp.

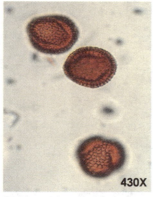

Pollen grains of a lilac, *Syringa* sp.

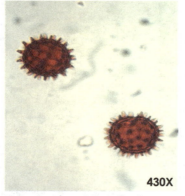

Pollen grains of the dicot arrowroot, *Balsamorhiza* sp.

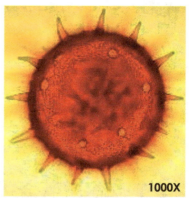

Pollen grain of hibiscus, *Hibiscus* sp.

Figure 27.18 Examples of pollen.

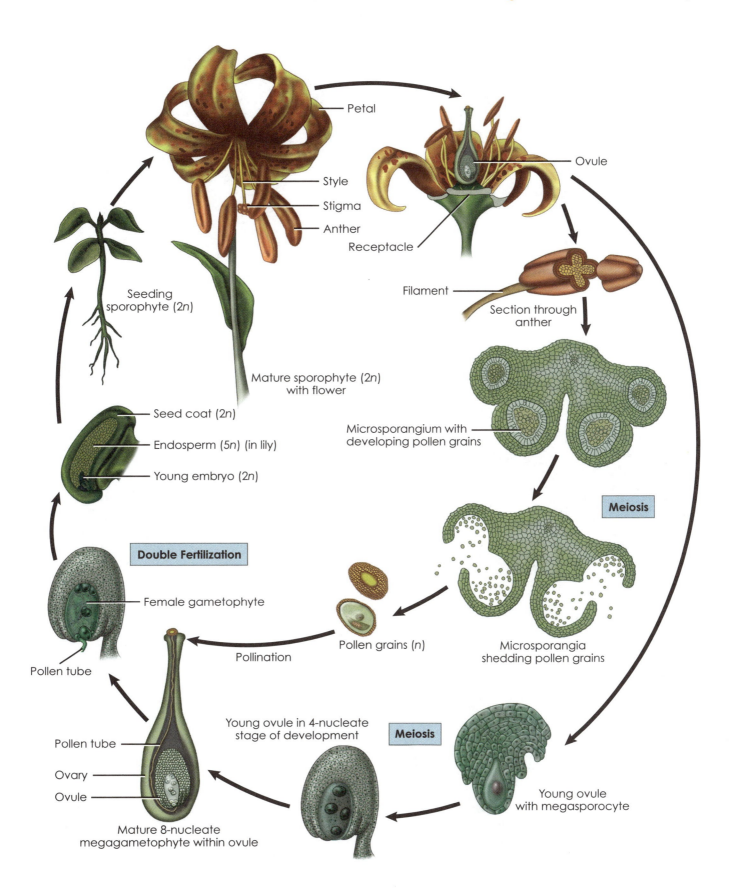

Petal

Ovule

Style

Stigma

Anther

Receptacle

Filament

Section through anther

Seeding sporophyte (2n)

Microsporangium with developing pollen grains

Meiosis

Mature sporophyte (2n) with flower

Seed coat (2n)

Endosperm (5n) (in lily)

Young embryo (2n)

Double Fertilization

Female gametophyte

Pollen tube

Pollen grains (n)

Pollination

Microsporangia shedding pollen grains

Young ovule in 4-nucleate stage of development

Meiosis

Pollen tube

Ovary

Ovule

Young ovule with megasporocyte

Mature 8-nucleate megagametophyte within ovule

Figure 27.19 The life cycle of an angiosperm.

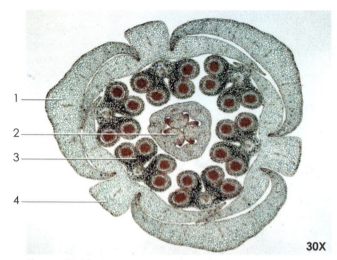

Figure 27.20 Transverse section of a flower bud from a lily, *Lilium* sp.
1. Sepal   3. Anther
2. Ovary   4. Petal

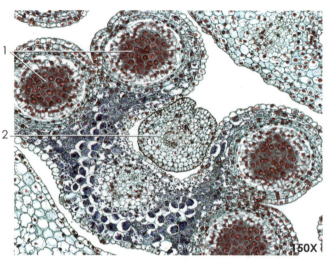

Figure 27.21 Transverse section of an anther from a lily, *Lilium* sp.
1. Sporogenous tissue   2. Filament

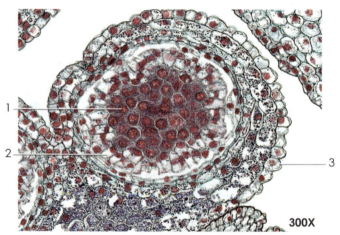

Figure 27.22 Transverse section of an anther from a lily, *Lilium* sp.
1. Young microsporocytes   2. Tapetum   3. Anther wall

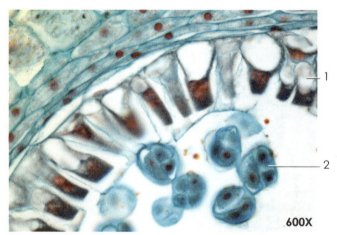

Figure 27.23 Transverse section of an anther from a lily, *Lilium* sp., magnified view.
1. Tapetum   2. Tetrad of microspores

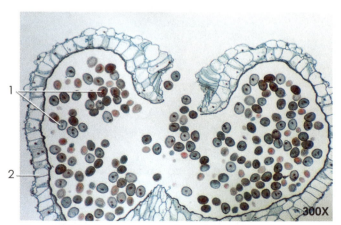

Figure 27.24 Transverse section of an anther from a lily, *Lilium* sp., showing mature pollen.
1. Pollen grains with two cells   2. Anther wall

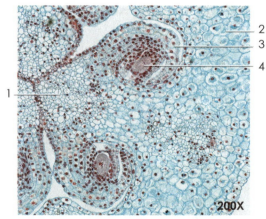

Figure 27.25 Transverse section of a lily, *Lilium* sp., ovary showing ovules.
1. Placenta   3. Ovule
2. Ovary wall   4. Megasporocyte (2n)

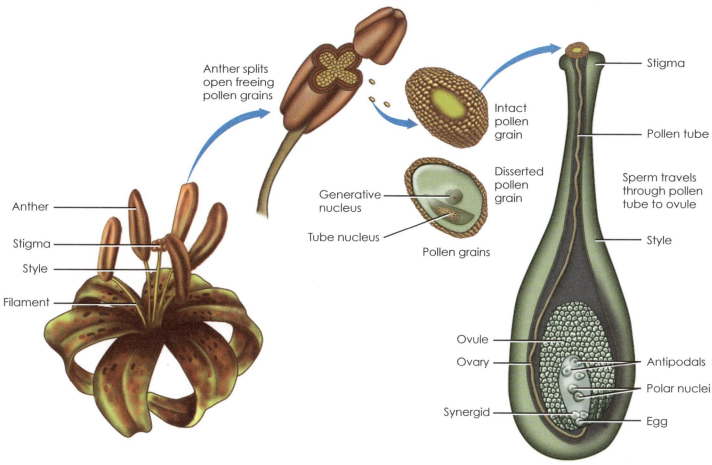

Figure 27.26 The process of pollination.

 Student Activity—Macroscopic Anatomy of Flowers

Availability of flowers is regional and seasonal. The instructor will provide a variety of basal dicot, eudicot, and monocot flowers for observation, and the student is encouraged to responsibly and legally collect flowers for observation as well. In addition, the student is encouraged to photograph flowers and bring the images to class for discussion. This is a good time to practice macrophotography.

Materials
- dissecting microscope
- hand lens
- compound microscope
- slides and coverslips
- scalpel
- colored pencils
- wet mount

 Procedure 27.1
Flower Dissection

1. Procure the specimens and equipment.
2. Describe the specimens in detail and, using a dissecting microscope or a hand lens, observe, draw, and label the anatomical features of your specimens on the next page. Be sure to include the common name and the scientific name of the plant in your description. Also include whether the flower is a basal dicot, eudicot, or monocot. In addition, include whether the flower is complete or incomplete, perfect or imperfect, solitary or an inflorescence, actinomorphic or zygomorphic, and if it has a hypogynous, epigynous, or perigynous ovary.

4. Based on the fragrance, color, and shape of the flower, infer the kind of pollination that the flower undergoes.

_____

_____

5. Using the point of the scalpel, remove some pollen from the anther. Make a wet mount of the pollen and observe it under the compound microscope. Record and illustrate your observations.

_____

_____

6. Carefully remove the sepals and petals from a flower designated by the instructor. Closely observe the male and female reproductive parts, using a dissecting microscope.

7. Carefully cut the base of the pistil and ovary longitudinally, and record your observations below.

3. Describe the smell of your flower.

_____

_____

_____

_____

 Student Activity—Microscopic Anatomy of Flowers

In this activity, slides should include, but not be limited to, eudicot and monocot flower bud, transverse section of a lily flower bud, transverse section of an anther from a lily, and a transverse section of an ovary to a lily.

Materials
- compound microscope
- colored pencils
- prepared slides of eudicot and monocot flowers

 Procedure 27.2
Microanatomy of the Flower

1. Procure prepared slides of various eudicot and monocot flowers from the instructor.
2. Observe, sketch, and label the flower slides on the next page.

3. Compare your observations to Figure 27.11.

Check Your Understanding

1. Compare and contrast eudicot and monocot flowers.

2. Sketch a typical eudicot flower, and label the parts.

3. What is a catkin?

FRUITS

Fruits are exclusive to angiosperms. All fruits are derivatives of the ovary or ovaries of a flower and associated structures such as the receptacle. The diversity of fruits is astonishing, ranging from acorns to zucchini. Fruits house, protect, and nourish a seed or seeds, and disperse seeds. Some fruits, such as the cultivated banana, wild parsnips, seedless watermelons, and seedless grapes, are **partheno-carpic**; they do not require fertilization to form a fruit.

Upon maturation, the ovary of a fleshy fruit usually has three regions—the exocarp, the mesocarp, and the endocarp (Fig. 27.27). Because these regions may merge, it sometimes is difficult to distinguish between the regions. The three regions are known collectively as the **pericarp**. In dry fruits, the pericarp may be thin, as in the hull of a peanut.

1. The **exocarp**, which forms the skin or peel of a fruit, is variable in color and texture. With its associated glands, it is called a **flavedo** in a citrus fruit. As an orange ripens, the outside of the flavedo changes from green (chlorophyll) to orange (mostly xanthrophyll) in color.
2. The **mesocarp** is the fleshy portion of the fruit between the exocarp and the endocarp. In citrus fruits, the whitish region just beneath the exocarp is actually the mesocarp, called the **albedo**.
3. The **endocarp** is the inside layer of the pericarp directly surrounding the seed. The endocarp may be papery as in apples, or stony as in a peach, or slimy as in a tomato, or a shell as in a pecan. In citrus fruits the endocarp is divided into distinct **segments**.

Juice vesicles provide the treasured juice in a citrus fruit.

Fruit classification is based on several features including whether the fruit is simple or compound, fleshy or dry, and whether other floral parts are present. This scheme is not exact, and arguments abound. In any case, Table 27.1 may be helpful in classifying fruits.

What is a Mexican Jumping Bean?

A Mexican jumping bean is not a bean at all! It is a carpel of a seed capsule from the Mexican shrub *Sebastiana pavoniana*, which houses the larva of a small gray moth called the jumping bean moth (*Laspeyresia saltitans*). While eating the nutritive material within the carpel, the larva wiggles, causing the jumping movements of the "bean."

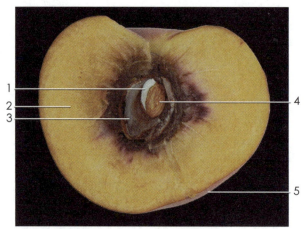

Longitudinal section of a peach fruit.
1. Endosperm   4. Seed coat
2. Mesocarp   5. Exocarp
3. Endocarp

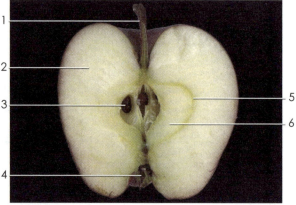

Longitudinal section of an apple fruit.
1. Pedicel   4. Remnants of floral parts
2. Mesocarp   5. Ovary wall
3. Seed   6. Endocarp

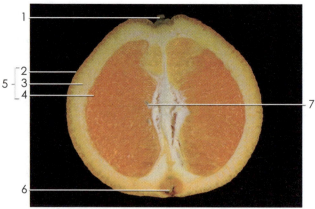

Longitudinal section of an orange fruit.
1. Pedicel   5. Pericarp
2. Exocarp   6. Remnants of floral
3. Mesocarp   parts
4. Endocarp   7. Seed

Figure 27.27 Anatomy of a peach, an apple, and an orange.

Table 27.1 Fruit Classification

Fruit Group	Type	Description	Examples
Simple Fruits			
Fleshy fruits	Berry	ovary compound, skin from exocarp, fleshy pericarp	tomato, grape, guava, kiwi, persimmon, papaya, pomegranate, avocado. *Note:* bananas, cranberries, and blueberries are false berries.
	Pome	accessory fruit, derived from several carpels, ovary (core) surrounded by fleshy receptacle tissue	apple, pear, quince
	Hip	accessory fruit, derived from several carpels, encloses achenes	rose
	Pepo	accessory fruit, berry with hard, thick rind, receptacle partially or completely enclosed the ovary	squash, watermelon, cantaloupe, gourd
	Drupe	derived from a single carpel, possesses one seed, endocarp a stony pit, exocarp a thin skin	peach, cherry, plum, olive, mango, pecan, walnut, coconut, almond, pistachio, cashew, macadamia
	Hesperidium	berry with a leathery rind and juice sacs	orange, lemon, lime, grapefruit, kumquat
Dry Fruits			
Dehiscent	Legume	single carpel, pod splits along two sides	peas, mimosa, bean, peanut, wisteria, redbud
	Follicle	single carpel, splits along one side	milkweed, oleander, columbine
	Silique	two carpels that separate at maturity, leaving a permanent partition between them	radish, mustard, cabbage, turnip
	Capsule	composed of several carpels, separates in several ways	okra, poppy, iris, yucca, cotton, orchid, sweet gum, agave, Mexican jumping bean, Brazil nut
Indehiscent	Achene	simple ovary with pericarp is dry and free from the internal seed, except at the placental attachment	sunflower, dandelion, buttercup, sycamore, buckwheat
	Samara	simple ovary with a winged pericarp, produced in clusters	maple, ash, elm
	Caryopsis (grain)	simple ovary, one seed with the pericarp fused to the seed coat	corn kernel, rice, oats, barley, wheat, Johnson grass, Bermuda grass
	Schizocarp	two or more sections break apart at maturity, each with one seed	carrot, fennel, celery, dill, puncture vine
	Nut	single seed with hard pericarp surrounded by bracts and/or a receptacle	oak (acorn), hickory, hazelnut, beech, chestnut
Compound Fruits			
Aggregate Fruits			
Fleshy Fruits	Achenes	consist of a number of matured ovaries from a single flower, arranged over the surface of a single receptacle; individual ovaries are called fruitlets.	strawberry, buttercup
	Drupes		dewberry, blackberry, raspberry, boysenberry
Dry Fruits	Follicles		southern magnolia
	Samaras		yellow (tulip) poplar
Multiple Fruits			
Fleshy Fruits	Achenes	collection of fruits produced by the grouping of many flowers crowded together in a single inflorescence, typically surrounding a fleshy stem axis	fig
	Drupes		breadfruit, mulberry, Osage orange
	Fused berries		pineapple
Dry Fruits	Achenes		sycamore
	Capsules		sweet gum
	Caryopsis		corn cob and kernels

Simple fruits such as grapes, beans, and hickory are derivatives of a single ovary. Many simple fruits, such as apples, oranges, and watermelons, are classified as **fleshy fruits**. Others are classified as **dry fruits.** In **dehiscent dry fruits,** the pericarp is dry and the fruit splits at maturity; these include peas, radishes, milkweed, and orchids. **Indehiscent dry fruits** do not split at maturity; examples are acorns, corn kernels, parsley, and rice.

Compound fruits, such as strawberries, blackberries, and figs, develop from several individual ovaries. **Aggregate compound fruits** are derived from a single flower with many pistils. In an aggregate fruit, the tiny fruitlets can be an achene or a drupe existing on a single receptacle. A strawberry is a fleshy fruit with achenes on the surface of the receptacle (the body). Look at a blackberry: Each tiny drupe came from an individual ovary. Strawberries and blackberries are **aggregate fleshy fruits.** Aggregate fruits can also be **dry**, such as the fruits of magnolia and yellow poplar. Multiple fruits such as figs, pineapples, and mulberries are derived from several individual flowers in an infloresence.

Fruits that develop from tissues surrounded by the ovary are called **accessory fruits.** These generally develop from flowers with inferior ovaries, and the receptacle becomes a part of the fruit. Accessory fruits can be simple, aggregate, or multiple (Fig. 27.28–Fig. 27.42).

Astonishing!

In the United States, 4,427,000 metric tons of apples are produced each year.

From a moth's point of view, flowers that reliably provide nectar are like docile, productive milch cows. From the flowers' point of view, moths that reliably transport their pollen to other flowers of the same species are like a well-paid Federal Express service, or like a well-trained homing pigeons."

—Richard Dawkins (1941–present)

How to Grow a Pineapple

A green thumb isn't necessary to grow a pineapple—just patience. This can be done by following these steps:

1. Obtain a fresh pineapple with healthy green leaves.
2. Remove several of the lower leaves to expose the stem. Cut off the crown about 3 inches below the stem. Trim any tissue from around the rim.
3. Place the crown upside down in a cool, dry, insect-free place for 1 week.
4. Plant the crown in an 8-inch clay pot filled with light garden soil with a 30% blend of organic matter. Be sure to form the soil around the crown up to the base of the stem.
5. Place the plant in a humid, sunny environment. Lightly water the plant weekly. (Pineapples don't like to get wet!)
6. During the summer, lightly fertilize the pineapple monthly.
7. After 18–24 months, inspect the plant's development. The pineapple will produce a red cone surrounded by blue flowers. The flowers will drop, and the fruit will begin to develop.
8. If desired, force the fruit to develop by covering the entire plant with a clear polyethylene bag and placing two ripe apples in the pot. Ethylene gas produced by the apples encourages fruit development.
9. When the pineapple matures, photograph it, then harvest it. *Bon appetit!*

Flower and fruit of the strawberry, *Fragaria* sp. The strawberry is an aggregate fruit.

Flower and fruit of a tomato, *Lycopersicon esculentum*. A tomato fruit is a berry.

Flower and fruit of the pear *Pyrus* sp. The pear fruit develops from the floral tube (fused perianth) as well as the ovary.

Figure 27.28 Flowers and fruits.

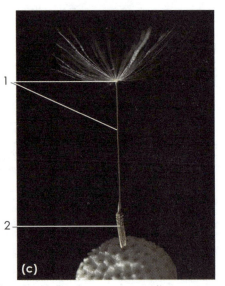

Figure 27.29 Flower (a) and the fruits (b and c) of the dandelion, *Taraxacum* sp. The dandelion has a composite flower. The wind-borne fruit (containing one seed) of a dandelion, and many other members of the family Asteraceae, develop a plumelike pappus, which enables the light fruit to float in the air.

1. Pappus   2. Ovary wall, with one seed inside

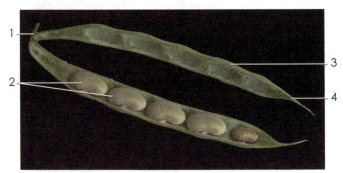

Figure 27.30 Dissected legume, garden bean, *Phaseolus* sp.

1. Pedicel   3. Fruit
2. Seeds   4. Style

Figure 27.31 Cob of corn from *Zea mays*. Corn was domesticated approximately 7,000 years ago from a Mexican grass, family Poaceae.

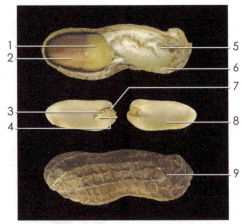

Figure 27.32 Fruit and seed of a peanut plant.

1. Cotyledon   6. Mesocarp
2. Integument   7. Radicle
   (seed coat)   8. Cotyledon
3. Plumule   9. Fruit wall (pericarp)
4. Embryo axis
5. Interior of fruit

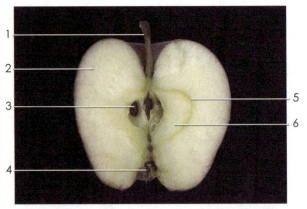

Figure 27.33 Longitudinal section of an apple fruit.

1. Pedicel   5. Ovary wall
2. Mesocarp   6. Endocarp (2 & 6
3. Seed   comprise the fruit)
4. Remnants of floral
   parts

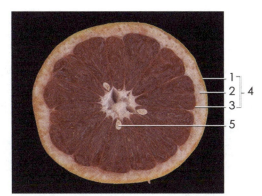

Figure 27.34 Transverse section through a grapefruit fruit.
1. Exocarp
2. Mesocarp
3. Endocarp
4. Pericarp
5. Seed

Figure 27.35 Longitudinal section of a pineapple.
1. Shoot apex
2. Central axis
3. Floral parts

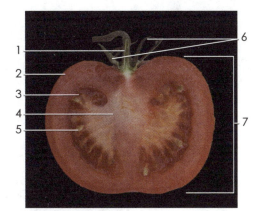

Figure 27.36 Longitudinal section of a tomato fruit (berry).
1. Pedicel
2. Pericarp
3. Locule
4. Placenta
5. Seed
6. Sepals
7. Mature ovary (fruit)

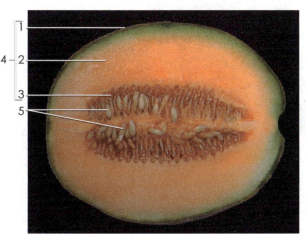

Figure 27.37 Longitudinal section of a cantaloupe.
1. Exocarp
2. Mesocarp
3. Endocarp
4. Pericarp
5. Seeds

Figure 27.38 Fruit and seed of a pecan.
1. Fruit wall (pericarp)
2. Cotyledon
3. Mesocarp
4. Mesocarp
5. Cotyledon

The Earth conceives by the sun and through him becomes pregnant with annual fruits.

—**Nicholas Copernicus (1473–1573)**

Figure 27.39 Examples of simple fruits: (a) peach, (b) grapes, (c) apple, (d) plum, and (e) a pea.

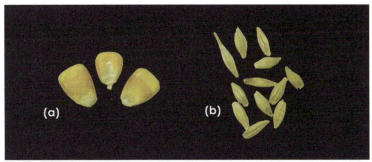

Anytime the perfume of orange and lemon groves wafts in the window, the human body has to feel suffused with a languorous well-being.

—**Frances Mayes (1940–present)**

Figure 27.40 Examples of dry fruits: (a) corn and (b) oats.

Figure 27.41 Example of an accessory fruit (a) strawberry and an aggregate fruit (b) blackberry.

Figure 27.42 Examples of multiple fruits: (a) pineapple and (b) fig.

Student Activity—Macroscopic Anatomy of Fruits

Fruit availability is regional and seasonal. The instructor will provide a variety of types of fruit for observation, preferably at least one fruit of each type. In addition, the student is encouraged to responsibly and legally collect fruits for observation. Further, the student is encouraged to photograph fruits and bring the images to class for discussion.

Materials
- dissecting microscope or hand lens
- scalpel
- dissecting tray
- forceps
- dissecting needle
- colored pencils
- fruits provided by the instructor

Procedure 27.3
Macroanatomy of Fruits

1. Procure the specimens and equipment.
2. Describe the specimens in detail and, using a dissecting microscope or a hand lens, observe, draw, and label the anatomical features of your specimens below. Be sure to include with your description both the common name and the scientific name of the plant, as well as the type of fruit and what you infer about seed dispersal.

3. Dissect the specific fruits provided (such as an apple, an orange, a peach, a bean, a strawberry, and others).
4. Record your observations and labeled illustrations below. In your labeling, include the anatomical information discussed here.

SEEDS

Remember the movie *Fern Gully*? A statement from this Disney classic is appropriate here: "All of the magic of creation exists within a single, tiny seed." Seeds link the historical development of a species with the present and the infinite possibilities of the future. The seed is a structure formed by maturation of the ovule of seed plants following fertilization. Seeds are the end products of sexual reproduction, and they house the embryo. Seeds protect, support, and nourish the embryonic plant until **germination** (resumption of growth and metabolic activity). Many plants have developed elaborate strategies to disperse the seeds into the environment. This explains the existence of yummy fruits, "beggar lice," and making a wish on a dandelion (Fig. 27.43).

All seeds are covered by a **testa** (seed coat), which protects the seed from drying out, extreme temperatures, bacteria, fungi, and predation. An opening called the **micropyle** is often visible on the seed coat as a small pore. The micropyle allows the pollen tube to enter the ovule to ensure fertilization. In addition, some seeds have a distinct scar, the **hilum**, which is left on the seed coat when the seed separates from the supportive **funiculus**, or stalk. The nutritive endosperm, beneath the testa, contains copious quantities of starch that nourish the seed after germination.

Angiosperm and gymnosperm seeds differ in the origin of their stored food. The female gametophyte in gymnosperms provides the food. In angiosperms, the food is supplied by **cotyledons**. In addition to the endosperm, an **embryo** can be found within the seed. The size of the embryo varies with the plant species. The mature embryo consists of a stem-like axis bearing one (monocots) or two (eudicots) cotyledons. The cotyledons, or "seed leaves," are the first leaves to appear in a new sporophyte. They serve as food-storage organs for the seedling plant.

Upon examination, a bean has two distinct halves, each a cotyledon, and a corn kernel has a single cotyledon. In monocots the cotyledon may be called the **scutellum**. In monocots the scutellum is highly absorptive. At the opposite ends of the plant **embryo** are the **apical meristem** of the shoot and the root. Many plants have a stem-like axis, the **epicotyl**, with one or more developing leaves above the cotyledon or cotyledons. The resulting embryonic shoot is called the **plumule**. The stem-like portion beneath the cotyledon or cotyledons is called the **hypocotyl**. The embryonic root, or **radicle**, exists at the lower end of the hypocotyl. The radicle and the plumule are enclosed in a sheath-like protective structure called the **coleorhiza** and the **coleoptile**, respectively. These structures protect the seed during germination.

Check Your Understanding

1. What is the function of a fruit?

2. Classify several fruits that can be seen while you're browsing in a market.

When environmental conditions are favorable, a seed breaks dormancy and germinates, forming a new generation of the plant. The embryos of different species remain viable (capable of germination) for varying periods of time. Seeds of one species of oriental lotus germinated after they were discovered in a 3,000-year-old tomb!

Among the variables influencing germination are temperature, light, water, and scarification, the latter of which can be brought about by bacterial action, stomach acid, freezing, or fire. Usually, seeds germinate while they are under the surface of the soil. Although soil is the ideal environment for roots, shoots are poorly designed for growth under abrasive soil conditions. Fortunately, nature has provided several mechanisms for protecting the young shoot during its emergence from the soil.

In beans, after development of the root and anchorage in the soil, the hypocotyl grows toward the surface in the form of a hook, gently pulling the cotyledons upward. When the hypocotyl hook reaches the surface, light induces the tissue of the hook to straighten, bringing the cotyledons and the young shoot to the surface. During this process, the plumule is protected between the cotyledons. In peas, the epicotyl elongates and forms a hook that otherwise is similar to the mechanisms of hypocotyl elongation in beans, but the cotyledons remain under the surface. In corn and in some grasses, the coleoptile protects the plumule. The coleoptile is a tough, protective sheath that completely surrounds the plumule. When the coleoptile reaches the surface, light induces it to split, and the plumule emerges (Fig. 27.44–27.49).

Burdock

Cocklebur

Dandelion

Poppy

Milkweed

Maple

Touch-me-not

Coconut

Double coconut, *Lodoicea maldivica*, the world's largest seed.

Figure 27.43 Diversity of seeds.

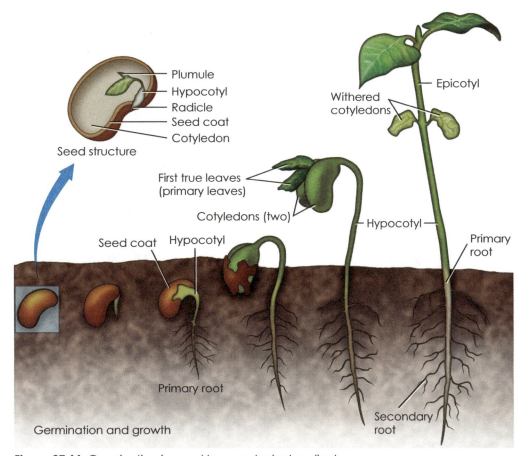

Plumule
Hypocotyl
Radicle
Seed coat
Cotyledon

Seed structure

Epicotyl

Withered
cotyledons

First true leaves
(primary leaves)

Cotyledons (two)

Hypocotyl

Primary
root

Seed coat   Hypocotyl

Primary root

Secondary
root

Germination and growth

Figure 27.44 Germination in a red bean, a typical eudicot.

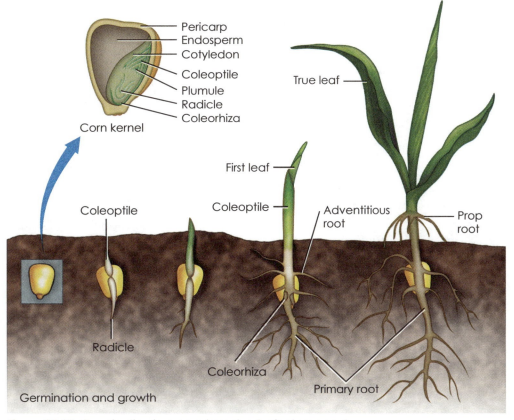

Pericarp
Endosperm
Cotyledon
Coleoptile
Plumule
Radicle
Coleorhiza

Corn kernel

True leaf

First leaf

Coleoptile

Coleoptile

Adventitious
root

Prop
root

Radicle

Coleorhiza

Primary root

Germination and growth

Figure 27.45 Germination in a corn kernel, a typical monocot.

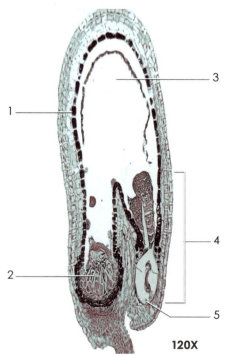

Figure 27.46 Photomicrograph of a developing dicot embryo from a shepherd's purse, *Capsella bursa-pastoris*.
1. Endothelium
2. Cellular endosperm
3. Endosperm
4. Developing embryo
5. Basal cell

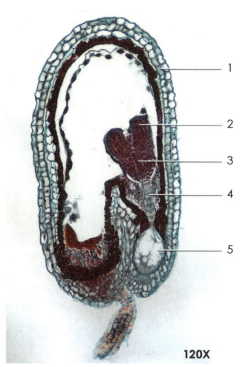

Figure 27.47 Photomicrograph of a developing dicot embryo from a shepherd's purse, *Capsella bursa-pastoris*, showing young embryo.
1. Seed coat
2. Cotyledon
3. Hypocotyl
4. Suspensor
5. Basal cell

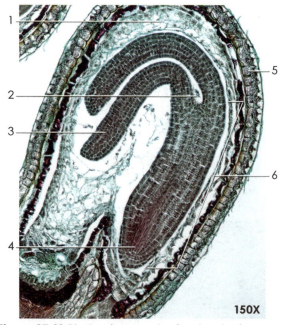

Figure 27.48 Photomicrograph of a developing dicot embryo from a shepherd's purse, *Capsella bursa-pastoris*, showing a nearly mature embryo.
1. Endosperm
2. Epicotyl
3. Cotyledon
4. Radicle
5. Seed coat
6. Hypocotyl

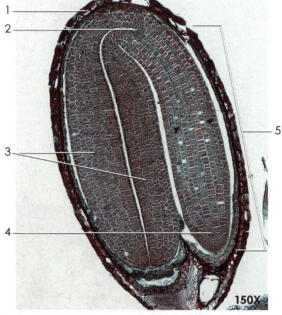

Figure 27.49 Photomicrograph of a developing dicot embryo from a shepherd's purse, *Capsella bursa-pastoris*, showing a mature embryo.
1. Seed coat
2. Epicotyl
3. Cotyledons
4. Radicle
5. Hypocotyl

Student Activity—Why Do They Do That?

Why does the clerk mark your $20 bill with that "magic pen?" As you learned in Chapter 2, paper money contains many chemical safeguards to protect it from counterfeiters, including removing starch (*amylose*) from the paper. Counterfeiters have not discovered how to remove the amylose and when counterfeit bills are marked with an iodine pen, the mark appears blue. Gotcha! Iodine reacts with starch, producing a bluish color. In this activity, the cut surface of a corn kernel is treated with a drop of iodine solution. The storage tissues turn blue because of the presence of amylose.

Materials
- iodine solution
- eyedropper
- scalpel
- plastic Petri dish
- paper towels
- paper money
- corn kernel soaked in water for 24 hours

Procedure 27.4
Iodine and Starch

1. Procure the material and equipment from the instructor.
2. Carefully cut the corn kernel in half longitudinally.
3. Place the kernel in the Petri dish, and place one drop of iodine solution on the kernel.
4. Wait 2 minutes, then describe what happened. What portion of the kernel is blue?

_____

_____

5. Drop a small drop iodine solution on the paper money and determine if your "bill" is counterfeit.

_____

_____

 Student Activity—Macroscopic Anatomy of Seeds

In this activity, the seeds should include, but not be limited to, a red bean, a peanut, a corn kernel, a true rice grain, and a watermelon seed. The student is encouraged to responsibly and legally collect additional seeds for observation. In addition, the student is encouraged to photograph seeds and bring the images to class for discussion. This is a good time to practice macrophotography.

Materials
- iodine solution
- dissecting microscope or hand lens
- scalpel
- colored pencils
- selected seeds provided by the instructor

Procedure 27.5
Macroanatomy of Seeds

1. Procure the equipment and seeds.

2. Describe the specimens in detail and, using a dissecting microscope or a hand lens, observe, draw, and label the anatomical features of your specimens below. In your description, be sure to include both the plant's common name and the scientific name.

3. Using a scalpel, carefully cut each seed longitudinally. Describe the specimens in detail and, using a dissecting microscope or a hand lens, observe, draw, and label the anatomical features of your specimens below.

Student Activity—Microscopic Anatomy of Seeds

Materials
- dissecting microscope
- compound microscope
- prepared slides of select eudicot and monocot seeds, such as longitudinal sections of a bean, shepherd's purse, and corn kernel
- colored pencils

Procedure 27.6
Microanatomy of Seeds

1. Procure the equipment and specimens from the instructor.
2. Observe, sketch, and label the seed slides. Include both the common name and the scientific name of the plant.

3. Compare your observations to Figures 27.46–27.49.

_____

_____

_____

_____

_____

A Digital Field Trip—Pictures of Plants

Think of the following activity as a digital field trip or a scavenger hunt. You may find these plants on campus, in your yard, in the woods, in a greenhouse, at an arboretum, or at a botanical garden. Locate the specimens in question, take your time to compose a photograph of the specimen, use good photographic technique, check out the quality of the image on the camera LCD, transfer the image to a computer, and develop a labeled presentation.

Materials
- digital camera
- notebook to record data
- computer
- presentation software
- storage device such as a jump drive or CD
- projector and screen

Procedure 27.7
Plant Pictures

1. Locate and photograph a specimen such as:

    ☐ a close-up of a flower showing basic floral anatomy

    ☐ a complete and an incomplete flower

    ☐ a perfect and an imperfect flower

    ☐ an actinomorphic and a zygomorphic flower

    ☐ a solitary flower

☐ an inflorescence

☐ a dissected parthenocarpic fruit

☐ a dissected apple showing basic internal anatomy

☐ a dissected orange showing basic internal anatomy

☐ a berry

☐ a drupe

☐ a pome

☐ a hip

☐ a hesperidium

☐ a legume

☐ a capsule

☐ an achene

☐ a nut

☐ a caryopsis

☐ an aggregate fruit

☐ a multiple fruit

☐ bean anatomy

☐ corn kernel anatomy

☐ peanut anatomy

2. Take photographs of the specimen. The photographs should include the entire specimen, as well as distinguishing features.

3. Using a computer and presentation software, develop a digital slide show of your project. Label your slides clearly. Include the title of your specimen, label important parts, and use both common and scientific names of the plant shown. And how about adding some music to your presentation?

4. Place your slide show on a storage device or CD. The instructor may ask you to present your slide show to the class.

NOTES

_____

_____

_____

_____

_____

_____

_____

_____

_____

_____

_____

_____

_____

_____

_____

_____

_____

Name: _____ Date: _____ Section: _____

?? Review Questions

1. Compare and contrast the flowers of eudicots and monocots.

_____

_____

_____

2. Label the following diagram.

a. _____ g. _____

b. _____ h. _____

c. _____ i. _____

d. _____ j. _____

e. _____ k. _____

f. _____ l. _____

3. Compare and contrast perfect and imperfect flowers, and provide an example of each.

_____

_____

_____

4. What is a catkin? Name several plants that produce catkins.

_____

_____

_____

5. Outline the life cycle of a typical angiosperm.

_____

_____

_____

6. Label the following micrograph.

a. _____

b. _____

c. _____

d. _____

7. Take an imaginary trip to a market. List 10 fruits that you will encounter, and indicate the group and type of fruit.

_____

_____

_____

_____

_____

_____

_____

_____

Name: _____ Date: _____ Section: _____

8. Compare the germination of a typical eudicot seed and a typical monocot seed.

_____

_____

_____

9. Sketch and label the anatomical features of a bean and a corn kernel below.

10. Draw a cross-section of an orange, and label the parts.

11. In relation to iodine, what do a counterfeit $20 bill and a corn kernel have in common?

_____

_____

_____

NOTES

Chapter 28
There's a Fungus Among Us: Understanding Fungi

Student Outcome Objectives

At the completion of this exercise, the student will be able to:

1. Explain the role of kingdom Fungi.
2. Describe the characteristics of kingdom Fungi.
3. Point out and identify the basic anatomical features of kingdom fungi.
4. Discuss the taxonomical organization of kingdom fungi.
5. Describe the characteristics of and biology of Phylum Chytridomycota.
6. Describe the characteristics of and biology of Phylum Zygomycota.
7. Trace the life cycle of *Rhizopus stolonifera*.
8. Describe and identify select examples of Phylum Zygomycota.
9. Describe the characteristics of and biology of Phylum Ascomycota.
10. Trace the generalized life cycle of a typical ascomycete.
11. Describe and identify select examples of Phylum Ascomycota.
12. Describe the characteristics of and biology of Phylum Basidomycota.
13. Trace the generalized life cycle of a typical basidiomycete.
14. Label the parts of a typical mushroom.
15. Describe and identify select examples of Phylum Basidiomycota.
16. Describe the characteristics of and biology of the artificial Phylum Deuteromycota.
17. Point out several significant deuteromycetes.
18. Describe the biology of lichens.
19. Identify and describe the three basic forms of lichens.

Overview

Sure, there are many species of fungi among us! (Fig. 28.1.) Unfortunately, the term *fungus* evokes images of molds, mildew, rotting organic matter, spoiled food, and various maladies of plants, animals, and humans. **Fungi** do not limit their enzymatic attack to living things or dead things. Species of fungi attack plastic, leather, paint, petroleum products, film, and even the multicoating of optical equipment. Millions of dollars are spent yearly trying to control fungal diseases in plants, including Dutch elm disease, wheat rust, and corn smut, and in humans, diseases such as ringworm, coccidiomycosis and aspergillosus. Some species of fungi produce powerful toxins, carcinogens, and hallucinogens.

Fungi play a vital role in ecosystems and are economically essential. Without certain species of fungi serving as decomposers, ecosystems would collapse. These decomposers break down dead organisms, leaves, feces, and organic matter and recycle their chemical components back into the environment. In addition, many species of plants depend upon mutualistic fungi to help their roots absorb minerals and water from the soil. Animals and humans eat many species of fungi. Truffles, morels, and some species of mushrooms are delicacies. Fungi also play a vital role in the bread, cheese, beer, and wine industries. Several species of fungi are used in the production of antibiotics including penicillin, cyclosporine, and other beneficial medicines.

For many years, the fungi were classified as imperfect plants. Today, it is clear that the fungi are not degenerate plants but, rather, unique eukaryotes deserving of their own kingdom. New classification schemes place fungi closer to the animal kingdom than to the plant kingdom. **Mycologists** (specialists in fungi) recognize nearly 100,000 species of fungi and predict that this number could increase to 2,000,000 species. Representative fungi

Figure 28.1 Examples of fungi, (a) yeast, (b) bread mold, (c) a mushroom, and (d) a morel.

include mushrooms, puffballs, bread mold, morels, truffles, smuts, rusts, blight, mildew, and yeasts.

Fungi are filamentous, spore-producing, heterotropic unicellular or multicellular eukaryotes. **Heterotropic fungi** release digestive enzymes onto a food source, partially dissolving the source to make the essential nutrients available. Most **coenocytic** (multinucleated or multicellular) fungi are composed of multiple filaments, known as **hyphae**, grouped together into a mass called a **mycelium**. Hyphae comprise the body of a mushroom. Hyphae and mycelia can grow rapidly. Under ideal conditions, a single fungus might produce a kilometer of hyphal growth in one day. That is why a mushroom can appear overnight.

Various modifications of hyphae can be found throughout the fungi. One modified hyphae, a **haustoria**, penetrates the tissues of a host. Other modified hyphae, called **rhizoids**, anchor fungi to a substrate. Some soil fungi have **snare**, or **loop hyphae**, to trap unsuspecting nematodes (roundworms) for future consumption. Fungi that have cross walls in their hyphae (**septate**) are connected to adjacent hyphae by tiny pores in the cross wall; in contrast, septa that separate reproductive cells have no pores. **Nonseptate** fungi are multinucleated and contain numerous nuclei in the cytoplasm. The majority of fungi are **saprobes**, which break down organic matter, but there are parasitic and mutualistic fungi as well. Many fungal species possess a cell wall composed of the polysaccharide chitin, and they store their energy in the form of glycogen. Fungi do not contain chlorophyll.

The majority of fungi are capable of undergoing asexual and sexual reproduction. Asexually, fungi can reproduce by budding, fragmentation, and spore formation. Spores can develop directly without uniting with another spore. Sexually, fungi produce gametes in specialized areas of the hyphae called **gametangia**. The gametes may be

Science is an integral part of culture. It's not this foreign thing, done by an arcane priesthood. It's one of the glories of the human intellectual tradition.

—**Stephen Jay Gould (1941–2002)**

Is It Mushroom or Toadstool?

In some circles, the common terms *mushroom* and *toadstool* cause confusion. Most Americans call any club-shaped fungus a mushroom. British people generally designate the term *mushroom* for edible fungi, and the inedible mushrooms as toadstools. Some people reserve the term *toadstool* for toxic fungi.

released to fuse into spores elsewhere, or the gametangia themselves may fuse. In the hyphae of some fungi, **dikaryons** (*Greek di = two, karyon = nucleus*) form as the result of unspecialized hyphae fusing. In this case, their two nuclei remain distinct for a portion of the life cycle. When the two nuclei finally fuse, the zygote undergoes meiosis prior to spore formation. Upon germination, the spores form haploid hyphae.

A large mushroom can produce billions of spores. One puffball in Canada measured more than 2.6 meters in girth and produced in excess of 7×10^{18} spores. When a spore lands on a suitable substrate, it germinates and grows. That is why bread mold can appear mysteriously on a slice of bread that you thought to be pristine.

Members of the kingdom Fungi are classified into five distinct phyla:

1. Phylum **Chytridiomycota** is the most ancient group of fungi. Most chytrids are either aquatic decomposers, feeding on dead plant or animal material in a pond, or parasites living on water molds, insects, or snakes.
2. Phylum **Zygomycota** includes bread molds and *Pilobus* spp., "the hat-throwing" fungus.
3. Phylum **Ascomycota**, the largest phylum of fungi, includes organisms such as truffles, morels, and yeast.
4. Phylum **Basidiomycota** includes mushrooms, shelf fungi, and puffballs.

Nature's Canon

Grazing animals such as cattle rarely graze near feces (dung pat). The zone of ungrazed grass around a dung pat is called the "ring of repugnance." Infective stages of several endoparasites are faced with the task of having to travel from the dung pat beyond the ring of repugnance to ungrazed grass.

The roundworm (nematode) *Dictyocaulus viviparous*, called the "cattle lungworm," has solved this problem in a unique way. The larvae of the cattle lungworm migrate up the sporangiophore of the saprobic fungus *Pilobolus* sp. and accumulate on the sporangium.

Pilobolus is known as the shotgun fungus or the hat-throwing fungus or the canon fungus because it has explosive sporangia that can shoot spores beyond the ring of repugnance up to 2.5 meters toward light. The lungworm larvae hitch a ride on the spores and land beyond the ring of repugnance. When ingested by a cow, the larvae penetrate the wall of the cow's intestine and are carried by the lymphatic and circulatory systems to the lungs, where the adult worms develop.

5. A new classification scheme places the ascomycetes and basidiomycetes as subphyla in the newly developed phylum **Dikaryomycota,** a heterogeneous group of unrelated species that includes *Penicillium* and the organisms responsible for athlete's foot and ringworm.

Fungal classification is based on anatomy, types of hyphae, means of reproduction, and molecular biology.

PHYLUM CHYTRIDIOMYCOTA

Perhaps the oldest and simplest fungi are the chytrids, members of phylum **Chytridiomycota.** The majority of these minute fungi are unicellular and can be found living in freshwater environments, moist environments, and leaf litter, and on animals such as insects and amphibians. The cell wall of chytrids is composed of chitin, and they display flagellated spores and gametes. One species, *Synchytrium endobioticum*, causes potato wart. Chytridiomycosis is a fungal infection of tadpoles and frogs caused by the chytrid *Batrachochytrium dendrobatidis*. It is a fatal disease and has caused the decline of amphibian population in several regions on Earth (Fig. 28.2–28.3).

Figure 28.2 A Panamanian golden frog, *Atelopus zeteki*, near death with chytridiomycosis. The small white specs are the fungal infection.

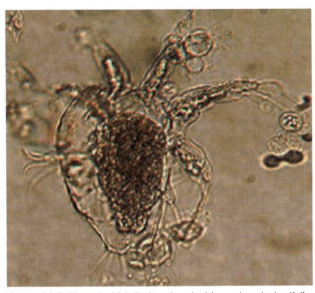

Figure 28.3 The chytrid, *Batrachochytrium dendrobatidis*.

 Student Activity—Macroscopic Anatomy of Selected Chyrids

Materials
- compound microscope
- colored pencils
- selected prepared slides of chyrids such as *Allomyces macrogynus, Rozella* sp., *Rhizophydium*, and *Batrachochytrium dendrobatidis*

 **Procedure 28.1
Macroanatomy of Chyrids**

1. Procure a microscope and selected slides from the instructor.
2. Observe and sketch select chytrid slides.
3. Record your observations and sketches.

Several other notable species of zygomycetes are medically and ecologically significant. *Rhizopus nigricans*, which also can grow on bread, and *Mucor* spp., which can grow on stored foods and seeds and is commonly found in house dust, cause fungal sinusitis and allergies. Other species of *Rhizopus* and *Mucor* can cause serious and sometimes deadly infections (mucormycosis) that affect the skin, digestive tract, facial region, lungs, and brain. *Pilobolus*, discussed in the *Nature's Canon* feature, is another zygomycete. One of the most important zygomycetes is *Glomus* sp., which grows in a symbiotic relationship with roots of certain plants. These organisms form mycorrhizae that aid in plant nutrition.

PHYLUM ZYGOMYCOTA

Phylum **Zygomycota** includes 1,000 species of primarily terrestrial fungi known as the coenocytic or conjugating fungi. Zygomycetes commonly occur in soil, decaying organic matter, and feces. The hyphae of zygomycetes lack septa and are called coenocytic. A representative example of phylum Zygomycota is *Rhizopus stolonifer*, the common black bread mold. Three types of hyphae are found in *Rhizopus*:

1. **rhizoids** (anchoring hyphae that penetrate the bread and have digestive enzymes);
2. **stolons** (horizontal surface hyphae); and
3. **sporangiophores** (reproductive hyphae).

Reproduction in *Rhizopus* can occur asexually or sexually. Asexually, when a **spore** (sporangiospore) lands on a suitable substrate such as a slice of bread, it germinates and forms hyphae that soon form a mycelium. After the mycelium develops, it produces sporangiophores, which rise above the surface and contain spore-containing **sporangia**. The sporangia release their spores and seek another supportive substrate (Fig. 28.4–Fig. 28.7).

Sexually, *Rhizupus* reproduces by **conjugation**. *Rhizopus* produces two different hyphae (starins), which develop a swollen **progametangia** on the ends facing each other. Eventually they touch, and a **cross wall** forms behind each tip. Next, a thick-walled **zygosporangium** forms, replacing the progametangia. The zygosporangium cracks open, forming sporangiophores and their associated sporangia. Mciosis occurs in the sporangia, producing **meiospores** that are released to seek another substrate.

Figure 28.4 *Rhizopus* sp.
1. Sporangia
2. Hyphae (stolon)

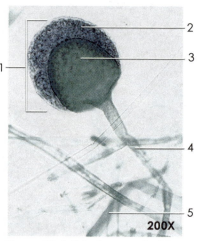

Figure 28.5 Whole mount of the bread mold, *Rhizopus* sp.
1. Sporangium
2. Spores
3. Columella
4. Sporangiophore
5. Hyphae

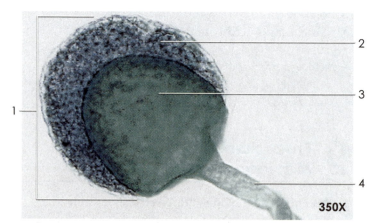

Figure 28.6 Mature sporangium in the asexual reproductive cycle of the bread mold, *Rhizopus* sp.
1. Sporangium   3. Columella
2. Spores   4. Sporangiophore

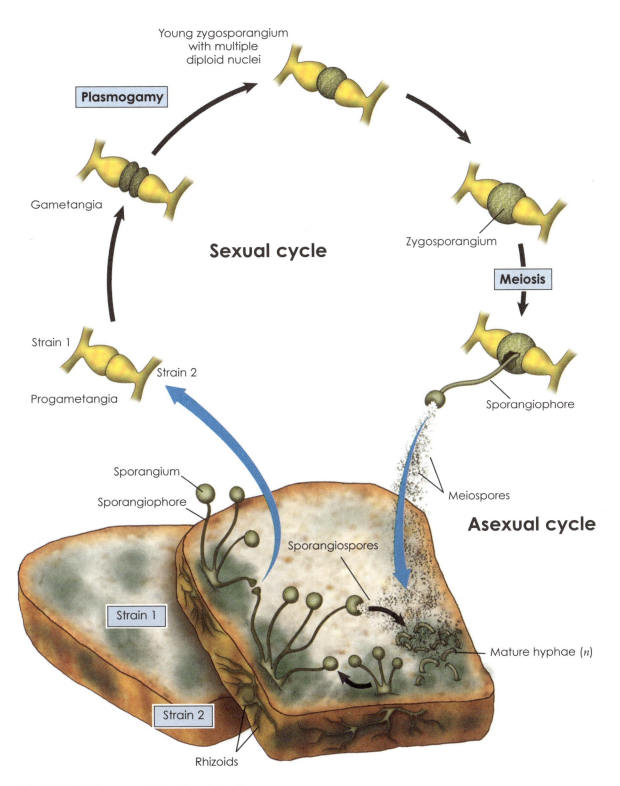

Figure 28.7 Life cycle of *Rhizopus stolonifera*.

Student Activity—Macroscopic Anatomy of *Rhizopus stolonifera*

Materials
- dissecting microscope or hand lens
- compound microscope
- slides and coverslips
- scalpel
- colored pencils
- paper towels
- water
- bread contaminated with *Rhizopus stolonifera*

Procedure 28.2
Macroanatomy of *Rhizopus*

1. Procure the equipment and the bread mold from your instructor.
2. Using a dissecting microscope or hand lens, observe your specimen. Attempt to find all of the anatomical structures. Record your observations, including a labeled sketch.

3. Using the scalpel, remove a sporangium from the sporangiophore. Place the sporangium on a microscope slide, and prepare a wet mount for observation. Record your observations.

Student Activity—Microscopic Anatomy of *Rhizopus stolonifera*

Materials
- compound microscope
- prepared slides of *Rhizopus stolonifera*
- colored pencils

 Procedure 28.3
Microanatomy of *Rhizopus*

1. Procure the microscope and prepared slides of *Rhizopus stolonifera*.
2. Examine the slides and compare them to the micrographs above.
3. Place your observations and labeled sketches below.

Student Activity—Target Practice

As noted in the *Nature's Canon* feature, *Pilobolus* becomes a canon or hat thrower, ejecting spores toward light. This activity tests the accuracy of *Pilobolus*. Wash your hands thoroughly during this activity.

Materials
- 2-day-old cow or horse dung that is growing *Pilobolus*
- glass dish with glass lid
- black construction paper
- paper towel

 Procedure 28.4
Horse Dung and *Pilobolus*

1. Procure the dung, glass dish with lid, and black construction paper.
2. Place the dung on a paper towel in the glass dish.
3. Cut out a star shape or a shape of your choice about 4.5 cm across in the center of the black construction paper.
4. Line the lid of the dish with the construction paper, and place it over the dish.
5. Place the apparatus in a place where it will receive light for 3–7 days.
6. After that time, examine the pattern of spores on the glass, and record your results.
7. Discard the material as instructed. Wash your hands and tabletop thoroughly.
8. Discuss your findings in the space below.

_____

_____

_____

_____

_____

PHYLUM ASCOMYCOTA

The majority of fungal species (50,000) belong to Phylum **Ascomycota**, the sac fungi (Fig. 28.8). Many species of ascomycetes are found in a symbiotic relationship with algae, forming **lichens**. Sac fungi live in a variety of terrestrial habitats. These organisms range from unicellular to elaborate multicellular forms. Ascomycetes are responsible for various serious plant diseases such as powdery mildew, chestnut blight, and Dutch elm disease.

The yeast *Saccharomyces cerevisia*, an ascomycete, plays important roles in the brewing industry and in genetic research (Fig. 28.9). Another yeast, *Canidida albicans*, causes a variety of fungus infections including oral thrush and vaginal infections. *Neurospora*, a type of bread mold, has contributed to genetic studies. Some ascomycetes are considered delicacies. Although plain to the sight, truffles have an exquisite flavor and rival the cost of gold per gram. True morels are common woodland ascomycetes featuring a convoluted cap. Several species of morels are prized for their flavor and consistency.

One of the most interesting ascomycetes, *Claviceps purpurea*, grows on rye and similar plants. *Claviceps* is responsible for **ergotism** in humans and other animals that consume infected food. In the Middle Ages, the dreaded Saint Anthony's Fire was caused by ergot. Ever since the Middle Ages, ergot has been used to induce abortions and to stop maternal bleeding following childbirth. Ergot produces a chemical that synthesizes lysergic acid, or LSD. Perhaps the witches of Salem and other "possessed" and "mad" people in the past were merely "tripping-out" as the result of ergotism.

Ascomycetes get their name from the **ascus**, a large sac-like cell that is responsible for producing reproductive **ascospores**. The hyphae in ascomycetes are septate, but the cross walls are not complete. Fruiting bodies in ascomycetes are well developed and are called **ascocarps**. Sexual reproduction in the ascomycetes starts when hyphae with one nucleus of opposite mating strains come into contact. Each female gametangia, called **ascogonia**, forms a **trichogyne**, which grows toward the male gametangia, called the **antheridia**. Once the trichogyne touches the antheridium, nuclei migrate from the antheridia to the female ascogonium. The ascogonium forms **ascogeneous hyphae**, which are dikaryotic. These hyphae form a **crozier**, or **hook**, and the nuclei fuse, forming a diploid nucleus. The nucleus undergoes meiosis, producing eight ascospores. Eventually, the **ascospores** forming in the ascocup are released. The asexual spores form singularly or in chains from **conidiophores** and are called **conidia** (Fig. 28.10–Fig. 28.15).

(a) 2X (b)

Checking Your Understanding

Q. Describe three distinct ascomycetes.

_____

_____

_____

Q. How do the ascomycetes get their name?

_____

_____

Q. How do zygomycetes get their name?

_____

_____

_____

Q. What is a chyrid?

_____

_____

_____

_____

_____

(c)

A Truffle Hound?

In France, where truffles are most plentiful, pigs are enlisted to find the fungal delight because truffles produce an aroma similar to a pig's sex pheromone. The truffle aroma is so powerful that a pig can find a truffle 15 meters away that is growing a meter below the surface. A muzzle is placed on the pig to preserve the truffle as it frantically digs for the prune-like fungus. The owner recovers the truffles and sells them at prices exceeding $300 a pound. In turn, the pig "truffle hound" is rewarded with a handful of acorns. Some truffle hunters now employ trained dogs for the task.

(d)

Figure 28.8 Representative ascomycetes: (a) *Peziza repanda* is a common woodland cup fungus. (b) *Scutellinia scutellata* is commonly called the eyelash cup fungus. (c) *Morchella esculenta* is a common edible morel. (d) *Helvella* is sometimes known as a saddle fungus since the fruiting body is thought by some to resemble a saddle.

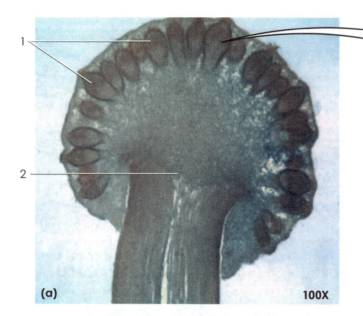

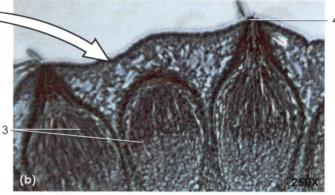

Figure 28.9 The ascomycete, *Claviceps purpurea*. (a) Longitudinal section through stoma showing ascocarps (ascoma). (b) Enlargement of three perithecia. This fungus causes serious plant diseases and is toxic to humans.

1. Perithecia
2. Stroma within multiple perithecia
3. Perithecia containing asci
4. Ostiole

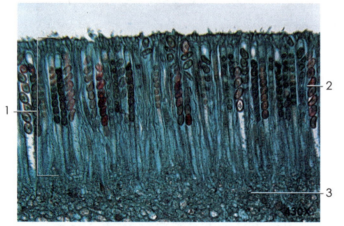

Figure 28.10 A section through the hymenial layer of the apothecium of *Peziza* sp., showing asci with ascospores.
1. Hymenial layer
2. Ascus with ascospores
3. Ascocarp (ascoma) mycelium

Figure 28.11 A section through an ascocarp (ascoma) of the morel, *Morchella* sp. True morels are prized for their excellent flavor.
1. Convoluted fruiting body
2. Hollow "stalk"
3. Hymenium

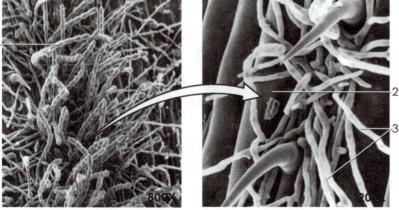

Figure 28.13 A scanning electron micrograph of a germinating spore (conidium) of the powdery mildew, *Erysiphe graminis*. The spore develops into a mycelium that penetrates the epidermis and then spreads over the host plant, producing a powdery appearance.

Figure 28.12 A scanning electron micrographs of the powdery mildew, *Erysiphe graminis*, on the surface of wheat. As the mycelium develops, it produces spores (conidia) that give a powdery appearance to the wheat.
1. Conidia
2. Wheat host
3. Hyphae of the fungus

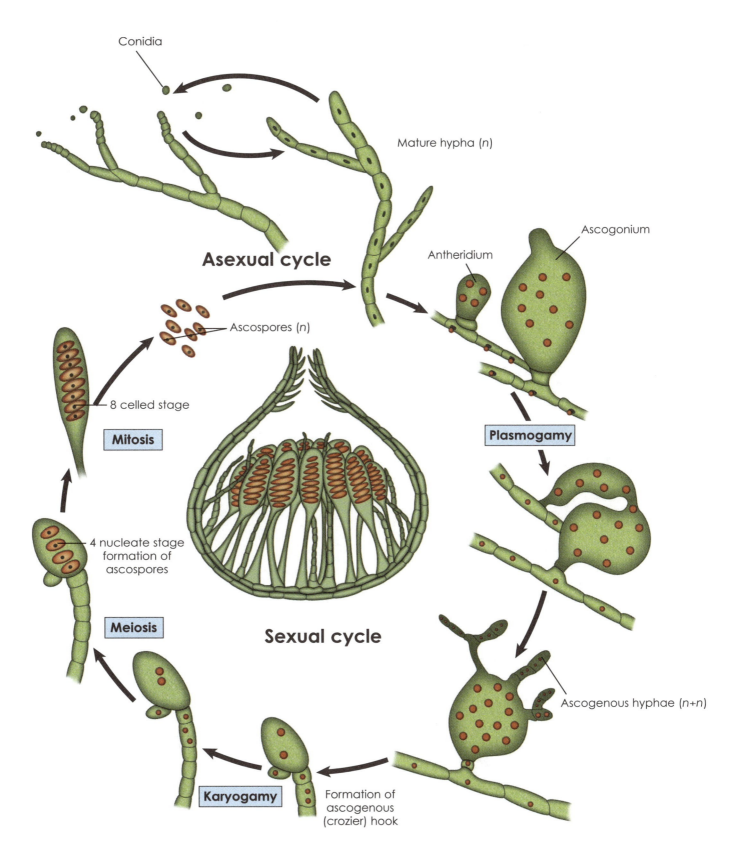

Figure 28.14 Life cycle of an ascomycete.

Student Activity—Macroscopic Anatomy of Select Ascomycetes

This activity is regional and seasonal. The instructor will provide examples of true and false morels, plants infected with ascocycetes, and other examples of this phylum.

Materials
- hand lens or dissecting microscope
- compound microscope
- slides and coverslips
- scalpel
- colored pencils
- paper towels
- water
- specimens provided by the instructor

Procedure 28.5
Macroanatomy of Ascomycetes

1. Procure the equipment and sample specimens.
2. Using a hand lens or dissecting microscope, observe your specimens. Record your observations and sketches below.

3. Carefully dissect your specimen, scraping away some of the tissue. Record your observations and sketches below.

Student Activity—Microscopic Anatomy of Select Ascomycetes

Materials
- compound microscope
- prepared slides of select Ascomycetes including a morel (*Morchella* sp.), *Claviceps* sp., *Peziza* sp., and other select specimens
- colored pencils

 Procedure 28.6
Microanatomy of Ascomycetes

1. Procure the compound microscope and select slides.
2. Observe the specimens on both low and high power.
3. Record your observations and sketches below.

 Student Activity—Observing Baker's Yeast, *Saccharomyces cerevisiae*

Materials
- compound microscope
- blank slides and coverslips
- colored pencils
- dropper
- prepared solution containing Baker's yeast
- methylene blue
- *Saccharomyces cerevisiae*

 Procedure 28.7
Baker's Yeast

1. Procure a compound microscope, blank slides, cover-slips, a dropper, and a sample of *Saccharomyces cerevisiae*.
2. Prepare a wet mount of the *Saccharomyces cerevisiae*.

Figure 28.15 Yeast, *Saccharomyces cerevisiae*

3. Carefully place a drop of methylene blue on your slide to view the yeasts more easily.

4. Record your observations below, and label the stages of the yeast life cycle.

PHYLUM BASIDIOMYCOTA

The best-known phylum of fungi is **Basidiomycota**, with more than 30,000 known species. Example basidiomycetes include mushrooms, puffballs, jelly fungi, earth stars, chanterelles, stinkhorns, rusts, and smuts (Fig. 28.16). Basidiomycetes are called "club fungi" because they produce spores, **basidiospores**, in a club-shaped structure, the **basidium**. The majority of basidiomycetes are saprobes living on dead or dying plants.

Mushrooms, the most obvious basidiomycetes, may be seen living in a circle, called a **fairy-ring**, on the forest floor. Several species of mushrooms, including portabella (*Agaricus bisporus*), oyster mushroom (*Pleurotus ostreatus*), shiitake (*Lentinula edodes*), and chanterelles (*Cantharellus cibarius*), are edible, known for their delectable taste.

One has to be careful when collecting mushrooms for consumption, as many poisonous mushrooms resemble edible species to the untrained eye. Some poisonous mushrooms such as *Amanita* spp. are colorful and appealing, yet this species is termed the "death angel" or "death cap" because of its deadly poison. Some mushrooms are hallucinogenic or psychedelic, such as the "magic mushroom" *Psilocybe* spp.

Puffballs are other common basidiomycetes. They literally release their spores into the wind (Fig. 28.18). Shelf or bracket fungi resemble small shelves growing on the trunk of a tree. Jelly fungi usually are colorful and feel rubbery or gelatinous to the touch. Stinkhorns are diverse, from orange finger-like structures erupting from the soil to resembling a whiffle ball. If this were a "scratch and sniff" manual, everyone would agree that the slimy covering of these fungi smells like rotting flesh. Rusts such as cedar-apple rust and wheat rust (*Puccinia triticina*) are parasitic fungi that are devastating to wheat and rye crops. Smuts are parasitic fungi, attacking sugar cane, corn (*Ustilago maydis*), and other cereal crops, resulting in much devastation.

The basidiomycetes reproduce primarily through sexual reproduction. The life cycle of a mushroom is typical of most basidiomycetes (Fig. 28.17). When a spore lands on a suitable substrate, it germinates into a network of hyphae that form a mycelium beneath the surface. Haploid hyphae exist in several reproductive types. When two compatible types unite, they form a new dikaryotic mycelium. These mycelia can live for perhaps a hundred years, and spread, forming the underground surface of a fairy ring. The mycelia eventually form a **button**, which emerges from the soil. The button develops into a typical mushroom sometimes called a **basidiocarp**, or **basidioma**.

A typical mushroom is composed of perhaps a cup-shaped **volva** at the base, a stalk-like structure called a **stipe**, a ring around the upper end of the stipe called an **annulus**, and a **cap**, or **pileus**. Beneath the cap are slit-like structures called **gills**, or they may be pore-like structures. The gills are composed of individual basidia. In immature mushrooms, a **veil** may cover the developing gills. The basidia mature, and the two nuclei fuse, forming a diploid nucleus that undergoes meiosis. The resulting four basidiospores can be found on peg-like **sterigma**. A large mushroom can produce several million basidiophores in a few days. The spores are released, and the cycle begins again (Fig. 28.18–Fig. 28.28).

This activity is regional and seasonal. The instructor will provide examples of basidiomycetes, such as mushrooms and puffballs. The student is encouraged to responsibly and legally collect specimens for observation. (Remember to wash your hands!) In addition, the student is encouraged to photograph basidiomycetes and bring the images to class for discussion.

I confess, that nothing frightens me more than the appearance of mushrooms on the table, especially in a small provincial town.

—**Alexandre Dumas (1802–1870)**

Pleurotus sp., oyster mushroom

Hericium sp., lion's mane

Chantarella sp., golden chanterelle

Coprinus sp., shaggy mane

Amanita sp., death angel

Coriolus sp., turkey tail

Astreus sp., earth star

Clathrus sp., orange stinkhorn

Nidularia sp., bird's nest

Boletus sp., penny cap

Puccinia podophylli, wheat rust

Ustilago maydis, smut

Figure 28.16 Some representative basidiomycetes.

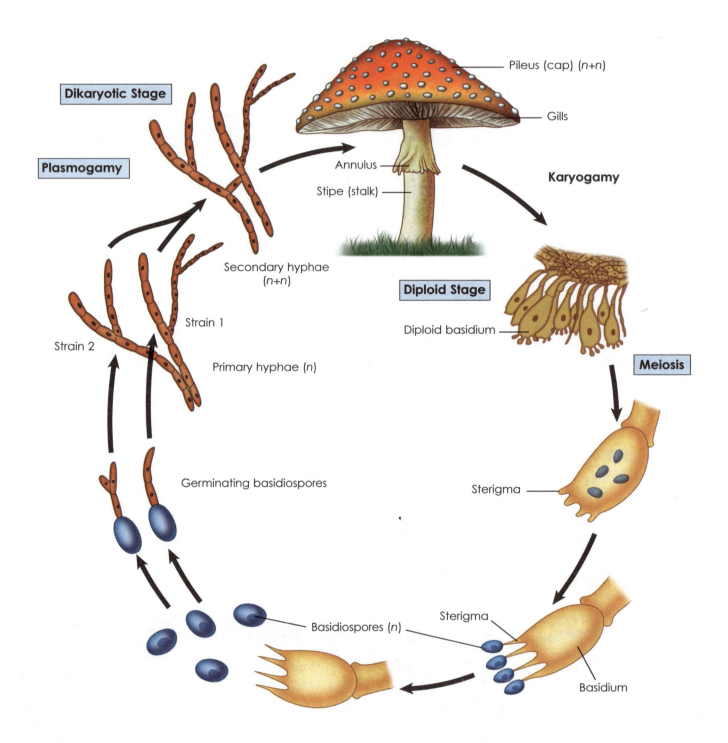

Figure 28.17 Life cycle of a "typical" basidiomycete (mushroom).

Figure 28.18 Basidiomycete puffballs.

Figure 28.19 The wood fungus, *Stropharia semiglobata*. Growing on decaying wood and other organic matter, basidiomycetes are important decomposers in forest communities.

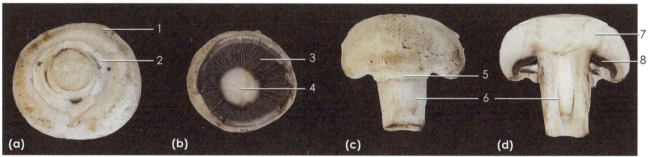

Figure 28.20 Structure of a mushroom. (a) An inferior view with the annulus intact, (b) an inferior view with the annulus removed to show the gills, (c) a lateral view, and (d) a longitudinal section.

1. Pileus (cap)
2. Veil
3. Gills

4. Stipe (stalk)
5. Annulus
6. Stipe (stalk)

7. Pileus (cap)
8. Gills

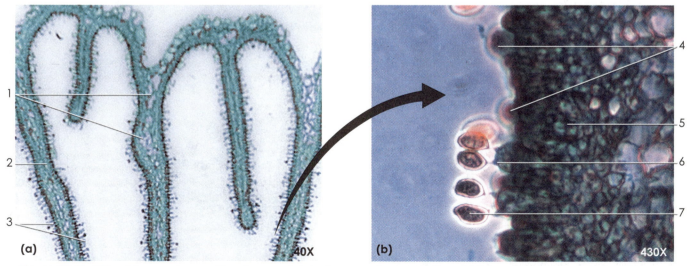

Figure 28.21 Gills of the mushroom *Coprinus* sp. (a) A close-up of several gills, and (b) a close-up of a single gill.

1. Hyphae comprising the gills
2. Gill
3. Basidiospores

4. Immature basidia
5. Gill (comprised of hyphae)
6. Sterigma

7. Basidiospore

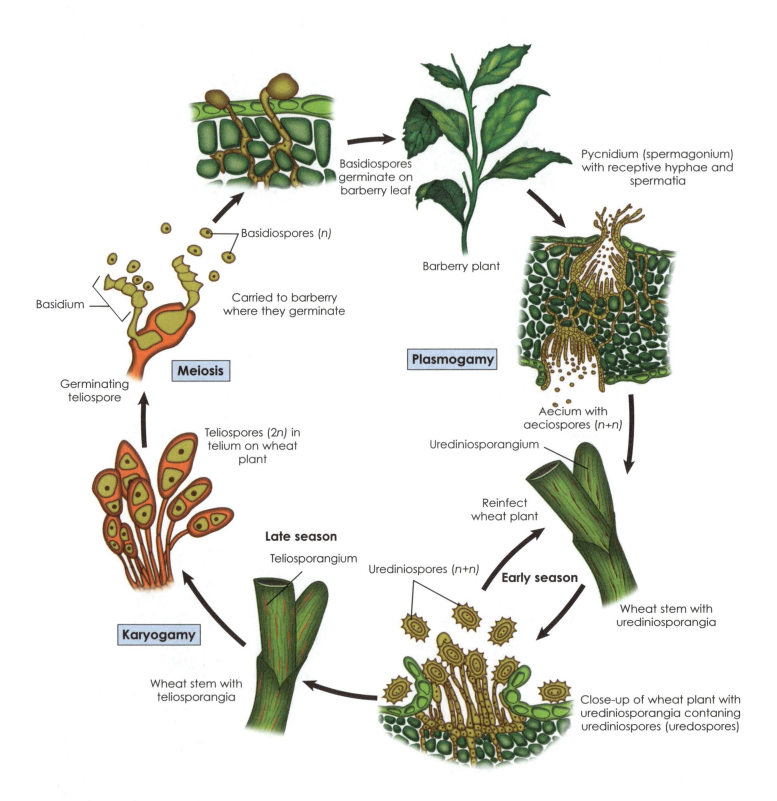

Figure 28.22 Life cycle of wheat rust, *Puccinnia graminis*.

Figure 28.23 Wheat rust, *Puccinia graminis*.

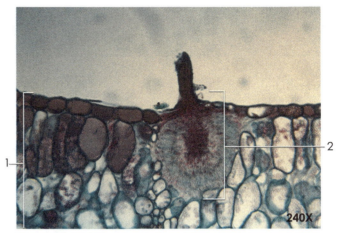

Figure 28.24 Wheat rust, *Puccinia graminis*, pycnidium on barberry leaf.
1. Barberry leaf   2. Pycnidium

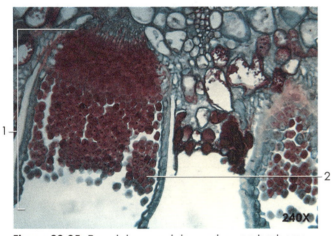

Figure 28.25 *Puccinia graminis*, aecium on barberry leaf.
1. Aecium   2. Aeciospores

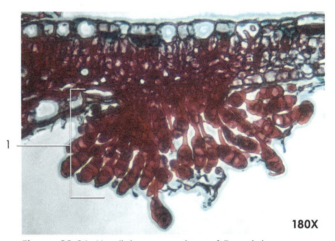

Figure 28.26 Urediniosporangium of *Puccinia* sp. on wheat leaf.
1. Urediniosporangium

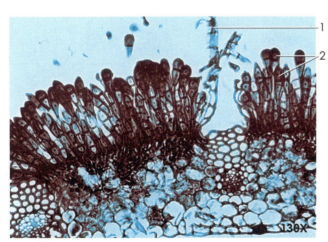

Figure 28.27 Close-up of a wheat leaf sheath showing telia of wheat rust, *Puccinia graminis*.
1. Epidermis of leaf   2. Teliospores

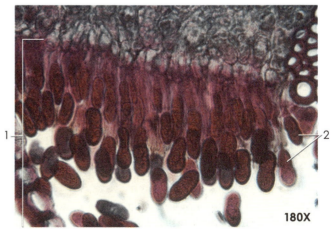

Figure 28.28 Close-up of telium of *Puccinia* sp. on wheat leaf.
1. Telium
2. Teliospores

Student Activity—Macroscopic Anatomy of Select Basidiomycetes

Materials
- hand lens or dissecting microscope
- compound microscope
- slides and coverslips
- scalpel
- colored pencils
- paper towels
- water

Procedure 28.8
Macroanatomy of Basidiomycetes

1. Procure the equipment and sample specimens.
2. Using a hand lens or dissecting microscope, observe your specimens. Record your observations and sketches below.

3. Carefully dissect your specimens, scraping away some of the tissue. Using the compound microscope, prepare slides of the spores if possible. Record your observations and sketches below.

In my first publication I might have claimed that I had come to the conclusion, as a result of serious study of the literature and deep thought, that valuable antibacterial substances were made by moulds and that I set out to investigate the problem. That would have been untrue and I preferred to tell the truth that penicillin started as a chance observation. My only merit is that I did not neglect the observation and that I pursued the subject as a bacteriologist.

—Sir Alexander Flemming (1881–1955)

Student Activity—Microscopic Anatomy of Select Basidiomycetes

Materials
- compound microscope
- prepared slides of select basidiomycetes, including a morel *Coprinus* and *Puccinia*
- colored pencils

 Procedure 28.9
Microanatomy of Basidiomycetes

1. Procure the compound microscope and select slides.
2. Observe the specimens on both low and high power.
3. Record your observations and sketches below.

PHYLUM DEUTEROMYCOTA

Today, some scientists place many of the members of **Phylum Deuteromycota** (imperfect fungi or fungi imperfecti) into phylum Ascomycota. Approximately 17,000 diverse fungi are considered in this artificial phylum. These fungi reproduce exclusively by asexual **conidia**. When a means of sexual reproduction is described, the organism is placed into its true phylum. Although classification is still complex and confusing, the majority of deuteromycetes are terrestrial and free-living. There are many interesting deuteromycetes. Some species of ants and termites literally farm these fungi deep within their mounds, providing them with leaf cuttings and reaping the fungi for food and liquids.

Aspergillus is a genus of green mold that can cause deadly respiratory infections. Some species of *Aspergillus* are used in producing soy sauce, inks, toothpaste, chewing gum, black ink, and photographic developers. The *Aspergillus flavus* species that may grow on improperly stored grain produces a potent carcinogenic substance that can cause liver cancer (Fig. 28.29–Fig. 28.31). *Stachybotrys chartarum* is a black mold responsible for "sick-building" syndrome, in which chronic exposure to the spores can cause chronic sickness. This mold presented a major problem in New Orleans, Louisiana, following hurricane Katrina.

The best known deuteromycete is *Penicillium chrysogenum*, formerly called *Penicillium notatum*. It is the source of penicillin, the antibiotic that was discovered fortuitously by Alexander Flemming in 1928. Penicilium also is used in the production of gourmet cheese. (What do you think comprises the "the blue stuff" in blue cheese?) (Fig. 28.32–Fig. 28.33.) But less friendly species of deuteromycetes, including *Trichophyton* sp., cause athlete's foot, ringworm, jock itch, and small non-pigmented splotches of skin (Fig. 28.34).

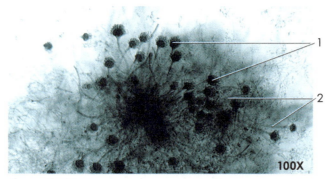

Figure 28.29 Common mold, *Aspergillus* sp.
1. Conidia   2. Conidiophores

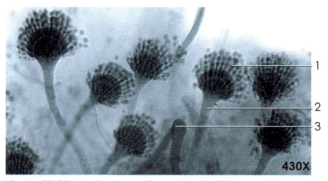

Figure 28.30 A close-up of sporangia of the mold, *Aspergillus* sp. The conidia, or spores, of this genus are produced in a characteristic radiate pattern.
1. Conidia (spores)   3. Developing conidiophore
2. Conidiophore

Figure 28.31 Electron micrograph of an *Aspergillus* sp. spore. Note the rodlet pattern on the spore wall.

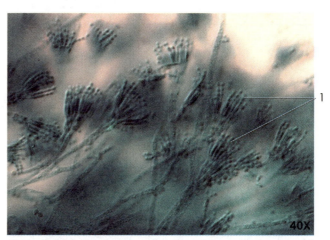

Figure 28.32 The fungus *Penicillium* sp. causes economic damage as a mold but is also the source of important antibiotics.
1. Conidia

Figure 28.33 Blue mold, *Penicillium expansum*, growing on a rotten pear.

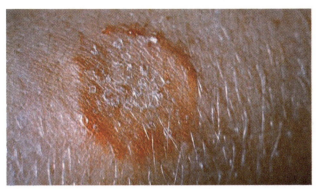

Figure 28.34 Ringworm, *Tinea corporis*, is a fungal infection of the skin.

 Student Activity—Macroscopic Anatomy of Deuteromycetes

Materials
- dissecting microscope or hand lens
- colored pencils
- tray
- several examples of deuteromycetes for observation, such as *Penicillium* and *Aspergillus*, provided by the instructor

Procedure 28.10
Macroanatomy of Deuteromycetes

Consider photographing this activity.

1. Procure the equipment and fungal specimens.
2. Observe the specimens with a dissecting microscope or hand lens.
3. Record your observations and sketches below.

Student Activity—Microscopic Anatomy of Deuteromycetes

Materials
- compound microscope
- colored pencils
- select slides of deuteromycetes including *Penicillium* and *Aspergillus*

Procedure 28.11
Microanatomy of Deuteromycetes

1. Procure a compound microscope and the selected slides.
2. Observe the microscope slides on both low and high power.
3. Record your observations and sketches below.

Checking Your Understanding

Q. What are five types of basidiomycetes?

Q. How did the basidiomycetes get their name?

Q. What are several significant deuteromycetes?

Equipped with his five senses, man explores the universe around him and calls the adventure Science.

—Edwin Powell Hubble (1889–1953)

LICHENS

Lichens are interesting symbionts consisting of a green algae or a cyanobacterium and, with the exception of a few species, ascomycetes. Algal cells or cyanobacteria are thought to provide food for both symbionts through photosynthesis, and the ascomycete retains water and minerals, anchors the organism, and protects the algae. Presently, nearly 25,000 species of lichen have been described. Lichens typically reside on trunks and branches of trees, bare rocks, and human-made structures such as walls and gravestones. They also can survive in extreme conditions such as the tundra (reindeer moss) and hot deserts. Lichens have been used to make dyes, litmus paper, bandages, antibiotics, packing material, decorations, and perfume. In the environment, lichens help to build soil, and they provide food and habitat for small animals. Some species of lichen serve as environmental indicators of pollution.

The body, or **thallus**, of a lichen is derived from an ascomycete surrounding algal cells and enclosing them within complex fungal tissues. The thallus ranges in size from less than 1 millimeter to more than 2 meters in diameter. Lichens are noted for their longevity, perhaps living 4,500 years. Lichens vary in color from dull gray to bright red, green, and orange.

Three basic types of lichens exist in nature:

1. **Crustose** lichens form brightly colored patches or crusts on rock or tree bark, without evident lower surfaces.
2. **Foliose** lichens appear to have leaf-like thalli that overlap, forming a scaly, lobed body. These lichens frequently are found on tree bark and on human-made structures.
3. **Fruiticose** lichens may appear shrub-like or hanging moss-like on trees. Their thalli are either highly branched or cylindrical. Many people think that lichens are parasites on trees but, with the exception of a few species, this is incorrect (Fig. 28.35–Fig. 28.41).

Figure 28.35 Lichens are often categorized informally by their form. (a) Crustose lichen, (b) foliose lichen, and (c) fruticose lichen.

Figure 28.36 Crustose lichen, *Lecanora* sp. growing on sandstone in Southern Utah.

Figure 28.37 Fruticose lichen, British soldier, *Cladonia cristatella*, growing in Alaska.

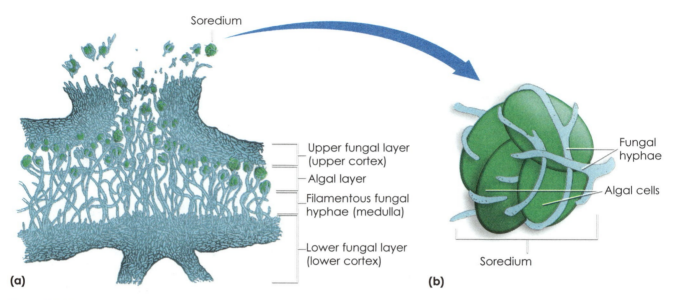

Figure 28.38 Many lichens reproduce by producing soredia, which are small bodies containing both algal and fungal cells. (a) Lichen thallus, and (b) soredium.

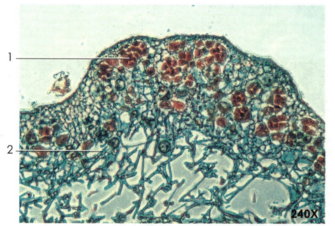

Figure 28.39 Transverse section through a lichen thallus.
1. Algal cells   2. Fungal hyphae

Figure 28.40 Ascomycete lichen thallus demonstrating a surface layer of asci.
1. Asci   2. Loose fungal filaments

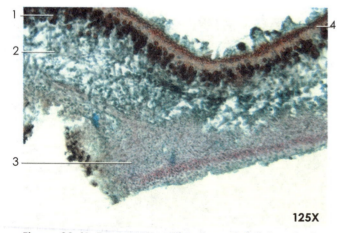

Figure 28.41 Transverse section through a lichen thallus.
1. Algal cells   3. Lower cortex
2. Medulla   4. Fungal layer (upper cortex)

Lichens in Space

In 2005, lichens were exposed to the harsh conditions of space aboard the BIOPAN 5 section of the European Space Agency facility for 16 days. Fungal and algal cells of lichens were found to survive in space after full exposure to massive UV and cosmic radiation—conditions proven to be lethal to bacteria and other microorganisms. In addition, after being dehydrated as the result of a vacuum, the lichens recovered within 24 hours.

Student Activity—Macroscopic Anatomy of Lichens

The instructor will provide several examples of crustose, fruticose, and foliose lichens for observation. The student is encouraged to photograph lichens and bring the images to class for discussion. This will be a good time to practice macrophotography.

Materials
- dissecting microscope or hand lens
- colored pencils
- tray

 Procedure 28.12
Macroanatomy of Lichens

Consider photographing this activity.

1. Procure the equipment and specimens.
2. Using a dissecting microscope or hand lens, observe, describe and sketch the lichens below.

A Digital Field Trip—Fungus and Lichens

Think of the following activity as a digital field trip or a scavenger hunt. Locate the specimens in question, take your time to compose a photograph of the specimen, use good photographic technique, check out the quality of the image on the camera LCD, transfer the image to a computer, and develop a labeled presentation.

Materials
- digital camera
- notebook to record data
- computer
- presentation software
- storage device such as a jump drive or CD
- projector and screen
- objects to photograph: students may find these specimens on campus, in the yard, in the woods, in a market, in a greenhouse, at an arboretum, or at a botanical garden.

Procedure 28.13
Taking Pictures

1. Locate and photograph the following:

   ☐ an ascomycete

   ☐ three different basidiomycetes

   ☐ a mushroom with the major anatomical features discussed in this chapter

   ☐ a deuteromycete

   ☐ three forms of lichens

2. Identify your specimens, using both common and scientific names.
3. Take photographs of the specimens. The photographs should include the entire specimen as well as distinguishing features.
4. Using a computer and presentation software, develop a digital slide show of your project. Clearly label your slides. Include the title of your specimen, and label the important parts, using both common and scientific name of the plant being shown. And how about adding some music to your presentation?
5. Place your slide show on a storage device or CD. The instructor may allow you to present your slide show to the class.

NOTES

Name: _____ Date: _____ Section: _____

??? Review Questions

1. What is the significance of kingdom Fungi to medicine, commerce, and the environment?

_____

_____

_____

_____

_____

2. What are chytrids, and where can they be found?

_____

_____

_____

_____

_____

3. What is the evolutionary significance of Phylum Chytridiomycota?

_____

_____

_____

_____

4. Name and describe five typical ascomycetes.

_____

_____

_____

_____

_____

5. Trace the generalized life cycle of an ascomycete.

6. What are three poisonous and three hallucinogenic fungi?

_____

_____

_____

_____

_____

_____

7. What are five typical basidiomycetes?

_____

_____

_____

_____

_____

Name: _____ Date: _____ Section: _____

8. Trace the generalized life cycle of a basidiomycete.

9. Draw and label a typical mushroom.

10. Why is Phylum Deuteromycota like a holding tank?

_____

_____

_____

_____

_____

11. What are three medically important deuteromycetes?

_____

_____

_____

_____

_____

12. What is a lichen? Name and describe three types of lichens.

_____

_____

_____

_____

_____

_____

_____

_____

_____

_____

_____

Chapter 29
From the Sea:
Understanding Animals—Part I

Student Outcome Objectives

At the completion of this exercise, the student will be able to:

1. Discuss the major characteristics of animals.
2. Compare and contrast acoelomates, pseudocoelomtes, and coelomates.
3. Compare and contrast protostomes and deuterostomes.
4. Define the basic terms of anatomical orientation and body planes.
5. Compare and contrast classical and modern classification schemes of animals.
6. Describe the fundamental characteristics and natural history of Phylum Porifera.
7. Name the three classes of Phylum Porifera.
8. Record and sketch macroscopic and microscopic observations of poriferans.
9. Identify several species of common poriferans.
10. Describe the fundamental characteristics and natural history of Phylum Cnidaria.
11. Name the four classes of Phylum Porifera.
12. Record and sketch macroscopic and microscopic observations of cnidarians.
13. Identify several species of common cnidarians.
14. Describe the fundamental characteristics and natural history of Phylum Ctenophora.
15. Record and sketch macroscopic and microscopic observations of ctenophorans.
16. Identify several species of common ctenophores.

Overview

From simple sponges to the great blue whale, and from trilobites etched in stone to the mighty *Tyrannosaurs rex*, the great diversity of animals captures our curiosity and imagination. Kingdom **Animalia** conservatively consists of approximately 1.5 million extant organisms and many more extinct forms. Animals are eukaryotic, multicellular, heterotrophs that exist in marine, aquatic, and terrestrial environments. Many animals—frogs and cats, as examples—are **free-living**, whereas several species, including barnacles and others, are **sessile**. A number of animal species exists in complex symbiotic relationships such as mutualism and parasitism.

Animal cells do not have a cell wall and are organized into complex tissues including muscle, nerve, and other tissue. The muscular system working with the nervous system results in most animals being motile. Sexual reproduction is the primary means of reproduction, and the diploid stage dominates the life cycle. Some animals, such as jellyfishes and adult starfishes, exhibit **radial symmetry**, and others, ranging from tapeworms to bees to and humans, exhibit **bilateral symmetry**. In radial symmetry the body parts are arranged around a central axis. In bilateral symmetry, the body can be divided into two mirror images (Fig. 29.2). In evolution, bilateral symmetry is tied closely to development of the head region, called **cephalization,** and forward movement such as swimming and running.

Although the shape and size of animals vary tremendously, only a few basic body plans exist in nature. Some animals, examples of which are jellyfishes, tapeworms, and flukes, have a sac-like body plan. These animals have an **incomplete digestive system**—"what goes in the mouth goes back out the mouth." Other animals, examples of which are squid and lions, have a tube-within-a-tube body plan, possessing a **complete digestive system**. Jellyfishes and their relatives are **diploblastic**, composed of two germ layers, the ectoderm and the endoderm. Animals such as earthworms, beetles, snakes, and eagles are **triploblastic**—having three germ layers, the ectoderm, mesoderm, and endoderm.

In relation to the body cavity, or **coelom**, three distinct arrangements are found in the animal world:

1. **Acoelomates** such as planarians do not have a coelom between the digestive system and the outer body wall.
2. In **pseudocoelomates** such as rotifers, pinworms, and horsehair worms, the coelom is derived from both the endoderm and the mesoderm.
3. The body cavity of **true coelomates** is derived from only mesodermal tissues.

In gastrulation, the fate of the **blastopore** has resulted into two major groups of animals: In **protostomates** such as tapeworms, hookworms, snails, leeches, and ants, the blastopore gives rise to the mouth. In **deuterostomates** such as sea urchins, sharks, turkeys, and monkeys, the blastopore gives rise to the anal opening.

Earthworms, insects, and vertebrates are **segmented** animals, in which body parts are repeated along the length of the animal's body. *Hox* **genes** influence the embryological patterning of the body plan, including the body

Figure 29.1 Kingdom Animalia is extremely diverse. (a) A jellyfish, (b) a sea urchin, (c) a butterfly, (d) a fish, (e) a bird, and (f) a bison.

axis and arrangement of parts on the body in animals. A mix-up in a *Hox* gene can cause a condition called antennapedia in fruit flies. In this case, legs replace the antenna that normally grow on the head. The study of homeotic, homeobox, and *Hox* genes are providing scientists with a better understanding of the organization and evolution of the animal kingdom.

The most likely candidate for the ancestor to animals is thought to be a colonial flagellated protist that probably was similar to today's choanoflagellates. Between 600 and 550 million years ago, complex, soft-bodied multicellular animals first appeared in the fossil record. These fossils were named Ediacaran fauna, after the site in Australia where they were first discovered.

Today, living animals, **Metazoans,** have been placed in approximately 35 distinct phyla. The **Parazoa** includes the sponges, and the other phyla are designated **Eumetazoans,** animals with radial symmetry, **Radiata**, and those with bilateral symmetry, **Bilateria**. The Bilateria consists of (a) the **Protostomia—Lophotrochozoa** and **Ecdysozoa,** and (b) the **Deuterostomia—Echinodermata** and **Chordata**.

Keep in mind that the classification of animals, as well as the classification of eukaryotes as a whole, is in a state of transition. Traditional classification was based primarily on body plans, whereas modern schemes are more molecular in origin. Because traditional views have been followed for more than a century with wide acceptance, coupled with the understanding that modern views are not complete, taxonomy is confusing to expert and novice alike. Many biologists and medical professionals are taking a more "user-friendly" approach to classification based upon a synthesis of classical and modern concepts until newer schemes are solidified.

For convenience, the animal kingdom will be divided here into four chapters. This chapter will cover the sponges, cnidarians (jellyfish and coral), and ctenphores (comb jellies). Chapter 30 will study the Lophotrochozoa including platyhelminths (flatworms), rotifers, mollusks (snails, octopi), and annelids (segmented worms). In Chapter 31, the Ecdysozoa, including the nematodes (roundworms) and arthropods (insects and arachnids), will be discussed. Chapter 32 will cover the echinoderms (starfishes) and chordates (fishes, mammals).

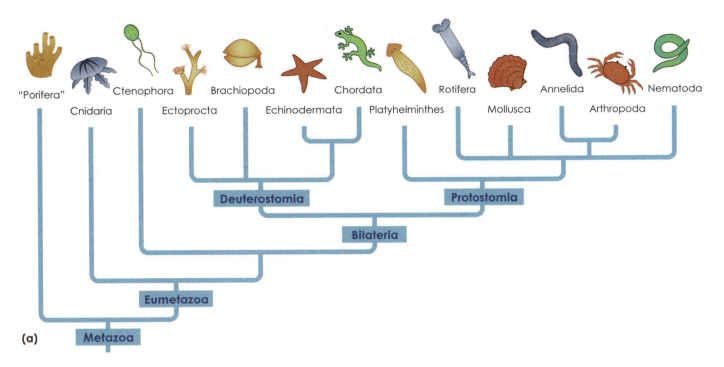

(a)

ANCESTRAL COLONIAL FLAGELLATE

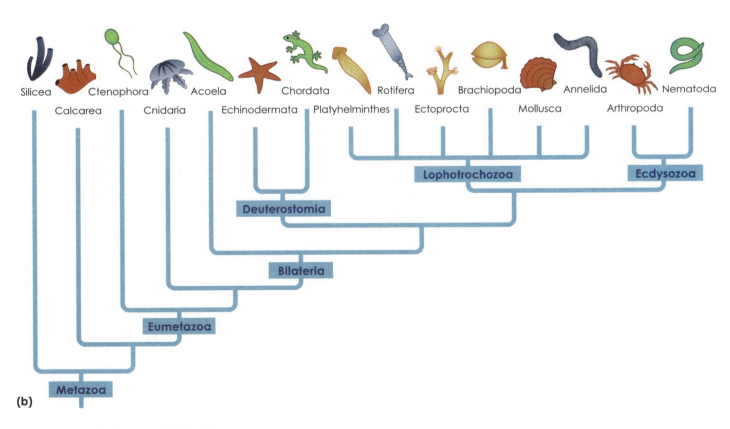

(b)

ANCESTRAL COLONIAL FLAGELLATE

Figure 29.2 (a) Traditional classification scheme of animals, and (b) modern classification scheme of animals.

Helpful Anatomical Orientation Terms

dorsal	pertaining to back	cephalic	head
ventral	pertaining to underside	celiac	abdomen
lateral	to the side	caudal	tail
anterior	front end	cural	leg
posterior	rear end	oral	mouth
superior	above another part or closer to head	pedal	foot
inferior	below another part or toward feet	body planes	
medial	toward imaginary midline	sagittal plane (medial)	lengthwise cut that divides the body into left and right halves.
central	middle	transverse plane	divides the body into superior and inferior sections
peripheral	nearest surface	frontal plane (coronal)	divides the body into anterior and posterior portions

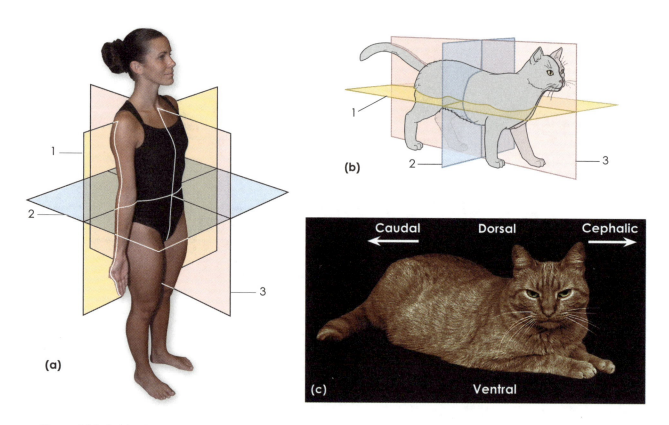

Figure 29.3 Body planes and basic anatomical orientation terms in (a) a bipedal vertebrate and (b) a quadrupedal vertebrate, and (c) directional terminology.

1. Frontal plane (coronal plane) divides anterior from posterior
2. Transverse plane (cross-sectional plane) divides superior from inferior
3. Sagittal plane divides body left and right

PHYLUM PORIFERA

Typically, when thinking of a sponge, visions of bathing or washing the car enter the mind. The term "sponge," however, has different connotations in biology. Upon examining a living or dried sponge, some people are amazed that sponges are actually the simplest of the multicellular animals. Many scientists think that sponges evolved from a group of flagellated aquatic eukaryotes known as the choanoflagellates (Fig. 29.4).

Phylum **Porifera** (Latin–*pore-bearer*) includes approximately 10,000 species of parazoan animals known as sponges. Although sponges are multicellular, they are phylogenetically distinct from other metazoans (multicellular animals) because they do not have tissues or organs. Fewer than 200 species of sponges live in freshwater environments. The vast majority of sponges are sessile marine organisms. Despite the adult sponge being sessile, larval sponges are free-swimming.

Sponges vary in size from a few millimeters to 2 meters across. The body of a sponge is organized around a system of water canals and chambers. Many species are brightly colored (red, yellow, orange, purple, or green). Sponges vary from radially symmetrical to irregularly shaped. Some sponges bore holes in shells and rocks, and others stand erect or form low masses on a substrate (Fig. 29.5). The skeletal structure of sponges consists of fibrous collagen and calcareous or siliceous crystalline **spicules** (Fig. 29.6). These structures are associated with **spongin** in many species.

Excretion and respiration in sponges occur through diffusion. Digestion in sponges is intracellular. Sponges reproduce asexually by budding or by forming **gemmules,** which are structures formed to survive harsh conditions. Under favorable conditions, gemmules form new sponges. Sexually, sponges produce sperm and egg, uniting to form free-swimming larvae. Most sponges are monoecious, having both sexes in the same organism.

Although the body of a sponge may vary in shape and size, the general anatomical features are similar. Many sponges are shaped like a porous vase. The pores, or **ostia**, allow water into the interior of the sponge. The central cavity, or **spongocoel**, is lined with flagellated collar-shaped cells known as **choanocytes**. Water is eliminated from the sponge by way of the **osculum**. The flow of water through the sponge allows food to be taken in and circulated within the sponge, and also enables the intake of sperm. The cells of sponges are arranged in a gelatinous matrix called **mesohyl**.

Figure 29.5 The diversity of sponges; sponges come in many colors and shapes.

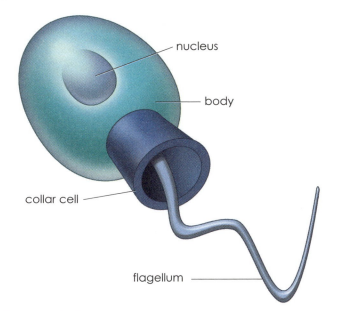

Figure 29.4 Choanocytes might be the ancestors of sponges and other animals.

nucleus

body

collar cell

flagellum

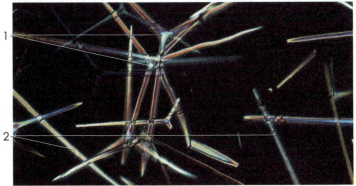

Figure 29.6 The branched silica spicules of a freshwater sponge.
1. Triaxonal   2. Monaxonal

One type of cell, the **pinacocytes,** consists of flat, thin cells covering the exterior and some interior surface of sponges, called **pinaderm**. In another type of cell, **amoebocytes,** or archaeocytes, move about in the mesohyl and absorb, digest, and transport food. Amoebacytes also are involved in the formation of spicules and spongin. **Spicules** are either siliconaceous or calcareous supportive (skeletal) structures. They can vary in shape and are important in sponge classification. Some sponges do not possess spicules. **Spongin** is a collagenous protein material that serves to support sponges.

Most species of sponge have one of three types of canal systems: asconoid, syconoid, or leuconoid (Fig. 29.7).

1. Sponges that have an **asconoid** canal system are generally small and tube-shaped. Water enters these sponges via tiny ostia in the dermis and makes its way to a large cavity called a **spongocoel**, which is lined with choanocytes. The water is filtered and exits via a large opening called an **osculum**. Asconoid sponges are placed in Class Calcarea.
2. **Syconoid** sponges resemble large versions of asconoid sponges. These sponges possess a tubular body with a single prominent osculum. Syconoid sponges, however, have a more complex canal system than asconoid sponges. The choanocytes are found in numerous radial canals that empty into the spongocoel, which is lined with epithelia-like cells in syconoid sponges. The water, with its nutrients, enters the sponge through a large number of ostia into an incurrent canal. The water then passes through **prosopyles** into the radial canals, where the food is ingested by the choanocytes. The flagella of the choanocytes force the water through **apopyles** into the spongocoel. Finally, filtered water exits the osculum. Syconoid bodies are found in Classes Calcarea and Hexactinellida.
3. **Leuconoid** sponges, the most common and complex of the types of sponge, generally form large masses, each member having its own osculum. Clusters of flagellated chambers receive water from incurrent canals, and discharged water exits via the excurrent canals and eventually to the osculum. One species of leuconoid sponge has been estimated to have several million flagellated chambers.

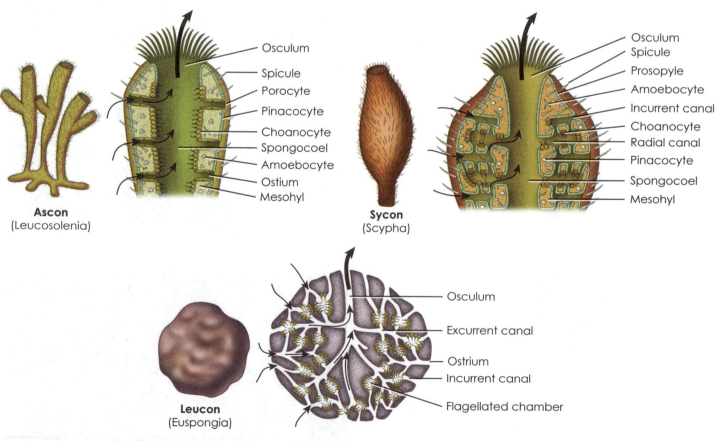

Ascon
(Leucosolenia)

- Osculum
- Spicule
- Porocyte
- Pinacocyte
- Choanocyte
- Spongocoel
- Amoebocyte
- Ostium
- Mesohyl

Sycon
(Scypha)

- Osculum
- Spicule
- Prosopyle
- Amoebocyte
- Incurrent canal
- Choanocyte
- Radial canal
- Pinacocyte
- Spongocoel
- Mesohyl

Leucon
(Euspongia)

- Osculum
- Excurrent canal
- Ostrium
- Incurrent canal
- Flagellated chamber

Figure 29.7 Examples of sponge body types. A diagrammatic representative of each of the three types depicts with arrows the flow of water through the body of the sponge.

Traditionally, zoologists have recognized three classes of sponge:

1. **Class Calcarea** consists of small marine sponges with spicules composed of calcium carbonate. The spicules vary from monaxonal (needle-shaped) to triaxonal (three rays), tetraxonal (four rays) and six-rayed. The majority of sponges in this phylum are vase-shaped and drab in color, although a few bright yellow, lavender, red, and green species exist. Asconoid, syconoid, and leuconoid body forms are found in Class Calcarea. *Sycon* (*Grantia*) is a vase-shaped syconoid sponge that can live in colonies. It is only 1 to 3 centimeters long and possesses a group of monaxonal spicules at the entrance of the osculum. *Leuconolenia* is a small, branched asconoid sponge (Fig. 29.9–29.14).

2. **Class Hexactinellida** is referred to as the "glass sponge" because of the six-rayed siliceous spicules that are fused into an intricate glass-like lattice. Members of this class of sponges are primarily deep-water marine forms. The body of these sponges is usually cylindrical or funnel-shaped. The flagellated chambers can be simple syconoid or leuconoid. Some attain lengths of 1.3 meters. The Venus flower basket (*Euplectella* sp.) is a beautiful member of this class (Fig. 29.8).

3. **Class Demospongiae** is the largest class of sponges, usually brilliantly colored with monaxonal or tetraxonal siliceous spicules, sometimes bound together by spongin. Members of this class have leuconoid canal systems. One family of this class lives in freshwater habitats. Examples are the bath sponge (*Spongilla* spp.) and the barrel sponge (*Xestospongia testudinaria*) (Fig. 29.15).

The affinities of all the beings of the same class have sometimes been represented by a great tree. . . As buds give rise by growth to fresh buds, and these if vigorous, branch out and overtop on all sides many a feebler branch, so by generation I believe it has been with the great Tree of Life, which fills with its dead and broken branches the crust of the earth, and covers the surface with its ever branching and beautiful ramifications.

—**Charles Darwin (1809–1882)**

Figure 29.8 *Euplectella*, the Venus flower basket.

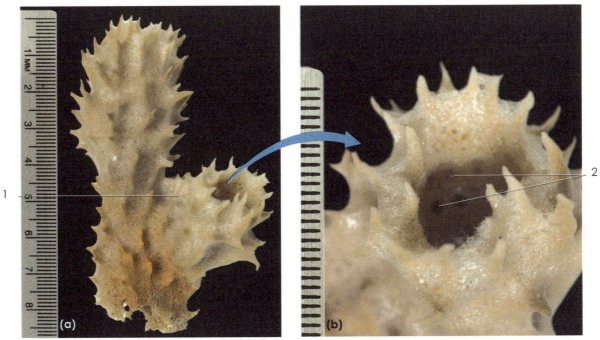

Figure 29.9 A member of class Calcarea: (a) a sponge with an ascon body type, and (b) a close-up view of osculum (scale in mm).

1. Osculum

2. Ostia (seen from inside osculum)

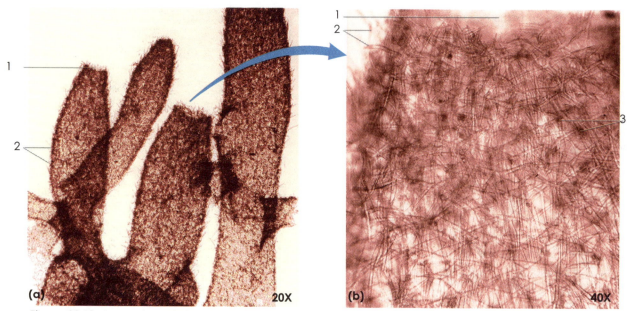

Figure 29.10 A member of class Calcarea: (a) *Leucosolenia* has an ascon body type, and (b) a high magnification of the spicules and ostia.

1. Osculum
2. Spicules
3. Ostia

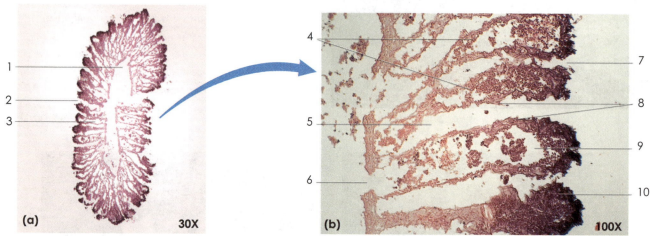

Figure 29.11 Transverse sections of the sponge, *Scypha* (*Grantia*), a member of the class Calcarea. (a) A low magnification and (b) high magnification.

1. Spongocoel
2. Ostium (incurrent canal)
3. Radial canal
4. Choanocytes (collar cells)

5. Incurrent canal
6. Apopyle
7. Ostium
8. Pinacocytes

9. Radial canal
10. Mesohyl

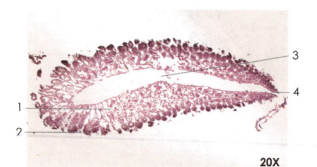

Figure 29.12 A longitudinal section of the sponge, *Scypha* (*Grantia*).
1. Radial canal
2. Ostium (incurrent canal)
3. Spongocoel
4. Osculum

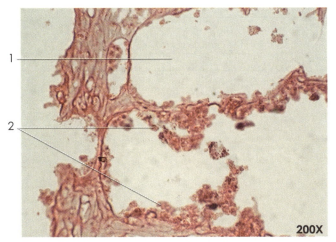

Figure 29.13 A transverse section of the sponge, *Scypha* (*Grantia*), showing collar cells.
1. Radial canal   2. Choanocytes (collar cells)

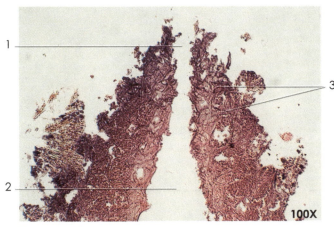

Figure 29.14 A longitudinal section of the sponge, *Scypha* (*Grantia*), showing magnified view of osculum.
1. Osculum   3. Spicules
2. Spongocoel

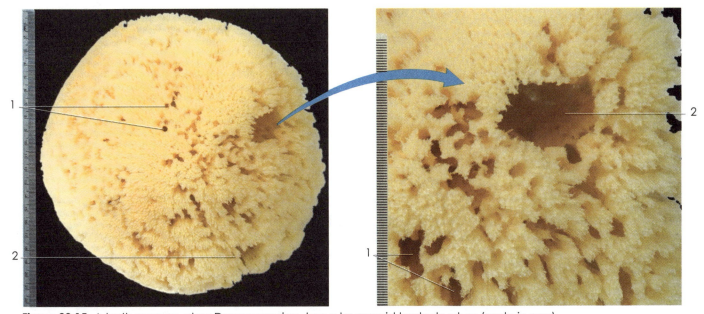

Figure 29.15 A bath sponge, class Demospongiae, has a leuconoid body structure (scale in mm).
1. Ostia   2. Osculum

Student Activity—Macroscopic Anatomy of Selected Poriferans

Materials
- dissecting microscope
- hand lens
- colored pencils
- selected specimens of sponges such as *Sycon* (*Grantia*), *Euplectella*, and *Spongilla*

Procedure 29.1
Macroanatomy of Poriferans

1. Procure the needed equipment and specimens.
2. Observe the specimens with a dissecting microscope and hand lens.
3. Record your observations and sketches on the next page.

Living animals are not ordinarily given the proper amount of credit for their part in forming the character of the Earth's surface.

—C.C. Furnas (1900–1969)

Student Activity—Microscopic Anatomy of Selected Poriferans

Materials
- compound microscope
- colored pencils
- specimen of *Sycon* (*Grantia*)
- microscope slides and coverslips
- teasing needle
- water
- bleach
- dropper
- selected prepared slides of spicules, *Leukosolenia*, *Sycon* (*Grantia*), *Euplectella*, and *Spongilla*

Procedure 29.2
Microanatomy of Poriferans

1. Procure the equipment and specimens.
2. Observe the slides with a compound microscope on both low and high power. Record your observations and sketches.

3. Tease a small section of *Sycon (Grantia)* onto a microscope slide and make a wet mount, being sure to tease and crush the sponge specimen. Record your observations and sketches below.

4. Follow the directions in step 3, but place two drops of bleach on the sponge specimen and gently stir. The bleach should dissolve the sponging and make the spicules easier to see. Record your observations and sketches.

Learn what is true in order to do what is right.

—**Thomas Henry Huxley (1825–1895)**

PHYLUM CNIDARIA

"Ouch! I was just taking a dip in the ocean when I suddenly felt something stinging my leg! When I looked down, I saw this blob of jelly floating near my aching leg." Sound familiar? You may have had similar encounters with one of its classical members, the jellyfish, at the beach. Welcome to the incredible world of **Phylum Cnidaria**!

The Radiata are eumetazoans that are represented by two distinct phyla, Cnidaria and Ctenophora. **Phylum Cnidaria** contains approximately 10,000 species of primarily marine invertebrates, including many bizarre and beautiful forms such as sea anemones, jellyfish, the Portugese man-of-war, coral, and the freshwater *Hydra* (Fig. 29.16–Fig. 29.17). The majority of species of cnidarians are sessile, although many floating or free-swimming forms exist. The colonial cnidarian the Portuguese man-of-war is an excellent example of a cnidarian that has its own sail used for wind locomotion. The cnidarians get their name from cells called **cnidoblasts** (cnidocytes), which contain the stinging cells, **nematocysts**, found in members of this phylum. Nematocysts aid in food gathering and, as you perhaps know, defense (Fig. 29.18).

Check Your Understanding

Q. What is a choanoflagellate, and why is it evolutionarily important?

Q. Describe three body forms seen in sponges.

Q. Describe three classes of sponges.

Figure 29.16 The sea nettle, *Chrysaora fuscescens*, often mass in large swarms off the Pacific coast, where they feed on zooplankton.

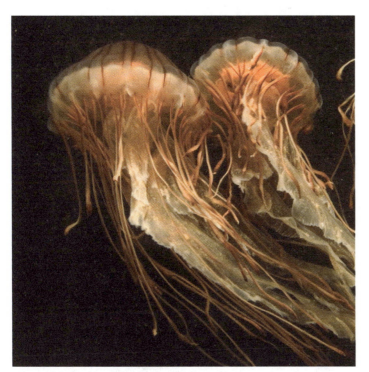

Figure 29.17 The red-striped jellyfish, *Chrysaora melanaster*, is common near the surface of the Bering Sea. Mature adults have a bell close to 30 cm across, and their tentacles can reach 3–6 meters.

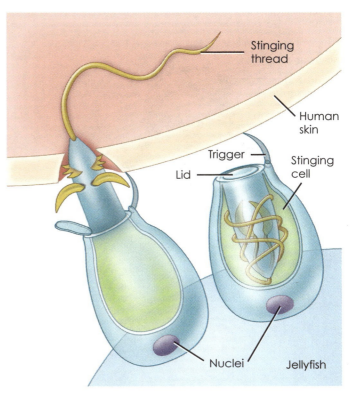

Figure 29.18 The nematocyst is capable of delivering a painful and sometimes fatal sting.

The cnidarians exhibit two obviously different body forms, termed **dimorphism**. The **medusa** form resembles a swimming upside-down cup with tentacles, and the **polyp** form consists of a tubular sessile body. Jellyfish represent the medusa form, and coral represent the polyp form. The life cycle of several species of cnidarians including *Obelia* spp. consists of both a medusa and a polyp generation. Cnidarians can vary in size from less than 1 millimeter to longer than 70 meters including the tentacles.

Cnidarians exhibit radial symmetry and are diploblastic. Because these organisms do not have mesoderm, the muscular system is made from contractile ectodermal and endodermal, or **epitheliomuscular,** cells. A gelatinous non-living substance called **mesoglea** exists between the **epidermis** and the **gastrodermis** or endodermis. (That's why a jellyfish is like a blob of jelly!) (Fig. 29.19.)

Amoebocytes within the mesoglea aid in digestion, transport, storage, repair, and defense against bacteria. Cnidarians have a single opening leading into their digestive system, and tentacles surround the mouth.

Nature's great and wonderful power is more demonstrated in the sea than on land.

—**Pliney the Elder (23–79 A.D.)**

Digestion is extracellular, and they have no coelom. They do not have a respiratory system; gas exchange occurs through diffusion. In these curious organisms, individual cells undergo excretion. Cnidarians possess a **nerve net** composed of **neurites** and sensory organs. In cnidarians, specialized ocelli serve as photosensitive organs, and **statocysts** serve as organs of balance.

Several species of cnidarians are **bioluminescent**; they can produce their own light. Interestingly, scientists have isolated this gene and inserted it into the embryos of animals such as mice and pigs to get "glow-in-the-dark" critters. Asexual reproduction in some members of this phylum can take place through **budding**, and in some colonial forms the life cycle includes an asexual part. Many cnidarians are **dioecious**. Male and female gametes are produced in separate individuals and unite, forming an embryo that eventually becomes a ciliated **planula larva**. The larva eventually attaches to a substrate and develops into the polyp form.

Four classes comprise phylum Cnidaria.

1. In **Class Hydrozoa**, the polyp form is dominant. Representative hydrozoans include *Hydra* spp., *Obelia* spp., and the Portugese man-of-war *Physalia* spp. (Fig. 29.20). The majority of hydrozoans are marine, but a few are freshwater species. The life cycle of many

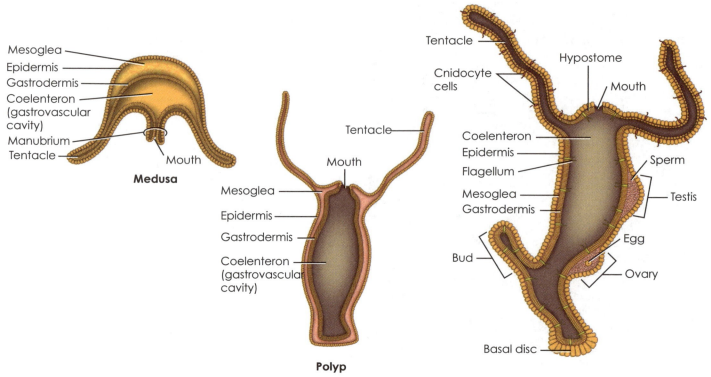

Figure 29.19 (a) Generalized body plans of cnidarians; (b) the basic anatomy of *Hydra*.

hydrozoans consists of an asexual polyp and a sexual medusa stage. Each summer on the East Coast of the United States, the Portuguese man-o-war (*Physalia physalis*) inflicts up to half a million stings. The purple gas-filled **pneumatophore** is characteristic of this colonial hydrozoan. A large *Physalia* may have tentacles exceeding 50 meters.

2. **Class Scyphozoa** (true jellyfish) are solitary organisms in which a polyp stage is reduced or absent, and a dominant bell-shaped medusa stage. The edge of the bell, the **umbrella**, has eight notches provided with sense organs. The umbrella of some jellyfish exceeds 2 meters across. Examples of true jellyfish are *Aurelia* spp., *Chrysaora* spp., and *Cassiopeia* spp.

3. **Class Cubozoa** once was considered an order of class Scyphozoa. The medusoid form is prominent, and the polyp is inconspicuous. The bell is cubical and bent inward. The tentacles are suspended from four flat **pedalia** at the corners of the umbrella. These organisms commonly are called box jellyfish. The venom of box jellyfish has extreme cardiotoxic, neurotoxic, and dermatonecrotic components. Examples of cubozoans are *Tripedalia Cystophora* and *Carybdea* spp. One species, *Chironex fleckeri,* found off the coast of Australia, can deliver lethal stings. In fact, its venom is the deadliest venom known to toxicology. The Irukandji jellyfish *Carukua barnesi* is a silent, mysterious, nearly invisible killer cubozoan found in the seas off northern Australia.

4. **Class Anthozoa** (flower animals) exists as polyps only. Anthozoans can be colonial or solitary marine organisms. The pharynx leads into a gastrovascular cavity divided by eight or more septa. Examples of anthozoans are sea anemones, sea fans, sea pens, sea pansies, and the corals.

Observing Class Hydrozoa

Class Hydrozoa is made up of the hydras and many colonial species collectively called hydroids. The life cycle of most hydrozoans consists of an asexual polyp and a sexual medusa stage. One of the most readily available hydrozoans is a small (25 millimeter) freshwater species, *Hydra*. This interesting organism is common in cool, clean, freshwater pools and streams throughout the world.

Hydra appears to be a cylindrical tube, with its **aboral** (away from the mouth) end forming a slender stalk ending in a **basal disc** for attachment. The basal disc contains specialized gland cells that allow it to attach to a substrate (perhaps a lily pad). In addition, it allows the organism to form a gas bubble for floating. The mouth of *Hydra* is located on an elevated portion of the oral end, the **hypostome**.

The hypostome is encircled by **hollow tentacles** (6–10 in number). The tentacles help to capture food such as small insect larvae, crustaceans, and worms. The mouth itself opens into a **gastrovascular cavity** that is continuous with the tentacles. Upon close examination, **testes** or **ovaries**, when present, appear as rounded structures on the body. Hydra can reproduce asexually by **budding**. Many times, a bud projects from the side of the animal. The epidermis of the Hydra contains specialized cells, including the **nematocysts** (Fig. 29.21–29.26).

The hydrozoan *Obelia* is a typical member of Class Hydrozoa. *Obelia* can exist in both the asexual polyp form and the sexual form of the medusa. This organism is one of many colonial hydroids that are attached to rocks, pilings, and shells in the brackish and marine environment.

Obelia attaches to the substrate via a root–like structure, the **stolon**, which gives rise to various **stalks**. Within the stalk, the tubular **coenosarc** (composed of three cnidarian layers) surrounds the gastrovascular cavity. The stalk is protected by a chitinous sheath, the **perisarc**. The individual polyps also are attached to the stalk. Most polyps, also called **zooids**, are used for feeding and are called **hydranths**. In addition to the feeding polyps, reproductive polyps individually called the **gonangium** are attached to the stalk. Dioecious medusa are budded from the gonangium. The free-swimming medusae mature and form gametes. After fertilization, a free-swimming **planula** larva develops and finds a new substrate. Upon settling, a new *Obelia* colony is established (Fig. 29.27–29.31).

(a) 40X (b) 15X (c)

Figure 29.20 (a) *Hydra* is a common freshwater hydrozoan; (b) *Obelia* is a colonial hydrozoan found in brackish and marine water; (c) the Portuguese man-of-war, *Physalia* sp., is actually a colony of medusae and polyps acting as a single organism. The tentacles are composed of three types of polyps: the gastrozooids (feeding polyps), the dactylozooids (stinging polyps), and the gonozooids (reproductive polyps) (scale in mm).

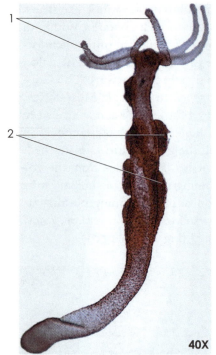

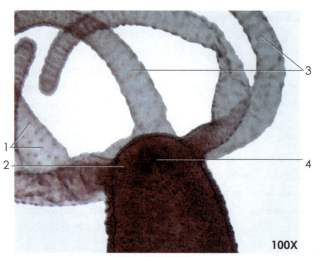

Figure 29.23 The anterior end of a *Hydra*.
1. Cnidocytes   3. Tentacles
2. Hypostome   4. Mouth

Figure 29.21 A male *Hydra*.
1. Tentacles
2. Testes

Figure 29.22 A female *Hydra*.
1. Tentacles
2. Ovary
3. Basal disc (foot)

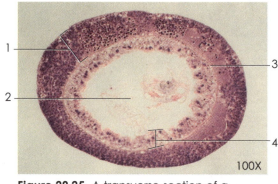

Figure 29.25 A transverse section of a female Hydra.
1. Epidermis (ectoderm)   4. Gastrodermis
2. Coelenteron   (endoderm)
3. Mesoglea

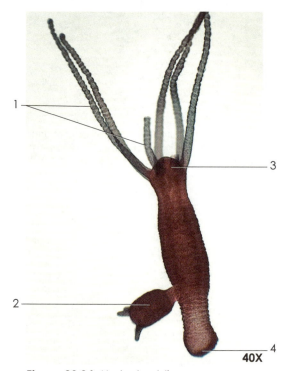

Figure 29.24 *Hydra* budding.
1. Tentacles   3. Hypostome
2. Bud   4. Basal disc (foot)

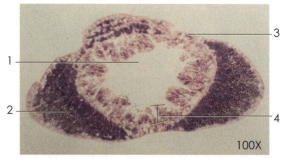

Figure 29.26 A transverse section of a male Hydra.
1. Coelenteron   4. Gastrodermis
2. Testes   (endoderm)
3. Epidermis (ectoderm)

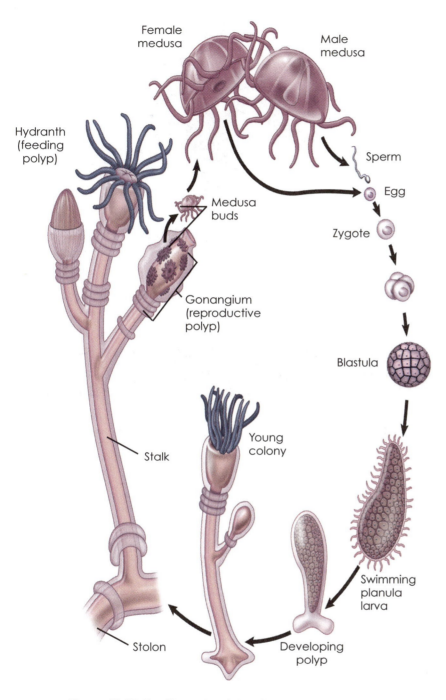

Female medusa

Male medusa

Hydranth (feeding polyp)

Sperm

Medusa buds

Egg

Zygote

Gonangium (reproductive polyp)

Stalk

Blastula

Young colony

Swimming planula larva

Stolon

Developing polyp

Figure 29.27 The life cycle of *Obelia*.

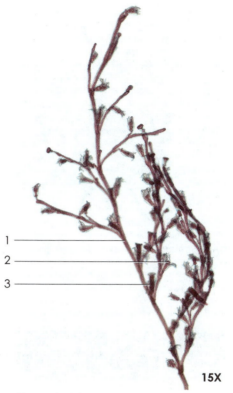

15X

Figure 29.28 An *Obelia* colony.
1. Coenosarc (soft tissue connecting polyps)
2. Hydranth (feeding polyp)
3. Gonangium (reproductive polyp)

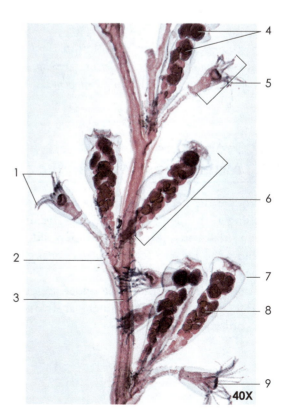

40X

Figure 29.29 A detailed view of an *Obelia* colony.

1. Tentacles
2. Perisarc (horny covering that encloses the polyp)
3. Coenosarc
4. Medusa buds
5. Hydranth (feeding polyp)
6. Gonangium (reproductive polyp)
7. Gonotheca
8. Blastostyle
9. Hypostome

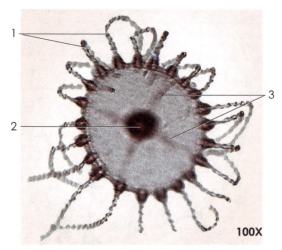

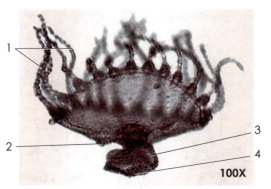

Figure 29.31 *Obelia* medusa in feeding position.
1. Tentacles   3. Manubrium
2. Gonad   4. Mouth

Figure 29.30 An aboral view of an *Obelia* medusa.
1. Tentacles   3. Radial canals
2. Manubrium

Student Activity—Macroscopic Anatomy of Select Hydrozoans

Materials
- dissecting microscope
- hand lens
- Petri dish or depression slide
- colored pencils
- probe
- 5% vinegar or congo red solution
- living *Hydra*, preserved *Obelia*, *Physalia*, and other select hydrozoans such as *Gonionemus*
- *Daphnia* or *Artenia*

Procedure 29.3
Macroanatomy of Hydrozoans

1. Procure the needed equipment and supplies.
2. Using a hand lens, obtain a living *Hydra* and place it in a small Petri dish or a depression slide.
3. Allow the *Hydra* a few minutes to acclimate to the new conditions.
4. Place the Petri dish or slide under the dissecting microscope. Sketch the specimen, label your sketch, and record your observations.

5. Tap the dish, or gently touch the *Hydra* with a probe. Record the response of the *Hydra* below.

6. Place some *Daphnia* or *Artemia* into the Petri dish, and observe the feeding behavior of *Hydra*. You may have to gently nudge them toward the tentacles of the *Hydra* with a probe.

7. If a solution of 5% vinegar or congo red is available, place a few drops into the Petri dish and record the reaction of the *Hydra*.

_____

_____

8. Follow your instructor's directions about clean-up of your station and return of the *Hydra*.

9. Procure specimens of *Obelia*, *Physalia*, and select hydrozoans. Sketch, label, and describe your specimens below.

_____

_____

_____

_____

Student Activity—Microscopic Anatomy of Select Hydrozoans

Materials
- compound microscope
- colored pencils
- select microscope slides of *Hydra* whole mount, *Hydra* longitudinal section, *Hydra* cross-section, *Hydra* budding, *Obelia* colony, and *Obelia* medusa

Procedure 29.4
Microanatomy of *Hydra*

1. Procure the compound microscope and select microscope slides.

2. Observe the *Hydra* whole mount slide on low power. Sketch and label a composite of the *Hydra* below.

3. Observe and scan the *Hydra* whole mount slide on high power. Sketch and label the anatomical features.

6. Observe the *Hydra* budding slide on low power. Sketch and label the it below.

4. Observe the *Hydra* longitudinal section slide on low and high power. Sketch and label it below.

7. Observe the *Obelia* colony slide on low and high power. Sketch and label the *Obelia* colony below.

5. Observe the *Hydra* cross-section slide on low and high power. Sketch and label it below.

8. Observe the *Obelia* medusa slide on low and high power. Sketch and label the *Obelia* medusa below.

Observing Class Scyphozoa

Class **Scyphozoa** is made up of the true jellyfish. The word is derived from the Greek, meaning a kind of drinking cup and referring to the cup shape of the organism. The jellyfish *Aurelia* is a classic example of a scyphozoan. In the life cycle of *Aurelia*, male and female **medusa** produce their respective gametes that undergo fertilization, forming a **zygote** and eventually a ciliated **planula** larva. The larva land on a suitable substrate and form a **scyphistoma** that grows perhaps asexually, buds and forms an asexual **strobila**. The strobila give rise to swimming **ephyra**, which eventually develop into a medusa (Fig. 29.32–Fig. 29.38).

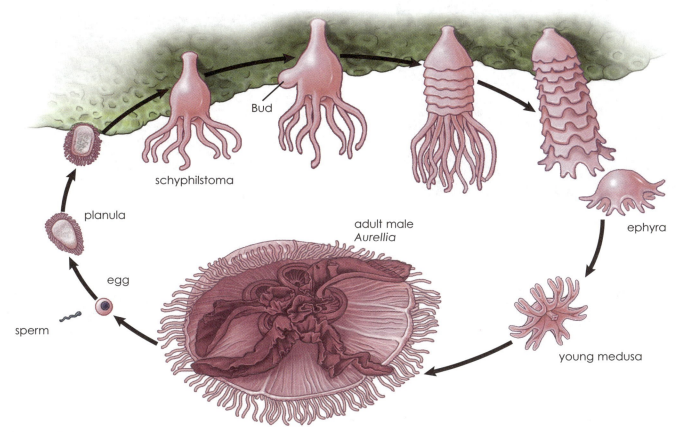

Figure 29.32 The life cycle of *Aurelia*.

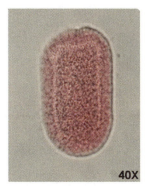

Figure 29.33 The *Aurelia* planula larva develops from a fertilized egg that may be retained on the oral arm of the medusa.

Figure 29.34 An *Aurelia* scyphistoma. The polyp is a developmental stage in the life cycle of the jellyfish.

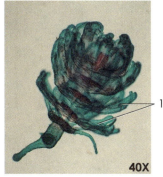

Figure 29.35 An *Aurelia* strobila. Under favorable conditions, the scyphistoma develops into the strobila.

1. Developing ephyrae

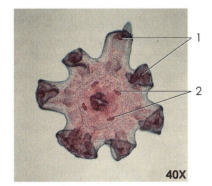

Figure 29.36 An *Aurelia* ephyra larva, which gradually develops into adult jellyfish.

1. Rhopalia (sense organs)
2. Gonads

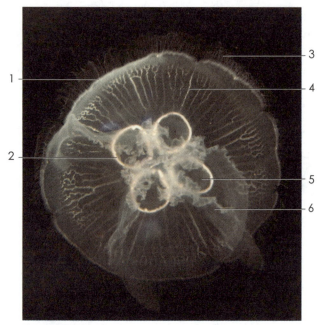

Figure 29.37 An oral view of *Aurelia* medusa.

1. Ring canal
2. Gonad
3. Marginal tentacles
4. Radial canal
5. Subgenital pit
6. Oral arm

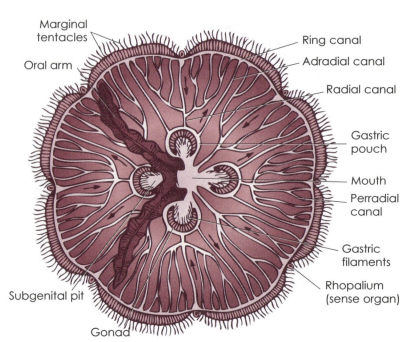

Figure 29.38 An oral view of *Aurelia* medusa. In this diagram, the right oral arms have been removed. The arrows depict circulation through the canal system.

Student Activity—Macroscopic Anatomy of Select Scyphozoans

Materials
- dissecting microscope or hand lens
- colored pencils
- preserved specimens of *Aurelia* and select scyphozoans

Procedure 29.5
Macroanatomy of Scyphozoans

1. Procure the equipment and select specimens.
2. Using the dissecting microscopie or hand lens, observe the anatomical features of *Aurelia* and selected specimens.

3. Record your observations and detailed sketches below.

_____

_____

_____

Student Activity—Microscopic Anatomy of Select Scyphozoans

Materials
- compound microscope
- colored pencils
- select microscope slides of *Aurelia* planula larva, scphistoma, strobila, and ephyra

Procedure 29.6
Microanatomy of Scyphozoans

1. Procure the compound microscope and slides.
2. Observe and sketch each stage of the development of *Aurelia* below.

Those who admire the massive, rigid bone structure of dinosaurs should remember that jellyfish still enjoy their secure ecological niche.

—**Beau Sheil**

Observing Class Cubozoa

Class **Cubozoa** includes the venomous box jellyfish. Stings from these organisms can be very painful and perhaps fatal. This class includes approximately 20 species of marine jellyfish, including the deadly *Chironex fleckeri* (Fig. 29.39–29.40).

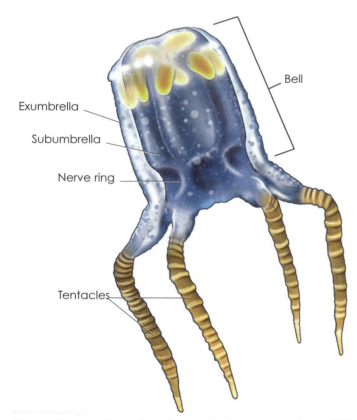

Bell
Exumbrella
Subumbrella
Nerve ring
Tentacles

Figure 29.39 An illustration of a box jellyfish, *Carybdea sivickisi*, showing basic external structures.

Figure 29.40 The box jellyfish, *Carybdea sivickisi*, is named from their cube-shaped bell. All cubozoans have four tentacles.

Student Activity—Macroscopic Anatomy of Select Cubozoans

Materials
- dissecting microscope or hand lens
- colored pencils
- preserved specimens of *Carybdea, Chironex*, and select cubozoans

 ## Procedure 29.7
Macroanatomy of Cubozoans

1. Procure the equipment and the select specimens.
2. Using a dissecting microscope or hand lens, observe and sketch a representative cubozoan.

Observing Class Anthozoa

Class **Anthozoa** is highly diverse, including sea anemones, sea fans, sea pens, and coral, among others. More than 6,000 species have been described, and great numbers of fossil forms have been found. Anthozoans occur from the intertidal zone of the ocean to the depths of the great marine trenches (6000 m). The anthozoans exit only in the polyp form.

Sea anemones occur in warm coastal waters worldwide. They are sessile, attaching by their **pedal disc** to a suitable substrate. Several species can burrow into the sand or mud. Sea anemones are cylindrically shaped, with a crown of **tentacles** surrounding the **mouth**. A **pharynx** leads to the **gastrovascular cavity**, compartmentalized by **septae**. Some anemones have separate sexes, and others are monoecious. Anemones feed primarily upon fishes. Asexual reproduction can occur as fragments of the pedal disc break off (pedal laceration), transverse fission, or budding. Some anemones have complex symbiotic relationships with other organisms such as algae and fishes (Fig. 29.41–Fig. 29.57).

True Love or Mutualism?

Some sea anemones have an interesting mutualistic relationship with hermit crabs. The anemone provides camouflage and protection for the crab and normally being sessile, this anemone "kinda joins the Navy" and gets to travel, meeting other anemones and sampling new foods!

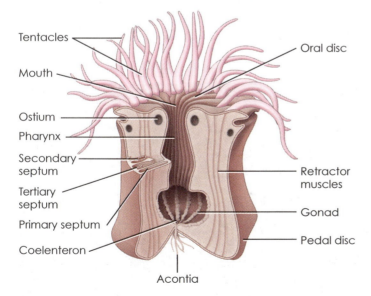

Figure 29.41 A diagram of a partially dissected sea anemone, *Metridium*.

Figure 29.42 A group of anemones, *Anthopleura*, in a tide pool with other tidal organisms.

Figure 29.43 The sunburst anemone, *Anthopleura sola*, gets its green coloration from symbiotic algae within it.

Figure 29.44 The firecracker coral, *Dendrophyllia,* a filter feeder actively feeds day and night.

Figure 29.45 The tube anemone, *Pachycerianthus fimbriatus*, makes a leathery tube and sinks it up to two feet into the sand.

Figure 29.46 The sea pen, *Ptilosarcus gurneyi*, is a colony of polyps that may reach two feet in height.

Figure 29.48 Anemones and clown fish have developed a symbiotic relationship.

Figure 29.47 Disk anemones, *Actinodiscus*, form large colonies.

Figure 29.49 Sunburst anemones, *Anthopleura sola*, at low tide.

Figure 29.50 Brain coral, *Goniastrea*.

Figure 29.51 The skeletal structure of brain coral, *Goniastrea*.

Figure 29.52 Mushroom coral, *Rhodactis*.

Figure 29.53 The skeletal structure of mushroom coral, *Rhodactis*.

Figure 29.54 Elkshorn coral, *Acropora*.

Figure 29.55 The skeletal structure of elkshorn coral, *Acropora*.

Figure 29.56 A detailed view of the polyps of candy cane coral, *Caulastrea furcata*.

Figure 29.57 A detailed view of the polyps of glove xenia, *Xenia umbellata*.

Student Activity—Macroscopic Anatomy of Select Anthozoans

Materials
- dissecting microscope or hand lens
- colored pencils
- scalpel
- dissecting tray
- preserved specimens of *Metridium,* corals, sea fans, and other select anthozoans

Procedure 29.8
Macroanatomy of Anthozoans

1. Procure the needed materials and specimens.
2. Place a specimen of the sea anemone *Metridium* on a dissecting tray and examine it with a dissecting microscope or hand lens. Record your observations, label, and sketch the organism below.

3. Using the scalpel, make a longitudinal cut through *Metridium.* Locate the structures found in Figure 29.41. Record your observations and labeled sketches.

4. Observe and sketch the various forms of coral, sea pens, and sea fans below.

5. Dispose of your dissected specimens as directed, and clean your equipment.

Check Your Understanding

Q. What are the major classes of cnidarians?

_____

_____

_____

Q. Compare and contrast a *hydra* and a *medusa*.

_____

_____

_____

Q. Sketch and label the Portuguese man-of-war.

PHYLUM CTENOPHORA

Any discussion of Phylum **Cnidaria** should briefly mention **Phylum Ctenophora**. Often, a small, clear, walnut-shaped blob washes up on the beach. Chances are that it is a ctenophore. Members of Phylum Ctenophora are known as "comb jellies." People sometimes call these solitary, harmless, marine, jellyfish-like animals "sea walnuts," or "sea gooseberries." Presently, there are approximately 150 described species of ctenophores, ranging in size from 1 centimeter to 1.5 meters. Ctenophores are exclusively marine, living in warm waters. They exist as a medusa only. Ctenophores have adhesive cells called **colloblasts** to capture food and do not possess nematocysts. The tentacles of ctenophores are solid, consisting of epidermis only.

Ctenophores swim by means of rows of fused cilia, **comb plates**. The majority of ctenophores are monoecious, reproducing only by sexual means. Many ctenophores are bioluminescent. Common examples of ctenophores are *Pleurobrachia* spp. and *Mnemiopsis* spp (Fig. 29.58).

Figure 29.58 The ctenophore, *Mnemiopsis* sp.

Student Activity—Macroscopic Anatomy of Select Ctenophores

Materials
- dissecting microscope or hand lens
- colored pencils
- preserved specimens of *Pleurobrachia* spp. and *Mnemiopsis* spp. and select cubozoans
- camera or camera phone (optional)

Procedure 29.9
Macroanatomy of Ctenophores

1. Procure the needed equipment, supplies, and selected specimens.

2. Using a dissecting microscope or hand lens, record your observations and sketches below and on the next page.

NOTES

Name: _____   Date: _____   Section: _____

Review Questions

1. Briefly describe the three classes of sponges and provide examples of each.

_____

_____

_____

_____

_____

2. Compare and contrast asconoid, syconoid, and leuconoid sponges.

_____

_____

_____

_____

3. Describe the skeletal elements of sponges.

_____

_____

_____

4. Name several ecological and commercial values of sponges.

_____

_____

_____

_____

5. Why are the sponges considered an evolutionary "dead-end?"

_____

_____

_____

6. What are five characteristics of cnidarians?

_____

_____

_____

_____

_____

7. Draw the life cycle of the jellyfish *Aurelia*.

8. What are the classes of Phylum Cnidaria?

_____

_____

_____

Name: _____   Date: _____   Section: _____

9. Describe the reproduction cycle in *Hydra*.

_____

_____

_____

10. Label the *Obelia* colony.

a. _____

b. _____

c. _____

d. _____

e. _____

f. _____

g. _____

h. _____

i. _____

11. Cite the dangers of several Cnidarians.

_____

_____

_____

12. Discuss the biology of coral.

_____

_____

_____

13. Sketch and label a longitudinal section of a sea anemone.

14. What are three characteristics of a ctenophore?

_____

_____

_____

_____

15. Why can clown fish live in harmony with a sea anemone?

_____

_____

_____

_____

Chapter 30
Lophotrochozoa, Diversity Abounds: Understanding Animals—Part II

Student Outcome Objectives

At the completion of this exercise, the student will be able to:

1. Describe the characteristics of Lophotrochozoa.
2. Describe the characteristics, natural history, and organization of Phylum Platyhelminthes.
3. Describe the characteristics of Class Turbellaria.
4. Identify basic anatomical structures of selected turbellarians.
5. Describe the characteristics of Class Cestoda.
6. Identify basic anatomical structures of selected cestodes.
7. Describe the characteristics of Class Trematoda.
8. Identify basic anatomical structures of selected trematodes.
9. Describe the characteristics, natural history, and organization of Phylum Rotifera.
10. Identify basic anatomical structures of selected rotifers.
11. Describe the characteristics, natural history, and organization of Phylum Mollusca.
12. Describe the characteristics of Class Polyplacophora.
13. Identify basic anatomical structures of selected polyplacophorans.
14. Describe the characteristics of Class Gastropoda.
15. Identify basic anatomical structures of selected gastropods.
16. Describe the characteristics of Class Scaphopoda.
17. Identify basic anatomical structures of selected scaphopods.
18. Describe the characteristics of Class Bivalvia.
19. Dissect and identify the anatomical features of a clam.
20. Describe the characteristics of Class Cephalopoda.
21. Dissect and identify the anatomical features of a squid.
22. Describe the characteristics, natural history, and organization of Phylum Annelida.
23. Describe the characteristics of Class Polychaeta.
24. Identify basic anatomical structures of selected polychaetes.
25. Describe the characteristics of Class Oligochaeta.
26. Dissect and identify the anatomical features of an earthworm.
27. Describe the characteristics of Class Hirudinia.
28. Describe the basic biology and natural history of Phylum Bryozoa (Ectoprocta).
29. Describe the basic biology and natural history of Phylum Brachiopoda.

Overview

In that great tree of life, where should flukes, snails, leeches, crabs, crinoids, and lemurs be placed? The debate on how to classify bilaterally symmetrical animals is on-going with no clear immediate resolution. Classically, the bilaterally symmetrical animals (protostomes and deutero-stomes) were divided into acoelomates, pseudocoelomates, and coelomates. In recent years, based upon molecular studies of 18S rRNA and Hox genes, scientists have reorganized the protostomes into two distinct clades and kept the deuterostomes in a separate clade. The two clades of protostomes are Lophotrochozoa and Ecdysozoa.

Lophotrochozoa is a large clade consisting of several interesting phyla. The best-known phyla within this clade are **Platyhelminthes** (flatworms), **Rotifera** (rotifers), **Mollusca** (snails, oysters, and squid), and **Annelida** (seg-mented worms). Several lesser known phyla are **Acanthocephala** (thorny-headed worms), **Gastrotricha** (spiny aquatic organisms), **Bryozoa** (ectoprocts and moss animals), **Entoprocta** (entoprocts), **Brachiopoda** (lamp shells), and **Nemertea** (ribbon worms). The two defining characteristics of Lophotrochozoa found in many members are (a) the presence of a horseshoe-shaped crown of ciliated tentacles (**lophophores**), and (b) a minute, translucent top-shaped ciliated larvae (**trochophores**) (Fig. 30.1).

PHYLUM PLATYHELMINTHES

Phylum **Platyhelminthes** consists of approximately 20,000 species of organisms collectively called the flat-worms. Common representatives of this phylum are planarians, flukes, and tapeworms (Fig. 30.2). Flatworms vary in size from shorter than 1 millimeter to longer than 10 meters (a species of tapeworm). The bodies of platy-helminths characteristically are flattened dorsoventrally

When assessing family connections, the anatomy of a larva is obviously as valid a piece of evidence as the adult.

—**David Attenborough (1926–present)**

Figure 30.1 Examples of Lophotrochozoans: (a) tapeworm, (b) rotifer, (c) snail, (d) leech and (e) a lamp shell *Lingula*.

and are ribbon-like, ensuring a large surface area. Flatworms are bilaterally symmetrical, triploblastic acoelomates. Some species of flatworms are dull in coloration, and others are brightly colored. Many species of platyhelminths are free-living and are found in terrestrial, aquatic, and marine environments. Others, such as tapeworms and flukes, are parasitic.

The platyhelminths lack specialized respiratory and circulatory systems. As a result, they exchange gases through diffusion. The digestive system of flatworms is incomplete, with only one opening to the exterior. Many flatworms possess a **mouth**, connected to the **gastrovascular** cavity by a **muscular pharynx**. In larger flatworms, the gastrovascular cavity is branched within the body. The main structures in the excretory system are the **protonephridia**, capped by **flame cells.** Parasitic flatworms are covered with a protective **syncytial tegument**. Platyhelminths exhibit cephalization. A pair of **cerebral ganglia** receives sensory information from the environment. **Eyespots** are present is some species. Each ganglia is connected to a **nerve cord** that runs the length of the body. Some flatworms can reproduce asexually through fission. Planaria have tremendous powers of **regeneration**. Many platyhelminths are monoecious but practice cross-fertilization. In addition, many parasitic flatworms have complex life cycles.

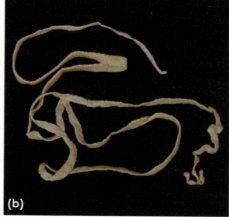

Figure 30.2 Examples of platyhelminthes: (a) a planarian, (b) a tapeworm, and (c) a liver fluke.

Phylum Platyhelminthes contains four classes of flatworms.

1. **Class Turbellaria** consists of more than 3,000 mostly free-living flatworms such as the planaria (*Dugesia* and *Bipalium*).
2. **Class Cestoda** includes approximately 3,500 species of parasitic tapeworms such as *Dipylidium caninum*, *Diphyllobothrium latum*, and *Taenia* spp.
3. **Class Trematoda** consists of flukes. Approximately 11,000 species of trematodes have been described, all of which are parasitic. *Fasciola* spp., *Clonorchis sinensis*, and *Schistosoma* spp. are typical trematodes.
4. **Class Monogenea** contains ectoparasitic flatworms, usually found on the skin and gills of fishes. Approximately 1,400 species have been identified, examples of which are *Gyrodactylus cylindriformis* and *Polystoma intergerrimum*.

Class Turbellaria

The **turbellarians** are mostly free-living flatworms. Although most turbellarians live in marine environments, both freshwater and a few terrestrial forms exist. Members of **Class Turbellaria** range in size from less than 5 millimeters to more than 60 centimeters. Turbellarians, like other members of Phylum Platyhelminthes, are dorsoventrally flattened. These organisms are covered by a ciliated epidermis that can range in color from shades of gray or brown to bright, rich colors. In many species, the anterior end bears eyespots, and a few species have tentacles. Beneath the epidermis is a muscle layer of specialized fibers. Turbellarians are considered acoelomates because they do not have a cavity between the body wall and the internal organs.

Turbellarians usually swim or crawl along the bottom of a water or terrestrial environment by ciliary propulsion. Some species move by means of undulating waves of muscle contractions. Glands in the epidermis and underlying tissue secrete a mucous film to help the organism glide over the environment. Most turbellarians are carnivores, feeding on other small invertebrates. The mouth is located along the mid-ventral line of the organism. The gut consists of the pharynx and the intestinal sac. These organisms do not have an anus; they eject food through the mouth.

In turbellarians, gas exchange occurs at the surface of the body. The flattened body increases the surface area, thus increasing the rate and efficiency of gas exchange. Movement of food materials in flatworms occurs primarily through diffusion. Paired protonephridia function in removing nitrogenous wastes and in osmoregulation. The ends of protonephridia are associated with specialized flame cells.

In many turbellarians, the neurons are found in longitudinal bundles located beneath the epidermis. In planarians, the connecting lateral nerve cords form a characteristic ladder-like pattern. The brain appears as a bilobed mass of ganglion cells at the anterior end of the organism. Some members of this class have light-sensitive eyespots, or **ocelli**. In many species, chemoreception and tactile reception are well-developed. The **auricles** (lobes) on the side of a planarian's head are associated with tactile reception and chemoreception.

Many turbellarians reproduce asexually through fission. Being monoecious, they also are capable of sexual reproduction. Although these animals are **hermaphroditic**, they do not exhibit self-fertilization. Turbellarians generally reproduce by mutual fertilization, eventually resulting in cocoons that are laid in jelly-like masses. Turbellarians exhibit amazing regenerative powers. A piece excised from the center of a planarian can develop into two new planarians.

A common turbellarian found in many gardens and greenhouses is *Bipalium kewense*. A native of Indo-China, it has been found in the United States for more than a century. *Bipalium* is photo-negative and can be found in dark, cool, moist areas under objects such as rocks or logs, in debris, or under shrubs. *Bipalium* is slender and usually brown in color, with dark longitudinal stripes. It can measure up to 25 cm in length. The head is shovel-shaped, and eyespots are obvious. The best-known member of Class Turbellaria is the planaria *Dugesia* sp., a common inhabitant of freshwater environments living on plants, under rocks, and in debris. *Dugesia* is 3–15 mm long and brown to gray in color. *Dugesia* feeds primarily upon other invertebrates. The large mouth and pharynx are in the middle of their body. The auricles and eyespots are prominent on their anterior end (Fig. 30.3–Fig. 30.8).

Like a lower form of life, like the cross-eyed planarian or squashed amoeba, the sort of creature that can't die even when it is cut to pieces.

—**Paul Edward Theroux (1941–present)**

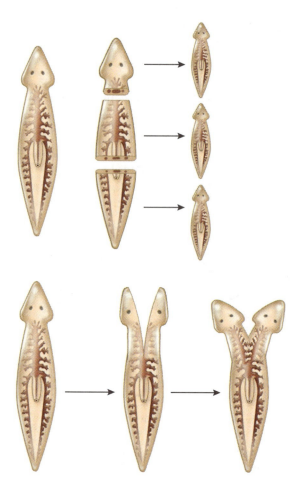

Figure 30.3 Planarians are capable of regeneration.

Figure 30.4 A Planarian (a) *Dugesia* sp. is aquatic, while (b) *Bipalium* sp. is a common inhabitant of gardens.

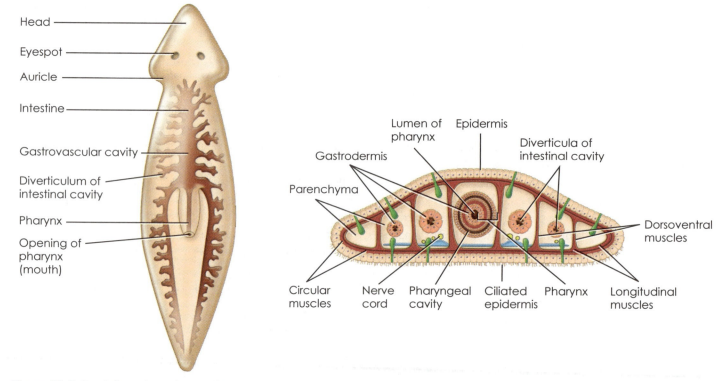

Head

Eyespot

Auricle

Intestine

Gastrovascular cavity

Diverticulum of intestinal cavity

Pharynx

Opening of pharynx (mouth)

Lumen of pharynx

Epidermis

Diverticula of intestinal cavity

Gastrodermis

Parenchyma

Dorsoventral muscles

Circular muscles

Nerve cord

Pharyngeal cavity

Ciliated epidermis

Pharynx

Longitudinal muscles

Figure 30.5 The internal anatomy of *Dugesia*: (a) A longitudinal section and (b) a transverse section through the pharyngeal region.

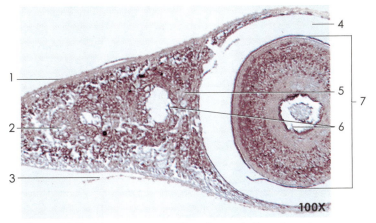

Figure 30.7 A transverse section through the pharyngeal region of *Dugesia*.
1. Epidermis
2. Testis
3. Cilia
4. Pharyngeal cavity
5. Dorsoventral muscles
6. Gastrodermis (endoderm)
7. Pharynx

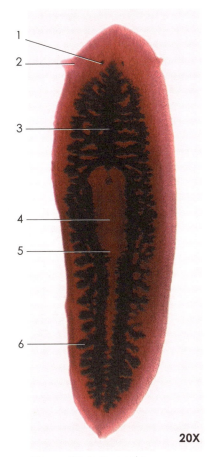

Figure 30.6 *Dugesia*.
1. Eyespot
2. Auricle
3. Gastrovascular cavity
4. Pharynx
5. Opening of pharynx (mouth)
6. Diverticulum of intestinal cavity

Figure 30.8 A transverse section through the posterior region of *Dugesia*.
1. Epidermis
2. Intestinal cavity
3. Mesenchyme
4. Dorsoventral muscles
5. Endoderm

Student Activity—Macroscopic Anatomy of *Bipalium* and *Dugesia*

Materials
- hand lens or dissecting microscope
- colored pencils
- Petri dish or watch glass
- water
- probe
- pipette
- living specimens of *Bipalium, Dugesia*, and earthworms (if available)
- egg yolk

Procedure 30.1
Macroanatomy of Turbellarians

1. Procure the needed equipment and specimens.
2. Transfer *Bipalium* to a Petri dish or watch glass with a probe. Be gentle, as this organism fragments easily.
3. Observe *Bipalium* with a hand lens or a dissecting microscope. Pay particular attention to the way it moves. Perhaps nudge the specimen with the probe. If an earthworm is available, place it near the *Bipalium* and observe the feeding behavior of the *Bipalium*. Record your observations and sketches on the following page.

4. *Dugesia* can be transferred to a Petri dish or watch glass with a pipette. Do not let the planarian sit in the pipette very long, as it will attach to the sides and be extremely hard to expel.

5. After placing the *Dugesia* into a Petri dish or watch glass, allow the organism a few minutes to acclimate. Using a hand lens or dissecting microscope, observe the specimen. Pay attention to locomotion, sticking ability, and behavior. Nudge the specimen with a probe, and record your results.

6. Using transmitted light, observe the internal structures. Record your observations and sketches and labels below.

7. Using a probe or pipette, place a small piece of egg yolk next to the planarian. If possible, turn down the lights on your scope and in the room. (Planaria do not like to eat in bright light.) Record your observations below.

8. Follow the instructor's directions regarding clean-up and storage.

Student Activity—Microscopic Anatomy of *Dugesia*

Materials
- compound microscope
- colored pencils
- selected prepared slides of *Dugesia*; these slides should include a whole mount, a transverse section through the pharyngeal region, and a transverse section through the posterior region

Procedure 30.2
Microanatomy of *Dugesia*

1. Procure the microscope and selected slides.
2. Using the compound microscope on scanning power or low power, observe the whole mount of *Dugesia*. Record your observation and labeled sketch below.

3. Using the microscope on low and high power, observe a slide of the transverse section through the pharyngeal region, and a transverse section through the posterior region of *Dugesia*. Place your observations and labeled sketches to the right.

Student Activity—Regeneration of *Dugesia*

Materials
- dissecting microscope or hand lens
- colored pencils
- living specimens of the planaria *Dugesia*
- Petri dishes and lids
- water
- scalpel
- Sharpie
- egg yolk

 Procedure 30.3
Regeneration of the Planarian

1. Procure the equipment and a specimen of *Dugesia*, and place it in a Petri dish with water.
2. Using a sharp scalpel, cut planarian into several cross sections.
3. Place the lid on the Petri dish, and label the dish with a Sharpie. Put your Petri dish in a designated area.
4. Each week for 5 weeks, observe your planarian. Ensure that the Petri dish does not dry out, using a pipette or probe, place a tiny piece of egg yolk in the Petri dish several times over the 5-week period.
5. Record your observations and sketches below.

Strange but True!

Tapeworms have been the subjects of lore and wives tales in the past. Tapeworm eggs actually have been given to patients to help them lose weight. Not good! In lore, if a long tapeworm were passed by a person, it was thought to be a "bosom serpent." Not really. Several other intestinal parasites also share this distinction.

Class Cestoda

Class Cestoda consists of approximately 3,500 species of flatworms known as tapeworms. The tapeworms are endoparasites of humans and other vertebrates. Most tapeworms require two hosts. Invertebrates are usually the intermediate host of the larval tapeworm. Adult tapeworms usually are present in the digestive tract of the final vertebrate host. Some tapeworms reach lengths of more than 10 m and perhaps up to 25 m in their final host. Other tapeworms live up to 20 years in their host. Perhaps at least 135 million people worldwide have a tapeworm infection at any given moment!

Tapeworms are highly adapted for the life of a parasite. They are totally dependent on the host for nutrition because they lack a digestive tract. The tegument of the tapeworm permits nutrients to enter the body while protecting the body against alkaline substances and digestive enzymes in the host.

Cestodes possess an anterior region called a **scolex** that contains **hooks** and **suckers** for attachment (Fig. 30.9). The hooks usually encircle a crown, or **rostellum**. Behind the scolex is a series of subunits called **proglottids**. The **strobiia**, the main mass of the tapeworm, is composed of proglottids. Immediately behind the scolex are **germinative proglottids**, which are in the process of asexually forming new proglottids. The proglottids mature as they are pushed posteriorly. Following the germinative proglottids, a series of mature proglottids can be found. Mature proglottids can contain numerous testes and ovaries. Each proglottid is a hermaphroditic individual, and any two proglottids, on either the same or different tapeworms, are capable of exchanging sperm. Usually, proglottids do not self-fertilize. Tapeworm eggs containing an embryo are surrounded by a protective shield. Some eggs exit the proglottid via the **gonopore** and enter the host's intestine, and others are stored in the **uterus**. **Gravid** (egg-bearing) **proglottids** often break away from the strobilia and rupture in the host's intestines, or they may exit the host with the feces (Fig. 30.10).

A gravid proglottid can contain more than 100,000 eggs in some species. Once the tapeworm's eggs are released, they must be ingested by **an intermediate host** to be able to hatch. In the intermediate host, the eggs hatch, forming **larvae** that bore through the intestinal wall, where they are picked up by the circulatory system and make their way to striated muscle. In the muscles, the larvae develop into a **cysticercus stage**. If the cysticercus is eaten in raw or poorly cooked meat, it develops into an adult tapeworm in the intestine of the final host. (Fig. 30.11–30.15)

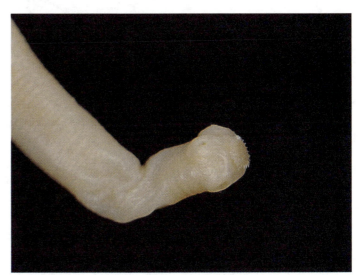

Figure 30.9 The scolex of a parasitic tapeworm, *Taenia pisiformis*.

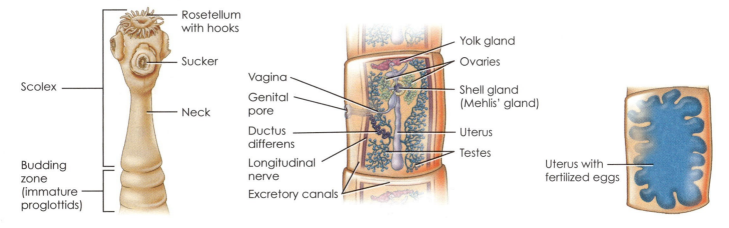

Rosetellum with hooks

Sucker

Scolex

Neck

Budding zone (immature proglottids)

Vagina

Genital pore

Ductus differens

Longitudinal nerve

Excretory canals

Yolk gland

Ovaries

Shell gland (Mehlis' gland)

Uterus

Testes

Uterus with fertilized eggs

Figure 30.10 Diagrams of a parasitic tapeworm, *Taenia pisiformis*. (a) The anterior end, (b) mature proglottids, and (c) a ripe proglottid.

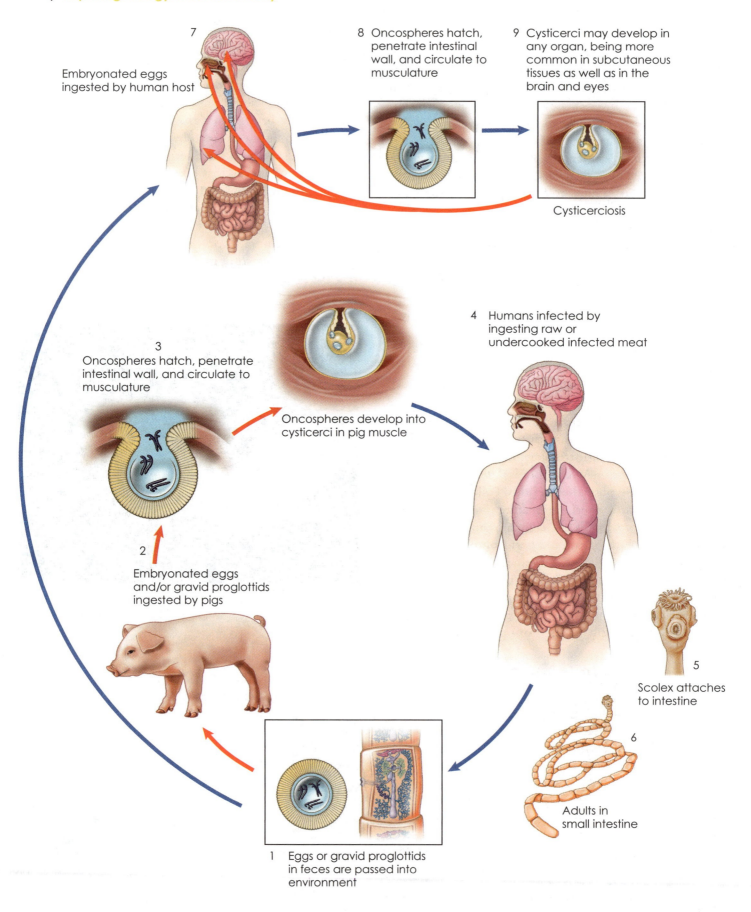

7 Embryonated eggs ingested by human host

8 Oncospheres hatch, penetrate intestinal wall, and circulate to musculature

9 Cysticerci may develop in any organ, being more common in subcutaneous tissues as well as in the brain and eyes

Cysticerciosis

3 Oncospheres hatch, penetrate intestinal wall, and circulate to musculature

Oncospheres develop into cysticerci in pig muscle

4 Humans infected by ingesting raw or undercooked infected meat

2 Embryonated eggs and/or gravid proglottids ingested by pigs

5 Scolex attaches to intestine

6 Adults in small intestine

1 Eggs or gravid proglottids in feces are passed into environment

Figure 30.11 The life cycle of the pork tapeworm, *Taenia saginata*.

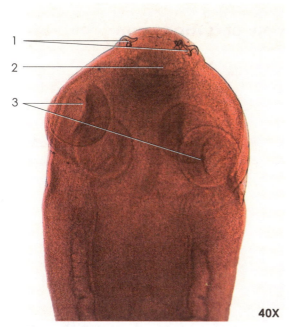

Figure 30.12 The scolex of *Taenia pisiformis*.
1. Hooks
2. Rostellum
3. Suckers

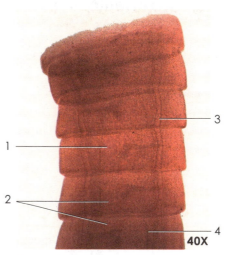

Figure 30.13 The immature proglottids of *Taenia pisiformis*,
1. Early ovary
2. Early testes
3. Excretory canal
4. Immature vagina and ductus deferens

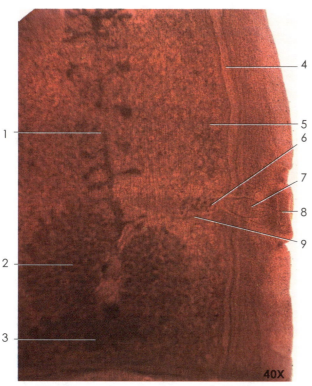

Figure 30.14 The mature proglottid of *Taenia pisiformis*.

1. Uterus	4. Excretory canal	7. Cirrus
2. Ovary	5. Testes	8. Genital pore
3. Yolk gland	6. Ductus deferens	9. Vagina

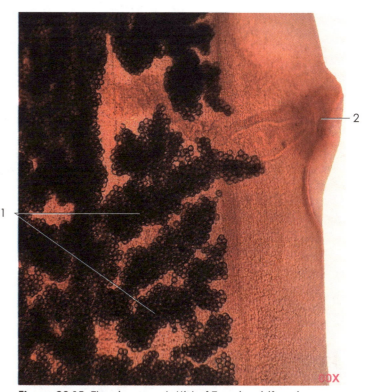

Figure 30.15 The ripe proglottid of *Taenia pisiformis*.
1. Zygotes in branched uterus   2. Genital pore

Student Activity—Macroscopic Anatomy of a Tapeworm

Materials
- dissecting microscope or hand lens
- colored pencils
- Petri dish or jar?
- Instructor's choice of preserved specimens of various tapeworms

 **Procedure 30.4
Macroanatomy of a Tapeworm**

1. Procure the needed equipment and specimens.
2. Observe the specimens in a jar or in a Petri dish.
3. Record your observations and labeled sketches below.

 Student Activity—Microscopic Anatomy of a Tapeworm

Materials
- compound microscope
- colored pencils
- selected prepared slides of various tapeworms such as *Taenia* spp., *Diphyllobothrium latum*, *Echinococcus granulosus*, or *Dipylidium caninum*

 **Procedure 30.5
Microanatomy of a Tapeworm**

1. Procure the microscope and selected slides.
2. Record your observations and labeled sketches of a composite tapeworm in the space provided.

Being parasitic, flukes have a variety of general and species-specific adaptations. General adaptations include

1. the presence of various specialty glands for penetration or cyst formation, suckers and hooks for attachment,
2. a mouth at the anterior end of the organism, and
3. an ability to produce a tremendous number of offspring.

The trematodes share with the turbellarians a well-developed alimentary canal, similar musculature, and similar body systems. In trematodes, the sense organs are poorly developed.

A representative fluke is *Clonorchis sinensis,* better known as the Chinese liver fluke. This parasite is common in many regions of the Orient, where it parasitizes dogs, cats, pigs, and humans. The adult worms vary in length from 10 to 70 millimeters. *Clonorchis* possesses an oral and a ventral sucker, and the digestive system consists of an esophagus and two long, unbranched intestinal **ceca**. Two pronephridial tubules unite to form a median bladder that points to the outside of the organism. The sense organs are degenerate, and the nervous system is similar to that of turbellarians (Fig. 30.18–Fig. 30.20).

The life cycle of this organism is interesting. Fertilized eggs exit the body of an infected host where, hopefully, they land in water and are ingested by a snail. In the snail, the egg hatches and forms a **miracidium larva**. At this point, the snail is the first intermediate host. The miracidium develops within the snail's body and forms a sporocyst. The **sporocyst** asexually produces thousands of larvae, the **redia**. Each redia in turn reproduces asexually to produce up to 50 **cercaria**, which emerge through the epidermis of the snail and enter the watery environment.

The cercariae are free-swimming and make their way to a fish, where they burrow into the muscle beneath the scales. The infected fish is the second intermediate host. The cercaria encyst in the fish and are called **metacercariae**. Next, the fish is eaten by the final host (perhaps a human) and the parasite makes its way to the bile ducts, where it parasitizes the host. It reaches sexual maturity and produces eggs to start the cycle over again (Fig. 30.21).

Class Trematoda

Members of **Class Trematoda** are all parasitic flukes. As adults, most trematodes are endoparasites of various vertebrates; however, there are some ectoparasitic species. Almost 11,000 species of flukes have been described, many of which are have great economic and medical importance. The flukes, belonging to the class Trematoda, are different from those of Class Monogenia in several ways, including the following:

1. The larvae of Trematoda are capable of reproduction.
2. The larvae of Trematoda are endoparasitic in intermediate hosts.
3. The adults of Trematoda are tissue endoparasites in a vertebrate host.

The majority of trematodes reside in the lung, liver, bile ducts, pancreatic ducts, intestines, and blood. The flukes are flattened and leaf-like in appearance. The body of a fluke is covered by a nonciliated tegument that lacks cell membranes between the nuclei (Fig. 30.16–Fig. 30.17).

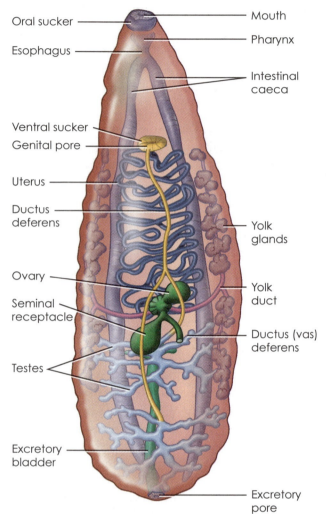

Oral sucker — Mouth
Esophagus — Pharynx
— Intestinal caeca
Ventral sucker
Genital pore
Uterus
Ductus deferens — Yolk glands
Ovary — Yolk duct
Seminal receptacle
Testes — Ductus (vas) deferens
Excretory bladder
— Excretory pore

Figure 30.16 A diagram of the human liver fluke, *Clonorchis sinensis*.

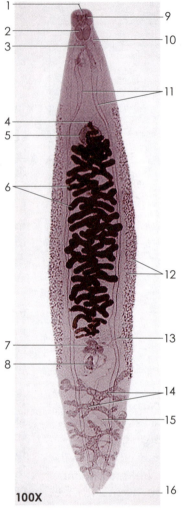

100X

Figure 30.17 The liver fluke, *Clonorchis*.
1. Mouth
2. Pharynx
3. Esophagus
4. Genital pore
5. Ventral sucker
6. Uterus
7. Ovary
8. Seminal receptacle
9. Oral sucker
10. Cerebral ganglion
11. Intestine
12. Yolk glands
13. Yolk duct
14. Testis
15. Ductus (vas) deferens
16. Excretory pore

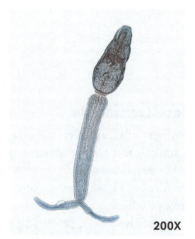

200X

Figure 30.18 The cercaria stage of a trematode species.

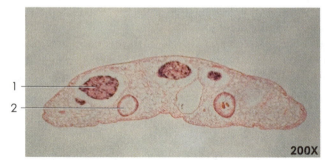

200X

Figure 30.19 A transverse section through the midbody body region of *Clonorchis*.
1. Uterus   2. Intestine

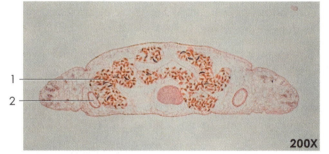

200X

Figure 30.20 A transverse section through the upper body region of *Clonorchis*.
1. Testes   2. Intestine

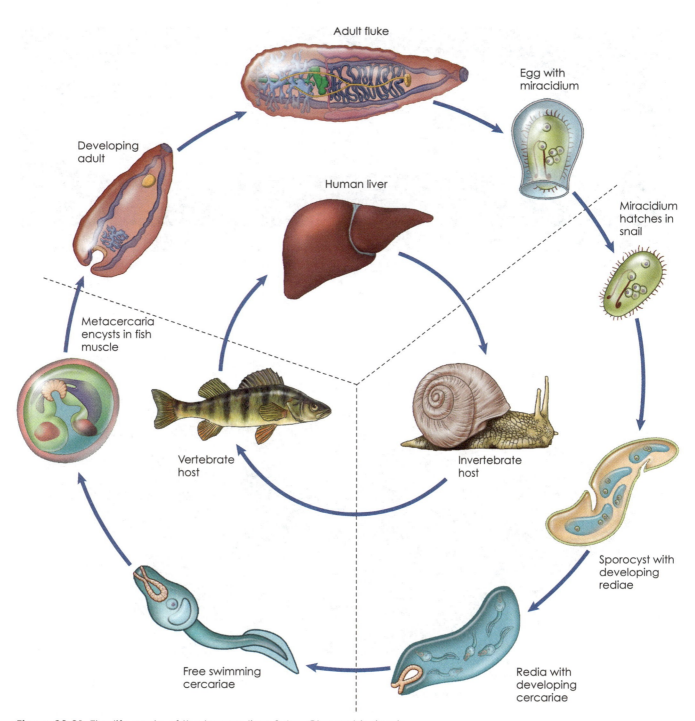

Figure 30.21 The life cycle of the human liver fluke, *Clonorchis sinesis*.

Fasciola hepatica, the sheep liver fluke, is one of the largest flukes, It has a leaf-shaped body measuring up to 3.5 cm by 1.5 cm. The fluke lives in the liver and bile duct of its host. Hosts include herbivorous mammals such as cattle and sheep, as well as humans. Its intermediate host is a snail which lives near standing water. This fluke's life stages are typical of other trematodes such as *Clonorchis* (Fig. 30.22–Fig. 30.24).

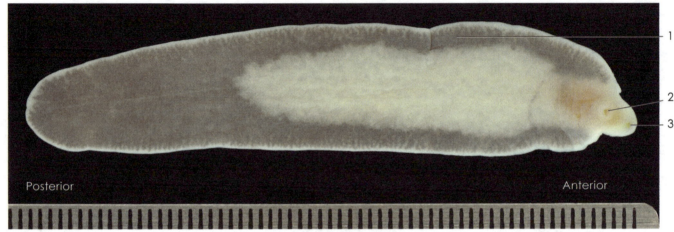

Posterior | Anterior

Figure 30.22 The cow liver fluke, *Fasciola magna*, is one of the largest flukes, measuring up to 7.75 cm long (scale in mm).

1. Yolk gland   2. Ventral sucker   3. Oral sucker

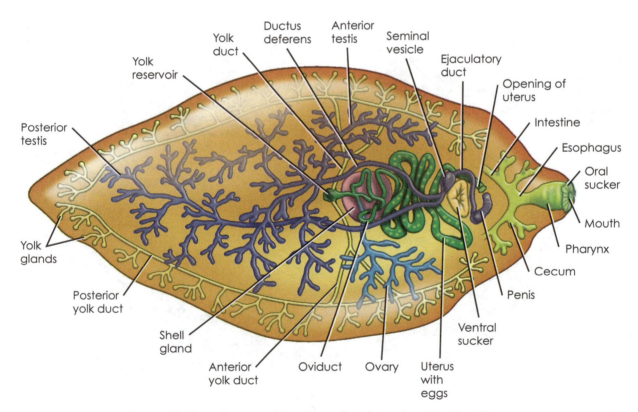

Figure 30.23 A diagram of the Sheep liver fluke, *Fasciola hepatica*.

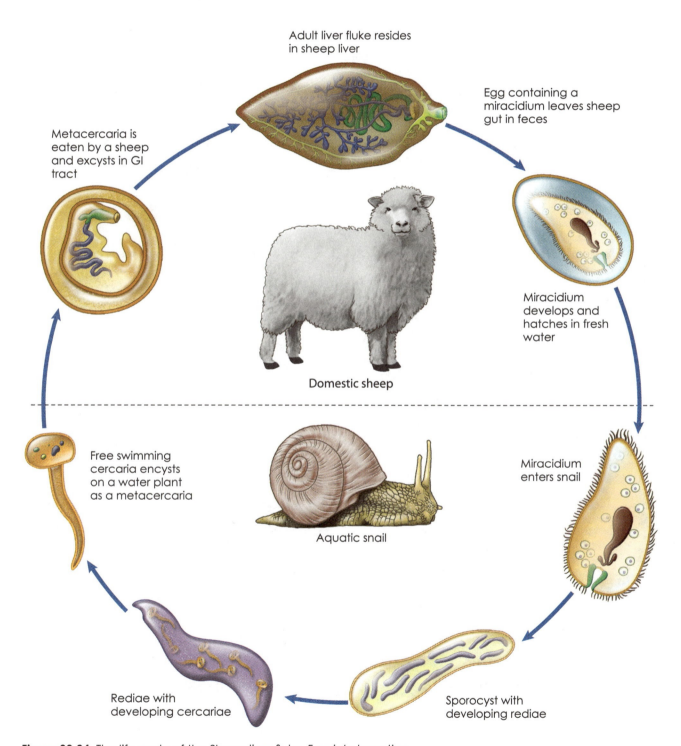

Adult liver fluke resides
in sheep liver

Egg containing a
miracidium leaves sheep
gut in feces

Metacercaria is
eaten by a sheep
and excysts in GI
tract

Miracidium
develops and
hatches in fresh
water

Domestic sheep

Free swimming
cercaria encysts
on a water plant
as a metacercaria

Miracidium
enters snail

Aquatic snail

Rediae with
developing cercariae

Sporocyst with
developing rediae

Figure 30.24 The life cycle of the Sheep liver fluke, *Fasciola hepatica*.

Schistosoma spp., commonly called blood flukes, are found mostly in Africa. Worldwide, approximately 200 million people are infected with schistosomiasis. Three species infect humans: *Schistosoma mansoni, S. haematobium,* and *S. japonicum.* Water becomes contaminated with *Schistosoma* spp. eggs when infected people urinate or defecate in the water. The eggs hatch in the water, grow, and develop inside snails. The parasite exits the infected snail and enters the water, where it can survive for as long as 48 hours. The larvae penetrate the skin of an individual who wades, swims, or bathes in the contaminated water. Within several weeks, blood flukes grow inside the blood vessels of the body and produce eggs. Eventually, some of the eggs travel to the bladder or intestines and are passed into the urine or stool. (Fig. 30.25) Interestingly, because it is difficult to find a mate in the miles of vessels during copulation, the female lives in a groove in the body of the male (Fig. 30.26).

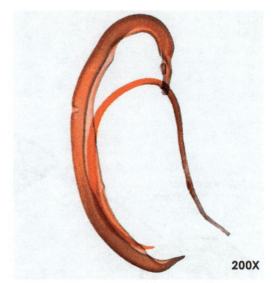

200X

Figure 30.26 Micrograph of a schistosome showing the female within the groove of the male.

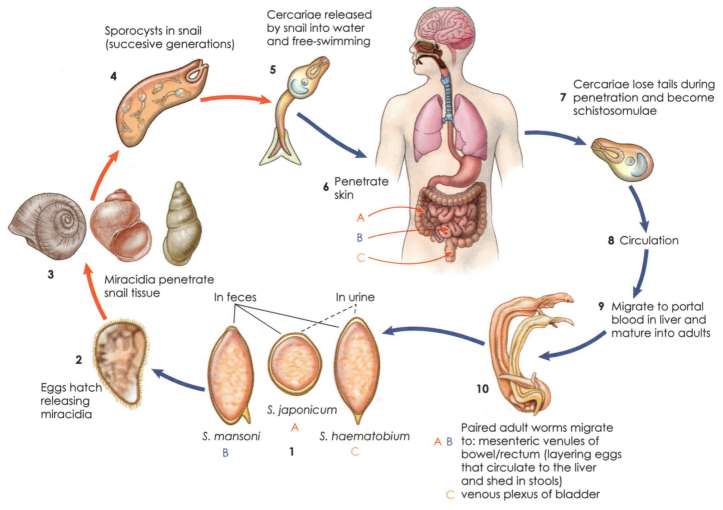

Figure 30.25 Generalized life cycle of *Schistosoma* spp.

Student Activity—Macroscopic Anatomy of *Fasciola hepatica*

Materials
- compound microscope
- colored pencils
- selected prepared slides of various trematodes such as a whole mount of *Clonorchis sinensis*, transverse sections of *Clonorchis, Fasciola* hepatica, *Schistosoma* spp., and a cercaria of a fluke

 Procedure 30.6
Macroanatomy of a Trematode

1. Procure the needed equipment and specimens.
2. In a jar or a Petri dish, observe the fluke with the dissecting microscope or hand lens. Be sure to observe the internal structures with transmitted light.
3. Record your observations and labeled sketch.

 Student Activity—Microscopic Anatomy of Select Trematodes

Materials
- compound microscope
- colored pencils
- selected prepared slides of various trematodes such as a whole mount of *Clonorchis sinensis*, transverse sections of *Clonorchis, Fasciola* hepatica, *Schistosoma* spp., and a cercaria of a fluke

Procedure 30.7
Microanatomy of a Trematode

1. Procure the microscope and selected slides. Using low power record your observations.

A man who dares to waste one hour of time has not discovered the value of life.

—Charles Darwin (1809–1882)

2. Observe a whole mount slide of *Clonorchis sinensis.* Record your observations and labeled sketch.

3. Observe a transverse section slide of *Clonorchis sinensis.* Record your observations and labeled sketch.

4. Observe a whole mount slide of *Fasciola hepatica.* Record your observations and labeled sketch below.

5. Observe a slide of *Schistosoma* spp. Record your observations and labeled sketch below.

6. Observe a slide of a cercaria of a fluke. Record your observations and labeled sketch below.

Check Your Understanding

Q. Compare and contrast turbellarians, cestodes, and trematodes.

Q. How are intestinal platyhelminthes protected by harsh digestive juices?

Q. Name and describe four parasitic platyhelminths.

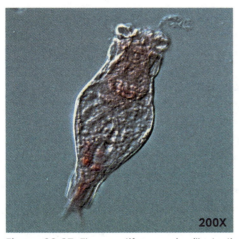 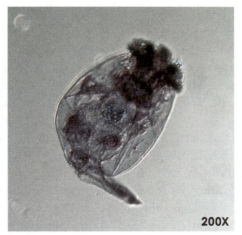 

Figure 30.27 Three rotifer species illustrating their diversity.

PHYLUM ROTIFERA

Antony van Leeuwenhoek termed a curious group of active aquatic creatures the "wheel animalcules." These tiny animals were so named because of a crown of cilia on their head. The cilia in motion resembled a spinning wheel. Leeuwenhoek's enthusiasm can be shared by examining a bottom sample of a nearby ditch or pond with a compound microscope. Today, Leeuwenhoek's wheel animacules are placed in **Phylum Rotifera.** The rotifers are considered to be pseudocoelomates. Ecologically, the rotifers are part of the food chain. Commercially and medically, rotifers are not significant. However, curious biologists study rotifers to gain a better understanding of evolution, diversity, physiology, ecology, and behavior (Fig. 30.27).

Phylum Rotifera consists of approximately 1,900 species of primarily freshwater organisms. Rotifers range in size from 0.1 to 1.0 millimeters and have an elongated sac-like body. The anterior portion of a rotifer bears a **corona**, or crown, with numerous cilia arranged on two **discs** which beat in a circular motion in opposite directions. These cilia are used in feeding and locomotion. Sensory bristles near the head region are called **papillae**. In some rotifers the **cuticle**, or covering on the **trunk** of the body, forms an armor-like girdle called a **lorica** that may bear **spines**. The posterior portion of a rotifer, the **foot**, contains **adhesive glands** that open to the exterior via **spurs**, or toes.

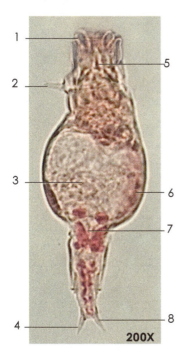

Figure 30.28 A rotifer.
1. Corona
2. Antenna
3. Stomach
4. Spur
5. Mastax
6. Vitellarium
7. Intestine
8. Toe

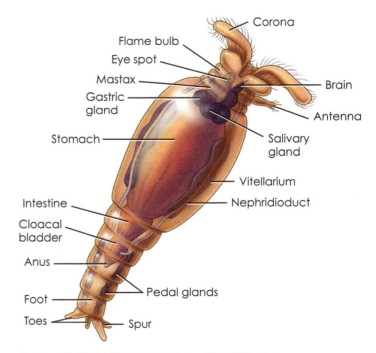

Figure 30.29 A diagram of the rotifer, *Philodina*.

Being transparent, rotifers provide the observer with an open window to their internal anatomy. A **mouth** can be seen beneath the ciliated corona. Behind the mouth is a unique pharyngeal apparatus called the **mastax**, which possesses grinding jaws, tropi, in several species. In living rotifers, the **tropi** easily can be seen grinding up algae and smaller invertebrates.

Rotifers have a complete digestive system starting with a mouth and ending in an **anus**. They have a large **stomach** and short **intestines**. The terminal portion of the intestines is called the **cloaca** because it receives solid wastes, liquid wastes from **protonephridia**, and sex cells such as eggs from the **oviducts**. The protonephridia are associated with several **flame cells** that empty into the **bladder**. Rotifers lack a blood circulatory system; they respire through their body surface. The nervous system of rotifers consists of a bilobed brain that sends paired nerves to various organs. Sensory organs in rotifers include papillae, ciliated pits, dorsal antennae, and paired eyespots in several species (Fig. 30.28–Fig. 30.29).

Rotifers are dioecious. Generally, female rotifers are larger than the males. Males either insert their **penis** into the female's cloaca or stab the female with the penis to insert sperm. This type of activity is called **hypodermic impregnation.** Rotifers can produce thin-shelled, fast-hatching eggs or thick-shelled, dormant eggs. Rotifers are tiny organisms that live in environments susceptible to drying up. To cope with these conditions, rotifers can enter an arrested state of biological activity termed **cryptobiosis**. So far, rotifers have been known to stay in this state for up to 4 years.

Student Activity—Macroscopic Anatomy of Rotifers

Materials
- compound microscope
- colored pencils
- microscope slides and coverslips
- teasing needle
- water
- dropper
- wet mount
- selected prepared slides of rotifers
- live culture of rotifers

Procedure 30.8
Macroanatomy of Rotifers

1. Procure the needed equipment, prepared slides, and culture.
2. Record your observations along with a labeled sketch of the prepared microscope slide below.

_____

_____

_____

3. Place a living rotifer on a microscope slide and prepare a wet mount. Observe the behavior, feeding activities, and anatomy of the rotifer.
4. Record your observations, prepare a sketch, and record the locomotion and activities of the rotifer below.

_____

_____

_____

Figure 30.30 Examples of molluscs: (a) mussels, (b) nautilus, (c) giant clam, (d) sea hare, (e) chiton, and (f) banana slug.

PHYLUM MOLLUSCA

Although we enjoy looking at the intricate patterns and beauty of seashells, few people take time to study the wonder of the natural history of the members of **Phylum Mollusca.** Scientists who study molluscs are called **malacologists.** Phylum Mollusca consists of 90,000 living species plus more than 70,000 fossil species (dating back to the Cambrian), making it the second largest phylum. The molluscs inhabit a variety of environments, including marine, freshwater, and terrestrial habitats. This highly diverse phylum encompasses chitins, limpets, tooth shells, slugs, snails, whelks, abalones, conchae, nudibranchs, clams, oysters, scallops, squids, octopuses, and nautiluses (Fig. 30.30). Molluscs vary in size from almost microscopic to gigantic. The giant squid can attain lengths of more than 20 meters and weigh more than 1,000 pounds. The giant clam can reach 1.5 meters in length and weigh more than 500 pounds. Phylum Mollusca includes herbivores, carnivores, filter feeders, detritus feeders, and even parasites.

Although the molluscs exhibit great diversity, they share a common body plan: They are bilaterally symmetrical. The molluscs are considered **coelomates** even though the coelom is limited to the space around the heart. The basic body plan of molluscs consists of two major portions: the head-foot region and the visceral-mass region. The **head-foot region** contains the cephalic portions of the organism, as well as the feeding and locomotor structures. The **visceral-mass** region contains the digestive, respiratory, circulatory, and reproductive systems.

Molluscs possess a heart, vessels, and sinuses, but the majority have an **open circulatory system** in which blood is not contained entirely in vessels. Most cephalopods have a **closed circulatory system,** in which the blood is contained in vessels. **Gills** and, in some species, **lungs** are responsible for gas exchange. The digestive tract is complete and highly specialized. The majority of molluscs possess a pair of **metanephridia**, or kidneys, which open into the coelom through a **nephrostome**. In many molluscs, the kidney ducts also discharge sperm and eggs.

The majority of molluscs have a well-developed head region that bears the mouth and sensory organs. Within the mouth of most molluscs is a unique structure called the radula, a protrusible tongue-like organ used for rasping. The radula can contain up to 250,000 "teeth," which serve to scrape, pierce, and cut. The next time you see a snail crawling up an aquarium glass, notice the radula rasping the algae. Also, while at the beach, notice perfectly round holes in a shell. These holes were done by a mollusc's radula such as that of an oyster drill.

The nervous system is composed of several pair of ganglia and their associated nerve cords. The nervous system in molluscs, especially the cephalopods (squid and octopi), is well-developed. Cephalopods even exhibit problem-solving ability. Sensory organs of vision, touch, smell, taste, and equilibrium vary in molluscs. The eyes of the cephalopods are particularly well-developed.

The **mantle** in molluscs is a sheath of skin extending from the visceral mass and hanging down each side of the body, protecting the soft parts of the organism. Between the soft parts of the organism is a **mantle cavity.** The outermost surface of the mantle is responsible for secreting and lining the shell of some molluscs. Typically, the shell has three distinct layers: The outer layer of the shell, the **periostracum**, serves to protect the inner layers. The middle portion, called the **prismatic layer,** is composed of calcium carbonate and a protein matrix. The innermost layer, the **nacreous layer** of the shell, is produced by the adjacent portions of the mantle surface. This is the iridescent "mother-of-pearl" layer that is visible in many shells. Many molluscs secrete nacre around foreign or induced particles, producing **pearls.** The mantle cavity houses the respiratory organs of molluscs.

Members of Phylum Mollusca reproduce sexually. Although some monoecious forms exist, the majority of molluscs are dioecious. Most molluscs undergo external fertilization, but a few forms have evolved internal fertilization. Many molluscs and the segmented worms (Annelids) have free-swimming, ciliated larvae called the **trophophore** larvae. In others, such as many gastropods and bivalves, the trophophore larvae form larvae with the beginnings of a foot, shell, and mantle, called the **veliger larvae.** The cephalopods and some other molluscs produce juveniles that hatch directly from the egg.

The molluscs (Fig. 30.32) are generally placed into six major classes: Monoplacophora, Polyplacophora,

How is a Pearl Made?

Buying pearl jewelry can damage a budget! Pearls can occur naturally when an irritant such as a parasite enters an oyster or mussel and the mantle tissue secretes calcium carbonate from the nacre over the irritant. In cultured marine species, a mother of pearl "seed" is placed in the oyster or mussel. In freshwater species, a tiny "seed" of mantle material is placed in the mussel to initiate pearl formation (Fig. 30.31).

Figure 30.31 The formation of pearls can be natural or artificially induced.

Gastropoda, Scaphopoda, Bivalva, and Cephalopoda. **Class Monoplacophora** consists of 25 described species of molluscs. Once thought to be extinct, living species of monoplacophpran *Neopilina* was discovered off the west coast of Costa Rica. These organisms resemble limpets but have serial repetition of organs.

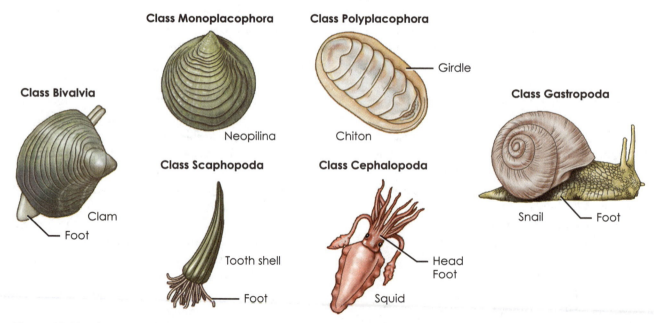

Figure 30.32 Diagrams of specimens representing several classes of molluscs.

Class Polyplacophora

Members of **Class Polyplacophora** are commonly referred to as chitins. Approximately 1,000 species have been identified. These bottom-dwelling marine molluscs vary in length from 2 mm to 40 cm. In chitins, the shell covers the dorsal side of the organism with a series of eight plates (Fig. 30.33). The body is elongated and flattened, and the foot is on ventral side, flattened, and extends the length and width of the chitin's body. Many species have beautiful, ornate shells. Examples include *Mopalia spp.*, *Lepidopleurus spp.*, and *Tonicella spp.*

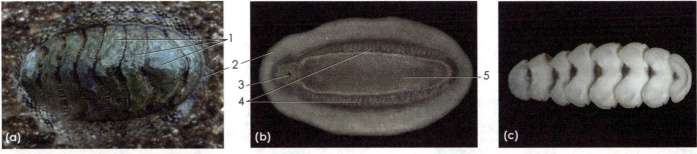

Figure 30.33 Chitons are easily recognized by their eight dorsal plates. (a) A dorsal view and (b) ventral view, and (c) A ventral view of a chiton skeleton showing the eight dorsal plates.

1. Dorsal plates   2. Girdle   3. Mouth   4. Gill filaments   5. Ventral foot

---

 Student Activity—Macroscopic Anatomy of Class Polyplacophora

Materials
- dissecting microscope or hand lens
- colored pencils
- Petri dish
- dissecting tray
- probe
- preserved specimens of selected chitins

 Procedure 30.9
Macroanatomy of Polyplacophora

1. Procure the needed equipment and specimens.
2. Using the dissecting microscope or hand lens, examine the specimens in their jars or, if the instructor permits, place them in a dissecting tray for examination.
3. Observe each specimen, paying close attention to the structures discussed above.

4. Record your observations and a detailed labeled sketch to the right.

Class Gastropoda

Class Gastropoda is the most diverse class of molluscs (Fig. 30.34–Fig. 30.37), with more than 70,000 identified species. Examples of gastropods are snails, abalone, whelks, conchae, limpets, periwinkles, sea hares, slugs, and nudibranchs. Gastropods live in marine, freshwater, and terrestrial environments. The shell in gastropods that have a shell is a one-piece **univalve**. Other gastropods, such as slugs, do not have a shell.

The end of the shell is called the **apex**. As the animal grows, the successive **whorls** increase in size. The central axis is called the **columnella**. Some shells coil to the left and are said to be left-handed, or **sinistral**; others are right-handed, or **dextral**, coiling to the right. The opening of the shell is called the **aperture**, and in many species an **operculum** covers the aperture.

Generally, gastropods have well-developed cephalic regions. The head usually is characterized by the presence of two tentacles with an eye at the end of each. Gastropods use a **muscular foot** for locomotion. Both monoecious and dioecious gastropods exist. Eggs of marine species are encased in **egg cases**, commonly found by beach combers.

Figure 30.34 Many gastropods have ornate shells, such as the Venus comb murex, *Murex pectin* (scale in mm).

Figure 30.35 A keyhole limpet, *Megathura crenulata*.
1. Shell   2. Mantle   3. Foot

Figure 30.36 A snail.
1. Shell   4. Head
2. Foot   5. Sensory tentacle
3. Occular tentacle

Figure 30.37 The locomotion of the slug, class Gastropoda, requires the production of mucus. Slugs differ from snails in that a shell is absent.
1. Foot   3. Mantle   5. Occular tentacle   7. Pneumostome
2. Mucus   4. Head   6. Sensory tentacle

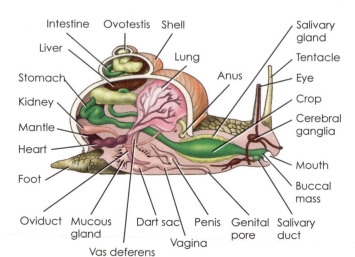

Figure 30.38 A diagram of pulmonate snail anatomy.

Intestine   Ovotestis   Shell
Liver
Stomach
Kidney
Mantle
Heart
Foot
Oviduct   Mucous gland   Dart sac   Penis   Genital pore   Salivary duct
Vas deferens   Vagina

Salivary gland
Tentacle
Eye
Crop
Cerebral ganglia
Anus
Lung
Mouth
Buccal mass

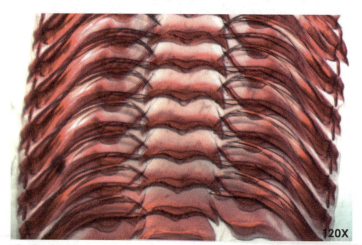

Figure 30.39 A snail radula is made up of small horny teeth made of chitin, called denticles.

Student Activity—Macroscopic Anatomy of Class Gastropoda

Materials
- dissecting microscope or hand lens
- colored pencils
- Petri dish
- dissecting tray or jars
- probe
- preserved specimens and shells of selected gastropods including a snail, a slug, a whelk, a conch, an abalone, a limpet, a nudibranch, and other specimens

Procedure 30.10
Macroanatomy of Gastropoda

1. Procure the needed equipment and specimens.
2. Examine the specimens in their jars or, if the instructor permits, place them in a dissecting tray for examination.
3. Observe each specimen, paying close attention to the structures discussed.
4. Record your observations and detailed labeled sketches below.

The life of wild animals is a struggle for existence. The full exertion of all their faculties and all their energies is required to preserve their own existence and provide for that of their infant offspring.

—Alfred Russel Wallace (1823–1913)

Student Activity—Microscopic Anatomy of a Snail Radula

Materials
- compound microscope
- colored pencils
- selected prepared slides of a snail radula

Procedure 30.11
Microanatomy of the Radula

1. Procure the equipment and prepared slide of the radula of a snail.
2. Observe the slide under both low and high power.
3. Record your observations and sketch.

Class Bivalvia

Class Bivalvia (Pelecypoda) (Fig. 30.40–Fig. 30.46) consists of nearly 25,000 species of molluscs that feature two separate shells (valves) joined by a ligament called the **hinge**. The oldest part of the shell, the umbo, looks like a large hump on the anterior end of the dorsal side of each valve. Powerful adductor muscles hold the valves of the shell together. Most bivalves have a posterior and an anterior **adductor muscle.** Scallops have just one.

Have you ever eaten fried clams or grilled scallops (*Pecten* spp.)? If so, you were eating adductor muscles! The bivalves also are called "hatchet-footed animals" because of their obvious hatchet-shaped muscular foot that is attached to the visceral mass of the organism. Examples of bivalves are clams, mussels, scallops, oysters, quahogs, and shipworms. They lack a head and radula. They vary in size from seed shells about 1 mm in length to the giant clam that can be longer than a meter and weigh 225 kg. Bivalves live in marine as well as freshwater environments. The majority are **filter feeders.** They take in water and nutrients in the **incurrent siphon** and eliminate by the **excurrent siphon**. Gaseous exchange occurs through the mantle and gills. Bivalves are mostly dioecious. In freshwater clams the trochopore larva develops into a specialized veliger larva known as a **glochidium larva,** which attaches to fishes to complete their development.

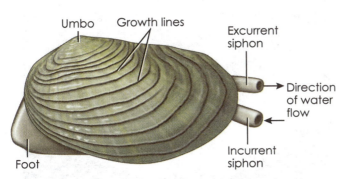

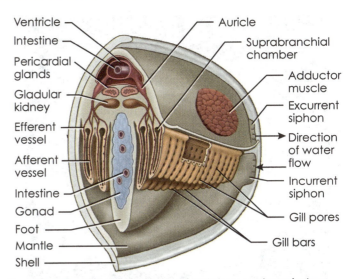

Figure 30.40 The surface anatomy of a freshwater clam (left valve).

Figure 30.41 A diagram of the circulatory and respiratory systems of a freshwater clam.

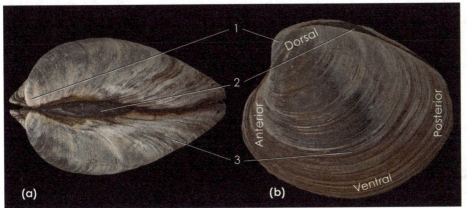

Figure 30.42 An external view of a clam shell: (a) dorsal view and (b) the left valve.
1. Umbo   2. Hinge ligament   3. Growth lines

Figure 30.43 Internal view of a clam shell showing the muscle scars where the adductor muscles attached to the shell.
1. Muscle scar

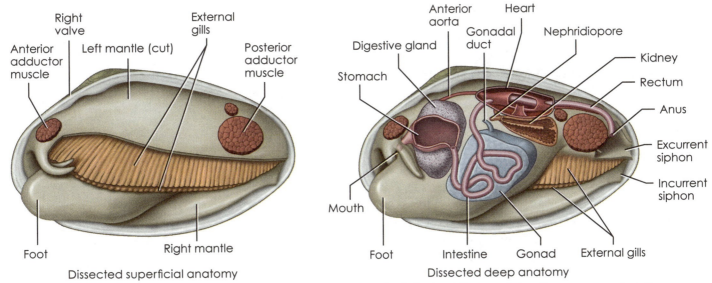

Dissected superficial anatomy

Dissected deep anatomy

Figure 30.44 The anatomy of a freshwater clam. Bivalves have two shells (valves) that are laterally compressed and dorsally hinged.

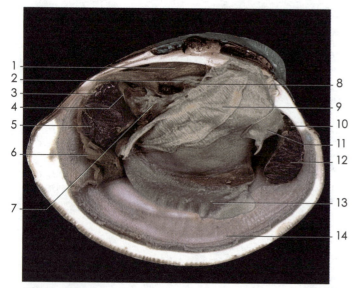

Figure 30.45 A lateral view of a clam.
1. Pericardium
2. Ventricle of heart
3. Anus
4. Posterior retractor muscle
5. Posterior adductor muscle
6. Excurrent siphon
7. Nephridium (kidney)
8. Atrium of heart
9. Gills
10. Anterior retractor muscle
11. Labial palps
12. Anterior adductor muscle
13. Foot
14. Mantle

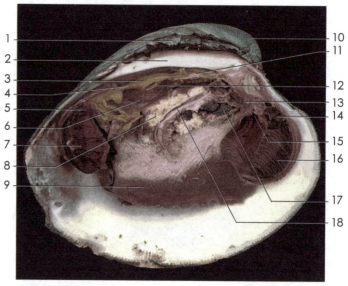

Figure 30.46 A lateral view of a clam, foot cut.
1. Hinge ligament
2. Hinge
3. Ventricle of heart
4. Posterior aorta
5. Posterior retractor muscle
6. Nephridium (kidney)
7. Posterior adductor muscle
8. Gonad
9. Foot
10. Umbo
11. Intestine
12. Opening between atrium and ventricle
13. Esophagus
14. Anterior retractor muscle
15. Mouth
16. Anterior adductor muscle
17. Digestive gland
18. Intestine

 Student Activity—Macroscopic Anatomy of Class Bivalvia

Materials
- dissecting microscope or hand lens
- colored pencils
- Petri dish
- dissecting tray
- probe
- preserved specimens and shells of select bivalves in jars, including an oyster, a freshwater clam or mussel, a scallop, a quahog, a shipworm, and other bivalves

 Procedure 30.12
Macro Anatomy of a Bivalve

1. Procure the needed equipment and specimens.
2. Examine the specimens in their jars or if the instructor permits, place them in a dissecting tray for examination. Preserved shells should be placed on a tray.
3. Observe each specimen closely paying particular attention to the structures discussed in the narrative

above. Record your observations and a detailed labeled sketch below.

Student Activity—Dissection of a Clam

Materials
- dissecting tray
- water
- gloves
- safety glasses
- lab coat or apron
- hand lens
- screwdriver
- dissection kit
- colored pencils
- clam provided by the instructor

Procedure 30.13
Clam Dissection

1. Procure the supplies and the clam. Wash your specimen thoroughly under running water. If preservative gets in your eyes, wash your eyes immediately and contact the instructor. Place the specimen on the dissecting tray.

2. Observe the external features of your specimen. Record your observations and labeled sketch below.

3. Place the clam with the dorsal side down, and carefully insert a screwdriver between the edges of the valves. Continue to move the tip of the screwdriver between the valves. Turn the screwdriver so the valves spread apart.

Q. Why are the valves so hard to separate?

4. Keeping the tip of the screwdriver between the valves, place the clam on the dissecting tray with one valve facing upward. Look inside the clam and find the adductor muscles. With a scalpel, cut the anterior adductor muscle and the posterior adductor muscle, cutting as close to the shell as possible.

5. Carefully and slowly bend the valves back so the clam lies flat on the dissecting tray. You may have to carefully separate the mantle from the valve.

6. Examine the specimen, noting textures. Locate the clam superficial structures shown in Figure 30.40 through 30.46. Record your observations and labeled sketch below. Pay particular attention to the umbo, hinge ligament, and growth lines.

7. Examine the interior surface of the valve without the mantle. Locate the anterior and posterior adductor muscle scars, nacreous layer, and other structures shown in Figure 30.42 and 30.43. Record your observations and labeled sketch below.

8. Using a pair of scissors and your fingers, carefully remove the half of the mantle that lined the valve with the visceral mass. After removing this part of the mantle, you can see the internal structures. You may have to use a probe to move structures within the visceral mass for better viewing. Compare your specimen to Figure 30.44 through 30.46, paying particular attention to the incurrent and excurrent siphons, foot, gills, heart, kidney, mouth, intestines, gonads, and stomach. Record your observations and labeled sketch.

9. Clean your equipment and desktop thoroughly. Return the equipment and discard your specimen as directed.

Class Cephalopoda

Octopi, squid, cuttlefish, and nautiluses are members of **Class Cephalopoda**. This unique class of marine predators is composed of approximately 800 living species. Ammonites and belemnites are fossil cephalopods that are significant in evolutionary studies and appealing to fossil collectors (Fig. 30.47). Cephalopods possess a modified foot in the head region and appears as a funnel-shaped structure, called a **siphon**, which is used to expel water from the mantle cavity. This structure is surrounded by **tentacles** with **suckers**. The siphon is used in a form of jet propulsion, expelling water during locomotion. Cephalopods range in size from 1 cm in length to *Architeuthis*, the giant squid, which may exceed 12 m in length.

Nautiloids such as the chambered nautilus and the paper nautilus have an external chambered shell and, as the organism grows, new chambers are added (Fig. 30.48). Cuttlefish have a small, curved shell enclosed by the mantle called the **cuttlebone**. In the cuttlefish it is a chambered, gas–filled shell that keeps the animal buoyant. Commercially, it is obtained as a source of calcium for birds, turtles, and hermit crabs. The shell of a squid is restricted to a thin strip called a **pen**. Octopi have no remnants of a shell (Fig. 30.49–Fig. 30.51).

Respiration in cephalopods occurs primarily through gills. Cephalopods have a closed circulatory system with a heart and vessels. The nervous system is highly developed. The brain is lobed and is the largest of any invertebrate. With the exception of nautiloids, the sense organs are well-developed. Octopi, as ambush hunters, are capable of learning, and they demonstrate complex behaviors. Many cephalopods have **ink glands** that secrete dark **sepia**, which serves as a "smoke screen" used in escape behaviors. Octopi, squid, and cuttlefish can change colors for sexual attraction, warning, camouflage, communication, and to attract and mesmerize prey. **Chromatophores**, which contain pigment granules, are responsible for color and pattern changes. Cephalopods are dioecious. Juveniles hatch directly from eggs (Fig. 30.49–Fig. 30.55).

Figure 30.47 Ammonites are fossilized cephalopods. (a) Ammonite fossil encrusted limestone, and (b) *Metoicoceras geslinianum*.

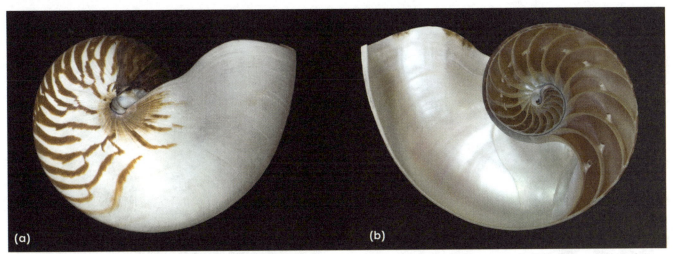

Figure 30.48 The Nautilus (a), a cephalopod, has gas-filled chambers within its shell, as seen in this cross-section of the shell. (b) These chambers regulate buoyancy.

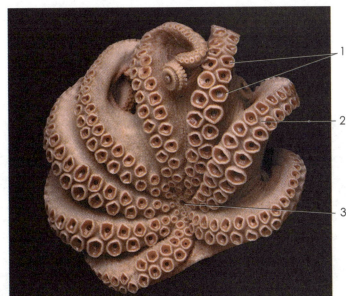

Figure 30.49 A dorsal view of an octopus collected in the Sea of Cortez, San Carlos, Mexico.

1. Mantle   2. Head   3. Arms

Figure 30.50 A ventral view of an octopus.

1. Suction cups   3. Mouth
2. Arm

Figure 30.51 (a) The giant octopus lives in the cooler waters of the North Pacific while (b) the cuttlefish lives in tropical waters.

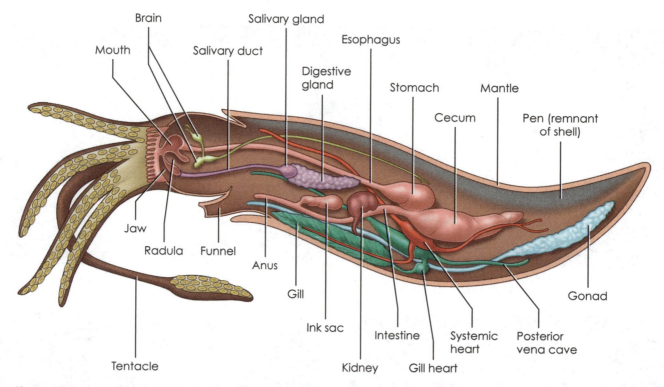

Brain

Mouth

Salivary duct

Salivary gland

Esophagus

Digestive gland

Stomach

Mantle

Cecum

Pen (remnant of shell)

Jaw

Radula

Funnel

Anus

Gill

Ink sac

Kidney

Intestine

Gill heart

Systemic heart

Posterior vena cave

Gonad

Tentacle

Figure 30.52 The internal anatomy of a squid.

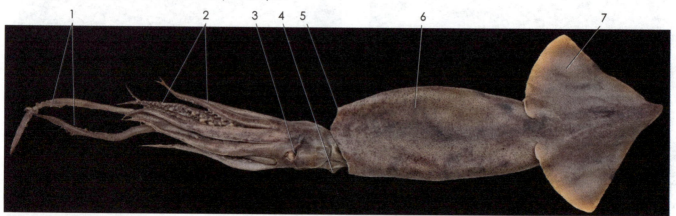

1   2   3   4   5   6   7

Figure 30.53 The external anatomy of the squid, *Loligo*.
1. Tentacles   2. Arms   3. Eye   4. Funnel   5. Collar   6. Mantle (body tube)   7. Fin

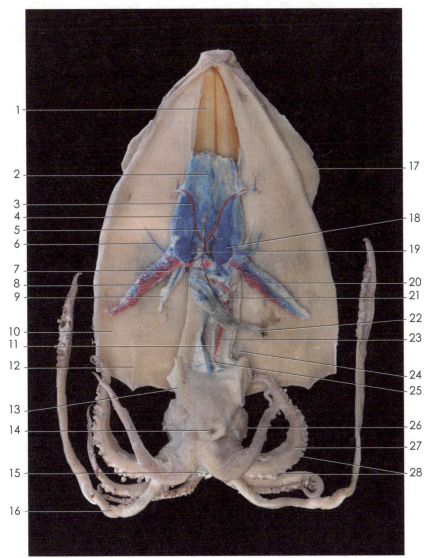

Figure 30.54 The internal anatomy of the squid.

1. Pen
2. Gonad
3. Lateral mantle artery
4. Posterior vena cava
5. Median mantle artery
6. Median mantle vein
7. Afferent branchial artery
8. Gill
9. Genital opening
10. Mantle
11. Esophagus
12. Articulating ridge
13. Articulating cartilage
14. Siphon
15. Mouth
16. Tentacle
17. Fin
18. Branchial heart
19. Systemic heart
20. Efferent branchial vein
21. Ink sac
22. Rectum
23. Cephalic aorta
24. Stellate ganglion
25. Cephalic vena cava
26. Eye
27. Arm
28. Suckers

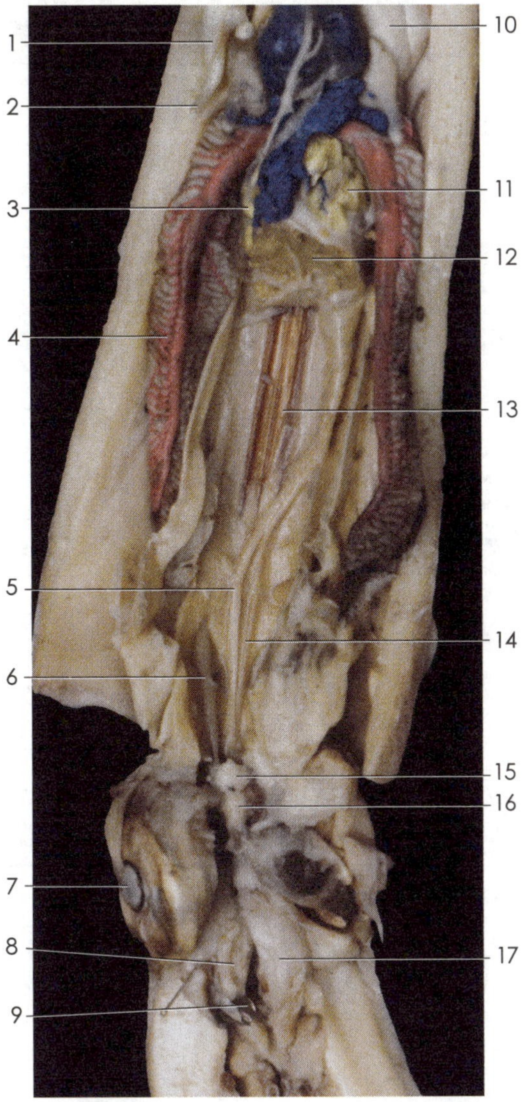

Figure 30.55 The internal anatomy of the squid including head region.

1. Spermatophoric duct
2. Penis
3. Kidney
4. Gill
5. Esophagus
6. Pleural nerve
7. Eye
8. Radula
9. Beak
10. Stomach
11. Pancreas
12. Digestive gland (cut)
13. Pen
14. Cephalic aorta
15. Visceral ganglion
16. Pedal ganglion
17. Buccal bulb

Student Activity—Macroscopic Anatomy of Class Cephalopoda

Materials
- dissecting microscope or hand lens
- colored pencils
- Petri dish
- dissecting tray or jars
- probe
- preserved specimens and shells of select cephalopods including: an octopus, a squid, cuttlefish, a chambered nautilus, a fossil ammonite, and other cephalopods

Procedure 30.14
Macroanatomy of a Cephalopod

1. Procure the necessary equipment and specimens.
2. Examine the specimens in their jars or if the instructor permits, place them in a dissecting tray for examination.
3. Observe each specimen, paying particular attention to the structures discussed above. Record your observations and detailed labeled sketches below.

Student Activity—Dissection of a Squid

Materials

- dissecting tray
- dissecting kit
- gloves
- safety glasses
- lab coat or apron
- hand lens
- water
- squid

Procedure 30.15
Squid Dissection

1. Procure supplies and the squid.
2. Wash your specimen thoroughly under running water. If preservative gets in your eyes wash your eyes immediately and contact the instructor. Place the specimen on the dissecting tray.
3. Using a hand lens, observe the external features of your specimen. Pay particular attention to the tentacles, mouth region, fin, anus, and eye. Record your observations and labeled sketch.

4. Keep the squid on the dissecting tray with the funnel facing upward. Carefully separate the two long tentacles from the eight shorter arms of the squid. Observe the tentacles and suckers.
5. Using a pair of scissors, cut through the mantle from the anterior end to the posterior end along the midline. Fold back the sides of the mantle and pin them down.

6. Feel the mantle of the squid and locate the pen. Remove the pen by carefully pulling it away from the mantle, examine it, and place it aside. Describe and sketch the pen.

7. Examine the squid and locate the structures found in Figure 30.32. Pay particular attention to the mouth region, gill, anus, ink sac, hearts, reproductive structures, stomach, kidney, and eye. Record your observations and labeled sketches below.

8. Carefully remove the gills and ink sac and examine them through a hand lens. Record your observations and sketches on the following page.

The relics of the squid's ancestral shell can be found deep within. This is that flat leaf of powdery chalk, the cuttle bone, that is often washed up on the seashore.

—David Attenborough (1926–present)

9. Clean your equipment and desktop thoroughly. Return the equipment, and discard your specimen as directed.

Student Activity—Microscopic Anatomy of Mollusc Larvae

Materials
- compound microscope
- colored pencils
- selected prepared slides of trophophore larvae, veliger larvae, and glochidium larvae
- culture

Procedure 30.16
Mollusc Larvae Anatomy

1. Procure the equipment, prepared slides, and culture.
2. Using a compound microscope, observe each slide under both low and high power.
3. Record your observations and sketches.

PHYLUM ANNELIDA

Perhaps you already have been introduced to a classic member of **Phylum Annelida** on a fishing trip. The earthworm or night crawler that you used for bait is a common representative of Phylum Annelida, which includes approximately 15,000 species of segmented worms. In contrast to your likely image of "worms," the annelids are highly diverse, and many species are quite attractive. In addition to earthworms, this large phylum includes a variety of other worms, such as leeches, tubiflex worms, sand worms, lug worms, parchment worms, clam worms, and blood worms. They live in terrestrial, freshwater, and marine environments. Annelids can vary in size from less than a millimeter to one species of tropical earthworm that is 4 meters long.

Annelids (Fig. 30.56) are **coelomates** and exhibit bilateral symmetry. The body of this phylum is divided into similar rings, or **segments**. The **annuli**, the grooves dividing each segment, are readily visible on most annelids. Each segment is termed a **metamere**, or a **somite**. The repeated pattern of these metameres is termed **metamerism**. Internally, the segments are delimited by structures called **septa**. Annelids, with the exception of leeches, possess tiny bristles known as setae, which vary in size, form, and function in different species of annelids. Some setae, such as those in earthworms, serve as anchor mechanisms. These structures can be felt along the sides of an earthworm. Other setae are used in locomotion and respiration.

The typical annelid has a two-part head consisting of the **prostomium** and the **periostomium**. The prostomium, found in front of the mouth, usually is a small, lip-like extension over the dorsal portion of the mouth. The periostomium contains the mouth. A segmented body follows the periostomium, and the posterior-most portion is called the **pygidium**. New segments form during development from the posterior end. Each segment typically contains the coelom and nervous, respiratory, circulatory,

and excretory structures. With the exception of leeches, the coelom is filled with fluids, providing the animal with a **hydrostatic skeleton.** The outer layer of annelids is covered by a protective, non-chitinous **cuticle**.

Annelids have extensive muscle systems consisting of both circular and longitudinal muscles. The muscles are involved in locomotion, as well as **peristolysis** (movement of food through the digestive system). The digestive system of these animals is complete, beginning with a mouth and ending with an anus. Annelids undergo cutaneous respiration. The circulatory system consists of ventral and dorsal longitudinal vessels in all annelids except leeches. Five pairs of pumping vessels serve as muscular hearts in annelids. The blood of many annelids contains the transport protein **hemoglobin**. The excretory system consists of a pair of **nephridia**, which remove waste from each segment and discharge the waste through external pores. The nervous system is fairly well-developed, consisting of cerebral ganglia and a ventral nerve cord. The sense organs in annelids include eyes, chemoreceptors, and statocysts. Annelids may be monoecious like earthworms and leeches, or dioecious like many sand worms.

Phylum Annelida consists of three classes: **Polychaeta, Oligochaeta,** and Hirudinea. Some biologists suggest that the oligochaetes and hirudineans should comprise a new class called **Clitellata**, because both classes possess a **clitellum**, a glandular structure used during reproduction. This proposed class is still under debate.

Class Polychaeta

The polychaetes are the largest class of annelids, consisting of more than 10,000 species of mostly marine worms. These worms are commonly called "paddle worms" because of their **parapodia** (fleshy structures bearing setae) on each body segment. Polychaetes range in size from less than 1 millimeter to 3 meters. Many species of polychaetes are brightly colored. Polychaetes are dioecious animals. Gamete production occurs in individual segments

Figure 30.56 Examples of annelids: (a) a leech, (b) an earthworm, and (c) a bloodworm.

or in specialized portions of the body. Most polychaetes do not copulate, instead releasing their gametes into the water for **external fertilization.**

Polychaetes most often are placed into two sub-classes, depending upon their mode of life. Errant polychaetes, **Subclass Errantia,** crawl around stones, shells, coral, and algae. These species usually have a well-developed head with eyes and antennae. The parapodia are large and function like legs. The pharynx commonly contains teeth and powerful jaws. *Nereis* spp., the clam worm, is a typical errant polychaete. Its body can contain 200 segments and exceed 40 cm in length. Sedentary polychaetes, **Subclass Sedentaria,** includes species that rarely expose more than their head from protective tubes and burrows. Sedentary burrowers construct a vertical burrow with only one or two openings at the surface (Fig. 30.57–30.60).

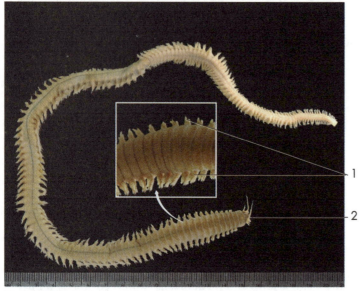

Figure 30.57 The sandworm, *Nereis* (scale in mm).
1. Parapodia
2. Mouth

Figure 30.58 The anterior end of the sandworm, *Nereis.* (a) A dorsal view and (b) a ventral view.
1. Palpi
2. Prostomium
3. Peristomial cirri
4. Peristome
5. Parapodia
6. Setae
7. Mouth
8. Everted pharynx

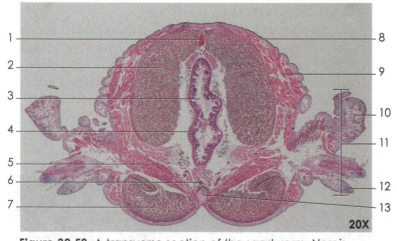

Figure 30.59 A transverse section of the sandworm, *Nereis.*
1. Dorsal blood vessel
2. Dorsal longitudinal muscle
3. Lumen of intestine
4. Intestine
5. Oblique muscle
6. Ventral blood vessel
7. Ventral longitudinal muscle
8. Integument
9. Circular muscle
10. Notopodium
11. Parapodium
12. Neuropodium
13. Ventral nerve cord

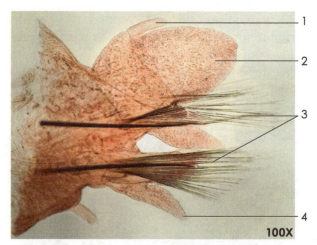

Figure 30.60 The parapodium of the sandworm, *Nereis.*
1. Dorsal cirrus
2. Notopodium
3. Setae
4. Neuropodium

Student Activity—Macroscopic Anatomy of Class Polychaeta

Materials

- dissecting microscope
- hand lens
- colored pencils
- Petri dish
- dissecting tray or jars
- probe
- preserved specimens of select polychaetes including *Nereis* spp., *Neanthes* spp., *Chaetopterus* spp., and other specimens

Procedure 30.17
Macroanatomy of a Polychaete

1. Procure the needed equipment and specimens.
2. Examine the specimens in their jars, or if the instructor permits, place them in a dissecting tray for examination.
3. Observe each specimen, paying close attention to the structures discussed above.
4. Record your observations and detailed, labeled sketches in the space provided.

It may be doubted whether there are many other animals which have played so important a part in the history of the world, as have these lowly organized creatures.

—Charles Darwin on earthworms (1809–1882)

Student Activity—Microscopic Anatomy of Class Polychaeta

Materials
- compound microscope
- colored pencils
- prepared slides of the transverse section of *Nereis* spp. and a parapodium of *Nereis* spp.

 Procedure 30.18
Microanatomy of a Polychaete

1. Procure the equipment and the prepared slides.
2. Using a compound microscope on low power, observe and record your observations, along with a labeled sketch of the prepared microscope slide, below.

CLASS OLIGOCHAETA

The best-known oligochaete is the earthworm or night crawler (*Lumbricus terrestris*). It burrows in moist, rich soil, and usually emerges at night. During dry weather, these worms can burrow several feet below the surface and become dormant. In wet weather, they stay near the surface, with their anus or mouth protruding through the burrow. When disturbed by too much water or certain stimuli, such as vibrations or chemicals, they will emerge from the burrow. Most earthworms are between 15 and 30 centimeters in length.

Earthworms are characterized by a prostomium on the anterior end and a pygostyle on the posterior end. In most earthworms, each segment contains four pairs of setae, projecting from small pores in the cuticle to the outside. Earthworms use circular and longitudinal muscles to produce peristolic movement. They feed primarily on decayed organic matter, bits of plant material, refuse, and animal matter. Food enters the mouth and, after leaving the esophagus, is stored in the crop. From the **crop**, food is passed to the **gizzard**, where it is ground into small pieces. Digestion and absorption take place in the **intestines**. Waste products are discharged through the **anus**. A pair of **metanephridia** are responsible for excretion.

Earthworms have no respiratory organs. Gaseous exchange takes place at the surface of the moist skin. The circulatory system is closed, and characterized by the presence of five pair of **aortic arches.** The nervous system consists of a **central nervous system** and **peripheral nerves.** Earthworms possess a brain and a ventral nerve cord that runs along the floor of the coelom to the last somite (Fig. 30.61–Fig. 30.62).

Earthworms are monoecious organisms, possessing both male and female organs in the same body, but they cannot mate with themselves. The gonads are restricted to the anterior portion of the body. A distinct swelling called the **clitellum** can be seen on certain segments behind the genital pores. Along with the genital setae, the clitellum secretes mucus, which holds mating earthworms together. Copulation involves the reciprocal transfer of sperm, which requires up to 3 hours. A few days after copulation, the clitellum secretes a dense mucus that eventually will form the cocoon. Eggs from the female gonopores and sperm from the seminal receptacles are collected enroute, and fertilization occurs within the cocoon. Embryonation takes place within the cocoon. The young worm hatches, appearing similar to an adult, in 2 or 3 weeks (Fig. 30.63–Fig. 30.70).

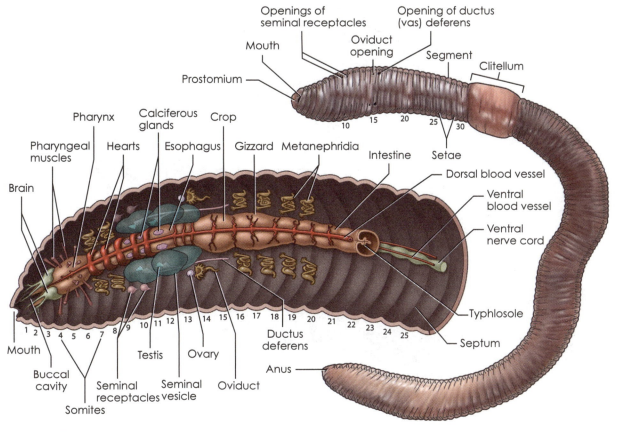

Figure 30.61 The anatomy of an earthworm, *Lumbricus*.

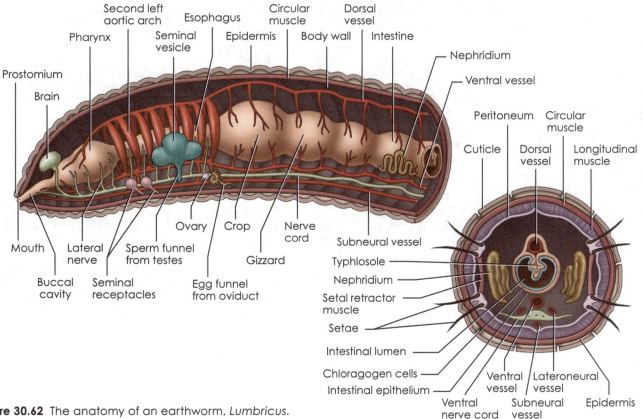

Figure 30.62 The anatomy of an earthworm, *Lumbricus*.

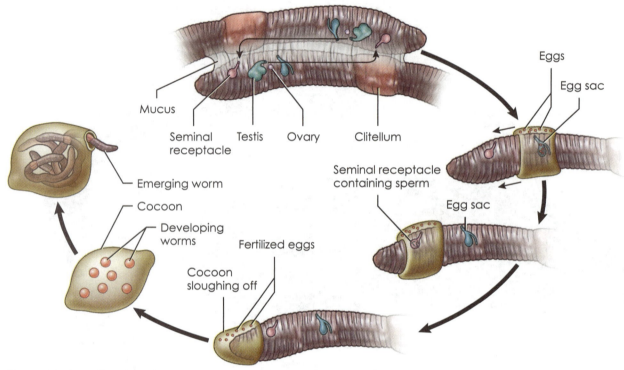

Mucus

Seminal receptacle

Testis

Ovary

Clitellum

Eggs

Egg sac

Seminal receptacle containing sperm

Egg sac

Emerging worm

Cocoon

Developing worms

Fertilized eggs

Cocoon sloughing off

Figure 30.63 A diagram of earthworm copulation and the formation of an egg cocoon.

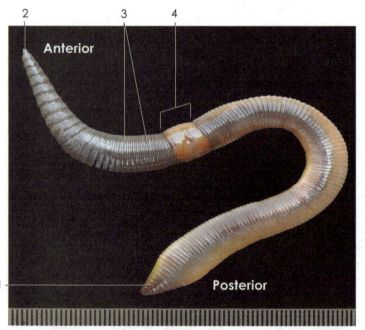

Figure 30.64 Dorsal view of an earthworm, *Lumbricus* (scale in mm).
1. Pygidium
2. Prostomium
3. Segments, or metameres
4. Clitellum

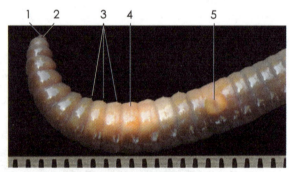

Figure 30.65 An anterior end of an earthworm, *Lumbricus* (scale in mm).
1. Prostomium   3. Setae   5. Opening of
2. Mouth   4. Segment 10   ductus (vas)
                                 deferens

Figure 30.66 Earthworm cocoons (scale in mm).

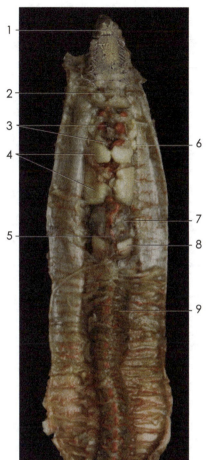

Figure 30.67 The internal anatomy of the anterior end of an earthworm, Lumbricus.

1. Brain
2. Pharynx
3. Hearts
4. Seminal vesicles
5. Dorsal blood vessel
6. Seminal receptacles
7. Crop
8. Gizzard
9. Intestine

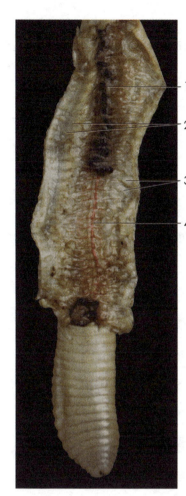

Figure 30.68 The internal anatomy of the posterior end of an earthworm with part of the intestine removed.

1. Intestine
2. Septae
3. Nephridia
4. Ventral blood vessel

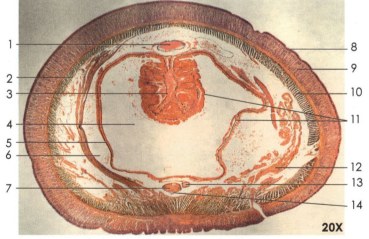

Figure 30.69 A transverse section of an earthworm posterior to the clitellum.

1. Dorsal blood vessel
2. Peritoneum
3. Typhlosole
4. Lumen of intestine
5. Intestine
6. Coelom
7. Ventral nerve cord
8. Epidermis
9. Circular muscles
10. Longitudinal muscles
11. Intestinal epithelium
12. Nephridium
13. Ventral blood vessel
14. Subneural blood vessel

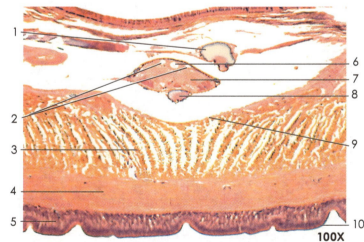

Figure 30.70 A magnified view of a transverse section of an earthworm posterior to the clitellum.

1. Ventral blood vessel
2. Lateral giant axons
3. Longitudinal muscles
4. Circular muscles
5. Epidermis
6. Medial giant axons
7. Ventral nerve cord
8. Subneural blood vessel
9. Peritoneum
10. Cuticle

Student Activity—Macroscopic Anatomy of Class Oligochaeta

Materials

- dissecting microscope
- hand lens
- colored pencils
- Petri dish
- dissecting tray
- probe
- preserved specimens of select *oligochaetes* including *Lumbricus terrestris, tubiflex worms,* and other specimens

Procedure 30.19
Macroanatomy of a Oligochaete

1. Procure equipment and specimens.
2. Examine the specimens in their jars or, if the instructor permits, place them in a dissecting tray for examination.
3. Using the dissecting microscope, observe each specimen, paying particular attention to the structures discussed above.
4. Record your observations and a detailed labeled sketch below.

Student Activity—Microscopic Anatomy of Class Oligochaeta

Materials
- compound microscope
- colored pencils
- prepared slide of a transverse section of *Lumbricus terrestris*

Procedure 30.20
Microanatomy of a Oligochaete

1. Procure the equipment and prepared slides.
2. Using a compound microscope, observe and record your observations along with a labeled sketch of the prepared microscope slide to the right.

Student Activity—Dissection of *Lumbricus terrestris*

Materials
- dissecting tray
- dissecting kit
- gloves
- safety glasses
- lab coat or apron
- hand lens
- water
- specimen of *Lumbricus terrestris,* provided by the instructor

Procedure 30.21
Earthworm Dissection

1. Procure supplies and the earthworm.
2. Wash your specimen thoroughly under running water. If preservative gets in your eyes, wash your eyes immediately and contact the instructor. Place the specimen on the dissecting tray.
3. Using a hand lens, observe the external features of your specimen. Notice that the dorsal side is more rounded and usually darker in color. Pay particular attention to the prostomium, clitellum, setae, genital pores, and anus. Record your observations and labeled sketch.

4. Lay the earthworm on the dissecting tray with its dorsal side facing up. Begin the dissection about an inch posterior to the clitellum. Carefully lift up the integument and snip an opening with a pair of sharp dissecting scissors. Insert the scissors into the opening and cut in a straight line all the way up through the mouth. Take your time and be sure to cut just the integument. If the cut is too deep, it may damage the internal organs.
5. Using forceps and dissection pins, carefully pull apart the two flaps of skin and pin them flat on the tray. The walls of some of the septa may have to be cut to pin down the earthworm properly.

6. Carefully examine the internal anatomy of the earthworm, comparing it to Figures 30.61 through 30.68. Pay particular attention to the pharynx, crop, gizzard, intestines, ventral nerve cord, reproductive structures, and aortic arches (hearts). Record your observations and a labeled sketch.

7. Clean your equipment and desktop thoroughly. Return the equipment and discard your specimen as directed.

Leeches should be kept a day before applying them. They should be squeezed to make them eject the contents of their stomachs.

—Avicenna (981–1037)

CLASS HIRUDINEA

The majority of leeches are inhabitants of freshwater habitats, with a few marine and terrestrial forms. Approximately 500 species of leeches have been described. The majority of leeches are between 2 and 6 centimeters in length, but one species of Amazonian leech, *Haementeria sp,*. can reach 30 centimeters. Leeches usually are dorsoventrally flattened, in a variety of colors. Leeches are monoecious, having an obvious clitellum only in breeding season. The head of a leech usually is reduced, and setae are absent. In addition, leeches have no internal septa.

Contrary to popular belief, not all leeches are monstrous blood suckers. Approximately 25% of leeches are predaceous, feeding upon oligochaetes, snails, and insect larvae. The other 75% of leeches are ectoparasites on a variety of invertebrates and vertebrates. Blood sucking leeches produce salivary secretions that contain an anesthetic, anticoagulants (**hirudin**), and a vasodilator. Many species of leeches can consume up to 11 times their weight in blood in a single 40-minute feeding period. After water is removed from the blood, the digestive process may take up to 6 months (Fig. 30.71–Fig. 30.73).

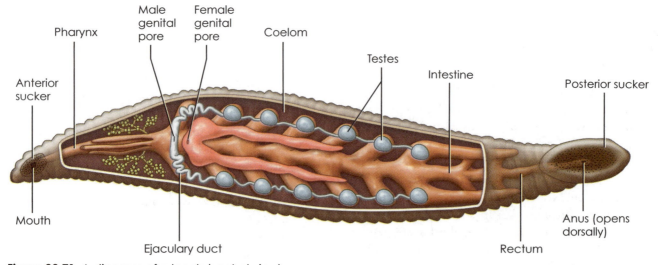

Figure 30.71 A diagram of a leech (ventral view).

Figure 30.72 A dorsal view of a leech. Leeches are more specialized than other annelids. They have lost their setae and developed suckers for attachment while sucking blood (scale in mm).

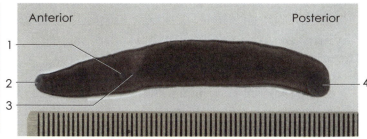

Figure 30.73 A ventral view of a leech (scale in mm).
1. Male genital pore
2. Anterior sucker
3. Female genital pore
4. Posterior sucker

 Student Activity—Macroscopic Anatomy of Class Hirudinia

Materials
- dissecting microscope
- hand lens
- colored pencils
- Petri dish
- dissecting tray
- probe
- preserved specimens of leeches

Procedure 30.22
Macroanatomy of a Leech

1. Procure the necessary equipment and specimens.
2. Examine the specimens in their jars or, if the instructor permits, place them in a dissecting tray for examination.
3. Observe each specimen, pay particular attention to the structures discussed in the figures above.
4. Record your observations and detailed labeled sketches below.

MISCELLANEOUS PHYLA

Phylum Bryozoa (Ectoprocta)

The majority of the members of **Phylum Bryozoa** (**Ectoprocta**) encrust themselves on hard surfaces such as seashells, ship hulls, and pilings. Some species, however, appear as jelly-like masses of eggs in a quiet pond or as a mossy covering on a submerged object. Bryozoans are colony builders, with each member of a colony about 0.5 mm in length. Individual members, called **zooids**, are encased in an exoskeleton called a **zoecium**, and extend their lophophores into the water to capture food. There are approximately 4,000 bryozoan species. *Pectinatella magnifica* is a common freshwater gelatinous bryozoan (Fig. 30.74–Fig. 30.75).

Figure 30.74 A freshwater bryozoan such as *Pectinatella magnifica* is often mistaken for a mass of eggs.

Figure 30.75 A preserved bryozoan exoskeleton or zoecium.

 Student Activity—Macroscopic Anatomy of Phylum Bryozoa (Ectoprocta)

Materials
- dissecting microscope or hand lens
- colored pencils
- Petri dish
- dissecting tray
- probe
- preserved specimens of select bryozoans including *Pectinatella* sp., *Plumatela* spp., a fossil bryozoan, and encrusting forms

 Procedure 30.23
Macroanatomy of a Bryozoan

1. Procure the needed equipment and specimens.
2. Examine the specimens in their jars or, if the instructor permits, place them in a dissecting tray for examination.
3. Observe each specimen.
4. Record your observations and detailed labeled sketches below and on the next page.

Student Activity—Microscopic Anatomy of Phylum Bryozoa (Ectoprocta)

Materials
- compound microscope
- colored pencils
- prepared slides of bryozoans, obtained from the instructor

Procedure 30.24
Microanatomy of a Bryozoan

1. Procure the needed equipment and prepared slides.
2. Observe and record your observations along with a labeled sketch of each prepared microscope slide below.

Student Activity—Macroscopic Anatomy of Phylum Brachiopoda

PHYLUM BRACHIOPODA

From the Paleozoic through the Mesozoic, **Phylum Brachiopoda** were abundant marine organisms. Nearly 12,000 species of fossil brachiopods have been identified. Today, only 325 species survive (Fig. 30.76–Fig. 30.77). Brachiopods are called "lamp shells" or, in ancient lore, "goose-neck barnacles." Modern brachiopods are not very different from their ancestors. *Lingula* sp. resembles its Ordovician ancestors and is considered to be a living fossil. Modern brachiopods range in length from 5 to 30 millimeters. Brachiopods attach to rocks or substrate by a **pedicile**, or they burrow in the sand or mud, feeding on protist, bacteria, and particulate matter that they draw in over their tentacle-covered **lophophore.** The lopophore also serves as a respiratory organ. At first glance, a brachiopod may resemble a bivalve mollusc, but the valves are oriented dorsally and ventrally in a brachiopod. The valves are held together by either a hinge or muscles. Two common species are *Lingula* and *Terebratella*.

Materials

- dissecting microscope
- hand lens
- colored pencils
- Petri dish
- dissecting tray
- probe
- preserved specimens of select brachiopods including a fossil brachiopod, *Lingula* spp. and others, if available from your instructor

Procedure 30.25
Macroanatomy of a Brachiopod

1. Procure the needed equipment and specimens.
2. Examine the specimens in their jars or, if the instructor permits, place them in a dissecting tray for examination.
3. Observe each specimen, paying close attention to the structures discussed earlier.
4. Record your observations and a detailed labeled sketch.

Figure 30.76 A fossil brachiopod,

Figure 30.77 (a) A fossil of the brachiopod, *Lingula* sp. and (b) a living example of a lamp shell, *Lingula* sp.

Name: _____ Date: _____ Section: _____

Review Questions

1. What are the basic characteristics of the Lophotrochozoa?

   _____

   _____

   _____

2. What are five characteristics of Phylum Platyhelminthes?

   _____

   _____

   _____

   _____

   _____

3. Where can one find turbellarians?

   _____

   _____

   _____

4. Describe reproduction in tapeworms.

   _____

   _____

   _____

5. What is the relationship between tapeworms and poorly cooked beef or pork?

   _____

   _____

   _____

6. Sketch and label the basic external anatomy of a tapeworm.

7. Discuss the diversity of trematodes.

_____

_____

_____

8. Draw the life cycle of *Clonorchis sinensis*.

9. Why were rotifers called "wheel animalcules?"

_____

_____

Name: _____ Date: _____ Section: _____

10. Name five characteristics of Phylum Mollusca.

_____

_____

_____

_____

11. List and give examples of the classes of Phylum Mollusca.

_____

_____

_____

12. Compare and contrast gastropods and bivalves.

_____

_____

_____

13. Why are cephalopods considered advanced molluscs?

_____

_____

_____

14. Describe the three layers of a clam shell.

_____

_____

_____

15. How do molluscs use their radula?

_____

_____

16. What are five characteristics of Phylum Annelida?

_____

_____

_____

_____

_____

17. Describe and give examples of the three classes of Phylum Annelida.

_____

_____

_____

18. Describe reproduction in the earthworms.

_____

_____

_____

Chapter 31
"Shedding Light" on the Ecdysozoans: Understanding Animals—Part III

Student Outcome Objectives

At the completion of this exercise, the student will be able to:

1. Describe the characteristics of ecdysozoans.
2. Describe the characteristics and provide examples of Phylum Nematoda and Phylum Nematomorpha.
3. Discuss the basic biology of *Ascaris lumbricoides* and other nematodes.
4. Describe the characteristics and classification of the panarthropods.
5. Describe the basic anatomy and trace the natural history of the tardigrades.
6. Discuss the fundamental characteristics and natural history of Phylum Arthropoda.
7. Provide specific examples of arthropods.
8. Describe the taxonomical organization of Phylum Arthropoda.
9. Discuss the fundamental characteristics and natural history of Subphylum Chelicerata.
10. Discuss the basic biology, natural history, and classification of Class Merostomata.
11. Discuss the fundamental characteristics, natural history and classification of Subphylum Myriapoda.
12. Discuss the fundamental characteristics, natural history, and classification of Subphylum Crustacea.
13. Dissect and describe the external and internal anatomy of a crayfish.
14. Discuss the fundamental characteristics, natural history, and classification of Subphylum Hexapoda.
15. Identify representatives of the major insect orders.
16. Dissect and describe the external and internal anatomy of a grasshopper.

Overview

If you are an ecdysozoan, it's a hassle to grow because you must shed your protective **cuticle**, or **exoskeleton**. If you aren't careful, you may be eaten by a predator, or perhaps end up on someone's soft-shell crab platter. Look around: spiders, cicadas, and other ecdysozoans are just about everywhere.

Ecdysis, or molting, is the basis for separation of the **Ecdysozoa** from the Lophotrochozoa. Evolutionarily, the development of ecdysis influenced the further development

of respiratory structures such as the trachea, gills, and lungs, as well as internal fertilization and metamorphosis. As a result of these incredible innovations, the ecdysozoans are a large and diverse group with a great impact upon ecology, commerce, and medicine.

The ecdysozoans are represented by eight very different phyla of protostomes: **Phylum Nematoda** (roundworms) and **Phylum Nematomorpha** (horse-hair worms) comprise a group called the **Nematoidea**. The following phyla are small and have only a few examples: **Phylum Kinrhyncha** (minute marine worms), **Phylum Priapulida** (about 16 species of cold-water marine worms), and **Phylum Loricifera** (fewer than 100 species of tiny marine animals). **Clade Panarthropoda** consists of three phyla: **Phylum Arthropoda** (insects and their relatives), **Phylum Tardigrada** (water bears), and **Phylum Onychophora** (velvet worms). The panarthropodans possess a true coelom (body cavity). The nematodes, nematomorphans, and kinorhynchs have a pseudocoelom, and the loriciferans can be a pseudocoelomate or an acoelomate depending upon the species (Fig. 31.1).

PHYLUM NEMATODA

Phylum Nematoda consists of a group of long, slender pseudocoelomate worms known as roundworms. Usually, when one thinks of roundworms, the following diabolical parasites come to mind: hookworms, pinworms, guinea worms, eye worms, heartworms, and whipworms (Fig. 31.2). It has been suggested that every animal on Earth has its own personal parasitic nematode, but just a small percentage of the 25,000 known species are "bad guys." Many species are harmless, existing in just about every environment on Earth. Rich topsoil can contain more than a billion nematodes per acre. Don't be surprised that the actual number of nematode species approaches between 500,000 and 1 million.

As a general rule, the bodies of nematodes are long, cylindrical, and tapered at both ends. Nematodes possess a thick, protective, non-living **cuticle** that grows between molts. The **epidermis and subepidermal muscle** can be found beneath the cuticle. In nematodes, the

Now as you can see, the molecular studies show the nematodes as relatives to the "panarthropods," which on anatomical grounds alone is surprising, for superficially at least they look nothing like each other.

—Colin Tudge (1943–present)

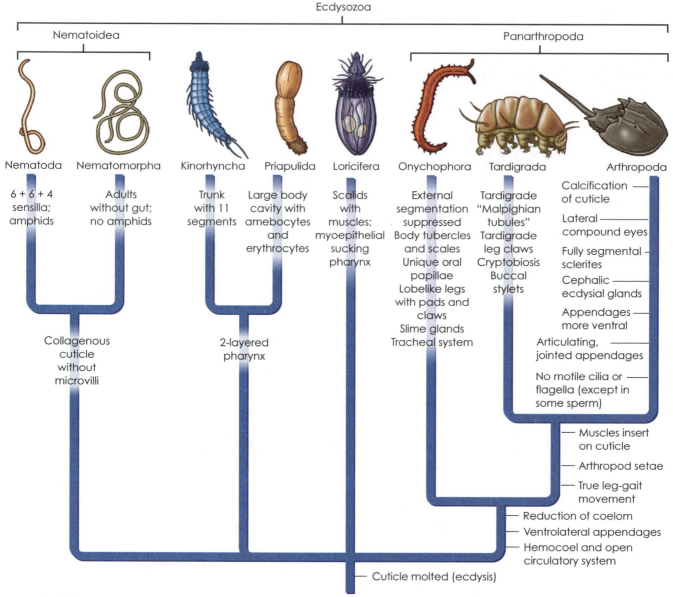

Ecdysozoa

Nematoidea

Panarthropoda

| Nematoda | Nematomorpha | Kinorhyncha | Priapulida | Loricifera | Onychophora | Tardigrada | Arthropoda |

6 + 6 + 4 sensilla; amphids

Adults without gut; no amphids

Trunk with 11 segments

Large body cavity with amebocytes and erythrocytes

Scalids with muscles; myoepithelial sucking pharynx

External segmentation suppressed
Body tubercles and scales
Unique oral papillae
Lobelike legs with pads and claws
Slime glands
Tracheal system

Tardigrade "Malpighian tubules"
Tardigrade leg claws
Cryptobiosis
Buccal stylets

Calcification of cuticle

Lateral compound eyes

Fully segmental sclerites

Cephalic ecdysial glands

Appendages more ventral

Collagenous cuticle without microvilli

2-layered pharynx

Articulating, jointed appendages

No motile cilia or flagella (except in some sperm)

Muscles insert on cuticle

Arthropod setae

True leg-gait movement

Reduction of coelom
Ventrolateral appendages
Hemocoel and open circulatory system

Cuticle molted (ecdysis)

Figure 31.1 The ecdysozoans are a highly diverse group of animals.

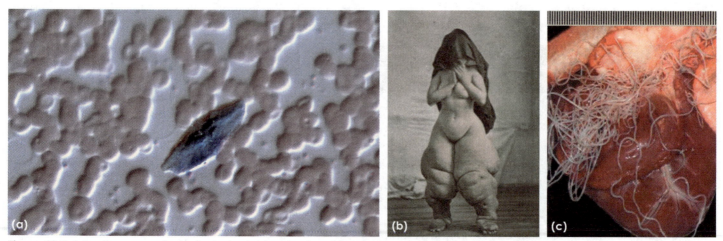

Figure 31.2 Infectious nematodes include: (a) Loa loa in a blood smear, (b) elephantitis, and (c) heartworm.

fluid-filled **pseudocoel** is well-developed and serves as a **hydroskeleton**. Nematodes move in a thrashing motion produced by a layer of longitudinal muscle. This motion allows them to move between spaces in algae, sand, and soil particles. Nematodes vary in size from microscopic to longer than 30 cm.

The mouth of a nematode opens into a **buccal cavity** with teeth or a **stylet** (spear-like structure). The buccal cavity connects to the pharynx. Nematodes have a long, straight intestine that serves as a site of digestion and absorption. Nematodes lack protonephridia but possess glands and tubules that open through a **mid-ventral pore**. The nervous system consists of a variety of **nerve rings**. Nematodes have sensory papillae, and non-parasitic species possess **amphids** (sensory organs) on each side of the head. Parasitic species have **phasmids** (sensory organs) near their posterior end. The majority of nematode species are **dioecious**, with the male being smaller than the female. Male nematodes usually possess **copulatory spicules** for internal fertilization. The fertilized eggs are stored in the **uterus** until deposition

Nematodes have developed a variety of lifestyles. Free-living nematodes feed on bacteria, algae, yeast, fungi, small invertebrates, and other nematodes. Some free-living nematodes, called **coprozoic** nematodes, feed on fecal material, and still others may be **saprobes**. Some species cause great agricultural damage by feeding on the juices of higher plants (usually roots). Perhaps the parasitic nematodes stimulate the most interest in humans, parasitic

forms are responsible for a variety of diseases in humans, as well as other animals. Nematodes are part of the food chain, eaten by insect larvae, mites, and even some species of nematode-capturing fungi. One species, *Caenorhabditis elegans*, has become a model organism in genetic and developmental research.

Ascaris lumbricoides is an excellent model of a typical nematode. Relatively common and large enough to dissect, it is one of the most common intestinal parasites in humans. Parasitologists estimate that nearly 1.27 billion people are stricken with *Ascaris* infections worldwide. Other species of *Ascaris* are found in cats, horses, pigs, and a number of other vertebrates.

A typical female *Ascaris* is prolific, producing more than 200,000 eggs daily. The eggs are eliminated with the host's feces and, given the proper soil conditions, can undergo embryonation. The eggs are capable of maintaining viability in the soil for many months to perhaps 10 years. Eggs enter the body via uncooked vegetables, soiled fingers, etc. Once the eggs enter the host, they hatch in the intestines. The juveniles burrow through the intestinal wall and into the veins and lymph vessels and are carried to the heart and lungs. Many times, their presence in the lungs initiates serious pneumonia. From the lungs, the larvae make their way up the trachea. When the larvae reach the pharynx, they are swallowed and pass to the stomach, and finally to the intestines, where the larvae mature and begin feeding on intestinal contents.

Symptoms of an *Ascaris* infection depend on the site and stage of the infection. In the intestines, *Ascaris* can cause malnutrition, blockage, and poor health. *Ascaris* has been known to block the bile duct, pancreatic duct, and appendix. Wandering worms can emerge from the throat and anus. The female is larger than the male. The male has a distinct crook in the tail. The female can attain a length of 30 cm or more (Fig. 31.3–Fig. 31.9).

Ultimate Survivor

In February 2006, the space shuttle Columbia was destroyed in a devastating accident. Aboard the shuttle were seven canisters containing the nematode *Caenorhabditis elegans*, used in zero-gravity experiments. Five of the canisters were recovered, and 7 weeks after the crash, four canisters contained living *C. elegans*. Life in space, the fiery reentry, and a 600-miles–per-hour impact had little effect on the ultimate survivors.

If all the matter in the universe except the nematodes were swept away, our world would still be dimly recognizable as a thin film of nematodes.

—N.A. Cobb (1859–1932)

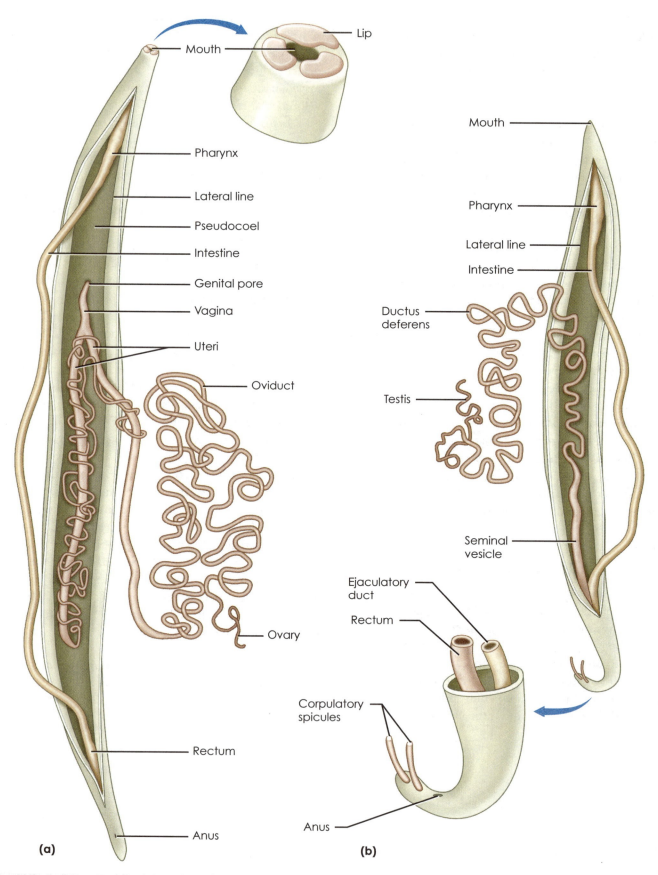

Figure 31.3 A diagram of the internal anatomy of (a) a female and (b) a male *Ascaris*.

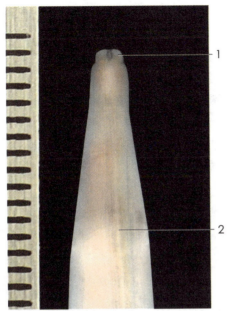

Figure 31.4 The head end of a male *Ascaris* (scale in mm).
1. Lip
2. Lateral line

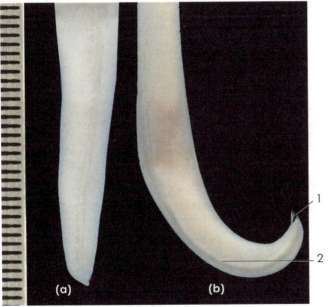

Figure 31.5 The posterior end of (a) a female and (b) a male *Ascaris* (scale in mm).

1. Copulatory spicules
2. Ejaculatory duct

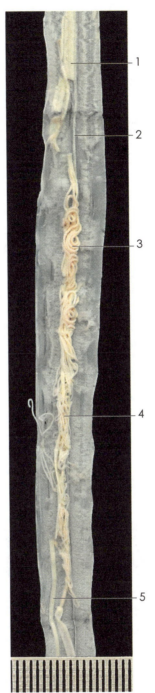

Figure 31.6 The internal anatomy of a male *Ascaris* (scale in mm).
1. Intestine
2. Lateral line
3. Ductus deferens
4. Testes
5. Seminal vesicle

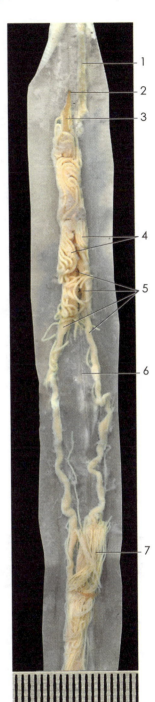

Figure 31.7 The internal anatomy of a female *Ascaris* (scale in mm).
1. Intestine
2. Genital pore
3. Vagina
4. Oviducts
5. Uteri (Y-shaped)
6. Lateral line
7. Ovary

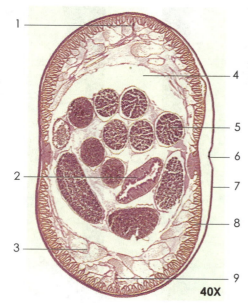

Figure 31.8 A transverse section of a male *Ascaris*.

1. Dorsal nerve cord
2. Intestine
3. Longitudinal muscle cell body
4. Pseudocoel
5. Testis
6. Lateral line
7. Cuticle
8. Contractile sheath of muscle cell
9. Ventral nerve cord

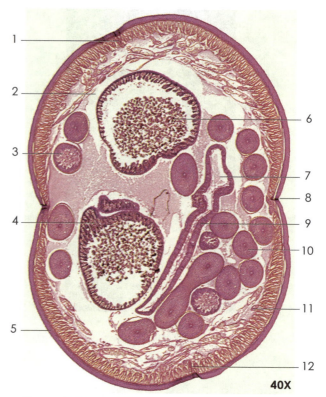

Figure 31.9 A transverse section of a female *Ascaris*.

1. Dorsal nerve cord
2. Pseudocoel
3. Oviduct
4. Uterus
5. Cuticle
6. Eggs
7. Lumen of intestine
8. Lateral line
9. Intestine
10. Ovary
11. Longitudinal muscles
12. Ventral nerve cord

Student Activity—Macroscopic Anatomy and Dissection of *Ascaris lumbricoides*

Materials

- hand lens or dissecting microscope
- colored pencils
- dissecting tray
- dissecting kit
- pins
- dissecting pins
- safety glasses
- gloves
- lab coat or apron
- water
- male and female specimen of *Ascaris lumbricoides*, obtained from your instructor

Procedure 31.1
Macroanatomy of *Ascaris*

1. Procure the needed equipment and specimens.
2. Carefully rinse a female and a male *Ascaris* and place them on a dissecting tray with water. (The water keeps the specimens from drying out.)
3. Observe female and male *Ascaris* with a hand lens or a dissecting microscope. Record your observations and sketches on the following page.

6. Using the scalpel or pin, continue your incision to the mouth of both specimens. Carefully pin the cuticle of both specimens with dissecting pins into the dissecting tray, exposing the internal structures.

7. Carefully examine the internal structures of both the female and the male *Ascaris* paying particular attention to those shown in Figure 31.6 and 31.7. Record your observations and labeled sketches below.

4. Position the specimens in the dissecting tray, ventral side down. Rotate the male to accommodate the curl of the tail.

5. Gently grasp the female *Ascaris* with your thumb and forefinger of one hand. Use a scalpel or a pin to scrape a longitudinal mid-dorsal incision in the anterior third of the body through the cuticle and longitudinal muscles of the body wall. Perform the same procedure on the male *Ascaris*.

8. Clean your equipment and desktop thoroughly. Return the equipment and discard your specimen as directed.

Student Activiy—Microscopic Anatomy of *Ascaris lumbricoides*

Materials
- compound microscope
- colored pencils
- prepared slides of transverse section of a female and a male *Ascaris*

Procedure 31.2
Microanatomy of *Ascaris*

1. Procure the needed equipment and prepared slides.
2. Using the compound microscope, observe the prepared slides.
3. Compare your slides to Figure 31.8 and 31.9.
4. Observe and record your observations, along with labeled sketches of the prepared microscope slides, below.

The painful and potentially deadly disease **trichinosis**, caused by *Trichinella spiralis*, can infect a variety of mammals, including humans. Trichinella is introduced to a host when the host eats muscle containing encysted juvenile parasites. The parasite develops in the host's intestine. Females produce living juveniles that penetrate blood vessels and eventually are carried to skeletal muscles, where they encyst. On occasion, the larvae encyst in the heart or brain. Mature females are approximately 3 mm in length, and males 1.5 mm (Fig. 31.10).

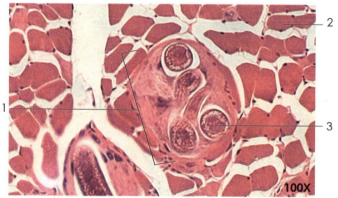

Figure 31.10 *Trichinella spirilis* can encyst in skeletal muscle.
1. Cyst   2. Muscle   3. Larva

Hookworms (*Necator americanus*) possess hook-like anterior ends. Females can attain a length of 11 mm, and males 9 mm. Hookworms infect a number of mammals including humans. Hookworm eggs pass in the feces of the infected host, and the juveniles hatch in the soil. Thus, going barefooted in an area where animals frequently defecate is not wise. The juveniles feed upon bacteria until they have the opportunity to burrow through the skin of a new host. The parasite travels to the lungs and, eventually, the intestines. The hookworm attaches to the intestinal mucosa, where it feeds on blood (Fig. 31.11).

Figure 31.11 Micrograph of a hookworm, *Necator americanus*.

Pinworms (*Enterobius vermicularis*) are the most common intestinal parasite in the United States. They have the uncanny ability to spread through an elementary school like wildfire. A female pinworm can exceed 12 mm in length. At night, the female pinworm migrates to the anal opening and lays her eggs. In the past, parents who suspected their child of having pinworms would place tape over the anus at bedtime, to capture pinworms and diagnose an infection. The eggs contaminate bedding and clothing, and the eggs can be easily spread. One of the symptoms of pinworms is scratching the anal region. The eggs can get on the fingers and, with kids, there's no telling where they can end up, be swallowed, and start the cycle again (Fig. 31.12).

Filarial worms get their name from their young, called **microfilariae**. In humans, a devastating filarial worm called *Wuchereria bancrofti* infects people in tropical countries such as Africa and South America. Females can exceed 10 mm in length and live in the lymphatic system. Females release microfilariae into the lymphatic system and blood. A mosquito bites the infected individual and is the vector for spreading the disease.

Wuchereria bancrofti is responsible for the disfiguring disease **elephantitis** (Fig. 31.13). The most common filarial disease in the United States is **heartworm**, which occurs primarily in dogs and is caused by *Dirofilaria immitis*. Occasionally, a cat, a sea lion, or a human can acquire the infection. This disease can be deadly to dogs if left untreated. Mosquitoes are the vector for heartworms.

Figure 31.12 *Micrograph of a pinworm, Enterobius vermicularis.*

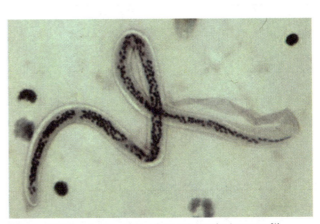

Figure 31.13 *Micrograph of Wuchereria bancrofti. Wuchereria causes elephantitis.*

 Student Activity—Microscopic Anatomy of Select Nematodes

Materials
- compound microscope
- colored pencils
- prepared slides of *Trichinella spirilis, Necator americanus, Enterobius vermicularis, Wuchereria bancrofti*, and *Dirofilaria immitis*

 **Procedure 31.3
Microanatomy of Nematodes**

1. Procure the needed equipment and prepared slides.
2. Compare your slides to Figure 31.10 to 31.13.
3. Record your observations, along with labeled sketches of the prepared microscope slides, below and on the next page.

Vinegar eels (*Turbatrix aceti*) are free-living nematodes commonly found in fermented fruit juices and unpasteurized vinegar. *Turbatrix* actively feeds upon bacteria living in these liquids. Fortunately, the vinegar in your kitchen is pasteurized. How would you like to see these little beasties thrashing about in your salad dressing? Under the microscope, the 2 mm nematodes are noted for their thrashing movements (Fig. 31.4).

We hope that, when the insects take over the world, they will remember with gratitude how we took them along on all our picnics.

—**Bill Vaughan (1915–1977)**

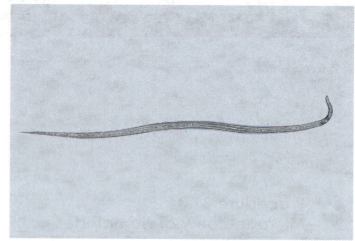

Figure 31.14 The vinegar eel, *Turbatrix aceti.*

Student Activity—Observations of Vinegar Eels, *Turbatrix aceti*

Materials
- compound microscope
- colored pencils
- microscope slides and coverslips
- dropper
- sand grains
- culture of vinegar eels, obtained from your instructor

Procedure 31.4
Vinegar Eels

1. Procure the necessary equipment, supplies, and specimens.
2. Place a drop of the culture containing vinegar eels on a microscope slide with a dropper.
3. Place several sand grains on the slide, and properly place a coverslip on top of the solution. The sand grains keep the slide above the vinegar eels and provide them with a medium to move.
4. Observe the anatomy, behavior, and movement of the vinegar eels and record your observations and sketches below.

Check Your Understanding

Q. Describe several characteristics of nematodes.

Q. Why do pinworm infections spread so easily around schools?

Q. How did vinegar eels get their name?

PHYLUM NEMATOMORPHA

An old-wives tale once warned that if a hair would fall from a horse's mane or tail, it would turn into a worm. This is an excellent example of spontaneous generation. Today, we know better and realize that the "worm" is actually a member of **Phylum Nematomorpha**. The nematomorphans are still known as horsehair worms, or Gordian worms. Superficially, the worm resembles a skinny horsehair, brownish in color and, in some cases, nearly a meter in length. Approximately 350 species of nematomorphans have been identified.

In horsehair worms the digestive system is **vestigial** (a structure that has lost its function), and digestion occurs through absorption. Adult horsehair worms are free-living and can be found in damp environments such as watering troughs around a barn and in puddles. The juvenile stage

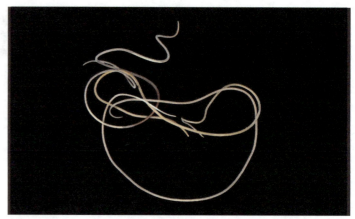

Figure 31.15 The horsehair worm, *Gordius aquaticus*.

usually encysts on a plant that is eaten by a cricket or grasshopper. Once eaten, the juvenile develops in the host and eventually emerges (Fig. 31.15).

PANARTHROPODA

The Clade Panarthropoda consists of two smaller phyla, **Onychophora** (velvet worms) and **Tardigrada** (water bears), and **Phylum Arthropoda**. These phyla possess a **hemocoel** that is lined by an extracellular matrix. Having an open circulatory system, blood enters the hemocoel and bathes the internal organs. A muscular heart and blood vessels also are present. Panarthropods possess a ventral nervous system, a segmented body, and appendages such as legs and claws (Fig. 31.6).

PHYLUM TARDIGRADA

Under the microscope, members of **Phylum Tardigrada**, commonly called water bears, are fascinating. Most of the 1,000 known species are microscopic. Tardigrades can be found living in a film of water associated with moss, lichens, detritus, or moist plants. Both marine and fresh water species exist. Tardigrades are considered **polyextremophyles**, surviving in both low and high temperatures, high radiation levels, dry conditions, and, as NASA determined, in the vacuum of space. Under extremely harsh conditions, tardigrades can enter a **cryptobiotic** state.

Figure 31.16 Examples of panarthropods include: (a) tardigrade, *Macrobiotus* sp., (b) tiger beetle, *Cicindela fulgida*, (c) shieldback katydid, *Neduba carinata*, (d) water beetle, *Lethocerus medius*, (e) stripped shore crab, *Pachygrapsus crassipes*, and (f) moon crab, *Gecarcinus quadratus*.

For all their toughness, evolution has left water bears very vulnerable to one thing—squashing! Despite being the only creatures able to survive having their photo taken by a Scanning Electron Microscope (which involves placing them in a vacuum and bombarding them with electrons), scientists have to be really careful not to crush them between the slides of a traditional optical microscope.

—**Rhiannon Buck,**
From *Tardigrades: The World's Toughest Critters* **(2009)**

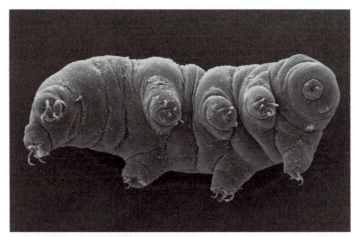

Figure 31.17 Tardigrades, *Macrobiotus* sp., are commonly called water bears.

Tardigrades have a stubby body with four pairs of short, unjointed legs terminating in four to eight claws. Tardigrades have a pair of **stylets** near their buccal tube, which allows them to pierce plant tissues or animal body walls. Several species of water bears eat entire rotifers, other tardigrades, and small invertebrates. The tardigrade brain is large, and the nerve cord is associated with four ganglia that control the legs. Tardigrades are dioecious (Fig. 31.7).

PHYLUM ARTHROPODA

Butterflies, fleas, centipedes, lobsters, barnacles, spiders, ticks, and fossil trilobites are a few examples of **Phylum Arthropoda**, the "joint-footed" animals. Presently, more than 100,0000 species have been identified, and this number will increase significantly through the years. No other phylum approaches their diversity or biomass (Fig. 31.8).

Arthropods have adapted successfully to every type of habitat and all modes of life. Living arthropods vary in size from 0.1 millimeters in length (*Stygotantulus stocki*, a

tiny crustacean) up to the Japanese spider crab *Macrocheira kaempferi*, which measures up to 4 meters in leg span. Fossil eurypterids reached lengths exceeding 2 meters, and a giant dragonfly (*Meganeura monyi*) had a wingspan of nearly a meter. A relative of centipedes and millipedes (*Arthropleura*) grew to nearly 3 meters in length during the Carboniferous. The arthropods are ecologically, medically, and commercially important. Without the arthropods, many ecosystems would literally collapse.

One of the most distinguishing characteristics of the arthropods is the presence of a **chitinous exoskeleton**. Some exoskeletons, such as those in a mosquito, are soft, and others are hard, such as those in crabs. The tough exoskeleton provides support and protection for the arthropods. Movement is possible because the exoskeleton

Figure 31.18 Example arthropods include: (a) flat rock scorpion, *Hadogenes troglodytes*, (b) American giant millipede, *Narceus americanus*, (c) fossil trilobite, *Modicia typicalis*, (d) fire shrimp, *Lysmata debelius*, and (e) leafhopper, *Homalodisca vitripennis*.

What a Buzz!

Ever wonder why a mosquito buzzes by your ear? A butterfly is capable of four wing beats per second; a housefly and a bee are capable of 100 wing beats per second; the tiny fruitfly is capable of 300 wing beats per second, and, last but not least, the mosquito and midge are capable of wing beats in excess of 1000 wing beats per second! That's why!

usually is divided into plates over the body and cylinders around the appendages. To grow, arthropods must undergo ecdysis, or molting. The interval between molts is termed **instars**.

Arthropods exhibit bilateral symmetry. Typically, they are divided into three body regions: **head, thorax,** and **abdomen**. Some species, however, possess a **cephalothorax** and abdomen, and others have a head and trunk. Arthropods also exhibit segmentation, **metamerism**, typically with each somite having a pair of jointed appendages. In arthropods, segments may fuse, forming functional groups called **tagmata**. These appendages vary throughout the phylum and perform a variety of functions such as locomotion, food gathering, copulation, gas exchange, and egg brooding. Arthropods have complex muscular systems. Typically, arthropods exhibit a great degree of **cephalization**. Also, they possess a variety of sense organs such as **compound eyes** and **antennae**. The senses of touch, smell, hearing, balance, and chemical reception are generally well developed in arthropods. Arthropods are coelomates, although the coelom is reduced to portions of the reproductive and excretory systems. The major body cavity, the **hemocoel**, consists of blood-filled spaces within the tissues.

Arthropods became adapted to a great number of dietary habits. Usually the appendages beside and just behind the mouth are used in feeding, and one pair of large appendages that flank the mouth are used in biting and tearing. The digestive system is complete, consisting of a **foregut, midgut,** and **hindgut**. Osmoregulation in arthropods occurs in **Malpighian tubes** and glands. Respiration in arthropods takes place through the gills, tracheae, book lungs, and the body surface.

Most arthropods have efficient **tracheae** (air tubes) that bring oxygen directly to the tissues and cells. The circulatory system is open; the heart consists of a pulsate vessel along the dorsal midline, and blood enters through the **ostia**. Generally, arthropods are dioecious and undergo internal fertilization. The females usually are **oviparous**

(egg-laying). Many arthropods undergo **metamorphic changes**, including a larval form that is very different from the adult.

The arthropods have been divided into the following five distinct subphyla based on current research:

1. **Subphylum Chelicerata**: horseshoe crabs, ticks, mites, sea spiders, spiders, scorpions, and extinct eurypterids
2. **Subphylum Myriapoda**: centipedes and millipedes
3. **Subphylum Trilobita**: extinct trilobites
4. **Subphylum Crustacea**: crabs, shrimp, lobsters, crayfish, barnacles, copepods, and pill bugs
5. **Subphylum Hexapoda**: insects, the largest subphylum.

Subphylum Chelicerata

Members of **Subphylum Chelicerata** are easily distinguished from other members of Phylum Arthropoda. They are characterized by six pair of appendages including four pairs of **walking legs**, a pair of **chelicerae** (appendages behind the mouth used for feeding), and a pair of **pedipalps** (appendages that aid in chewing). Chelicerates have no antennae and no mandibles. The body has two regions: a cephalothorax and an abdomen. Chelicerates include horseshoe crabs, scorpions, spiders, mites, ticks, sea spiders, and extinct eurypterids. Three classes of chelicerates have been described: Merostomata, Pycnogonida, and Arachnida.

Class Merostomata

Members of **Class Merostomata**, commonly called the horseshoe crabs, have an ancient lineage that has not changed since the Triassic Period. These animals are the only gill-bearing chelicerates. The five species that remain today are *Limulus polyphemus,* living in the Gulf of Mexico and the Atlantic Ocean.

The dorsal side of the horseshoe crab is distinguished by a heavy **carapace**, a pair of **compound eyes** composed of receptors called **ommatidia**, and a long spike **telson**. The ventral side is characterized by the presence of a pair of **chelicerae**, a pair of **pedipalps**, and four pair of **walking legs**, all except the last pair having **pincers**. **Book gills** are visible on the abdomen. Horseshoe crab blood does not contain hemoglobin. Instead, it contains a copper-rich protein, **hemocyanin**. Despite their formidable appearance, horseshoe crabs are harmless. They are omnivores and scavengers, feeding on a variety of invertebrates and algae. Their sheds are commonly found on the beach. The sexes are separate, and fertilization is external. Horseshoe crabs produce trilobite-like larvae (Fig. 31.19).

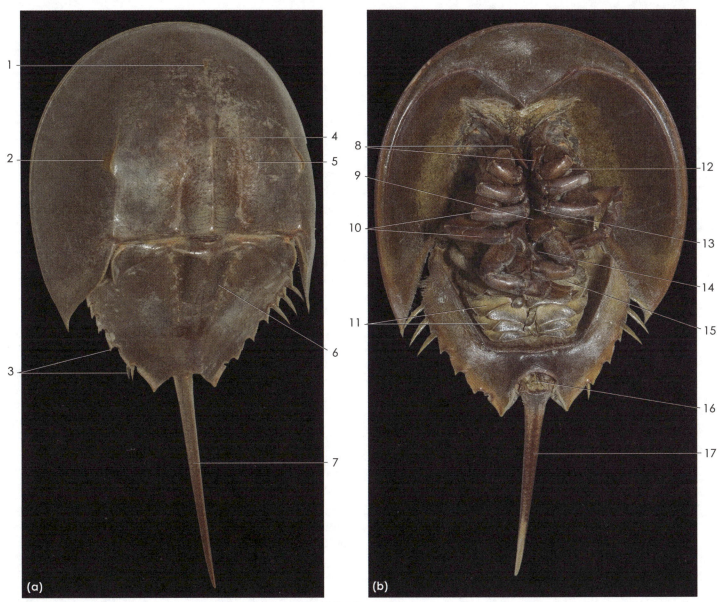

Figure 31.19 (a) A dorsal view and (b) a ventral view of the horseshoe crab, *Limulus*. This animal is commonly found in shallow waters along the Atlantic coast from Canada to Mexico.

1. Simple eye
2. Compound eye
3. Abdominal spines
4. Anterior spine
5. Cephalothorax (prosoma)
6. Abdomen (opisthosoma)
7. Telson
8. Chelicerae
9. Gnathobase
10. Chelate legs
11. Book gills
12. Pedipalp
13. Mouth
14. Chilarium
15. Genital operculum
16. Anus
17. Telson

Student Activity—External Anatomy of *Limulus polyphemus*

Materials
- hand lens
- colored pencils
- specimen of a horseshoe crab, provided by the instructor
- dissecting tray
- probe

 **Procedure 31.5
Horseshoe Crab**

1. Procure the needed equipment and specimen.
2. Place the horseshoe crab in a dissecting tray.
3. Using a hand lens, observe the horseshoe crab and record your observations and labeled sketch.

[blank box]

_____

_____

Class Arachnida

Class Arachnida consists of 80,000 species of arthropods such as spiders, scorpions, mites, ticks, and harvestmen ("daddy longlegs") (Fig. 31.21). The majority of arachnids are terrestrial. Many are predators and may possess fangs, claws, poison glands, and stingers. Generally, arachnids have **sucking mouth parts** with which they suck the fluids and soft tissues from the bodies of their prey.

Defining characteristics of arachnids are two **tagmata,** the **cephalothorax,** and **abdomen.** The cephalothorax consists of the head region and thorax. The head usually has a pair of **pedipalps**, a pair of **chelicerae**, and four pair of **walking legs**. If you look closely at the head region of a jumping spider crawling across the window sill, you will see the obvious metallic green fangs, as well as the

pedipalps, which seem to be moving constantly. The fangs actually are modified **chelicerae**. The claws, or **chelae**, of scorpions are modified pedipalps.

The vast majority of arachnids are harmless and even helpful to the environment. Still, one should be careful in handling these animals. Some species of spiders, such as the black widow and the brown recluse, are relatively common and can deliver a potentially dangerous and sometimes fatal bite. Some species of scorpions also are capable of producing a sting that is painful, dangerous, and potentially deadly to humans. Other species of arachnids, such as ticks, mites and chiggers (red bugs), can cause irritation. Many diseases of plants, animals, and humans are associated with various species of ticks and mites.

Do you have arachnophobia? Many people are terrified by arachnids, particularly spiders. Despite their fearsome and perhaps creepy appearance, the majority of spiders are harmless to humans. Presently, more than 40,000 species of spiders belong to **Order Araneae**. Spiders can be found on every continent with the exception of Antarctica, and they live in a variety of habitats. These arachnids vary in size from 0.37 mm to the Goliath birdeater spider (*Theraphosa leblondi*), with a leg span of more than 30 cm. By the way, it does not eat birds!

The body of a spider is compact, consisting of a cephalothorax joined to the abdomen by a **pedicel**. Spiders possess chelicerae, modified to form fangs complete with **venom sacs.** They also have a pair of sensory pedipalps. Spiders do not have antennae. They have eight pair of walking legs. Most of the spiders liquefy the tissue of their prey, then suck the liquid into the digestive system. Silk is produced by **silk glands**, opening into two or three pair of **spinnerets**. Some spiders weave complex and intricate webs. One interesting spider, *Mastophora dizzydeani* (named after the hall of fame pitcher Dizzy Dean), throws a sticky ball-like web on the end of a thread to capture prey. Book lungs, tracheae, or both structures are used in breathing. The heart of a spider is a long, slender tube in the abdomen, which pumps the blood throughout the body. The blood returns to the heart through **ostia** from the hemocoel. In the excretory system, **Malpigian tubules** extract nitrogenous waste and uric acid from the hemocoel and release it into the **cloaca**. The nervous system is well-developed. The majority of spiders have eight simple eyes, located on the top anterior region of the cephalothorax. Spiders also have **slit sensillae** in the joints of their limbs, which detect vibrations, especially when prey become entangled in a web.

Spiders are noted for their elaborate courtship behavior. Fertilized eggs develop in a cocoon that can be hidden or carried around by the female. In several species, the female devours the male shortly after mating. As a defense, tarantulas can release **urticating hairs**, which are irritating, and deliver a painful bite (Fig. 31.20–Fig. 31.21)).

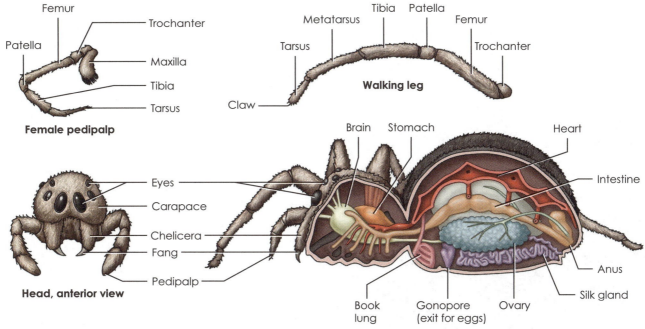

Figure 31.20 A diagram displaying the anatomy of a spider.

Figure 31.21 Example arachnids: (a) jumping spider, *Phidippus regius,* (b) green lynx spider, *Peucetia viridans,* (c) brown recluse, *Loxosceles apache,* (d) black widow, *Latrodectus hesperus,* (e) orb weaver, *Argiope trifasciata,* (f) cobalt blue tarantula, *Haplopelma lividum,* (g) solpugid, *Eremobates pallipes,* and (h) wolf spider, *Lycosa* sp.

Order Scorpiones

Approximately 2,000 species of scorpions have been classified and placed in **Order Scorpiones**. Most scorpions are found in tropical and subtropical environments. The largest living species measures more than 22 cm in length. Many species can deliver a painful sting, and approximately 40 species of scorpions have a venomous sting that is potent enough to kill a human. Interestingly, some scorpions have fluorescent chemicals in their cuticle, which fluoresce when exposed to black light. The majority of scorpions are secretive and nocturnal, living as ambush hunters and stalkers, and consume insects (Fig. 31.22).

The body of a scorpion is composed of a short **cephalothorax**, a **preabdomen** called a **mesosoma** with seven segments, and a slender **postabdomen**, or **metasoma**, with five segments, and a **stinger**. The bulb-like **telson** holds the venom gland. Scorpions have four pairs of eight-jointed walking legs. The pedipalps are large and pincer-like. Scorpions have a large pair of median eyes and up to five pairs of lateral eyes but do not have good vision. **Pectines** are comb-like tactile organs on the ventral side of the scorpion. The female scorpion is usually larger and more robust than the male. Scorpions give live birth to their young, termed **viviparous**, and the brood is carried on the back of the mother scorpion until they have undergone their first molt (Fig. 31.23).

Figure 31.22 Examples of scorpions: (a) three-lined, *Hottentotta trilineatus*, (b) bark, *Centruroides hentzi*, (c) tri-colored, *Opistophthalmus ecristatus*, and (d) emperor, *Pandinus imperator*.

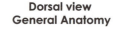

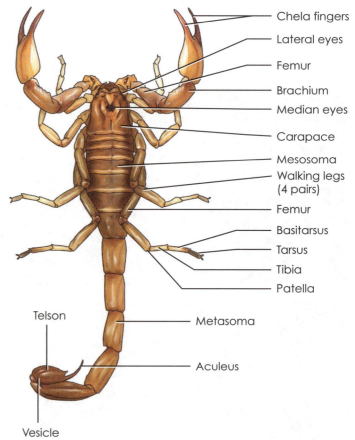

Figure 31.23 External anatomy of a scorpion.

Because we focused on the snake, we missed the scorpion.

—**Egyptian Proverb**

Order Acari

Order Acari consists of ticks, mites, and chiggers (red bugs). This order is large, with 40,000 described species and perhaps nearly a million total species. The mites are especially hard to find because many are tiny and others live in habitats that have not been studied thoroughly. Members of Order Acari can be found in a variety of habitats including freshwater, marine, and many terrestrial habitats. Many species are free-living, but several diabolical parasitic species exist (Fig. 31.24).

In acarines, the cephalothorax and abdomen are completely fused. The mouth parts exist on an anterior projection called the **capitulum**. The **chelicerae** on each side of the mouth are used for holding, ripping, or piercing food. Adult acarines generally possess four pair of legs.

This order is medically significant because many of its members are responsible for a host of diseases or serve as a vector for other diseases. Several species of minute mites known as house mites or dust mites including *Dermatophagoides farinae* share our dwellings and cause allergies in many people. These tiny mites consume dead skin cells that are a major component of house dust. Larval chiggers cause irritating dermatitis. The next time you fall victim to a chigger (*Trombicula* spp.), examine it with a dissecting microscope or a hand lens—they are strange looking creatures! In humans, *Sarcoptes scabiei* can cause severe itching as they burrow beneath the skin, and they cause the dreaded disorder scabies. In dogs, *Demodex canis* is responsible for demodectic mange. Ticks can be an ectoparasite on a number of animals including humans. Not only are ticks irritating, but they can be a vector for several diseases including Lyme disease and Rocky Mountain spotted fever.

Harvestmen

Surprisingly, the **harvestmen**, or "daddy longlegs," which may reside in a clump under the eaves of a house, are not spiders. Harvestmen belong to **Order Opiliones**, which contains approximately 5,500 species. The majority of nocturnal harvestmen are dull-colored, and the diurnal species are brightly colored. Most harvestmen have exceptionally long walking legs and a body consisting of a joined cephalothorax and abdomen. The feeding mechanism allows harvestmen to devour pieces of food (fungi, insects) instead of liquids. Harvestmen have a single pair of eyes in the middle of their head, oriented laterally (Fig. 31.25).

Myth Buster!

You've undoubtedly heard the urban legend claiming that daddy longlegs are the most venomous animal known to humans. But its fangs are too short and its mouth is too small to bite a human. None of the known species of harvestmen has venom glands. They're harmless!

Figure 31.25 Harvestmen, *Phalangium opilio*, commonly called daddy long legs, are not really spiders.

Figure 31.24 Examples of mites and ticks include: (a) a living red mite, *Dermanyssus gallinae*, on a lizard, (b) a micrograph of a mite, (c) living ticks on a lizard, and (d) a close-up of a wood tick, *Dermacentor* sp.

Student Activity—Macroscopic Anatomy of Harvestmen

Materials
- hand lens
- colored pencils
- dissecting tray
- probe
- specimen of harvestmen

Procedure 31.6
Macroanatomy of Harvestmen

Consider photographing this activity.

1. Procure the needed equipment and specimen.
2. Examine your specimen in a jar, or place it on a dissecting tray for study.
3. Using a hand lens, record your observations and labeled sketch below.

_____

_____

_____

Subphylum Myriapoda

Subphylum Myriapoda contains approximately 13,000 species of centipedes, millipedes, and their relatives. All members of this subphylum are terrestrial. The myriapods possess one pair of antennae and a pair of simple eyes. Similar to that of insects, myriapods breathe through their spiracles, connected to the tracheal system. Malpigian tubules are the sites of excretion in the myriapods. The brain is poorly developed in this subphylum. Females lay eggs that hatch into young resembling the adults but with fewer segments and legs. The four classes of myriapods are:

1. **Chilopoda**: centipedes
2. **Diplopoda**: millipedes
3. **Symphyla**: a soft-bodied myriapode
4. **Pauropoda**: a soft-bodied myriapode

Class Chilopoda

Perhaps as a child you heard of a "hundred-legger." This curious creature actually is a centipede, a member of **Class Chilopoda**. Of course, the majority of centipedes do not have a hundred legs, and most species have only 30 pair of legs. One species, however, does have 382 legs. Centipedes feature one pair of legs per segment. These flattened, elongate arthropods have a chitinous exoskeleton and range in length from 4 mm to 30 cm. The majority of centipedes are dull brown and can be found in the underbrush, or perhaps visiting your house! Approximately 3,000 species of centipedes have been described (Fig. 31.26).

Centipedes use their venom claws, which actually are the first pair of legs, to capture and kill their prey. Centipedes are dioecious. Despite the poisonous bite of some species such as the poisonous black centipede of North America, centipedes as a whole are largely beneficial in gardens.

Figure 31.26 Examples of centipedes: (a) giant Sonoran, *Scolopendra heros*, (b) Florida blue, *Hemiscolopendra marginata*, and (c) Vietnamese centipede, *Scolopendra subspinipes*.

Student Activity—Macroscopic Anatomy of Select Members of Class Chilopoda

Materials
- hand lens or dissecting microscope
- colored pencils
- Petri dish
- probe
- specimens of several species of centipedes

Procedure 31.7
Macroanatomy of Centipedes

Consider photographing this activity.

1. Procure the needed equipment and specimens.
2. Observe your specimen in a jar, or place it on a Petri dish. Thoroughly examine your specimens, using a hand lens or dissecting microscope, pay particular attention to the mouthparts and legs.
3. Record your observations and labeled sketches below and to the right.

Class Diplopoda

Millipedes, a member of **Class Diplopoda**, classically have been called "thousand-leggers" because, as they busily scoot across the ground, they may look like they have a thousand legs! Approximately 10,000 species have been identified. Usually, millipedes hide under logs and stones. Millipedes are herbivorous or **detritovores,** compared to the carnivorous centipede. When disturbed, millipedes may roll into a ball and emit a foul-smelling odor from their **repugnant glands**.

The millipede body is more dome-shaped than the centipede body and consists of 25 to more than 100 segments. One species, *Illacme plenipes*, can have 750 legs. Millipedes range in size from 2 mm to 28 cm. They have two pair of legs per segment, with the exception of some anterior and posterior segments. Millipedes reproduce sexually, although several parthenogenic species have been described. Initially, larval forms may be mistaken for centipedes because they have one pair of legs per segment (Fig. 31.27).

Figure 31.27 Examples of millipedes: (a) American giant millipede, *Narceus americanus*, (b) Sonoran desert, *Orthoporus ornatus*, and (c) African giant millipede, *Archispirostreptus gigas*.

Student Activity—Macroscopic Anatomy of Select Members of Class Diplopoda

Materials
- hand lens or dissecting microscope
- colored pencils
- Petri dish
- probe
- specimens of several species of millipedes

Procedure 31.8
Macroanatomy of Millipedes

Consider photographing this activity.

1. Procure the needed equipment and specimens.
2. Observe your specimens in a jar, or place them on a Petri dish. Thoroughly examine your specimens, using a hand lens or dissecting microscope.
3. Record your observations and labeled sketches.

Subphylum Trilobita

Like the dinosaurs, the trilobites, **Suborder Trilobita**, are icons of geologic time. Trilobite-like organisms are perhaps the ancestors of all living arthropods. Trilobites first appeared in the Cambrian Period approximately 540 million years ago, and eventually went extinct during the Permian Period about 280 million years ago. During this time, trilobites were the most common arthropods roaming the ancient seas, serving mostly as bottom-dwelling scavengers and detritivores. Presently, the fossil record has yielded more than 5,000 known species, with many more yet to be named. Some fossil trilobites are actually sheds, from ecdysis. Trilobites ranged in size from less than a centimeter to more than 70 cm (Fig. 31.28).

Trilobites at first showed little variation between the segments. In later arthropods, the segments tended to fuse and specialize. The term "trilobite" is derived from the pair of **longitudinal grooves** on the dorsal side, forming the **left and right pleural lobes** and the **axial lobe**. The body consisted of three **tagmata**: the **cephalon** (head region), the **thorax** (main body), and the **pygidium** (posterior fused segments). The appendages were **biramous** (having two distinct branches). Trilobites possessed a **hypostome** (hardened structure near the mouth used in feeding) instead of true mouth parts. Gills served as respiratory structures. Many trilobites could roll up similar to pill bugs today (Fig. 31.29).

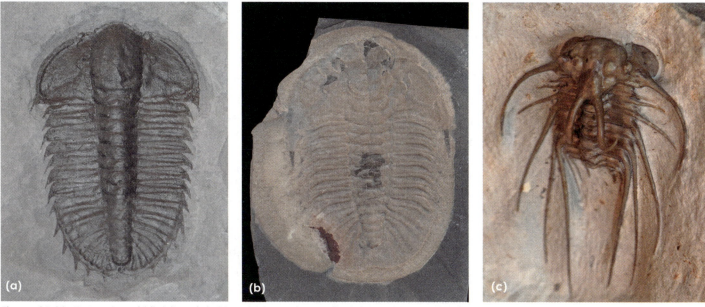

Figure 31.28 Examples of trilobites: (a) *Olenoides marjumensis*, (b) *Hemirhodon amplipyge*, and (c) *Dicranurus elegans*.

Check Your Understanding

Q. Compare and contrast centipedes and millipedes.

_____

_____

_____

Q. What is a harvestman?

_____

_____

_____

Q. Compare and contrast a spider and a scorpion.

_____

_____

_____

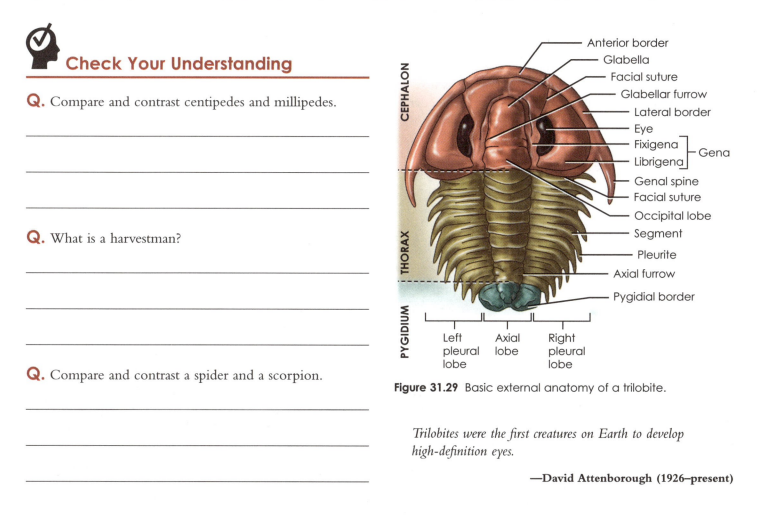

Figure 31.29 Basic external anatomy of a trilobite.

Trilobites were the first creatures on Earth to develop high-definition eyes.

—**David Attenborough (1926–present)**

Subphylum Crustacea

Fried shrimp, boiled crawfish, crab gumbo, steamed lobster—let's hear it for the crustaceans! The vast majority of the 68,000 members of **Subphylum Crustacea** are marine, but there are several freshwater and terrestrial forms. Example crustaceans are shrimp, lobsters, hermit crabs, water fleas, barnacles, copepods, and pill bugs (Fig. 31.30). The majority of crustaceans are free-living, although some parasitic species exist. Crustaceans possess a hard chitinous and calcified **cuticle** that must be shed (**ecdysis**) for the animal to grow. Most crustaceans have two **tagmata**—the **cephalothorax** and the **abdomen**. The number of body segments ranges from 6 to 32 segments, each with a pair of appendages. The more advanced forms are thought to have fewer segments.

Many crustaceans have a hardened **carapace** that covers the cephalothorax and may extend past the head in the form of a **rostrum**. Crustaceans are different from the other arthropods because they have two pairs of anterior antennae. The first pair is called the **antennules** and the second pair the **antennae**. The antennules are shorter and usually are associated with the sense of smell. Antennae are longer and are also sensory structures. Crustaceans have modified mouth parts, such as **mandibles, maxillae,** and **maxillipeds**, with important feeding and sensory functions. In some species, such as crabs, crayfish, and lobsters, the first pair of walking legs, **chelipeds**, are modified to form powerful claws. Other walking legs are **biramous** (having two branches). In addition to walking legs, some crustaceans have feathery **swimmerets**, or **pleopods**, in the abdominal region. The tail of a crustacean consists of a nonsegmented **telson** flanked by fan-like **uropods**.

Crustaceans have an open circulatory system. If present, the heart resides in a blood-filled sinus and communicates by paired **ostia** (valves) to the body. In many crustaceans, food is broken down physically in the large **cardiac stomach** and also in the smaller **pyloric stomach** as part of the **foregut**. In the majority of crustaceans, excretion of nitrogenous wastes occurs by diffusion across the gills. **Green glands**, located near the base of the antennae, are excretory organs ridding the organism of water in some species. Although several monoecious and parthenogenic species exist, the majority of crustaceans are dioecious. Fertilized eggs may be released into water or carried by the female crustacean. Larval crustaceans are variable and may include the **nauplius**, **zoea**, and **megalops** stages.

Biologists recognize six classes of crustaceans. The four large classes are:

1. **Class Branchiopoda**: brine shrimp, tadpole shrimp, and fairy shrimp
2. **Class Ostracoda**: ostracods and seed shrimp
3. **Class Maxillopoda**: barnacles, tongue worms, fish lice, and copepods
4. **Class Malacostraca**: lobsters, crabs, shrimp, isopods, and pill bugs

Classes **Remipedia** and **Cephalocardia** include rather obscure crustaceans.

Figure 31.30 Example crustaceans: (a) cyclops copepod, *Abyssorum tatricus*, (b) pill bug, *Armadillidium* sp., (c) hermit crab, *Coenobita clypeatus*, (d) gooseneck barnacles, *Pollicipes polymerus*, and (e) lobster, *Panulirus interruptus*.

Class Branchiopoda

About 10,000 species of **Class Branchiopoda** have been identified, of which fairy shrimp and brine shrimp are common members. These shrimp do not have a carapace. Brine shrimp (*Artemia salina*) are sold as novelty items, packaged as sea monkeys. Brine shrimp eggs are metabolically inactive and in a state of cryptobiosis when they are purchased. After being placed in salt water, the brine shrimp hatch within a few hours, forming larvae called **nauplii**. The larvae soon mature and develop into those frolicking sea monkeys. The best-known branchiopodan is the tiny water flea *Daphnia* spp., commonly found in ponds and other freshwater environments (Fig. 31.31–Fig. 31.32).

Every species has come into existence coincident both in time and space with a pre-existing closely related species.

—**Alfred Russel Wallace (1823–1913)**

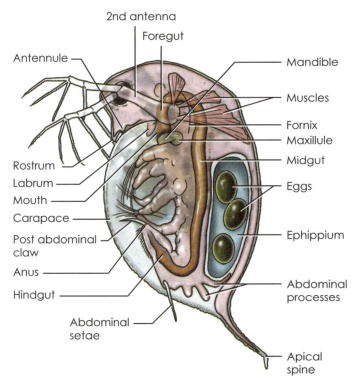

Figure 31.31 Internal anatomy of the water flea, *Daphnia* sp.

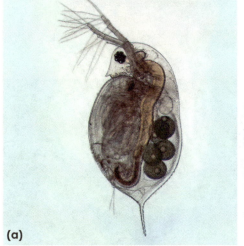

Figure 31.32 (a) A water flea, *Daphnia* sp., (b) brine shrimp, *Artemia salina*, and (c) tadpole shrimp, *Triops longicaudatus*.

Class Maxillopoda

Approximately 12,000 species of **Class Maxillopoda** have been described. This diverse class includes the subclasses copepods and barnacles. As a general rule, its members have five cephalic segments, six thoracic segments, four abdominal segments, and a telson. The majority of maxillopodans lack appendages. Some large barnacles (*Balanus nubilus*) can exceed 7.5 cm in length.

Copepods (**Subclass Copepoda**) are some of the most common animals on Earth, living in both freshwater and marine environments. Perhaps they comprise the largest animal biomass on Earth. Most copepods are less than a millimeter in length and have an elongated body that tapers toward the posterior. Early scientists referred to some copepods as a cyclops because they have a single compound median eye in the adult form. In addition, copepods lack a carapace. In copepods, the antennules may be longer than the appendages. Copepods are a major component of zooplankton, important in the food chain. Several species of copepods are parasites on invertebtrates, fishes, sharks, and marine mammals (Fig. 31.33–Fig. 31.34).

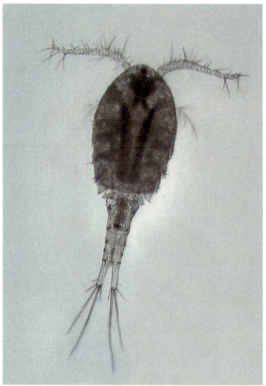

Figure 31.33 A cyclops copepod, *Abyssorum tatricus.*

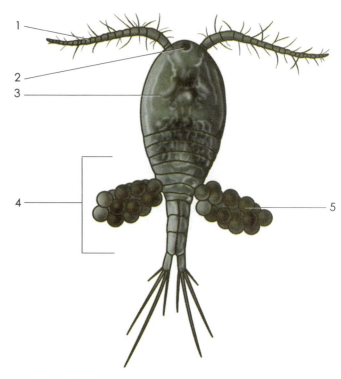

Figure 31.34 Generalized copepod anatomy.
1. Antennule   4. Abdomen
2. Compound eye   5. Egg sac
3. Cephalosome

Barnacles

Charles Darwin devoted much of his life to studying barnacles, **Subclass Cirripedia**. Biologists have described approximately 1,200 species of marine barnacles (Fig. 31.35). Barnacles have two larval stages. The **nauplius larvae** have a single median eye and are free-swimming. The **cyprid larvae** attach to a suitable substrate head-first and develop into juvenile barnacles.

As adults, barnacles are **sessile**, and some species may be attached to a substrate by a stalk (gooseneck barnacle) or cemented to a substrate (bay barnacle). The majority of barnacles are enclosed in calcareous plates. Barnacles have no abdomen and a reduced head. When feeding, the top plates open and feathery **cirri** filter food from the water. Some barnacles are parasitic on crabs. If barnacles grow in great numbers on a ship's hull, they can slow the ship, and these **fouling organisms** have to be removed periodically.

Figure 31.35 Gooseneck barnacles, *Pollicipes polymerus.*

No naturalist has devoted more painstaking attention to the structure of the barnacles than Mr. Darwin.

—**Richard Owen** (1804–1892)

Check Your Understanding

Q. What is the function of the green gland in crustaceans?

_____

_____

_____

Q. Describe five common crustaceans.

_____

_____

_____

_____

_____

Q. Where do brine shrimp live?

_____

_____

_____

Class Malacostraca

More than 22,000 members of **Class Malacostraca** have been described. They are found worldwide, inhabiting terrestrial, freshwater, and marine environments. The malacostracans usually possess three tagmata: a **head** with five segments, a **thorax** with approximately eight segments, and an **abdomen** with usually six segments. Members of Class Malacostraca are diverse, and their classification is lengthy and presently undergoing reorganization. For convenience, the two best known orders will be discussed here.

Order Isopoda includes terrestrial, freshwater, and marine species. Isopods typically are flattened dorsoventrally and have seven pair of legs. They range in size from 0.03 cm to more than 50 cm in length. Eyes, when present, are attached to the head. Abdominal **pleiopods** bear gill-like or lung-like structures called **pseudotrachae**. One of the best-known isopods is the pill bug, or sow bug (*Armadillium vulgare*), a common backyard resident living under stones, bricks, and leaves. When touched, it rolls up into a ball.

Another common isopod is *Ligia* spp., sometimes called a wharf roach. It can be seen scurrying along rocky shores and manmade structures. Several species of isopods are ectoparasites of the gill region of marine fishes. They resemble terrestrial pill bugs and possess massive claws on the tips of each appendage that serve as holdfast structures (Fig. 31.36).

Figure 31.36 Example isopods, (a) Pill bug, *Armadillidium* sp., and (b) sea slater, *Ligia italica*.

Order Decapoda

Order Decapoda contains the most familiar crustaceans, including true shrimp, prawn, hermit crabs, fiddler crabs, snow crabs, blue crabs, lobsters, and crayfish. Approximately 18,000 species of decapods have been described. All decapods have 10 legs. The anterior three pairs of legs function as mouthparts and are called **maxillipeds.** In many species, one pair of legs has enlarged pincers called **chelae,** or claws. More appendages exist on the abdomen, each segment with a pair of **biramous pleopods.** The last pair, which forms part of the tail fan along with the **telson,** comprises the **uropods** (Fig. 31.37).

The crayfish is commonly used as a representative crustacean in biology laboratories. Observe the drawings and photographs in Figures 31.38–31.46. These will be helpful in your dissection of the crayfish.

Figure 31.37 Example crustaceans: (a) peppermint shrimp, *Lysmata wurdemanni,* (b) fiddler crab, *Uca* sp., (c) hermit crab, *Coenobita clypeatus,* (d) ghost crab, *Ocypode ceratophthalmus,* (e) blue crab, *Callinectes sapidus,* and (f) red reef lobster, *Enoplometopus* sp.

Figure 31.38 A lateral view of the crayfish.

1. Carapace
2. Abdomen
3. Uropod
4. Swimmeret (pleopod)
5. Rostrum
6. Compound eye
7. Maxilliped
8. Cheliped
9. Walking legs

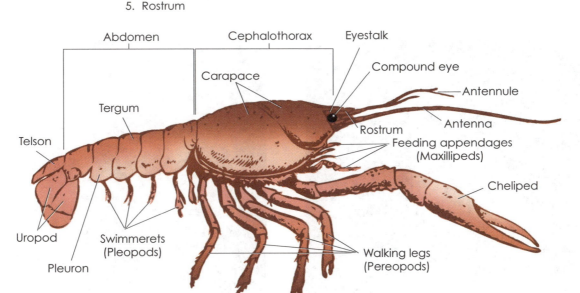

Figure 31.39 A diagram of the crayfish, *Cambarus*.

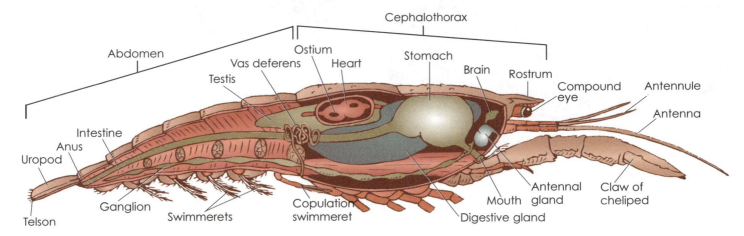

Figure 31.40 The anatomy of a crayfish. A sagittal section of an adult male.

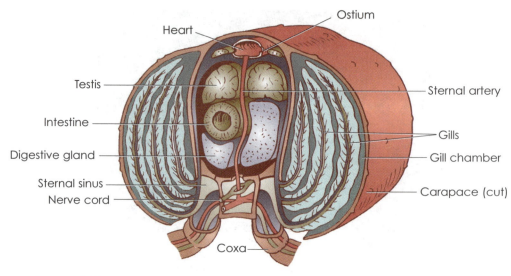

Figure 31.41 The anatomy of a crayfish. A transverse section of an adult male.

Figure 31.42 A ventral view of the oral region of the crayfish.

1. Third maxilliped
2. Second maxilla
3. First maxilliped
4. Green gland duct
5. Mandible

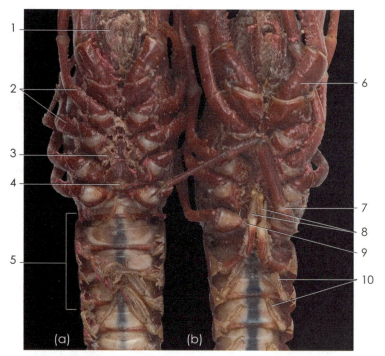

Figure 31.43 A ventral views of (a) a female and (b) a male crayfish. The first pair of swimmerets are greatly enlarged in the male for the depositing of sperm in the female's seminal receptacle.

1. Third maxilliped
2. Walking legs
3. Disk covering oviduct
4. Seminal receptacle
5. Abdomen
6. Base of cheliped
7. Base of last walking leg
8. Copulatory swimmerets (pleopods)
9. Sperm ducts (genital pores)
10. Swimmerets (pleopods)

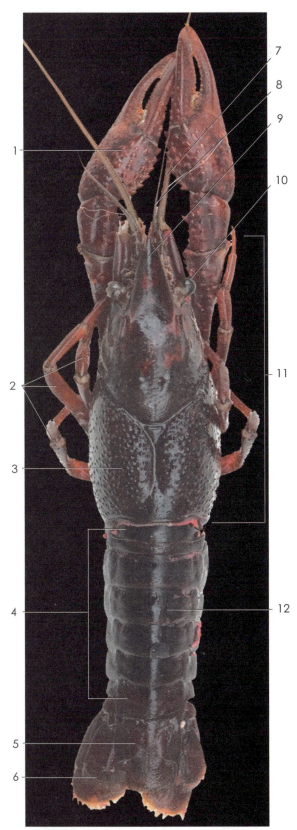

Figure 31.44 A dorsal view of the crayfish.

1. Cheliped
2. Walking legs
3. Carapace
4. Abdomen
5. Telson
6. Uropod
7. Antenna
8. Antennule
9. Rostrum
10. Compound eye
11. Cephalothorax
12. Tergum

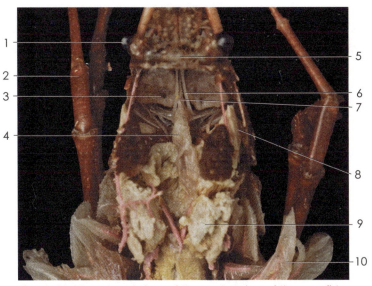

Figure 31.45 A dorsal view of the oral region of the crayfish.

1. Compound eye
2. Walking leg
3. Green gland
4. Cardiac chamber of stomach
5. Brain
6. Circumesophageal connection (of ventral nerve cord)
7. Esophagus
8. Region of gastric mill
9. Digestive gland
10. Gill

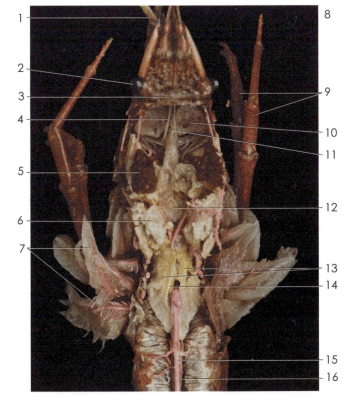

Figure 31.46 A dorsal view of the anatomy of a crayfish.

1. Antenna
2. Compound eye
3. Brain
4. Circumesophageal connection (of ventral nerve cord)
5. Mandibular muscle
6. Digestive gland
7. Gills
8. Antennules
9. Walking legs
10. Green gland
11. Esophagus
12. Pyloric stomach
13. Testis
14. Ductus deferens
15. Aorta
16. Intestine

Student Activity—Macroscopic Anatomy of Select Decapods

Materials
- dissecting microscope
- hand lens
- colored pencils
- dissecting tray
- preserved specimens of shrimp, prawn, hermit crabs, blue crabs, fiddler crabs, lobsters, and crayfish

Procedure 31.9
Macroanatomy of Decapods

Consider photographing this activity.

1. Procure the needed equipment and specimens.
2. Examine the preserved specimens of decapods in their jars and, if possible, place them on a dissecting tray for closer examination.
3. After examining them closely with a dissecting microscope and hand lens, place your observations and sketches below.

Student Activity—Crayfish Dissection

Materials
- dissecting tray
- dissecting kit
- hand lens
- safety glasses
- gloves
- lab coat or apron
- scalpel
- forceps
- water
- preserved crayfish

Procedure 31.10
Crayfish Dissection

Consider photographing this activity.

1. Procure the specimen and the needed equipment.
2. Rinse the preservative off of the crayfish with running water. Place the crayfish on its ventral side in the dissecting tray.
3. Locate the external features of the crayfish shown in Figures 31.40 through 31.46. Record your observations and detailed sketch.

4. With your hand lens, closely examine the mouthparts and appendages. Include your descriptions and detailed sketch.

5. Determine the sex of the crayfish. Compare it to the other sex, which may be found at another lab station. Compare your male and female crayfish to Figure 31.43. Record your observations and sketch below.

6. With a scalpel and forceps, carefully remove the carapace from the rostrum to the abdomen. Also remove the dorsal portion of three abdominal segments. Using Figure 31.46, find the major internal structures of the crayfish. Place your observations and sketch below.

7. Detach one of the walking legs of the crayfish, and notice the small gill attached to the leg.
8. Dispose of all crayfish body parts in the waste container provided by the instructor.
9. Clean the equipment and station thoroughly.
10. Return the clean and dry equipment to the proper location.

Subphylum Hexapoda

Nearly one-third of all animals on the Earth are members of **Subphylum Hexapoda**, which has two classes:

1. Class Entognatha, which includes Order Collembolla (springtails), and
2. Class Insecta (ants, moths, fleas, etc.).
3. Hexapods have six pair of uniramous legs. They also have three tagmata: a head, a thorax, and an abdomen.

Class Insecta

The study of insects is called **entomology**. **Class Insecta** is the most diverse and abundant of all of the groups of arthropods. This class also represents the largest group of animals on Earth, with more than a million recognized species, and is estimated to increase many-fold in the future (Fig. 31.47). Beetle species alone number more than 350,000. The insects have undergone great adaptive radiation and have occupied virtually every type of habitat. Basically, insects differ from other arthropods by having three pairs of legs and, in the majority of species, two pair of wings in the thoracic region. Insects vary in body shape, color, and size. The smallest insect is shorter than a millimeter, and the largest measures more than 20 centimeters (Fig. 31.48).

The insect **tagmata** consists of the head, thorax, and abdomen. The cuticle of each body segment is made of four plates, **sclerites**, connected to each other by a flexible hinge joint. The head of insects usually has a pair of large **compound eyes**, three **ocelli**, and two **antennae**. The mouth-parts are highly variable in insects. The thorax usually is divided into the **prothorax, mesothorax**, and **metathorax**, each having a pair of legs, which vary among insects. In the majority of insects, the mesothorax and metathorax have a pair of wings. The abdomen has 9 to 11 segments. The end of the abdomen bears the external sex organs (Fig. 31.49 and Fig. 31.53).

The digestive system consists of a **foregut** (mouth, salivary gland, esophagus, crop, and gizzard), a **midgut** (stomach and gastric ceca), and a **hindgut** (intestine, rectum, and anal opening). The feeding habits of insects vary from predaceous (praying mantis), to phytophagous (grasshopper), to saprophagous (dung beetles) to parasitic (fleas). In insects, a **tubular heart** located in the pericardial cavity pumps **hemolymph** through the **dorsal aorta**. The hemolymph has little to do with oxygen transport. The **tracheal system** in insects is an efficient respiratory apparatus used primarily for breathing air. Insects, like spiders, possess a unique excretory system consisting of **Malpighian tubules**, which operate in conjunction with specialized glands in the wall of the rectum for excretion (Fig. 31.54–Fig. 31.57).

The nervous system of insects is similar to the nervous systems of crustaceans. Insects have a variety of keen sense organs, which are primarily microscopic and located in the body wall. **Sensilla** are responsible for auditory reception. Interestingly, some insects such as grasshoppers have **tympanic organs** in their legs. Many insects have keen **chemoreception**. Although chemoreceptors usually are located on mouthparts, they may appear on antennae (ants, bees) or even legs (butterflies, moths). Insect eyes follow two general plans: **Simple eyes** are found in larvae, nymphs, and some adult insects. **Compound eyes** are found in the majority of species of insects. The compound eye aids insects to see simultaneously in nearly all directions. Although a typical insect could not read the newspaper, its vision can detect movement. Insects exhibit a variety of complex behaviors and communication skills.

Insects are **dioecious**. Many species practice internal fertilization. The majority of insects are **oviparous** (egg-laying), but some **viviparous** (live-bearing) species exist. Female insects may possess an **ovipositor**, which lays eggs. The **stinger** of insects is a modified ovipositor associated with venom glands. The insect egg exhibits early development. The hatching young insect escapes the egg through a variety of means. Approximately 88% of insects undergo **complete metamorphosis**, which consists of three stages: the **larvae**, the **pupa**, and the **adult** (Fig. 31.49–Fig. 31.51).

Check Your Understanding

Q. What are three characteristics of insects?

_____

_____

_____

Q. How are pill bugs classified?

_____

_____

_____

There are more than 30 orders of insects, and entomologists have developed several means of classification. The orders listed in Table 31.1 are considered the "user-friendly" versions of some of the common orders, along with examples of each.

Table 31.1 Major Orders of Insects	
Order	**Examples**
Collembolla	springtail
Diptera	house fly, crane fly, deer fly, fruit fly, gnat, mosquito, love bug, robber fly
Orthoptera	grasshopper, cricket, mole cricket, locust
Coleoptera	beetle, ladybug
Siphonoptera	flea
Lepidoptera	butterfly, moth
Dermoptera	earwig
Odonata	dragonfly, damselfly
Isoptera	termite
Hymenoptera	bee, wasp, hornet, ant, cicada killer
Blatodea	roach
Mantodea	praying mantis
Phasmatodea	walking stick
Neuroptera	lacewing, antlion, dobsonfly
Homoptera	cicadas, leafhopper, planthopper, scale insects, aphids, mealybug, spittlebug
Hemiptera	stink bug, giant water bug, assassin bug, water strider
Thysanura	silverfish, bristletails
Ephemeroptera	mayfly
Phthiraptera	head louse, body louse, pubic louse, bird lice, poultry lice
Plecoptera	stonefly
Psocoptera	barklouse, booklouse

Figure 31.47 Example insects: (a) greater arid-land katydid, *Neobarrettia spinosa*, (b) Eastern lubber grasshopper, *Romalea microptera*, (c) giant cockroach, *Blaberus giganteus*, (d) cicada, *Diceroprocta apache*, (e) flame skimmer dragonfly, *Libellula saturata*, (f) cynthia moth, *Samia cynthia*, (g) Carolina mantis, *Stagmomantis carolina*. and (h) milkweed beetle, *Tetraopes tetraophthalmus*.

Orthoptera
(cricket)

Dermaptera
(earwig)

Thysanura
(silverfish)

Neuroptera
(lacewing)

Odonata
(damselfly)

Hemiptera
(stink bug)

Strepsiptera
(stylops)

Diptera
(fly)

Isoptera
(termite)

Coleoptera
(beetle)

Homoptera
(cicada)

Siphonaptera
(flea)

Lepidoptera
(butterfly)

Hymenoptera
(wasp)

Phthiraptera
(louse)

Blattaria or Blattodea
(cockroach)

Plecoptera
(stonfly)

Figure 31.48 Representatives from some of the orders of insects.

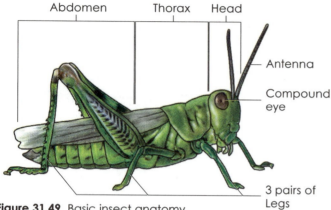

Figure 31.49 Basic insect anatomy.

A Game of Deception

The flickering of fireflies on a quiet summer evening seems so peaceful. Well—maybe not. A game of deception may be going on beneath your eyes. Male fireflies fly around at night using specific blinking patterns to gain the attention of female fireflies residing on the ground. Females return a sequence of flashes that signal acceptance of the male. Unfortunately, some species of female fireflies can imitate the acceptance pattern of other females. When the unsuspecting male approaches the deceptive female, he is eaten by that female.

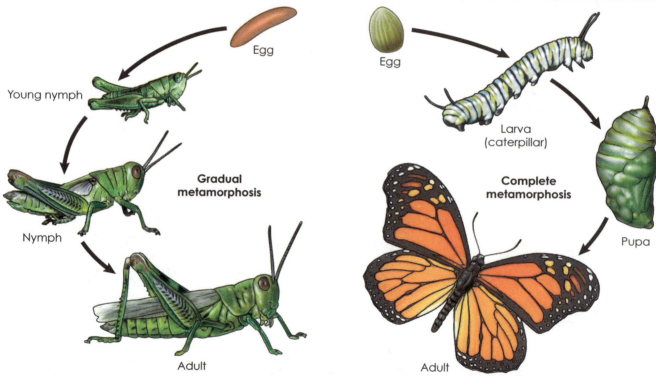

Figure 31.50 A diagram showing Insect development. In gradual (incomplete) metamorphosis the young resemble the adults, but they are smaller and have different body proportions. In complete metamorphosis, the larvae look different from the adult and generally have different food requirements.

Figure 31.51 The developmental stages of the monarch butterfly, *Danaus plexippus*, include (a) egg, (b) larval stage, (c) chrysalis, and (d) adult.

Figure 31.52 The developmental stages of the common honeybee, *Apis mellifera*, include (a) larval stage, (b) pupa, and (c) adult (scale in mm).

Figure 31.53 Anatomy of the grasshopper. (a) Male and (b) female.

1. Antenna	9. Mandible	17. Tarsus
2. Ocelli	10. Labrum	18. Tegmen
3. Compound eye	11. Labial palp	19. Wing
4. Prothorax	12. Metathorax	20. Abdomen
5. Mesothorax	13. Tibia	21. Dorsal valve
6. Tympanum	14. Cercus	22. Ventral valve
7. Femur	15. Subgenital plate	23. Ovipositor
8. Pronotum	16. Spiracle	

From the first dawn of life, all organic beings are found to resemble each other in descending degrees, so they can be classed in groups under groups.

—**Charles Darwin (1809–1882)**

So That's the Answer!

Have you ever been called a nitpicker? What's a nitpicker anyway?

A "nit" is the egg of a louse that is glued to the shaft of the hair of the host mammal. The term "nitpicking" commonly refers to the detailed and meticulous effort that has to be undertaken to locate the nit and remove it from the hair.

Tiny combs were found in the wreckage of the Civil War ironclad, the *USS Cairo*, in Vicksburg, Mississippi. Archaeologists determined that these combs were used to remove lice from the beards of sailors.

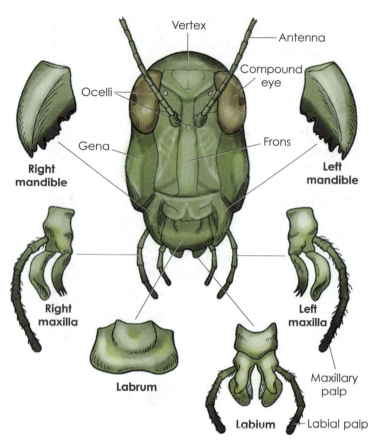

Figure 31.54 A diagram of the head and mouthparts of a grasshopper.

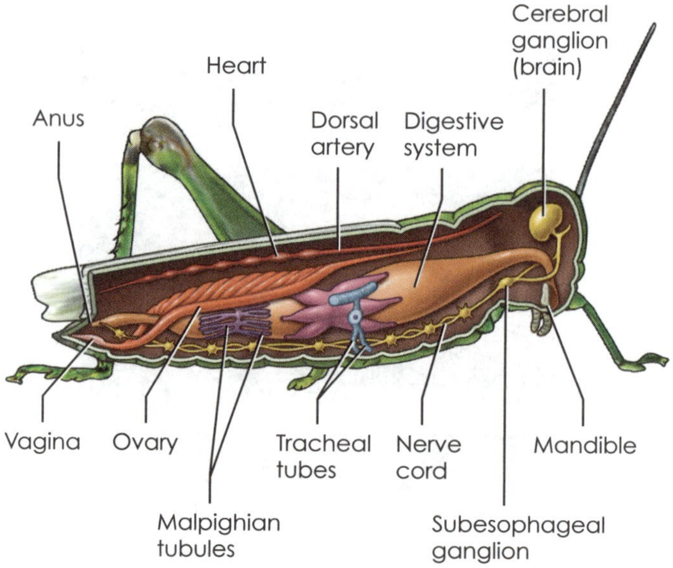

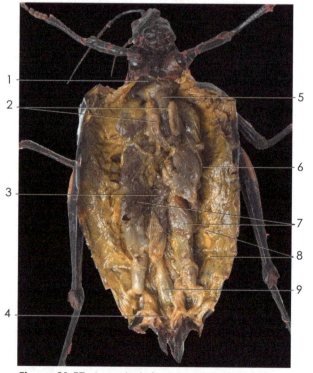

Figure 31.55 The internal anatomy of a grasshopper.

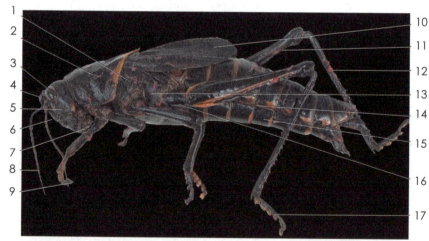

Figure 31.56 A preserved specimen of a grasshopper, order Orthoptera.

1. Mesothorax
2. Pronotum
3. Compound eye
4. Vertex
5. Gena
6. Frons
7. Maxilla
8. Antenna
9. Claw
10. Wing
11. Femur
12. Tibia
13. Mesothorax
14. Spiracle
15. Abdomen
16. Trochanter
17. Tarsus

Figure 31.57 A ventral view showing the internal anatomy of a grasshopper.

1. Esophagus
2. Gastric caecum
3. Stomach
4. Rectum
5. Crop
6. Malpighian tubule
7. Ovaries
8. Trachea
9. Intestine

Student Activity—Grasshopper (*Romalea microptera*) Dissection

Materials
- dissecting tray
- dissecting kit
- hand lens
- safety glasses
- gloves
- lab coat or apron
- scalpel
- forceps
- scissors
- preserved grasshopper

Procedure 31.11
Grasshopper Dissection

Consider photographing this activity.

1. Procure the needed equipment and specimen.
2. Rinse the preservative off of the grasshopper with running water. Place the specimen on its ventral side in the dissecting tray.
3. Using a hand lens, locate the external features of the grasshopper shown in Figure 31.53. Record your observations and sketch below.

4. Using a hand lens, closely examine the head, mouthparts, and appendages of the grasshopper. Compare your specimen to the head region illustrated in Figure 31.54. Include descriptions and sketch.

5. Referring to Figure 31.53, determine the sex of the grasshopper. Compare it to the opposite sex, which may be at another lab station. Record your observations and sketches below.

6. With a scalpel and forceps, carefully remove the carapace from the rostrum to the abdomen. Also remove the dorsal portion of three abdominal segments. Referring to Figures 31.55 and 31.57, find the major internal structures of the grasshopper. Place your observations and sketch below.

8. Using the foreceps, pull the sides apart. Pin each side of the insect to the dissecting pan. Locate the internal structures shown in Figure 31.57. Record your observations and sketch below.

9. Dispose of all grasshopper body parts in the waste container provided by the instructor.
10. Clean the equipment and station thoroughly.
11. Return the clean and dry equipment to the proper location.

7. Place the grasshopper in the dissecting tray, ventral side up. Using scissors, cut through the exoskeleton on the ventral side from the head to the most posterior end of the abdomen.

 ## A Digital Field Trip—Ecdysozoans

Think of the following activity as a digital field trip or a scavenger hunt. Locate the specimens listed in Procedure 31.12, take your time to compose a photograph of the specimen, use good photographic technique, check the quality of the image on the camera LCD, transfer the image to a computer, and develop a labeled presentation.

Materials
- digital camera
- notebook to record data
- computer
- presentation software
- storage device such as a jump drive or CD
- projector and screen

 ### Procedure 31.12
Digital Field Trip

1. Locate and photograph the following: 3 crustaceans, 3 arachnids, and 15 other nematodes or arthropods of your choice.
2. Take photographs of the specimens. The photographs should include the entire specimen, as well as distinguishing features.
3. Using a computer and presentation software, develop a digital slide show of your project. Include the title of your specimen, and label the important parts, using both the common and scientific names of the organisms. And how about adding some music to your presentation?
4. Place your slide show on a storage device or CD. The instructor may allow you to present your slide show to the class.

Name: _____ Date: _____ Section: _____

Review Questions

1. List several characteristics of ecdysozoans.

_____

_____

_____

2. Explain the taxonomical organization of the ecdysozoans.

_____

_____

_____

3. Name five characteristics of Phylum Nematoda.

_____

_____

_____

_____

4. Trace the natural history of five sample nematodes.

_____

_____

_____

5. What is a nematomorphan, and where does the adult live?

_____

_____

_____

6. What are tardigrades and onycophorans?

_____

_____

_____

7. Name five characteristics of Phylum Arthropoda.

_____

_____

_____

8. Describe and provide specific examples of the organization Subphylum Chelicerata.

_____

_____

_____

9. What is a harvestman?

_____

_____

10. Compare and contrast chilopodans and diplopodans.

_____

_____

11. Why are trilobites icons of geologic time?

_____

_____

12. Describe the orders of arachnids.

_____

_____

13. What is the difference between an isopod, an amphipod, and an ostracod?

_____

_____

_____

14. Explain the organization of the crustaceans.

_____

_____

15. Why are insects successful?

_____

_____

Chapter 32
The Deuterostomes: Understanding Animals—Part IV

Student Outcome Objectives

At the completion of this exercise, the student will be able to:

1. Provide examples of and describe the characteristics of deuterostomes.
2. Provide examples of and describe the basic characteristics of Phylum Echinodermata.
3. Provide examples of and describe the basic characteristics of Class Crinoidea.
4. Provide examples of and describe the basic characteristics of Class Asteroidea.
5. Dissect and identify external and internal anatomical features of a starfish.
6. Provide examples of and describe the basic characteristics of Class Ophiuroidea.
7. Provide examples of and describe the basic characteristics of Class Echinoidea.
8. Provide examples of and describe the basic characteristics of Class Holothuroidea.
9. Provide examples of and describe the basic characteristics of Phylum Hemichordata.
10. Provide examples of and describe the basic characteristics of Phylum Chordata.
11. Provide examples of and describe the basic characteristics of Subphylum Urochordata.
12. Provide examples of and describe the basic characteristics of Subphylum Cephalochordata.
13. Identify the external anatomical features of an amphioxus.
14. Provide examples of and describe the basic characteristics of Subphylum Vertebrata (Craniata).
15. Provide examples of and describe the basic characteristics of Superclass Agnatha.
16. Provide examples of and describe the basic characteristics of Superclass Gnathostomata.
17. Provide examples of and describe the basic characteristics of Class Chondrichthyes.
18. Provide examples of and describe the basic characteristics of Class Osteichthyes.
19. Perform a perch dissection and identify specific external and internal structures.
20. Provide examples of and describe the basic characteristics of Class Amphibia.
21. Perform a bullfrog dissection and identify specific external and internal structures.
22. Provide examples of and describe the basic characteristics of Class Reptilia.
23. Compare, contrast, and identify anapsid, diapsid, and synapsid skulls.
24. Provide examples of and describe the basic characteristics of Class Aves.
25. Identify the anatomy of a feather and the skeletal features of a bird.
26. Provide examples of and describe the basic characteristics of Class Mammalia.
27. Perform a fetal pig dissection and identify specific external and internal structures.
28. Identify the skeletal features of a cat.

Overview

Perhaps we are more familiar with the **deuterostomes** than any other group of animals. They include, among others, starfish, sand dollars, sharks, salamanders, turtles, eagles, and humans (Fig. 32.1). All deuterostomes are triploblastic coelomates that undergo radial cleavage, and during embryological development, their blastopore forms the anal opening. The three phyla of deuterostomes are the **Phylum Echinodermata** (starfish), **Phylum Hemichordata** (acorn worms), and **Phylum Chordata** (fishes and tetrapods).

PHYLUM ECHINODERMATA

Phylum Echinodermata consists of nearly 7,000 species of organisms called the "spiny-skinned" animals. This phylum includes organisms such as starfish, brittle stars, sea daisies, sea urchins, sand dollars, crinoids, and sea cucumbers. Typically, echinoderms range in size from small sea cucumbers and brittle stars measuring less than a centimeter in length or diameter, to large sea cucumbers exceeding 2 meters in length. Except for some rare brackish water species, echinoderms inhabit marine environments.

I profess to learn and to teach anatomy not from books but from dissections, not from the tenets of Philosophers but from the fabric of Nature.

—**William Harvey (1578–1657)**

Figure 32.1 The deuterstomes are a diverse group of animals: (a) starfish, *Pisaster ochraceus*, (b) a sea squirt, *Polycarpa aurata*, (c) a amphioxus, *Branchiostoma* sp., (d) stone fish, *Synanceia verrucosa*, (e) a big-eye tree frog, *Leptopelis vermiculatus*, (f) American alligator, *Alligator mississippiensis*, (g) kingfisher, *Todiramphus chloris*, and (h) a bobcat, *Lynx rufus*.

The echinoderms are an unusual group of organisms distinct from the remainder of the animal kingdom. Echinoderms have **pentamerous** (five-pointed) radial symmetry. They have a body wall containing an endoskeleton of small, calcareous **ossicles** that include surface **spines**. The spines can have jaw-like pincers, **pedicellarie**, which discourage barnacles and other fouling organisms from settling on their surface. In some species, such as sea urchins, the spines are associated with **poison glands**. The echinoderms possess a unique **water vascular system** of canals and appendages that function in locomotion, feeding, sensory reception, and gas exchange. The **tube feet** of starfish are powered by this hydraulic system.

Adult echinoderms lack heads, brains, and segmentation. The digestive system of most echinoderms is **complete**. The circulatory system is reduced and radiates in five directions, circulating colorless blood. Minute **gills** that protrude from the coelom are responsible for respiration in some species. Some echinoderms, such as the sea cucumbers, use cloacal structures called **cloacal trees** for respiration. The nervous system basically consists of nerves in a ring around the mouth extending radially outward from inside the body. Echinoderms do not have excretory organs. Many species, such as starfish, have great regenerative powers. Some species can purposely detach a limb to escape a predator, called **autonomy**. Echinoderms are dioecious. After fertilization, the zygote becomes a bilaterally symmetrical ciliated larva, **bipinnaria**, which passes through several stages before becoming a radially symmetrical adult.

Phylum echinodermata consists of five distinct living classes. **Class Crinoidia** is composed of numerous fossil species and living sea lilies and feather stars. The majority of starfishes belong **to Class Asteroidea. Class Ophiuroidea,** the largest class of echinoderms, contains brittle stars. **Class Echinoidea** includes the sea urchins, sand dollars, and sea biscuits. **Class Holothuroidea** is home to the unique sea cucumbers (Fig. 32.2).

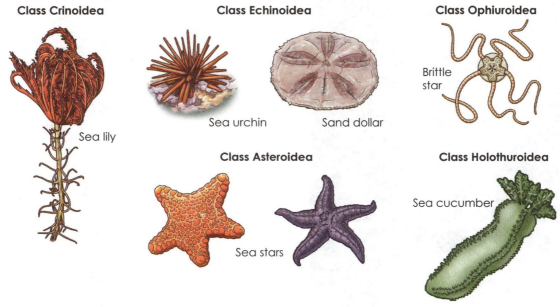

Figure 32.2 A diagram representing each class of echinoderms.

Taxonomy (the science of classification) is often undervalued as a glorified form of filing—with each species in its folder, like a stamp in its prescribed place in an album; but taxonomy is a fundamental and dynamic science, dedicated to exploring the causes of relationships and similarities among organisms. Classifications are theories about the basis of natural order, not dull catalogues compiled only to avoid chaos.

—**Steven Jay Gould (1941–2002)**

Class Crinoidea

Members of **Class Crinoidea** have been present in the Earth's oceans since Cambrian times, and about 700 species still thrive today (Fig. 32.3). These mostly sessile, bottom-dwelling sea creatures are commonly called sea lilies or feather stars. The oral region of a sea lily faces upward and resembles an upside-down starfish. The opposite side is attached to a stalk that attaches to a substrate.

The body of a crinoids, called the **calyx**, is covered by a leathery **tegmen**. The five **arms** are composed of feathery branches called **pinnules**. The arms and the calyx comprise the **crown** of a crinoid. Sessile species possess a **stalk** of ring-shaped **ossicles** and branches called **cirri**. **Holdfast**s anchor the crinoid to a substrate. Crinoids are dioecious (Fig. 32.4).

Figure 32.3 Crinoids have basically remained unchanged since the fossil record. (a) A fossil crinoid (b) a partial crown, or aboral cup, and (c) the stalk.

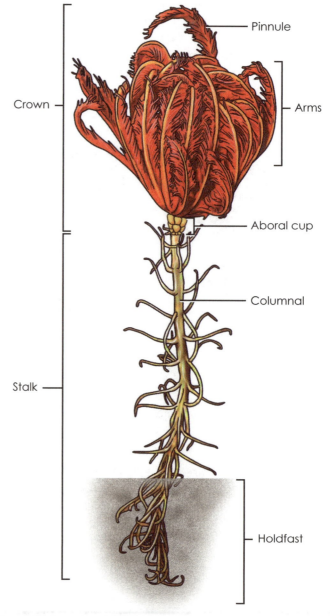

Figure 32.4 Basic crinoid anatomy.

Student Activity—Macroscopic Anatomy of a Crinoid

Materials
- dissecting microscope or hand lens
- colored pencils
- dissecting tray
- probe
- select specimens of crinoids, including fossil crinoids, provided by the instructor

 ### Procedure 32.1
Macroanatomy of a Crinoid

Consider photographing this activity.

1. Procure the needed equipment and specimens.
2. Using the hand lens or dissecting microscope, observe the specimens in the jars. If permitted, remove them from the jars and place them on a dissecting tray for closer examination. Note their external anatomy. Record your observations and labeled sketches below.

3. Using the hand lens or dissecting microscope, observe the fossil crinoids. Record your observations and labeled sketches below. Is there much anatomical difference between the living and fossil crinoids?

4. Clean the equipment and station thoroughly.
5. Return the clean and dry equipment to the proper location.

It is that range of biodiversity that we must care for—the whole thing rather than just one or two stars.

—David Attenborough (1926–present)

Class Asteroidea

The starfishes or sea stars are the representative members of **Class Asteroidea**. Approximately 1,800 living species have been identified. Starfishes range in size from a few centimeters up to a meter in diameter (Fig. 32.5–Fig. 32.6). Starfish are distinguished from other echinoderms because their bodies are gradually drawn out into five **arms** (rays), radiating from a central disc. Some species, however, can have six or more arms. The body of starfish appears to be flattened and is covered by a thin epidermis that may contain spines. The mouth is located on the oral side (ventral) of the starfish. An **ambulacral area** runs along the oral side of each arm to the tip of the arm. The ambulacral groove is found in the center of this area. The groove is bordered by **tube feet**, or podia. Protective, movable spines appear near the tube edges of the tube feet. A distinct **radial nerve** is found in the center of each ambulacral groove.

The **aboral** (dorsal) surface may be smooth, granular, or covered with spines. Beneath the epidermis, a thick dermis layer secretes the endoskeleton of small **ossicles**, perforated by irregular canals filled with cells. A well-developed muscle layer allows the ossicle lattice to be flexible. Starfishes possess a variety of shapes of calcareous spines. In most species of starfish, the spines are blunt, but in a few species such as the Crown-of-Thorns starfish, the spines are sharp. Near the base of the spines are groups of minute **pedicellarie** (pincer-like structures) used to free the surface of debris and encrusting organisms. Small finger-like projections of the coelomic cavity, **papulae**, cover the epidermis of the starfish. The papulae function in gas exchange and excretion (Fig. 32.7).

In most starfish, a distinct **madreporite** (external opening of the water vascular system) can be seen on one side of the central disc. In starfish, a water vascular system is responsible for movement, food gathering, respiration, and excretion.

Interestingly, starfish that feed on bivalve molluscs actually can open the shell of the prey through forces exerted over a period of time with the tube feet. When the bivalve becomes weak, the starfish exerts its stomach into the opening between the shells. After feeding, the starfish redraws its stomach. Primarily, the digestive system consists of a short esophagus, a large stomach, small intestines, and an inconspicuous anus (Fig. 32.8).

In starfish, the sense organs are not well-developed. Tactile organs and other sensory structures are scattered over the surface of the animal. Light-sensitive **ocelli** are present at the tip of each arm. Most starfish are dioecious.

Figure 32.5 A group of sea stars (starfish), *Asterias*, in a tide pool in Oregon.

Figure 32.6 A sea star (starfish), *Asterias* sp.

Figure 32.7 An aboral view of the internal anatomy of a sea star.

1. Ambulacral ridge 3. Spines   5. Pyloric caecum
2. Gonad   4. Ring canal   (digestive gland)

Fertilization is external and usually occurs in early summer when the sperm and eggs are shed from the adults. Regeneration of lost parts is common in starfish. Embryological development is interesting in starfish and important in evolutionary studies. Fertilized eggs develop into planktonic **bipinnaria** larvae and, in some species, into **brachiolaria** larvae. The larvae are bilaterally symmetrical, unlike adult starfishes.

Figure 32.8 An oral view of a sea star (a) showing the cardiac stomach extended through mouth and (b) after retracting the stomach.
1. Cardiac stomach

 Student Activity—Macroscopic Anatomy of a Starfish

Materials
- dissecting microscope
- hand lens
- colored pencils
- dissecting tray
- probe
- starfish

 Procedure 32.2
Macranatomy of Starfish

Consider photographing this activity.

1. Procure the needed equipment and specimens.
2. Using the hand lens or dissecting microscope, observe the external anatomy of a starfish and compare it to Figure 32.7. Record your observations and labeled sketch.

3. Clean the equipment and station thoroughly.
4. Return the clean and dry equipment to the proper location.

Figure 32.9 A brittle star, *Ophioderma* sp.

Figure 32.10 A green brittle star, *Ophiarachna incrassata.*

Student Activity—Dissection of a Starfish

Materials
- dissecting microscope
- hand lens
- colored pencils
- dissecting tray
- dissecting kit
- safety glasses
- lab coat or lab apron
- gloves
- water
- dissecting pins
- starfish provided by your instructor

 Procedure 32.3
Starfish Dissection

Consider photographing this activity.

1. Procure the needed equipment and specimen.
2. Thoroughly rinse the starfish in running water.
3. Place the starfish in a dissecting tray with its dorsal side facing upward. Pin the starfish down using dissecting pins.
4. Cut off the tip of one arm with a scalpel. Proceed to make two long, parallel incisions from the tip of the arm to the central disc. Carefully peel back the top layer of tissue and locate the anatomical features featured in the figures above. Record your observations and labeled sketch.

5. Gently cut the tissue away from the central disc to find the internal structures featured in Figure 32.7. Record your observations and labeled sketch.

6. Thoroughly clean your laboratory station and equipment. Return the dry equipment and discard the starfish as indicated by the instructor.

Class Ophiuroidea consists of over 2000 known species. The members of this class are commonly called brittle stars or basket stars. The arms of brittle stars are slender and sharply set off from the central disc. Their ambulacral grooves are covered with ossicles and they do not possess papulae and pedicellariae. The tube feet do not have suckers in brittle stars. In these animals, the madreporite is located on the oral surface. Five movable plates that serve as jaws surround the mouth of brittle stars. These animals do not possess intestines or an anus. Indigestible

Science is not perfect. It can be misused. It is only a tool. But it is by far the best tool we have, self-correcting, ongoing, applicable to everything. It has two rules. First: there are no sacred truths; all assumptions must be critically examined; arguments from authority are worthless. Second: whatever is inconsistent with the facts must be discarded or revised. . . . The obvious is sometimes false; the unexpected is sometimes true.

—**Carl Sagan (1934–1996)**

material is cast out the mouth. The visceral organs are confined to the central disc. Brittle stars are capable of autonomy and regeneration (Fig. 32.9 and 32.10).

Class Echinoidea

Nearly 1,000 members of **Class Echinoidea** have been identified. The echinoides include sea urchins, sand dollars, and sea biscuits. Many beach-combers enjoy collecting

the **test** formed by the fused **ossicles** of a sand dollar. Echinoids lack arms but their pentamerous arrangement of parts is apparent on the dorsal side of the test. The arm-like extensions are called **petalloids**. Echinoids are marine organisms living in the deep ocean as well as the intertidal zone. Sea urchins seem to prefer rocky areas and sand dollars prefer to burrow into sediment. Sea urchins feature a skeletal structure known as **Aristotle's lantern**. It has five

Figure 32.11 Examples of sea urchins: (a) A green sea urchin, *Strongylocentrotus droebachiensis*, (b) A red slate sea urchin, *Heterocentrotus mammillatus*, (c) pencil sea urchin, *Eucidaris*, sp. (d) helmet sea urchin, *Colobocentrotus atratus*, (e) common sand dollars, *Echinarachnius parma*, and (f) sea biscuits skeleton.

Student Activity—Macroscopic Anatomy of a Representative Member of the Class Echinoidea

hard "teeth" that are moved by complex struts and muscles and used for grinding food. Sea urchins possess large spines that can penetrate human skin. Echinoids are dioecious and may produce planktonic larvae (Fig. 32.11).

Materials
- dissecting microscope
- hand lens
- colored pencils
- dissecting tray
- probe
- select specimens of sea urchins, sand dollars, and sea biscuits

Procedure 32.4
Macroanatomy of Echinoderm

Consider photographing this activity.

1. Procure the needed equipment and specimens.
2. Using a hand lens or dissecting microscope, observe the specimens in the jars. If permitted, remove them from the jars and place them on the dissecting tray for closer examination. Note their external anatomy. Pay close attention to the tests.
3. Record your observations and labeled sketches to the right.
4. Thoroughly clean your laboratory station and equipment. Return the dry equipment.

Class Holothuroidea
Class Holothuroidea includes about 1,200 species of sea cucumbers. Sea cucumbers are shaped like a cucumber or link of sausage and live on the surface of various substrates or burrow into sediments (Fig. 32.12). These organisms have fleshy bodies, and their skeleton is reduced to isolated **ossicles** embedded in a muscular body wall. Sea cucumbers lie on their side by means of three ambulacra called the **sole**. **Tentacles** surround the mouth and vary in number from 8 to 30. In sea cucumbers, the tube feet aid in locomotion. The fluid-filled coelom serves as a hydrostatic skeleton. The **hemal** system is well developed. The complete digestive system empties into a muscular **cloaca**.

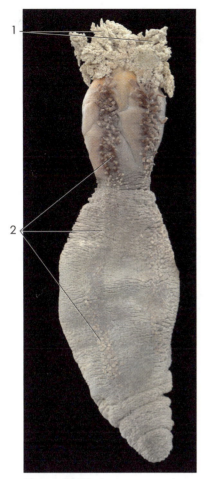

Figure 32.12 A sea cucumber, *Cucumaria*.
1. Tentacles
2. Tube feet

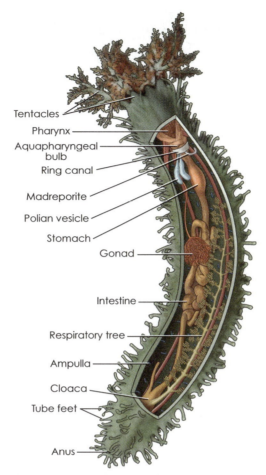

Figure 32.13 A diagram of the internal anatomy of a sea cucumber.

Tentacles
Pharynx
Aquapharyngeal bulb
Ring canal
Madreporite
Polian vesicle
Stomach
Gonad
Intestine
Respiratory tree
Ampulla
Cloaca
Tube feet
Anus

Figure 32.14 The internal anatomy of a sea cucumber.

1. Tentacles	7. Aquapharyngeal bulb
2. Mouth	
3. Polian vesicle	8. Esophagus
4. Respiratory tree	9. Retractor muscle
5. Cloaca	10. Intestine
6. Anus	11. Ampulla
	12. Gonad

In sea cucumbers, the unique respiratory tree is both a respiratory and an excretory structure. The majority of holothurians are dioecious, although a few hermaphroditic species exist. Fertilization is external, and the zygote develops into planktonic larvae. When threatened, sea cucumbers can undergo **evisceration**, in which the entire digestive system and other organs, including the gonads, can be shot out of the mouth or anus. When threatened, sea cucumbers can rupture the hindgut, extruding **Cuverian threads**, which can wrap around a threatening adversary (Fig. 32.13–Fig. 32.15).

Figure 32.15 California sea cucumber, *Parastichopus californicus*.

Student Activity—Macroscopic Anatomy of Representative Members of Class Holothuroidea

Materials
- dissecting microscope
- hand lens
- colored pencils
- dissecting tray
- probe
- select specimens of sea cucumbers

Procedure 32.5
Macroanatomy of Holothuroidea

Consider photographing this activity.

1. Procure the needed equipment and specimens.
2. Using a hand lens or dissecting microscope, observe the specimens in the jars. If permitted, remove them from the jars place them on the dissecting tray for closer examination. Note their external anatomy.
3. Record your observations and labeled sketches below.

_____

_____

4. Thoroughly clean your laboratory station and equipment. Return the dry equipment.

Check Your Understanding

Q. Describe three characteristics of echinoderms.

_____

_____

_____

Q. Compare and contrast members of Class Asteroidea and Class Ophiuredia.

_____

_____

Q. Sketch and label a sand dollar.

Biophilia, if it exists, and I believe it exists, is the innately emotional affiliation of human beings to other living organisms.

—**Edward O. Wilson (1929–present)**

PHYLUM HEMICHORDATA

Any study of the deuterostomes should mention **Phylum Hemichordata**. The hemichordates (half chordates) are a group of marine organisms that once were placed in Phylum Chordata and now reside in their own phylum. Hemichordates have gill slits like chordates. The most common hemichordates are the 80 species of acorn worms, belonging to **Class Enteropneusta**. Acorn worms vary in size from a few centimeters to 2.5 meters. Their body is covered in mucus and has three regions; a **proboscis**, a **collar**, and a **trunk** (Fig. 32.16).

I look into the jar;
I don't know what they are;
Some have scales and some lay eggs;
And some look pretty bizarre;
I still don't know what they are!
They must be vertebrates;
With the same subphylum traits;
The farther I move from left;
The closer I get to myself.

—**Robert Everett,**
from "Bio-Notes Vertebrate Songs" (1988)

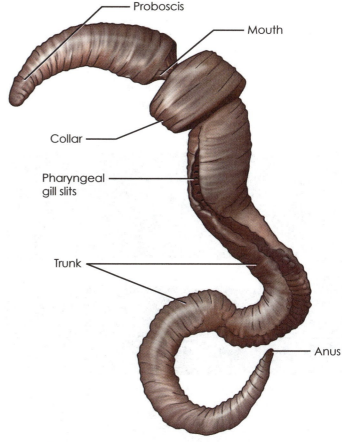

Figure 32.16 An acorn worm.

 Student Activity—Macroscopic Anatomy of Representative Members of Phylum Hemichordata

Materials
- dissecting microscope
- hand lens
- colored pencils
- dissecting tray
- probe
- selected specimens of an acorn worm

 Procedure 32.6
Macroanatomy of Hemichordata

Consider photographing this activity.

1. Procure the needed equipment and specimens.
2. Observe the specimens in the jars. If permitted, remove them from the jars and place them on the dissecting tray for closer examination and note their external anatomy.

3. Record your observation and labeled sketch below.

4. Thoroughly clean your laboratory station and equipment. Return the dry equipment.

PHYLUM CHORDATA

Phylum Chordata consists of nearly 57,000 species, including a variety of animals, among them sea lancelets, lampreys, sharks, bass, frogs, turtles, mockingbirds, giraffes, whales, and humans. Fundamental characteristics of chordates are a **notochord**, a **dorsal hollow nerve cord**, **pharyngeal pouches**, and a **post-anal tail.**

These common features can exist throughout an organism's lifetime or only during embryological development. Additional chordate traits include bilateral symmetry, segmentation, and radial cleavage. Chordates are also triploblastic deuterostomes (Fig. 32.17).

Figure 32.17 Images of chordates: (a) a lancelet, *Branchiostoma* sp. (b) giant grouper, *Epinephelus lanceolatus,* (c) red-eyed tree frog, *Agalychnis callidryas,* (d) snake-necked turtle, *Chelodina parkeri,* (e) black-crowned night heron, *Nycticorax nycticorax,* and (f) a chimpanzee, *Pan troglodytes.*

Table 32.1 Representatives of the Phylum Chordata	
Subphyla and Representative Kinds	**Characteristics**
Urochordata — tunicates	Marine, larvae are free-swimming and have notochord, gill slits, and dorsal hollow nerve cord; most adults are sessile (attached), filter-feeders, saclike animals
Cephalochordata — lancelets, amphioxus	Marine, segmented, elongated body with notochord extending the length of the body; cirri surrounding the mouth for obtaining food
Vertebrata — agnathans (lampreys and hagfishes), fishes (cartilaginous and bony), amphibians, reptiles, birds, mammals	Aquatic and terrestrial forms; distinct head and trunk supported by a series of cartilaginous or bony vertebrae in the adult; closed circulatory system and ventral heart; well-developed brain and sensory organs

The notochord is a flexible, supportive, rod-like structure found along the dorsal midline of all chordates during some part of their lifespan. In most protochordates, such as tunicates and lancelets, the notochord persists throughout the organism's life. In the vertebrates (except for some fishes), the notochord is found only in embryos, where it guides development of the vertebrae. Another vertebrate characteristic is the presence of a dorsal hollow (tubular) nerve cord, found along the midline dorsal to the vertebrae. The spinal cord forms from the dorsal hollow nerve cord. In vertebrates, it is encased in the neural arches of the vertebrae. The brain forms at the anterior end of the nerve cord and is encased in the cranium.

In chordates, pharyngeal pouches and gill slits form in the embryo as pockets of ectoderm. Eventually they grow inward, fusing with pockets of endoderm and lining the pharynx. Pharyngeal gill slits form when two pockets break through, forming a slit. In some chordates, such as fishes and amphibians, these structures are the precursors to gills. In tetrapod vertebrates (amphibians, reptiles, birds, and mammals), the pharyngeal pouches give rise to structures such as the middle ear cavity, Eustachian tubes, parathyroid glands, and the palatine tonsil of mammals. A post-anal tail appears during some portion of the lifespan of all chordates. The post-anal tail, along with muscles, and the notochord, are an important locomotor device in protochordates. In other vertebrates, the post-anal tail performs a variety of functions.

Three distinct subphyla of Phylum Chordata have been recognized. **Subphylum Urochordata** is a non-vertebrate subphylum that includes the tunicates (sea squirts) and salps. Another non-vertebrate subphylum is **Subphylum Cephalochordata**, which includes the lancelets. The largest subphylum of chordates is **Subphylum Vertebrata (Craniata)**, which includes fishes, amphibians, reptiles, birds, and mammals.

Subphylum Urochordata

At first glance, the brightly colored bags of jelly attached to a marine substrate (rocks, eelgrass, pilings), filtering the water for morsels of food, do not resemble a typical chordate. Upon closer examination, however, adult organisms possess gill slits and an examination of their larval form yields the presence of the other traits fundamental to chordates. **Subphylum Urochordata** is composed of nearly 3,000 species of marine animals—tunicates, sea squirts, salps, ascidians, and larvaceans. The three classes of urochordates are **Class Ascidiacea** (sea squirts), **Class Thaliacea** (salps), and **Class Appendicularia** (larvaceans). These organisms are enclosed in a non-living **tunic** made of protein and tunicin.

Tunicates take in water through an **incurrent siphon**, filter the water for food, and eliminate water and wastes through an **excurrent siphon**. The digestive system of tunicates is complete. The simple circulatory system consists of a heart and two large vessels. The nervous system is underdeveloped. Tunicates are monoecious, with a single ovary and testes. The larvae are free-swimming and resemble a tadpole in form (Fig. 32.18–Fig. 32.19).

Figure 32.18 An adult tunicate, *Ciona intestinalis*.

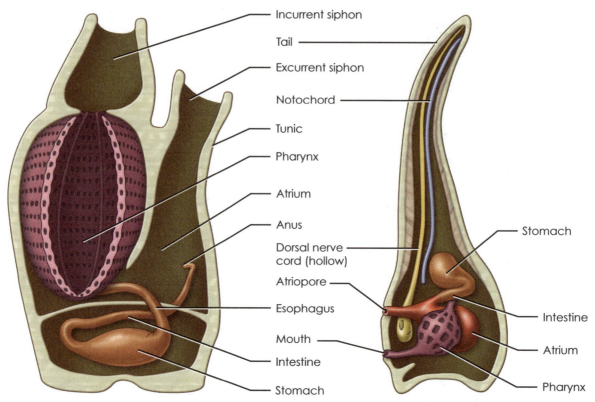

Figure 32.19 A diagram of a tunicate, (a) adult and (b) a larva.

 Student Activity—Macroscopic Anatomy of Representative Members of Subphylum Urochordata

Materials
- dissecting microscope
- hand lens
- colored pencils
- dissecting tray or Petri dish
- probe
- select specimens of *Ciona intestinalis* or other urochordate

 **Procedure 32.7
Macroanatomy of Urochordata**

Consider photographing this activity.

1. Procure the needed equipment and specimens.
2. Using the dissecting microscope or hand lens, observe the specimens in the jars. If permitted, remove them from the jars and place them on a dissecting tray for closer examination, and note their external anatomy.

3. Record your observations and labeled sketch below.

4. Thoroughly clean your laboratory station and equipment. Return the dry equipment.

Student Activity—Microscopic Anatomy of a Larval Urochordate

Materials
- compound microscope
- colored pencils
- microscope slide of a larval urochordate such as *Ciona intestinalis*

Procedure 32.8
Microanatomy of Urochordata

Consider photographing this activity.

1. Procure microscope and slide of a larval urochordate.
2. Using the compound microscope, observe the slide under both low and high power. Closely examine the anatomical features illustrated in Figure 32.19.
3. Record your observations and labeled sketches.

4. Thoroughly clean your laboratory station and equipment. Return the dry equipment.

Subphylum Cephalochordata

Members of **Subphylum Cephalochordata** are commonly called sea lancelets because of their slender, laterally compressed body, which may resemble a willow leaf. There are approximately 25 known species of cephalochordates. These animals are generally translucent, and adults vary in length from 3 to 7 centimeters. The sea lancelets usually inhabit sandy coastal waters, and to an untrained eye may look like a worm or a larval fish. Sea lancelets bury their posterior end into the sand and stick their anterior end above the sand to filter-feed. The term "amphioxus" is sometimes used interchangeably with sea lancelets. The best-known species is *Branchiostoma lanceolatus*.

In cephalochordates, the notochord and nerve cord persist along the entire length of the organism. Sea lancelets have obvious segmented **myomeres**. During feeding, water enters the mouth and passes to the **endostyle**, where it is trapped by mucus and moved to the **hepatic cecum**, where it is digested. Filtered water exits from the **atriopore**, and waste material from the **anus**. These animals possess a **diverticulum** that resembles a vertebrate pancreas and liver in function. Sea lancelets have a complex, closed circulatory system. Gas exchange occurs at the surface of the body. Sea lancelets have a small brain and simple sense organs such as an ocelli. Sea lancelets are dioecious, and their gametes are released from the gonads to the environment by the atriopore. Fertilization in sea lancelets is external (Fig. 32.20–Fig. 32.21).

It's a long way from amphioxus
It's a long way to us…
It's a long way from amphioxus
To the meanest human cuss.
It's good-bye, fins and gill slits,
Hello, lungs and hair!
It's a long, long way from amphioxus,
But we all came from there!

—**Sewell H. Hopkins (1906–1984)**
(sung to the tune of "It's a Long Way to Tipperary")

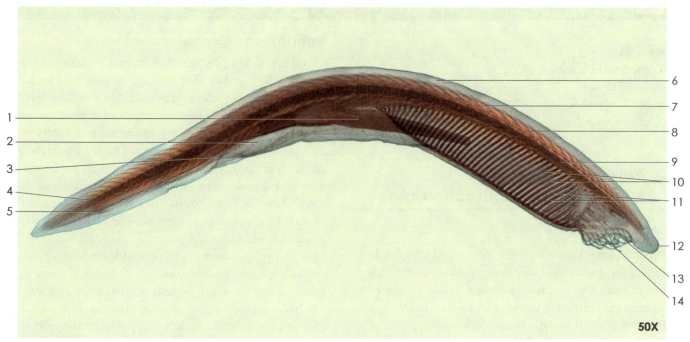

50X

Figure 32.20 A whole mount of Amphioxus.

1. Esophagus
2. Atrium
3. Atriopore
4. Caudal fin
5. Anus
6. Fin rays
7. Myomeres
8. Dorsal nerve cord
9. Notochord
10. Gill slits
11. Gill bars
12. Rostrum
13. Wheel organ
14. Oral cirri

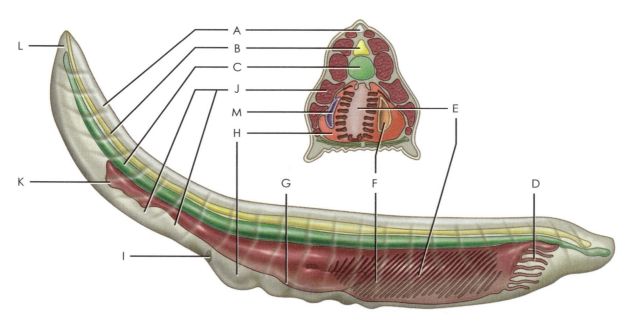

(A) Fin ray
(B) Nerve cord
(C) Notochord
(D) Tentacles
(E) Pharynx with gill slits
(F) Digestive caecum
(G) Intestine
(H) Atrium
(I) Atriopore
(J) Myonemes
(K) Anus
(L) Tail
(M) Gonad

Figure 32.21 Lancelet, Amphioxus.

Student Activity—Macroscopic Anatomy of Representative Members of Subphylum Cephalochordata

Materials
- dissecting microscope
- hand lens
- colored pencils
- dissecting tray or Petri dish
- probe
- select specimen of *Branchiostoma lanceolatus*
- whole mount slides of *Branchiostoma lanceolatus*

 **Procedure 32.9
Macroanatomy of Amphioxus**

Consider photographing this activity.

1. Procure the needed equipment and specimens.
2. Observe the specimen in a jar. If permitted, remove it from the jar and place it on a dissecting tray for closer examination, noting the external anatomy. Record your observations and labeled sketches below.

3. Using the dissecting microscope, thoroughly examine a whole mount of a sea lancelet. Locate the features illustrated in Figures 32.20. Record your observations and labeled sketches below.

4. Thoroughly clean your laboratory station and equipment. Return the dry equipment.

 Student Activity—Microscopic Anatomy of Representative Members of Subphylum Cephalochordata

Materials
- compound microscope
- colored pencils
- microscope slides of a whole mount and a cross-section of a male and female sea lancelet

 **Procedure 32.10
Microanatomy of Amphioxus**

1. Procure microscope and slides.
2. Using the compound microscope, observe the slide of the whole mount under both low and high power. Closely observe the anatomical features illustrated in Figures 32.21. Pay close attention to the cephalic and posterior regions.

3. Record your observations and labeled sketch below.

[sketch box]

_____

_____

4. Now observe the slide of the cross-section of a male and female sea lancet under both low and high power.
5. Record your observations and labeled sketches below.

[sketch box]

_____

_____

[sketch box]

_____

_____

6. Thoroughly clean your laboratory station and equipment. Return the dry equipment.

Subphylum Vertebrata (Craniata)

Subphylum Vertebrata (Craniata) is the largest and most dominant assemblage of chordates, with more than 55,000 species. Vertebrates occupy the majority of Earth's habitats, from deep underwater trenches to the sky above. Examples of vertebrates range from the anchovy on your pizza to the alluring giant panda. The smallest known vertebrate is *Paedocypris progenetica*, a tiny fish measuring 7.9 mm in length from the swamps of Sumatra, and the largest vertebrate is the blue whale, *Balaenoptera musculus*, measuring 30 m in length and weighing up to 180,000 kg (Fig. 32.22).

All vertebrates follow the general body plan of the chordates. The vertebrates are characterized by a vertebral column, which is a chain of structures that pass along the dorsal side from the head to the tail. In the majority of vertebrates, the vertebral column surrounds or replaces the notochord. **Phylum Craniata** perhaps more accurately reflects vertebrata because all members of this group have a skull or cranium. Technically, some jawless fish, such as hagfish, lack vertebrae and keep their notocord. The craniates also have a **neural crest**, which exists as a group of embryonic cells that contribute to forming the cranium, jaws, teeth, and some nerves.

The vertebrates or craniates are metabolically more active than the protochordates and possess a more complex muscular system. Vertebrates feature a bony or cartilaginous endoskeleton consisting of a cranium, limb girdles, and two pairs of appendages. To keep the metabolic rate higher, all members of this group have multi-chambered heart (two to four chambers) and hemoglobin-rich blood. Vertebrates also possess a number of specialized organs including the liver and kidneys. The endocrine system is also well-developed in vertebrates.

Subphylum Vertebrata consists of two superclasses: **Agnatha** (jawless fishes) and **Gnathostomata** (jawed vertebrates). Superclass Agnatha consists of **Class Myxini** (hagfishes) and **Class Cephalaspidomorphi (Petromyzontida)** (lampreys). Superclass Gnathostomata consists of six classes: **Chondrichthyes** (cartilaginous fishes), **Osteichthyes** (bony fishes), **Amphibia** (amphibians), **Reptilia** (reptiles), **Aves** (birds), and **Mammalia** (mammals). Class Osteichthyes is further divided into Subclass **Sarcopterygii** (lobe-finned fishes) and **Actinopterygii** (ray-finned fishes). In recent years, with the reorganization of classical taxonomy, many variations of the above theme may exist.

Did You Know?

Blue whales are the largest animals ever known to have lived on earth. The heart of a blue whale is about the size of a Volkswagen beetle and the tongue is as big as an elephant. A human actually can crawl through the aorta of a blue whale. A blue whale baby when born ranks as one of the largest animals on earth weighing in at over three tons and over 25 feet in length. It gains over 200 pounds a day for the first year of it's life.

Figure 32.22 Examples of vertebrates include: (a) a lamprey, *Lampetra tridentata,* (b) hammer head shark, *Sphyrna tiburo,* (c) leafy sea dragon, *Phycodurus eques,* (d) tiger-Legged monkey frog, *Phyllomedusa hypochondrialis,* (e) tokay gecko, *Gekko gecko,* (f) Clark's nutcracker, *Nucifraga columbiana,* and (g) an American bison, *Bison bison.*

Table 32.2 Representatives of the Subphylum Vertebrata

Taxa and Representative Kinds	Characteristics
Superclass Agnatha	Eel-like and aquatic; sucking mouth (some parasitic); lack jaws and paired appendages
Class Myxini — hagfishes	Terminal mouth with buccal funnel absent; nasal sac connected to pharynx; 4 pairs of tentacles; 5 to 10 pairs pharyngeal pouches
Class Cephalaspidomorphi (Petromyzontida) — lampreys	Suctorial mouth with rasping teeth; nasal sac not connected to buccal cavity; 7 pairs of pharyngeal pouches
Superclass Gnathostomata	Jawed vertebrates; most with paired appendages
Class Chondrichthyes — sharks, rays, and skates	Cartilaginous skeleton; placoid scales; most have spiracle; spiral valve in digestive tract
Class Osteichthyes	Bony fishes; Gills covered by bony operculum; most have swim bladder
Class Sarcopterygii	Bony skeleton; lobe-finned; paired pectoral and pelvic fins;
Class Actinopterygii	Bony skeleton; most have dermal scales; ray-finned
Class Amphibia — salamanders, frogs, and toads	Larvae have gills and adults have lungs; scaleless skin (except apoda); an incomplete double circulation; three–chambered heart
Class Reptilia (=Sauropsida) — turtles, snakes, and lizards	Amniotic egg; epidermal scales; three- or four-chambered heart; lungs
Class Aves — birds	Homeothermous (warm-blooded); feathers; toothless; air sacs; four-chambered heart with right aortic arch
Class Mammalia — mammals	Homeothermous; hair; mammary glands; most have seven cervical vertebrae; muscular diaphragm; three auditory ossicles; four–chambered heart with left aortic arch

Superclass Agnatha

Superclass Agnatha consists of approximately 70 species of jawless fishes. Members of this group lack scales, internal ossification, and paired fins. Agnathans possess eel-like bodies with paired pore-like gill openings.

Class Myxini includes approximately 30 species of bottom-dwelling marine scavengers known as the hagfishes. Hagfishes range in size from 18 cm to 1 m in length. The mouth of a hagfish contains two keratinized plates with tooth-like structures. Hagfishes have a small brain and eyes, and highly developed senses of smell and taste. Lateral **slime glands** produce copious amounts of slime that is used to repulse other organisms and in self-defense.

Sit down before fact as a little child, be prepared to give up every preconceived notion, follow humbly wherever and to what ever abyss nature leads, or you shall learn nothing.

—**T. H. Huxley (1825–1895)**

The best-known hagfish is the Atlantic hagfish, *Myxine glutinosa*. **Class Cephalospidomorphi (Petromyzontida)** includes approximately 41 species of freshwater and marine organisms called lampreys. Lampreys can vary in size from 15 cm to 1 meter in length. Some lampreys are non-parasitic, and others are parasites of fishes. The non-parasitic lampreys do not feed after emerging as adults, because the alimentary canal degenerates. They usually spawn after reaching the adult stage and soon die. Marine lampreys are parasitic as adults. The parasitic forms attach to a fish with their sucker-like mouth and sharp horny teeth. They suck out body fluids, many times causing the death of the host. Marine lampreys such as *Petromyzon marinus* are **anadromous**, living in the ocean most of their lives and returning to freshwater to spawn (Fig. 32.23–Fig. 32.30).

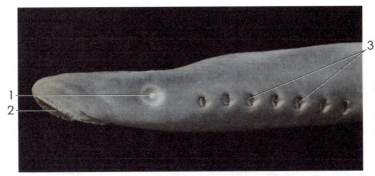

Figure 32.23 A lateral view of the anterior anatomy of a marine lamprey.
1. Eye
2. Buccal funnel
3. External gill slits

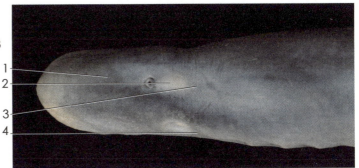

Figure 32.24 A dorsal view of the anterior anatomy of a marine lamprey.
1. Head
2. Nostril
3. Pineal body
4. Eye

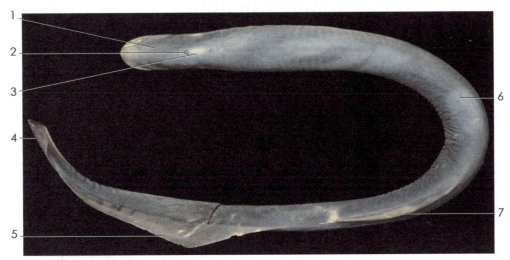

Figure 32.25 A dorsal view of the external anatomy of a marine lamprey, *Petromyzon marinus*.

1. Head
2. Nostril
3. Pineal body
4. Caudal fin
5. Posterior dorsal fin
6. Trunk
7. Anterior dorsal fin

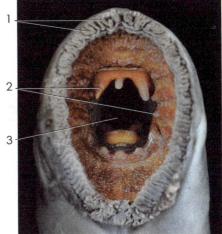

Figure 32.26 The oral region of a marine lamprey.
1. Buccal papillae
2. Horny teeth
3. Mouth

Figure 32.27 A sagittal section through the anterior region of a lamprey.

1. Pineal organ
2. Nostril
3. Brain
4. Pharynx
5. Mouth
6. Annular cartilage
7. Lingual cartilage
8. Internal gill slit
9. Buccal muscle
10. Myomeres
11. Dorsal nerve cord
12. Notochord
13. Dorsal aorta
14. Atrium
15. Ventricle
16. Liver

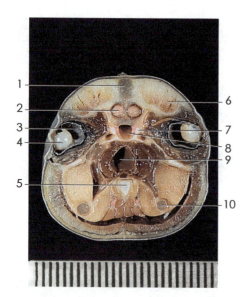

Figure 32.28
A transverse section through the head at the level of the eyes of a lamprey.
1. Pineal organ
2. Brain
3. Lens of eye
4. Retina of eye
5. Lingual cartilage
6. Myomere
7. Cranial cartilage
8. Nasopharyngeal pouch
9. Pharynx
10. Pharyngeal gland

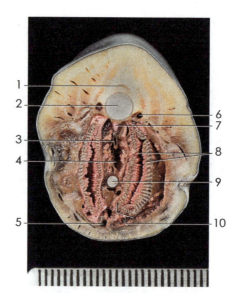

Figure 32.29
A transverse section through the body at the level of the fourth gill slit of a lamprey.
1. Spinal cord
2. Notochord
3. Esophagus
4. Respiratory tube
5. Ventral jugular vein
6. Anterior cardinal vein
7. Dorsal aorta
8. Gill filaments
9. Ventral aorta
10. Branchial pouch

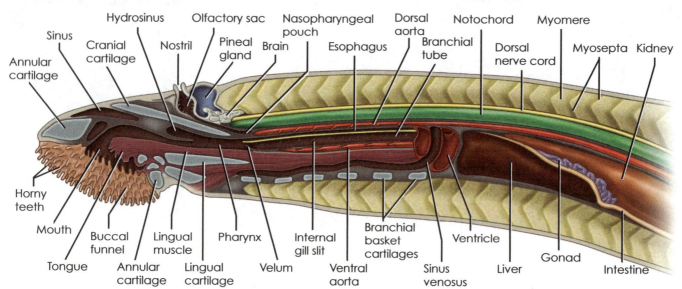

Figure 32.30 A diagram of a sagittal section of a marine lamprey.

 Student Activity—Macroscopic Anatomy of Representative Members of Superclass Agnatha

Materials
- dissecting microscope
- hand lens
- colored pencils
- dissecting tray
- probe
- select specimens of *Myxine glutinosa* and *Petromyzon marinus*

 Procedure 32.11
Macroanatomy of the Lamprey

Consider photographing this activity.

1. Procure equipment and specimens.
2. Using a dissecting microscope or a hand lens, observe a specimen of a hagfish (*Myxine glutinosa*) and a sea lamprey (*Petromyzon marinus*) on a dissecting tray.
3. Compare and contrast the external features of the two agnathans below, including observations and sketches. Pay attention to the anatomical features of the lamprey in Figures 32.23 through 32.26.

As gnathostomes, members of Class Chondrichthyes possess a ventrally oriented mouth. The entire skeleton of members of Class Chondrichthyes, including the skull, is cartilaginous. These animals feature paired **pectoral** and **pelvic** fins and two **dorsal** median fins. In males, the pelvic fins are modified to aid in sperm transfer and are called **claspers**. The skin of cartilaginous fishes has **placoid scales** (denticles) and mucous glands in the majority of species. The teeth of sharks and other members of the class are actually placoid scales (Fig. 32.32–Fig. 32.33).

Members of Class Chondrichthyes have a **complete digestive system** and a **closed circulatory system** with a **two-chambered heart**. Cartilaginous fishes also possess five to seven pair of **gill slits**. **Spiracles**, found just below the eyes in some species such as sawfishes and rays, aid in drawing water into the gills. Many members of class Chondrichthyes have well-developed senses of smell and hearing.

Sharks, rays, and skates possess an electric sensory system known as the **ampullae of Lorenzini**. This system enables them to detect the weak electric fields produced by their prey. They also may have a **lateral-line** system to detect movements and vibrations in the water. Skates are **oviparous** (egg-laying), producing a hard rectangular egg case called a **"mermaid's purse."** The majority of rays are **ovoviviparous** species. The young develop in the mother and are nourished by a yolk sac until birth. Some sharks are ovoviviparousas well, but several species are **viviparous**, in which the embryo receives nourishment from the mother's bloodstream from a placenta-like structure until birth (Fig. 32.34).

Class Chondrichthyes consists of many intriguing organisms. Some of the most bizarre members of this class are the marine sawfishes. These unique organisms possess a saw-like **rostrum** covered with motion-sensitive and electro-sensitive pores (ampullae of Lorenzini) that allow the sawfish to detect movements and perhaps the heartbeat of prey buried in the ocean floor. The rostrum may be used to dig up prey or as a deadly slashing instrument.

Another interesting group of cartilaginous fishes are the **skates**. Presently, 200 species of skates have been named. Skates have a flattened body with a fleshy tail that lacks spines. Skates can be distinguished from rays by the presence of a prominent dorsal fin. They possess small teeth, as opposed to the grinding plates of rays. Rays feature a slender, whip-like tail with serrated spines with associated venom glands at the base. These bottom-dwelling fishes are dorso-ventrally flattened and possess grinding plates for eating.

4. Thoroughly clean your laboratory station and equipment. Return the dry equipment.

Superclass Gnathostomata

The majority of animals with which we are familiar are members of **Superclass Gnathostomata**. These organisms differ from the agnathans in that they possess jaws and a variety of feeding devices designed to grasp, crush, shear and chew. The gnathostomes include sharks, rays, bony fishes, amphibians, reptiles, birds, and mammals.

Class Chondrichthyes

Some members of **Class Chondrichthyes** have a bad reputation as the result of their depiction in movies and some scattered deadly incidents. Of nearly 1,000 living species of cartilaginous fishes, however, only a few species are dangerous to humans. Examples of cartilaginous fishes are sawfishes, skates, rays, chimaeras, and sharks (Fig. 32.31).

Some species of rays, such as the torpedo ray, have paired electric organs on each side of the head. These organs are used to stun prey and in defense. Chimaeras, sometimes called ratfish or ghostfish, are ancient cartilaginous fishes that usually inhabit deep water. Sharks are the best-known members of class chondrichthyes. Approximately 450 species of sharks have been identified (Fig. 32.36).

A Shocking Fact!

The torpedo ray *Torpedo spp.* can produce an impressive electrical charge ranging from 8 to more than 200 volts! Luckily, the amps are usually not high enough to kill an average person. Ancient Greeks and Romans used the shock from these rays to numb the pain of childbirth, and for headaches, gout, and other medical maladies. The term "torpedo" in Latin means "stiff." This is the reaction to a shock by one of these rays.

Figure 32.31 Examples of chondrichthyes: (a) black tip reef shark, *Carcharhinus melanopterus*, (b) gray reef shark, *Carcharhinus amblyrhynchos*, (c) gray smoothhound shark, *Mustelus californicus*, (d) nurse shark, *Ginglymostoma cirratum*, (e) blue spotted stingray, *Taeniura lymma*, and (f) chimaera, *Hydrolagus colliei.*

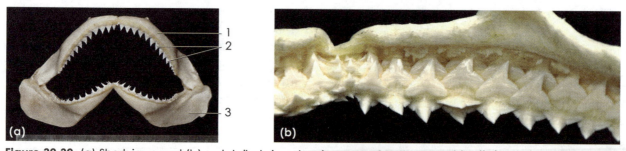

Figure 32.32 (a) Shark jaws and (b) a detailed view showing rows of replacement teeth (scale in mm).
1. Palatopterygoquadrate cartilage (upper jaw)   2. Placoid teeth   3. Meckel's cartilage (lower jaw)

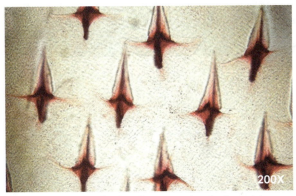

Figure 32.33 A photomicrograph of placoid scales.

Figure 32.34 Many members of Class Chondrichthyes produce an egg case is known as a mermaid's purse. Here a young shark can be seen in silhouette, the round shape on the right is the yoke.

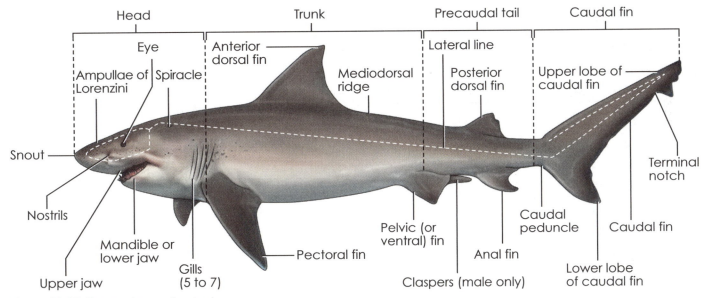

Figure 32.35 The anatomy of a shark.

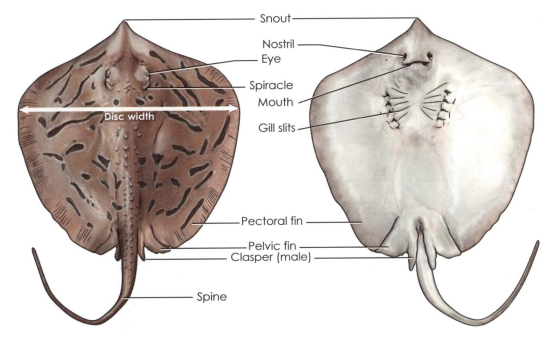

Figure 32.36 The external anatomy of a stingray.

Student Activity—Macroscopic Anatomy of Representative Members of Class Chondrichthyes

Materials

- dissecting microscope or hand lens
- colored pencils
- dissecting tray
- probe
- select specimens of available members of Class Chondrichthyes, including sharks, rays, skates, and a chimaera
- examples of shark integument, shark teeth, a shark jaw, a sawfish blade, and a mermaid's purse

Procedure 32.12
Macroanatomy of Chondrichthyes

Consider photographing this activity.

1. Procure the needed equipment and specimens.
2. Using a dissecting microscope or hand lens, observe the available specimens, paying particular attention to the shark and stingray anatomy.
3. Record your observations and labeled sketches

4. Thoroughly clean your laboratory station and equipment. Return the dry equipment.

Why does Sea World have a seafood restaurant? I'm halfway through my fish burger and I realize, oh my God... I could be eating a slow learner.

—**Lynda Montgomery**

Class Osteichthyes

The next time you eat an anchovy pizza, check out the tiny anchovy bones. Welcome to **Class Osteichthyes**! This taxa classically includes approximately 27,000 species of bony fishes, the largest and most diverse chordate group. In recent years, taxonomists have divided the osteichthyes into two distinct classes: **Class Actinopterygii**, or ray-finned fishes; and **Class Sarcopterygii**, the lobe finned fishes.

The ray-finned fishes belonging to class Actinopterygii comprise nearly 95% of vertebrate species. Examples of this diverse class are goldfish, catfish, gars, tuna, seahorses, puffer-fish, swordfish, and trout. They are known as ray-finned fishes because their fins consist of webs of skin supported by bony spines. If you have ever been "finned" by a fish, you'll understand (Fig. 32.37).

Actinopterygians feature paired pectoral and pelvic fins, and skin with mucous glands usually embedded with dermal scales. Three types of scales can exist in these fishes:

1. **ganoid** scales: flat, heavy scales shaped like an arrow-head, characteristic of gars (Fig. 32.38);
2. **cycloid** scales: thin scales featuring growth rings; common in fish such as carp and salmon, and (Fig. 32.39)
3. **ctenoid** scales: resemble cycloid scales but have spine-like structures on the free edge; include bass and sunfish (Fig. 32.40).

The majority of fishes have a **fusiform** body tapered at both ends, but a variety of unusual shapes can be seen in scorpionfish, flounders, and ocean sunfish. Evident segmentation of the muscles is present in the zigzag **myomeres**. Respiration in the actinopterygians occurs in the gills, which are covered with a protective flap, or **operculum**. Many species have a **swim bladder**, which serves as a floatation device. Fishes possess a two-chambered heart and a closed circulatory system. Fishes are **ectothermic**, controlling their body heat through external sources. They have a **complete digestive system** supported by **accessory organs** such as the pancreas and liver. The **kidneys** filter the wastes contained in the blood.

Freshwater fishes have well-developed kidneys, and saltwater fishes have poorly developed kidneys. The nervous system is well-developed. The **brain** consists of a small cerebrum, olfactory lobes, a large cerebellum, and optic lobes. Fishes have 10 pairs of **cranial nerves**. In many species, the eyes are acute, sounds can be detected in the inner ear, and **olfaction** (smell) is well-developed. Also, in many fishes a **lateral line** system serves to detect vibrations. Fishes are dioecious. The majority of fishes reproduce by external fertilization and are oviparous (Fig. 32.41–Fig. 32.45).

In 1938, the worlds of ichthyology and evolution were stunned when Mary Latimer observed and described a strange fish that appeared to be a living fossil, captured of the coast of South Africa. It turned out to be a coelacanth (*Latimeria chaumnae*). Today, the coelacanth is one of eight living species of Class Sarcopterygii, the lobe-finned fishes. This class includes the coelocanths and lungfishes, found in Africa, South America, and Australia (Fig. 32.46–Fig. 32.47).

Sarcopterygian fins are fleshy, and their bones equate to the limb bones of land vertebrates. Their fins are considered to have evolved into the legs of the first land vertebrates, the **tetrapods**. In 2006, scientists discovered a fossil sarcopterygian, named *Tiktaalik roseae*, which is thought to be an intermediate between fishes and amphibians.

Figure 32.37 Ray-finned fishes: (a) lookdown fish, *Seiene vomer*, (b) piranha, *Pygocentrus nattereri*, (c) copper band butterfly fish, *Chelmon rostratus*.

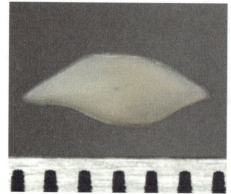

Figure 32.38 Ganoid scales, present in primitive fishes like the gar, are composed of silvery ganoin on the top surface and bone on the bottom. (scale in mm).

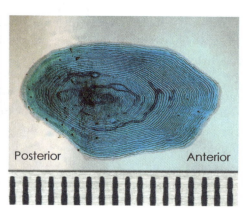

Posterior   Anterior

Figure 32.39 Cycloid scales, along with ctenoid scales, are found on advanced bony fishes. They are much thinner and more flexible than ganoid scales and overlap each other (scale in mm).

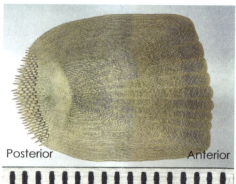

Posterior   Anterior

Figure 32.40 Ctenoid scales differ from cycloid scales in that they have comblike ridges on the exposed edge. This is thought to improve swimming efficiency (scale in mm).

DORSAL

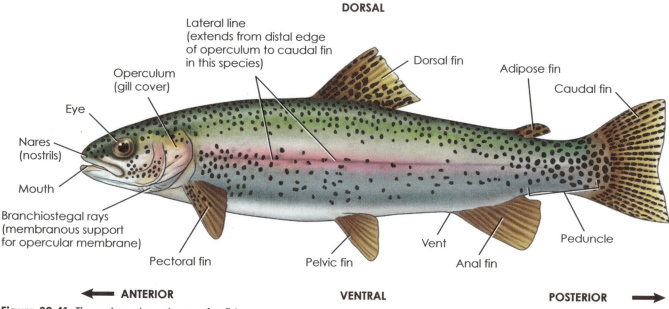

Lateral line (extends from distal edge of operculum to caudal fin in this species)

Operculum (gill cover)

Eye

Nares (nostrils)

Mouth

Branchiostegal rays (membranous support for opercular membrane)

Pectoral fin

Pelvic fin

Vent

Anal fin

Dorsal fin

Adipose fin

Caudal fin

Peduncle

◄── **ANTERIOR**   **VENTRAL**   **POSTERIOR** ──►

Figure 32.41 The external anatomy of a fish.

Figure 32.42 The external anatomy of a perch.

1. Anterior dorsal fin
2. Eye
3. Nostrils
4. Mandible
5. Dentary
6. Operculum
7. Pectoral fin
8. Pelvic fin
9. Lateral line
10. Posterior dorsal fin
11. Caudal fin
12. Anal fin
13. Anus

Figure 32.43 The skeleton of a perch.

1. Anterior dorsal fin
2. Fin spines
3. Neurocranium
4. Premaxilla
5. Maxilla
6. Dentary
7. Opercular bones
8. Pectoral fin
9. Pelvic fin
10. Vertebral column
11. Posterior dorsal fin
12. Soft rays
13. Caudal fin
14. Neural spine
15. Haemal spine
16. Anal fin
17. Ribs

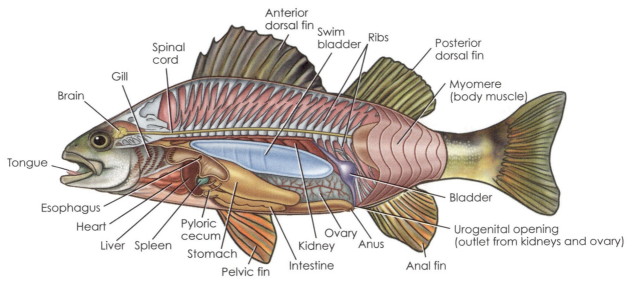

Figure 32.44 The anatomy of a female perch.

Figure 32.45 The viscera of a perch.

1. Epaxial muscles
2. Stomach
3. Gill
4. Heart
5. Pyloric cecum
6. Liver (cut)
7. Pancreas
8. Vertebrae
9. Urinary bladder
10. Gonad
11. Anus
12. Intestine

Figure 32.46 The coelacanth, *Latimeria chalumnae*, a lobe-fin fish, was once thought to be extinct.

Figure 32.47 The African lungfish, *Neoceratodus forsteri*.

 Student Activity—Macroscopic Anatomy of Representative Members of Class Actinopterygii

Materials
- dissecting microscope or compound microscope
- hand lens
- colored pencils
- dissecting tray
- probe
- select specimens of available members of Class Actinopterygii
- examples of scale types and fish skeleton

 Procedure 32.13
Macroanatomy of Actinopterygii

Consider photographing this activity.

1. Procure the needed equipment and specimens.
2. Using a hand lens, observe the available specimens on the dissecting tray. Pay particular attention to their external anatomy and their skeleton. Record your observations and labeled sketches below.

3. Observe the types of fish scales using the dissecting microscope. Or you may need a compound microscope for these observations.
4. Record your observations and sketches below.

5. Thoroughly clean your laboratory station and equipment. Return the dry equipment.

Student Activity—Dissection of a Perch

Materials

- dissecting microscope
- compound microscope
- hand lens
- colored pencils
- dissecting tray
- dissecting kit
- safety glasses
- lab coat or lab apron
- gloves
- preserved perch

Procedure 32.14
Perch Dissection

1. Procure the needed equipment and specimen.
2. Thoroughly rinse the perch in running water.
3. Place the perch in a dissecting tray, laying it on its left side.
4. Locate the external anatomical features of the perch, and examine the eyes. Using a probe, lift up the operculum and observe the gill filaments. Remove a single scale from the fish.
5. Place the scale on a slide, and observe it with a compound microscope. From your observation, perch have _____ scales.
6. Record your observations and labeled sketch below.

7. Using sharp scissors, remove the operculum by cutting it just behind the orbit. Be careful not to damage the underlying gills. The perch has _____ gills on each side. Note the bony gill arches. How many gill arches are present? _____
8. Record your observations and sketches below.

9. Using the scissors, carefully cut a large section out of the body of the perch, beginning at the posterior edge of the last gill and extending dorsally to anus. Cut around the anus, and proceed back to the edge of the opercular region.
10. Carefully lift out the large section and view the internal structures illustrated in Figures 32.44. Record your observations and sketches below.

11. Thoroughly clean your laboratory station and equipment. Return the dry equipment, and discard the perch as indicated by the instructor.

Check Your Understanding

Q. Describe the defining characteristics of the chordates.

Q. Why would one think that a sea lancelet were a worm at first glance?

Q. Compare and contrast actinopterygian and sarcopterygian fishes.

Q. How many gill arches are present on fish? _____

Class Amphibia

Class Amphibia is composed of nearly 6,000 species. Modern amphibians include frogs, toads, salamanders, amphiumas, sirens, newts, and caecilians. The amphibians evolved from sarcopterygian fishes during the Devonian Period. To venture away from the waters and colonize the land, amphibians developed a protective integument, a means to breathe in the terrestrial environment and specialized limbs. Despite these changes, the amphibians did not evolve the ability to live their entire lives on land. They remained dependent on water to lay their eggs.

Members of class Amphibia are characterized by their bony skeletons with vertebrae, gills during development in the majority of species, and moist, glandular integument without external scales. In some amphibians the integument allows for taking in air through the skin, called **cutaneous respiration**. Pigment cells, **chromatophores**, in the skin are responsible for a variety of colors. Specialized integumentary glands in several species, such as marine toads and poison arrow frogs, produce potent and deadly toxins.

The skull of amphibians is short, broad, and incompletely ossified. The mouth is usually large, with small teeth. Two internal **nares** open into the mouth cavity. Many species possess a protrusable muscular **tongue** attached to the front of the mouth. The circulatory system is closed, and the heart has three chambers with two atria and one ventricle. Amphibians eliminate nitrogenous waste primarily in the form of urea. Their nervous system is well-developed. The brain consists of a **forebrain** a **midbrain**, and a **hindbrain**). In many species the eyes have adapted to a terrestrial lifestyle. The eyes are protected by an eyelid, the **nicitating membrane**. The auditory system contains a **tympanic membrane** for hearing.

Amphibians are dioecious. Frogs and toads undergo external fertilization, and salamanders primarily undergo internal fertilization. The amphibians are primarily oviparous. The eggs of amphibians are **mesolecithal**, having a large yolk and jelly-like membranes. Larval forms such as tadpoles are aquatic and possess gills. Several species of salamanders are **neotenic**, retaining their gills in the adult form.

Class Amphibia is composed of three orders: Gymnophiona, Caudata, and Anura (Fig. 32.48). **Order Gymnophiona** consists of 173 species of elongated limbless apodans known as caecilians. At first glance, they may be mistaken for earthworms or snakes. Caecilians are typically blind and possess sensory tentacles between their eyes and nostrils. The body of a caecilian is arranged in rings,

Figure 32.48 Representatives from the three orders of amphibians: (a) blue-webbed gliding tree frog, *Rhacophorus reinwardtii*, from the order Anura, (b) spotted salamander, *Ambystoma maculatum*, from the order Caudata, and (c) Cameroon caecilian, *Crotaphatrema bornmuelleri*, from the order Gymnophiona.

called **annuli**, giving caecilians an earthworm-like appearance (Fig. 32.49).

Salamanders, newts, amphiumas, and sirens are members of **Order Caudata** (Urodela), consisting of nearly 560 described species. Salamanders and newts (more aquatic salamanders) can be found in almost all of the northern temperate regions of the world and in Central America and South America. The body of a typical salamander is lizard-like, characterized by a slender body, a short nose, and an elongate tail. The majority of salamanders have four toes on their front legs and five toes on their hindlegs, and they lack claws. Amphihumas and sirens have degenerate legs and an eel-like appearance. Some adult terrestrial salamanders utilize lungs, and some terrestrial species lack both lungs and gills. These salamanders undergo cutaneous respiration. Gills are used in larval salamanders and some **neotinic** (paedomorphic) forms such as sirens and axolotl (Fig. 32.50–Fig. 32.56).

Order Anura consists of 5,290 species of toads and frogs. Generally, toads possess a dry, bumpy integument, parotid glands behind the tympanic membrane, a blunt nose, no teeth, and they do not have webs on their hind digits. Frogs have smooth, moist skin, a longer body, long legs, teeth in the upper jaw, and webs between the hind toes. They possess extremely long, muscular hindlegs and front legs that serve as shock absorbers. Adult frogs and toads are carnivorous, and the tadpoles are herbivorous. Some anurans are poisonous. The marine toad *Bufo marinus* has enlarged **parotoid glands** behind their eyes, and other glands on the back. When toads are in danger, their glands secrete a milky white fluid, **bufotenin**, which can be deadly to animals such as dogs and even humans. Poison arrow frogs and poison dart frogs produce a number of deadly chemicals that are being studied for their potential use in medicine such as muscle relaxants, heart stimulants, and appetite suppressants (Fig. 32.57–32.79).

Figure 32.49 Cameroon caecilian, *Crotaphatrema bornmuelleri*. The rings or annuli can clearly be seen.

Figure 32.50 A lesser siren, *Siren intermedia*.

Figure 32.51 A tiger salamander, *Ambystoma tigrinum*.

Figure 32.52 An Eastern newt, *Notophthalmus viridescens*.

Figure 32.53 A blue-spotted salamander, *Ambystoma laterale*.

Figure 32.54 An axolotl, *Ambystoma mexicanum*. This individual is leucistic.

Figure 32.55 An amphiuma, *Amphiuma means*.

Figure 32.56 A mudpuppy, *Necturus maculosus*.

 Student Activity—Macroscopic Anatomy of Representative Members of Class Amphibia

Materials
- dissecting microscope (you may need a compound microscope as well)
- hand lens
- colored pencils
- dissecting tray
- probe
- select specimens of the following amphibians: caecilian, salamanders, newts, amphiumas, sirens, toads, frogs

Consider photographing this activity.

 Procedure 32.15 Macroanatomy of Amphibia

1. Procure the needed equipment and specimens.
2. Using a hand lens or dissecting microscope, observe the available specimens. If permitted, remove the specimen from the jar and place them on a dissecting tray for closer examination.
3. Record your observations and labeled sketches on the following page. When possible, provide the scientific name of the specimen.
4. Thoroughly clean your laboratory station and equipment. Return the dry equipment.

Caecilian

Amphiumas and sirens

Salamanders and newts

Toads and frogs

Figure 32.57 A White's tree frog, *Litoria caerulea.*

Figure 32.58 A marine or Cane toad, *Bufo marinus.*

Figure 32.59 A bumble bee poison-dart frog, *Dendrobates leucomelas*.

Figure 32.60 An African clawed frog, *Xenopus laevis*.

Figure 32.61 A Woodhouse's toad, *Bufo woodhousii*.

Figure 32.62 The leopard frog, *Rana pipiens*.

Figure 32.63 An American bullfrog, *Rana catesbeiana*.

 Student Activity—External Anatomy of the Bullfrog

Materials
- dissecting microscope or hand lens
- colored pencils
- dissecting tray
- dissecting kit
- safety glasses
- lab coat or lab apron
- gloves
- doubly injected preserved bullfrog (*Rana catesbeiana*)

Consider photographing this activity.

 **Procedure 32.16
External Anatomy of the Bullfrog**

1. Procure the needed equipment and specimen.
2. Thoroughly rinse the bullfrog in running water.
3. Place the bullfrog in a dissecting tray. Locate the dorsal and ventral external anatomical features of the bullfrog labeled in Figure 32.64.
4. Record your observations below and label the external features of the bullfrog in the frog outline on the next page.

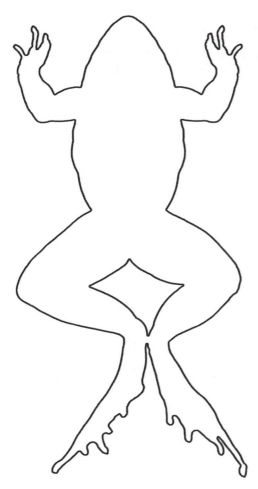

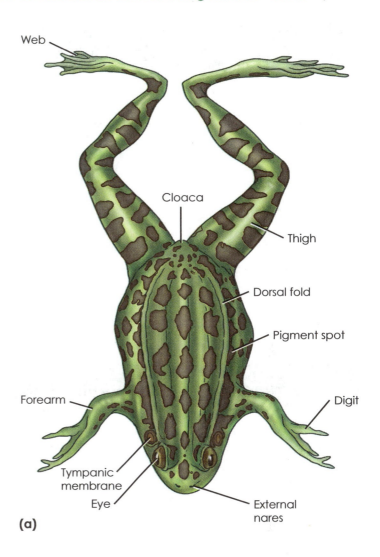

Web

Cloaca

Thigh

Dorsal fold

Pigment spot

Forearm

Digit

Tympanic membrane

Eye

External nares

(a)

5. Using a probe, pry open the mouth of the bullfrog and open the mouth wide. Locate the anatomical features labeled in Figure 32.64. Press on the eyes and notice the movement of the vacuities. When a frog swallows, it blinks its eyes and the vacuities help push the food into the esophagus. Record your observations and labeled sketch below.

6. Thoroughly clean your laboratory station and equipment. Return the dry equipment.

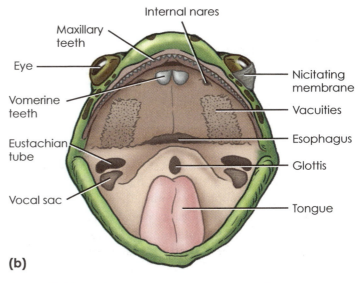

Internal nares

Maxillary teeth

Eye

Nicitating membrane

Vomerine teeth

Vacuities

Eustachian tube

Esophagus

Glottis

Vocal sac

Tongue

(b)

Figure 32.64 (a) The external anatomy of a bullfrog, *Rana catesbeiana*, and (b) the anatomy of a bullfrog's mouth.

Student Activity—Skeleton of the Bullfrog (*Rana catesbeiana*)

Materials
- hand lens
- colored pencils
- skeleton of a bullfrog (*Rana catesbeiana*)

 ### Procedure 32.17
Bullfrog Bones

Consider photographing this activity.

1. Procure the needed equipment and specimen.
2. Using the hand lens, observe the skeleton of the bull-frog, comparing the bones to Figure 32.65 and 32.66.

3. Record your observations below, and draw and label the bones of the bullfrog in the frog outline on page 694.

_____

_____

_____

_____

_____

4. Thoroughly clean your laboratory station and equipment. Return the dry equipment.

 Student Activity—Muscles of the Bullfrog

Materials
- dissecting microscope or hand lens
- colored pencils
- dissecting tray
- dissecting kit
- safety glasses
- lab coat or lab apron
- gloves
- doubly injected preserved bullfrog (*Rana catesbeiana*)

 ### Procedure 32.18
Muscles of the Bullfrog

Consider photographing this activity.

1. Procure the needed equipment and the specimen. (Use the bullfrog from Procedure 32.18).
2. There's more than one way to skin a bullfrog! With a pair of sharp scissors, carefully make a cut to include just the skin completely around the bullfrog's waist (about where a pair of pants would fit).
3. Remove the top half of the skin (the sweater) and the bottom half of the skin (the pants). Use a scalpel or a pair of scissors as an aid.

4. Once the skin is completely removed, begin comparing the musculature of the specimen with Figures 32.67 to 32.62.
5. Record your observations below and draw and label the muscles of the bullfrog in the frog outline on page 694 and 695.

_____

_____

_____

_____

6. Thoroughly clean your laboratory station and equipment. Return the dry equipment.

The important thing in science is not so much to obtain new facts as to discover new ways of thinking about them.

—**Sir Lawrence Bragg (1890–1971)**

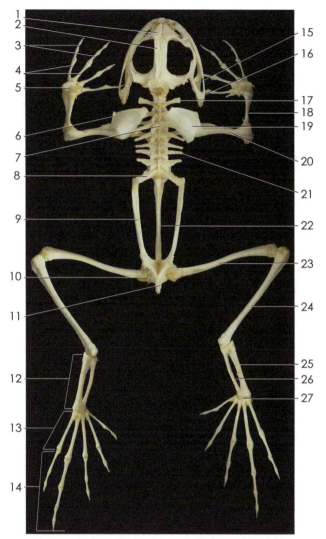

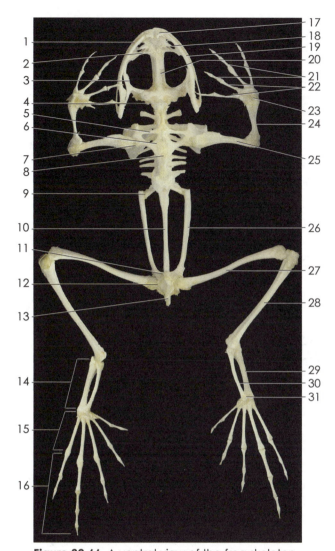

Figure 32.65 A dorsal view of the frog skeleton.

1. Nasal bone	14. Phalanges of digits
2. Frontoparietal bone	15. Squamosal bone
3. Phalanges of digits	16. Quadratojugal bone
4. Metacarpal bones	17. Transverse process
5. Carpal bones	18. Radioulna
6. Scapula	19. Suprascapula
7. Vertebra	20. Humerus
8. Transverse process of sacral (ninth) vertebra	21. Transverse process
	22. Urostyle
9. Ilium	23. Femur
10. Acetabulum	24. Tibiofibula
11. Ischium	25. Fibulare (calcaneum)
12. Tarsal bones	26. Tibiale (astragalus)
13. Metatarsal bones	27. Distal tarsal bones

Figure 32.66 A ventral view of the frog skeleton.

1. Maxilla	16. Phalanges of digits
2. Palatine	17. Premaxilla
3. Pterygoid bone	18. Vomer
4. Exoccipital bone	19. Dentary
5. Clavicle	20. Parasphenoid bone
6. Coracoid	21. Phalange of digits
7. Glenoid fossa	22. Metacarpal bone
8. Sternum	23. Carpal bones
9. Transverse process of sacral (ninth) vertebra	24. Humerus
	25. Radioulna
10. Urostyle	26. Ilium
11. Pubis	27. Femur
12. Acetabulum	28. Tibiofibula
13. Ischium	29. Fibulare (calcaneum)
14. Tarsal bones	30. Tibiale (astragalus)
15. Metatarsal bones	31. Distal tarsal bones

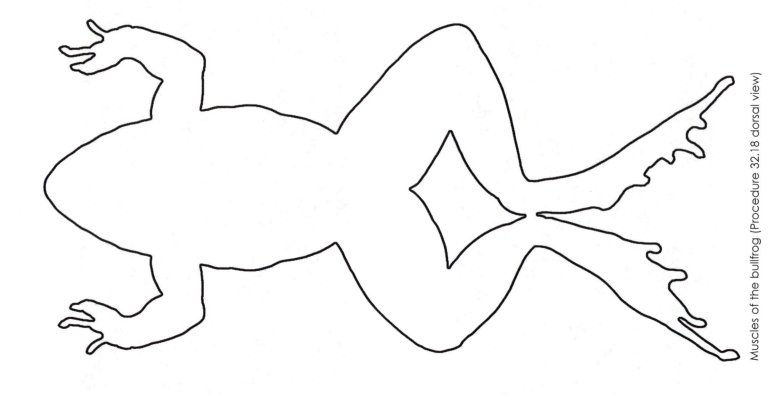

Muscles of the bullfrog (Procedure 32.18 dorsal view)

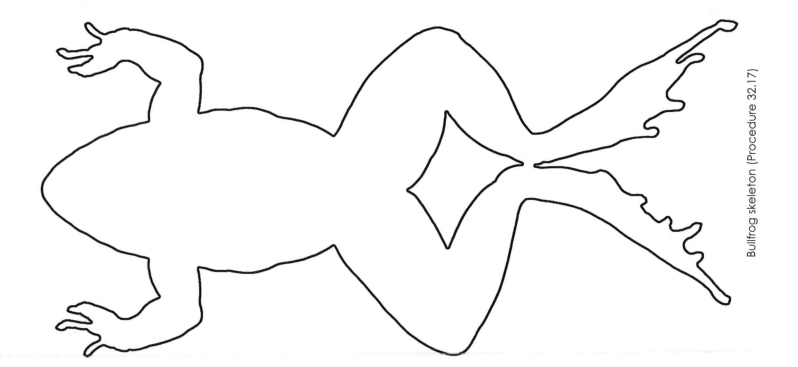

Bullfrog skeleton (Procedure 32.17)

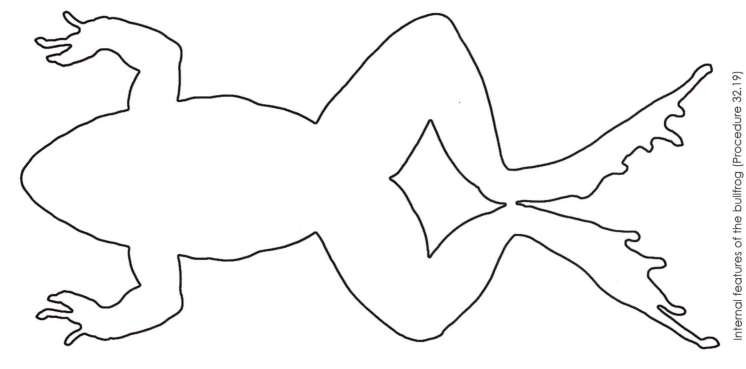

Internal features of the bullfrog (Procedure 32.19)

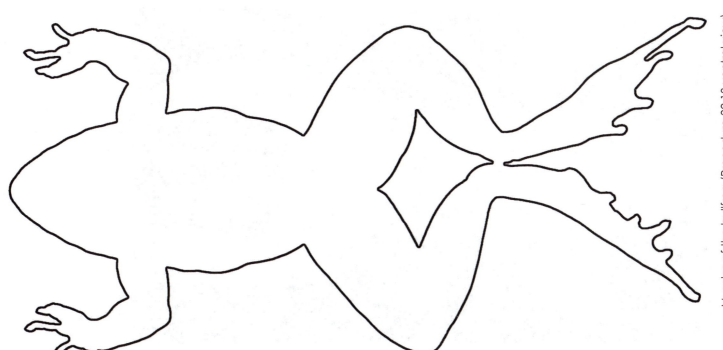

Muscles of the bullfrog (Procedure 32.18 ventral view)

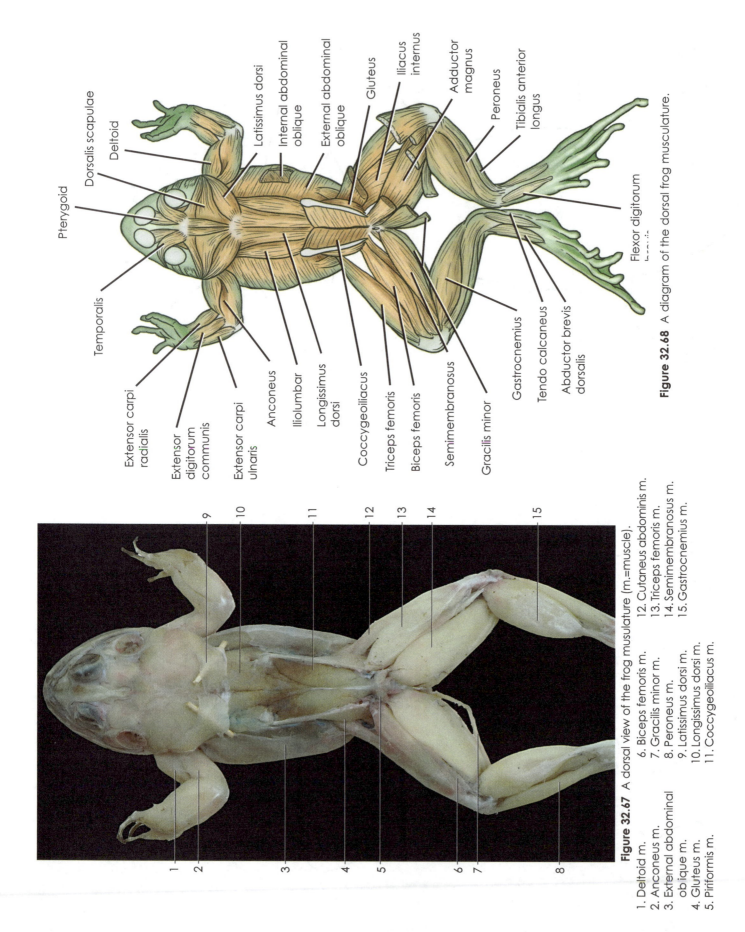

Pterygoid

Temporalis

Dorsalis scapulae

Deltoid

Latissimus dorsi

Internal abdominal oblique

External abdominal oblique

Gluteus

Iliacus internus

Adductor magnus

Peroneus

Tibialis anterior longus

Flexor digitorum

Extensor carpi radialis

Extensor digitorum communis

Extensor carpi ulnaris

Anconeus

Iliolumbar

Longissimus dorsi

Coccygeoiliacus

Triceps femoris

Biceps femoris

Semimembranosus

Gracilis minor

Gastrocnemius

Tendo calcaneus

Abductor brevis dorsalis

Figure 32.68 A diagram of the dorsal frog musculature.

Figure 32.67 A dorsal view of the frog musulature (m.=muscle).

1. Deltoid m.
2. Anconeus m.
3. External abdominal oblique m.
4. Gluteus m.
5. Piriformis m.
6. Biceps femoris m.
7. Gracilis minor m.
8. Peroneus m.
9. Latissimus dorsi m.
10. Longissimus dorsi m.
11. Coccygeoiliacus m.
12. Cutaneus abdominis m.
13. Triceps femoris m.
14. Semimembranosus m.
15. Gastrocnemius m.

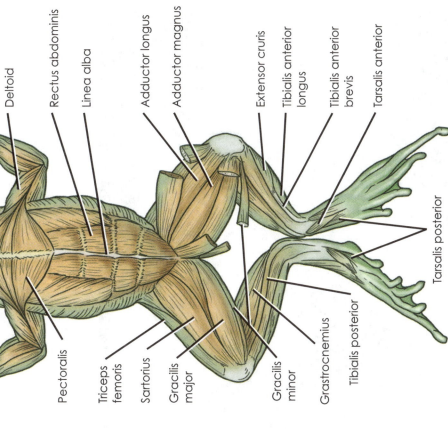

Figure 32.70 A diagram of the ventral frog musculature.

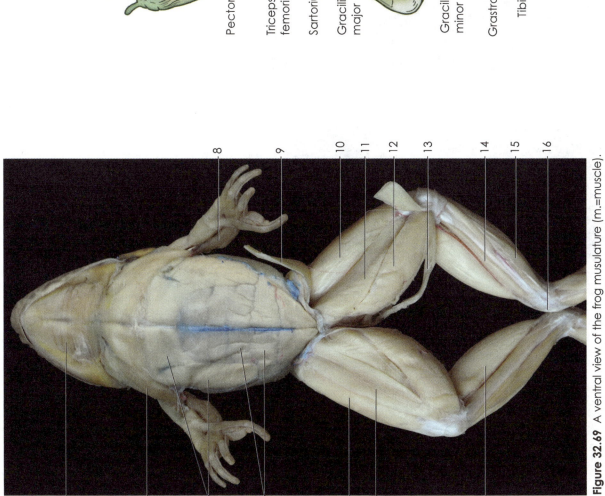

Figure 32.69 A ventral view of the frog musculature (m.=muscle).

1. Mylohyoid m.
2. Deltoid m.
3. Pectoralis m.
4. Rectus abdominis m.
5. Triceps femoris m.
6. Sartorius m.
7. Gastrocnemius m.
8. Palmarir longus m.
9. Sartorius m.
   (cut and reflected)
10. Adductor longus m.
11. Adductor magnus m.
12. Gracilis major m.
13. Gracilis minor m.
14. Tibialis posterior m.
15. Tibialis anterior m.
16. Tendo calcaneus

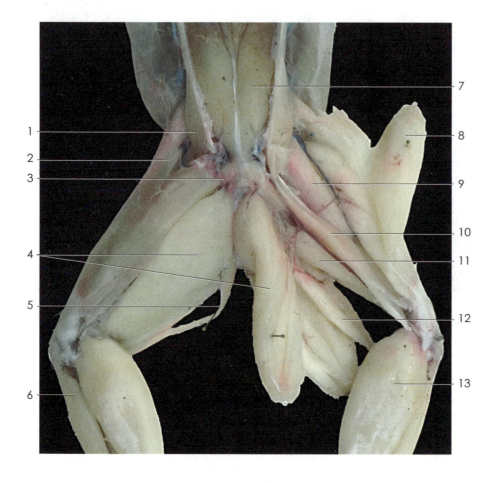

Figure 32.71 A dorsal view of the leg muscles of a frog (m.=muscle).

1. Gluteus m.
2. Cutaneus abdominis m.
3. Piriformis m.
4. Semimembranosus m. (cut and reflected)
5. Gracilis minor m.
6. Peroneus m.
7. Coccygeoiliacus m.
8. Triceps femoris m. (cut and reflected)
9. Iliacus internus m.
10. Biceps femoris m.
11. Adductor magnus m.
12. Semitendinosus m. (cut and reflected)
13. Gastrocnemius m.

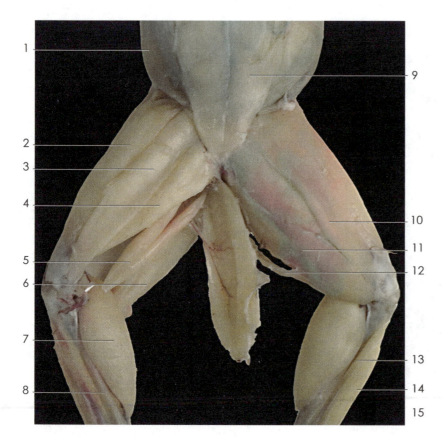

Figure 32.72 A ventral view of the leg muscles of a frog (m.=muscle).

1. External abdominal oblique m.
2. Triceps femoris m.
3. Adductor longus m.
4. Adductor magnus m.
5. Semitendinosus m. (cut)
6. Semimembranosus m.
7. Gastrocnemius m.
8. Tibialis posterior m.
9. Rectus abdominis m.
10. Sartorius m.
11. Gracilis major m.
12. Gracilis minor m.
13. Extensor cruris m.
14. Tibialis anterior longus m.
15. Tibialis anterior brevis m.

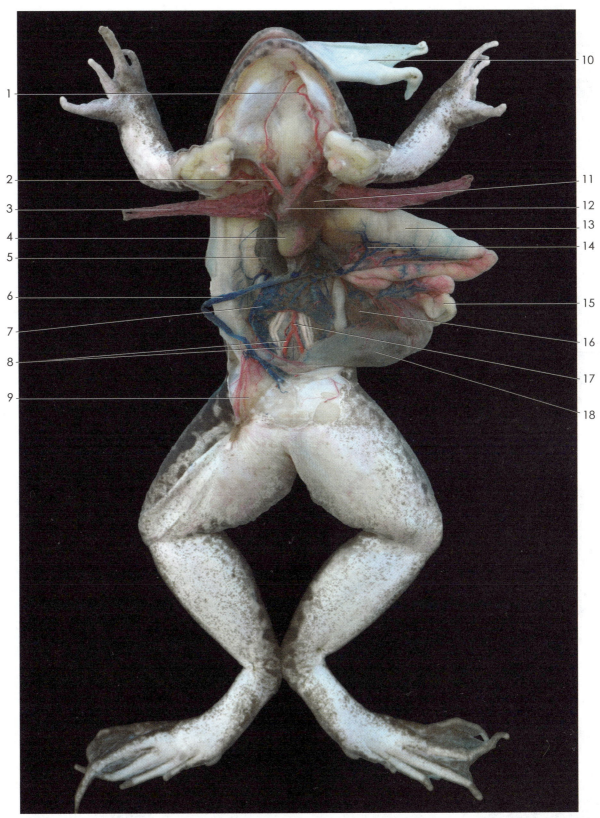

Figure 32.73 The internal anatomy of the frog.

1. External carotid artery
2. Truncus arteriosus
3. Lung (reflected)
4. Ventricle of heart
5. Liver (cut)
6. Ventral abdominal vein
7. Kidney
8. Sciatic arteries
9. Bladder
10. Tongue
11. Atrium of heart
12. Conus arteriosus
13. Stomach (reflected)
14. Gastric vein
15. Small intestine
16. Spleen
17. Dorsal aorta
18. Large intestine

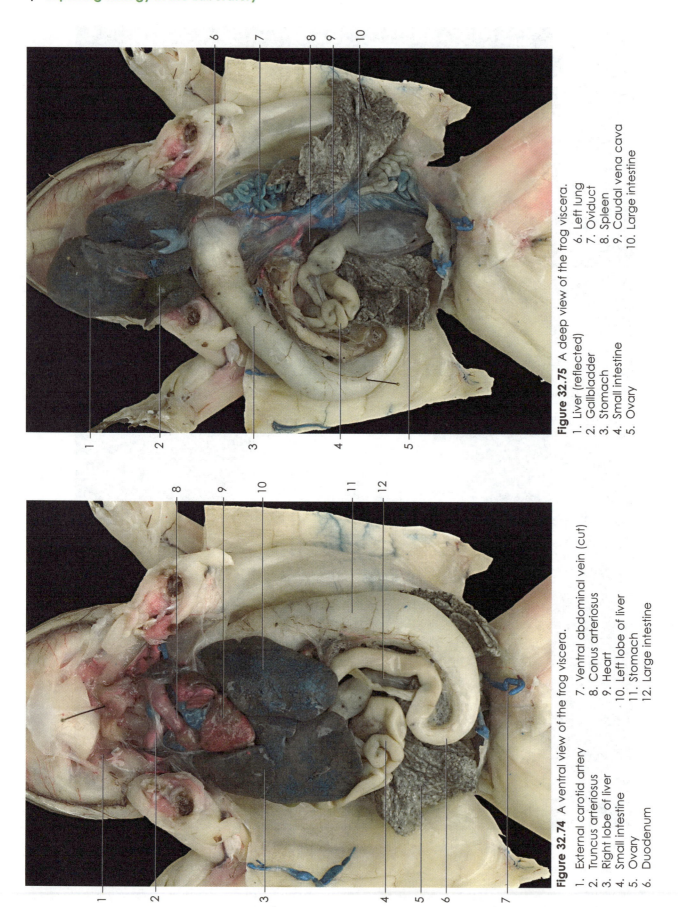

Figure 32.75 A deep view of the frog viscera.

1. Liver (reflected)
2. Gallbladder
3. Stomach
4. Small intestine
5. Ovary
6. Left lung
7. Oviduct
8. Spleen
9. Caudal vena cava
10. Large intestine

Figure 32.74 A ventral view of the frog viscera.

1. External carotid artery
2. Truncus arteriosus
3. Right lobe of liver
4. Small intestine
5. Ovary
6. Duodenum
7. Ventral abdominal vein (cut)
8. Conus arteriosus
9. Heart
10. Left lobe of liver
11. Stomach
12. Large intestine

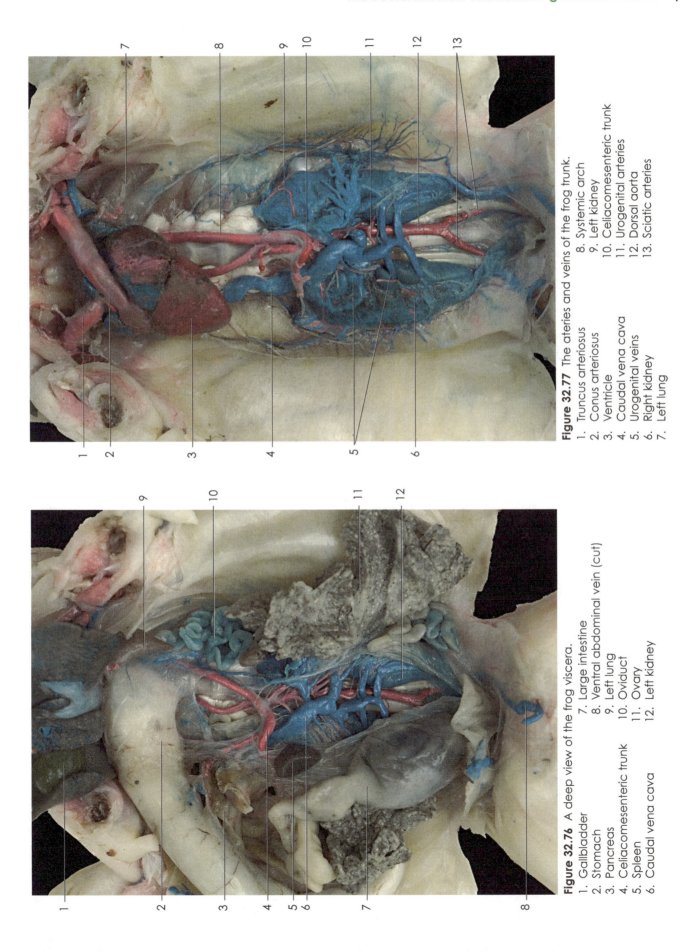

Figure 32.77 The arteries and veins of the frog trunk.

1. Truncus arteriosus
2. Conus arteriosus
3. Ventricle
4. Caudal vena cava
5. Urogenital veins
6. Right kidney
7. Left lung
8. Systemic arch
9. Left kidney
10. Celiacomesenteric trunk
11. Urogenital arteries
12. Dorsal aorta
13. Sciatic arteries

Figure 32.76 A deep view of the frog viscera.

1. Gallbladder
2. Stomach
3. Pancreas
4. Celiacomesenteric trunk
5. Spleen
6. Caudal vena cava
7. Large intestine
8. Ventral abdominal vein (cut)
9. Left lung
10. Oviduct
11. Ovary
12. Left kidney

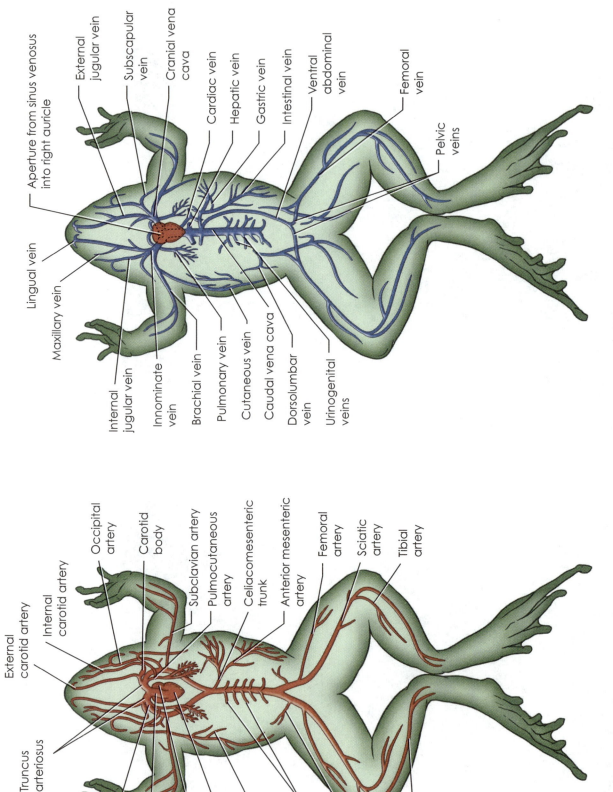

Figure 32.79 A diagram showing the veins of the frog.

Aperture from sinus venosus into right auricle

External jugular vein

Subscapular vein

Cranial vena cava

Cardiac vein

Hepatic vein

Gastric vein

Intestinal vein

Ventral abdominal vein

Femoral vein

Pelvic veins

Lingual vein

Maxillary vein

Internal jugular vein

Innominate vein

Brachial vein

Pulmonary vein

Cutaneous vein

Caudal vena cava

Dorsolumbar vein

Urinogenital veins

Figure 32.78 A diagram showing the arteries of the frog.

External carotid artery

Internal carotid artery

Occipital artery

Carotid body

Subclavian artery

Pulmocutaneous artery

Celiacomesenteric trunk

Anterior mesenteric artery

Femoral artery

Sciatic artery

Tibial artery

Truncus arteriosus

Systemic arch

Conus arteriosus

Atrium

Ventricle

Cutaneous artery

Urinogenital arteries

Epigastric artery

Peroneal artery

Student Activity—Dissection of the Bullfrog

Materials
- dissecting microscope or hand lens
- colored pencils
- dissecting tray
- dissecting kit
- safety glasses
- lab coat or lab apron
- gloves
- doubly-injected preserved bullfrog (*Rana catesbeiana*)

 ## Procedure 32.19
Bullfrog Dissection

Consider photographing this activity.

1. Procure the equipment and specimen. (Use the bull-frog from Procedure 32.18.)
2. Using sharp scissors, make two shallow incisions just through the muscles from the junction of the hind legs upward to the bottom of the jaw just a few millimeters on each side of the linea alba. You will have to cut through the pectoral girdle.
3. Using the scalpel, separate the blue ventral abdominal vein from the muscles. Proceed to move the muscular flap. Using the scissors, cut laterally through the external oblique muscle on each side from the "shoulder" to the "waist." Remove the muscle or pin it in place.
4. Determine the sex of the bullfrog. Locate the anatomical features and circulatory system features found in Figures 32.73 through 32.75. Observe the anatomy of a frog of the opposite sex from another laboratory station.
5. Record your observations below and draw and label the internal features of the bullfrog in the outline on page 695.

_____

_____

_____

_____

6. Thoroughly clean your laboratory station and equipment. Return the dry equipment and discard the frog as indicated by the instructor.

CLASS REPTILIA

With the evolution of the reptiles, the vertebrates finally conquered the land. Today, approximately 7,500 species of unique animals have been placed in **Class Reptilia**. Living reptiles include turtles, lizards, snakes, tuataras, and crocodiles. Presently, the classification of the reptiles is undergoing major revision and several themes have become popular. In the future, terms such as "non-avian reptiles" may be used to describe turtles, lizards, snakes, tuataras, and crocodiles (Fig. 32.80).

Modern reptiles share several fundamental traits in common. Whether oviparous or ovoviviparous, reptiles produce **amniotic eggs** that are ideal for the transition to land. These eggs possess a **yolk** for nourishment and four distinct membranes that are important in development: the yolk sac, the amnion, the chorion, and the allantois. The **yolk sac** provides food for the embryo, the **amnion** encases and cushions the developing embryo in a fluid-filled cavity, the **chorion** allows for the exchange of vital respiratory gases, and the **allantois** collects metabolic waste.

In addition to reptiles, the birds and mammals have amniotic eggs. The water-tight **shell** of the reptilian egg is highly protective. Fertilization occurs **internally** before the egg forms. The tough, dry, and scaly **integument** of reptiles protects them from desiccation and injury. The skeleton of reptiles is **well-ossified** and strong. The limbs are paired with five toes, with the exception of snakes, amphisbaenians, and some lizards. Usually two sacral vertebrae support the pelvic girdle in reptiles. They possess well-developed **lungs** and undergo **thoracic breathing**, in which specialized muscles and ribs provide for the transport of copious amounts of air into and out of the lungs. With the exception of crocodilians, which have a four-chambered heart, reptiles have a three-chambered heart.

As **ectotherms,** reptiles depend upon the environment for thermoregulation. Reptilians have **complete digestive systems** with accessory structures. Reptiles have well-developed **kidneys** and urine is voided in a semi-solid mass containing uric acid crystals. Reptiles possess a well-developed nervous system with twelve pair of cranial nerves. Various sensory structures exist in reptiles. The reptiles are dioecious and no larval stage appears in reptiles.

In studying amniotes, three types of skull patterns become apparent. **Anapsid skulls** do not have openings in the temporal region of the skull behind the orbit. Turtles possess anapsid skulls. The remainder of the reptiles and birds possess **diapsid skulls**, in which two temporal openings are evident on each side of the skull. Fossil mammal-like reptiles and modern mammals possess a **synapsid skull** featuring a single pair of temporal openings. These features have been important in classifying reptiles (Fig. 32.81). Despite a rich history, only four living orders of reptiles exist today. **Order Chelonia** consists of turtles and tortoises, **Order Squamata** includes lizards, snakes, and amphisbaenians, **Order Sphenodonta** consists of tuataras, and **Order Crocodillia** includes alligators and crocodiles.

While in the Galapagos Islands, young Charles Darwin was fascinated by and took the opportunity to ride one of the large Galapagos tortoises. These tortoises, along with the common box turtle, gopher tortoise, sliders, soft-shelled turtles, snapping turtles, and marine turtle,

Figure 32.80 Members of the Class Reptilia: (a) a star tortoise, *Geochelone elegans*, (b) a green basilisk, *Basiliscus plumifrons*, (c) kingsnake, *Lampropeltis getulus*, (d) a tuatara, *Sphenodon punctatus*, and (e) American alligator, *Alligator mississippiensis*.

have been placed in **Order Chelonia (Testudines)**. Approximately 300 species of turtles have been described. Although the majority of turtles are aquatic, terrestrial and marine species exist (Fig. 32.82).

The most distinguishing feature of turtles is the **protective shell,** which provides turtles with an adequate defense. Many species can retract their head and appendages into the shell. The box turtle, *Terrepene Carolina*, possesses a **hinge** that actually closes up the head region. The dorsal portion of the shell is called the **carapace**, and the ventral portion is the **plastron.** The shell of a turtle is composed of hard, bony plates covered by corresponding keratinized **scutes**. Soft-shelled turtles have a pliable, leathery shell. The ribs and body vertebrae are fused to the interior of the carapace in turtles. The shoulder and hip girdles of turtles are located within the rib cage, instead of outside the rib cage. Instead of teeth, turtles have a sharp, keratinized covering over their maxilla and mandible (Fig. 32.83–Fig. 32.84).

Although the brain of a turtle is rather small, the sense of sight is well developed. Turtles are dioecious and oviparous. In several turtle families, such as crocodiles and some lizards, the sex of turtles is determined by temperature. In **temperature–dependent sex determination**, high nest temperature results in females and low nest temperature results in males.

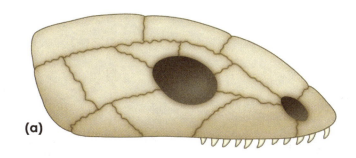

(a)

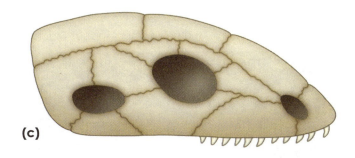

(b)

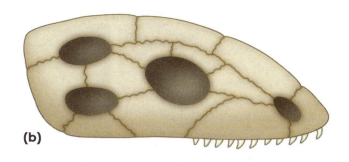

(c)

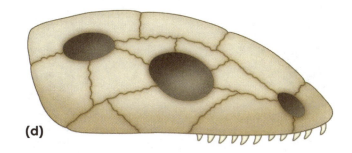

(d)

Time Traveler

In June 2006, the biological world was saddened by the death of Harriet the giant Galapagos land tortoise (*Geochelone nigra porteri*). It is believed that Harriet was collected by Charles Darwin in 1835 while visiting the Galapagos Islands on board the *HMS Beagle*. Records indicate that Harriet was brought to Australia by a former Captain of the Beagle, John Wickham in 1841. Harriet died in Australia Zoo in 2006, the same year its co-owner and TV personality Steve Irwin (Crocodile Hunter) died.

I have destroyed almost the whole race of frogs, which does not happen in that savage Batrachomyomachia of Homer.... For in this (frog anatomy) owing to the simplicity of the structure, and the almost complete transparency of the vessels which admits the eye into the interior, things are more clearly shown so that they will bring the light to other more obscure matters.

—**Marcello Malpighii (1628–1694)**

Figure 32.81 Amniote skulls: (a) An anapsid skull, (b) a diapsid skull, and (c) and (d) synapsid skulls.

Figure 32.82 Examples from the Order Chelonia: (a) Hawaiian green sea turtle, *Chelonia mydas*, (b) desert tortoise, *Gopherus agassizii*, (c) red-eared slider, *Trachemys scripta elegans*, (d) Aldabra giant tortoise, *Dipsochelys dussumieri*, (e) spiny soft shell turtle, *Apalone spinifera*, and (f) Alligator snapper, *Macrochelys temminckii*.

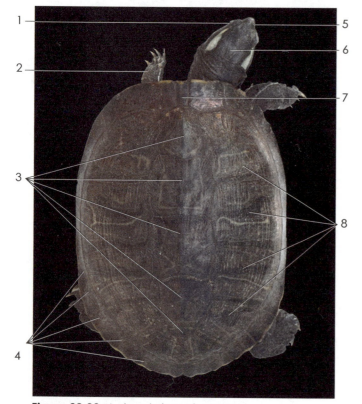

Figure 32.83 A dorsal view of a turtle.

1. Eye
2. Pentadactyl foot
3. Vertebral scales
4. Marginal scales
   (encircle the carapace)
5. Nostril
6. Head
7. Nuchal scale
8. Costal scales

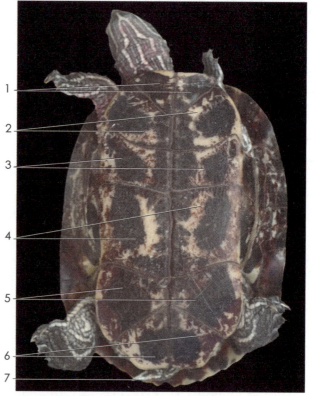

Figure 32.84 A ventral view of a turtle.

1. Gular scales
2. Humeral scales
3. Pectoral scales
4. Abdominal scales
5. Femoral scales
6. Anal scales
7. Tail

Order Squamata, the largest order of reptiles, encompasses nearly 7,000 species of lizards, snakes, and amphisbaenians. Squamates possess a diapsid skull with movable joints, called a **kinetic skull,** because of the development of the **quadrate bones.** The modified skull allows squamates to seize, hold, and swallow prey efficiently. Members of this order also possess distinct scales. Male squamates have a **hemipenis,** used in copulation. Squamates also possess a **Jacobson's organ** (vomeronasal organ) that enhances the sense of smell. A snake flicking its tongue is carrying back molecules to the Jacobson's organ on the roof of the mouth for evaluation. Several members of this order, such as the Gila monster and

rattlesnake, are venomous. Two living suborders of squamates are **Suborder Sauria** (lizards and amphisbaenians) and **Suborder Serpentes** (snakes) (Fig. 32.85).

The lizards and amphisbaenians, **Suborder Sauria,** are a diverse group of reptiles. The majority of the nearly 4,000 species of lizards are terrestrial, but aquatic and marine forms do exist. Lizards have an elongated body and, with the exception of the legless lizards, have paired appendages. In contrast to snakes, lizards have external auditory openings and a pair of eyelids. All lizards have the ability to lose their tail, and many can regenerate the lost tail. Some lizards have a colorful flag-like structure, the **dewlap,** in the throat region. Well-known lizards include

Figure 32.85 Images of squamates: (a) Jackson's chameleon, *Chamaeleo jacksonii*, (b) Komodo dragon, *Varanus komodoensis*, (c) emerald swift, *Sceloporus malachiticus*, (d) Eastern glass lizard, *Ophisaurus ventralis*, is a legless lizard (e) anole, *Anolis carolinensis*, (f) gila monster, *Heloderma suspectum*, and (g) a fire skink, *Riopa fernandi*.

anole lizards, geckos, skinks, chameleons, iguanas, monitor lizards, Komodo dragons, Gila monsters, and bearded dragons.

Amphisbaenians also are known as "worm-lizards" because they lack limbs and auditory openings and have underdeveloped eyes. The worm-like body has numerous independently moving rings that are used in locomotion. Approximately 160 species of amphisbaenians have been identified. One species found in Florida is known as the "graveyard snake."

Unfortunately, snakes, **Suborder Serpentes**, have a bad reputation because of the several dangerous species among them. Of the 3,000 species of snakes, the majority are harmless, but rattlesnakes, moccasins, copperheads, coral snakes, cobras, mambas, kraits, sea snakes, taipans, the African boomslang, Russell's viper, and the Fer de Lance keep the reputation going. Snakes first appeared in the early Cretaceous Period, evolving from burrowing lizards. Constricting snakes such as boas and pythons appear to be the oldest group and still retain **vestigial** hindlimbs, called **spurs**. Today, snakes are found on every continent except Antarctica (Fig. 32.86).

Snakes possess an elongate body with numerous vertebrae. The vertebrae (perhaps as many as 400) and their associated ribs are essential in the locomotion of snakes. Snakes have a complete digestive system. They can detect vibrations but do not have a good sense of hearing. Snakes lack eyelids and do not blink. In addition, they have a protective transparent membrane, the **spectacle**, over their eye. The **Jacobson's organ** is well-developed in these reptiles. Snakes known as **pit-vipers** possess heat-sensitive pits in the head region. Pit vipers include rattlesnakes, copperheads, and water moccasins. Vipers have fangs hidden within a sheath. When a viper strikes its prey, venom is injected into the prey. Most snakes are oviparous, although some ovoviviparous species exist (Fig. 32.87–Fig. 32.89).

Figure 32.86 Examples of snakes: (a) water moccasin, *Agkistrodon piscivorus*, (b) ball python, *Python regius*, (c) Arizona coral snake, *Micruroides euryxanthus*, (d) Western coachwhip, *Masticophis flagellum*, (e) sidewinder rattlesnake, *Crotalus cerastes*, (f) scarlet kingsnake, *Lampropeltis triangulum elapsoides*, and (g) great basin rattlesnake, *Crotalus viridis lutosis*.

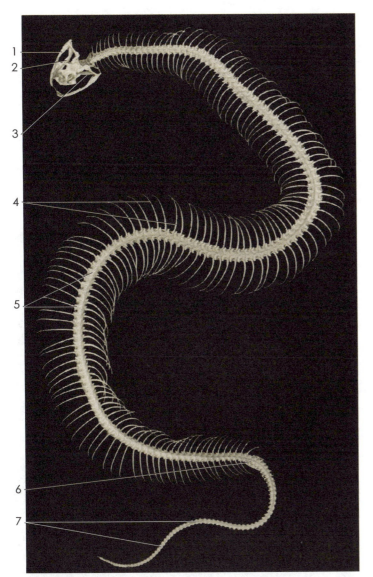

Figure 32.87 The skeleton of a snake (python).

1. Dentary
2. Quadrate bone
3. Supratemporal bone
4. Ribs
5. Trunk vertebrae
6. Vestigial pelvic girdle
7. Caudal vertebrae

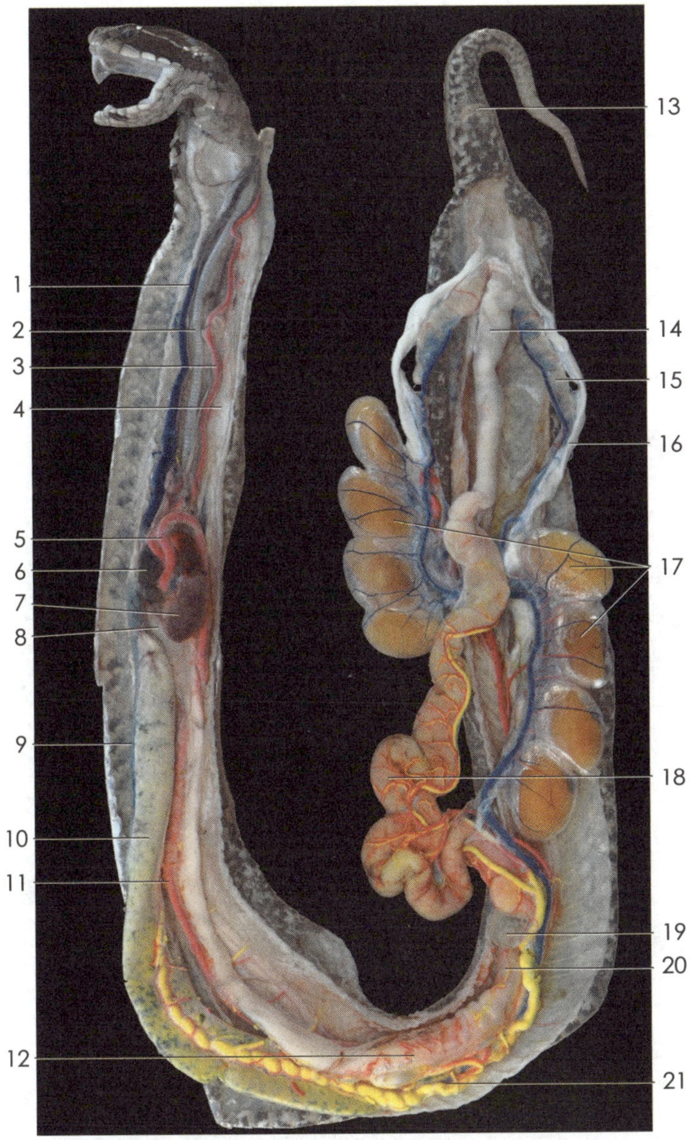

Figure 32.88 A ventral view of the internal anatomy of a female water moccasin, *Agkistrodon piscivorus*.

1. Jugular vein
2. Trachea
3. Common carotid artery
4. Esophagus
5. Aortic arch
6. Auricle of heart
7. Ventricle of heart
8. Lung
9. Hepatic portal vein
10. Liver
11. Dorsal aorta
12. Stomach
13. Anus
14. Colon
15. Kidney
16. Oviduct
17. Eggs
18. Small intestine
19. Pancreas
20. Duodenum
21. Abdominal vein

Figure 32.89 A garter snake, *Thamnophis sirtalis*, extending its tongue. Reptiles use their tongue in conjunction with the Jacobson's organ or vomeronasal organ, an auxiliary olfactory organ, to aid with smell.

Members of **Order Crocodilia** such as alligators, crocodiles, caimans, and gharials, along with the birds, are the only living descendents of archosaurian linage (Fig. 32.90). Crocodilians possess an elongate, heavy skull with a robust mandible and maxilla housing large teeth set into bony sockets (**thecodont**). Members of order Crocodilia possess a secondary bony palate in the mouth that allows them to breathe air when they are partially submerged and when their mouth is full of water. Crocodilians have a four-chambered heart, a complete digestive system and a well-developed nervous system.

Crocodilians have excellent vision, including color vision. The presence of a light-reflecting layer, the **tapetum**, behind the retina greatly increases their ability to see at night. The tapetum also is responsible for their eyes seeming to glow at night. A **nicitating membrane** covers the eye when the animal is underwater. Numerous **sensory pits** line the jaws of crocodilians and serve as a type of lateral line system. Crocodilians are oviparous, and some species show parental care. Crocodilians undergo temperature-dependent sex determination.

Figure 32.90 Examples of crocodilians: (a) American alligator, *Alligator mississippiensis*, (b) Johnston's fresh-water crocodile, *Crocodylus johnstoni*, (c) Cuvier's dwarf caiman, *Paleosuchus palpebrosus*, and (d) a gharial, *Gavialis gangeticus*.

Student Activity—Observations of Representative Members of Class Reptilia

Materials

- dissecting microscope
- hand lens
- colored pencils
- dissecting tray
- probe
- select specimens of the following reptiles: turtles, lizards, amphisbaenians, snakes, and a crocodilian

Procedure 32.20
Observations of Class Reptilia

Consider photographing this activity.

1. Procure the needed equipment and specimens.
2. Using a dissecting microscope or hand lens, observe the available specimens. If permitted, remove the specimen from the jar for closer examination. Pay close attention to the previous description of the organisms.
3. Record your observations and labeled sketches below. When possible, provide the scientific name of the specimen.
4. Thoroughly clean your laboratory station and equipment. Return the dry equipment.

Lizards

Amphisbaenians

Turtles

Snakes

Crocodilians

Check Your Understanding

Q. To invade new niches, what fundamental characteristics were necessary in amphibians and reptiles?

Q. Describe three skull types found in amniotes.

Q. Why does a snake flick its tongue?

The zoologist is delighted by the differences between animals, whereas the physiologist would like all animals to work in fundamentally the same way.

—Sir Alan Lloyd Hodgkin (1914–1998)

CLASS AVES

The next time you feed your canary, you are feeding a distant relative of a velociraptor. The age-old question, "Where have all the dinosaurs gone?' can be answered today as, "They are the birds!" Members of **Cass Aves,** birds are among the most colorful and intriguing animals on the planet. Approximately 9,700 extant species of birds have been described (Fig. 32.91 and Fig. 32.100).

Birds are bipedal, oviparous, warm-blooded (endothermic) winged tetrapods. Their most distinguishing characteristic is the presence of feathers. The specific arrangement and appearance of feathers is called the bird's **plumage**. Plumage may vary within a bird species with respect to age and sex. Feathers are arranged on the bird's body in specific tracts, the **pterylae**. A typical bird has several types of feathers:

- **Contour feathers** are typical feathers and provide the bird with shape and aid in flight.
- **Down feathers** are soft feathers that help to insulate a bird.
- **Natal down feathers** are the typical feathers that make a duckling or chick appear fluffy.
- **Filoplumes** are simple, hair-like feathers associated with contour feathers and provide sensory feedback on contour feather activity.
- **Bristles** are sensory and protective feathers found near the mouth of some birds such as flycatchers.

Feathers require constant maintenance, and birds preen or groom them every day. Birds apply an oily substance from the **uropygial gland**, located near the base of the tail, to discourage microbial growth and for waterproofing. A typical contour feather consists of a smooth, nonpigmented base extending beneath the skin, called the **calamus** (quill). The **inferior umbilicus** at the base of the calamus appears as a small hole. The shaft above the skin is called the **rachis**. The **vane** of a feather is the flat structure on each side of the feather, composed of filaments called **barbs**. The barbs, in turn, consist of **barbules** connected together by **hooklets** (Fig. 32.92–Fig. 32.95).

Birds have a light but sturdy skeleton. Bird bones have numerous **air cavities**, which makes them lightweight. To make flight possible, birds' vertebrae are fused, with the exception of the neck (cervical) vertebrae. The tail, or caudal vertebrae, are fused, forming a **pygostyle**. And the trunk vertebrae form a **synsacrum**. The sturdy pelvic girdle allows birds to walk and perch. In many birds, the sternum is keeled, allowing for the attachment of flight muscles. The "wishbone," or **furculum**, of a bird consists of fused clavicles that allow for flight. Bird skulls are highly fused and feature large **orbits** to accommodate the eyes. Birds have a single occipital condyle. A ring of bones called **sclerotic bones** encircles and supports the eye of birds. Flight muscles are massive in birds of flight. Bird **talons**, such as those of eagles, can produce great force (Fig. 32.96–Fig. 32.97).

The circulatory system of birds features a large, strong, four-chambered heart. Metabolism and heart rate are heightened. A chicken may have a resting heart rate of 250 beats per minute (bpm), and an active hummingbird more than 1,200 bpm. Birds have an advanced immune system. Their respiratory system is complex. Much of the air inhaled goes into a system of nine **air sacs**, located primarily in the thorax and the abdomen. The air sacs, coupled with the lungs, keep the bird supplied with rich air to fuel their demanding oxygen requirement. A **syrinx**, located at the junction where the trachea forks into the lungs, allows birds to produce a myriad of sounds and songs.

Members of Class Aves are voracious eaters. Birds exhibit a variety of feeding preferences and habits. The keratinized, toothless **beaks** of birds are highly specialized for specific diets (Fig. 32.98–Fig. 32.99). The digestive system is complete and features some interesting structures:

The **crop** is an enlargement of the esophagus that serves as a storage chamber. The crop of a pigeon or duck is distended after feeding. The stomach has two unique compartments. Because birds do not have teeth, a muscular **gizzard** helps to grind food. The gizzard holds pebbles that help in the mechanical breakdown of food. The **proventriculus** secretes gastric juices to help chemically break down food. Solid wastes, urine, and reproductive cells exit the body by means of the **cloaca**. Birds excrete nitrogenous wastes as uric acid crystals. Marine birds have **salt glands** above each eye to eliminate excess salt.

In birds, the nervous system is well-developed. The three major portions of the brain are:

1. the **cerebrum**, which primarily controls complex behavior patterns, migration, navigation, mating behavior, and nest building;
2. a fine-tuned **cerebellum**, which controls muscular coordination and flight-related matters; and
3. large **optic lobes**, responsible for acute vision and association.

Birds have good color vision and excellent monocular and binocular vision. A **nicitating membrane** lubricates and protects the eye. With the exception of flightless birds, waterfowl, and several other groups, the senses of smell and taste are poorly developed in birds. Hearing in birds, like vision, is highly developed.

Figure 32.91 Examples of birds: (a) brown pelican, *Pelecanus occidentalis*, (b) roseate spoonbill, *Ajaja ajaja*, (c) snow goose, *Chen caerulescens*, and (d) magellanic penguin, *Spheniscus magellanicus*.

714

All birds are oviparous. In most male species, the **testes** becomes active only during breeding season. The majority of male bird species lack a penis. In females, only the left **ovary** and **oviduct** develop. To reproduce, most birds must bring their **cloacal surfaces** together. Fertilization occurs in the upper region of the oviduct before **albumin** (egg white) and the **shell** are added to the egg. Eggs usually are laid in a nest and are incubated by one or both parents. Upon hatching, some young, such as chicks and ducklings, are **precocial**—ready to run or swim. Other birds, such as the mockingbird and canary, are **altricial** (helpless), requiring parental care for a period of time.

Extant birds have been divided into two convenient Superorders. The **Superorder Paleognathae** includes flightless birds (raties) such as the ostrich, emu, cassowary, rhea, kiwi, and tinamous. The ratites have no keel on their sternum, thus making flight impossible. **Superorder Neognathae** is composed of 27 orders of modern birds. These birds possess a keeled sternum, but some are flightless. Example neognathans are ducks, crows, eagles, sparrows robins, woodpeckers, hummingbirds, penguins, parrots, pelicans, and owls.

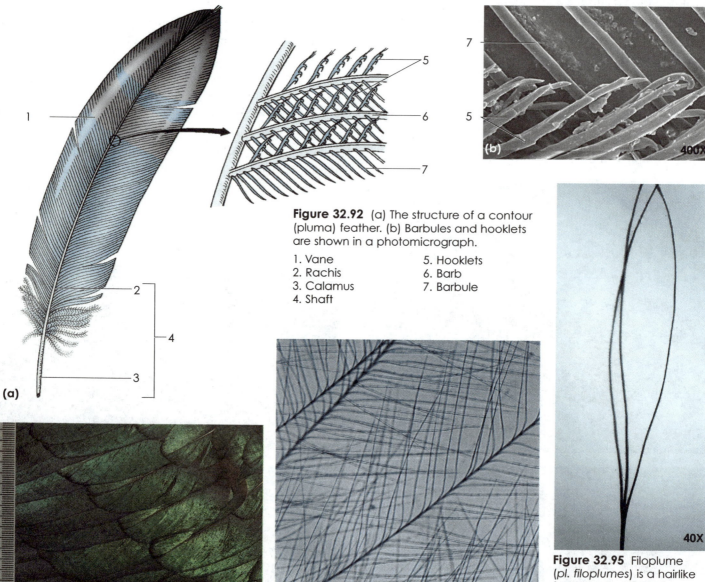

Figure 32.92 (a) The structure of a contour (pluma) feather. (b) Barbules and hooklets are shown in a photomicrograph.

1. Vane
2. Rachis
3. Calamus
4. Shaft
5. Hooklets
6. Barb
7. Barbule

Figure 32.93 Pluma (*pl. plumae*) are contour feathers that include the quill feathers and flight feathers. Quill feathers cover most of the body. Flight feathers are confined to the wings as remiges and to the tail as retrices (scale in mm).

Figure 32.94 Plumula (*pl. plumulae*) is a soft insulating feather that lacks a rachis and hooklets. Also called down feathers, plumulae are abundant on the breast and abdomen of a bird. Plumulae are shown in this micrograph.

Figure 32.95 Filoplume (*pl. filoplumes*) is a hairlike feather consisting of a single rachis and either no barbs or just a few at its distal tip. Filoplumes are scattered throughout the plumae. A filoplume is shown in this micrograph.

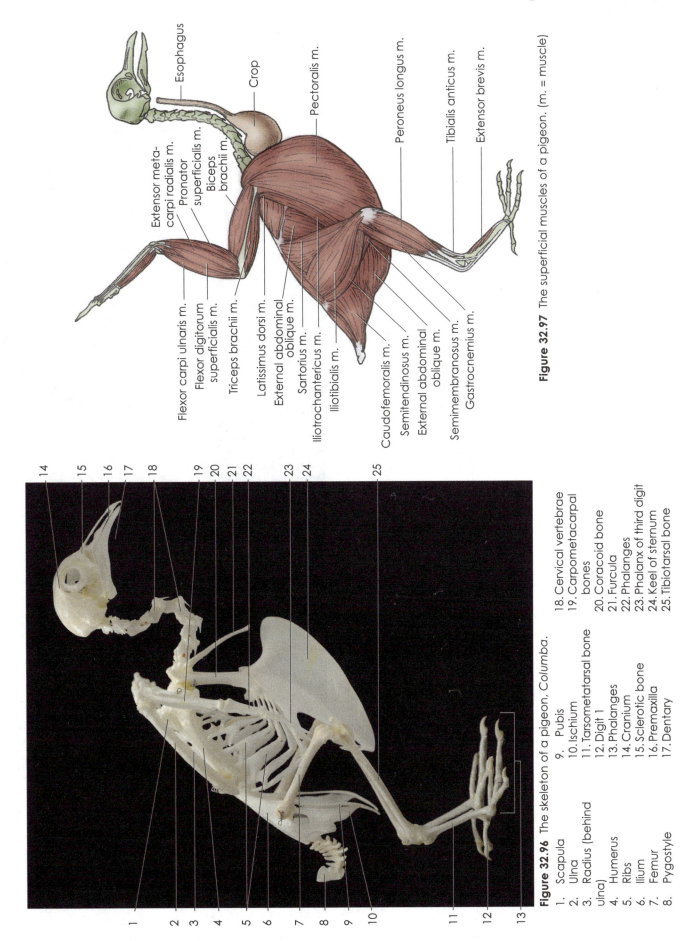

Figure 32.97 The superficial muscles of a pigeon. (m. = muscle)

Esophagus
Crop
Pectoralis m.
Peroneus longus m.
Tibialis anticus m.
Extensor brevis m.

Extensor meta-carpi radialis m.
Pronator superficialis m.
Biceps brachii m.

Flexor carpi ulnaris m.
Flexor digitorum superficialis m.
Triceps brachii m.
Latissimus dorsi m.
External abdominal oblique m.
Sartorius m.
Iliotrochantericus m.
Iliotibialis m.
Caudofemoralis m.
Semitendinosus m.
External abdominal oblique m.
Semimembranosus m.
Gastrocnemius m.

Figure 32.96 The skeleton of a pigeon, *Columba*.

1. Scapula
2. Ulna
3. Radius (behind ulna)
4. Humerus
5. Ribs
6. Ilium
7. Femur
8. Pygostyle
9. Pubis
10. Ischium
11. Tarsometatarsal bone
12. Digit 1
13. Phalanges
14. Cranium
15. Sclerotic bone
16. Premaxilla
17. Dentary
18. Cervical vertebrae
19. Carpometacarpal bones
20. Coracoid bone
21. Furcula
22. Phalanges
23. Phalanx of third digit
24. Keel of sternum
25. Tibiotarsal bone

Generalist

Insect catching

Grain eating

Nut cracking

Nectar feeding

Fruit eating

Chiseling

Dip netting

Surface skimming

Filter feeding

Probing

Pursuit fishing

Scavenging

Flesh eating

Figure 32.98 Variation in bird beaks.

Over increasingly large areas of the United States, spring now comes unheralded by the return of the birds, and the early mornings are strangely silent where once they were filled with the beauty of bird song.

—Rachel Carson (1907–1964)

(a)

(b)

Figure 32.99 The function of beaks determines the size and shape. (a) The great blue heron, *Ardea herodias*, spears its prey, while (b) the flamingo, *Phoenicopterus chilensis*, has evolved a beak that allows it to filter feed with its head pointed down.

Coraciiformes
(kingfisher)

Sphenisciformes
(penguin)

Strigiformes
(owl)

Columbiformes
(dove)

Falconiformes
(falcon)

Anseriformes
(duck)

Gaviiformes
(loon)

Charadriiformes
(gull)

Pelecaniformes
(pelican)

Ciconiiformes
(heron)

Podicipediformes
(grebe)

Struthioniformes
(ostrich)

Galliformes
(pheasant)

Piciformes
(toucan)

Passeriformes
(bluebird)

Apodiformes
(hummingbird)

Phoenicopteriformes
(flamingo)

Psittaciformes
(parrot)

Figure 32.100 Representative bird orders.

Student Activity—Observations of Bird Feathers

Materials
- dissecting microscope or hand lens
- colored pencils
- compound microscope
- dissecting tray
- probe
- a variety of bird feathers
- microscope slides of bird feathers

 Procedure 32.21
Feathers

Consider photographing this activity.

1. Procure the needed equipment and specimens.
2. Observe the available feather specimens with a dissecting microscope or hand lens. Record your observations and labeled sketch below.

3. Observe a microscope slide of a feather, noting the ultra structure of the feather.
4. Record your observations and labeled sketches below.

5. Thoroughly clean your laboratory station and equipment. Return the dry equipment.

 Student Activity—Observations of a Pigeon Skeleton

Materials
- hand lens
- colored pencils
- skeleton of a pigeon (*Columba* spp.)

Procedure 32.22
Pigeon Skeleton

Consider photographing this activity.

1. Procure the equipment and specimen.
2. Using the hand lens, observe the skeleton of the pigeon, and compare the bones to Figure 32.96. Pay particular attention to the special features discussed in the narrative.
3. Record your observations, and sketch and label the bones of the pigeon on the next page.

4. Thoroughly clean your laboratory station and equipment. Return the dry equipment.

CLASS MAMMALIA

Members of **Class Mammalia** are extremely diverse and remarkable animals; having conquered marine, freshwater, aerial, and terrestrial habitats. Presently, nearly 5,000 species of mammals have been described, including the duck-billed platypus, koala bears, squirrels, mountain lions, dolphins, aardvarks, zebras, and orangutans (Fig. 32.101–Fig. 32.102).

One group of extant mammals, the **monotremes**, are egg-laying mammals. The duck-billed platypus and echidna are the final vestiges of this ancient group. About 120 million years ago, the **marsupial** (pouched) mammals appeared. Living marsupials include wombats, kangaroos, sugar gliders, Tasmanian devils, and the Virginia Opossum. Shortly after the marsupials, another branch, the **placental** mammals, appeared. In placental mammals the embryo remained in the uterus, receiving vital nutrients and oxygen from the mother for an extended time, allowing for further development of the brain. Representative placental mammals include shrews, bats, dogs, manatees, and humans.

The most obvious characteristic of mammals is the presence of hair. Hair can provide insulation and protection, serve as camouflage or warning, and give sensory feedback. The coat of a mammal is called its **pelage**.

Vibrissae, or whiskers, are long, coarse hairs used in gathering tactile information. In porcupines, modified hairs called **quills** are used for defense, and rhinoceros horns are made of keratinized hair-like filaments. Some mammals, such as cows and sheep, have **true horns**, consisting of keratin that covers a bony core. Male deer and moose have **antlers**, which are actually bony structures covered in epidermis, or **velvet**, at the beginning of each breeding season. Antlers are shed yearly.

Mammals also possess a number of glands in the integument, including sweat glands, scent glands, sebaceous (oil), and mammary glands. Members of class Mammalia are **endothermic** (warm-blooded), maintaining their own body temperature. Mammals have a four-chambered heart and a respiratory system driven by a muscular diaphragm. Mammals can be carnivorous, herbivorous, or omnivorous, each with specific modifications to the complete digestive system. Mammals exhibit a number of types of **dentition** based on their diet, from toothless anteaters to the conical, fish-eating **homodont** teeth of dolphins. Most mammals exhibit a number of tooth types (**heterodont)** including incisors, canines, premolars, and molars.

In mammals, the lower jaw is a single bone. The middle ear of mammals contains three **ossicles** (bones): the stapes (stirrup), incus (anvil), and malleus (hammer). With the exception of several mammals, all mammals from mice to giraffes have seven **cervical** (neck) **vertebrae**. The nervous system of mammals is well-developed and features a large cerebrum in the brain, responsible for the majority of complex behaviors and learning in mammals. The senses are developed in various mammals depending upon their role in nature. The special sense of echo-location is extremely well-developed in bats and dolphins.

Mammals are dioecious animals and undergo internal fertilization. The duck-billed platypus and the echidna are oviparous mammals. In marsupials, pregnant females develop a modified yolk sac in the womb, which provides the embryo with nutrients. Marsupials give birth to their altricial young at an early stage of embryological development. After birth, the newborn marsupial crawls to the pouch (marsupium) and attaches itself to a nipple for nourishment. The majority of mammals are placentals. The placenta serves as the interface between the embryo and the mother. The young are born **altricial** (such as kittens and mice) or **precocial** (such as calves, and dolphins).

Presently, 26 orders of mammals are recognized. The monotremes, or **protoherians,** are represented by one order; the marsupials, or **metatherians**, are represented by seven orders; and the placentals, or **eutherians**, are represented by 18 orders. Figure 32.102 illustrates some representative orders of mammals.

Figure 32.101 Examples of mammals: (a) echidna, *Tachyglossus aculeatus*, (b) Eastern grey kangaroo, *Macropus giganteus*, (c) bottlenose dolphin, *Tursiops truncatus*, (d) lion, *Panthera leo*, (e) meerkat, *Suricata suricatta*, and (f) gorilla, *Gorilla gorilla gorilla*.

Figure 32.102 Representatives from some of the orders of mammals.

Student Activity—Observations of Representative Members of Class Mammalia

Materials
- dissecting microscope or hand lens
- colored pencils
- dissecting tray
- probe
- select specimens of representative mammals

Procedure 32.23
Mammal Observations

Consider photographing this activity.

1. Procure the needed equipment and specimens.
2. Observe the available specimens using a dissecting microscope or hand lens. If permitted, remove the specimens from their jars and place them on the dissecting trays for closer examination. Pay close attention to the previous description of the organisms.
3. Record your observations and labeled sketches below. Along with the common name, when possible, provide the order and the scientific name of the specimen.

4. Thoroughly clean your laboratory station and equipment. Return the dry equipment.

Student Activity—Dissection of the Fetal Pig (*Sus scrofa*)

Materials
- dissecting microscope
- hand lens
- colored pencils
- dissecting tray
- dissecting kit [is scalpel included?]
- safety glasses
- lab coat or lab apron
- gloves
- paper towel
- preserved fetal pig (*Sus scrofa*)

 Procedure 32.24
Fetal Pig Dissection

Consider photographing this activity.

1. Procure the needed equipment and specimen.
2. Thoroughly rinse the fetal pig in running water.
3. Examine the external anatomy of the fetal pig, comparing it to Figure 32.103 and 32.104.
4. Place your specimen on a dissecting tray, ventral side up. Using a sharp scalpel, make a shallow incision through the skin, extending from the chin caudally to the umbilical cord. Carefully continue your cut around one side of the umbilical cord. If your specimen is a male, make a diagonal cut from the umbilical cord to the scrotum. If a female, continue a mid-ventral incision from the umbilical cord to the genital papilla. Make an incision around the genitalia and tail.

5. From the mid-ventral incision, extend an incision down the medial surfaces of the front legs to the hoofs, then do the same for the skin of the hindlegs. Make circular incisions around each of the hooves. Following the ventral borders of the lower jaws, make extended cuts from the chin dorsolaterally to just below the ears.
6. Grasp the cut edge of the skin, and carefully remove it from your specimen. If the skin is difficult to remove, grasp the cut edge of the skin with one hand and push on the muscle with the thumb of the other hand.
7. After the specimen is skinned, the muscles can be seen more easily if the moisture on it is sponged away with a paper towel. The muscles of a fetal pig are extremely delicate, and as you proceed to dissect your specimen, make certain that you separate the muscles along their natural boundaries. When transection of a muscle is necessary, carefully isolate the muscle from its attached connective tissue and make a clean cut across the belly of the muscle, leaving the origin and insertion intact.
8. Record your observations and sketch and label the external and internal anatomy of the pig fetus on the next page (See Fig. 32.105–Fig 32.119 for reference).
9. Thoroughly clean your laboratory station and equipment. Return the dry equipment and discard the fetal pig as indicated by the instructor.

Internal anatomy of a pig fetus (Procedure 32.28)

External anatomy of a pig fetus (Procedure 32.28)

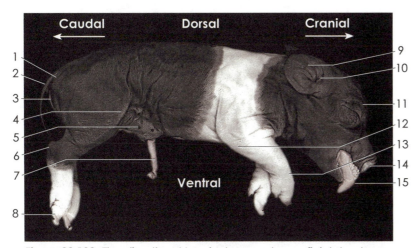

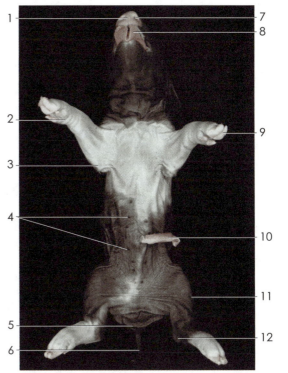

Figure 32.103 The directional terminology and superficial structures in a fetal pig (quadrupedal vertebrate).

1. Anus
2. Tail
3. Scrotum
4. Knee
5. Teat
6. Ankle
7. Umbilical cord
8. Hoof
9. Auricle (pinna)
10. External auditory canal
11. Superior palpebra (superior eyelid)
12. Elbow
13. Wrist
14. Naris (nostril)
15. Tongue

Figure 32.104 A ventral view of the surface anatomy of the fetal pig.

1. Nose
2. Wrist
3. Elbow
4. Teats
5. Scrotum
6. Tail
7. Nostril
8. Tongue
9. Digit
10. Umbilical cord
11. Knee
12. Ankle

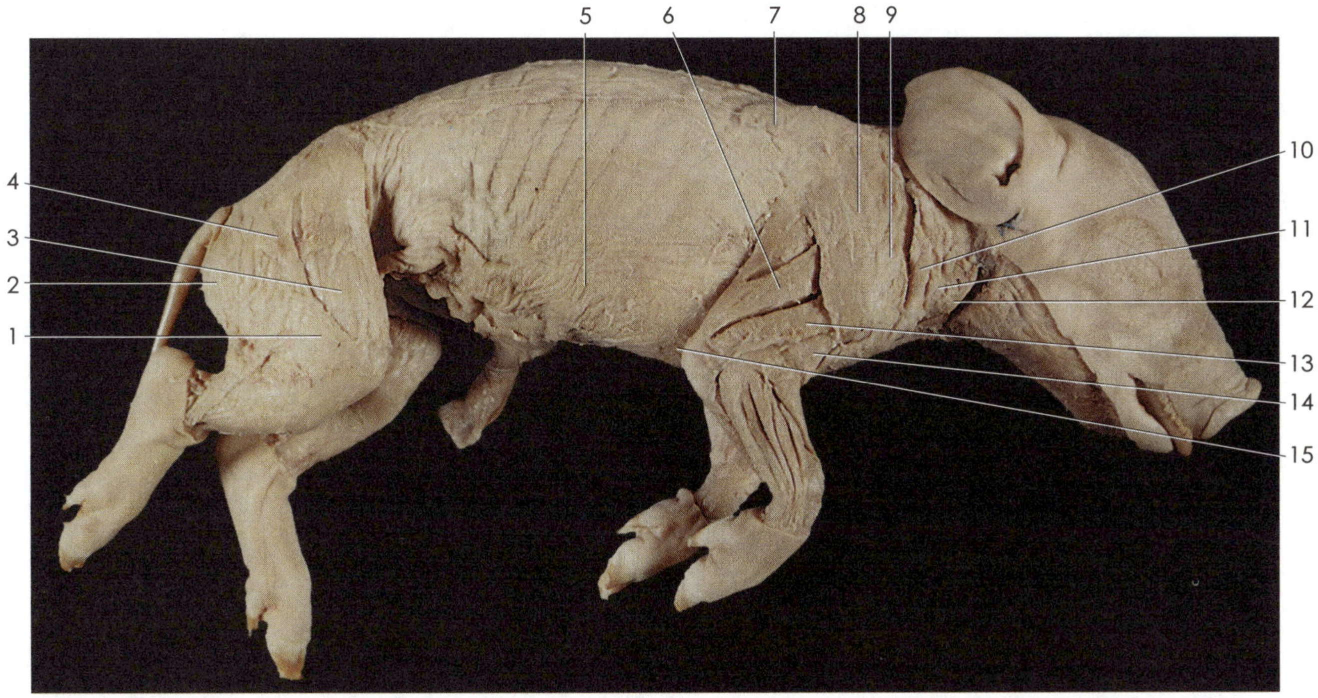

Figure 32.105 A lateral view of superficial musculature of the fetal pig.

1. Biceps femoris m.
2. Semitendinosus m.
3. Tensor fasciae latae m.
4. Gluteus medius m.
5. External abdominal oblique m.
6. Triceps brachii m. (long head)
7. Trapezius m.
8. Deltoid m.
9. Supraspinatus m.
10. Cleidooccipitalis m.
11. Cleidomastoid m.
12. Sternocephalicus m.
13. Triceps brachii m. (lateral head)
14. Brachialis m.
15. Pectoralis profundus m.

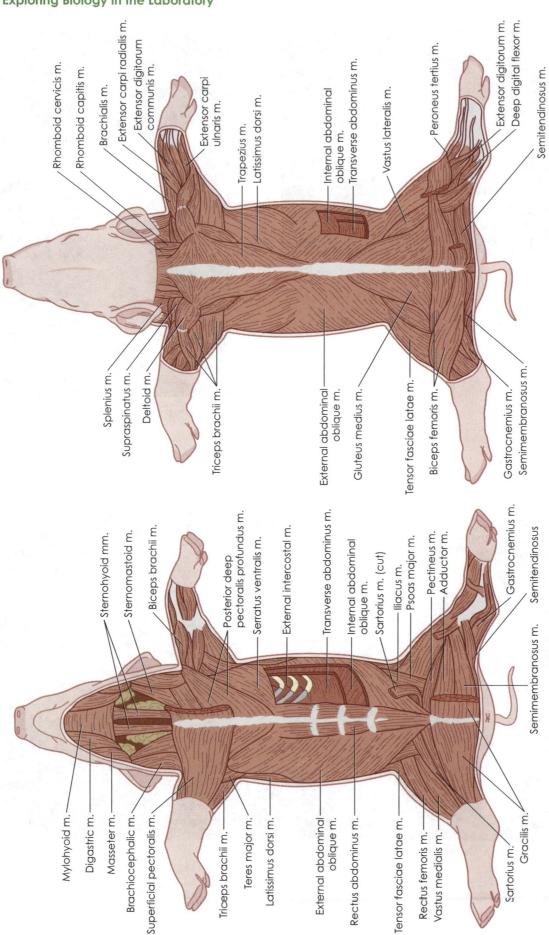

Rhomboid cervicis m.
Rhomboid capitis m.
Brachialis m.
Extensor carpi radialis m.
Extensor digitorum communis m.
Extensor carpi ulnaris m.
Trapezius m.
Latissimus dorsi m.
Internal abdominal oblique m.
Transverse abdominus m.
Vastus lateralis m.
Peroneus tertius m.
Extensor digitorum m.
Deep digital flexor m.
Semitendinosus m.

Splenius m.
Supraspinatus m.
Deltoid m.
Triceps brachii m.
External abdominal oblique m.
Gluteus medius m.
Tensor fasciae latae m.
Biceps femoris m.
Gastrocnemius m.
Semimembranosus m.

Figure 32.107 A dorsal view of the muscles of the fetal pig.

Sternohyoid mm.
Sternomastoid m.
Biceps brachii m.
Posterior deep pectoralis profundus m.
Serratus ventralis m.
External intercostal m.
Transverse abdominus m.
Internal abdominal oblique m.
Sartorius m. (cut)
Iliacus m.
Psoas major m.
Pectineus m.
Adductor m.
Gastrocnemius m.
Semitendinosus

Mylohyoid m.
Digastric m.
Masseter m.
Brachiocephalic m.
Superficial pectoralis m.
Triceps brachii m.
Teres major m.
Latissimus dorsi m.
External abdominal oblique m.
Rectus abdominus m.
Tensor fasciae latae m.
Rectus femoris m.
Vastus medialis m.
Sartorius m.
Gracilis m.
Semimembranosus m.

Figure 32.106 A ventral view of the muscles of the fetal pig.

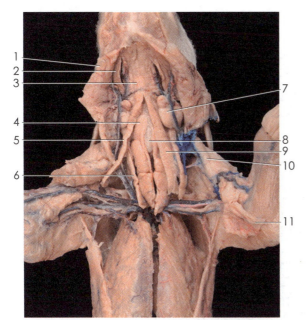

Figure 32.108 A ventral view of superficial muscles of neck and upper torso.

1. Platysma m. (reflected)
2. Digastric m.
3. Mylohyoid m.
4. Sternohyoid m.
5. Omohyoid m.
6. Sternomastoid m.
7. Mandibular gland
8. Larynx
9. Sternothyroid m.
10. Brachiocephalic m.
11. Pectoralis superficialis m. (cut and reflected)

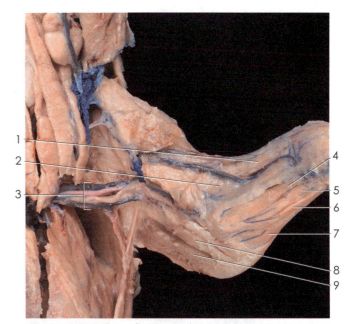

Figure 32.109 Superficial medial muscles of the forelimb.

1. Extensor carpi radialis m.
2. Biceps brachii m.
3. Axillary artery and vein, brachial plexus
4. Flexor carpi radialis m.
5. Flexor digitorum profundus m.
6. Flexor digitorum superficialis m.
7. Flexor carpi ulnaris m.
8. Triceps brachii m. (lateral head)
9. Triceps brachii m. (long head)

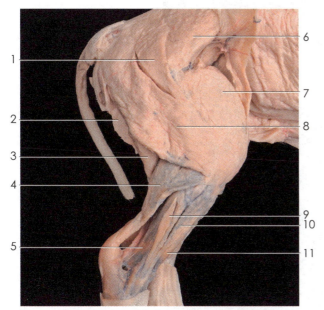

Figure 32.110 A lateral view of the superficial thigh and leg.

1. Gluteus superficialis m.
2. Semitendinosus m.
3. Semimembranosus m.
4. Gastrocnemius m.
5. Extensor digitorum quarti and quinti mm.
6. Gluteus medius m.
7. Tensor fasciae latae m.
8. Biceps femoris m.
9. Fibulais (peroneus) longus m.
10. Fibulais (peroneus) tertius m.
11. Tibialis anterior m.

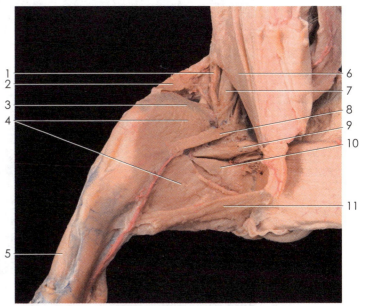

Figure 32.111 Medial muscles of thigh and leg.

1. Iliacus m.
2. Tensor fasciae latae m.
3. Rectus femoris m.
4. Semimembranosus m.
5. Tibialis anterior m.
6. External abdominal oblique m.
7. Psoas major m.
8. Sartorius m.
9. Pectineus m.
10. Adductor m.
11. Semitendinosus m.

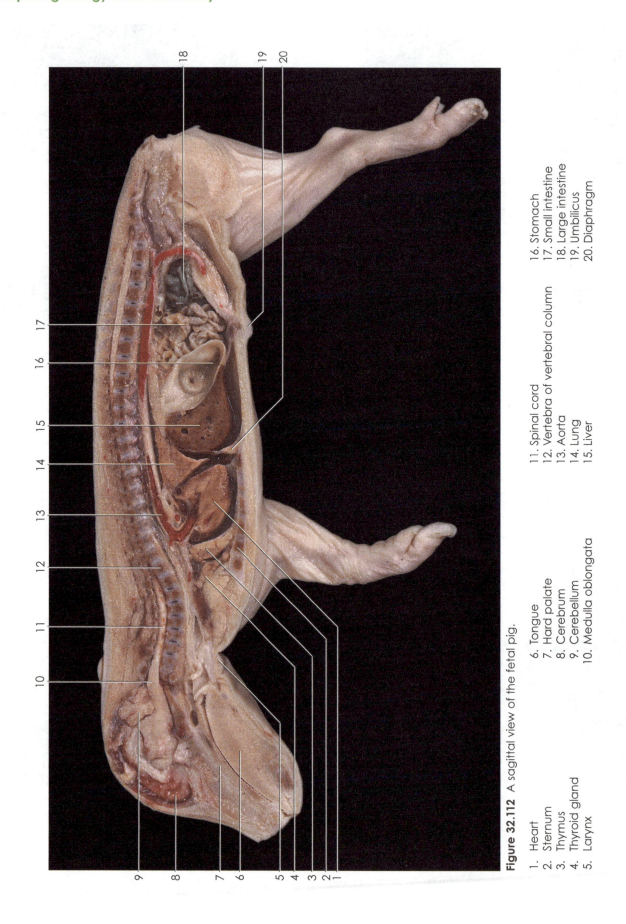

Figure 32.112 A sagittal view of the fetal pig.

1. Heart
2. Sternum
3. Thymus
4. Thyroid gland
5. Larynx

6. Tongue
7. Hard palate
8. Cerebrum
9. Cerebellum
10. Medulla oblongata

11. Spinal cord
12. Vertebra of vertebral column
13. Aorta
14. Lung
15. Liver

16. Stomach
17. Small intestine
18. Large intestine
19. Umbilicus
20. Diaphragm

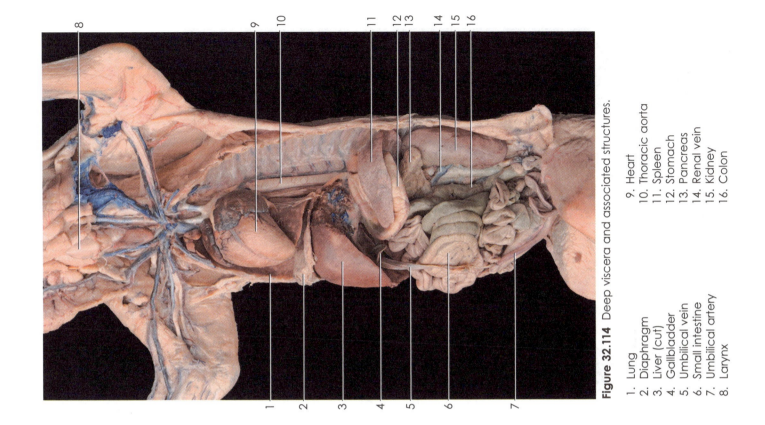

Figure 32.114 Deep viscera and associated structures.

1. Lung
2. Diaphragm
3. Liver (cut)
4. Gallbladder
5. Umbilical vein
6. Small intestine
7. Umbilical artery
8. Larynx

9. Heart
10. Thoracic aorta
11. Spleen
12. Stomach
13. Pancreas
14. Renal vein
15. Kidney
16. Colon

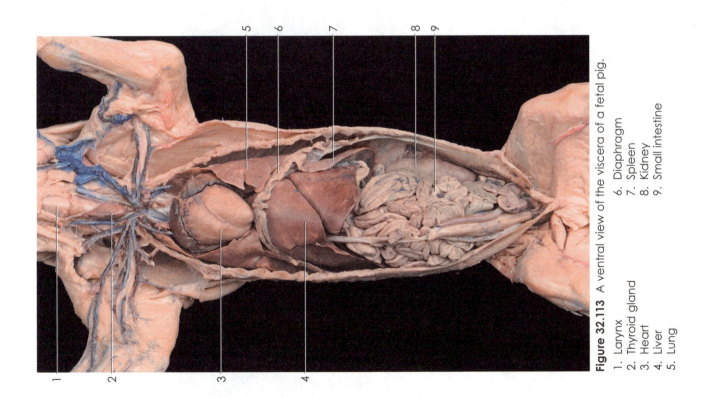

Figure 32.113 A ventral view of the viscera of a fetal pig.

1. Larynx
2. Thyroid gland
3. Heart
4. Liver
5. Lung

6. Diaphragm
7. Spleen
8. Kidney
9. Small intestine

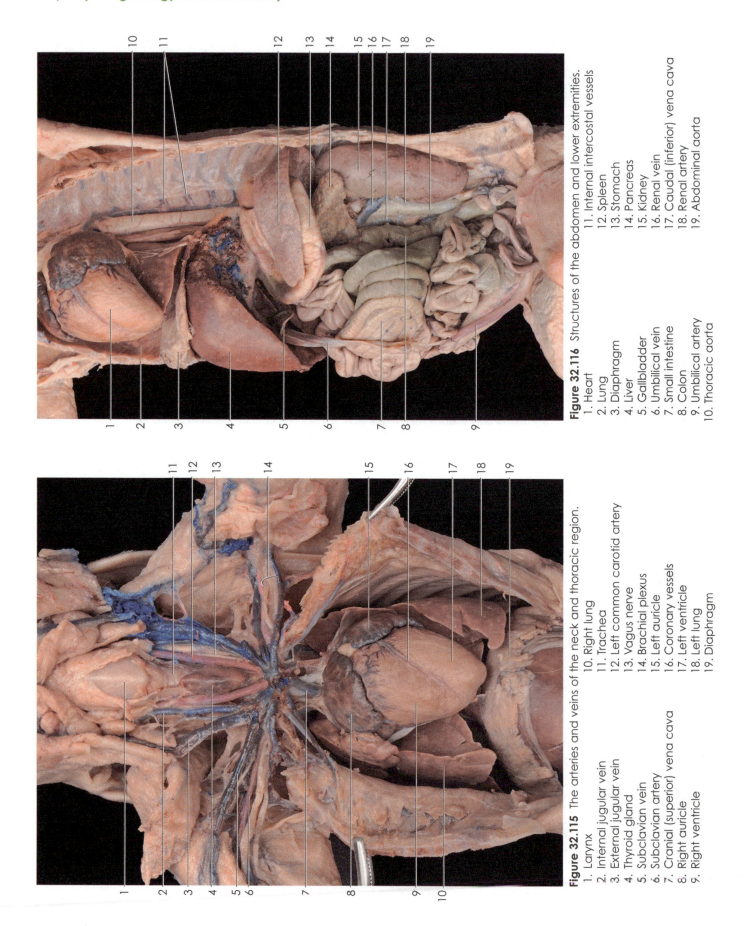

Figure 32.116 Structures of the abdomen and lower extremities.

1. Heart
2. Lung
3. Diaphragm
4. Liver
5. Gallbladder
6. Umbilical vein
7. Small intestine
8. Colon
9. Umbilical artery
10. Thoracic aorta
11. Internal intercostal vessels
12. Spleen
13. Stomach
14. Pancreas
15. Kidney
16. Renal vein
17. Caudal (inferior) vena cava
18. Renal artery
19. Abdominal aorta

Figure 32.115 The arteries and veins of the neck and thoracic region.

1. Larynx
2. Internal jugular vein
3. External jugular vein
4. Thyroid gland
5. Subclavian vein
6. Subclavian artery
7. Cranial (superior) vena cava
8. Right auricle
9. Right ventricle
10. Right lung
11. Trachea
12. Left common carotid artery
13. Vagus nerve
14. Brachial plexus
15. Left auricle
16. Coronary vessels
17. Left ventricle
18. Left lung
19. Diaphragm

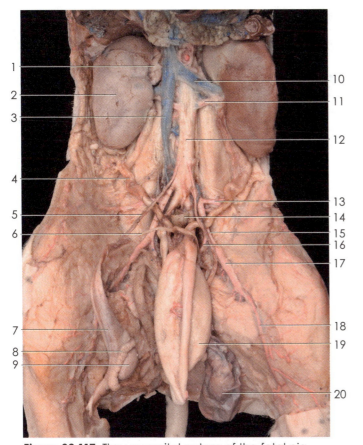

Figure 32.117 The urogenital system of the fetal pig.

1. Adrenal gland
2. Right kidney
3. Caudal (inferior) vena cava
4. Ureter
5. Genital vessels
6. Vas (ductus) deferens
7. Spermatic cord
8. Epididymis
9. Testis
10. Renal vein
11. Renal artery
12. Descending aorta
13. Iliolumbar artery
14. Rectum (cut)
15. Common iliac artery
16. Internal iliac artery
17. External iliac artery
18. Femoral artery
19. Urinary bladder
20. Testis

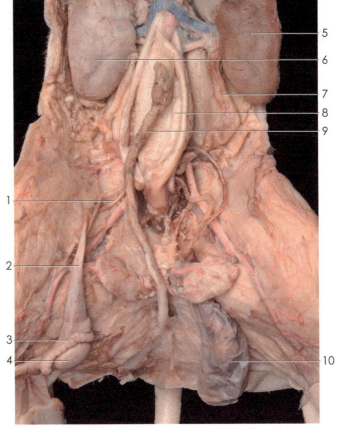

Figure 32.118 The urogenital system of the fetal pig.

1. Vas (ductus) deferens
2. Spermatic cord
3. Epididymis
4. Right testis
5. Left kidney
6. Right kidney
7. Ureter
8. Urinary bladder
9. Penis
10. Left testis

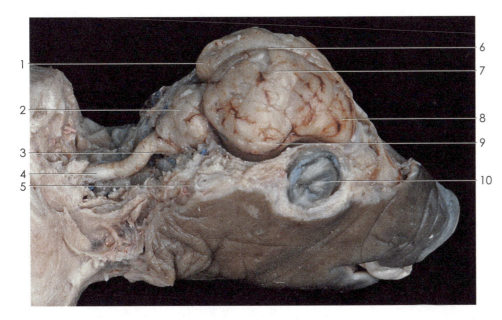

Figure 32.119 The general structures of the fetal pig brain. Because the cerebrum is less defined in pigs, the regions are not known as lobes as they are in humans.

1. Occipital region of cerebrum
2. Cerebellum
3. Medulla oblongata
4. Spinal cord
5. External acoustic meatus
6. Longitudinal fissure
7. Parietal region of cerebrum
8. Frontal region of cerebrum
9. Temporal region of cerebrum
10. Eye

Student Activity—The Cat (*Felis domesticus*) Skeleton

Materials
- hand lens
- colored pencils
- skeleton of a cat (*Felis domesticus*)

 Procedure 32.25
Cat Skeleton

Consider photographing this activity.

1. Procure the needed equipment and specimen.
2. Using a hand lens, observe the skeleton of the cat, comparing the bones to Figures 32.120 through 32.122.
3. Record your observations, and sketch and label the bones of the cat below.
4. Thoroughly clean your laboratory station and equipment. Return the dry equipment and discard the cat skeleton as indicated by the instructor.

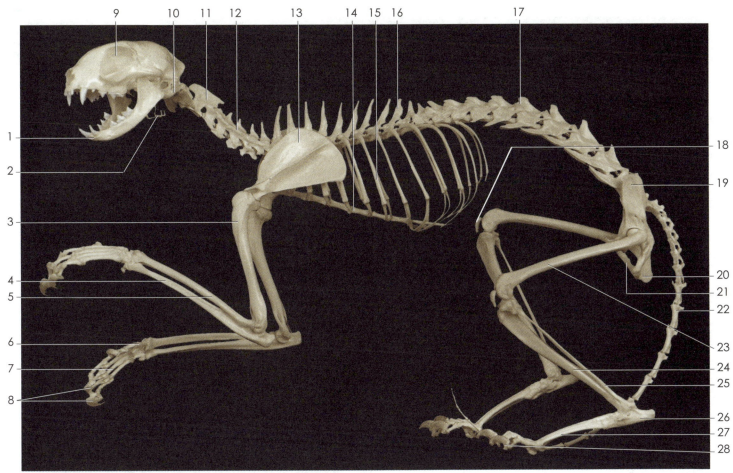

Figure 32.120 The cat skeleton.

1. Mandible
2. Hyoid bone
3. Humerus
4. Ulna
5. Radius
6. Carpal bones
7. Metacarpal bones
8. Phalanges
9. Skull
10. Atlas
11. Axis
12. Cervical vertebra
13. Scapula
14. Sternum
15. Rib
16. Thoracic vertebra
17. Lumbar vertebra
18. Patella
19. Ilium
20. Ischium
21. Pubis
22. Caudal vertebra
23. Femur
24. Tibia
25. Fibula
26. Tarsal bones
27. Metatarsal bones
28. Phalanges

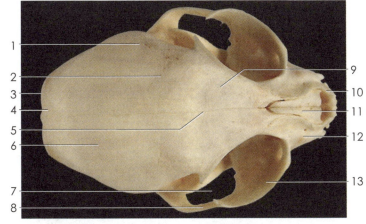

Figure 32.121 A dorsal view of a cat skull.

1. Temporal bone
2. Coronal suture
3. Nuchal crest
4. Interparietal bone
5. Sagittal suture
6. Parietal bone
7. Orbit
8. Zygomatic arch
9. Frontal bone
10. Premaxilla
11. Nasal bone
12. Maxilla
13. Zygomatic (malar) bone

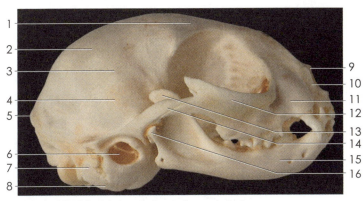

Figure 32.122 A lateral view of a cat skull.

1. Frontal bone
2. Parietal bone
3. Squamosal suture
4. Temporal bone
5. Nuchal crest
6. External acoustic meatus
7. Mastoid process
8. Tympanic bulla
9. Nasal bone
10. Premaxilla bone
11. Maxilla
12. Zygomatic (malar) bone
13. Coronoid process of mandible
14. Zygomatic arch
15. Mandible
16. Condylar process of mandible

A Digital Field Trip—Pictures of Vertebrates

Photography is a fascinating hobby and profession. Besides, taking photographs of family, friends, pets, and vacation is fun. Why not try some biophotography with your digital camera? You do not have to have an expensive digital single lens reflex (DSLR) camera to complete this assignment.

Think of the following activity as a digital field trip or a scavenger hunt (Fig. 32.123–Fig. 124). It is green, too: No animals will be harmed in this activity. Find the animals in question, take your time to compose a photograph of the animal, use good photographic technique, check out the quality of the image on the camera LCD, transfer the image to a computer, and develop a labeled presentation.

Materials
- digital camera
- notebook to record data
- computer
- presentation software
- storage device such as a jump drive or CD
- projector and screen
- various vertebrates as subjects for your photographs

Procedure 32.26
Pictures of Vertebrates

1. Using the examples of vertebrates discussed in this chapter, as well as your own resourcefulness, take photographs of at least 30 different vertebrate species. Visit a zoo, a wildlife park, a local park, an aquarium, or your neighborhood to acquire the images.
2. Take photographs of the animals. The photographs should include the entire organism, as well as distinguishing features of the organism.
3. With a computer and presentation software, develop a digital slide show of your vertebrates. Label your slides clearly. Include the classification of the animal. And how about adding some music to your presentation?
4. Place your slide show on a storage device or CD. The instructor may allow you to present your slide show to the class.

No one knows the diversity in the world, not even to the nearest order of magnitude. ... We don't know for sure how many species there are, where they can be found or how fast they're disappearing. It's like having astronomy without knowing where the stars are.

—**Edward O. Wilson (1929–present)**

Figure 32.123 Panther chameleon, *Furcifer pardalis*.

Figure 32.124 Mandrill, *Mandrillus sphinx*.

Name: _____ Date: _____ Section: _____

1. What is the difference between a protostome and a deuterostome?

_____

_____

_____

2. Describe the general characteristics of Phylum Echinodermata.

_____

_____

_____

3. Draw and label the external anatomy of a typical starfish.

4. Describe the classes of Phylum Echinodermata.

_____

_____

_____

5. List the general characteristics of a chordate.

_____

_____

_____

6. Compare and contrast urochordates and cephalochordates.

_____

_____

_____

7. Why are scientists interested in the Phylum Hemichordata?

_____

_____

_____

8. Compare and contrast a lamprey, an eel, and an amphuma.

_____

_____

_____

9. Compare and contrast Class Chondrichthyes and Class Osteichthyes.

_____

_____

_____

10. How did the amphibians become adapted for life on land?

_____

_____

_____

11. What features allowed reptiles to successfully radiate into terrestrial environments?

_____

_____

_____

12. List the adaptations for flight in birds.

_____

_____

_____

Name: _____ Date: _____ Section: _____

13. Discuss how mammals have successfully radiated into a variety of environments.

_____

_____

_____

14. Compare and contrast "cold-bloodedness" and "warm-bloodedness."

_____

_____

_____

15. What evidence links echinoderms, hemichordates, and chordates?

_____

_____

_____

16. List the major classes of Phylum Chordata, and give an example of each.

_____

_____

_____

_____

_____

_____

_____

_____

17. Where have all the dinosaurs gone?

_____

_____

_____

NOTES

Chapter 33
Homo sapiens: Understanding Human Biology—Part I

Student Outcome Objectives

At the completion of this exercise, the student will be able to:

1. List and describe characteristics that make us human.
2. List and describe the basic organ systems of a human.
3. Discuss the basic components of the integumentary system and their function.
4. Identify basic fingerprint types.
5. Discuss the functions of the skeletal system.
6. Identify the basic components of the skeletal system.
7. Discuss the function of the articulations.
8. Discuss the classification of joint types and provide examples of each.
9. Define and conduct articular movements.
10. Describe the function of muscle.
11. Identify common superficial muscles of the human body.
12. Compare and contrast the central nervous system and the peripheral nervous system.
13. Describe the basic anatomy and function of the olfactory system.
14. Describe the basic anatomy and function of the gustatory system.
15. Discuss basic smell and taste sensations.
16. Describe the major functions of the ear.
17. Describe the basic anatomy of the ear.
18. Discuss the basic anatomy of the eye.
19. Describe the endocrine glands, their associated hormones, and the effect of the hormones.

Overview

In 1871 when Charles Darwin speculated about the origins of humans in his work *The Descent of Man and Selection in Relation to Sex*, he had no idea that just over a century later, the fossil record would yield a plethora of hominid fossils that would substantiate his ideas. Since the humble emergence of our species, we have come a long way in understanding our world and ourselves.

Like all other organisms, humans possess certain traits that make them a unique member of the living world. Humans are **generalist species**, being highly adaptable to a number of environments and ways of life. Human body type varies tremendously. The average size of a human is 1.5 to 1.8 m tall and 54 to 83 kg in mass. Being **bipedal** (walking on two legs) frees up the upper limbs to engage in a number of activities from operating a computer to catching a baseball. The **manipulative hands,** coupled with an **opposable thumb,** allow humans to display a **power grip** for grasping and a **technical grip** for fine activities.

Humans are **omnivorous,** consuming both plant and animal products. Humans have **stereoscopic color vision.** The **brain** is highly developed, capable of abstract reasoning, problem solving, tool-making, consequential thinking, and introspection. **Broca's area** of the brain in humans is well-developed, allowing humans to have complex **language** skills. Humans are eutherian that give birth usually to one offspring. The young have an extended **pre-adult** period in which to grow both physically and intellectually.

The population of humans is approaching 7 billion, more than 300 million of whom reside in the United States. These numbers are alarming because as the human population continues to exponentially increase, increased demands are placed on natural resources and planet Earth.

The human body is an incredible machine consisting of nearly 100 trillion cells working together in order to sustain life. These cells are organized into tissues such as those discussed in Chapter 7. Tissues comprise organs, and the organs comprise organ systems. Humans consist of several organ systems, outlined in Table 33.1.

THE INTEGUMENTARY SYSTEM

Surprisingly, the **integumentary system,** including the skin, hair, nails, and associated glands, is the largest organ system in the body, covering nearly 2 square meters. The integumentary system serves as a protective layer against ultraviolet (UV) light and harmful chemicals. It is the first line of defense against potentially harmful microbes. The system is involved in vitamin D synthesis, prevention of desiccation, regulation of body heat, excretion, and sensory reception.

The skin consists of a superficial layer, known as the **epidermis,** a second layer called the **dermis,** and a deep subcutaneous layer called the **hypodermis.** The **dermis** is found beneath the epidermis. It may be thick in some regions of the body such as the palms of the hands. The surface appears uneven because of the **dermal papillae,** which in the fingers can form fingerprints.

Table 33.1 Organ Systems of the Human Body

Organ System	Principle Organs	Basic Functions
Integumentary System	Skin, hair, nails, cutaneous glands	Protection, defense, thermoregulation, vitamin D synthesis, electrolytic balance, tactile information
Skeletal System	Bones, cartilage, ligaments	Movement, protection, support, blood formation, calcium storage
Muscular System	Muscles, tendons, aponeuroses (tendinous sheets)	Movement, support, production of heat
Nervous System	Brain, spinal cord, nerves, sense organs	Coordination, control, communications, sensory input, response, reflexes, cognition
Endocrine System	Pituitary gland, thyroid gland, parathyroid glands, hypothalamus, thalamus, thymus, adrenal glands, pancreas, ovaries, testes	Coordination, control, communications, metabolic activities, sex cell production
Circulatory System	Heart, arteries, veins, blood	Distribution of oxygen, nutrients and hormones, removal of wastes
Lymphatic System	Lymph, lymphatic vessels, lymph nodes, spleen, thymus, tonsils	Defense against disease, immunity, returns fluids to bloodstream
Respiratory System	Nose, sinuses, pharynx, larynx, trachea, bronchi, lungs, alveoli	Delivery of air, provides oxygen to blood cells, removal of carbon dioxide, acid-base balance, generation of sounds
Digestive System	Teeth, tongue, pharynx, esophagus, stomach, small intestines, large intestines, liver, gall bladder, pancreas	Physical and chemical breakdown of food, digestion, absorption of nutrients, removal of wastes, metabolism
Urinary System	Kidneys, ureters, urinary bladder, urethra	Storage and removal of urine and other wastes, regulation of blood pressure and blood volume, electrolytic balance, water balance, acid-base balance
Female Reproductive System	Ovaries, uterine tubes, uterus, vagina, mammary glands	Production of the egg, site of fertilization, site of fetal development, nourishment of fetus, production of sex hormones, production of milk
Male Reproductive System	Testes, epididymis, seminal vesicles, prostate gland, penis	Production and delivery of sperm, production of sex hormones

 Student Activity—Fingerprint Impressions

Fingerprint identification, or **dermatoglyphics,** is a method of personal identification using the impressions of fingerprints. No two persons have exactly the same fingerprint patterns, and the patterns remain unchanged throughout life. This activity is designed to introduce the fascinating world of dermatoglyphics. The content and fingerprint types have been simplified. The true science of dermatoglyphics is tedious and requires experience, a keen eye, and patience. This activity should be conducted in groups of two to four students.

Materials
- printer's ink
- ink roller, if necessary
- stamp pad
- hand lens
- colored pencils
- soap
- paper towels
- students' fingers for fingerprints

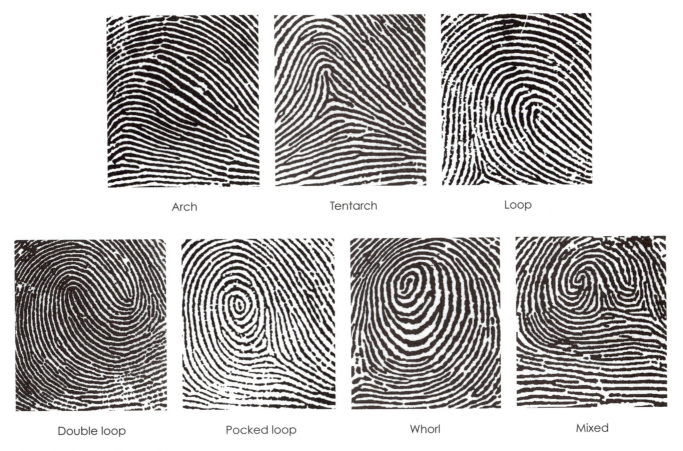

Figure 33.1 Basic fingerprint types.

 **Procedure 33.1**
Fingerprints

1. Procure the needed equipment and supplies.
2. Make sure the ink is evenly distributed on the stamp pad. It can be recharged with printer's ink and spread evenly with an ink roller. Each member of the team should insert their fingerprints in their manual.
3. Carefully place the medial edge of the thumb of the right hand against the stamp pad. Roll the thumb slowly and naturally across the pad to the lateral edge of the thumb, covering the thumb in ink.
4. Using the same movement, roll the thumb across the space below.

Right Thumb

5. One at a time, repeat Steps 3 and 4 for each finger.

Index Finger   Middle Finger

Ring Finger   Little Finger

6. Wash your hand thoroughly, and allow the ink enough time to dry.

7. Compare your fingerprints to Figure 33.1, and determine the specific pattern of your fingerprints. Record your observations below, including any unusual patterns that were noted. More details on fingerprints can be found on the Internet, such as http://www.policensw.com/info/fingerprints/finger07.html.

_____

_____

_____

8. Compare your fingerprints to other members of the group. Record your observations below.

_____

_____

_____

THE SKELETAL SYSTEM

The vertebrates, including humans, have an **endoskeleton** composed of cartilage or bone, as opposed to the chitinous **exoskeleton** of arthropods. In humans, the **skeletal system** consists of bones, cartilage, ligaments, and associated connective tissues. The bones protect internal organs, allow for movement and leverage, support the body, provide sites of attachment for muscles, ligaments, and tendons, allow for growth, store energy, provide a reservoir for calcium, aid in electrolytic and acid-base balance, and give rise to blood cells in the activity called **hemopoiesis**.

Amazingly, at birth, a human may have over 270 bones. The average adult human has approximately 206 bones, depending upon genetic and developmental factors. Distinct anatomical variation within individuals may result in the formation of more or fewer bones. The skeleton is divided into the axial skeleton and the appendicular skeleton.

The **axial skeleton** consists of 80 bones comprising the skull, ossicles of the ear, hyoid bone, ribs, vertebrae, and sternum. These bones form the axis of the body, providing support to the body, attachment for muscles, stabilization of the appendicular skeleton, movements associated with respiration, and protection to the brain and internal organs.

The **appendicular skeleton** consists of 126 bones that comprise the pectoral girdle, the arms, the hands, the pelvic girdle, the legs, and the feet. The pectoral and pelvic girdle attach the appendicular skeleton to the axial skeleton, aid in locomotion, and allow for manipulation of objects.

Each one of you has something no one else has, or has ever had: your fingerprints, your brain, your heart. Be an individual. Be unique. Stand out. Make noise. Make someone notice. That's the power of individuals.

Jon Bon Jovi (1962-present)

Student Activity—The Human Skeleton

Materials
- colored pencils
- articulated human skeletons
- skull

Procedure 33.2
The Skeleton

Consider photographing this activity.

1. Procure skeleton and skull.
2. Observe the skeleton and skull, noting anatomical features found in Figures 33.2–33.5.

Somewhere, something incredible is waiting to be known.

Carl Sagan (1934-1996)

3. Place your observations in the space below.

_____

_____

_____

_____

_____

_____

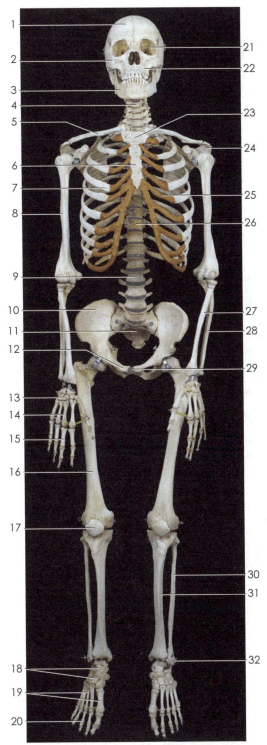

Figure 33.2 An anterior view of the skeleton.

1. Frontal bone
2. Zygomatic bone
3. Mandible
4. Cervical vertebra
5. Clavicle
6. Body of sternum
7. Rib
8. Humerus
9. Lumbar vertebra
10. Ilium
11. Sacrum
12. Pubis
13. Carpal bones
14. Metacarpal bones
15. Phalanges
16. Femur
17. Patella
18. Tarsal bones
19. Metatarsal bones
20. Phalanges
21. Orbit
22. Maxilla
23. Manubrium
24. Scapula
25. Costal cartilage
26. Thoracic vertebra
27. Radius
28. Ulna
29. Symphysis pubis
30. Fibula
31. Tibia
32. Lateral malleolus

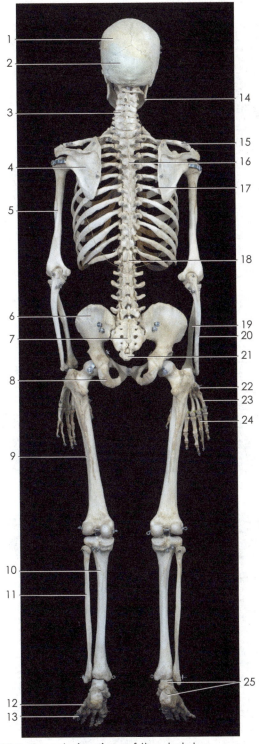

Figure 33.3 A posterior view of the skeleton.

1. Parietal bone
2. Occipital bone
3. Cervical vertebra
4. Scapula
5. Humerus
6. Ilium
7. Sacrum
8. Ischium
9. Femur
10. Tibia
11. Fibula
12. Metatarsal bones
13. Phalanges
14. Mandible
15. Clavicle
16. Thoracic vertebra
17. Rib
18. Lumbar vertebra
19. Radius
20. Ulna
21. Coccyx
22. Carpal bones
23. Metacarpal bones
24. Phalanges
25. Tarsal bones

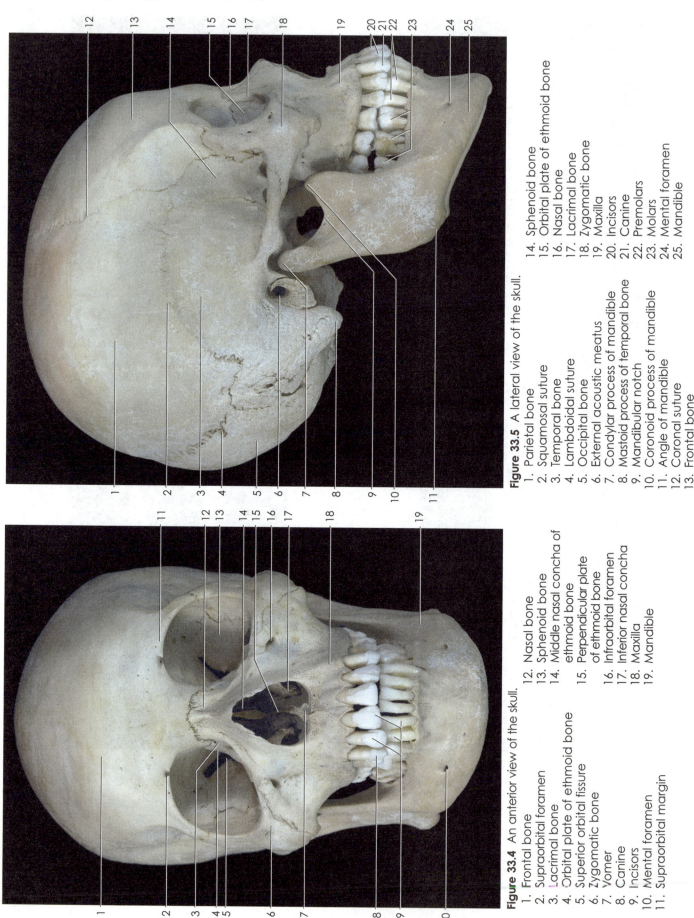

Figure 33.5 A lateral view of the skull.

1. Parietal bone
2. Squamosal suture
3. Temporal bone
4. Lambdoidal suture
5. Occipital bone
6. External acoustic meatus
7. Condylar process of mandible
8. Mastoid process of temporal bone
9. Mandibular notch
10. Coronoid process of mandible
11. Angle of mandible
12. Coronal suture
13. Frontal bone
14. Sphenoid bone
15. Orbital plate of ethmoid bone
16. Nasal bone
17. Lacrimal bone
18. Zygomatic bone
19. Maxilla
20. Incisors
21. Canine
22. Premolars
23. Molars
24. Mental foramen
25. Mandible

Figure 33.4 An anterior view of the skull.

1. Frontal bone
2. Supraorbital foramen
3. Lacrimal bone
4. Orbital plate of ethmoid bone
5. Superior orbital fissure
6. Zygomatic bone
7. Vomer
8. Canine
9. Incisors
10. Mental foramen
11. Supraorbital margin
12. Nasal bone
13. Sphenoid bone
14. Middle nasal concha of ethmoid bone
15. Perpendicular plate of ethmoid bone
16. Infraorbital foramen
17. Inferior nasal concha
18. Maxilla
19. Mandible

ARTICULATIONS

An **articulation** is a joint and refers to a point where two bones meet. Humans have approximately 147 joints. Articulations allow for movement and provide strength. Obvious examples of articulations include movable joints (**diarthrosis**) such as the knee, elbow, and shoulder. Immovable joints (**synarthrosis**) include the sutures of the skull and the gomphosis between the teeth and jaw. An **amphiarthrosis** is a slightly movable fibrous or cartilaginous joint. Examples include the junction between the pubic bones at the pubic symphysis and the articulation between the vertebrae at the intervertebral discs. The joints allow for a number of movements as indicated in Figure 33.6.

Table 33.2 Movements Permitted at Synovial Joints

Type of movement	Description
Angular movement	Increase or decrease the joint angle
Flexion	Decreasing the angle between two bones
Extension	Increasing the angle between two bones
Hyperextension	Excessive extension beyond 180° (angle of anatomical position)
Dorsiflexion	Bending the foot toward the tibia
Plantar flexion	Bending the foot away from the tibia
Abduction	Movement of a body part away from the axis of the body, or away from the midsagittal plane, in a lateral direction
Adduction	Movement of a body part toward the axis of the body, or toward the midsagittal plane, in a mesial direction
Inversion	Movement of the sole of the foot inward, or medially
Eversion	Movement of the sole of the foot outward, or laterally
Opposition	Movement of the thumb toward the surface of the palm
Protraction	Moving a part of the body anteriorly in the horizontal plane
Retraction	Moving a part of the body posteriorly in the horizontal plane
Elevation	Moving a part of the body in the superior direction
Depression	Moving a part of the body in the inferior direction
Circular movement	Rotation on an axis
Rotation	Turning of a bone at joint axis
Pronation	Rotation of the forearm causing the wrist and hand to go from palm facing front to palm facing back
Supination	Rotation of the forearm causing the wrist and hand to go from palm facing back to palm facing front
Circumduction	Movement of a body segment in a circular, conelike motion

Flexion at the joints of the vertebral column.

Hyperextension at the joints of the vertebral column.

Flexion at the shoulder, hip, and knee joints on left side of body; extension at the elbow and wrist joints; plantar flexion at the left ankle joint.

Hyperextension at the shoulder and hip joints on left side of body. Extension at the wrist and ankle.

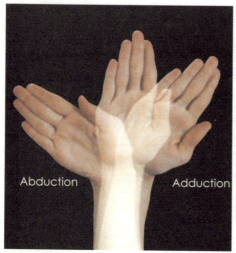

Abduction and adduction of the right hand at the wrist.

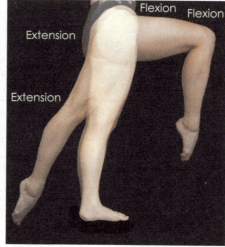

Flexion and extension at the right hip and knee joints.

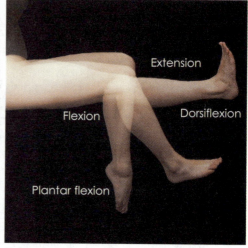

Flexion and extension at the knee joint and plantar flexion and dorsiflexion at the ankle joint.

Flexion at left shoulder, elbow, and knee joints. Extension of right shoulder, elbow, and knee joints.

Maximal flexion at each of the principal body joints.

Rotation at the joints of the neck; elevation at the shoulder joint; and flexion at the right elbow and right wrist joints.

Figure 33.6 A photographic summary of joint movements.

Student Activity—Joint Movements

Materials
- just students!

 ## Procedure 33.3
Movement of Joints

1. Observe Table 33.2 and Figure 33.6.
2. Using these as guides, review the joint movements illustrated in each.
3. Record your observations below.

_____

_____

_____

_____

MUSCULAR SYSTEM

The three types of muscle tissue were discussed in Chapter 7. Review this information, particularly the information related to skeletal muscle. **Skeletal muscle** is responsible for movement, the maintenance of posture, the generation of body heat, support of soft tissues, and it serves to guard and regulate bodily entrances and exits. A human has approximately 600 skeletal muscles, accounting for about 40% of body mass.

The majority of skeletal muscles are attached to bone at both ends by tendons. A **tendon** attaches muscle to bone and a **ligament** attaches bone to bone. **Fascia** is composed of fibrous connective tissue and serves to cover muscles and attaches muscle to skin. The **origin** of a muscle refers to the more stationary attachment, and the **insertion** refers to the more moveable attachment. The movement provided by a muscle is called **muscle action**. Skeletal muscles usually work in groups to provide movement.

 Student Activity—Superficial Muscles of the Human Body

Materials
- Figures 33.7 and 33.8, showing human muscles

 ## Procedure 33.4
Muscles

1. Locate and identify the muscles on the muscle model shown in Figures 33.7 and 33.8.
2. Record your observations below.

_____

_____

_____

THE NERVOUS SYSTEM

The nervous system consists of two major anatomical divisions: the **central nervous system (CNS)** and the **peripheral nervous system (PNS)**. The CNS consists of the brain and spinal cord, and the PNS consists of all nervous tissue outside of the brain and spinal cord, including cranial nerves, spinal nerves, and ganglia.

The Brain

The **brain** (Fig. 33.9–33.11) is a fascinating organ consisting of billions of neurons and containing nearly 98% of the body's neural tissue. Although the brain varies among individuals, the typical adult human brain weighs approximately 1.4 kg and has a volume of approximately 1200 cc. The largest portion of the brain is the **cerebrum,** divided into left and right **cerebral hemispheres** by a **longitudinal fissure**. The **corpus callosum** is a thick collection of nerve fibers connecting the hemispheres. A superficial layer of **gray matter,** or the **cerebral cortex** (5 cm thick), covers the surface. Obvious ridges, the **gyri,** and shallow depressions, **sulci,** cover the surface. The deep depressions are called **fissures**. Gray matter contains neural cell bodies and is not myelinated like the **white matter** that exists below it. The cerebrum is the seat of conscience thought, memory, sensations, intellect, and higher mental functions.

The brain consists of three distinct regions: the hindbrain, midbrain, and forebrain.

1. The **hindbrain** consists of the cerebellum, pons, and medulla oblongata.

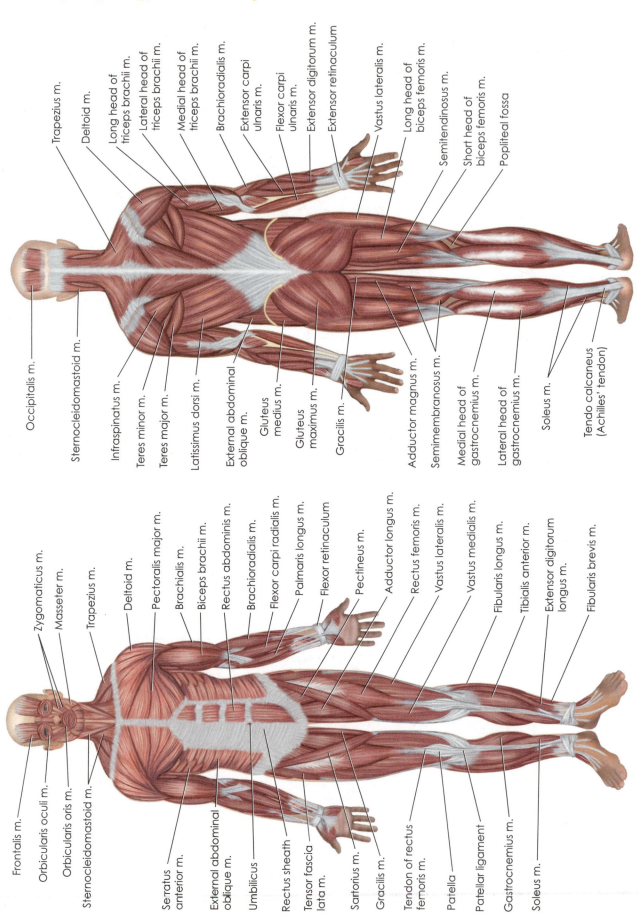

Occipitalis m.

Trapezius m.

Sternocleidomastoid m.

Deltoid m.

Long head of triceps brachii m.

Lateral head of triceps brachii m.

Medial head of triceps brachii m.

Brachioradialis m.

Extensor carpi ulnaris m.

Flexor carpi ulnaris m.

Extensor digitorum m.

Extensor retinaculum

Vastus lateralis m.

Long head of biceps femoris m.

Semitendinosus m.

Short head of biceps femoris m.

Popliteal fossa

Infraspinatus m.

Teres minor m.

Teres major m.

Latissimus dorsi m.

External abdominal oblique m.

Gluteus medius m.

Gluteus maximus m.

Gracilis m.

Adductor magnus m.

Semimembranosus m.

Medial head of gastrocnemius m.

Lateral head of gastrocnemius m.

Soleus m.

Tendo calcaneus (Achilles' tendon)

Figure 33.8 A posterior view of human musculature (m=muscle).

Frontalis m.

Orbicularis oculi m.

Orbicularis oris m.

Sternocleidomastoid m.

Zygomaticus m.

Masseter m.

Trapezius m.

Deltoid m.

Pectoralis major m.

Brachialis m.

Biceps brachii m.

Rectus abdominis m.

Brachioradialis m.

Flexor carpi radialis m.

Palmaris longus m.

Flexor retinaculum

Pectineus m.

Adductor longus m.

Rectus femoris m.

Vastus lateralis m.

Vastus medialis m.

Fibularis longus m.

Tibialis anterior m.

Extensor digitorum longus m.

Fibularis brevis m.

Serratus anterior m.

External abdominal oblique m.

Umbilicus

Rectus sheath

Tensor fascia lata m.

Sartorius m.

Gracilis m.

Tendon of rectus femoris m.

Patella

Patellar ligament

Gastrocnemius m.

Soleus m.

Figure 33.7 An anterior view of human musculature (m=muscle).

- The **cerebellum** can be found inferior to the cerebrum. The cerebellum functions in motor coordination, monitoring sensory input, muscle movements, and muscle tone.

- The **pons** connects the cerebellum to other parts of the brain and spinal cord and appears as a bulge in the **brainstem**. The pons contains sensory and motor nuclei of several cranial nerves, and nuclei that involve sleep, respiration, swallowing, hearing, equilibrium, taste, eye movements, bladder control, and movements of the head.

- The **medulla oblongata,** of *Waterboy* fame, is a portion of the brainstem that is continuous with the spinal cord. It is the center for autonomic functions such as heartbeat, breathing, blood pressure, sneezing, coughing, gagging, vomiting, hiccupping, and swallowing.

2. The **midbrain** connects the hindbrain to the forebrain. Components of the midbrain include the **superior colliculi** (vision), **inferior colliculi** (hearing), **red nucleus** (muscle tone and upper limb positioning), **substantia nigra** (relays inhibitory signals to the thalamus), and **reticular formation** (cardiovascular control, pain modulation visual tracking, consciousness, habituation). Habituation of the reticular formation allows many students to study with the iPod blaring or to ignore a noisy air conditioner in a lecture hall.

3. The **forebrain** consists of the diencephalon and cerebrum.

- The **diencephalon** consists of epithalamus, thalamus, and hypothalamus. The **epithalamus** contains the **pineal gland**, an endocrine structure that secretes melatonin. It functions in circadian rhythms, day-night cycles, and regulation of reproductive functions. The **thalamus** is the final relay point for ascending sensory information going to the primary sensory cortex. Thalamic nuclei are involved in emotion, motivation, touch, pain, temperature, position, and visual and auditory signals. The **hypothalamus** controls thirst, eating, body temperature, and circadian rhythms. The **limbic system,** found along the border of the diencephalon and cerebrum, facilitates memory storage and retrieval, learning, establishing emotional states, and linking the conscious and unconscious functions of the cerebral cortex.

- The **cerebrum,** the largest portion of the brain, communicates with and interacts with all other portions of the brain. It also contains the centers for higher-level thought, learning, speech, and memory. The **left hemisphere** is more adept at analytical thought, mathematics, and language. **The right hemisphere** is more adept at artistic and musical skills, emotions, spatial relationships, and pattern recognition. Sulci divide each hemisphere into distinct lobes.

- The **frontal lobe** is associated with memory, planning, emotion, speech, judgment, mood, voluntary control of skeletal muscle, and aggression.
- The **parietal lobe** is concerned with perception of touch, pressure, pain, taste, and temperature.
- The **temporal lobe** is involved in hearing, smell, memory, emotional behavior, and visual recognition.
- The **occipital lobe** contains the visual centers of the brain.

The brain is like a muscle. When it is in use we feel very good. Understanding is joyous.

Carl Sagan (1934–1996)

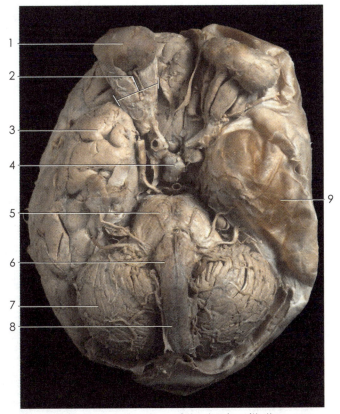

Figure 33.9 An inferior view of the brain with the eyes and part of the meninges still intact.

1. Eyeball
2. Muscles of the eye
3. Temporal lobe of cerebrum
4. Pituitary gland
5. Pons
6. Medulla oblongata
7. Cerebellum
8. Spinal cord
9. Dura mater

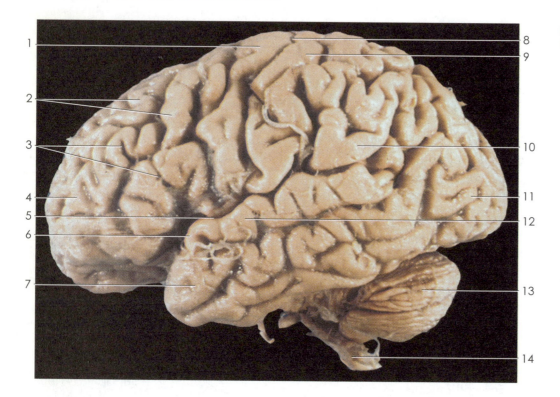

Figure 33.10 A lateral view of the brain.
1. Primary motor cerebral cortex
2. Gyri
3. Sulci
4. Frontal lobe of cerebrum
5. Lateral sulcus
6. Olfactory cerebral cortex
7. Temporal lobe of cerebrum
8. Central sulcus
9. Primary sensory cerebral cortex
10. Parietal lobe of cerebrum
11. Occipital lobe of cerebrum
12. Auditory cerebral cortex
13. Cerebellum
14. Medulla oblongata

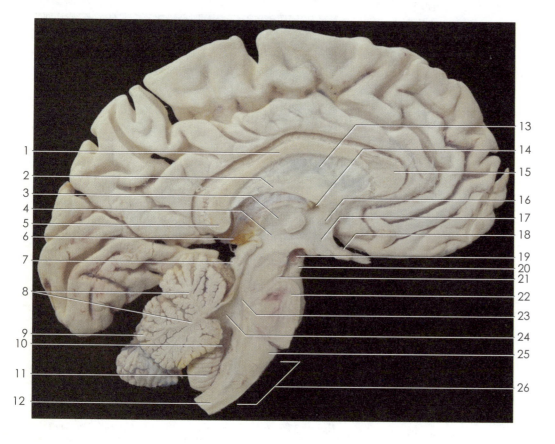

Figure 33.11 A sagittal view of the brain.
1. Body of corpus callosum
2. Crus of fornix
3. Third ventricle
4. Posterior commissure
5. Splenium of corpus callosum
6. Pineal body
7. Inferior colliculus
8. Arbor vitae of cerebellum
9. Vermis of cerebellum
10. Choroid plexus of fourth ventricle
11. Tonsilla of cerebellum
12. Spinal cord
13. Septum pellucidum
14. Intraventricular foramen
15. Genu of corpus callosum
16. Anterior commissure
17. Hypothalmus
18. Optic chiasma
19. Oculomotor nerve
20. Cerebral peduncle
21. Midbrain
22. Pons
23. Mesencephalic (cerebral) aqueduct
24. Fourth ventricle
25. Pyramid of medulla oblongata
26. Medulla oblongata

Student Activity—Examination of Human Brain Model

Materials
- model of human brain
- colored pencils

 **Procedure 33.5
Human Brain**

Consider photographing this activity.

1. Procure the model.
2. Observe the human brain model, and locate the structures in Figures 33.9–33.11.
3. Record your observations below.

_____

_____

_____

_____

 Student Activity—Dissection of the Sheep Brain

In this activity, the sheep brain will be observed and dissected. Locating important structures in the brain of a sheep will greatly aid in understanding how the brains of other animals, including humans, are organized.

Materials
- dissecting tray
- dissecting kit
- gloves
- safety glasses
- lab coat or apron
- hand lens
- colored pencils
- preserved sheep brain

 **Procedure 33.6
Sheep Brain**

Consider photographing this activity.

1. Procure the needed supplies and the sheep brain.
2. Wash your specimen thoroughly under running water. *Caution: If preservative gets in your eyes, wash your eyes immediately and contact the instructor.*
3. Place the specimen on the dissecting tray, and locate the anatomical features labeled in Figures 33.12–33.17. Record your observations and labeled sketch to the right.

_____

_____

4. Before proceeding with identification of the structures on the dorsal and ventral surfaces of the sheep brain, carefully remove the meninges with a pair of scissors.
5. Locate and observe the anatomical features of the dorsal and ventral surfaces of the sheep brain labeled in Figures 33.14–33.15. A hand lens may be handy. Record your observations and labeled sketch on the following page.

7. Referring to Figures 33.16 and 33.17, locate the anatomical structures labeled in each half of the mid-sagittal view. A hand lens may be handy. Record your observations and labeled sketch below.

6. Hold the brain level in your hand with the cerebellum facing away from you. Carefully cut along the longitudinal fissure, eventually separating the brain into left and right halves. This procedure is known as a **mid-sagittal cut**.

8. Clean your equipment and desktop thoroughly. Return the equipment and discard your specimen as directed.

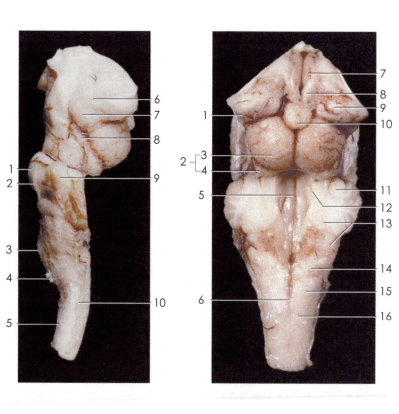

Figure 33.12 A lateral view of the brainstem.
1. Pons
2. Abducens nerve
3. Medulla oblongata
4. Hypoglossal nerve
5. Spinal cord
6. Lateral geniculate body
7. Medial geniculate body
8. Trochlear nerve
9. Trigeminal nerve
10. Accessory nerve

Figure 33.13 A dorsal view of the brainstem.
1. Medial geniculate body
2. Corpora quadrigemina
3. Superior colliculus
4. Inferior colliculus
5. Fourth ventricle
6. Dorsal median sulcus
7. Intermediate mass
8. Habenular trigone
9. Thalmus
10. Pineal gland
11. Middle cerebellar peduncle
12. Anterior cerebellar penduncle
13. Posterior cerebellar peduncle
14. Tuberculum cuneatum
15. Fasciculus gracilis
16. Fasciculus cuneatus

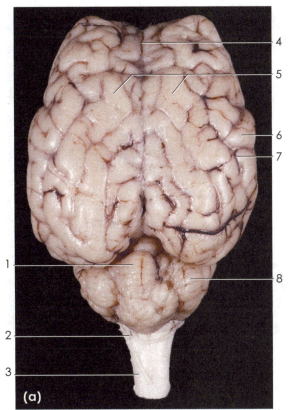

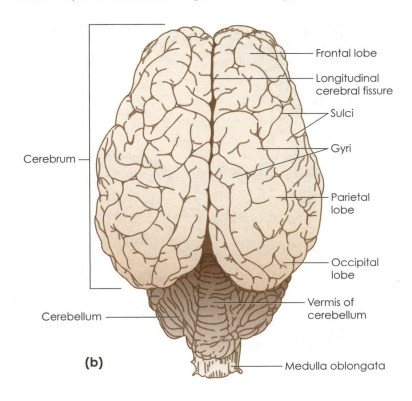

(b)

Figure 33.14 A dorsal view of the sheep brain, (a) photograph; (b) diagram.

1. Vermis
2. Vermis of cerebellum
3. Spinal Cord
4. Longitudinal cerebral fissure
5. Cerebral hemispheres
6. Gyrus
7. Sulcus
8. Cerebellar hemisphere

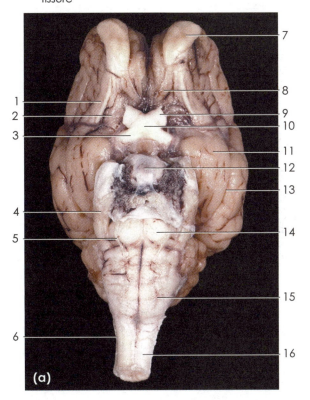

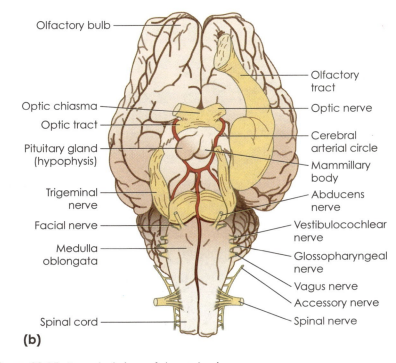

(b)

Figure 33.15 A ventral view of sheep brain, (a) photograph and (b) diagram.

1. Lateral olfactory band
2. Olfactory trigone
3. Optic tract
4. Trigeminal nerve
5. Abducens nerve
6. Accessory nerve
7. Olfactory bulb
8. Medial olfactory band
9. Optic nerve
10. Optic chiasma
11. Pyriform lobe
12. Pituitary gland (hypophysis)
13. Rhinal sulcus
14. Pons
15. Medulla oblongata
16. Spinal cord

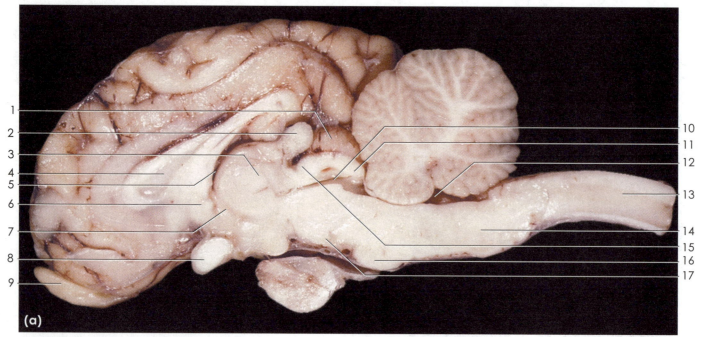

Figure 33.16 A right sagittal view of the sheep brain, (a) photograph and (b) diagram.

1. Superior colliculus
2. Pineal body (gland)
3. Intermediate mass
4. Septum pellucidum
5. Interventicular foramen
   (foramen of Monro)
6. Anterior commissure
7. Third ventricle
8. Optic chiasma
9. Olfactory bulb
10. Mesencephalic
    (cerebral) aqueduct
11. Inferior colliculus
12. Fourth ventricle
13. Spinal cord
14. Medulla oblongata
15. Posterior commissure
16. Pons
17. Cerebral peduncle

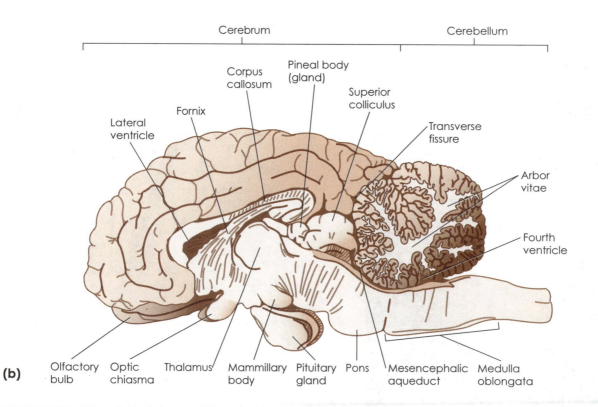

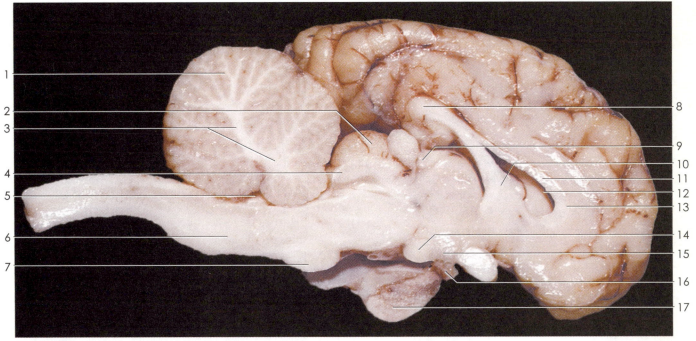

Figure 33.17 A left sagittal view of the sheep brain.

1. Cerebellum
2. Superior colliculus
3. Arbor vitae
4. Inferior colliculus
5. Fourth ventricle
6. Medulla oblongata
7. Pons
8. Splenium of corpus callosum
9. Habenular trigone
10. Fornix
11. Body of corpus callosum
12. Lateral ventricle
13. Genu of corpus callosum
14. Mammillary body
15. Tuber cinereum
16. Pituitary stalk
17. Pituitary gland (hypophysis)

THE SPECIAL SENSES

The smell of a rose, the taste of your favorite candy, the sound of a soothing voice, and the sight of a majestic waterfall stimulates your senses. We interpret the world through a number of receptors including those of the special senses. A **receptor** is a specialized structure that detects a given stimulus. Receptors are not limited to the special senses of smell (**olfaction**), taste (**gustation**), hearing, sound, and vision; receptors can also detect touch (**tactile**), pressure (**baroreceptors**), temperature (**thermoreceptors**), pain (**nociceptors**), and position (**proprioreceptors**).

Smell and Taste

The senses of smell (olfaction) and taste (gustation) involve special receptors classified as **chemoreceptors**. **Olfactory receptors** are located in the roof of the **nasal cavity**. The

average human can distinguish between 2,000 and 4,000 distinct odors. One common classification of odors recognizes eight primary odors that humans can detect:

1. **putrid** (rotten eggs, vomit)
2. **pungent** (vinegar, acetone)
3. **minty** (peppermint)
4. **floral** (rose)
5. **musky** (cat musk, ferret musk)
6. **ethereal** (ether, solvents)
7. **camphoraceous** (turpentine, eucalyptus oil), and
8. the newly added **fishy** (fish).

These eight basic odors can be combined to produce a myriad of unique odors. As examples, the smell of a jasmine employs more than 90 odoriferous compounds and the smell of cow manure is composed of 75 odoriferous compounds.

Student Activity—Olfaction Activity

Materials

Small labeled bottles with removable caps containing the following substances, to be divided among the groups:

- perfume
- strawberries or a freshly cut fruit
- freshly ground coffee
- lemon or orange peels
- garlic powder
- oil of peppermint
- oil of wintergreen
- oregano
- vinegar
- vanilla extract

- chocolate
- distilled water
- milk of magnesia
- Dr. Pepper®
- eucalyptus oil
- rose oil
- cinnamon
- fresh roll bits
- fingernail polish remover
- pungent cheese

Procedure 33.7
The Nose Knows—or Does It?

This activity demonstrates the sense of olfaction using common substances. Students are to work in groups of two to four.

1. Procure the bottles.
2. Instruct your partner to sit comfortably and close his or her eyes and keep them closed when testing a substance.
3. Remove the cap from the first bottle, and hold it approximately 5 cm from your partner's nose for about 3 seconds. Place the cap back on the bottle.
4. Have your partner describe or identify the odor, and record his or her response in the chart below.
5. Allow ample time for your partner to recover.
6. Repeat Steps 4 and 5.

Substance	Identify or Description of Substance Trial I	Identify or Description of Substance Trial II
perfume		
strawberries or a freshly cut fruit		
freshly ground coffee		
lemon or orange peels		
garlic powder		
oil of peppermint		
oil of wintergreen		
oregano		
vinegar		
vanilla extract		
chocolate		
distilled water		
milk of magnesia		
Dr. Pepper®		
eucalyptus oil		
rose oil		
cinnamon		
fresh roll bits		
fingernail polish remover		
pungent cheese		

7. Repeat the following procedure using the same bottles in a different order. Record the results in the chart.

8. Return the bottles, and discuss your results.

9. Tally the number of correct and incorrect responses offered by your partner for each trial.

10. Compare your group's results with those of other groups. Did you find any evidence that individuals vary in their ability to recognize odors?

Q. Did you note any discrepancies between trials I and II?

_____

_____

_____

Q. Which odors were the hardest to detect?

_____

_____

_____

Taste receptors are associated with taste buds located on the tongue (Fig. 33.18), inside the cheeks, and on the soft palate, pharynx, and epiglottis. An average human has between 3,000 and 4,000 taste buds, the majority of which are on the tongue. The tongue has four kinds of observable bumps, or **linguinal papillae**. The most common papillae are called **filiform papillae**. These structures are generally rough to the feel (as observed more easily in a cat's tongue) which create friction and sense the texture of food. **Foliate papillae,** more common in young children, are found on the sides of the tongue. These papillae are not involved in taste. **Circumvallate papillae** are large and located in a V near the back of the tongue. An average person possesses from 7 to 12 of these structures, which contain about half of all taste buds. **Fungiform papillae** are shaped like mushrooms and are located all over the tongue, especially at the tip and on the sides. These papillae house approximately three taste buds each. Physiologists recognize five primary tastes:

1. **sweet** (sugars),
2. **salty** (salt),
3. **sour** (citrus fruits, and acids),
4. **bitter** (alkaline substances, caffeine),
5. and **umami** (meaty broth, amino acids such as glutamic and aspartic acid).

Recently, the presence of water receptors has been described primarily in the pharynx by physiologists. The senses of smell and taste are important in helping us enjoy favorable stimulants and help us detect dangerous substances.

The Nose with a Dog Attached

The bloodhound is such a good and dependable tracker that the results of a bloodhound hunt are admissible evidence in court. The bloodhound's ultra-sensitive nose allows it to distinguish smells over a thousand times better than humans.

 ## Student Activity—Gustation Activity

The primary taste sensations can be detected all over the tongue; however, specific regions of the tongue seem to be more sensitive than others to these sensations. The tip of the tongue is generally more sensitive to sweet tastes. The frontal lateral margins of the tongue are generally more sensitive to salty tastes, and the mid-lateral margins are more sensitive to sour tastes. Generally, the rear of the tongue is more sensitive to bitter tastes.

Materials
- paper cups
- drinking water
- clean paper towels
- clean cotton swabs or Q-Tips®
- small capped bottles with the following solutions: 5% sucrose, vinegar, 5% NaCl solution and 0.5% quinine-sulfate solution

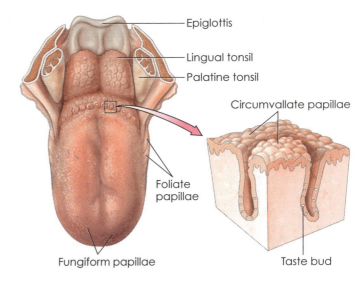

Figure 33.18 Generalized overview of the tongue.

Procedure 33.8
Mapping the Tongue

1. Procure the needed materials.
2. Have one member of the group thoroughly rinse out his or her mouth with water.
3. Using a clean, dry paper towel, have the lab member blot his or her tongue. Dispose of the paper towel.
4. Dip a clean cotton swab into the 5% sucrose solution. Remove as much excess liquid from the swab as possible by pressing it against the side of the bottle.
5. If your lab partner permits (he or she may want to do this independently, as it can trigger the gag reflex), gently touch specific regions of the tongue, cheek, gums, and roof of the mouth with the moistened tip of the swab.

Q. What regions of the tongue seemed more sensitive to sweet sensations?

6. Wait 5 minutes, then repeat Steps 2 through 5, using a vinegar solution.

Q. What regions of the tongue seemed more sensitive to sour sensations?

7. Wait 5 minutes, then repeat Steps 2 through 5, using a 5% NaCl solution.

Q. What regions of the tongue seemed more sensitive to salty sensations?

8. Wait 5 minutes, then repeat Steps 2 through 5, using a 0.5% quinine-sulfate solution.

Q. What regions of the tongue seemed more sensitive to bitter sensations?

9. Properly dispose of the cotton swabs and paper towels.

10. Repeat the same series of steps with another member of the group.
11. Based upon your results, briefly sketch the tongue and denote specific regions that were sensitive to certain tastes.

Hearing and Equilibrium

The two major functions of the ear are **equilibrium** and **hearing**. Equilibrium serves to monitor rotation, linear acceleration, and gravity, allowing us to determine the position of the head with respect to the environment. Hearing (audition) detects and interprets sounds in the environment. The range of frequencies detected by various animals varies. In humans, the general audible range of frequencies is from 20 Hz (cycles per second) to 20,000 Hz, although individuals vary considerably with respect to genetics, sex, age, and occupation. (To gain a better understanding of frequencies, play with the equalizer on a stereo.)

The ear has three major regions:

1. The **outer ear** consists of the conspicuous **auricle** or **pinna,** and the **external auditory meatus** or canal. The canal is lined with **ceruminous glands** that produce earwax, or **cerumin**.
2. The middle ear is primarily a cavity filled with air. The middle ear contains the three **auditory ossicles**—the **malleus, stapes,** and **incus**—as well as the **tympanic membrane** (eardrum) and the **Eustacean tube.**
3. The inner ear consists of the **vestibule** and the **semicircular canals** (both involved in equilibrium and hearing), and the **cochlea** (organ of hearing).

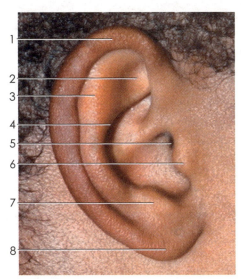

Figure 33.19 The surface anatomy of the auricle.

1. Helix
2. Triangular fossa
3. Antihelix
4. Concha
5. External auditory canal
6. Tragus
7. Antitragus
8. Earlobe

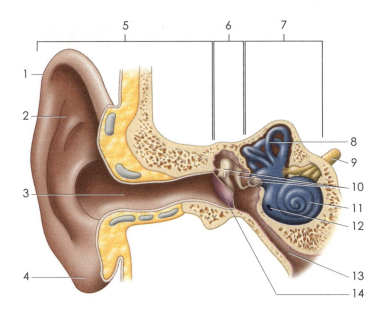

Figure 33.20 Generalized overview of the ear.

1. Helix
2. Auricle
3. External auditory canal
4. Earlobe
5. Outer ear
6. Middle ear
7. Inner ear
8. Semicircular canals
9. Vestibulo-cochlear nerve
10. Auditory ossicles
11. Cochlea
12. Round window
13. Auditory tube
14. Tympanic membrane

 Student Activity—Examination of Human Ear Model

Materials
- model of human ear
- colored pencils

 **Procedure 33.9
Human Ear**

1. Procure model.
2. Observe the human ear model and locate the structures in Figures 33.19 and 33.20.
3. Record your observations below.

_____

_____

_____

_____

_____

Oh, That Terrible Feeling!

Have you ever been seasick? If so, it's a miserable feeling. Seasickness happens when the brain receives conflicting messages about motion and body position. The conflicting information is delivered from the inner ear, eyes, and perhaps skin and muscle receptors. The symptoms of seasickness may include dizziness, profuse sweating, nausea, and vomiting.

Student Activity—Hearing Acuity Test

Hearing acuity refers to the sharpness of hearing or the ability to detect a sound with respect to intensity and distance. This simple test measures hearing acuity.

Materials
- analog clock (one that ticks)
- meter stick
- tuning fork (500–1000 Hz)
- sterile cotton
- blindfold
- stopwatch or watch with secondhand

 Procedure 33.10
Hearing Acuity

1. Have your lab partner sit comfortably on a lab stool or chair.
2. Have your partner keep his or her eyes closed, or blindfold your partner during testing. Instruct your partner to keep his or her head stationary during testing.
3. Gently place a sterile cotton plug into your partner's left ear.
4. Hold a ticking clock approximately 0.5 meters from your partner's ear, and slowly move the sound source away from the ear in a straight line. Hold a meter stick in the other hand. Have your partner indicate when he or she can no longer hear the ticking sound.

5. Using the meter stick, measure the distance to where your partner lost the sound of the ticking clock. Convert the distance to centimeters, and record it in the chart below.
6. Repeat the procedure, using a new sterile cotton plug, but place it in your partner's right ear. Record your results in the chart above.

Hearing Acuity Test		
Student	**Distance (Right Ear)**	**Distance (Left Ear)**

7. Exchange places with your partner and repeat the above steps.

Q. Discuss your findings from the hearing acuity test. Was the hearing acuity the same for the right and left ears of you and your partner?

_____

_____

 Student Activity—Equilibrium Test

Materials
- no special materials are needed for this activity

 Procedure 33.11
Testing Equilibrium

1. Students should work in groups of four.
2. Have your laboratory partner stand erect and as still as possible with both feet placed closely together for 2 minutes. Observe how he or she maintains balance. Record your observations in the chart provided on the next page.

3. Have your laboratory partner repeat the procedure with his or her eyes closed. For safety during this activity, a lab partner should stand on each side of the subject. Observe how the test subject maintained his or her balance, and record your observations in the chart provided.
4. Have your partner stand erect on one foot for 1 minute with his or her eyes open. For safety during this activity, a lab partner should stand on each side of the subject. Observe how the test subject maintained balance, and record your observations in the chart.
5. Have your laboratory partner repeat the procedure with their eyes closed. For safety during this activity, a lab partner should stand on each side of the subject.

Equilibrium Test

Procedure	Observations of Response
Standing on two feet with both eyes open	
Standing on two feet with both eyes closed	
Standing on one foot with both eyes open	
Standing on one foot with both eyes closed	

Observe how the test subject maintained balance, and record your observations in the chart.

Q. How important was the relationship of vision to equilibrium during these activities?

Vision

Humans are animals that are highly dependent upon vision, and the eyes are the structure responsible for vision. The eye refracts and focuses incoming light waves onto sensitive **photoreceptors,** the **rods** and **cones,** which are located in the **retina**. Nerve impulses from the stimulated photoreceptors are conveyed along visual pathways to the **occipital lobe** of the cerebrum of the brain, where vision is perceived.

The human eyeball (Fig. 33.20–33.21) is an irregular, sphere-shaped structure about the size of a ping-pong ball, consisting of three layers, called the **tunics**. The **fibrous** or **outer tunic** consists of the **sclera** and the **cornea**. The sclera, the "white" of the eye, is comprised of fibrous tissue. A thin layer called the **conjunctiva** covers the sclera. The cornea is transparent and is continuous with the sclera. The **limbus** serves as the border between the sclera and cornea. The fibrous tunic provides support, is a site of attachment for muscles, and aids in focusing. The **vascular or middle tunic** contains vessels, lymphatics, and some eye muscles; it regulates light entering the eye, secretes and absorbs aqueous humor, and controls the shape of the eye.

The iris, choroid, and ciliary body make up the vascular tunic. The **iris,** the colored portion of the eye, regulates the diameter of the **pupil**. The pupil, like the aperture of a camera, allows light into the eye. Behind the retina is the **choroid,** a highly vascularized and pigmented portion of the eye. The ciliary body is a thickened portion of the choroid that forms a ring around the **lens**.

The lens helps to form images and divides the eye into two chambers or compartments. The **anterior chamber,** between the cornea and lens, is filled with **aqueous humor,** a liquid that provides nutrients, carries away wastes, and cushions the eye. The **posterior chamber** behind the eye contains the gelatinous **vitreous humor,** which helps to give the eyeball its shape and supports the retina. The neural, or inner, tunic contains the retina, and the retina has photoreceptors. **Rods** are responsible for vision in dim light, such as twilight, and cones are responsible for color vision. **Cones** provide sharper images than rods do. The highest concentration of cones is in the **fovea centralis** of the **macula lutea** (a region with no rods). The fovea centralis is the region of sharpest vision. The **optic disc,** the blind spot, is a region where the **optic nerve** connects to the wall of the eye.

Figure 33.21 The surface anatomy of the eye.

1. Eyebrow
2. Superior eyelid (palpebra)
3. Palpebra commissure
4. Lateral canthus
5. Sclera
6. Eyelashes
7. Pupil
8. Iris
9. Lacrimal caruncle
10. Medial commisure
11. Conjunctiva
12. Inferior eyelid (palpebra)

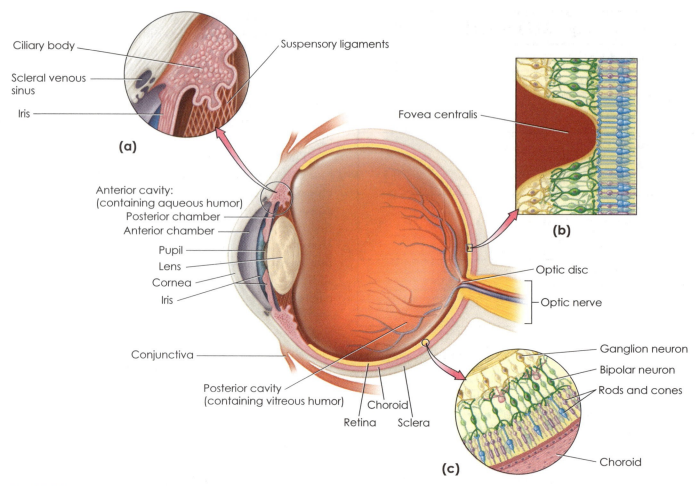

Fig. 33.22 The structure of the eye. (a) Ciliary body, (b) Fovea centralis, and (c) Retina and choroid.

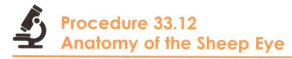

Student Activity—Dissection of the Sheep Eye

Materials
- hand lens
- colored pencils
- dissecting tray
- dissecting kit
- safety glasses
- lab coat or lab apron
- gloves
- preserved eye of a sheep (*Ovis aries*)

Procedure 33.12
Anatomy of the Sheep Eye

Consider photographing this activity.

1. Procure the needed equipment and specimen.
2. Thoroughly rinse the sheep eye in running water.

3. Use your scissors to carefully remove the eyelid. Examine the anterior and posterior external anatomy of the sheep eye, comparing it to Figures 32.23 and 33.24. You may have to use scissors to remove the adipose tissue and four extrinsic muscles in order to see the optic nerve in the posterior view.

4. Place the sheep eye in the dissection tray. Turn the specimen so the cornea is facing left and the optic nerve is on your right. Make an incision through the tough sclera midway between the cornea and optic nerve with a sharp scalpel, cutting the eye into anterior and posterior portions. During this process, aqueous and vitreous humor will leak from the cut.

5. Observe the tunics of the eye as illustrated in Figure 32.25.

6. Removal of the vitreous humor reveals the lens, ora serrata, iris, and ciliary body as illustrated in Figure 32.26.

7. Record your observations and sketch your specimen below.

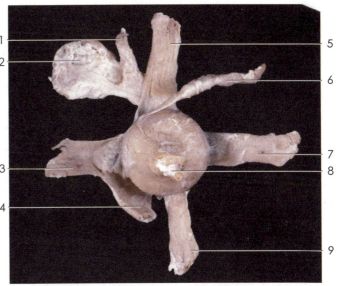

Figure 33.24 Example of posterior view of the sheep eyeball.

1. Levator palpebrae superioris m.
2. Lacrimal gland
3. Lateral rectus m.
4. Inferior oblique m.
5. Superior rectus m.
6. Superior oblique m.
7. Medial rectus m.
8. Optic nerve
9. Inferior rectus m.

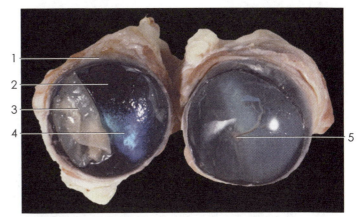

Figure 33.25 Tunics of the sheep eyeball.

1. Sclera
2. Choroid
3. Retina
4. Tapetum lucidum
5. Optic disc

8. Thoroughly clean your laboratory station and equipment. Return the dry equipment and discard the eye parts as indicated by the instructor.

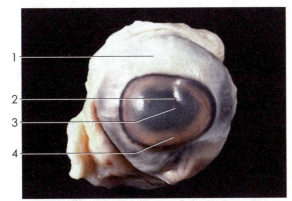

Figure 33.23 A superficial view of the anterior view of the eyeball of a sheep.

1. Sclera
2. Cornea
3. Pupil (dark opening)
4. Iris

Figure 33.26 The internal anatomy of sheep eye.

1. Ciliary body
2. Ora serrata
3. Iris
4. Pupil
5. Lens (removed)

Student Activity—The Snellen Test of Visual Acuity

The **Snellen eye chart** is commonly used to measure visual acuity (Fig. 33.27). The chart has several sets of letters in different sizes printed on a white poster. The larger letters are at the top of the chart and the smallest letters at the bottom. Each set of letters is associated with an acuity value such as 20/20. Normal eyes can see these letters at a distance of 20 feet. If a person has a visual acuity of 20/50, that person can see detail from 20 feet away the same as a person with normal eyesight would see it from 50 feet away. People with a visual acuity of 20/50 have less than normal acuity.

Materials
- masking tape
- white index cards
- Snellen chart
- meterstick or yardstick

Procedure 33.13
Visual Acuity

1. Hang a Snellen chart in a quiet, well-lighted area in the laboratory.
2. Measure a distance of 20 feet from the chart on the floor, and place a piece of masking tape on the floor at this spot.
3. Remove your eyeglasses or contact lenses if you wear them.
4. Face the Snellen chart and place an index card over your left eye. Read the letters on the fifth line of the chart aloud so your partner can hear your response. If you cannot read this level, go up one level at a time until you can accurately read the line. If you can accurately read the fifth line, proceed to the line with the smallest letters that you can read accurately.

Q. Which line could you accurately read with your right eye?

_____

Q. What was the visual acuity value?

_____

5. Repeat Steps 3 and 4, testing your left eye.

Q. Which line could you accurately read with your left eye?

_____

Q. What was the visual acuity value?

_____

6. If you removed your eyeglasses or contact lenses, use them and repeat Steps 3 and 4.

Q. Which line could you accurately read with your right eye?

_____

Q. What was the visual acuity value?

_____

Q. Which line could you accurately read with your left eye?

_____

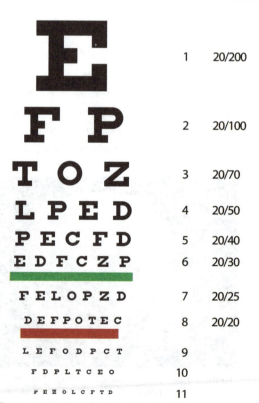

Figure 33.27 An example of a Snellen chart.

Student Activity—Color Vision Test

The term **colorblind** is commonly used to describe individuals that cannot distinguish certain colors; however, the term **color-deficient** is more appropriate. Although the primary cause of color deficiency is genetic, certain types of eye, nerve, or brain damage can cause color deficiency. The most common form of color deficiency is red/green color deficiency. The test for evaluating color deficiency is based on color plates developed by Shinobu Ishihara (1879–1963).

Materials
• *Ishihara's Tests for Colour Blindness* book

 **Procedure 33.14
Color Vision**

1. If available, examine the plates in *Ishihara's Tests for Colour Blindness*. If not, the three plates in Figure 34.28 will serve as example plates.
2. Administer this test in bright light. Place the plate about 76 cm (30 in.) in front of your lab partner. Your partner should respond by telling you a number within the plate or not describe a number at all within 5 seconds. Record his or her response.
3. Test other members of the group as well as yourself.

   Lab partner's response _____

   Lab partner's response _____

Did You Know?

The English scientist John Dalton (1766–1844) famous for his work on atomic theory and gas laws, was colorblind. He attempted to study colorblindness, and for many years colorblindness was called Daltonism. He requested that, after his death, his eyes be removed and studied. His physician, Joseph Ransome, removed the eyes, and they were scientifically examined. The examination was normal. Today, one of his eyes is on display in Dalton Hall in Manchester, England.

Lab partner's response _____

Your response _____

4. Compare your group's responses with the correct response, and discuss the results below. If anyone exhibited color deficiency, record their sex below.

_____

_____

_____

_____

5. Fill in the chart on the next page based upon the class responses.

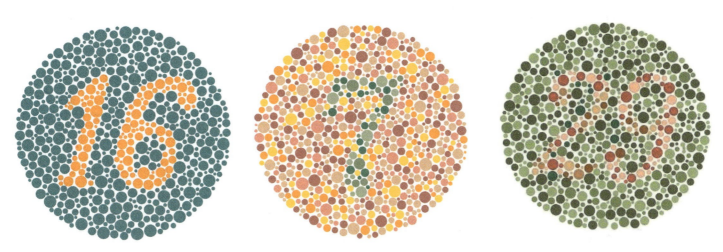

Figure 33.28 Color blindness tests. These are only examples and not meant for diagnosis. (a) Everyone should see the number 16. (b) A person with normal color vision would see a 7 here. (c) A person with normal color vision would see a 29 here. A person with red/green color defeciency would see the number 70 or no number.

Q. Did the class results for the males reflect the expected results? Why?

Q. Did the class results for the females reflect the expected results? Why?

_____

_____

_____

Class Results for Color Vision Test (Males)					
Test	Total Number of Students	Number of Males	Percent of Males	Expected Percent for Males	Discussion
A. Normal				100%	
B. Deficient				93%	
C. Red/Green				7%	
Other					
Other					

Class Results for Color Vision Test (Females)					
Test	Total Number of Students	Number of Females	Percent of Females	Expected Percent for Females	Discussion
A. Normal					
B. Deficient					
C. Red/Green					
Other					
Other					

Student Activity—Afterimage Test

You have seen an **afterimage** after staring at a light or seeing a ghost image on a TV screen when it goes blank. An afterimage results when an image continues to appear after exposure to the original image has stopped. Negative afterimages are caused when photoreceptors in the eyes are overstimulated and lose their sensitivity. When the eyes are diverted to a blank wall or closed, the colors are muted and the brain interprets it as the opposite color.

Materials
- Figure 33.29

 ## Procedure 33.15
Afterimage

1. Hold the page with the flag about 30 cm from your face.
2. Intensely fixate on the bottom right star for 1 minute.

3. After 1 minute, look away toward the ceiling or blank wall, or close your eyes.
4. Describe the flag.

_____

_____

Figure 33.29 Afterimage test.

 ## Student Activity—Blind-spot Test

The eye has a **blind spot,** where the optic nerve exits the rear of the eye. This region, termed the **optic disc,** contains no receptor cells. Your blind spot can be easily detected by the following test.

Materials
- Figure 33.30

Procedure 33.16
Blind Spot Test

1. Close your left eye, and keep it closed.
2. Hold the page with the blind-spot test about 30 cm from your face.
3. Fixate on the plus sign with your right eye, and start moving the page towards your eye until the dot disappears. When the image of the dot disappears, have your lab partner measure the distance from the eye to the dot, in centimeters. The dot disappeared because the image of the dot passed over the optic disc.

Distance _____

4. Repeat Steps 1 through 3 closing the right eye and testing the left eye.

Distance _____

Q. Did you note any differences between the right and left eye?

_____

_____

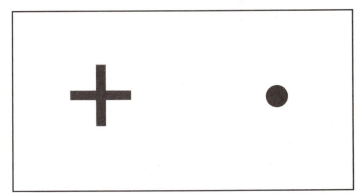

Figure 33.30 Blind spot test.

THE ENDOCRINE SYSTEM

The **endocrine system** is a major control system of the body, and like the nervous system, helps maintain homeostasis. The endocrine system is responsible for the production of a variety of **hormones**. The hormones are released into the bloodstream and are carried throughout the body, where they affect **target tissues**. In contrast to rapidly acting nervous controls, the majority of endocrine controls are slower and long-acting.

The endocrine system is comprised of a number of ductless glands that produce and release hormones. In addition to these glands, structures such as the heart, liver, and kidneys also have endocrine functions. Also, adipose tissue has some endocrine function. Refer to Figure 33.31 for the location of the primary hormone-producing structures in the human body. In addition, Table 33.3 provides an overview of several endocrine structures, the hormones they produce, and the primary function or action of these hormones.

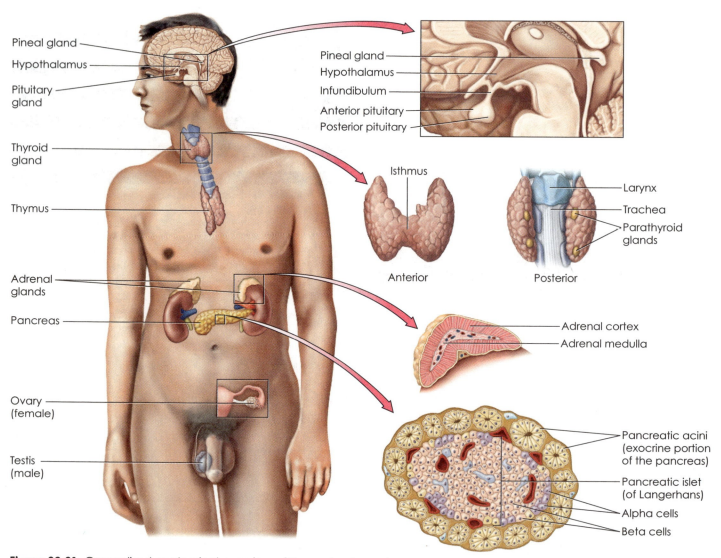

Figure 33.31 Generalized anatomical overview of the endocrine system.

I think that science changes the way your mind works, to think a little more deeply about things.

—P.Z. Myers (1957–present)

Student Activity—Observing a Model of the Human Endocrine System

Materials
- model of the endocrine structures

 Procedure 33.17
Endocrine System

1. Procure the model of the endocrine structures.
2. Locate and be able to briefly describe the function of the endocrine organs included in Figure 33.31.
3. Sketch and label the model in the space provided. Write the basic function to each structure adjacent to the label.

Table 33.3 Endocrine Glands, Associated Hormones and their Functions

Endocrine Gland or Structure	Hormone	Target	Action
Hypothalamus	Releasing Hormones	Anterior pituitary	Stimulates release of hormones of the anterior pituitary
	Inhibiting hormones	Anterior pituitary	Inhibits release of hormones of the anterior pituitary
Anterior Pituitary (Adenohypophysis)	Follicle Stimulating Hormone (FSH)	Ovaries and testes	Stimulates spermatogenesis and development of ovarian follicle
	Lutenizing Hormone (LH)	Ovaries and testes	Stimulates ovulation and testosterone secretion
	Thyroid Stimulating Hormone (TSH)	Thyroid gland	Stimulates the thyroid gland and metabolic rate
	Adrenocorticotrophic Hormone (ACTH)	Adrenal cortex	Stimulates adrenal cortex
	Prolactin (PRL)	Mammary glands	Production of milk
	Growth Hormone (Somatotropin or GH)	Connective tissue, internal organs, soft tissues	Cell division, protein synthesis, and growth
Posterior Pituitary (Neurohypophysis)	Antidiuretic hormone (ADH)	Kidneys	Stimulates water resorption
	Oxytosin (OT)	Uterus and mammary glands	Stimulates uterine contraction and release of milk

Table continued on page 770

Table 33.3 Endocrine Glands, Associated Hormones and their Functions

Table continued from page 769

Endocrine Gland or Structure	Hormone	Target	Action
Thyroid	Thyroxine (T_4) and Triiodothyronine (T_3)	All tissues	Stimulates metabolic rate and regulates growth and development
	Calcitonin	Bone	Reduces blood calcium level, promotes ossification
Parathyroid	Parathyroid hormone (PTH)	Kidneys, bone, and digestive tract	Raises blood calcium levels and activates vitamin D
Adrenal Glands Adrenal Cortex	Glucocorticoids (cortisol)	All tissues	Raises blood glucose level, stimulates protein breakdown, and mobilizes fat
	Mineralocorticoids (aldosterone)	Kidneys	Excretes potassium and reabsorb sodium maintaining proper balance between potassium and sodium
	Sex steroids (androgen, estrogen)	Gonads, skin, bones, and muscles	Stimulates the development of sex characteristics and reproductive organs
Adrenal Glands Adrenal Medulla	Epinephrine (adrenaline) and norepinephrine (noradrenaline)	Muscles, heart, and most tissues	Raises heart rate, blood glucose levels, and metabolic rate. Dilates blood vessels and mobilizes fat reserves
Pancreas	Insulin	Liver, skeletal muscle, adipose tissue	Lowers blood glucose levels. Promotes fat, protein, and glycogen synthesis
	Glucagon	Liver, skeletal muscle, adipose tissue	Raises blood glucose levels. Stimulates the livers breakdown of glycogen
Thymus	Thymosins	T lymphocytes	Benefits the immune system by promoting the development of T lymphocytes.
Pineal Gland	Melatonin	Gonads, brain, pigment cells	Controls biological rhythms such as circadian rhythms. Perhaps controls sex organ maturation
Testes	Androgens (testosterone)	Gonads, muscles, bones, skin	Stimulates the development of male secondary sex characteristics. Stimulates spermatogenesis
	Inhibin	Anterior pituitary	Inhibits secretion of FSH
Ovaries	Estrogen	Female reproductive tract, skin, muscles, and bones	Stimulates the development of female secondary sex characteristics. Prepares uterus for pregnancy
	Progesterone	Uterus and mammary glands	Stimulates development of mammary glands and completes preparation for pregnancy
	Inhibin	Anterior pituitary	Inhibits secretion of FSH

Name: _____ Date: _____ Section: _____

Review Questions

1. Describe three characteristics that make us human.

_____

_____

_____

2. Describe the functions of the integumentary system.

_____

_____

_____

3. What are the major functions of bone?

_____

_____

_____

4. Describe the basic functions of muscle.

_____

_____

_____

5. Compare and contrast the central and peripheral nervous systems as to composition and function.

_____

_____

_____

6. Describe the basic types of odors.

_____

_____

_____

7. Describe the basic types of taste receptors.

_____

_____

_____

8. What are the two major functions of the ear?

_____

_____

_____

9. Describe the hormones secreted by the anterior and posterior pituitary.

_____

_____

_____

10. Label the following:

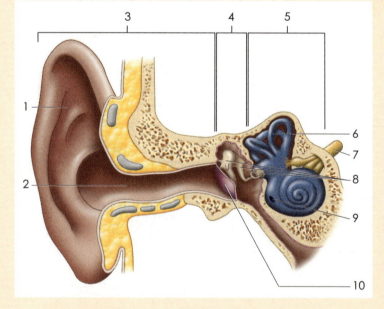

1. _____

2. _____

3. _____

4. _____

5. _____

6. _____

7. _____

8. _____

9. _____

10. _____

Name: _____   Date: _____   Section: _____

11. Label the following:

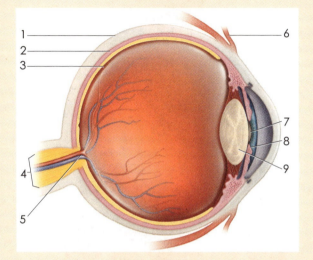

1. _____

2. _____

3. _____

4. _____

5. _____

6. _____

7. _____

8. _____

9. _____

12. Label the following:

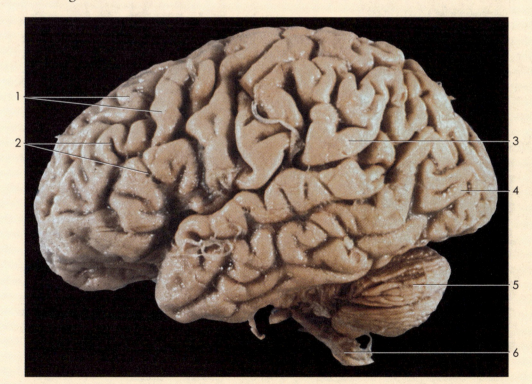

1. _____   4. _____

2. _____   5. _____

3. _____   6. _____

13. Label the following:

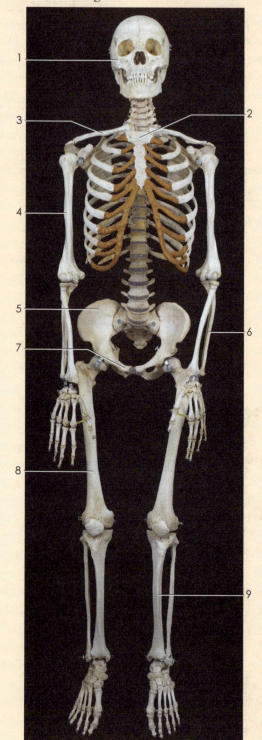

14. Label the following:

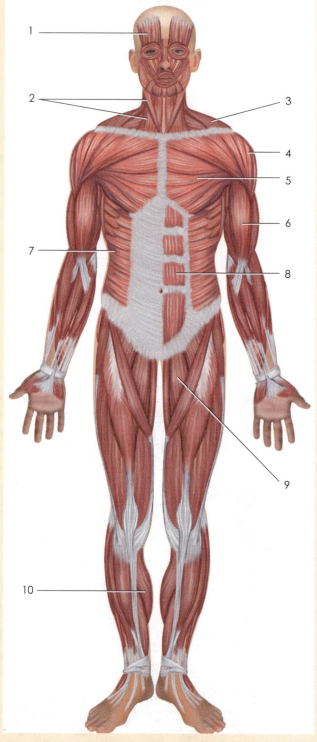

1. _____   6. _____

2. _____   7. _____

3. _____   8. _____

4. _____   9. _____

5. _____

1. _____   6. _____

2. _____   7. _____

3. _____   8. _____

4. _____   9. _____

5. _____   10. _____

Chapter 34
Homo sapiens:
Understanding Human Biology—Part II

Student Outcome Objectives

At the completion of this exercise, the student will be able to:

1. Describe the major components of the circulatory, lymphatic, and respiratory systems and their basic functions.
2. Identify and discuss basic heart anatomy.
3. Trace the route of blood through the heart.
4. Describe heartbeat, pulse, and blood pressure.
5. Describe the functions of the respiratory system.
6. Describe and locate the major structures of the respiratory system.
7. List and describe the major components of the digestive system.
8. Discuss the basic functions of the digestive system.
9. List the major components of the urinary system.
10. List the major components of the female and male reproductive system.

Overview

The circulatory, lymphatic, and respiratory systems are essential components of the human body, serving primarily as delivery and defense systems. The **circulatory system**, or cardiovascular system, consists of the blood, heart, and a variety of vessels including arteries and veins. The **lymphatic system** consists of lymph, lymphatic vessels, lymph nodes, and lymphatic organs such as the thymus and spleen. The **respiratory** system is made up of the airways that carry air to and from the lungs.

To survive as an individual, the digestive and urinary systems must function properly. And to perpetuate the species, the female and male reproductive systems must function efficiently. The **digestive system** is composed of the **gastrointestinal tract** (oral cavity, pharynx, esophagus, stomach, small intestine, and large intestine) and the **accessory digestive organs and structures** (teeth, tongue, gums, salivary glands, liver, gallbladder, and pancreas). The **urinary system** consists of the kidneys, ureters, urinary bladder, and the urethra. The **female reproductive system** contains the ovaries, uterus, oviducts, vagina, external genitalia, and mammary glands. The **male reproductive system** includes the testes, scrotum, seminal vesicles, prostrate gland, bulbourethral gland, and penis.

The Heart

The **heart** is a muscular organ about the size of a clenched fist (Fig. 34.1–34.6). The heart lies in the mediastinum, resting upon the diaphragm surrounded by the sternum, lungs, and thoracic vertebrae. The distal end, or **apex**, of the heart is blunt-shaped and points to the left. Several large vessels attach to the top (base) of the heart. The vessels comprise the **pulmonary circuit,** which carries blood to and from the lungs, and the **systemic circuit,** which carries blood to and from the rest of the body. The heart is enclosed in the **pericardium**, a double-walled sac. The fibrous outer portion of the pericardium is called the **parietal pericardium,** and the inner serous layer is called the **visceral pericardium,** or **epicardium**. The epicardium covers the surface of the heart.

A **pericardial cavity** filled with **pericardial fluid** exists between the parietal and visceral pericardia. The fluid reduces friction within the beating heart. The **endocardium** is a thin layer of tissues that covers the inside of the heart and vessels. The muscular **myocardium** exists between the epicardium and endocardium. The myocardium is composed of cardiac muscle, the most massive portion of the heart.

Within the heart are four chambers consisting of the **atria** and the **ventricles**. The left and right atria receive blood, and the left and right ventricles pump blood. The atria are separated by a thin-walled **interatrial septum** and the ventricles are separated by a muscular **interventricular septum**. The **right atrium,** a part of the systemic circuit, receives deoxygenated blood from the superior vena cava, the inferior vena cava, and the coronary sinus.

Deoxygenated blood flows from the right atrium to the **right ventricle** through the **right atrioventricular** or **tricuspid valve**. As part of the pulmonary circuit, the right ventricle pumps blood into the **pulmonary artery** or trunk, which divides into the left and right pulmonary arteries. These arteries carry deoxygenated blood to the lungs for oxygenation. Oxygenated blood returns to the **left atrium,** a part of the pulmonary circuit, by means of the left and right pulmonary veins.

The opening between the left atrium and ventricle is guarded by the **left atrioventricular,** or **bicuspid** or **mitral valve**. **Chordae tendon,** also called the "strings of the heart," connect the muscles to the tricuspid and bicuspid valves. The **left ventricle** has a thick wall because it is responsible for pumping oxygenated blood through the systemic circuit to the rest of the body.

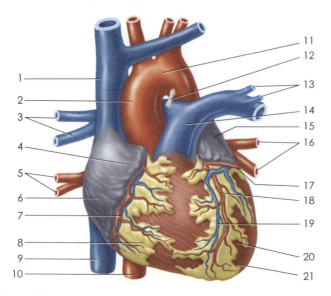

Figure 34.1 An anterior view of the structure of the heart.

1. Superior vena cava
2. Ascending aorta
3. Branches of right pulmonary artery
4. Auricle of right atrium
5. Right pulmonary veins
6. Right atrium
7. Right coronary artery and vein
8. Right ventricle
9. Inferior vena cava
10. Thoracic aorta
11. Aortic arch
12. Ligamentum arteriosum
13. Branches of left pulmonary artery
14. Pulmonary trunk
15. left atrium
16. Left pulmonary veins
17. Circumflex artery
18. Anterior interventricular artery
19. Anterior interventricular vein
20. Left ventricle
21. Apex of heart

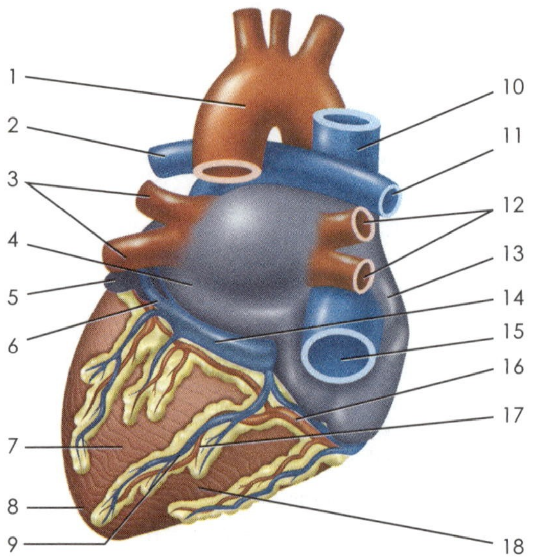

Figure 34.2 A posterior view of the structure of the heart.

1. Aortic arch
2. Left pulmonary artery
3. Left pulmonary veins
4. Left atrium
5. Auricle of left atrium
6. Great cardiac vein
7. Left ventricle
8. Apex
9. Middle cardiac vein
10. Superior vena cava
11. Right pulmonary artery
12. Right pulmonary veins
13. Right atrium
14. Coronary sinus
15. Inferior vena cava
16. Right coronary artery
17. Posterior interventricular artery
18. Right ventricle

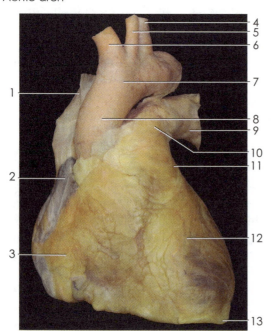

Figure 34.3 An anterior view of the heart.

1. Superior vena cava
2. Right atrium
3. Right ventricle
4. Left subclavian artery
5. Left common carotid artery
6. Brachiocephalic trunk
7. Aortic arch
8. Ascending aorta
9. Pulmonary arteries
10. Pulmonary trunk
11. Left atrium
12. Left ventricle
13. Apex of heart

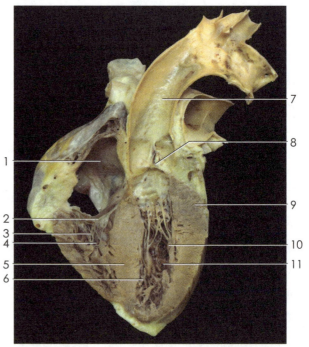

Figure 34.4 The internal structure of the heart.

1. Right atrium
2. Right atrioventricular valve
3. Chordae tendinae
4. Left ventricle
5. Interventricular septum
6. Trabeculae carneae
7. Ascending aorta
8. Aortic valve
9. Myocardium
10. Papillary muscle
11. Left ventricle

Blood leaves the left ventricle by way of the **aortic valve,** or **semilunar valve,** to the **aorta** (the largest artery in the body). From the aorta, blood is distributed throughout the body. For example, the **coronary arteries** supply the cells of the heart with blood.

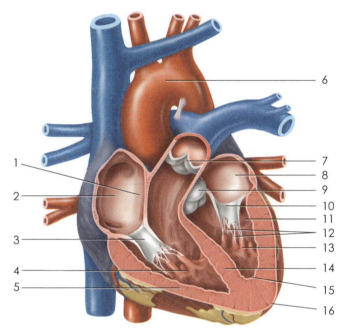

Figure 34.5 An internal view of the structure of the heart.

1. Interatrial septum
2. Right atrium
3. Tricuspid valve
4. Right ventricle
5. Myocardium
6. Aortic arch
7. Pulmonary valve
8. Left atrium
9. Aortic valve
10. Bicuspid valve
11. Left ventricle
12. Chordae tendinae
13. Papillary muscle
14. Interventricular septum
15. Endocardium
16. Visceral pericardium

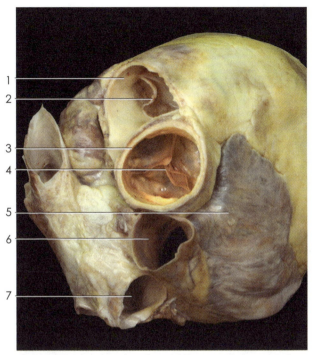

Figure 34.6 A superior view of the great vessels of the heart and aortic valve.

1. Pulmonary trunk
2. Pulmonary valve
3. Ascending aorta
4. Aortic valve
5. Right atrium
6. Superior vena cava
7. Pulmonary vein

 Student Activity—Examination of a Human Heart Model

Materials
- model of heart
- colored pencils

 **Procedure 34.1
Human Heart**

1. Procure the model of the human heart.
2. Examine the external anatomy of the model heart and locate the structures labeled in Figure 34.1 through Figure 34.6.

3. Sketch and label the model below.

Student Activity—Dissection of the Sheep Heart

Materials
- sheep heart
- dissecting tray
- dissecting kit
- safety glasses
- lab coat or apron
- gloves

Procedure 34.2
Sheep Heart

1. Procure the needed equipment and the specimen.
2. Thoroughly rinse the sheep heart in running water. Run water through the large blood vessels to remove preservative and blood clots.
3. Place the heart in the dissecting tray with the ventral surface facing you.
4. Locate the visceral pericardium, which appears as a thin, transparent layer on the outer surface of the heart. Use a scalpel to remove a portion of the visceral pericardium and expose the myocardium. Notice the layers of fat around the heart.
5. Pick up the heart and locate the external features of the heart as illustrated in Figures 34.7 and 34.8. Sketch and record your observations in the space below.

6. Using a scalpel, carefully make a coronal cut though the heart from the base to the apex and separate the heart as illustrated in Figure 34.9.
7. Observe the anatomical features illustrated in Figures 34.9, 34.10, and 34.11. Sketch and record your observations below.

8. Discard the heart and thoroughly clean your equipment and station as instructed.

The body of man has in itself blood, phlegm, yellow bile and black bile; these make up the nature of this body, and through these he feels pain or enjoys health.

Hippocrates (460 BC–370BC)

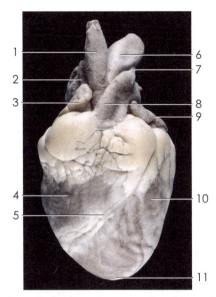

Figure 34.7 A ventral view of mammalian (sheep) heart.

1. Brachiocephalic artery
2. Cranial vena cava
3. Right auricle of right atrium
4. Right ventricle
5. Interventicular groove
6. Aortic arch
7. Ligamentum arteriosum
8. Pulmonary trunk
9. Left auricle of left atrium
10. Left ventricle
11. Apex of heart

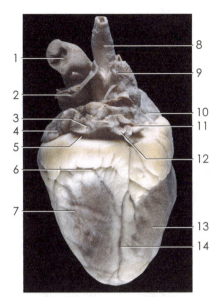

Figure 34.8 A dorsal view of mammalian (sheep) heart.

1. Aorta
2. Pulmonary artery
3. Pulmonary vein
4. Left auricle
5. Left atrium
6. Atrioventricular groove
7. Left ventricle
8. Brachiocephalic artery
9. Cranial vena cava
10. Right auricle
11. Right atrium
12. Pulmonary vein
13. Right ventricle
14. Interventricular groove

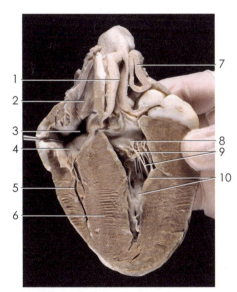

Figure 34.9 A coronal section of the mammalian (sheep) heart.

1. Aorta
2. Cranial vena cava
3. Right atrium
4. Right atrioventricular (tricuspid) valve
5. Right ventricle
6. Interventricular septum
7. Pulmonary artery
8. Left atrioventricular (bicuspid) valve
9. Chordae tendineae
10. Papillary muscles

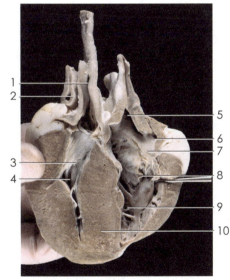

Figure 34.10 A coronal section of the mammalian (sheep) heart showing the valves.

1. Opening of the brachiocephalic artery
2. Pulmonary artery
3. Left atrioventricular (bicuspid) valve
4. Left ventricle
5. Opening of cranial vena cava
6. Opening of coronary sinus
7. Right atrium
8. Right atrioventricular (tricuspid) valve
9. Right ventricle
10. Interventricular septum

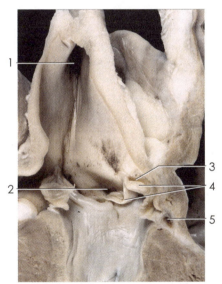

Figure 34.11 A coronal section of the mammalian (sheep) heart showing openings of coronary arteries.

1. Opening of brachiocephalic artery
2. Opening of left coronary artery
3. Opening of right coronary artery
4. Aortic valve
5. Coronary vessel

THE CIRCULATORY SYSTEM IN ACTION

Auscultation is the process of listening to sounds generated by the body. Auscultation of the heart can provide a medical professional with vital information about the heart and its health. An instrument called the **stethoscope** is used to listen to heart sounds. Each cardiac cycle generates two, or perhaps three, sounds that are can be heard with a stethoscope. When listening to heart sounds, the first S_1 and second S_2 are responsible for the typical "lubb" "dupp" sounds. The S_1 sound "lubb" is louder and longer than S_2 and represents the closing of the bicuspid and bicuspid valves and the beginning of ventricular contraction. The S_2 sound occurs when the ventricles start to fill after the semilunar valves close. In individuals under age 30, a third heart sound S_3 sometimes can be heard. If so, it is called a "gallop" or "triple rhythm" and is not caused by valves closing.

Heart sounds are best heard in four regions, indicated in Figure 34.12. The normal resting heart rate is about 65 to 80 times per minute. Athletes may have slower heart rates.

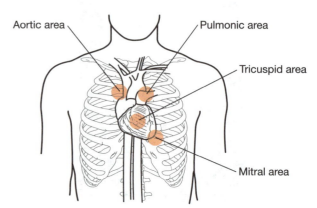

Figure 34.12 Common auscultation areas.

 Student Activity—Listening to the Heart

Materials
- stethoscope
- alcohol pads
- stopwatch or watch with a secondhand

 **Procedure 34.3
Heart Sounds**

Do not exercise if you have heart or lung problems.
1. Procure the stethoscope and alcohol pads.
2. Thoroughly clean the earpieces and diaphragm of the stethoscope with an alcohol pad.
3. Place the earpieces in your ears and gently tap on the diaphragm to ensure that you have the diaphragm oriented properly.
4. You may listen to your heart sounds or your partner's. You or your partner sit quietly and place the diaphragm in the aortic area under the shirt or blouse. Your partner may wish to place the diaphragm of the stethoscope on their own chest.
5. Listen for cardiac cycles and define the characteristic "lubb" "dupp."
6. Count the heart rate for 15 seconds, and multiply this number by 4 to determine the number of beats per minute. Record the number of beats below.

Number of heartbeats in 15 seconds: _____

Number of heartbeats in 1 minute: _____

7. Move your stethoscope to the pulmonic, tricuspid, and mitral areas of the heart.
8. If you or your partner does not have heart or lung problems, jog in place for 2 minutes and measure the heart rate immediately afterward. Record the number of heartbeats below.

Number of heartbeats in 15 seconds: _____

Number of heartbeats in 1 minute: _____

9. Clean the earpiece and diaphragm of the stethoscope with another alcohol pad, and return it to the proper place.

Q. Did you notice any differences between the four areas?

_____

_____

_____

Q. Did you witness an extra heartbeat or extra sounds?

_____

_____

_____

10. Record and discuss the difference in heartbeat at rest and after exercise.

Student Activity—Measuring the Pulse

As blood surges through the arteries, it creates the **pulse**. Monitoring the pulse rate with the fingertips is common and is called **pulse palpitation**. **Pulse rate** represents the heartbeats per minute. Typical pulse rates range from 65 to 80 beats per minute; however, athletes may have much slower rates because they possess stronger hearts. The body has several common pulse points, (Fig. 34.13) but the most common one used by medical professionals is **the radial pulse**.

Materials
• sanitizing soap

 Procedure 34.4
The Pulse

Do not exercise if you have heart or lung problems.
1. Thoroughly wash your hands and have your partner thoroughly wash their hands and wrist area with soap.
2. Have the person being measured sit down. Lightly place your index and middle finger on the thumb side of the inner wrist to locate the radial artery of your partner, and press down slightly.
3. Once you have located the pulse, count the pulse rate for 15 seconds. Multiply this number by 4 to determine the pulse rate per minute.

Pulse rate in 15 seconds: _____

Pulse rate for 1 minute: _____

4. Repeat the procedure, but this time monitor the carotid artery.

Pulse rate in 15 seconds: _____

Pulse rate for 1 minute: _____

5. Have your partner hold his or her breath for 20 seconds and measure the radial pulse.

Pulse rate in 15 seconds: _____

Pulse rate for 1 minute: _____

6. If your partner does not have heart or lung problems, ask him or her to vigorously jog in place for 2 minutes, then sit down. Measure the radial pulse immediately.

Pulse rate in 15 seconds: _____

Pulse rate for 1 minute: _____

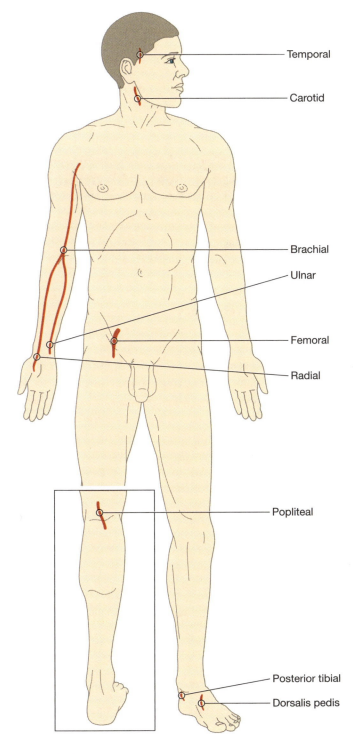

Figure 34.13 Common pulse points.

7. Measure the radial pulse of your partner 2 and 5 minutes after exercise.

2 minutes after exercise:

Pulse rate in 15 seconds: _____

Pulse rate for 1 minute: _____

5 minutes after exercise:

Pulse rate in 15 seconds: _____

Pulse rate for 1 minute: _____

8. Thoroughly wash your hands, and have your partner thoroughly wash his or her hands and wrist area with soap.

9. Fill in the chart below. Account for deviation of pulse rate from rest for each measurement.

	Pulse rate per minute
At rest	
Holding breath	
Immediately after exercise	
2 minutes after exercise	
5 minutes after exercise	

Student Activity—Measuring Blood Pressure

Basically, **blood pressure** is the measure of the pressure exerted upon the surface of the vessels by the blood. This pressure serves to circulate the blood throughout the body. The blood leaving the heart generates the **systolic pressure,** and the pressure when the heart relaxes is the **diastolic pressure**. Blood pressure is measured in millimeters of mercury (mmHg). The systolic pressure is the larger pressure and should average between 100 and 120 mm HG. Diastolic pressure averages between 60 and 80 mm Hg.

An instrument called a **sphygomanometer** is used to measure blood pressure. *Note: A number of types of sphygomanometers are available. Listen to your instructor to learn how to use the sphygomanometer used in the following laboratory.* Partners may be switched after each major measurement, but be sure to denote this in the chart. *Caution:* Do not exercise if you have heart or lung problems.

Materials
- sphygomanometer
- stethoscope
- alcohol pads
- watch
- bucket of ice water

Procedure 34.5
Blood Pressure

1. Procure the sphygomanometer, stethoscope, and alcohol pads.
2. Thoroughly clean the earpieces and diaphragm of the stethoscope with an alcohol pad.

3. Have your partner sit comfortably in a chair next to a lab table.
4. If your partner is not wearing short sleeves, have him or her roll up one sleeve. Wrap the cuff of the sphygomanometer apparatus around the arm (your choice) of your partner. The cuff should sit about 3.5 cm above the anticubital fossa. Do not make it tight, just tight enough to stay in place. If you are uncomfortable with the degree of tightness, have your instructor check your setup. Have your partner rest his or her arm on the table.
5. Clean the arm to be tested near the brachial artery with an alcohol pad.
6. Familarize yourself with the gauge of the sphygomanometer.
7. Place the diaphragm of the stethoscope over the brachial artery. Place the earpieces in your ears. No sounds should be heard.
8. Close the screw on the bulb of the sphygomanometer by tightening it in a clockwise direction. Watch the pressure gauge, and begin to gently pump the bulb. Pump the bulb up to about 190 mm Hg. Your partner may feel some discomfort.
9. Keeping the stethoscope above the brachial artery and listening closely, watch the pressure gauge and slowly release the pressure on the cuff by turning the screw in a counter-clockwise direction.
10. Eventually you will see the needle on the gauge jump and hear the pulse in the brachial artery. Note the value when you hear the pulse; this is the systolic pressure. Continue releasing the pressure with the

screw until you can no longer hear the pulse. Note the gauge at this point. This is the diastolic pressure.

Systolic pressure: _____

Diastolic pressure: _____

11. Record the blood pressure as a fraction with the systolic pressure on top.

Blood pressure: _____

12. Repeat the procedure now that you understand the process.

Blood pressure: _____

13. Repeat the process with your partner standing. Take a reading immediately after he or she stands. Then take readings 5 minutes and 10 minutes after your partner stands up.

Immediately after standing blood pressure: _____

Standing 5 minutes blood pressure: _____

Standing 10 minutes blood pressure: _____

14. Have your partner sit and relax for 5 minutes.
15. Measure your partner's blood pressure at rest.

Blood pressure: _____

16. Have your partner immerse their free hand in a bucket of ice water. With the hand in the water, quickly measure their blood pressure.

Blood pressure during ice immersion: _____

17. Have your partner remove his or her hand from the bucket and relax for 10 minutes.
18. Measure the blood pressure again 5 minutes after water immersion.

Blood pressure after ice immersion: _____

19. Take a resting blood pressure of your partner.

Blood pressure resting: _____

20. Have your partner rigorously jog in place for 3 minutes. Immediately take the blood pressure. Take the blood pressure while sitting 2 and 5 minutes after exercising.

Blood pressure immediately after exercise: _____

Blood pressure two minutes after exercise: _____

Blood pressure five minutes after exercise: _____

21. Clean the earpiece and diaphragm of the stethoscope with another alcohol swab, and return the materials to the proper place.
22. Fill in the following chart below.

	Blood Pressure
At rest	
At rest (2nd reading)	
Immediately after standing	
5 minutes while standing	
10 minutes while standing	
After resting 3 minutes	
During water immersion	
After water immersion	
Prior to exercise at rest	
Immediately after exercise	
2 minutes after exercise	
5 minutes after exercise	

Q. Discuss any difficulties that you encountered while taking blood pressure.

_____

_____

_____

Q. Discuss and account for differences in blood pressure readings throughout this activity.

_____

_____

THE RESPIRATORY SYSTEM

The **respiratory system** (Fig. 34.14), working in conjunction with the cardiovascular system, supplies oxygen to the cells of the body so they can undergo aerobic respiration and remove the waste product of cellular respiration—carbon dioxide—from the body. In addition, the respiratory system is involved in vocalizations, the sense of smell, and controlling the pH of the body.

The respiratory tract is divided into two distinct regions.

1. The **upper respiratory tract** consists of the structures found outside of the thorax and includes the nose, nasal cavity, paranasal sinuses, and pharynx (nasopharynx, oropharynx, and laryngopharynx). These structures filter, warm, and humidify the incoming air and protect the respiratory tract.

2. The **lower respiratory tract** (passages within the thoracic cavity) consists of the larynx (voice box), trachea (windpipe), bronchi, bronchioles, alveoli, and lungs.

The **conducting portion** of the respiratory system consists of the structures essential in the movement of air. The **respiratory portion** consists of structures involved in gas exchange, such as the alveoli.

The **lungs** are the largest organ in the respiratory tract. In humans, the right lung has three lobes and the left lung has two lobes. Air enters the lungs by way of the left and right **primary bronchi**. The **bronchial tree** has several divisions within the lungs, ending in **terminal bronchioles**. The **alveoli** are sacs that are responsible for gas exchange and are located at the end of respiratory bronchioles. The respiratory rate in an average adult human is 12 times per minute, or nearly 17,500 times a day.

I wish I could breathe
Like you, oh, breathe
I know the answer's out there.
Won't you help me...please?
Cause one day I'll breathe
I'll breathe, I'll breathe
I'll breathe like you.

Matt Scales (1979–2007) and Barnaby Pinny.
Breathe, **a song for cystic fibrosis.**

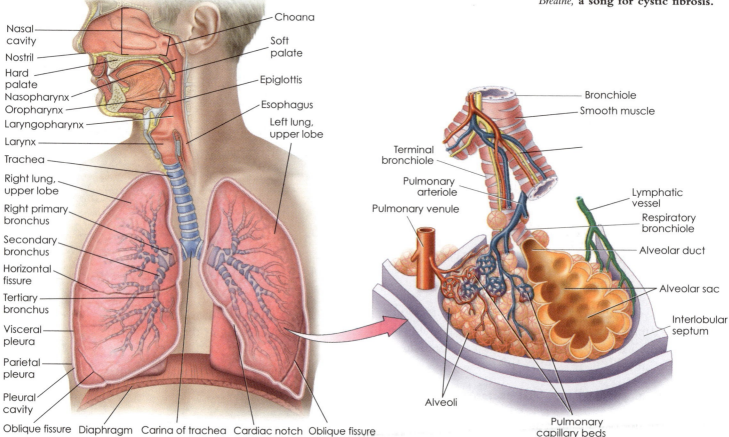

Figure 34.14 The structure of the respiratory system.

Student Activity—Examination of the Respiratory System Model

Materials
- model of respiratory system
- colored pencils

Procedure 34.6
Respiratory System

1. Procure the respiratory system model.
2. Examine the respiratory system model, locating the structures labeled in Figure 34.14.
3. Sketch and label the model below.

THE DIGESTIVE SYSTEM

The gastrointestinal tract, or alimentary canal, is basically a long tube extending from the mouth to the anal opening (Fig. 34.15). Accessory digestive structures are found along the tract to aid in the breakdown of food. The digestive system begins with the oral cavity with the opening of the mouth, where the food is chewed, called **mastication**, and mechanically broken down by the **teeth**. The **lips, cheeks, palate,** and **tongue** help to contain and manipulate the **bolus,** or food mass. The **salivary glands** release **saliva** (containing salivary amylase), which initiates the digestion of carbohydrates. Once the bolus is prepared, **deglutition** (swallowing) occurs.

The **pharynx** receives the bolus from the oral cavity and passes it to the esophagus. The pharynx is shared by the respiratory and digestive systems. The **esophagus** is a tube approximately 25 cm in length and 2 cm at its widest point. Its function is to allow the passage of the bolus to the stomach through peristalsis (a wave of smooth muscle contractions).

The stomach is a J-shaped organ consisting of a lesser and a greater curvature.

The stomach has four duties:

1. to serve as a storage structure for ingested food,
2. to continue the mechanical breakdown of food,
3. to chemically break down food, and
4. to produce a mixture of partially digested food called **chyme,** which is passed to the small intestine.

From the stomach, the chyme is passed to the **duodenum** of the **small intestine**. The next two sections of the small intestines include the **jejunum** and the **ileum**. The small intestine is responsible for most of the chemical digestion and nutrient absorption in the body. The small

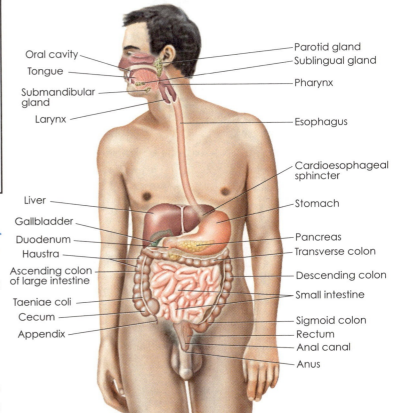

Figure 34.15 The basic anatomy of the digestive system.

intestine averages 6 meters in length and has a diameter of about 1 centimeter.

Secretions from the liver and pancreas are essential in the proper functioning of the small intestine. The **liver,** an accessory digestive structure, consists of four lobes: the right and left lobes are larger than the caudate and quadrate lobes. The liver is the largest organ, weighing about 1.5 kg. Major functions of the liver include:

- production of bile,
- storage of iron and copper,
- conversion of glucose to glycogen,
- storage of glycogen,
- synthesis of vitamins,
- storage of vitamins (A, D, E, and K), and
- detoxification of harmful substances.

The **bile** produced in the liver is stored in the **gallbladder,** a pear-shaped structure located behind the right lobe of the liver. Bile is released into the duodenum where it acts as an emulsifier of fats.

The **pancreas**, another accessory structure of the digestive system, is a pinkish, elongate structure about 15 cm in length, extending laterally from the duodenum toward the spleen. It is responsible for the secretion of pancreatic juice containing digestive enzymes and the production of the hormones insulin and glucagon.

The **large intestine,** or large bowel, receives undigested wastes from the small intestine, absorbs water, salts, and vitamins, and stores indigestible material until it is eliminated. The large intestine does not undergo digestion. The large intestine is about 1.5 meters in length and has a diameter of 7.5 cm.

The ileum of the small intestine empties into the **cecum** of the large intestine. The **vermiform appendix,** attached to the cecum, has some lymphatic function. The **colon,** the largest section of the large intestine, is shaped like a horseshoe and consists of the **ascending colon**, the **transverse colon**, the **descending colon,** and the **sigmoid colon**. The **rectum,** which lies below the sigmoid colon, forms the last 15 cm of the digestive tract. It serves as a temporary storage structure for feces. The last portion of the rectum terminates in the **anus**.

Student Activity—Examination of a Model of the Human Digestive System

Did You Know?

William Beaumont (1785–1853) is known as the Father of Gastric Physiology partially as the result of a lucky or not so lucky event. On June 6, 1822 Beaumont, an Army physician, treated the stomach wound of Alexis St. Martin (1794–1880) on Mackinac Island in Michigan. St. Martin had been shot with a musket that opened a hole in his stomach the size of a fist. Beaumont expected St. Martin to die but he survived. After St. Martin recovered; Beaumont began to perform digestive experiments on St. Martin. The majority of the digestive experiments were conducted by tying a piece of food to a string and inserting it through the hole into St. Martin's stomach. Periodically, Beaumont would remove the food and observe how well it had been digested. Beaumont also extracted and analyzed samples St. Martin's gastric juices. As the result of these observations and others Beaumont became an expert on digestion. Interestingly, St. Martin lived to 86 years old. After Beaumont died, St. Martin joined a traveling medical show selling snake-root oil.

Materials
- model of the human digestive system
- colored pencils

Procedure 34.7
Digestive System

1. Procure the model of the human digestive system.
2. Examine and locate the structures labeled in Fig. 34.15.
3. Sketch and label the model in the space provided.

Student Activity—Understanding Amylase

Salivary amylase is the digestive enzyme found in saliva. It initiates the process of carbohydrate digestion. This activity is designed to investigate the action of the enzyme amylase.

Materials
- clean test tubes
- test tube rack
- 0.5% amylase solution
- 0.5% starch solution
- wax pencil
- hot plate
- 500 ml beaker
- thermometer
- watch or stopwatch
- six medicine droppers
- three test tube clamps
- reaction wells
- blank typing paper
- toothpicks
- iodine-potassium-iodide solution
- Benedict's solution

Procedure 34.8
Amylase

1. Procure the test tubes and needed supplies.
2. With a wax pencil write "amylase," "starch," and "amylase + starch" on three separate test tubes.
3. Add 6 ml of amylase solution to the test tube marked "amylase," add 6 ml of starch solution to the test tube marked "starch," and add 1 ml of amylase solution and 5 ml of starch solution to the test tube marked "amylase + starch."
4. Gently shake the test tubes for 10 seconds.
5. Prepare a water bath using the hot plate and a 500 ml beaker filled ¾ full with water. Stabilize the temperature to 37 degrees Celsius (same as body temperature 98.6 degrees Fahrenheit).
6. Carefully place the three test tubes in the beaker of water, and let them sit in the water bath for 10 minutes.
7. Place a reaction well over a piece of white typing paper. Using a medicine dropper, drop 1 ml of the amylase solution in one well and label its position.

Using another medicine dropper, drop 1 ml of the starch solution in another well and label its position. Using a third medicine dropper, drop 1 ml of the amylase + starch solution in one well and label its position.

8. Add one drop of the iodine-potasium-iodide solution with a medicine dropper to each well, and stir with a toothpick.
9. Note any changes in color below. If the solution turns a bluish black, starch is present. Record and label your results in the reaction wells below.

amylase: _____

starch: _____

amylase + starch: _____

10. With a wax pencil, write "amylase," "starch," and "amylase + starch" on another set of three separate test tubes.
11. Using the previously prepared test tubes, add 1 ml of amylase solution to the test tube marked "amylase," add 1 ml of starch solution to the test tube marked "starch," and add 1 ml of amylase/starch solution to the test tube marked "amylase + starch."
12. Add 1 ml of Benedict's solution to each of the three test tubes.
13. Place a 500 ml beaker filled ¾ with water on the hot plate, and bring the water to boiling.
14. Using a test tube clamp, place one test tube at a time in the boiling water for 2 minutes. Record any changes in color. Keep in mind that the color blue indicates no sugar was present, red indicates the greatest amount of sugar, green indicates a minimal amount of sugar, and yellow or orange represents a significant amount of sugar.

amylase: _____

starch: _____

amylase + starch: _____

15. Discard the solutions as directed, clean the glassware thoroughly, and return the materials.

16. Complete the following chart.

Container	Starch Test	Sugar Test
Amylase solution		
Starch solution		
Amylase + Starch Solution		

17. Interpret the results of the tests below.

_____

_____

_____

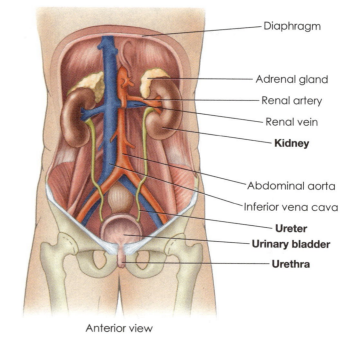

Anterior view

Figure 34.16 The organs of the urinary system.

THE URINARY SYSTEM

The **urinary system** (Fig. 34.16) is perhaps one of the most under-appreciated systems in the human body. The system consists of a pair of kidneys, two ureters, the urinary bladder, and the urethra. In the male, the urinary system is associated with the reproductive system because the urethra serves as a passage for both urine and sperm. The term **urogenital system** refers to both the urinary and reproductive system.

The **kidneys** are a pair of organs about 10 cm long, 5 cm wide, and 2.5 cm thick (each about the size of a bar of soap) that lie against the dorsal abdominal wall about the height of the 12th rib. The **adrenal glands** sit atop the kidneys. A single kidney is composed of 1.2 million functional units called **nephrons**. The primary function of a nephron is to filter the blood to regulate the concentration of water and several soluble substances (sodium

salts). Needed substances are reabsorbed, and waste is excreted as **urine**.

Urine is a pale yellow to amber-colored solution composed of water, metabolic waste such as urea, salts, and other organic compounds. The characteristic smell of urine is from ammonia, (Did you ever smell a kitty liter box that has been ignored for a while?), a product from the breakdown of urea. Other smells, such as those from ketones, also may be detected in urine. For centuries, medical practitioners have analyzed urine (odor, color, consistency) to determine the health of individuals. At one time, urine was even tasted to determine if the person had diabetes. As in the past, diagnostically helpful attributes of urine, such as pH, specific gravity, the presence of nitrates, billirubin, glucose, proteins, ketones, blood, and leukocytes, are ascertained through **urinalysis**.

 Student Activity—Examination of a Model of the Human Urinary System

Materials
- model of the human urinary system
- colored pencils

 **Procedure 34.9
Urinary System**

1. Procure the model of the human urinary system.
2. Examine and locate the structures labeled in Figure 34.16.
3. Sketch and label the model on the next page.

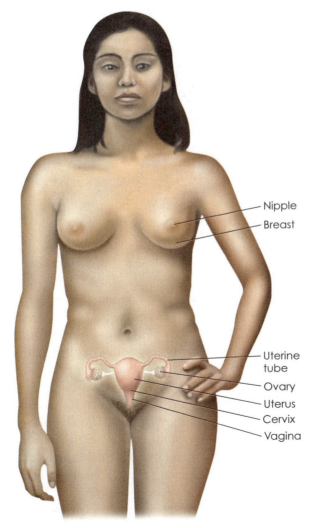

- Nipple
- Breast

- Uterine tube
- Ovary
- Uterus
- Cervix
- Vagina

Figure 34.17 The organs of the female reproductive system.

THE REPRODUCTIVE SYSTEM

In humans, the reproductive system consists of **primary sex structures** and **secondary sex structures**. The primary sex structures are responsible for the production of sex cells, or **gametes**. In females the **ovaries** produce ova, or eggs, and in males the **testes** produce sperm. Secondary sex structures are structures other than those involved in the direct production of gametes. In the female they include the uterine tubes, uterus, vagina, external genitalia, and mammary glands. In the male, the secondary sex structures include the seminal vesicles, prostrate gland, bulbourethral gland, urethra, and penis.

The **female reproductive system** is much more complex than the male reproductive system, simply because the former has many more duties (Fig. 34.17). With the exception of the ovaries, the female reproductive system exists in the pelvic cavity. The **ovaries** are a pair of almond-shaped organs that lie in the **ovarian fossa** of the dorsal portion of the pelvic wall. Each ovary is about 5 cm long, 1.5 cm wide, and 1 cm thick. The ovaries are responsible for the production of the egg, or ova, and several hormones including estrogen, progesterone, and inhibin. The ovaries are held in place by ligaments.

Oocytes develop within several kinds of follicles within the ovary. During **ovulation,** a **secondary oocyte** is release from the **Graafian follicle** into the pelvic cavity, where finger-like projections called **fimbriae** direct the oocyte to the **uterine** or **Fallopian tube** (the oviduct). The uterine tube is about 10 cm long.

Fertilization occurs in the uterine tube, and if fertilization does not occur, unfertilized eggs degenerate in the uterine tubes or the **uterus**. About 6 days after fertilization, the blastocyst enters the uterus. If the embryo continues to develop in the uterine tube, it is called an **ectopic** (tubal) **pregnancy**.

The uterus is a muscular chamber about 7.5 cm long, 5 cm in diameter, and 2.5 cm thick. In pregnancy, the uterus can exceed 30 cm in length. Anatomically, the uterus consists of the upper **fundus**, the large middle region known as the **body** (where the fetus is housed), and the lower **cervix** extending into the **vagina**. Prior to birth, the cervix dilates. The uterus consists of three distinct layers: an outer layer called the **perimetrium**, a middle muscular layer called the **myometrium**, and an inner layer known as the **endometrium**. After it leaves the uterine tubes, the blastocyst implants into the endometrial lining. The endometrial lining becomes part of the **placenta** called the **decidua**. If there is no implantation of the blastocyst, the endometrium is shed during menstruation.

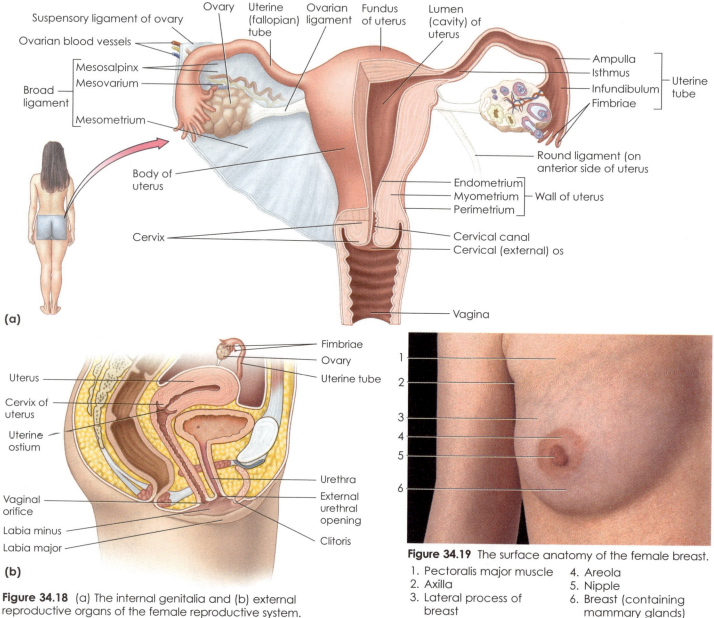

Figure 34.18 (a) The internal genitalia and (b) external reproductive organs of the female reproductive system.

Figure 34.19 The surface anatomy of the female breast.
1. Pectoralis major muscle
2. Axilla
3. Lateral process of breast
4. Areola
5. Nipple
6. Breast (containing mammary glands)

The **vagina** is a muscular tube that extends from the cervix to the **vaginal orifice** (opening). The external genitalia of the female, collectively called the **vulva**, consist primarily of the **urethral orifice**, **clitoris**, a number of glands, the **labia majora**, and the **labia minora**. The **mons pubis** is a bulge of adipose tissue that lies superior to the vulva (Fig. 34.19).

The paired **breasts** are mounds of tissue that lie above the pectoralis major muscle. The breasts consist of the **body, adipose tissue, axillary tail** (extension towards the armpit), **nipple**, and **areola.** Unfortunately, the lymphatics of the axillary tail can serve as a route for metastasis in breast cancer. The nipple is a conical structure where the ducts from the mammary glands open to the surface. A darker, circular region, the **areola**, surrounds

the nipple. Although not true reproductive organs, the **mammary glands** are an essential part of the female reproductive system. The function of the mammary glands is to produce milk, called **lactation**, to nourish the infant. The mammary glands are controlled by hormones of the reproductive system and placenta.

During pregnancy, the mammary glands develop 15 to 20 **lobes** that contain milk-producing **alveoli** within **lobules**. The milk travels from the alveoli to an **alveolar duct,** and eventually to a chamber called the **lactiferous sinus**. Milk is released from 15 to 20 **lactiferous ducts** that open at the surface of the nipple (Fig. 34.20).

The major functions of the **male reproductive system** are to produce and store spermatozoa or sperm, produce male sex hormones (androgens such as testosterone)

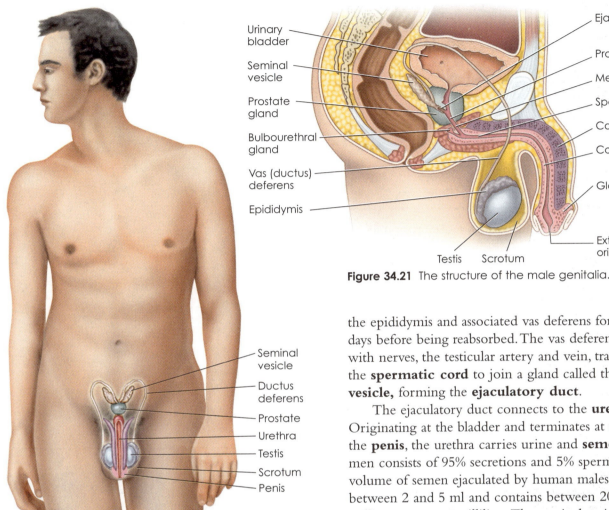

Urinary bladder

Seminal vesicle

Prostate gland

Bulbourethral gland

Vas (ductus) deferens

Epididymis

Ejaculatory duct

Prostatic urethra

Membranous urethra

Spongy urethra

Corpora cavernosa

Corpus spongiosum

Glans penis

External urethral orifice

Testis   Scrotum

Figure 34.21 The structure of the male genitalia.

Seminal vesicle

Ductus deferens

Prostate

Urethra

Testis

Scrotum

Penis

Figure 34.20 The organs of the male reproductive system.

and ejaculate semen. The gamete-producing structures in the male are the **testes,** housed in an external sac, the **scrotum**. Sperm cannot develop at 37 degrees (body temperature) Celsius. Thus, the scrotum is located outside the body cavity to allow sperm to develop in about a 2-degree cooler environment (Fig. 34.18).

A single testis is an oval structure that is approximately 4 cm long and 2.5 cm wide. The **tunica albuginea**, a sheath within each testis, divides it into over 250 **lobules**. Each lobule is composed of a several tightly packed **seminiferous tubules** that are the sites of sperm production. **Sertoli cells** in the seminiferous tubules protect and aid the development of sperm.

Located between the seminiferous tubules are **Leydig cells,** which produce testosterone. The seminiferous tubules lead to the **rete testes**, where sperm partially mature. **Efferent ductules** lead to the **epididymis**, where sperm finish maturation. Mature sperm can be stored in

the epididymis and associated vas deferens for up to 60 days before being reabsorbed. The vas deferens, along with nerves, the testicular artery and vein, travel though the **spermatic cord** to join a gland called the **seminal vesicle,** forming the **ejaculatory duct**.

The ejaculatory duct connects to the **urethra**. Originating at the bladder and terminates at the end of the **penis**, the urethra carries urine and **semen**. The semen consists of 95% secretions and 5% sperm cells. The volume of semen ejaculated by human males averages between 2 and 5 ml and contains between 20 and 130 million sperm per milliliter. The seminal vesicles secrete fructose to nourish the sperm and factors to enhance the motility of sperm. The **prostate gland** releases an alkaline substance that buffers seminal and vaginal acidity in order to activate sperm. The **bulbourethral gland** adds more alkaline substances to the semen.

The penis (Fig. 34.21) has the dual function of eliminating urine and depositing semen in the vagina of the female during intercourse. The **corpora cavernosa** (anterior) and the **corpus spongiosum** (surrounding the penile urethra) serve as erectile tissue during penile erection. The **root** of the penis is interior and connects the penis to the body wall. The **shaft** of the penis is the tubular external portion of the penis. The **glans penis** is the expanded distal end of the penis. A fold of skin called the **prepuce**, or foreskin, surrounds the tip of the penis. **Prepuce** or coronal glands secrete a waxy substance called **smegma**, which can serve as a lubricant. If the prepuce is present, bacteria can thrive on the smegma, causing infections. **Circumcision**, removal of the prepuce, is commonly done to reduce this risk. The **external urethral meatus** is the opening of the penis (Fig. 34.21).

Student Activity—Examination of a Model of the Female Human Reproductive System

Materials
- model of the human female reproductive system
- colored pencils

 Procedure 34.10
Female Reproductive System

1. Procure the model of the human female reproductive system.
2. Examine and locate the structures labeled in Figure 34.17 and Figures 34.19–34.20.

3. Sketch and label the model below.

 Student Activity—Examination of a Model of the Male Human Reproductive System

Materials
- model of the human male reproductive system
- colored pencils

 Procedure 34.11
Male Reproductive System

1. Procure the model of the human male reproductive system.
2. Examine and locate the structures labeled in Figure 34.18 and Figure 34.21.
3. Sketch and label the model.

Name: _____ Date: _____ Section: _____

1. Compare and contrast the pulmonary and systemic portions of the circulatory system.

_____

_____

_____

2. Trace the route of blood through the heart.

_____

_____

_____

3. Label the components of the respiratory system.

1. _____

2. _____

3. _____

4. _____

5. _____

6. _____

7. _____

8. _____

9. _____

10. _____

11. _____

4. Describe the upper and lower respiratory tract.

_____

_____

_____

5. Label the heart.

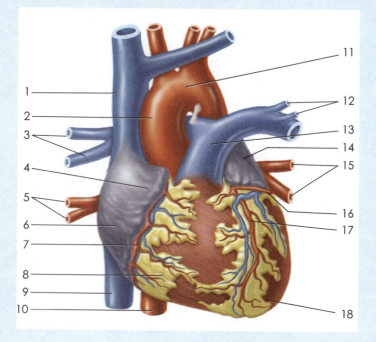

1. _____

2. _____

3. _____

4. _____

5. _____

6. _____

7. _____

8. _____

9. _____

10. _____

11. _____

12. _____

13. _____

14. _____

15. _____

16. _____

17. _____

18. _____

6. Label the following diagram.

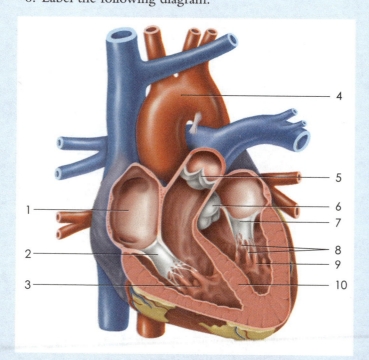

1. _____

2. _____

3. _____

4. _____

5. _____

6. _____

7. _____

8. _____

9. _____

10. _____

Name: _____ Date: _____ Section: _____

7. What is the function of salivary amylase?

_____

_____

_____

8. What is the function of bile?

_____

_____

_____

9. Discuss the role of the urethra in the urogenital system.

_____

_____

_____

10. Label the following diagram.

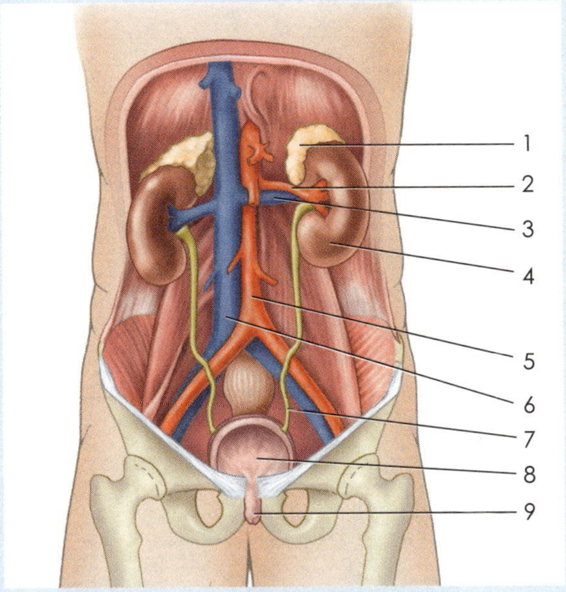

1. _____

2. _____

3. _____

4. _____

5. _____

6. _____

7. _____

8. _____

9. _____

11. What is the difference between primary and secondary sex structures?

_____

_____

_____

NOTES

Chapter 35
Acting It Out:
Understanding Animal Behavior

Student Outcome Objectives

At the completion of this exercise, the student will be able to:

1. Define behavior and ethology.
2. Discuss the factors that influence animal behavior.
3. Define and provide examples of innate and learned behaviors.
4. Compare and contrast habituation, classical conditioning, operant conditioning and cognitive learning.
5. Define an ethogram, and discuss how to develop and construct an ethogram.
6. Describe orientation behavior.
7. Compare, contrast, and give examples of taxis and kinesis.
8. Define and give examples of agnostic behavior.

Overview

Why do cats purr? Why do butterflies migrate? What is that dolphin thinking? Why do baby chicks follow me? Why are those fish putting on such a display? Why do birds sing? Animals display a variety of complex behaviors within their physical environments, within their own species, and with other species. **Behavior** is the response of an organism to environmental stimuli and includes the actions of protists, plants, and fungi, as well as animals. **Ethology** is the study of animal behavior in relation to evolution, ecology, and social interactions between and within species.

Studying animal behavior encompasses a broad range of disciplines including evolution, comparative anatomy, physiology, genetics, ecology, and sometimes parasitology. To understand the behavior of animals, a wide variety of variables must be considered. Factors influencing animal behavior are related to successful survival of the individual and, ultimately, the species. The two major factors influencing how an animal behaves are **genetic** or **innate behaviors** and **learned behaviors**. Examples of innate behaviors are withdrawing when touching a hot object, the scratching behavior of dogs and cats, and aggressive behaviors in male stickleback fish. Learned behaviors are differentiated as habituation, classical conditioning, operant conditioning, and cognitive learning.

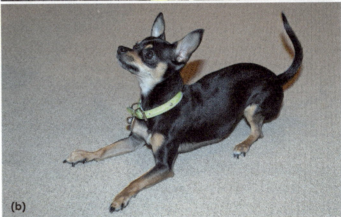

Figure 35.1 What are these animals "saying"? (a) A cottonmouth, *Agkistrodon piscivorus*, and (b) a chihuahua.

1. In **habituation**, an organism learns to ignore repeated stimuli. For example, pigeons in a city learn to ignore the hustle and bustle of a crowded street, birds ignore insects that are camouflaged like twigs, and animals in a wildlife preserve learn to ignore visitors.
2. **In classical conditioning**, a positive or negative involuntary response becomes associated with a given stimulus. The best-known example of classical conditioning is Ivan Pavlov's work involving the salivation of dogs.
3. In **operant conditioning**, an animal learns to associate a given behavior with a reward or punishment. A classic example is Skinner's box, in which rats are rewarded with food if they press certain buttons. An example of a negative stimulus is the fenceless dog yard to prevent dogs from roaming.
4. **Cognitive learning** involves thought, perception, judgment, and the ability to solve problems, such as use of tools by chimps, octopi, and certain birds, and problem-solving behavior in dolphins.

Behavior is a function of genetics and learning (nature and nurture). A form of this relationship is termed **imprinting**. During development, certain behaviors are established during a **critical period**, when the young imprint on the parents and learn fundamental species-specific behaviors. One of the pioneers in understanding imprinting behavior was Konrad Lorenz. The images of Lorenz and his imprinted goslings are often included in biology and psychology books.

Many topics are under investigation in animal behavior. **Migratory behavior** in animals is fascinating. How do monarch butterflies coordinate their migration from North America to the forests of central Mexico? How do ducks and other birds maintain their orientation and bearings during long winter migrations? **Communications behavior**, too, is fascinating, whether the organism uses chemical, visual, or auditory prompts. Insects such as ants and termites produce pheromones that serve as chemical signals. Many animals have developed elaborate visual behaviors, such as mating displays in the bird of paradise, flashing displays in fireflies, the dance of the honeybee, and the warning behavior of a puff adder. Examples of auditory communications are the chirps of crickets and the distinctive frog sounds. Dolphin and elephant acoustic communications are complex and intricate.

Pay closer attention to animal behavior. It's fascinating! A trip around the yard or a nearby park may yield intriguing results, such as a bird song, dewlap displays in anole lizards, predatory behaviors of spiders, playing dead or injured (feigning) in some birds and the opossum, and communications behavior in ants and bees. A trip to an animal preserve, marine life park, or zoo can yield even more incredible results, such as pacing behavior in great cats in a zoo (they are telling you that they are bored), courtship displays in peacocks, playfulness in seals, and grooming behavior in monkeys. Keep your eyes and mind open. There's a fascinating world to discover.

Figure 35.2 A male anole lizard, *Anolis carolinensis*, displaying a dewlap, which can be used to establish territory or attract a mate.

And when the day comes that we can communicate intelligently with dolphins, they may introduce us to the concept of survival without aggression, and the true joy of living, which at present eludes us. In that circumstance what they have to teach us would be infinitely more valuable than anything we could offer them in exchange.

—**Horace Dobbs**

THE ETHOGRAM

In studying the behavior of a given animal, an ethologist first develops an **ethogram**. The ethogram is an inventory and quantitative description of the behavior of the animal(s) under investigation. Constructing a meaningful ethogram demands much time spent watching an animal or animals, taking careful notes, perhaps photographing, videoing, or recording the subject, and making sense of the observed behaviors.

A good ethogram depends upon dedicated observations and accurate note taking. Notes should be taken with ink in a notebook, not on loose pages. Specialty equipment such as binoculars, stopwatches, camouflage, and measuring devices may be needed, and the observer must not interfere with the subject's routine. A catalog "language" of observed behaviors is imperative. These terms can be **broad categories**, such as: resting, eating, playing, courtship, mating, communicating, foraging, fighting, hunting, hiding, etc. Or they may be **specific categories** within a broad category; for example, within the broad category of courtship, the animal maybe engaged in vocalizations, displays, dancing, etc. A map or sketched map of the study area should be placed in the notebook. At the beginning of each observation, the physical environment (temperature, humidity, wind velocity and direction, percent cloud cover) should be noted, along with time and descriptors of the environment. Photographs, video, and recordings should accompany the notebook. Physical and behavioral data must be presented accurately (in metric measures), truthfully, and in detail.

The ethogram includes a time budget of the animal's activities over the study time. For example, "The subject was engaged in foraging behavior for 3 hours and 24 minutes, comprising 32% of the study time." This is accompanied by a detailed narrative and charts. Accurate ethograms can be the starting point for more detailed studies. Once the ethogram is presented, a hypothesis can be generated and plans made to perform an experiment based on the ethogram.

Student Activity—Developing an Ethogram

For this activity, students will build an ethogram for an animal of the instructor's choice (not someone's pet!) as part of a 4-week project. The animal should be studied at least 4 hours per week. The animal must be found and identified easily and will be in a specific location for the duration of the project. Ideally, wild animals would be studied in their natural habitat, such as a squirrel on campus. Or wild animals could be studied in enclosures, such as fish or a hermit crab in an aquarium, crickets in a terrarium, or an animal at a wildlife preserve, zoo, or aquarium. Or the study could utilize domestic animals in a large area, such as ducks in a park. Or domestic animals in an enclosure could be studied, such as mice in a cage.

Figure 35.3 The gray squirrel, *Sciurus carolinensis*, makes a good subject for an ethogram.

Materials
- notebook and pen
- specialty equipment specific for your study
- camera, video camera, or recording devices
- animal as subject for the ethogram

 Procedure 35.1 Ethogram

1. Following the above guidelines, prepare an ethogram to be turned in to the instructor. Your instructor may request a PowerPoint® presentation. Include a narrative, data charts, time budget, and other information requested by the instructor.
2. Answer the following questions.
   a. What animal did you choose for this study? Why?

   _____

   _____

   _____

   b. Describe the environment associated with the study.

   _____

   c. What problems were encountered in the study? How did you overcome these problems?

   _____

   _____

   _____

   d. What interesting behaviors did you observe?

   _____

   _____

   e. If time permits, how would you extend this study, and what sort of experiment might you design based upon the study?

   _____

   _____

   _____

There isn't a sharp line dividing humans from the rest of the animal kingdom. It's a very wuzzie line and it's getting wuzzier all the time. We find animals doing things that are in our arrogance, used to be thought that were just human.

—Jane Goodall (1934–present)

ORIENTATION

The way an organism positions its body in relation to an external stimulus is termed its **orientation**. Usually the organism moves toward or away from the stimulus. This basic type of behavior is essential to survival, whether the organism is seeking food or a mate or avoiding a predator. Ethologists note two types of orientation: taxis and kinesis. Taxis refers to movement directly toward or away from a stimulus. A **positive taxis** moves toward the stimulus, and a **negative taxis** moves away from the stimulus. To address a given taxis, prefixes such as photo, thermo, and chemo are added to taxis. For example, a plant exhibits positive phototaxis.

Kinesis refers to a random movement of the subject's body in response to a stimulus. The subject does not move toward or away from the stimulus. The rate of kinesis increases with the intensity of the stimulus. Positive kinesis denotes sudden movement, and negative kinesis denotes slower movement. A loud sound may bring about positive kinesis.

 Student Activity—Taxis in Brine Shrimp, *Artemia salina*

Brine shrimp (*Artemia salina*), typically sold as sea monkeys in pet stores, are small crustaceans that live in salt lakes and date back to Triassic times. Brine shrimp tend to swim upside down and appear highly active under a hand lens. They possess 11 pairs of appendages, two pairs of short antennae, and two relatively large compound eyes.

Materials
- dissecting microscope
- Petri dish
- black construction paper
- two test tubes
- saltwater
- lamp
- brine shrimp food
- 250 milliliter beaker
- brine shrimp (*Artemia salina*)
- colored pencils

 **Procedure 35.2
Brine Shrimp**

1. Procure the needed materials and specimens, including two test tubes filled 2/3 full with saltwater containing six to eight brine shrimp and a Petri dish with three or four brine shrimp in saltwater.
2. Stand the test tubes upright in the beaker for later use.
3. Using the dissecting microscope, observe the anatomy and movements of the brine shrimp in the Petri dish.
4. Sketch your observations below, and describe their movements.

5. To get a better view of the brine shrimp in the test tubes, place a sheet of black construction paper behind the beaker.
6. Describe the activity, distribution, and position of the brine shrimp in the test tubes in detail below.

Figure 35.4 Brine shrimp, *Artemia salina*.

Check Your Understanding

7. Place a lamp above the test tubes, and describe the activity, distribution, and position of the brine shrimp in the test tubes in detail below.

_____

_____

_____

8. Place a lamp below the test tubes, and describe the activity, distribution, and position of the brine shrimp in the test tubes in detail below.

_____

_____

_____

9. Drop several pieces of brine shrimp food in the test tubes, and describe the activity, distribution, and position of the brine shrimp in the test tubes in detail below.

_____

_____

_____

10. Return the equipment, and dispose of the brine shrimp and solutions as indicated by your instructor.

Q. Did the shrimp exhibit phototaxis? Describe their actions.

_____

_____

_____

Q. Did the brine shrimp exhibit a type of taxis toward food? How did they react?

_____

_____

_____

Q. Describe several other variables that could be tested in extensions of this experiment.

_____

_____

_____

_____

Student Activity—Kinesis in Pill Bugs, *Armadillidium* spp.

The pill bug (*Armadillidium* spp.) is a common terrestrial crustacean living in moist areas under logs, leaf litter, flower pots, and wet areas under bricks. These small animals have been nicknamed "roly polys" because, when threatened, they roll up into a ball. They feed upon moss, algae, bark, and other decaying organic matter. Their dorsal side consists of a number of overlapping, articulating plates. Pill bugs have seven pairs of legs and a pair of antennae. Most pill bugs are slate gray or brown in color and reach a length of about 13 mm and a width of about 7 mm. The kinesis involving water is called **hygrokinesis**.

Materials
- dissecting microscope
- three Petri dishes
- filter paper
- squirt bottle with water
- pill bugs (*Armadillidium* spp.)
- colored pencils

Procedure 35.3
Kinesis in Pill Bugs

1. Procure the materials listed.
2. Place two pill bugs in a Petri dish with no filter paper. Place six pill bugs in a Petri dish with wet filter paper, and place six pill bugs in a Petri dish with dry filter paper.
3. Place the Petri dishes with filter paper and their pill bugs in a dark drawer.
4. Using the dissecting microscope, observe the anatomy and movements of the pill bugs in the Petri dish with no filter paper.
5. Sketch these pill bugs below, and describe their movements.

Figure 35.5 (a) A pill bug, *Armadillidium* spp. and (b) rolled into a ball when threatened.

_____

_____

_____

7. Remove the Petri dish with wet filter paper from the drawer. The instructor will assign group members to specifically observe how many pill bugs are moving and the speed in which they move. With just one pill bug, describe how it moves around the dish, how many laps it makes in 1 minute, and how many times it changes directions in 1 minute. Place your numbers in the chart (above). Write a synopsis of your observations below.

6. Remove the Petri dish with dry filter paper from the drawer. The instructor will assign group members to specifically observe how many pill bugs are moving, their speed, and how many laps they make in 1 minute. With just one pill bug, describe how many times it changes directions in 1 minute. Place your numbers in the chart (on following page), and write a synopsis of your observations below.

_____

_____

_____

8. Return the equipment and pill bugs according to your instructor's directions.

Petri Dish	Number of pill bugs moving and description of movement activity	Number of laps by a specific pill bug in one minute	Number of changes of direction by a specific pill bug in one minute
Dry Petri Dish			
Wet Petri Dish			

Check Your Understanding

Q. Did the pill bugs exhibit positive or negative hygro-kinesis? Describe their actions.

_____

_____

Q. Suggest several other variables that could be tested in extensions of this experiment.

_____

_____

_____

AGNOSTIC BEHAVIORS

Animals are faced with a number of threatening situations in their environment. **Agnostic behaviors** (from the Greek *agonistes*, meaning "champion") result when an organism perceives that it is being threatened and reacts by threatening, attacking, or withdrawing. Members of the same species, called **intraspecies**, can exhibit agonistic behaviors, such as two male elephant seals battling for a dominance. Or agnostic behaviors can be exhibited by different species, **interspecies,** such as a mother cat protecting her litter of kittens from a dog. In many cases, the agnostic behavior results in a **ritualistic threat** that may make the animal look larger and more threatening.

In general, the majority of animals avoid fighting unless they have a good chance of triumphing without incurring serious injury or death. A successful bluff is more logical than an all-out battle. A male silverback gorilla threatens a foe by chest pounding, stomping, throwing vegetation, and roaring before it takes the battle up a notch by using its enormous strength and slashing canines. **Aggressive behaviors** are agnostic behaviors manifested by the use of force. At first, a grizzly bear may try to bluff its adversary by a series of vocalizations, popping the jaw, staring into the adversary's eyes with its ears back, and making a bluff charge before knocking the victim down and using force. Submissive behavior results in retreating from a conflict, such as a smaller ram fleeing from a larger, more mature ram.

Student Activity—Agnostic Behavior in Male Siamese Fighting Fish, *Betta splendens*

The Siamese fighting fish (*Betta splendens*) is a breed of fish found in the waters around the Malay Archipelago of Southeast Asia. The brilliant coloration and long, ornate, flowing fins of the Siamese fighting fish make it one of the most popular freshwater aquarium fishes. The Betta's colors range from red to blue to even white. The female Siamese fighting fish are not as brilliantly colored and have much shorter and less elaborate fins than the male. Selectively bred *Betta splendens* are much more ornate than those in the wild. Siamese fighting fish usually grow to an overall length of about 5 cm. Bettas can gulp air to take in oxygen and can live in water with low oxygen levels. *Bettas* can survive in very small bowls (16 oz or 475 ml), although a larger container is preferred.

The male *Betta splendens* is known for its aggressive behavior toward other males and its own reflection in a mirror. When encounters between male Siamese fighting fish occur, both undergo a ritualistic set of agnostic threatening behaviors. If the threats do not work, the fish, when evenly matched, will fight ferociously. Eventually, one fish is defeated and will withdraw.

When faced with an intruder, male Siamese fighting fish display a number of innate agnostic **fixed action patterns**. These patterns include broadside maneuvers, dorsal fin flare, gill flare exposing brachiostegal membrane, pelvic fin shudder, caudal fin shudder, body undulation, head oriented downward, head-on approach, change in color, and charging. In this activity, the fish will be exposed to their image in a mirror, different models of fighting fish, construction paper cutouts, and another fish in an adjacent container. Working in groups of four, the students, will closely observe the behavior of these fish.

You are capable of more than you know. Choose a goal that seems right for you and strive to be the best, however hard the path. Aim high. Behave honorably. Prepare to be alone at times, and to endure failure. Persist! The world needs all you can give.

—**E. O. Wilson (1929–present)**

Figure 35.6 A male Siamese fighting fish, *Betta splendens*.

Figure 35.7 A male Siamese fighting fish, *Betta splendens*, displaying aggressive behavior.

Figure 35.8 Model cutout of a male Siamese fighting fish.

Materials
- mirror
- Model cutout of fish (Figure 35.8)
- *Betta*-colored construction paper
- scissors
- applicator sticks
- transparent tape
- lab coat
- male Siamese fighting fish (*Betta splendens*) in (16 oz or 475 ml) bowl
- goldfish (*Carassius auratus*) in a bowl

 Procedure 35.4
Agnostic Behavior in *Betta* Fish

Consider photographing or shooting video of this activity.

1. Prior to beginning the activity, as a group, discuss the expectations of agnostic behaviors in Siamese fighting fish. The instructor will assign specific tasks to members of the group.
2. Procure the required equipment and fish. *(Reminder: Treat your fish with respect.)*
3. Cut out the model fish from Figure 35.8, and cut a round piece of colored construction paper about the size of your fish.
4. Attach the model fish and construction paper to an applicator stick, using transparent tape. Once complete, place the model fish out of sight of your fish.
5. Wear a lab coat, especially if you are wearing brightly colored clothes, and speak softly.
6. Keep all experimental images and other fish out of sight of your fish.
7. Avoid sudden movements or bumping the bowl, and do not tap on the bowl.
8. In this activity, record the subjective behavioral activities of the fish. Use 0 to indicate no response, 1 to indicate a weak response, 2 to represent a medium response, and 3 to represent a strong response.

9. Place a mirror in front of the bowl so the fish can see its reflection. Hold the mirror in front of the fish for 1 minute to avoid habituation. If habituation occurs, the fish may not respond during latter parts of the experiment. During this time, record the behavior of the fish to the right, and fill in the chart on the following page. If your fish fails to display a behavior or

the response is not intense, notify your laboratory instructor; you may need a more aggressive fish before continuing. Wait at least 5 minutes before performing the next step.

_____

_____

10. Place the round piece of colored construction paper in front of the bowl so the fish can see the paper. Hold the paper in front of the fish for 1 minute to avoid habituation. During this time, record the behavior of the fish below, and fill in the chart on the following page. Wait at least 5 minutes before performing the next procedure.

_____

_____

11. Place the plain model of the fish in front of the bowl so the fish can see the paper. Hold the paper in front of the fish for 1 minute to avoid habituation. During this time, record the behavior of the fish below, and fill in the chart on the following page. Wait at least 5 minutes before performing the next procedure.

_____

_____

12. Place the detailed model of the fish in front of the bowl so the fish can see the paper. Hold the paper in front of the fish for 1 minute to avoid habituation. During this time, record the behavior of the fish below, and fill in the chart on the following page. Wait at least 5 minutes before performing the next procedure.

_____

_____

13. Now procure another Siamese fighting fish in a separate bowl (perhaps this lab group will be combined with another one). Place the bowls next to each other so the fish can see each other. Allow the fish to "face-off" for 1 minute to avoid habituation. During this time, record the behavior of the fish below, and fill in the chart on the following page.

_____

_____

14. Procure a goldfish in a separate bowl. Place the bowls next to each other so the fish can see each other. Allow the fish to "face-off" for 1 minute to avoid habituation. During this time, record the behavior of the fish below, and fill in the chart on the following page.

_____

_____

15. Return the materials and fish to the proper place, and fill in the chart on the following page. In the chart, use 0 to indicate no response, 1 to indicate a weak response, 2 to represent a medium response, and 3 to represent a strong response.

Check Your Understanding

Q. Why do male Siamese fighting fish attempt to display? Why not just start fighting?

_____

_____

_____

Q. What is the evolutionary and ecological significance of the agnostic behavior in the Siamese fighting fish?

_____

_____

_____

Q. Describe the orientation of the fish during the display.

_____

_____

_____

In the long history of humankind (and animal kind too) those who learned to collaborate and improvise most effectively have prevailed.

—**Charles Darwin (1809–1882)**

Q. Describe the action of the fins and gills during the display.

_____

_____

_____

Q. Discuss any other displays peculiar to your fish that you observed.

_____

_____

_____

Q. Which stimuli did the fish respond to the most and the least?

_____

_____

_____

_____

_____

_____

Observations of Agnostic Behaviors in Male Siamese Fighting Fish (*Betta splendens*)						
	Mirror	Cutout	Plain Fish Image	Fish Picture	Real Male Siamese Fighting Fish	Goldfish
Observed Behavioral Response						
Broadside maneuvers						
Dorsal fin flare						
Gill flare exposing brachiostegal membrane						
Pelvic fin shudder						
Caudal fin shudder						
Body undulation						
Head oriented downward						
Head-on approach						
Charging						

Name: _____   Date: _____   Section: _____

1. Why is ethology a valuable discipline within the biological sciences?

_____

_____

_____

2. What is an ethogram, and when is it used?

_____

_____

_____

3. Define *learned behavior*, and provide several examples of learned behavior.

_____

_____

_____

4. Define *innate behavior*, and provide several examples of innate behavior.

_____

_____

_____

5. A Galapagos finch learning to dig for insects with a twig is an example of _____ learning.

6. Baby ducks following their mother, you, or even a remote hovercraft provide an example of _____ learning.

7. A tiger displaying _____ behavior at the zoo is essentially bored.

8. An opossum or a hognose snake playing dead is an example of _____ learning.

9. Using examples, distinguish between intraspecies and interspecies behavior.

_____

_____

_____

10. Using examples, distinguish between taxis and kinesis.

_____

_____

_____

11. Give several examples of positive taxis.

_____

_____

_____

_____

12. Name several types of agnostic behavior.

_____

_____

_____

_____

_____

_____

Chapter 36
But One Earth: Understanding Basic Ecology

Student Outcome Objectives

At the completion of this exercise, the student will be able to:

1. Define and differentiate ecology and environmental science.
2. Differentiate population, community, and ecosystem.
3. Define and discuss the biotic and abiotic factors of an ecosystem.
4. Describe different types of ecosystems.
5. List and describe several biomes and aquatic life zones.
6. Define and discuss the significance of keystone species.
7. Define and provide examples of tropic levels.
8. Construct and describe a food chain and a food web.
9. Define and describe the importance of biomass.
10. Construct and describe an ecological pyramid.
11. Define succession, and distinguish between primary and secondary succession.
12. Provide definitions and examples of pioneer species and climax communities.
13. Discuss the meaning of the acronym HIPPO.
14. Define and describe several endangered and threatened species.
15. Name several invasive species and tell why they are harmful.
16. Define pollution, and describe several types of pollution.
17. Construct and conduct an experiment addressing the effect of oil on chosen organisms.

Overview

John Muir (1838–1914) wisely stated, "When one tugs at a single thing in nature, he finds it attached to the rest of the world." We must develop an understanding and appreciation of the Earth, its composition, cycles, life, and connections that enable us to survive. **Ecology**, the study of the interrelationship between organisms and their environment, is a broad discipline encompassing principles of the physical sciences and the biological sciences. Among the many subdisciplines of ecology are molecular, evolutionary, organismal (physiological and behavioral), population, community, ecosystem, conservation, and quantitative ecology. Many times, the terms *ecology* and *environmental science* are used interchangeably. **Environmental science** is a multidisciplinary study including the scientific as well as social impact of humans upon the Earth. Environmental science includes topics such as philosophy, politics, ethics, economics, sustainability, and stewardship.

A **population** of organisms consists of a group of individual organisms of the same species living in a given area. Their home is called the **habitat**, and their **niche** is their role in the environment. These organisms interact with other species, forming a **community**. An **ecosystem** consists of the biological, or **biotic**, community and the non-living, or **abiotic**, environment. Abiotic components include factors such as moisture or water, sunlight, weather or climate, nutrients, substrate, soil, pH, salinity, and oxygen availability. An ecosystem can consist of a small ditch or an entire swamp. Ecosystems can be natural, such as a forest, or artificial, such as an aquarium.

All of the Earth's ecosystems comprise the **biosphere**. The terrestrial portions of the biosphere, **biomes**, consist of large regions of land with distinct climates and specific species of organisms. Examples of biomes are deserts, coniferous forests, deciduous forests, evergreen forests, tropical forests, grassland, savannas, and the tundra. The parts of the biosphere dominated by water comprise the **aquatic life zones** and include freshwater systems, marine environments, estuaries, and wetlands. Within a given ecosystem, **keystone species** provide a critical role in maintaining the ecosystem. Examples of keystone species are bison in a grassland, alligators in a swamp, gopher tortoises in a desert, sea otters in a kelp forest, beavers near a lake, starfish in a sound, and a killer whale in the ocean. Loss of these species would have a major impact on an ecosystem.

Q. Describe several ecosystems unique to your region, and name significant species in each.

_____

_____

_____

_____

_____

Figure 36.1 Despite the apparent differences between these ecosystems, they have many characteristics in common.

Q. Describe the major biome or biomes associated with your region.

_____

_____

_____

The organisms residing within an ecosystem can be assigned **tropic levels** based on their source of energy acquisition. **Producers** make up the fundamental tropic level and consist of **autotrophs** such as **photosynthetic** organisms (some bacteria, algae, and plants) and **chemosynthetic** organisms (some bacteria, such as those living in hydrothermal vents). Photosynthetic organisms directly capture the radiant energy of the sun and turn it into energy-rich molecules such as glucose, which they and other organisms can use. **Consumers** such as rabbits and hawks are **heterotrophs**, organisms that receive their energy from acquiring food. **Primary consumers** such as rabbits and deer eat plant materials and are called **herbivores**. **Secondary consumers** are organisms that eat primary consumers. In turn, they may be eaten by **tertiary consumers**. Consumers that eat flesh are called **carnivores**, of which bass, bobcats, dolphins, and hawks are examples. **Omnivores** consume both plant and animal matter and include, among many others, bears, chickens, raccoons, and humans. **Scavengers** are animals that primarily consume the carcasses of dead animals, such as vultures and occasionally an opossum or coyote. **Detritivores** feed on decomposing organic matter such as earthworms. **Decomposers** comprise the final tropic level, the ultimate recyclers. These organisms use

However fragmented the world, however intense the national rivalries, it is an inexorable fact that we become more interdependent every day. I believe that national sovereignties will shrink in the face of universal interdependence. The sea, the great unifier, is man's only hope. Now, as never before, the old phrase has a literal meaning: We are all in the same boat.

—**Jacques Yves Cousteau (1910–1997)**

non-living organic matter as a source of energy and return inorganic material to the environment.

Tropic levels can be appreciated by constructing a simple **food chain**. A **food web** is a more complete representation of the relationships in an ecosystem, illustrating the many tropic-level interactions in a community of organisms.

Energy flows through each tropic level. With each level, the amount of energy decreases. The percentage of usable energy that is transferred through the tropic levels is termed the **ecological efficiency** of a system. Only about 10% of the energy is passed from one level to the next. As a result, the **biomass** (total weight of all organisms at a given tropic level) decreases from one level to the next. This explains why there are more grazing animals than top carnivores in a grassland. An **ecological pyramid** illustrates energy relationships in an ecosystem. The most biologically productive ecosystems on Earth are the swamps and marshes, estuaries, and tropical rainforests.

Over time, communities undergo a series of predictable and recognizable changes known as **succession**. Succession can be brought about by changes in climate, changes in the physical characteristics of an area such as

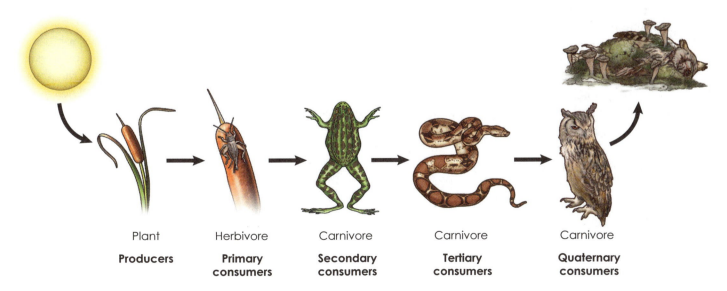

| Plant | Herbivore | Carnivore | Carnivore | Carnivore |
| **Producers** | **Primary consumers** | **Secondary consumers** | **Tertiary consumers** | **Quaternary consumers** |

Figure 36.2 A simple food chain.

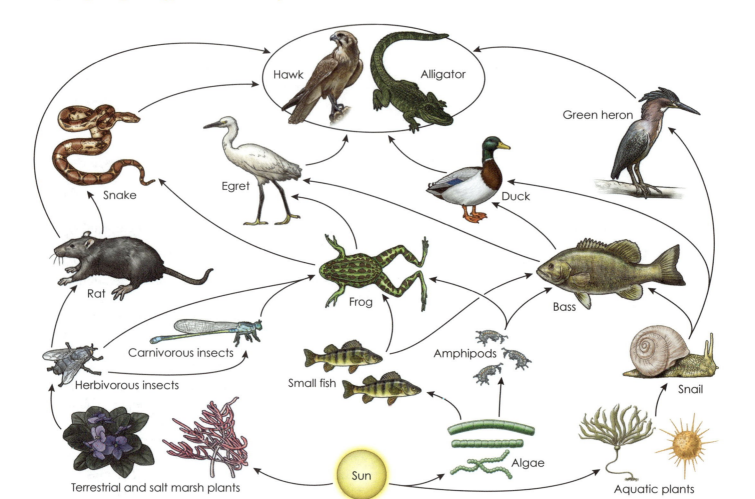

Figure 36.3 A simplified food web.

drop or rise in elevation, introduction of invasive species, and destruction of an ecosystem by factors such as volcanos and the impact of humans.

1. **Primary succession** involves literally starting from scratch with no life in an area. This may take place on lifeless, barren land left after a volcano. The first species, called the **pioneer species**, may consist of lichens, mosses, and algae. Once abiotic factors are more favorable, more complex organisms enter an area. Over time, the region becomes more biologically complex and a **climax community** may develop.

2. In **secondary succession**, the soil and some of the organisms remain after an event such as a fire or a flood. The progression resembles that of primary succession leading to a climax community. Many times, the new climax community will not be the same as the previous climax community.

Just For Thought!

How much biomass is there in a well-trimmed lawn? Calculating the biomass of grass clippings in a large lawn would be quite a task. Provided that the grass in the lawn is uniform, you could manage this task taking a square meter to study over the course of a year. On a regular schedule, you would clip all of the grass to a uniform height and save the clippings. Then you would determine the wet and dry weight of the clippings during the year.

What does most of the weight of grass consist of? For a rough estimate of the biomass of the clippings, multiply the dry weight by the number of square meters of grass in the yard. Unless you have the task of cutting the lawn, this number may be surprising. It does not include the biomass of the remainder of the plant and other living things in the yard such as trees, insects, and other animals. WOW!

OBSERVING ECOSYSTEMS AND SUCCESSION

One of the things that many people like about biology is that what is learned in the classroom is best illustrated outdoors. Students of biology can apply their knowledge of ecosystems and succession by taking a local field trip.

There are no words that can tell the hidden spirit of the wilderness that can reveal its mystery, its melancholy, and its charm.

—**Theodore Roosevelt (1858–1919)**

 Student Activity—Investigating Local Ecosystems

This activity introduces the student to several ecosystems near the campus. Preferably, at least two natural ecosystems (a terrestrial ecosystem and an aquatic ecosystem, if possible) and one artificial ecosystem will be observed. The instructor will gain permission to explore the environments and take photographs.

Materials
- appropriate clothes
- camera
- binoculars
- field thermometer
- small rake
- net
- field guides to common plants and animals
- compound microscope
- microscope slides and coverslips
- eyedropper
- collection jar
- hand lens

 ## Procedure 36.1
Local Ecosystems

1. Discuss various ecosystems near your campus. Determine which ecosystems would be interesting to visit and then take a field trip to these locations.
2. Answer the questions below for each ecosystem, and take photographs during the field trip.
3. If the instructor requests, compile a PowerPoint® presentation for each of the three ecosystems.

Terrestrial Ecosystem

Q. Where is the ecosystem located?

_____

_____

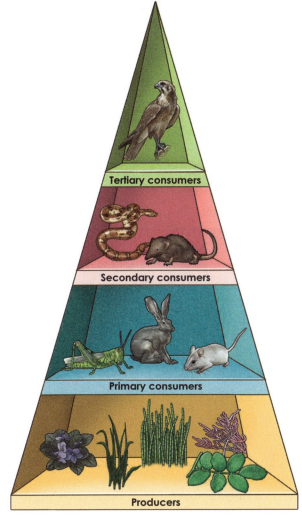

Figure 36.4 A simplified ecological pyramid.

Q. When are you conducting this study (time of year and day)?

_____

_____

_____

Q. Describe the weather conditions. Take the temperature 2 meters above the ground and at the surface of the soil in both shade and sunlight. Also, if there is cement nearby, take the temperature 2 meters above the cement and at the cement level in both shade and in sunlight. Record your results below.

Terrestrial Ecosystem		
	Shade Temperature	Sunlight Temperature
2 meters above ground		
Soil Surface		
2 meters above cement		
Cement surface		

Q. Describe a general overview of the study area, and take several general photographs.

_____

_____

Q. Describe the abiotic conditions of the study area.

_____

_____

Q. State the dominant soil type (sand, clay, loam).

_____

Q. What is the dominant plant ground cover?

_____

Q. Is there leaf litter? If so, how deep and rich is the litter? Do any animals inhabit the litter? If so, what are they? Do any animals live in the soil? If so, what are they?

_____

_____

Q. Describe the most obvious animals, both invertebrate and vertebrate, in the study area.

_____

_____

_____

Q. Using scientific and common names, identify several organisms and their place in the food web.

_____

_____

Aquatic Ecosystem

Q. Where is the ecosystem located?

_____

_____

Q. When are you conducting this study (time of year and day)?

_____

Q. Describe the weather conditions. Take the temperature at the shoreline 2 meters above the ground and at the surface of the ground in both shade and sunlight. What is the temperature at the surface of the water? If possible, take a temperature reading at various depths beneath the surface of the water and 2 meters above the surface of the water. Record these temperatures below.

Aquatic Ecosystem		
	Shade Temperature	Sunlight Temperature
2 meters above ground		
Soil Surface		
2 meters above water		
below water surface		

Q. Provide a general overview of the study area, and take several general photographs. Is the water flowing, still, stagnant, etc.?

_____

_____

Q. Describe the abiotic conditions of this aquatic ecosystem.

_____

_____

Q. Describe the type of bottom, if possible (sand, clay, silt).

_____

_____

Q. Describe the condition of the water as related to light penetrance (muddy, murky, clear).

_____

_____

Q. What are the dominant plants on the shoreline? Are any plants living in the water?

_____

_____

Q. Take a sample of the water, if possible, with the collection jar. Bring the jar to the lab as soon as possible, and examine the water with a compound microscope. Below, describe and identify, if possible, the organisms found in the water. Discard the water and slide as instructed.

_____

_____

Q. Name the most obvious animals, both invertebrate and vertebrate in your study area.

_____

_____

Q. Using both scientific and common names, identify several organisms and their place in the food web.

_____

_____

_____

Artificial Ecosystem

Q. Where is the ecosystem located (aquarium, wildlife park, etc.)?

_____

_____

Q. When are you conducting this study (time of year and day)?

_____

_____

Q. Describe the abiotic conditions or weather conditions. If possible, take temperature readings and record them below.

_____

_____

Q. Provide a general overview of the study area, and take several general photographs.

_____

_____

Q. Describe the abiotic conditions of the ecosystem.

_____

_____

Q. Describe the environment.

_____

_____

Q. Describe the most obvious plants and animals in the ecosystem.

_____

Q. Using both the scientific and the common names, identify several organisms and their place in the food web.

_____

_____

Q. Describe the apparent health of each ecosystem.

_____

_____

 Student Activity—Observing Ecological Succession

This activity introduces the student to ecological succession. Preferably, the student will be observing secondary succession in an empty lot or in an area recently ravaged by a fire or flood. The instructor will gain permission to explore the environment and take photographs.

Materials
- appropriate clothes
- camera
- field guides to common plants and animals
- hand lens

Procedure 36.2
Ecological Succession

1. With your classmates, discuss various areas where you have noticed that the environment has been disturbed by clearing, fire, or flood.
2. Specifically discuss the area to be visited in the field trip.
3. Answer the questions below.

Succession Study Area

Q. Where is the study area located?

_____

_____

Q. When are you conducting this study (time of year and day)?

_____

_____

Q. What event triggered the succession? Is it primary or secondary succession?

_____

_____

Q. Describe the abiotic conditions or weather conditions. If possible, take temperature readings, and record them below.

Succession Study Area		
	Shade Temperature	Sunlight Temperature
2 meters above ground		
Soil Surface		
2 meters above water		
below water surface		

Q. Provide a general overview of the study area, and take several general photographs.

_____

_____

Q. Describe the abiotic conditions of the ecosystem (soil, slope, runoff, etc.).

_____

_____

e. Condition of the species and how the organisms were housed or displayed:

_____

_____

3. Prepare a brief report on the organism, and provide photographs. In addition, your instructor may wish you to develop a PowerPoint® presentation.

 Check Your Understanding

Q. What is the difference between an endangered species and a threatened species?

_____

_____

Q. Describe several endangered and threatened species in your region.

_____

_____

_____

_____

Q. When saving species, why must we abandon the "it is so cute" attitude?

_____

_____

 Student Activity—Invaders!

Approximately 50,000 **non-native** or **exotic species** of plants and animals live in the United States. Through the years many non-native plant and animal species have been a source of food, medicine, aesthetics, and other beneficial applications. One in seven non-native species, however, are considered to be an **invasive species** because it is harmful to the ecosystem.

One of the most significant threats to ecosystems is the introduction of invasive species. Many of these organisms have no natural predators, competitors, pathogens, or parasites that can keep their populations in check. As a result, they overcompete with the native species, disrupting the natural ecosystem. In the United States alone, more than 7,000 invasive species have been introduced either accidentally or deliberately into various ecosystems.

This activity introduces students to non-native and invasive species. Before engaging in this activity, students should conduct some research on non-native and invasive species in the region. The instructor will gain permission for the group to explore targeted environments and photograph the species.

Materials
• appropriate clothes
• camera
• field guides to common plants and animals
• hand lens

 **Procedure 36.4
Invasive Species**

1. After determining several non-native and invasive species in your geographic region, list and describe several species and discuss their impact on local life on the next page.
2. Develop a plan of action to find several of these organisms.

In the area of species protection, we should concern ourselves with what is right as opposed to what might be easier, or popular in the short term.

—Richard Leakey (1944–present)

3. As arranged by your instructor, visit areas where you would expect to find these organisms, try to photograph them, and discuss your observations below.

_____

_____

_____

_____

Q. What can be done to discourage the spread of invasive species?

_____

_____

_____

POLLUTION

Pollution is defined as any physical or chemical entity that degrades the environment. When we think about pollution, some sources that come to mind are foul air, toxic wastes, acid rain, radiation leaks, oil spills, and intolerable noise. Also, pollution conjures up images of death, filth, and despair that haunt our very thoughts. Most pollution results from human activities such as burning fossil fuels; however, natural sources of pollution such as volcanoes also affect the environment. The two main sources of human pollutants are:

1. **point source pollutants**, such as oil spills, exhaust from vehicles, and noise from a jet engine; single-source, localized, and easily recognized forms of pollution;
2. **non-point source pollutants**, such as urban runoff, pesticide runoff, and fertilizer runoff are more difficult to identify.

Pollutants disrupt life-support systems such as food chains, they damage and destroy living things such as the destruction of wildlife in an oil spill, and they create a

Figure 36.6 Several invasive species: (a) nutria, *Myocaster coypus*, (b) water hyacinth, *Eichornia crassipes*, (c) marine toad, *Bufo marinus*, and (d) tamarisk (salt cedar), *Tamarix* sp.

Invasive Organisms

Describe the following invasive organisms, their point of origin, and how and where they were introduced.

Kudzu:

_____

_____

_____

Water hyacinth:

_____

_____

_____

Zebra mussels:

_____

_____

_____

Argentina fire ants:

_____

_____

_____

Hydrilla:

_____

_____

_____

Marine toad:

_____

_____

_____

Common pigeon:

_____

_____

_____

European wild boars:

_____

_____

_____

Nutria:

_____

_____

_____

Formosan termite:

_____

_____

_____

nuisance for the lifestyle of living things such as the unpleasant noise, odors, and sights surrounding some industrial areas.

On April 20, 2010, history was written as the Deepwater Horizon oil rig exploded in the Gulf of Mexico approximately 50 miles off the coast of Louisiana. The resulting spill released tens of thousands of barrels of crude oil a day into the waters of the Gulf of Mexico, threatening the fragile wetlands off the coast of Louisiana, Mississippi, Alabama, and Florida. In addition, the oil spill threatened the pristine beaches of Mississippi, Alabama, and Florida. The spill was responsible for the horrific destruction of wildlife and coastal plants, disruption of the tourist industry, devastation of the seafood industry, and it jeopardized a way of life and psyche of a region rich in tradition.

Initial efforts using contemporary strategies and unproven methods to stop the oil leak at its source failed, allowing unprecedented amounts of oil to leak into the Gulf of Mexico. This spreading "monster" threatened all in its path, including biologically rich estuaries, wetlands, and beautiful beaches. Several strategies were utilized to contain and clean up the spill. These methods included: letting the spill break down by natural means, using booms to channel and collect the oil, blocking the spill with barges, building sand berms, using biological agents to degrade the oil, burning off the oil, and using dispersants to break up the oil.

A common method to clean up oil spills is to use dispersants to break up the oil and speed the natural biodegradation process. **Dispersants** are similar to emulsifying or surfactant agents acting to reduce the surface tension of water that prevents oil and water from mixing. In turn, small droplets of oil are formed, which can break down more rapidly.

Unfortunately, many dispersants are toxic to living things. Dispersants are not appropriate for all oils and all locations. If the oil is dispersed through the water column, it can affect marine organisms that form the basis of the food chain. It also can affect organisms that are important in the seafood industry. The following activity has been designed to safely simulate the effect of a dispersant upon oil, using dishwashing liquid and cooking oil. (If laboratory safety permits, outboard motor oil can be used to replace the cooking oil.)

A common dispersant that is safe for classroom use is dishwashing liquid. The detergent loosens grease and oil from dirty dishes by acting as a **surfactant**, which stands for "surface active agent." Surfactants work at the interface between the oil or grease and the water. Surfactant molecules consist of a **hydrophilic face**, which is attracted to water, and a **hydrophobic face**, which is

repelled by water. In a cleaning solution, the hydrophobic face of the surfactant molecule orients itself toward the oil and grease. Then surfactant molecules attack the oil and grease, breaking them up into small pieces similar to the way that a dispersant treats oil in an aquatic environment. In addition, the hydrophilic face of the surfactant molecules project into the water, causing the oil and grease to break up further and become suspended in the cleaning solution.

Figure 36.7 The Deepwater Horizon incident was devastating to the environment, economy, and way of life in the Northern Gulf of Mexico.

Figure 36.8 Spraying dispersant in the Gulf of Mexico.

Figure 36.9 Tar balls washing up off the coast of Louisiana as the result of dispersant.

Student Activity—Simulating the Effects of Dispersant in an Aquatic System

In this activity, dishwashing liquid will be used as the dispersant. The solution will contain common soil sediments found in an aquatic ecosystem (sand, silt, and clay). The difference between sand, silt, and clay is determined primarily by particle size. Sand ranges in size from 2 mm for very large grains to 0.05 mm for the finest sand particles. Silt particles range in size from 0.05 mm to 0.002 mm. Clay particles are smaller than 0.002 mm in diameter. Roughly 500 clay particles can wrap around a typical grain of sand.

The chemical nature and percentage of the particles vary by locality. The sediment particles will settle in calm water according to their particle size. The largest particles, the sand, settle first, followed by silt and clay. In this experiment, all of the cylinders will contain the same amount of sand, silt, and clay mixture. In addition, all of the cylinders will contain water from the same source. (If silt samples are not available, fine sand can be substituted for silt, and coarse sand for sand samples.) Within groups, the instructor assigns specific tasks to each student.

Materials
- four 250 ml clear, graduated cylinders
- water
- vegetable/cooking oil
- sand
- silt
- clay
- Dawn® dishwashing liquid
- eight 25 ml plastic measuring cups
- wooden applicator sticks
- eyedropper
- plastic wrap
- lab coat or apron
- gloves
- eye protectors
- wax pencil

Procedure 36.5
Oil Dispersant in Water

1. Procure the needed materials.
2. With a wax pencil, label four 250 ml graduated cylinders 1 through 4, and label four 25 ml plastic measuring cups 1 through 4.

3. Prepare the dispersant solutions: Fill the four labeled 25 ml measuring cups with water. Cup 1 should contain water only, cup 2 should receive one drop of dishwashing liquid, cup 3 should receive two drops of dishwashing liquid, and cup 4 should receive three drops of dishwashing liquid. Thoroughly stir each cup with a different applicator stick.
4. Fill one 25 ml plastic measuring cup with sand. Scrape the excess sand from the top of the cup with an applicator stick. Pour the sand into a dry 250 ml graduated cylinder.
5. Fill one 25 ml plastic measuring cup with silt. Scrape the excess silt from the top of the cup with an applicator stick. Pour the silt on top of the sand into the same 250 ml graduated cylinder.
6. Fill one 25 ml plastic measuring cup with clay. Scrape the excess clay from the top of the cup with an applicator stick. Pour the clay on top of the silt and sand into the same 250 ml graduated cylinder.
7. Add faucet-temperature tap water to the graduated cylinder labeled number 1 to reach the 200 ml mark.
8. Place plastic wrap over the top of the graduated cylinder labeled number 1. Use one hand to hold the plastic on top of the cylinder and the other hand to hold the base of the cylinder. Thoroughly mix the water and sediments into a uniform solution. More water may have to be added to bring the solution up to 200 ml.
9. Add one plastic cup (25 ml) of cooking oil (vegetable) to the graduated cylinder.
10. Repeat steps 3 through 9 for the remaining three cylinders labeled 2, 3, and 4.
11. Let the cylinders sit undisturbed for 10 minutes.
12. Pour the contents of measuring cup number 1 of the dispersant solution into the graduated cylinder labeled number 1. The first graduated cylinder should receive 25 ml of water only from measuring cup number 1 as a control.
13. Place plastic wrap over the top of the graduated cylinder. Use one hand to hold the plastic on top of the cylinder and the other hand to hold the base of the cylinder, and shake the solution for 1 minute. Immediately place the graduated cylinder on the top of the lab table, start the stopwatch, remove the plastic wrap, and do not move the graduated cylinder for the remainder of the activity.

14. After 15 minutes, mark and label, using a wax pencil, cylinder number 1 with the lower and upper limits for layers of sand, silt, clay, water, and oil. Determine the thickness of each, and record your measurements and observations in the chart and in the space below.

_____

_____

_____

15. Repeat steps 11–14 with measuring cups 2, 3, and 4 and graduated cylinders 2, 3, and 4.
16. Dispose of the contents of the graduated cylinders as directed by the instructor. Thoroughly clean the materials and return them to their proper places.
17. Complete the graph below, using the data from the table.

Check Your Understanding

Q. Why is dispersant used on oil spills?

_____

_____

_____

Q. What are the advantages and disadvantages of using dispersant?

_____

_____

_____

Q. Describe the findings of your experiment.

_____

_____

_____

_____

Q. Using the graph and your observations on the previous page, draw a conclusion about the effect of the dishwashing liquid on the distribution of vegetable cooking oil.

_____

_____

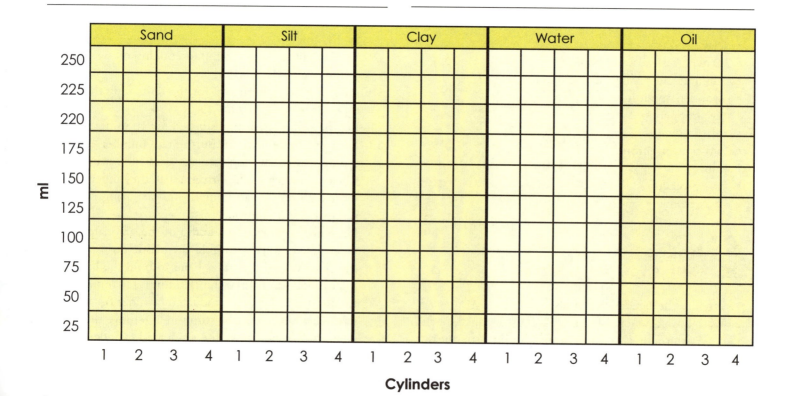

| | Sand | | | | Silt | | | | Clay | | | | Water | | | | Oil | | | |

Lower and Upper Limit Measurements

		Sand		Silt		Clay		Water		Oil	
Cylinders	1										
	2										
	3										
	4										

Student Activity—Cleaning Oil from Bird Feathers

One of the most heartbreaking scenes from an oil spill is the oil-covered wildlife. Birds such as pelicans are particularly vulnerable to the effects of oil. This experiment is designed to illustrate how dishwashing liquid can be used to clean bird feathers. Keep in mind that the actual process is highly coordinated and much more complex. Oiled animals should not be treated or cleaned without proper facilities and trained personnel.

A note of concern: Although valiant efforts have been taken to clean oil-soaked birds, recent studies indicate that only a small percentage of the birds actually survive, especially those inhabiting colder waters. Many die from hypothermia and shock, as well as kidney- and liver-related complications. Also, detergents may damage bird's natural waterproofing oil mechanisms. In addition, surviving and released birds may become disoriented and not adjust to new environments.

Materials
- six feathers soaked in outboard motor oil
- 1% Dawn® dishwashing liquid solution
- six small plastic trays
- cotton balls
- gloves
- lab coat or apron
- eye protection
- large plastic container
- paper towels
- disposal bin for oil and oily water
- stopwatch

Procedure 36.6
Oil and Feathers

1. Procure the equipment for this experiment. Place the oil-soaked feathers in one of the small plastic trays.

2. Place three feathers in a small plastic tray containing only water.
3. With gloves, cotton balls, a lab coat, and eye protection, gently scrub the feathers until the water becomes dirty with oil. When the water is dirty, move the feathers to another container, and so on until the feathers appear clean, or for a maximum of 5 minutes. Record your observations below.

_____

_____

_____

4. Place three feathers in a small plastic tray containing 1% dishwashing solution.
5. With gloves, cotton balls, a lab coat, and eye protection, gently scrub the feathers in the 1% dishwashing solution until the water becomes dirty with oil. When the water is dirty, move the feathers to another container, and so on, until the feathers appear clean, or for a maximum of 5 minutes.
6. When the feathers appear clean, rinse them with water. Record your observations below.

_____

_____

7. Dispose of the feathers, water, and oil as indicated by your instructor.
8. Discuss your findings and answer the questions below.

_____

_____

Check Your Understanding

Q. Did the water or the dishwashing solution work the best? Why?

_____

_____

_____

Q. Was the dishwashing liquid effective in cleaning the feathers?

_____

_____

_____

_____

Q. Describe several problems that a professional or a volunteer may encounter in cleaning birds.

_____

_____

_____

_____

_____

Q. Even though the feathers may become clean, what other problems does the bird face?

_____

_____

_____

_____

Student Activity—The Effects of Oil upon Radish Seeds: A Self-designed Experiment

Along the marshes of the northern coast of the Gulf of Mexico, wetland plants were exposed to oil. The oil proved lethal to many of the plants, their seeds, and seedlings, increasing the chances of erosion and the loss of valuable wetlands.

This activity will allow students to simulate the effect of oil on either the germination of radish seeds or the survival of radish seedlings. Working in groups, the students will design and conduct a valid experiment that tests the effect of oil on radish seeds. Communication within the group is important. The students also will prepare a detailed lab report, including photographs and a PowerPoint® presentation.

I think the environment should be put in the category of our national security. Defense of our resources is just as important as defense abroad. Otherwise what is there to defend?

—Robert Redford (1936–present)

Materials (some materials are optional)
- radish seeds
- radish seedlings growing on a moist paper towel in a Petri dish
- Petri dishes
- paper towels
- shoebox or plastic container
- ruler
- hand lens
- stopwatch
- pipette
- variety of graduated cylinders
- outboard motor oil
- mineral oil
- water
- lab coat or apron
- gloves
- eye protection
- food coloring
- variety of beakers
- thermometer
- camera
- colored pencils
- applicator sticks
- other materials that you decide upon

Procedure 36.7
Oil and Radish Seeds

1. Review the basic and integrated process skills discussed in Chapter 1 and seed germination discussed in Chapter 27. As always, stress safety.
2. Decide to test either seed germination or seedlings.
3. Decide to test either mineral oil or outboard motor oil.
4. If you decide to use a solution of mineral oil and water, consider adding a bit of food coloring to the water.
5. Answer the following questions:
   a. Discuss some background material.

   _____

   _____

   b. List the control, the independent, and dependent variables in your experiment.

   _____

   _____

   _____

   c. State your hypothesis.

   _____

   _____

   d. List the materials you decided to use.

   _____

   _____

   _____

   _____

   e. Describe the procedure.

   _____

   _____

   _____

   f. What basic and integrated process skills will you use in this experiment?

   _____

   _____

   _____

   _____

6. Do you have a means of safely disposing of the material at the end of the experiment?
7. Decide how your lab report and presentation will be done.
8. Develop a lab report and PowerPoint® presentation.

 Student Activity—The Effects of Oil upon Brine Shrimp: A Self-designed Experiment

Small invertebrates in the waters and along the shores of the northern coast of the Gulf of Mexico are a major part of the food chain. The oil has been lethal to many invertebrates, including crustaceans. This activity allows students to simulate the effect of oil on either the hatching of brine shrimp eggs or the survivability of brine shrimp in oil. Working in groups, the students will design an experiment that tests the effect of oil on brine shrimp. They will have to design and conduct a valid experiment using select materials. Finally, the students will prepare a detailed lab report including photographs and a PowerPoint® presentation.

Figure 36.10 Oil is detrimental to the survival of crustaceans such as the brine shrimp, *Artemia salina*.

Materials (some materials are optional)
- brine shrimp eggs
- adult brine shrimp
- Petri dishes
- hand lens
- dissecting microscope
- pipette
- applicator sticks
- variety of graduated cylinders
- outboard motor oil
- mineral oil
- prepared saltwater solution for brine shrimp
- water
- lab coat or apron
- gloves
- eye protection
- paper towels
- variety of beakers
- thermometer
- camera
- colored pencils
- materials that you decide upon

Procedure 36.8
Oil and Brine Shrimp

1. Review the basic and integrated process skills discussed in Chapter 1. Stress safety as always.
2. Decide if you want to test brine shrimp eggs or adults.

   Your choice: _____

3. Decide if you want to test mineral oil or outboard motor oil.

   Your choice: _____

4. If you decide to use a solution of mineral oil and water, consider adding a bit of food coloring to the water.
5. Answer the following questions to help get you started:

   a. Discuss some background material.

   _____

   _____

   _____

b. List the control, independent and dependent variables in your experiment.

_____

_____

_____

c. State your hypothesis.

_____

_____

_____

d. List the materials you decided to use.

_____

_____

_____

e. Describe the procedure.

_____

_____

_____

_____

f. What basic and integrated process skills will you use in this experiment?

_____

_____

_____

6. Do you have a means of safely disposing of the material at the end of the experiment?
7. Decide how the lab report and presentation will be done.
8. Develop a lab report and PowerPoint presentation.

The love for all living creatures is the most noble attribute of man.

—**Charles Darwin (1809–1882)**

Name: _____   Date: _____   Section: _____

Review Questions

1. What is the difference between ecology and environmental science? Provide several examples of each.

_____

_____

_____

_____

2. Name several keystone species in the marine ecosystem or marsh ecosystem disrupted by the Deepwater Horizon incident.

_____

_____

_____

_____

3. Describe several of the most bioproductive environments on Earth.

_____

_____

_____

_____

4. Construct a simple aquatic food chain.

5. Construct a simple marine food web.

6. What is the difference between primary succession and secondary succession?

_____

_____

_____

7. Why are scientists concerned about intrusive species?

_____

_____

_____

_____

8. What does HIPPO stand for?

_____

_____

_____

_____

9. Discuss several types of pollution in your region.

_____

_____

_____

_____

Index